**Books are to be returned on or before
the last date below.**

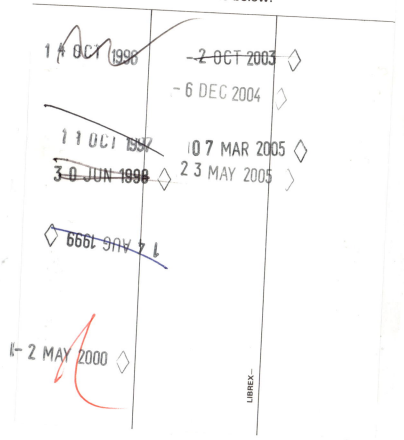

1 4 OCT 1996   2 OCT 2003

                     - 6 DEC 2004

11 OCT 1997   0 7 MAR 2005

3 0 JUN 1998   2 3 MAY 2005

1 4 AUG 1999

- 2 MAY 2000

LIBREX—

Fundamentals of Industrial Control

C. L. Albert, Editor
D. A. Coggan, Editor

Fundamentals of Industrial Control

C. L. Albert, Editor
D. A. Coggan, Editor

Practical Guides for Measurement and Control

Instrument Society of America

FUNDAMENTALS OF INDUSTRIAL CONTROL

Copyright © 1992, Instrument Society of America

INSTRUMENT SOCIETY OF AMERICA
67 Alexander Drive
P.O. Box 12277
Research Triangle Park, NC 27709

Library of Congress Cataloging in Publication Data

Fundamentals of industrial control / D.A. Coggan, editor, C.L. Albert, editor.
    p. cm.
    Part of the Practical guides series.—Pref.
    Includes bibliographical references and index
    ISBN 1-55617-335-0
    1. Process control. 2. Automatic control. I. Coggan, Donald A. II. Albert, C.L. III.
    Instrument Society of America
    TS156.8.F85 1992   92-29680
    629.8—dc20   CIP

About This Series

This volume is part of the Practical Guide Series developed and published by the Instrument Society of America (ISA).

The Practical Guides were conceived because of a shortage of published material in the field of measurement and control that bridges the gap between theory and actual industrial practice. Many books in the field have catered to the needs of technical students, who need to be oriented to basic control theory and concepts, or college-level readers, who are interested in engineering mainly from a classroom perspective. There are handbooks for practicing engineers that cover measurement and control, but these handbooks often devote only a chapter or two to topics that merit more attention. Within the Practical Guides Series, separate volumes address each of the important topics and give them comprehensive, book-length treatments. Each book in the series can be understood and used by technical students, sales engineers, sales personnel, and managers, and relied upon by those who have "real-life" industrial concerns such as correct application, safety, installation, and maintenance.

Another unique feature of the Practical Guides is the stress placed on the actual experience of measurement and control practitioners. The Practical Guides are overseen by three Series Editors and one Series Technical Editor, who have extensive experience in measurement and control. The Series Editors guide the Volume Editors, who have been selected for their specific expertise in the volume topics and who bring together numerous Contributing Writers with even more specialized knowledge.

The Practical Guides capture the hard-earned experience of the writers and, by employing examples and recording anecdotal observations, make that experience as applicable for the reader as possible. Case studies, either hypothetical or based on real case histories, are used to illustrate typical situations and show how good planning and practical applications made the difference between success and failure. Some of this information has never been documented before.

This volume is designed to be at home in a library, in a classroom, or on the plant floor. The comfortable reading style, large pages, and frequent illustrations will contribute to ease of use. The page design uses graphics to "call out" some of the major points of the text, such as crucial safety checks and important examples. Each Practical Guide gathers widely scattered information in a single text, with bibliographies directing the reader to other sources.

Providing editorial guidance for the Practical Guides Series are some of the most distinguished names in the field of measurement and control.

Paul W. Murrill, Ph. D., Series Editor

Paul W. Murrill has authored or co-authored ten textbooks and over 70 articles on process control, computers, and mathematical models. Formerly Chancellor of Louisiana State University, he currently serves as Special Advisor to the Chairman of the Board of the Gulf States Utilities, after having served as chairman and CEO of the company for five years.

Thomas M. Stout, Ph. D., Series Editor

Thomas M. Stout is the author or co-author of more than 125 technical papers and holds four patents as co-inventor of computer and process controls. Dr. Stout is a pioneer in the application of computers in process control, particularly in petroleum refining.

Robert H. Zielske, Series Editor

Robert H. Zielske is Chief Instrument Engineer for Georgia Pacific Corporation, where he oversees design, testing, installation, and start-up of new facilities and major expansions. During his 40-year career, he has worked in design and applications engineering, marketing and sales, and training.

John W. Bernard, Series Technical Editor

John W. Bernard has worked on the leading edge of industrial automation systems for 35 years. He has extensive experience in process management and research. Among his many achievements is installing the first digital computers for direct process control. Bernard has written numerous papers on computer control and systems engineering and has been deeply involved in standards activities.

Call for Participation

The major purpose of this series is to collect in one volume most of the existing practical knowledge about specific measurement and control subjects.

Additional material for inclusion in subsequent updates of this volume is most welcome.

If you wish to contribute any helpful hints, case studies, or other material, please contact:

Manager, Publication Services
Instrument Society of America
P.O. Box 12277
Research Triangle Park, NC 27709
(919) 549-8411

Table of Contents

Preface

Readers will wonder how a book on fundamentals can be published when there already seems to be so much similar material on the market. This one truly is different. As the introductory volume to the entire Practical Guide Series, it was written in the PGS spirit—with emphasis on the practical. This isn't always easy to do when dealing with fundamental concepts. The contributors to this book have succeeded, however, in finding the right balance between requisite theory and recommended application.

Another way in which this book is different is that it has a proven track record. Before becoming the introductory volume to the Practical Guide Series, it was used in a preliminary version as reference notes for an introductory course in process instrumentation given at McGill University in Montréal. This course has run continuously for 40 years, and the lecturers have always been practitioners. Their backgrounds have rubbed off on the course presentations and—fortunately—in this book.

Finally, this book was written by a dedicated group of professionals who, with the exception of one person, are all members of the Montréal chapter of ISA. Their enthusiasm was a major factor in assembling all the material needed to produce the words as they are printed here. The reader will undoubtedly be affected by this contagious enthusiasm.

Donald A. Coggan
Charles Albert
Montréal, Québec
July, 1992

About the Volume Editors

Donald A. Coggan, author of ISA's book and accompanying software, *Preparing for Instrumentation Technician Evaluation*, is an independent consulting engineer. He is owner and principal engineer of a consulting engineering firm specializing in control and automation systems. He is also the founder of Lab-Experts, which offers specialized engineering services for the control of laboratory ventilation systems. From 1981 to 1988, he carried on his varied consulting business activities under the umbrella company, Coggan Consulting Corporation.

Born and raised in Winnipeg, Manitoba, Mr. Coggan later moved to Montréal where he obtained his Bachelor of Electrical Engineering degree from McGill University. Before starting up his own consulting engineering business at the end of 1980, Mr. Coggan had previously worked for Johnson Controls and MCC Powers in positions of increasing responsibility. In addition, he was a part-time instructor from 1972 to 1986 at Vanier College where he taught courses in instrumentation, HVAC controls, energy conservation, and computer-aided drafting.

Mr. Coggan is the author of over 60 articles and technical papers, which he has presented throughout North America and in Europe and Asia. As founder and Editor-in-Chief of *Gaining Control*, his own technical publications company serving the control and automation industry, he has written a number of technical reports and software programs.

An avid reader and amateur health buff, Mr. Coggan lives in Outremont, Québec, with his wife, Huguette, and children, Rebecca, Christopher, and Melanie.

Charles Albert is a former president of the Montréal Section of ISA. He has been a member of the Executive Committee as well as a very active member of the Education Committee.

Mr. Albert has also carried out many functions at an international level out of ISA's head office in Research Triangle Park in North Carolina. These include participation in the Executive Committee on Education and the Computer Division. He has also played an active role as Official Delegate of the Montréal Section to international meetings and conferences in Canada, in the United States, and abroad.

Mr. Albert is employed by Canadian Pacific Forest Products where he is Senior Process Automation Engineer with Corporate Engineering Group.

Contributors

Biographies of each of the following contributors are included at the end of their respective chapter(s).

Sudhendu N. Banerjee
Instrumentation, Ltd.

Zdislaw Victor Barski
Consultant

Helen Beecroft
Sandwell, Inc.

Daniel Bellefontaine
Rosemount, Inc.

Diana C. Bouchard
*Pulp and Paper Research Institute
of Canada*

Gilles J.P. Bouchard
Sandwell, Inc.

James E. Bouchard
Johnson & Johnson, Inc.

Donald A. Coggan
Consultant

Alberto J. Dufau
SNC-Lavalin, Inc.

Eddie Marquis
Bechtel Canada

Lowell E. McCaw
*Professor Emeritus, Monroe
Community College*

Jean-Claude Moisan
Petromont, Inc.

Norman Peters
Consultant

Maurice L. Pyndus
Public Works, Canada

Michel Spilman
SOMIS, Inc.

1

Sensors

This chapter discusses the principles involved in the sensing of the most commonly encountered variables used in process control in an industrial facility. Sensors may be used for both monitoring and control.

Applications of Instrumentation

Everyday examples of instruments used for monitoring are the thermometers, barometers, and anemometers used by government weather services to indicate the condition of the environment. Similarly, water, gas, and electric meters are used to keep track of the consumption and cost of such commodities. Closer to the subject of this book, sensors are used to monitor and record important variables in a process.

The other and extremely important application of sensors is that in which the instrument serves as a component of an automatic control system. The role of the sensor in an automatic control system is clearly seen in the traditional functional block diagram (see Figure 1-1).

An application example close to home is the typical thermostatically controlled forced-air heating system. In this case, a sensor measures the room temperature and provides the information necessary for proper functioning of the control sys-

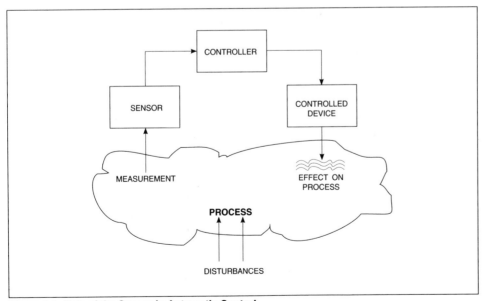

Figure 1-1. Role of the Sensor in Automatic Control

tem. Many more examples of automatic control will be found throughout this book.

Whatever the nature of the application, intelligent selection and use of measurement instrumentation depends on a broad knowledge of what is available in the market and how it will perform in a specific application.

In the following paragraphs, some of the uses of sensing instruments in process control applications are summarized.

Collecting and sending information about a measured variable. Examples include pressure sensors (such as bellows, diaphragms, manometers, and Bourdon tubes), temperature sensors (such as thermometers, thermal bulbs, thermocouples, thermistors, and resistance temperature detectors), level sensors (such as floats, level switches, and displacers), and flow sensors (such as orifice plates and Venturi tubes when used with a differential pressure sensor, and rotameters). Some instruments, called transmitters, combine the sensing and sending functions in one package.

Displaying and/or recording information about a measured variable. Instruments that display information include thermostats, speedometers, indicating lights on a control panel, and meters of all sorts. Instruments that record information include lie detectors, electrocardiograms, plotters, and chart recorders.

Comparing what is happening (value of the measured variable) to what should be happening (set point). Instruments that compare what is to what should be include thermostats, controllers, and microcomputers.

Making a decision about what action should be taken to adjust for deviation from the set point. Instruments that make decisions include thermostats, controllers, and microcomputers. This may also include taking action by adjusting the manipulated variable by means of control valves, fans, dampers, motors, and pumps. Note that the comparison and decision modes are often combined.

Initiating an alarm when the measured variable is either too high or too low. Instruments that actuate an alarm include smoke detectors and home security systems.

Introduction to Sensor Fundamentals

Transducers and Sensors

A transducer is a device that converts one form of energy into another. This conversion may be pressure to movement, electric current to pressure, liquid level to a twisting movement on a shaft, or any number of other combinations. Although the final output of a sensor may be electrical or pneumatic, there may also be one or more intermediate transducing stages.

There are two basic types of sensors: analog, which produces an output proportional to a change in a parameter, and digital, which produces an on/off type of output. Sensors that provide digital outputs (for example, pulses) proportional to changes in the parameter are regarded as digital sensors.

A sensor may also be looked upon as an active or a passive transducer. A sensor whose output energy is supplied entirely, or almost entirely, by its input signal is commonly called a passive transducer. The output and input signals may involve energy conversion from one form to another (for example, mechanical to electrical). An active transducer, on the other hand, has an auxiliary source of power that supplies a major part of the output power, while the input signal supplies only an insignificant portion. Again, there may or may not be a conversion of energy from one form to another (see Table 1-1).

Table 1-1. Some Physical Effects Used in Instrument Transducers

Energy Conversion Principles	Energy Controlling Principles
Electromagnetic	Resistance
Piezoelectric	Inductance
Magnetostrictive (as a generator)	Capacitance
Thermoelectric	Mechanoresistance (strain)
Photoelectric	Magnetoresistance
Photovoltaic	Thermoresistance
Electrokinetic	Photoresistance
Pyroelectric	Piezoresistance
	Magnetostrictive (as a variable inductance)
	Hall effect
	Radioactive ionization
	Radioactive screening
	Ionization (humidity in solids)

The Functional Elements of an Instrument

Examination of sensor elements reveals recurring similarities with regard to functional operation. Instruments can, therefore, be categorized into a limited number of types of elements according to the generalized function performed by the element.

Consider the diagram of Figure 1-2, which includes the basic elements needed for a description of an instrument. The primary sensing element receives energy from the measured medium and produces an output that depends on the measured quantity. Note that an instrument generally extracts some energy from the measured medium; thus the measured quantity may be disturbed by the act of measurement, making a perfect measurement theoretically impossible. Good instruments are designed to minimize this effect.

The output signal of the primary sensing element is a physical variable, such as displacement, voltage, or current. For the instrument to perform the desired function, it may be necessary to convert this variable to another, more suitable variable while preserving the information content of the original signal. An element

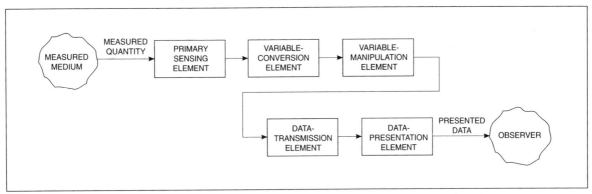

Figure 1-2. Generalized Description of an Instrument

that performs such a function is called a variable-conversion element. Note that while not every instrument needs a variable-conversion element, some require several. Also, the "elements" referred to here are functional elements, not physical elements.

In performing its intended task, an instrument may require that a signal represented by some physical variable be manipulated in some way. Manipulation means a change in numerical value according to some definite rule but a preservation of the physical nature of the variable. Thus, an electronic amplifier accepts a small voltage signal as input and produces an output signal that is also a voltage but is some constant times the input. An element that performs such a function will be called a variable-manipulation element.

When the functional elements of an instrument are physically separated, the data must be transmitted from one to another. The data transmission element performs this function. It may be as simple as a shaft and bearing assembly, or it may be as complex as a complete telemetry system.

If the information about the measured quantity is to be communicated to a human being for purposes of monitoring, control, or analysis, it must be put into a form that is recognizable by one of the human senses. The data presentation element performs this "translation" function. It may involve a simple pointer moving over an indicating scale or a pen moving over a recording chart.

These elements are present in the pressure-type thermometer (see Figure 1-3). The liquid-filled bulb acts as primary sensor and variable-conversion element because a temperature change results in a pressure change within the bulb due to the constrained thermal activity of the filling fluid. This pressure converts pressure to displacement. This displacement is manipulated by the linkage and gear-

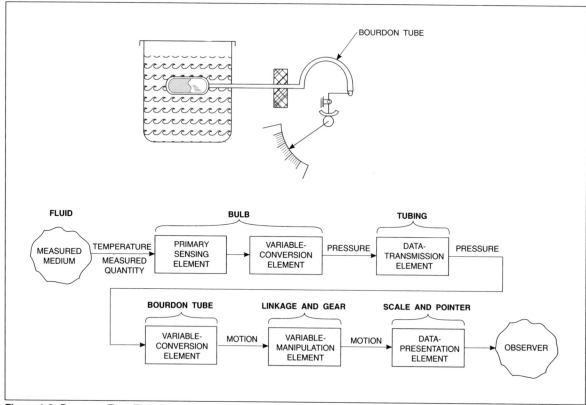

Figure 1-3. Pressure-Type Thermometer

ing to give a larger pointer motion. A scale and pointer again serve for data presentation.

Terminology

Instrument engineering has its own terminology. Some of the terms have subtle meanings, and a misunderstanding can lead to a completely wrong impression of the performance of a system. The following definitions are intended as an introduction to the use of the terminology. For more complete and precise definitions refer to the ISA standard ANSI/ISA-S51.1.

RANGE

Every sensor is designed to work over a specified workable range. While an electrical output may be adjusted to suit the application, this is not usually practical with mechanical transducing elements. The design ranges of these mechanisms are usually fixed and, if exceeded, can result in permanent damage to a sensor. Transducing elements must be used over the part of their range in which they provide predictable performance and often truer linearity.

ZERO

A measurement must be made with respect to a known datum. Often, it is convenient to adjust the output of the instrument to zero at the datum. For example, the output of a Celsius thermometer is zero at the freezing point of water; the output of a pressure gage may be zero at atmospheric pressure. Zero, therefore, is a value ascribed to some defined point in the measured range.

ZERO DRIFT

One of the problems experienced with sensors occurs when the value of the zero signal varies from its set value. This introduces an error into the measurement equal to the amount of variation, or drift, as it is usually termed. All sensors are affected by drift to some extent, and it is sometimes specified in terms of short-term and long-term drift. Short-term drift is usually associated with changes in temperature or electronics stabilizing. Long-term drift is usually associated with aging of the transducer or electronic components.

SENSITIVITY

Sensitivity of a sensor is defined as the change in output of the sensor per unit change in the parameter being measured. Sensors may have constant or variable sensitivities, in which cases they are described as having a linear or a nonlinear output, respectively. Clearly, the greater the output signal change for a given input change, the greater the sensitivity of the measuring element. Sensitivity depends on a number of variable factors. The mechanical properties of a transducer may vary with temperature and cause a variation in sensitivity, but often it is the electrical part of the sensor that is responsible for the greatest changes. An amplifier may change its gain because of temperature effects on components or variations in power supplies or even faulty operation. See Figure 1-4 for an illustration of the effect of varying sensitivity.

An example of when sensitivity would be critical is in a blending process that requires a certain mix. The load change that occurs every time an ingredient is added may cause a sharp change in the temperature. The mix could be ruined if the change in temperature were not measured and controlled immediately. High sensitivity of the measuring element increases the chances of a quick response.

RESOLUTION

Resolution is defined as the smallest change that can be detected by a sensor. Although it is evident that sensors using wire-wound potentiometers or digital

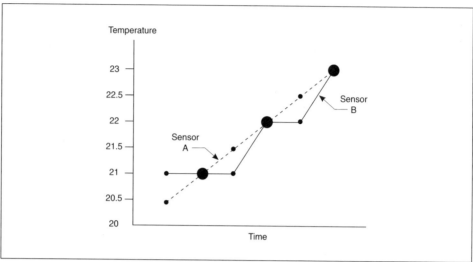

Figure 1-4. Effect of Different Sensitivities

techniques to provide their electrical output have finite resolutions, no known device has an infinitely small resolution.

RESPONSE

The time taken by a sensor to approach its true output when subjected to a step input is sometimes referred to as its response time. It is common to state the performance of a sensor in terms of having a flat response between specified limits of frequency. This is known as the frequency response, and it indicates that if the sensor is subjected to sinusoidally oscillating input of constant amplitude, the output will faithfully reproduce a signal proportional to the input. Fast sensors make it possible for controllers to function in a timely manner. Sensors with large time constants are slow and may degrade the overall operation of the feedback loop.

Sensor response is best understood by examining a bare bulb-type expansion thermometer (see Figure 1-5). For analysis purposes, the bare bulb is immersed into an agitated constant temperature bath as shown.

As the bare bulb reacts to the bath temperature, which is assumed to be higher than ambient, the thermometer needle rises (see Figure 1-6). The bath temperature is approached gradually, as shown, by the exponential response curve, which gives some experimental insight into the dynamics of this particular measuring device.

The time constant is defined as the time necessary for the response curve to reach 63.2% of its final value. In the illustration shown, the time constant for the bulb is approximately five seconds.

LINEARITY

A sensor described as having a linear transfer function is one whose output is directly proportional to the input over its entire range. This relationship appears as a straight line on a graph of output versus input. In practice, exact linearity is never quite achieved, although most transducers exhibit only small changes of slope over their working range. In such cases, the manufacturer fits a "best" straight line whose error is usually well within the tolerance of the measurement. Some sensors, particularly those using inductive transducing principles, demonstrate considerable changes in the slope of their output versus input graph and may even reach a point at which, regardless of change of input, there is no change of output. The working range of such a sensor is restricted and must be

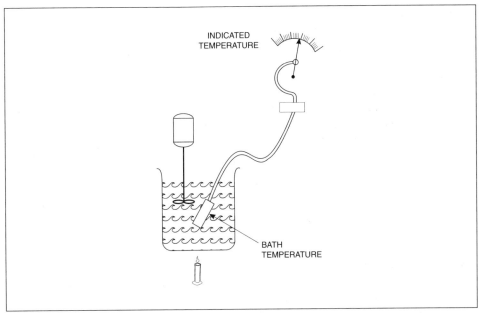

Figure 1-5. Thermometer Experiment

limited to where the graph is most linear; alternatively, a different factor must be applied to each reading.

HYSTERESIS

Hysteresis becomes apparent when the input to a sensor is applied in a cyclic manner. If the input is increased incrementally to the sensor's maximum and returned to its zero datum in a similar manner, the calibration may be seen to describe two output curves that meet at the maximum. In returning to zero input, the instrument has not returned to its original datum. If the calibration is continued in the negative direction of input, two further curves that are a mirror image of the previous ones will be produced. Further cycling will eventually link these two halves into one complete loop, which will then be repeatable with every cycle. This loop is normally referred to as the hysteresis loop of the sensor,

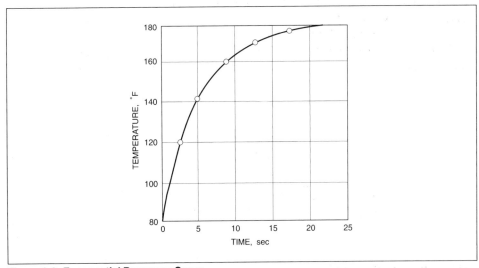

Figure 1-6. Exponential Response Curve

although it also contains any of the other nonlinearity effects that may be present. Consequently, it is usual when specifying a sensor to quote nonlinearity and hysteresis as one parameter.

CALIBRATION

To be meaningful, the measurement of the output of a sensor must be in response to an accurately known input. This process is known as calibration, and the devices that produce the inputs are described as calibration standards. It is usual to provide measurements at a number of points of the working range of the sensor, so that a ratio of output to input may be determined from the measured points by calculation. Such a ratio is described as a calibration factor. The ratio of output to input is not always a constant over the range of a sensor, and the calibration graph may describe a curve. In these instances, a best straight line may be fitted through the points and the errors accepted, or a different calibration factor must be provided for every measurement.

ACCURACY AND PRECISION

Accuracy of a measurement is the term used to describe the closeness with which the measurement approaches the true value of the variable being measured. Precision is the reproducibility with which repeated measurements of the same variable can be made under identical conditions. In matters of process control, the latter characteristic is more important than accuracy; it is normally more desirable to measure a variable precisely than it is to have a high degree of absolute accuracy. The distinction between these two properties of measurement is best illustrated graphically (see Figure 1-7).

The meaning of accuracy, when quoted as a percentage of full-scale output, is that of a value of uncertainty that is applied to converted sensor outputs throughout the entire range of measurement. For example, a measurement with an accuracy of $\pm1\%$ full scale and with a range of 0–100 units has a value of uncertainty of ±1 part in 100 or ±1 unit, which applies to every measurement. A measurement of 50 units would be made with a value of uncertainty of ±1 unit or $\pm2\%$ of the value.

Sensors are designed to be both accurate and precise. A sensor that is accurate but imprecise may come very close to measuring the actual value of the controlled variable, but it will not be consistent in its measurements.

A sensor that is precise but inaccurate may not come as close to measuring the actual value of the controlled variable, but its measurements will differ from the actual value by nearly the same amount every time. This consistency makes it possible to compensate for the sensor error.

Practitioners make a distinction between two types of accuracy: static or steady-state accuracy and dynamic accuracy. Static accuracy is the closeness of

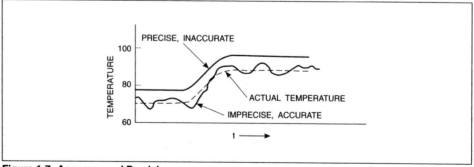

Figure 1-7. Accuracy and Precision

approach to the true value of the variable when that true value is constant. Dynamic accuracy, on the other hand, is the closeness of approach of the measurement when the true value is changing. These terms may be illustrated graphically (see Figure 1-8).

Transmitters

A transmitter carries a signal of the value of the measured variable from the sensor to a controller. Transmitters are necessary because the sensor and the controller are often physically far apart. The transmitter picks up the measurement provided by the sensor, converts it to a standard signal, which can be easily sent and read, and conveys the signal to the controller. Sensors and transmitters are often combined into one device. The two most common types of transmission used in industry are pneumatic and electronic.

A pneumatic transmitter converts the value of the measurement to an air pressure signal that is sent through tubing to the controller. The connecting tubing carries the transmitted pressure to a receiver, which is a component located in the controller housing. The tubing is almost always one-quarter inch in outside diameter and may be copper, aluminum, or plastic. The receiver is simply a pressure-gage element, and the transmitted air pressure is converted into the movement of a bellows or diaphragm (that is, pressure is transduced into a position or force that is used by the controller).

As distances increase, the speed of response of pneumatic transmission systems becomes a problem, and alternative solutions, including electronic transmission, are necessary.

An electronic transmitter converts the measurement to an electric signal, usually a voltage or a current, then transmits the signal by wire or radio linkage. Electronic signals can be transmitted over very long distances while still providing a virtually instantaneous response; therefore, their dynamics do not become serious problems in process control applications.

Other types of transmission sometimes used are hydraulic, telemetering, and optical.

Hydraulic. The transmitter converts the value of the measurement to an equivalent value of fluid pressure and sends the signal through tubing. Hydraulic transmitters can be more accurate than pneumatic ones because, unlike air, a liquid does not compress. However, hydraulic transmitters are temperature sensitive and a small leak will destroy the transmission.

Telemetering. The transmitter converts the value of the measured variable to certain frequencies or amplitudes of radio signals and sends the signal by radio linkage. An example of a telemetered transmitter is a microwave transmitter. A

The accuracy or uncertainty of power supplies, amplifiers, and recorders also contributes to the overall measurement value. Some instrumentation engineers treat all these quantities as a "measuring chain" and do not attempt to break them down, arguing that the accuracy of the measurement can be only the accuracy of the chain.

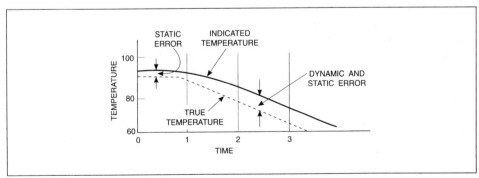

Figure 1-8. Static and Dynamic Accuracy

microwave transmitter converts the value of the measured variable to microwave radio signals. These transmitters require no wiring or cabling but are subject to interference.

Optical. The transmitter converts the value of the measured variable to light frequencies and sends the signal through optical fibers. Optical transmitters are not subject to electrical noise or interference.

Signals are usually transmitted within standard ranges. For electronic transmitters the most common standard is 4–20 mA DC. The most common standard pneumatic signal is 3–15 psig (20–100 kPa).

The information transmitted by the transmitter has to cover the entire range of information on the measured variable. For example, if the range of a process temperature is 100–500°F and the output signal range of the transmitter is 4–20 mA, the transmitter is calibrated so that 4 mA corresponds to 100°F, and 20 mA corresponds to 500°F.

Standards of Measurement

Standards of measurement should be reviewed before going further; accurate measurement is the basis of all science. There are six base units of measurement: length, mass, time, temperature, electric current, and light. All other units of measurement are expressed in terms of the fundamental units and are known as derived units of measurement. Measurements are expressed in both English and metric units. The latter are now properly known as SI units from the French "Système International" or International System of Units.

Length. Length is expressed in meters (m). The meter is defined as 1,650,763.73 wavelengths in vacuum of the orange-red line of the spectrum of krypton 86. The former standard meter was the distance between two engraved marks at a temperature of 0°C (32°F) on a metal bar retained at the International Bureau of Weights and Measures near Paris, France. An English unit for length is the standard yard, which is the length equal to 0.914 of the former standard meter.

Mass. Mass is expressed in kilograms (kg). The standard of mass is a cylinder of platinum-iridium alloy kept by the International Bureau of Weights and Measures in Paris, France.

Time. Time is expressed in terms of the second, which is defined as the duration of 9,192,631,770 cycles of the radiation associated with a specified transition of the cesium atom. This duration is realized by tuning an oscillator to the resonance frequency of the cesium atoms as they pass through a system of magnets and a resonant cavity into a detector.

Temperature. Temperature is expressed in terms of the kelvin (K). The thermodynamic or Kelvin scale of temperature used in the SI has its origin or zero point at absolute zero and has a fixed point at the triple point of water defined as 273.16 kelvins. The Celsius scale is derived from the Kelvin scale. The triple point is defined as 0.01°C on the Celsius scale, which is about 32.02°F on the Fahrenheit scale. The triple-point cell, an evacuated glass cylinder filled with pure water, is used to define a known fixed temperature. When the cell is cooled until a mantle of ice forms, the temperature at the interface of solid, liquid, and vapor is 0.01°C.

Electric Current. Electric current is expressed in terms of the ampere (A). The ampere is defined as the magnitude of the current, which, when flowing through each of two long parallel wires separated by one meter in free space, results in a force between the two wires (due to their magnetic fields) of 2×10^{-7} newtons for each meter of length. (The newton (N) is the SI unit of force. One newton will give one kilogram of mass a speed of one meter per second. One newton equals 0.22 lb of force.)

Light. Light, or luminous intensity, is expressed in terms of the candela (cd). It is defined as the luminous intensity of 1/600,000 of a square meter of a radiating cavity at the temperature of freezing platinum (2,042 K).

Level Measurement

Everyday examples of liquid level measurement devices are the engine oil and gas tank gages of a car. Another simple device is the level gage or level glass on a tank or boiler. The measurement and control of liquid level is essential in a process plant, where a wide variety of liquids are handled in both batch and continuous processes. The accurate measurement of level is important for environmental protection (for example, tank overflow to drains), plant safety, product quality, and inventory control.

Almost all liquid level devices measure by way of the position or height of the liquid above a zero or lowest point, or the hydrostatic pressure or head.

Direct and Indirect Measurements

The level measurement may be expressed either in units of length or volume, or in percentage of total volume. There are two methods to measure a liquid level: the direct method and the indirect or inferential method.

The direct method measures the liquid height above the zero point by any of the following techniques:

(1) Direct visual observation of the height by means of sight glass, level gage, or dip stick

(2) A float, which is mechanically linked or electrically connected to an indicator or alarm device

(3) An electrical probe in the liquid

(4) Reflection of sonic waves from the liquid surface or from the bottom

The indirect or inferential method of measurement uses the changing position of the liquid surface to determine level. Techniques involve:

(1) The buoyant force on a float or displacer, which is partially or completely immersed in liquid

(2) Hydrostatic pressure of the liquid

(3) The amount of radiation passing through the liquid

(4) Electric systems by which liquid level may be inferred

Visual Level Sensors

SIGHT GLASS

A sight glass is a device that is connected to a tank in such a way that the liquid level in the tank can be seen through the glass. They are very common in the process industry. There are two types of sight glass gages: tubular and flat glass. Sight glasses are usually installed with shutoff valves and a drain valve, mainly for purposes of maintenance, repair, and replacement. Graduations engraved on the glass or housing help one compare the level with a certain value (for example, between 0% and 100%). Refer to Figure 1-9.

For an open vessel, a simple, open-end tubular level gage is used. For pressure and vacuum vessels, the upper end of the tube is connected to the vessel to maintain an equilibrium.

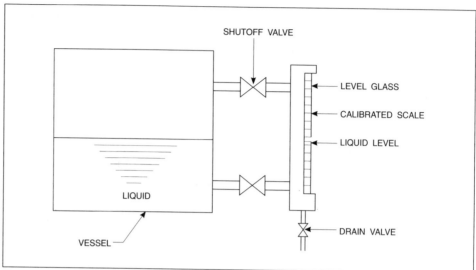

Figure 1-9. Sight Glass

Flat gages are used in industry for a wide range of pressure and temperature applications. There are two basic designs: reflex and transparent. The reflex-type design is chosen for nonviscous, colorless, liquid. The transparent gage is used for colored, viscous, and corrosive liquid.

FLOAT DEVICES

Another method of direct level measurement is the float-cable-pulley-weight arrangement. It operates by having a cable attached to a float pass over several pulleys. The movement of the float raises or lowers a counterweight on a scale attached to the side of the tank. This is usually installed on storage tanks for inventory purposes. See Figure 1-10.

Liquid level can also be measured using a float or displacer along with a mercury switch or microswitch. Figure 1-11 shows a ball float switch used to detect an increase (or decrease) in level beyond a set value.

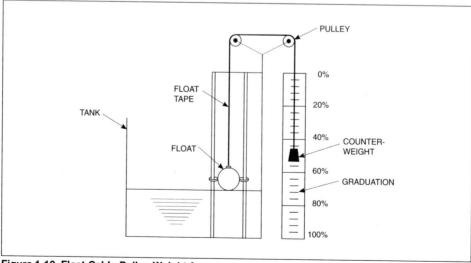

Figure 1-10. Float-Cable-Pulley-Weight Arrangement

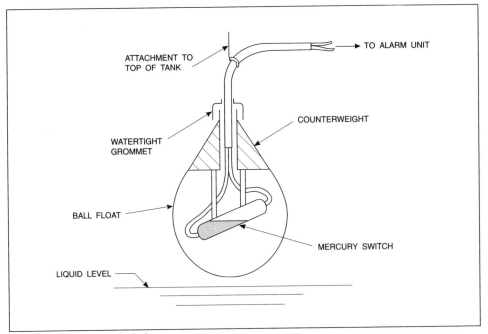

Figure 1-11. Ball Float Switch

The ball float of Figure 1-11 has an internal mercury switch and a counter-weight in the top part of the float. When the float is not lifted by the liquid, the mercury switch is tilted, and the mercury remains in the bottom of the glass tube (both ends of the glass tube have an electrical contact point). When the liquid rises, the float is lifted and tilted such that the mercury in the tube spreads and connects with both contact points on the end of the glass tube, thus completing an electrical circuit.

MAGNETIC DEVICES

The magnetic level switch of Figure 1-12 is more versatile, because it can detect both high and low level conditions if needed. This capability is a result of a

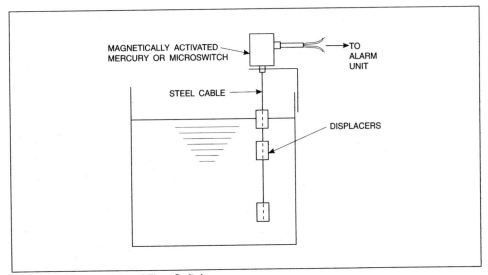

Figure 1-12. Magnetic Level Float Switch

permanent magnet being attached to the mercury switch, along with an attracting sleeve attached to the displacers to tilt the mercury switch. If the level falls, the sleeve drops below the magnetic position, the magnet is pulled away from the sleeve area, and the mercury switch tilts in the opposite angle, activating the contacts to sound an alarm or start a pump. The displacers may be stainless steel or porcelain, depending on the type of material sensed.

So far, the discussion has dealt only with level switches, which indicate when a maximum or minimum level condition exists. This is called point level measurement. It is also important to know values of level between the maximum and minimum values. There are a number of ways of making a continuous level measurement.

Displacement Devices

The displacement level transmitter is commonly used for continuous level measurement. It works on the buoyancy principle of Archimedes, which states that a body immersed in a liquid will be buoyed up by a force equal to the weight of the water displaced. The displacer body has a cylindrical shape. As a result, for each equal increment of submersion depth, an equal increment of buoyancy change will result. This gives a linear, proportional relationship, which is desirable.

Some displacers allow the addition or removal of weights when dealing with a liquid that has a higher or lower density than water. In some instances, the displacer body is filled with a predetermined amount of lead shot for specific applications. There are a variety of ways of conveying the buoyancy force change into a usable signal.

TORQUE TUBE

In this method, the displacer body is connected to a torque tube, which twists a specific amount for each increment of buoyancy change. It is insensitive to pressure changes in the vessel. The twisting force can drive a pointer on an indicator or be transferred to a pneumatic system. The pneumatic system is designed such that there is a different air pressure to the indicator or transmitter for each shaft position. The torque tube is an inferential or indirect method because, although the displacer is in direct contact with the liquid, it is the torque rather than the position of the displacer that is measured. Refer to Figure 1-13.

MAGNETIC LEVEL GAGE

The magnetic level gage is a unique measuring method employed when glass cannot be used due to the presence of corrosive, toxic, pyrophoric, very high-temperature, or otherwise dangerous material. A float containing a magnet (that has a greater force than the edge magnetized bicolor indicator wafers in the scale) is placed inside a sealed chamber. This chamber is connected to the vessel with nozzles and pipe flanges. The float is free to move and rise and fall with the level in the adjoining vessel. The scale or indicator is attached outside the sealed chamber, and it contains small bicolor wafers, which are free to rotate 180 degrees. As the float moves, its magnet causes the wafers to rotate. One side of the wafers is black and the other side is yellow. Black denotes liquid and yellow indicates the vapor space.

Electrical Level Sensors

CAPACITANCE

One of the most desirable characteristics of level measuring apparatus is that there be no moving parts in the tank or vessel. This requirement is met by a system that is based on changes in the capacitance of the probe as the level changes.

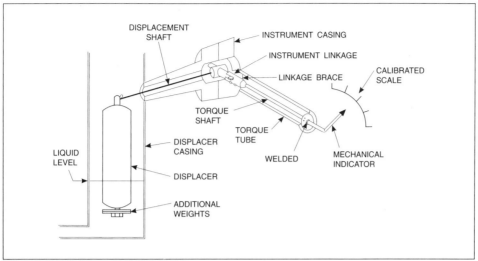

Figure 1-13. Displacer Level Sensor

This type of probe is suitable for the measurement of liquid level, interface level (an interface liquid level has two or more liquids of different specific gravity), and granular solids level. The advantages of a capacitance probe are that (1) it has no moving parts; (2) construction is simple and rugged; and (3) the probe can be anticorrosive and easily cleaned and able to withstand excessive temperature and pressure.

A capacitor consists of two conducting plates, separated by an insulator called the dielectric. The electrical characteristic that expresses the dielectric quality of a material is called the "dielectric constant," (k) (for example, k for air = 1, pure

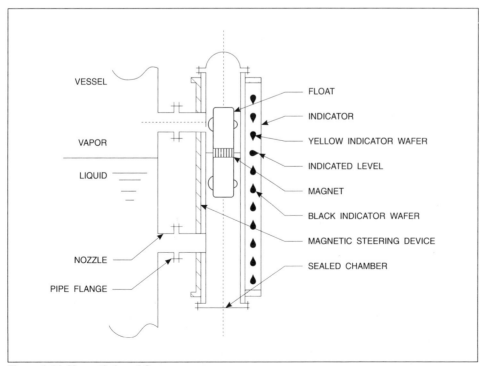

Figure 1-14. Magnetic Level Gage

water = 80, salt = 6, oil = 2.6). In Figure 1-15(a), the tubular metal shell and inner rod serve as the capacitor plates. The length of the probe may vary.

When the liquid level rises, more liquid acts as the dielectric of the capacitor. When the level drops and exposes the probe, more air or vapor above the liquid acts as the dielectric. This alters the capacitance of the probe, thus affecting the output of the electronic bridge circuit in which it is used.

Another application for level measurement using capacitance is shown in Figure 1-15 (b) where the probe has only one electrode. The tank wall is the outer conductor and is connected to the electronic circuit through a ground connection. Therefore, the probe serves as one plate and the vessel wall as the other plate. The material between the probe and the vessel is the dielectric. The insulation around the probe is required to prevent any electrical conduction between the plates. Capacitance probes have an active and an inactive length. The active length is the sensitive portion of the probe. Table 1-2 gives the dielectric constant for a variety of solids and liquids.

Table 1-2. Dielectric Constants for Various Materials

Solids		Granular and Powdery Materials			Liquids		
Material	**K**	**Material**	**K$_{LOOSE}$**	**K$_{PACKED}$**	**Material**	**°F**	**K**
Asbestos	4.8	Ash (Fly)	1.7	2.0	Butane	30	1.4
Bakelite	5.0	Coke	65.3	70.0	Castor oil	60	4.7
Calcium Carbonate	9.1	Gerber® oatmeal	1.47	Not tested	Ethanol	77	24.3
Cereals	3.0–5.0	Linde® 5A molecular sieve, dry 20% moisture	1.8 10.1	Not tested Not tested	Glycerine	68	47.0
Glass	3.7	Polyethylene	2.2	Not tested	Kerosene	70	11.8
Mica	7.0	Polyethylene powder	1.25	Not tested	Methyl alcohol	68	33.1
Nylon	45.0	Sand-reclaimed foundry	4.8	4.8	Sulphur	752	3.4
Paper	45.0	Cheer®	1.7	Not tested	Sulphuric acid	68	84.0
Rice	3.5	Fab® (10.9% moisture)	1.3+	1.3+	Water	32	88.0
Sugar	3.0	Tide®	1.55		Water	212	48.0

CONDUCTIVITY

Another method of detecting liquid level is the conductivity level probe, which is often used as a single-point device (detects only a single predetermined level in the tank). They are used only with liquids that conduct electricity (for example, acids, bases, and beverages).

The electrodes, supplied with a DC voltage, are mounted on top of the vessel and extend into the vessel. Figure 1-16 shows an application that measures high and low conditions in the vessel. When the liquid makes contact with an electrode, an electric current flows between the electrode and the ground (the metal of the vessel itself). The current energizes a relay coil, which causes the relay contacts to open or close, depending on the application.

To avoid frequent nuisance signals caused by momentary level fluctuations, a conductivity probe should not be mounted in a horizontal position. Also, conductivity adjustment must be made when the material in the vessel is changed.

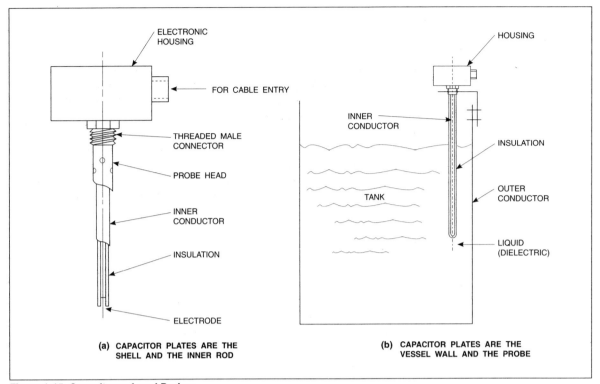

Figure 1-15. Capacitance Level Probes

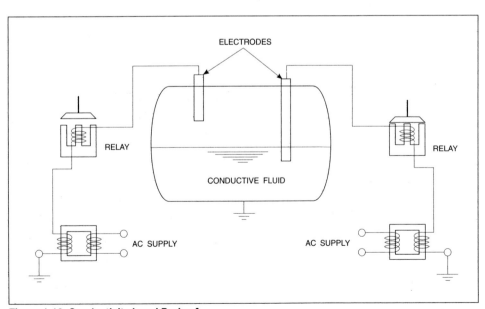

Figure 1-16. Conductivity Level Probes]

RESISTANCE

The resistance-type level sensor uses electrical level resistance as an analog to level. Refer to Figure 1-17. Two elements, one made of 301 stainless steel and one of helically wound Nichrome™, are separated by a polyester insulation open on one side. Precision wrapping of the Nichrome over the slot gives uniform separation of the conductors. An outside insulating sheath, sealed at the bottom and vented at the top, covers the sensor.

As the measured material rises, its weight shorts out part of the helically wound Nichrome, reducing overall sensor resistance. Resistance change is 100 ohms per linear meter (30.5 ohms/foot), and the readout device or controller can be located as far as 1,000 meters from the sensor. This method of level sensing can be used for both continuous and point measurements, and it is suitable for solids and liquids.

Sonic and Ultrasonic Level Sensors

In applications when it is not acceptable for the level measuring instrument to contact the process material, a sonic or ultrasonic device may be used. These devices measure the distance from a reference point in the vessel to the level interface, using sonic or ultrasonic (sound) waves.

A point measurement is made by detecting the interruption of a transmitted signal. A continuous measurement is made by measuring the elapsed time between emission and reception of a signal reflected from a surface in a vessel. Refer to Figure 1-18. The system shown in Figure 1-19 may be used for both continuous and point level measurement.

The amount of sound wave energy reflected from a surface depends on the process material and the sound frequency. Both liquids and solids are highly reflective, although solids are affected by surface porosity, material thickness, and rigidity. Some devices operate at frequencies of about 10 kHz, whereas ultrasonic devices operate in the 20 – 40 kHz range.

The two-element system shown in Figure 1-20 may be used only for point measurement (low- or high-level detection). In this system, piezoelectric transducers separated by a 4-in. (10-cm) gap are used in the sensor. When the

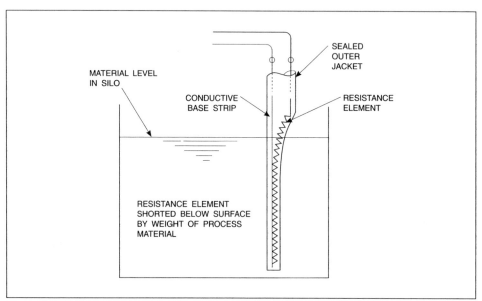

Figure 1-17. Resistance-Type Level Sensor

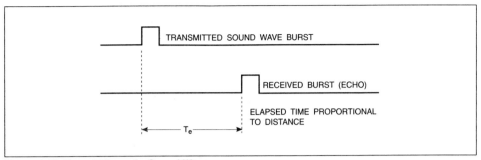

Figure 1-18. Transit Time of a Sound Wave

low-energy ultrasonic beam is interrupted by the presence of material, a control relay is de-energized to activate an alarm or control action.

NONINVASIVE ULTRASONIC SENSORS

Sometimes it is unacceptable to have contact between the sensing elements and the process material. For such applications, one can use noninvasive sensors, which are designed to be mounted on the outside walls of the vessel.

The transmitter is usually located on the wall opposite the receiver. The ultrasonic signal will reach the receiver only when the process material is present in the wave path. The signal will not transmit through gas.

To penetrate the vessel walls, the signal of a noninvasive ultrasonic sensor must be considerably higher than those of the other ultrasonic devices. Noninvasive probes cannot be used when the walls are constructed of materials that would absorb the ultrasonic signal.

Hydrostatic Pressure Level Sensors

This method uses the measurement of pressure exerted by the liquid. Consider a tank filled to a depth of one foot. There will be a pressure on the bottom because of the height of the column of liquid being supported by each unit area. If the tank is filled to a depth of two feet, there will be a column twice as high to be supported by each square inch or square foot, so the pressure will have been doubled.

A pressure gage connected at the bottom of the tank will register pressures that are directly proportional to the weight of the column of water liquid above it and, thus, proportional to the depth. Therefore, the pressure indicated by the instru-

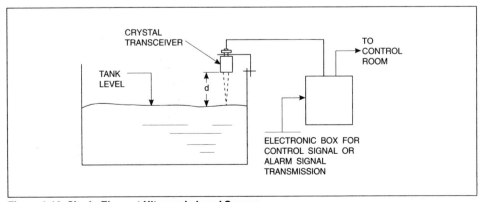

Figure 1-19. Single-Element Ultrasonic Level Sensor

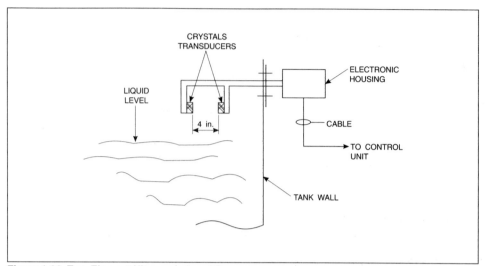

Figure 1-20. Two-Element Ultrasonic Level Sensor

ment is directly proportional to the height of the column of liquid above the gage. The relationship for open-vessel pressure gage level systems is:

$$P = SG \times D \times H$$

where:

P = hydrostatic pressure (head), psi (kPa)
SG = specific gravity of vessel liquid
D = density of water lb/in.^{3} (g/cm^3)
H = height of liquid level from reference point, in. (cm)

As shown in Figure 1-21, the dial of the pressure gage can be calibrated in units of length. This application is the simplest of all hydrostatic pressure methods.

In some applications there cannot be a direct connection of the gage to the vessel because of the nature of the liquid (for example, it could be viscous or corrosive). One then resorts to other alternatives:

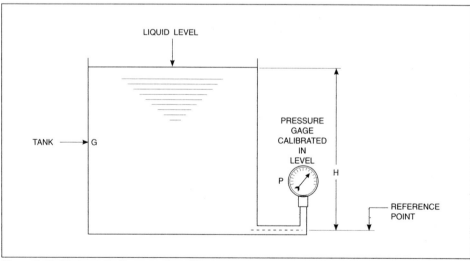

Figure 1-21. Hydrostatic Pressure Level Sensor

(1) Diaphragm-box system

(2) Purge system or bubble tube

(3) Force-balance system.

DIAPHRAGM-BOX SYSTEM

This system is totally submerged in the liquid and is applied whenever protection is required against plugging in slurry services or where dead-end cavities have a tendency to solidify material.

The diaphragms of these units are usually made of special alloys or coated with plastic materials that inhibit corrosion and prevent process contamination. Diaphragm units are furnished with either pneumatic or electronic transmission.

Figure 1-22 shows a pneumatic level sensing arrangement. The hydrostatic pressure on the bottom of the diaphragm (shown as detail in Figure 1-22) is converted into an air pressure indication by admitting a constant air flow into the chamber above the diaphragm. The movement of the diaphragm, caused by the hydrostatic pressure, will create a back pressure in the air tubing. This air pressure is proportional to the liquid level and is indicated on the pressure gage. (The gage dial may indicate percentage of level.) The pneumatic signal can also be transmitted to a controller.

AIR BUBBLE TUBE OR PURGE SYSTEM

If an open-ended tube is submerged in the liquid (refer to Figure 1-23), a constant air flow is admitted into it, and its pressure is regulated at a value slightly greater than the maximum head of the liquid in the tank, air pressure in the system will be equal due to the hydrostatic head of the tank liquid at any level because any excess pressure will bubble air out of the bottom of the tube.

Because air or some other gas, such as nitrogen, is continuously bubbling from the bottom of the pipe keeping liquid out, this system is also called a purge system. It is very well suited to measuring the level of corrosive liquids, viscous liquids, or liquids that contain entrained solids. An advantage of this system is

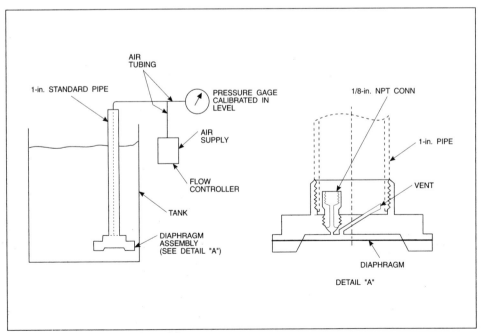

Figure 1-22. Diaphragm-Box Level Sensor

that it can be measured as far as 1,000 feet (approximately 300 m) from the indicator or controller.

In the case of a closed pressurized vessel, some kind of compensation device is required and is connected to the vessel with an equalizing line controlled by a differential pressure regulator.

DIFFERENTIAL PRESSURE TYPES

This system is based upon the same principle as discussed in hydrostatic pressure method. However, instead of using a pressure gage, a differential pressure cell transmitter (known as a dP cell or ΔP cell transmitter) is installed. This transmitter will transmit either a pneumatic or an electronic signal to a remote indicator or controller.

The hydrostatic pressure exerts its force against the steel diaphragm in the high pressure chamber (H), and it is balanced against the atmospheric pressure in the low pressure chamber (L). Any imbalance is detected by the transmitter, which contains an amplifier that will send out a signal in direct proportion to the level in the tank. Refer to Figure 1-24.

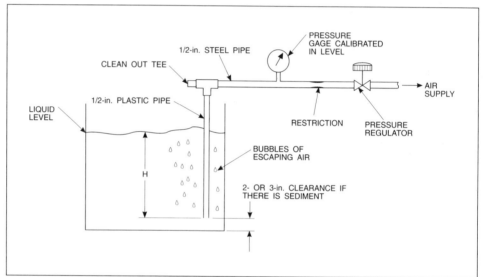

Figure 1-23. Bubble Tube Level Sensor

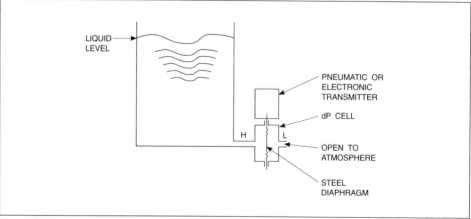

Figure 1-24. Differential Pressure Level Sensor

In the case of a closed, pressurized vessel, the low side (L) of the dP cell will be connected to a point above the maximum liquid level in order to cancel the pressure of the vapor.

Other Level Sensors

NUCLEAR LEVEL GAGES

Nuclear radiation from cesium 137 or cobalt 60 can be related to the liquid or solids level in a vessel. Since the nuclear gage does not come in contact with the product being measured, it is well suited to deal with applications where adverse conditions exist (such as like high or low temperature, abrasive, sticky, or corrosive materials) and to measure the level in highly pressurized vessels.

Radioisotopes used for level gaging emit energy in a random fashion but at a rate constant enough to use an average value for calibration. As the radioactive sources disintegrate, alpha, beta, and gamma particles are emitted. Gamma radiation is the phenomenon used for level gaging; it is a high-energy, short-length wave source of energy, having no substance but great penetrating power (alpha and beta particles have less penetrating power).

The basic unit of measurement for radiation intensity is the curie. A one-curie source undergoes 3.7×10^{10} disintegrations per second. The practical unit is the millicurie.

The radiation intensity is measured with a Geiger counter, also called a Geiger-Mueller tube. The components needed to detect or measure the level in the tank consist of the radioactive source and a gamma ray detector and amplifier, both of which are externally mounted so as not to be in touch with the measured medium. A source decay compensator is also built into the amplifier unit. A source decay is the depletion of radioactive material as the atoms of the material disintegrate to produce radiation. The rate of decay is logarithmic and is expressed in units of half-life, which is the time required for the material to lose half its strength. For example, the half-life of cesium 137 is 33 years; cobalt 60 is five years. This type of level detection may be applied to both point and continuous measurement.

The point measurement system shown in Figure 1-25(a) has two separate detectors: one high-level alarm switch and one low-level alarm switch. These switches may also be utilized to start and stop pumps. Therefore, the output of the detec-

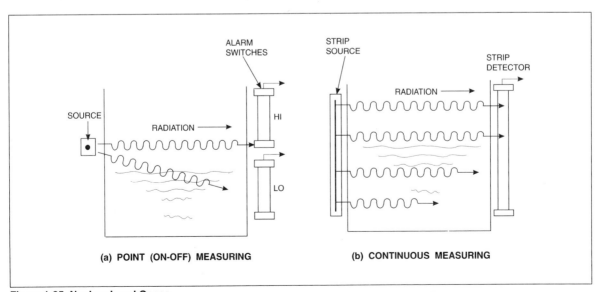

(a) POINT (ON-OFF) MEASURING   (b) CONTINUOUS MEASURING

Figure 1-25. Nuclear Level Gages

tors is an on or off signal. Continuous level measurement is somewhat more sophisticated than point measurement, but it is an extension of the same principle. In many cases instead of using a Geiger-Mueller tube, an ion chamber or electronic cell is used. Whereas with a Geiger-Mueller tube radioactivity has to be converted to electrical current, an electronic measuring cell generates an electrical current by ionization of a filling gas in the measuring cell as it is exposed to radiation. Electrodes within the cell attract the ionized electrons, causing a small current flow that is amplified and provides a signal output that is proportional to level. This signal is transmitted to a remote indicator or controller.

Electric Systems for Indirect Level Measurement

OSCILLATOR

This type of level detector is generally used to detect a level point, either a high or a low point.

An electronic oscillator generates a high-frequency alternating current, which is conducted to a pickup unit clamped to a clean gage glass, as shown in Figure 1-26. Any change of the liquid level in the field of the pickup detunes the oscillator. The resultant change in current, when amplified, can operate relays or contacts to activate alarms, pumps, solenoids, or motorized control valves. The output signal is therefore on or off. This method of level detection is very accurate.

PHOTOELECTRIC CELL

A typical photoelectric cell liquid measurement unit is shown in Figure 1-27. It uses the same type of level glass as the oscillator sensor. On one side of the level glass a light source is installed, and on the opposite side, a photocell (such as a phototransistor of photodiode) is used. A photocell converts light into an electrical signal.

The light rays from the light source are concentrated by the magnifying lens and pass through the gage glass when it is empty, reaching the base of the phototransistor. This transistor acts as a switch and is activated by the movement of electrons in the base material. This movement is caused by absorption of light rays. When the phototransistor is turned on, it activates the electronic control relay, which can be connected to a solenoid valve, alarm contacts, or motor control contacts to start or stop pumps. When the liquid level is above the phototransistor, no light rays reach the phototransistor. Therefore, no activation occurs in the circuit. This system is suitable for level point measuring.

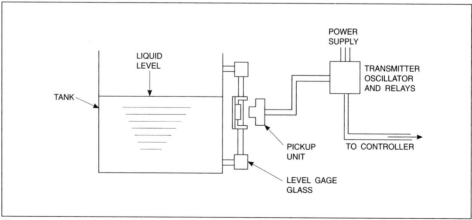

Figure 1-26. Oscillator Level Sensor

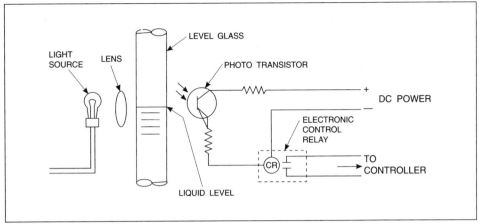

Figure 1-27. Photoelectric Cell Level Sensor

The Effect of Tank Shapes on the Change of Volume

With a vertical cylindrical tank, the change of volume is linear with respect to the change of level. However, when this cylindrical tank is placed in a horizontal position, the change of volume is nonlinear with respect to the change of level.

Figure 1-28 shows the difference between the volume change in a vertical cylindrical tank and a horizontal cylindrical tank. The example of the horizontal tank shows that even though height A is equal to height B ($H_A = H_B$), the volume

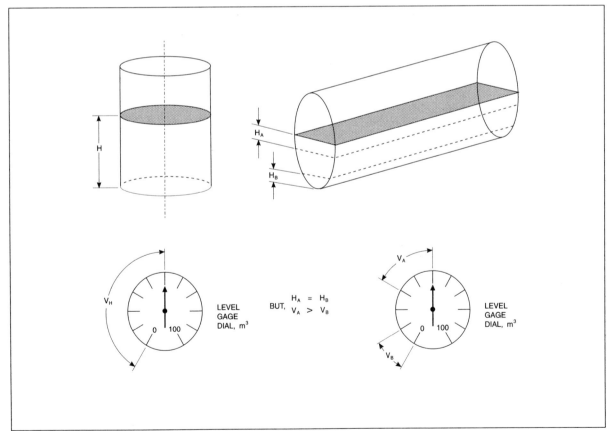

Figure 1-28. Vertical and Horizontal Cylindrical Storage Tanks

change A is greater than volume change B. Thus, it would be more effective in this application to indicate percentage of level instead of volumetric units. However, tables are available to relate the partial volume of horizontal cylinders at any height H to the total volume V_T by the formula:

$$V_H = K V_T$$

(Partial Volume in Height (H) = Cylindrical Coefficient for $H/D \times$ Total Volume)

where K is the cylindrical coefficient for height to diameter, as found in the table "Coefficients for Partial Volumes of Horizontal Cylinders" by the Natural Gas Processors Suppliers Association.

Solids Level Measurement Systems

Bulk solids can be in many forms: flakes, powders, granules, lumps, and rocks. These require a different approach to measurement, although some of the hardware used to measure liquid levels could be modified for this purpose.

The loading and discharging of bulk solids storage bins is done mechanically by various conveying systems. Level monitoring is required to prevent any overfill or total runouts. An automatic materials handling system requires that signals be provided for each critical level change. Generally, these are electric signals that actuate alarms and solenoid valves, or start and stop conveyer belts and conveyer screw motors.

ROTATING PADDLE

A low-torque, slow-speed motor is connected to a rotary paddle that is mounted inside the tank or storage bin. The power requirement of the motor when the paddle is rotating in the air is low. The power consumption will increase as the bulk level rises and starts to cover the paddle. In both cases the motor will actuate or deactuate a control circuit switch, and in some cases the motor will stop when the paddle gets buried underneath the solid. This system is for level point measuring only.

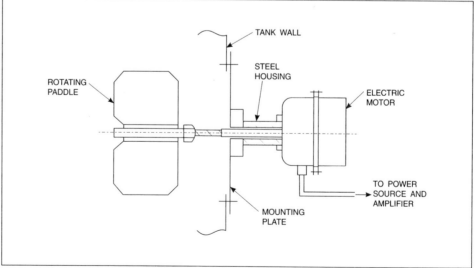

Figure 1-29. Rotating Paddle Level Sensor

Level Tilt Switches

Tilt switches are of the mercury or micro-switch type, mounted so as to hang from the top of the storage bin. The principle is similar to that of the level float switch. When the tilt switch hangs freely, there is no contact between the tilt switch and the control relay. As soon as the solid level reaches the tilt switch, the vertical angle changes, causing the contact to close. This creates a closed circuit with the control relay, which activates a solenoid valve, an alarm relay, or a motor control start/stop command. This type of level detector is used for point measuring only, usually for high- or very high-level alarms.

PRESSURE SENSITIVE DIAPHRAGM

The presence or absence of bulk solids can be determined by a pressure-sensitive instrument, such as a flexible diaphragm. A linear motion is derived for purposes of actuating or deactuating an electric control circuit switch or to transmit a pneumatic signal in the case of a pneumatic control system.

Figure 1-31 shows a pressure-sensitive diaphragm level switch. Its technique is that the force of two magnets, positioned with like poles repelling one another, is used to cause the tripping of a switch, with the magnet fields operating through the wall of the switch housing. The switch compartment may be completely sealed with no moving parts necessary to cause activation.

This level detector operates as follows:

(1) When the diaphragm is forced inward by the bulk solids, the driving magnet moves inward.

(2) The field of the driving magnet repels the field of the driven magnet and the micro-switch is tripped.

(3) Deactivation occurs when the amount of bulk solids is reduced, and the diaphragm and driving magnet return to their original position.

The micro-switch has a sensitivity adjustment to calibrate the level at which the switch should trip. This level switch is used only for point measurement, to activate audible or visual alarms, solenoid valves, and motors or feeders.

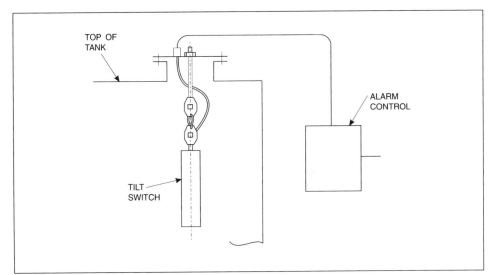

Figure 1-30. Level Tilt Switch

VIBRATION SENSOR

Figure 1-32 shows a vibration-type level sensor for solids application. In some cases, a paddle extends to the vibrating rod for better contact with the process medium. Two types of vibrating level sensors are available: mechanical and solid-state. In the mechanical type vibrations are induced by a driver coil and detected by or compared with the vibration of a second coil, which has a permanent magnet located in the pickup end of the device. In the solid-state type there is an integrated frequency vibration, which operates as follows:

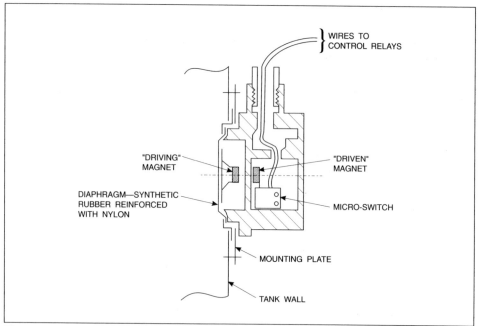

Figure 1-31. Pressure Sensitive Diaphragm

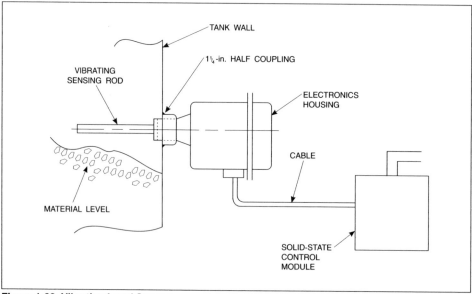

Figure 1-32. Vibration Level Sensor

(1) The control module sends a pulse to the sensor head.

(2) This pulse applies a force to the sensor rod, causing it to vibrate.

(3) If solid material restricts the end of the probe inside the bin, it is sensed and a signal is given.

(4) After pulsing the sensor electronically, the control module "listens."

(5) If vibrations at both ends of the sensor are equal, it indicates no restrictions, and thus the material is not present.

(6) If the vibrations are unequal, it indicates material is present at the outer sensor rod, and a relay is triggered.

The contacts of the control module can be used to actuate alarms, indicator lights, or process equipment such as conveyers and feeders.

THERMAL LEVEL SENSOR

These sensors employ DC circuits with low heat sources at the bulk-solids point of contact. The bulk material must provide a greater heat transfer coefficient than does the air above. When the material surrounds the sensor, it causes a lowering of the temperature of the contact element. The resistance then changes and affects the balancing of a Wheatstone bridge circuit, which can be used to activate a control relay circuit (refer to Figure 1-33).

LEVEL DETECTION BY WEIGHING

Level in a storage bin can be directly related to the weight of the bulk solids inside the tank. Mechanical, pneumatic, hydraulic, and electronic force-sensing systems are used to indicate vessel gross weight changes. The storage bin may be suspended or supported by the force-sensitive components, often involving some form of lever system. Relays are included to provide electric signals at specific weights (hence, levels). Actually, the true mass and not the actual bulk solids level determine when the signals will be given. The system has to be calibrated and adjusted in such a way that the sensor ignores the weight of the bin. Recently, load cells (refer to Figure 1-34) have been applied to measure the weight.

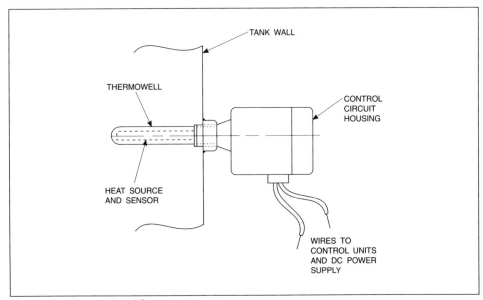

Figure 1-33. Thermal Level Sensor

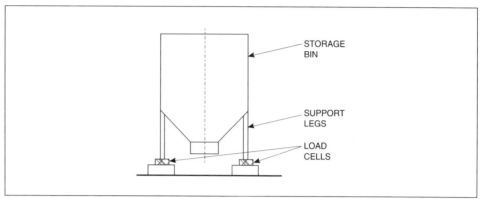

Figure 1-34. Load Cell Level Sensor

Load cells are specially constructed mechanical units that contain strain gages. A strain gage provides a measurable electrical output that is proportional to the stress applied by the weight of the vessel on the load cells. As the pressure on the cell changes, the electrical resistance of the strain gage changes. This causes an imbalance in a Wheatstone bridge circuit and either energizes a control relay circuit or provides a proportional electric signal to an indicator that is graduated in units of length.

SOUNDING LEVEL SENSOR

This system is also called a "plumb bob" remote inventory gage. A push button on a control panel activates the unit, releasing a weighted probe into a storage bin or tank. During downward travel, a counter in the control panel provides a reading of the amount or depth of material in the bin. When the probe reaches the actual material level, a slight slack in the cable is automatically sensed, reversing the motor action, stopping the counter instantly, and returning the sounding probe to its original position.

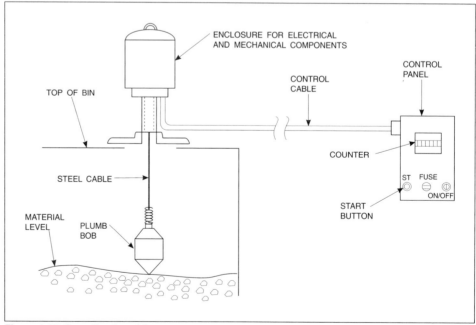

Figure 1-35. Sounding Level Sensor

Pressure Measurement

Pressure is one of the primary measurements used in instrumentation and control. Its meaning, units, and measuring devices are important to a proper understanding of not only what is occurring in a process but also how instruments and controls are used to monitor the process.

The atmosphere consists of gases and liquids. These gases and liquids have weight and exert a pressure on the surface of the earth. Pressure is defined as force exerted over a unit area. Mathematically, pressure is expressed as $P = F/A$ where P = pressure, F = force, A = area. Therefore, the amount of force exerted by a substance directly affects the amount of pressure. Note that pressure is not a fundamental quantity but is derived from force and area, which in turn are derived from mass, length, and time, the later three being fundamental quantities.

Force Exerted by Liquids

A liquid occupies the space in and conforms to the shape of the tank or vessel. The forces exerted on the walls and the bottom of the tank have to be determined in order to measure pressure. The amount of force a specific volume of liquid in an open tank exerts over a specific area depends on (1) the height of liquid above the measurement point, (2) the specific gravity or density of the liquid, and (3) the temperature of the liquid.

Consider the following example, which addresses these three factors. The first is height. If the water level in a tank is 30 centimeters and the weight of water at 16°C is 1 kilogram per cubic decimeter, than the water column exerts a force of 0.3 kg per square centimeter or 29.42 kilopascals at the bottom. If the level rises to 90 cm, the force exerted will be 88.26 kilopascals.

To measure the force on the side of the tank, the measurable force the water exerts on the side of the tank will depend on the height of water above the measuring instrument. If the instrument is placed 30 centimeters above the bottom level of the tank, the force exerted by the water would be 0.2 kilogram per square centimeter. This is because there is a 20-centimeter water column on top of the measuring instrument.

The second factor is weight, which must be determined in order to determine the pressure of the liquid. One usually works with specific gravity, which is a reference number that compares the weight of a specific volume of liquid to the same volume of water, and a specific volume of gas to air at the same temperature. Water and air are assigned specific gravity constants of one. These values serve as references for determining the specific gravity of other liquids and gases. Refer to Table 1-3.

A cubic decimeter of mercury weighs 13.57 kg, while the same volume of water weighs 1 kg at 4°C. Thus, mercury, weighing 13.57 times as much as water, has a specific gravity of 13.57 (see Figure 1-36).

The third factor is temperature, which affects the pressure of a liquid in an open tank or vessel. If the temperature of the liquid changes, its specific gravity also changes. This is because changes in the temperature will cause the liquid to expand or contract, depending on whether the temperature rises above or falls below the specific standard. When the temperature falls, the liquids expand, which causes the force per unit area to increase.

EFFECT OF PRESSURE ON CLOSED VESSELS

In many process systems, liquids are confined inside closed vessels or piping. The liquid will be subject to any forces that may be exerted on it by external sour-

Table 1-3. Specific Gravity of Selected Liquids

Liquid	Specific Gravity
Ethyl alcohol, C_2H_6O	0.7939
Kerosene 41, API at 60°F	0.8200
Ellison gage oil	0.8340
Benzene (benzol), C_6h_6	0.8794
Water	1.000
Ethylene glycol (glycol), $C_2H_6O_2$	1.1155
Carbitol, $C_6H_{14}O_3$	1.024
Glycerine (glycerol), $C_3H_8O_3$	1.260
Acetylene tetrabromide	2.964
Mercury	13.570

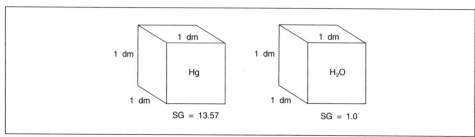

Figure 1-36.   Comparison of Specific Gravities of Water and Mercury

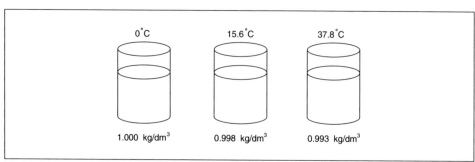

Figure 1-37. Effect of Temperature Changes on Water

ces. Therefore, these external sources will have to be taken into consideration, in addition to the three attributes discussed in the example.

Pascal's law states that when a force is applied to a confined fluid, the force will be transferred undiminished throughout the fluid to all surfaces of the containing vessel. This means that the force exerted on the liquid will be consistently indicated regardless of where the measurement is taken within the fluid.

Force Exerted by Gases

Gases have no definite shape. They expand to occupy the full volume of the vessel that contains them. As a result, a gas will exert an equal amount of pressure on all surfaces of the vessel that contains it.

BOYLE'S LAW

The force a gas exerts depends on the volume of the vessel and the temperature of the gas. The relationship of pressure exerted by the gas and the volume of the vessel is expressed by Boyle's law.

Boyle's law states that if the temperature is held constant, the force exerted by the gas on the walls of a containing vessel varies inversely with the volume of the vessel, provided the mass remains unchanged. This means that if the volume of the vessel increases, the force exerted by the gas decreases and vice versa. Thus, $P_1V_1 = P_2V_2$ (temperature kept constant), where P_1 and V_1 are the initial pressure and volume, respectively, and P_2 and V_2 are the final steady-state pressure and volume, respectively.

CHARLES' LAW

Charles' law states that the volume of gas will vary directly as the absolute temperature, if the pressure remains constant. This fact has also been determined by Gay-Lussac. Thus, $V_1/T_1 = V_2/T_2$ (pressure kept constant), where V_1 and T_1 are the initial volume and temperature, respectively, and V_2 and T_2 are the final volume and temperature. This indicates that if the absolute temperature is doubled, the volume will be doubled, and if the absolute temperature is reduced to 50%, the volume will be reduced to 50%.

IDEAL GAS LAW

In practice, it often happens that there is a change in all three variables — pressure, temperature, and volume. For instance, an increase in temperature may cause an increase in both volume and pressure. Therefore, another formula is needed to solve more complex applications. Boyle's law and Charles' law are combined on the following ideal gas law equation:

$$P_1 V_1 / T_1 = P_2 V_2 / T_2$$

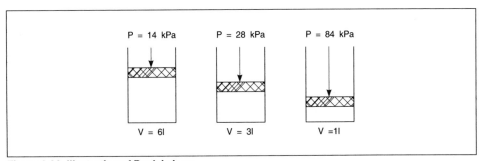

P = 14 kPa   P = 28 kPa   P = 84 kPa

V = 6l   V = 3l   V =1l

Figure 1-38. Illustration of Boyle's Law

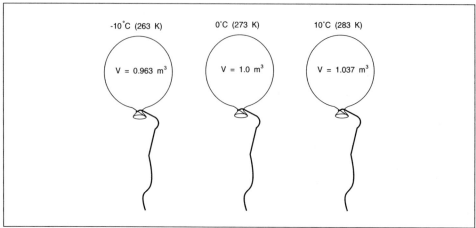

Figure 1-39. Illustration of Charles' Law

This equation can be manipulated to solve for any of the quantities, for example,

$$P_1 = P_2 V_2 T_1 / V_1 T_2$$

However, most gases deviate somewhat from Boyle's and Charles' laws because they do not behave like perfect gases. Thus, these laws give only close approximations.

DALTON'S LAW

In some processes, one must deal with a mixture of gases. A mixture of gases, provided they do not combine chemically, exerts a pressure equal to the sum of the pressures that would be exerted by each of the gases separately if allowed to fill the containing vessel at the same temperature. This is Dalton's law, which implies that the pressure exerted by one gas does not interfere with the pressure exerted by another gas. It is expressed by the equation

$$P_f = T_c / V_c \, (P_1 v_1 / T_1 = P_2 v_2 / T_2 + \ldots)$$

where P_f = final pressure, T_c = constant temperature, V_c = constant volume.

The pressure exerted by each gas is called its partial pressure; partial pressure is proportional to the concentration of any gas in a mixture.

In all calculations dealing with pressure, temperature, and volume changes in gases, one uses absolute pressure and absolute temperature.

Absolute pressure = gage pressure + atmospheric pressure

Absolute temperature (kelvins) = thermometer reading °C + 273

Absolute temperature (°Rankine) = thermometer reading + 459.69

Basic Methods of Pressure Measurement

Since pressure can be transduced to force by allowing it to act on a known area, the methods of measuring force and pressure are essentially the same, except for high vacuums for which a variety of special methods are used.

Pressure measuring instruments can be broadly categorized as mechanical or electrical/electronic. This includes manometers, which are widely used in laboratories for the calibration of secondary instruments.

MANOMETERS

Manometers are simple and accurate pressure measuring devices. They operate on the principle that change in pressure will cause a liquid to rise or fall in a tube or column.

The liquids most commonly used in a manometer tube are water, mercury, and red oil. Unlike mercury, red oil is not toxic. Water and mercury have been used because their specific gravity, thermal expansion, and mass are known. The specific gravity of water is 1.0 and that of mercury is 13.6. This means that it takes 13.6 times as much pressure to raise mercury by one inch. A pressure of one pound per square inch (psi) will support a 27.7-inch column of water and a two-inch column of mercury. Mercury manometers are used to measure higher pressure ranges than water manometers. The other considerations of the filling fluid are its chemical inertness to the process materials and its compatibility with the process temperature so that it will not freeze or vaporize under operating conditions. Manometers with sealed and evacuated columns are used to detect and indicate vacuum.

A manometer's accuracy depends on various factors, for example, the method used to attach the scale to the column, the type of graduation, and the shape of the liquid at the interface of liquid and air in the column. Water produces a concave meniscus because of the adhesive forces at the glass surface. For a mercury-air interface, there is a convex meniscus because of the cohesive force between glass and mercury.

A basic manometer consists of a reservoir filled with a liquid. The reservoir is connected to a source whose pressure must be measured. A transparent column is attached to the reservoir. The top of the column may be open, exposing it to the atmosphere, or it may sealed and evacuated. Manometers with sealed columns are used to measure absolute pressure (pressure in reference to absolute zero). Manometers with open columns are used to measure gage pressure in reference to atmospheric pressure.

Other types of manometers are the well-type, the inclined, the McLeod gage, the micromanometer, and the capacitance type.

THE U-TUBE MANOMETER

The U-tube manometer is a simple device used to accurately measure the difference of two pressures, P_1 and P_2, by the difference in height, h, of liquid columns. Refer to Figure 1-41, from which one can write the following equation:

$$P_1 - P_2 = h \times d \quad \text{or} \quad P_1 = P_2 + h \times d$$

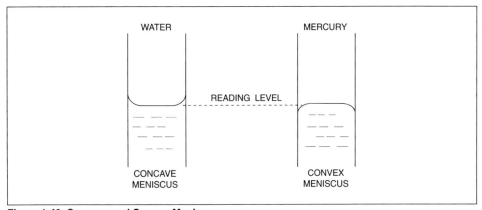

WATER

MERCURY

READING LEVEL

CONCAVE
MENISCUS

CONVEX
MENISCUS

Figure 1-40. Concave and Convex Meniscus

where:

P_1 = process pressure
P_2 = atmospheric pressure
d = density of the liquid
h = height of liquid

The process pressure is referred to as a gage pressure, which is the difference between atmospheric pressure and the pressure being measured. Therefore, gage pressure uses atmospheric pressure as a "zero" reference.

Atmospheric pressure is the pressure that surrounds everything on earth. It varies with altitude. At sea level the atmospheric pressure is 14.7 psi or 0 psig, 1 atmosphere, 101.325 kilopascals, 1.0133 bars, or 29.921 inches of mercury. Absolute pressure is defined as the difference in pressure between a perfect vacuum (0 psia, 0 kPa abs) and the pressure being measured. In other words absolute pressure = gage pressure + atmospheric pressure.

The well manometer shown in Figure 1-42 is widely used because of the convenience of reading only a single leg. The well area is made very large compared with the cross-sectional area of the tube, thus the zero level moves very little when pressure is applied. The cross-sectional areas of the well and the tube are not the same; this has to be considered in calculating the process pressure.

THE INCLINED MANOMETER

The inclined manometer is a variation of the well manometer (see Figure 1-43) where the tube is at an angle. This increases the sensitivity, giving a greater movement of liquid for small pressure differentials. The height, h, of the column, equals h' sin α, where h' is the distance measured along the scale and is the angle of inclination of the tube.

For statue balance: $P_1 - P_2 = $ SG $(1 + A_1/A_2)L$ sin α where A_1 = area of leg; A_2 = area of well; L = length of scale corresponding to height h; α = angle of inclination of smaller diameter leg.

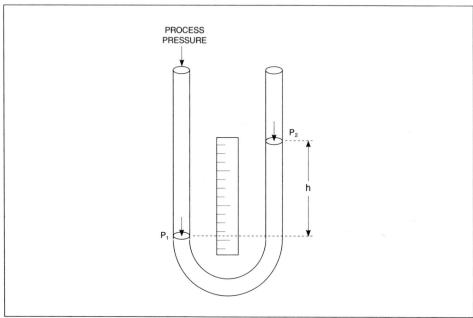

Figure 1-41. The U-tube Manometer

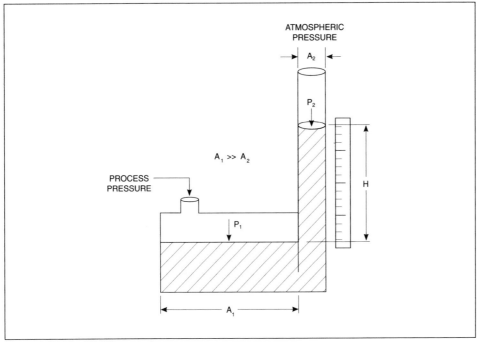

Figure 1-42. The Well Manometer

This instrument is for measuring very low pressures such as the draft in a fur-nace, a chimney, or a ventilation duct. The ratio of diameters should be as great as possible to reduce the error that results from the change in level in the large diameter well. Reducing the angle increases the scale length. The ratio of h' to h should not exceed 10:1 to maintain accuracy. The liquid used is often colored water.

BAROMETER
A barometer is a well-type manometer that indicates changes in atmospheric pressure. In its simplest form it is a glass tube with a sealed end and is filled with mercury; the tube is inverted and the open end is placed in a reservoir (cistern) of

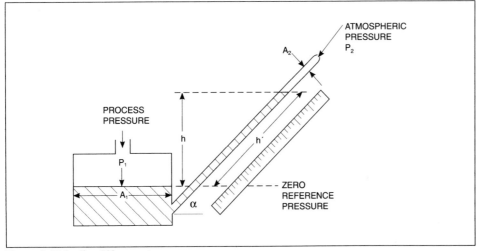

Figure 1-43. Inclined Manometer

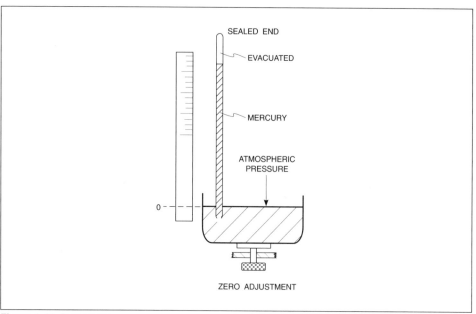

Figure 1-44. Barometer

mercury. The pressure of the atmosphere on the mercury surface in the reservoir holds up the mercury in the tube, while above the mercury there is a vacuum (that is, zero absolute pressure).

At standard atmospheric pressure (sea level, 0°C) the height of the mercury column is 760 mm. The pressure is then indicated as 760 mm Hg = 101.325 kPa. Note that the height of the mercury column is independent of the tube's cross-sectional area; the tube's core is usually and should be of high precision. The column's height is read from a scale that may be moved up or down. Corrections are made for temperature and altitude, since barometric readings in laboratories have to be at standard conditions, i.e., sea level and 0°C.

The well may be moved up or down, so that the liquid level will be made to coincide with the zero mark on the fixed scale prior to reading.

Factors that influence the accuracy of manometers are the method used for fastening the scale to the column, which affects its position; the spacing of the graduations; and the quality of the fill liquid of known specific gravity.

MCLEOD GAGE
Another type of manometer is the McLeod gage, which is used as a standard device for measuring vacuum by trapping and compressing a volume of gas.

A specific volume of mercury is contained in the gage reservoir and a secure connection is made to the vacuum source. After this, the gage is rotated 90 degrees, as shown in Figure 1-45. The mercury then traps and compresses the gas in the measuring tube. The gas volume in the container decreases and the pressure increases, according to Boyle's law. When the gas pressure equals the pressure exerted by the mercury, the mercury level is read on the scale.

CAPACITANCE MANOMETERS
The capacitance manometer, another variant, uses two cisterns with a interconnecting line. One cistern is fixed; the other, which is moveable, is positioned by means of a precision adjusting screw until capacitor plate equilibrium is reached. The mercury surfaces in both cisterns act as plates of a capacitor. Pressure is applied to the fixed cistern, while the moveable one is at some reference pressure (usually atmosphere). When the process pressure changes, mercury flows from

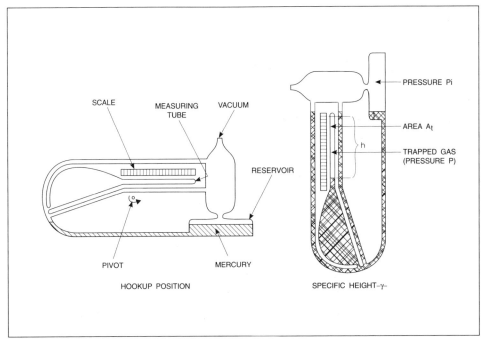

Figure 1-45. McLeod Gage

one cistern to the other. This changes the amount of mercury in both cisterns and also changes the distance between the mercury and the other capacitor plate. In both cisterns there is a change in capacitance, which can be measured as pressure using an electrical circuit.

MICROMANOMETER

The micromanometer is widely used to measure small pressure differences accurately. It is based on a variation of the indirect manometer principle.

The instrument is adjusted so that when $P_1 = P_2$, the meniscus is located at a reference point given by a fixed hairline. The reading of the micrometer is noted.

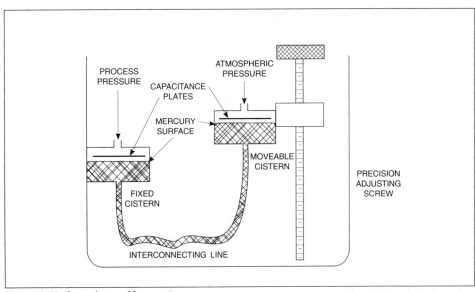

Figure 1-46. Capacitance Manometer

When an unknown pressure is applied, it causes the meniscus to move off the hairline. It can be restored to its initial position by raising or lowering the well with the micrometer. The difference between initial and final micrometer reading gives the height and, thus, the pressure.

Table 1-4.   Manometer Reference Chart

Class	Type	Optimum Range	Accuracy
U-type	Glass - mercury	0.5 – 50 in. Hg	0.02 in.
	Glass - water	0.5 – 100 in. H_2O	0.02 in.
			0.5%
	Glass - two-liquid	0.01 – 1 in. H_2O	0.05%
			0.5%
	Float, electric, etc.	0.5 – 50 in. Hg	
	Multiple-tube	0 – 100 in. Hg	
Single tube	Vertical - open top	2 – 100 in.	0.02 in.
	Compression manometer	vac. – 200 psi	1%
	Open-cistern, vacuum gage	0 – 30 in. Hg	0.02 in.
	Mercurial barometer	Atmospheric	0.01 in.
	do., photocell servo	Atmospheric	0.005 in.
	Absolute pressure gage	0 – 100 in. Hg	0.005 in.
	Inclined-tube	abs.	0.005 in.
		0 – 4 in. H_2O	
Micromanometer	Null method, etc.	0.001 – 20 in. H_2O	0.05%
Liquid column and plunger	Mercury	50 – 5000 psi	
	Vacuum	**Microns**	
Liquid column (compression manom)	Laboratory McLeod	0.01 – 2000	2%
	Industrial McLeod	0.1 – 100,000	

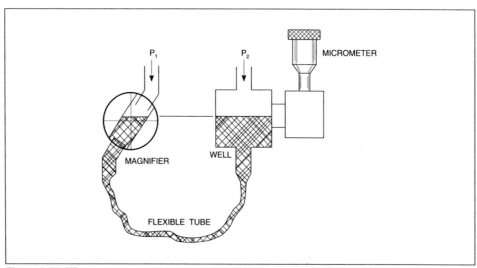

Figure 1-47. Micromanometer

Some types of liquids used in manometers are toxic and can be damaging to the environment. Therefore, when using manometers to measure or indicate pressure, do not connect any manometer to a pressure that has the potential to exceed the range of the manometer. This could cause the liquid to be forced out of the tube. In addition, since the tubes in many manometers are made of glass and can be easily broken, it is important to use care in handling these manometers. If the liquid is accidently spilled from a manometer, follow your facility's procedures for containing and cleaning hazardous materials.

Elastic Deformation Elements

There are several mechanical devices whose shape alters when pressure is applied. These devices, designed to respond to different pressure ranges, include the:

(1) diaphragm,

(2) diaphragm capsule,

(3) Bourdon tube pressure sensor,

(4) spiral pressure sensor,

(5) helical pressure sensor, and

(6) bellows pressure sensor.

DIAPHRAGM

The diaphragm is best suited for low-pressure applications. It is sensitive to small changes in pressure, since pressure is exerted over a larger area. A mechanical device such as a pin, rod, or bar, is connected to the diaphragm. Strain gages are bonded to diaphragm surfaces.

The element may be metallic or nonmetallic. The nonmetallic diaphragms are usually large and noncircular. Their pressure range is 0 to 10 inches of water. They are generally applied for indicating, recording, or controlling draft gages. An example of this type of diaphragm is shown in Figure 1- 48. It is made of neoprene rubber, Teflon™, polythene, silk, Koroseal™, silicone, or leather. The diaphragm is opposed by a light spring.

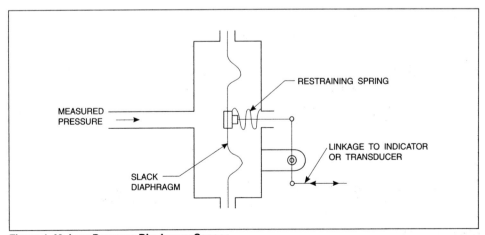

Figure 1-48. Low-Pressure Diaphragm Sensor

METALLIC DIAPHRAGMS

Metallic diaphragms are generally chosen to function within pressure transmitters on both pneumatic and electronic transmitters. The movement of this type of diaphragm is small, depending on its: (1) metal thickness, (2) diameter, (3) shape of the corrugations, (4) number of corrugations, (5) modulus of elasticity and (6) pressure applied. Refer to Figure 1- 49.

A diaphragm shell is designed so that the deflection approaches a linear relation to the pressure applied over as wide a range as possible. The depth and number of corrugations and the angle of formation of the diaphragm face determine the sensitivity (deflection per unit pressure) and the linearity of the diaphragm. This type of diaphragm does not require a restraining spring.

The application of single metallic diaphragms is similar to the "slack" non-metallic diaphragm. They are made of phosphate, bronze, beryllium, copper, trumpet brass, stainless steel, or Monel™.

DIAPHRAGM CAPSULES

A capsule is composed of two single metallic diaphragm shells bonded by soldering, brazing, or welding. The shells are either flat or corrugated. Two types of diaphragm capsules are available: convex and nested (see Figure 1-50). The area between the diaphragms is filled with a fluid that has a low forcing point, a high boiling point, a low viscosity, and a low coefficient of thermal expansion. Kerosene and toluene are typical fluids.

Capsules can be stacked and connected together to form one pressure sensor so that each capsule deflects upon pressure application. The total deflection is the sum of the deflections of all capsules. This type of diaphragm can be used for both motion-balance and force-balance instruments. These two terms are derived

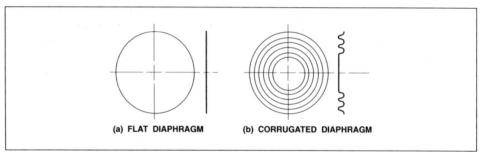

(a) FLAT DIAPHRAGM   (b) CORRUGATED DIAPHRAGM

Figure 1-49. Metallic Diaphragms

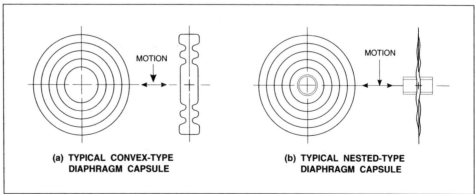

(a) TYPICAL CONVEX-TYPE DIAPHRAGM CAPSULE   (b) TYPICAL NESTED-TYPE DIAPHRAGM CAPSULE

Figure 1-50. Diaphragm Capsules

from the balancing systems applied to pressure indication or transmission methods.

BOURDON TUBE PRESSURE SENSOR

This type of pressure gage is the most frequently used in the process industry, especially for purposes of local pressure indication. The Bourdon tube was patented in 1852 by Eugene Bourdon, and it has changed little since then. It is described as a curved or twisted tube whose transverse section is not circular. The application of internal pressure causes the tube to unwind or straighten out. The movement of the free end is transmitted to an indicating element or transducer by way of a gear mechanism (see Figure 1-51).

The term "C-Bourdon tube" is often applied to this element because its form in a 250 degree arc makes it look like the letter C. The Bourdon tube is also suitable for pressure transmission and control mechanism applications and is easily adapted to transducer designs in order to obtain electrical outputs.

SPIRAL PRESSURE SENSOR

The spiral pressure sensor is a variation of a Bourdon tube. It is used when the free-end movement of the C-type is not great enough to provide the needed motion. This type of tube is formed by winding more than one turn in the shape of a spiral about a common axis. As pressure is applied to the spiral, it tends to uncoil and straighten out, producing the relatively long movement of the tip end (sealed) whose motion can be used for indication or transmission.

Since greater movement of the free end is attained with the spiral element, no gear mechanism is required (see to Figure 1-52).

HELICAL PRESSURE SENSOR

The helical pressure sensor is similar to the spiral element, except that it is wound in the form of a helix (see to Figure 1-53). It increases the tip travel considerably, producing even greater amplification than the spiral element. A central shaft is installed within the helical element, and the pointer is driven from this

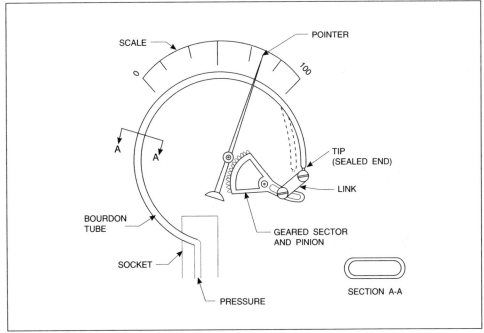

Figure 1-51. Bourdon Tube Pressure Sensor

shaft by connecting links. This system transmits only the circular motion of the tip to the pointer and, hence, is directly proportional to changes in pressure. Advantages over the other Bourdon tube sensors are high overrange capabilities (a ratio of 10:1), stability in fluctuating pressure services, and high-pressure service adaptability.

The number of coils used varies according to the pressure range (for example, three coils for low-pressure spans and 16 or more for higher spans).

Materials for Bourdon tubes will vary depending upon process medium, corrosiveness, pressure range, and temperature. The usual materials are phosphor bronze, beryllium-copper, 4130 alloy steel, 316 and 403 stainless steel, k-Monel™, and Monel™.

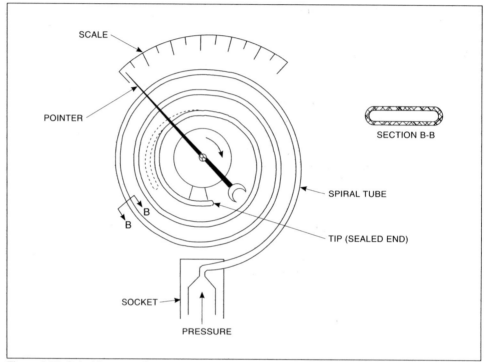

Figure 1-52. Spiral Pressure Sensor

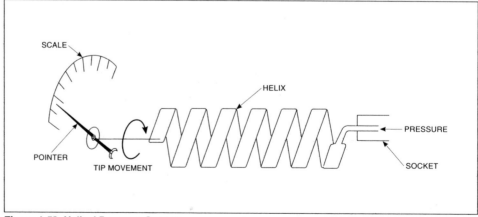

Figure 1-53. Helical Pressure Sensor

BELLOWS PRESSURE GAGE

A metallic bellows is a series of circular parts so formed or joined that they can be expanded axially by the introduction of pressure but not appreciably in any other direction. Because the materials can be of light gage and may have large effective areas, they can develop far larger forces than are needed to straighten through a desirable distance. Usually, a calibrated spring is employed to oppose its movement so that only part of the maximum stroke is used. In practice, the useful stroke is limited to 5 to 10% of the length of the bellows. The number of convolutions in the bellows determines the stroke. It should be noted that atmospheric pressure acts on the outside of the bellows, so the bellows responds only to pressure in excess of atmospheric.

Besides pressure gages, bellows elements are also used in pressure transmitters, receivers, and controllers for pneumatic instruments. Figure 1-54 shows a simplified schematic drawing of a bellows-type pressure gage. The materials used for bellows are similar to the ones used in Bourdon tube elements.

PRESSURE AND VACUUM GAGE ACCESSORIES

In certain processes, pressure and vacuum gages have to be protected from corrosive media, pulsating flows, high temperature, and so on. As a result, devices

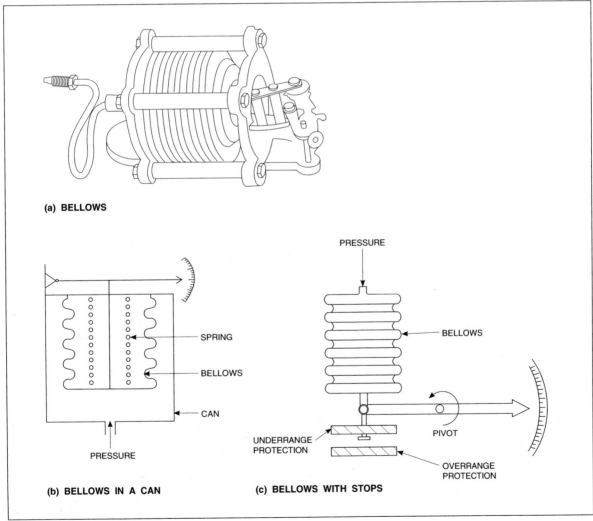

(a) BELLOWS

(b) BELLOWS IN A CAN

(c) BELLOWS WITH STOPS

Figure 1-54. Bellows Pressure Gage

such as the following are needed: chemical seals, pulsation dampeners, pigtail siphons, bleeders, and blowout disks. Sealed glycerin-filled cases are also available to reduce shock in rough or pulsating services. Figure 1-55 illustrates a typical screwed-type diaphragm chemical seal.

The space between the chemical seal and the pressure gage is filled with a liquid to transfer the deflection of the diaphragm to the pressure gage. This fill fluid should provide:

(1) minimum thermal expansion,

(2) no contamination of process fluid in case of a diaphragm rupture, and

(3) low viscosity and chemical stability under all exposed temperatures.

Fills may be silicone oils or a mixture of water and glycerin. Diaphragms and bodies are available in a wide variety of materials. Diaphragm seals are classified according to the method of attaching them to the process (that is, flanged, threaded, or saddle connection).

Pulsation dampeners come in several types:

(1) Needle valve

(2) Fine bore plug screwed into the gage socket

(3) Filled-bulb pulsation dampener

(4) Porous metal pulsation dampener

(5) Flow check pulsation dampener

(6) Moving pin pulsation dampener

These devices are needed when pressure cycling is frequent enough to damage measuring sensors or when pressure surges are great enough to cause damage. Therefore, a restricting device is included in the dampener to offer a sufficient resistance to flow that the pressure in the gage changes slowly enough for the element to respond (see Figure 1-56).

Pigtail siphons are used primarily for hot vapor services such as steam lines, boilers, and thermochemical processes. The two purposes for installing a pigtail siphon are: (1) to isolate hot vapors from the measuring element, and (2) to form

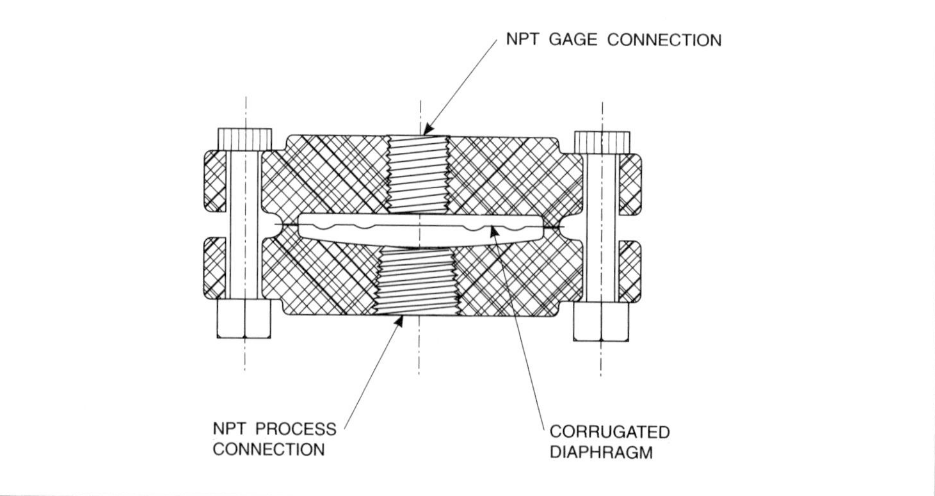

To preserve the liquid seal in the siphon, it should never be insulated.

Figure 1-55. Typical Screwed-Type Diaphragm Chemical Seal

a liquid seal in condensable vapor services to prevent instability due to mixed-phase conditions. Refer to Figure 1-57.

Vacuum Measurements

Most of the pressure gages previously described function satisfactorily in both the positive pressure range and the negative pressure range (vacuum), especially in the process industry. However, for high-vacuum measurements in laboratories, space simulation, and cryogenics, other gages are utilized. Types of vacuum sensors discussed here are thermal, Pirani, hot-cathode ionization, cold-cathode ionization, and Knudsen.

THERMOCOUPLE GAGE

A linear relation exists between pressure and thermal conductivity as predicted by the kinetic theory of gases for low-pressure measurement. One of the reasons is that the mean free path of molecules is large compared with the pertinent dimensions of the apparatus. It also depends on the spacing between the hot and cold surfaces. For high pressures, conductivity becomes independent of gas pressure.

The application of the thermal conductivity principle depends on the loss due to radiation of heat between the hot and cold surfaces. If the radiation losses are more pressure induced, conductivity changes will cause only a slight temperature change. Radiation losses can be minimized by using surfaces of low emissivity . The cold surface must be maintained at a constant temperature to ensure overall accuracy.

The element in this type of gage is a fine wire or ribbon heated electrically and immersed in a gas whose pressure is to be measured (see Figure 1-58). The hot surface is a thin metal strip of wire. The cold surface is the glass tube at room temperature. The steady-state temperature attained by the heated wire depends, among other things, on the loss of heat by conduction through the surrounding

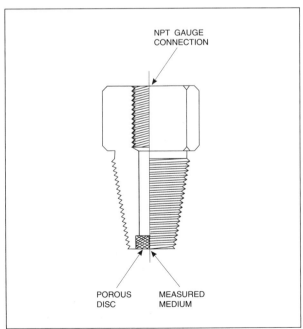

Figure 1-56. Typical Porous Metal Pulsation Dampener

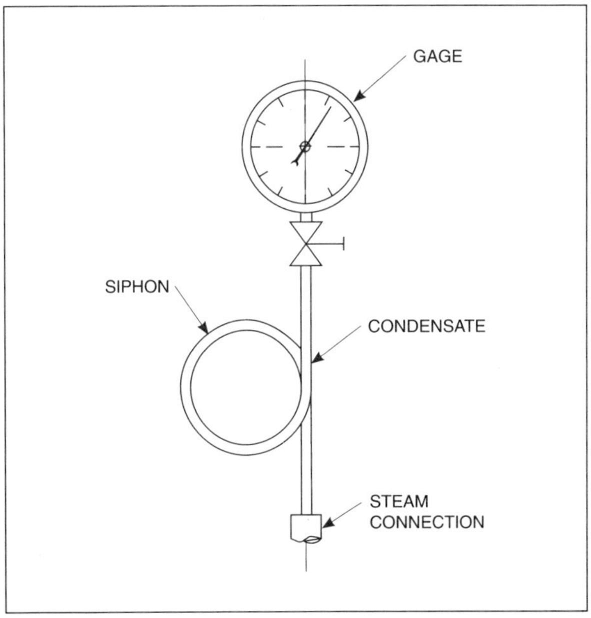

Figure 1-57. Pigtail Siphon

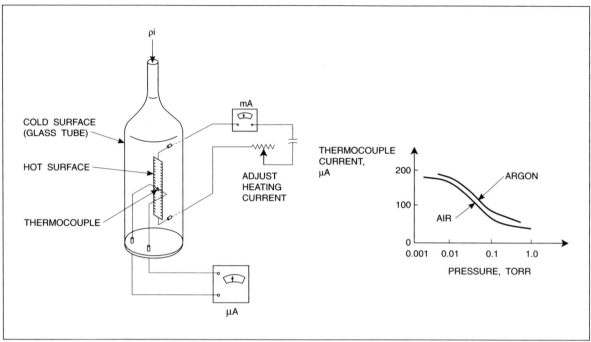

Figure 1-58. Thermocouple Gage

gas. The thermal conductivity of the gas varies with pressure in a range of vacuum conditions. Thus, the temperature of the ribbon will vary with gas pressure. The thermocouple gage measures the temperature of the strip with a thermocouple. Since thermal conductivity varies among gases, the gage must be individually calibrated to the gas for good accuracy. They cover the range from 10^{-4} to 1 torr.

PIRANI GAGE

In the Pirani gage, the wire element is made into an electrical resistance with two wire elements, one of which is sealed in a vacuum as a reference (see Figure 1-59). The evacuated tube acts as a compensator to reduce the effect of bridge excitation voltage changes and temperature changes on the output reading. The two elements are two electrical resistances that form two arms of a Wheatstone bridge. The cooling effect on the element exposed to the system unbalances the bridge to provide an output signal. Sensitivity is higher than with a thermocouple gage, and accuracy is some ±5 percent. Calibration is nonlinear and varies from one gas to another. Pirani gages cover the range from 10^{-5} to 1 torr.

HOT-CATHODE IONIZATION GAGE

In hot-cathode ionization gages, electrons emitted from a cathode move towards a grid. The electron acquires a kinetic energy that is proportional to the potential difference when the energy is large enough. Some of the electrons collide with molecules of the gas whose pressure is to be measured. The gas molecules lose electrons as a result of the collisions, producing positive ions. The remaining electrons are collected on the grid. The positive ions, however, are attracted to the negatively charged collector. Each ion so collected causes a pulse on current to flow in the collector circuit. The number of ions produced depends of the molecular density of the gas. This means that collector current is proportional to gas molecular density or pressure. Thus, the main advantage of this method is its linearity. Disadvantages of the cathode are decomposition of gases

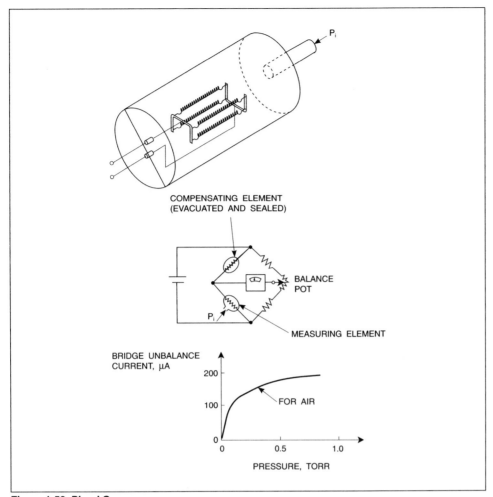

Figure 1-59. Pirani Gage

by the hot filament and contamination of the pressured gas by gases forced out of the hot filament. They cover the range from 10^{-10} to 1 torr. Refer to Figure 1-60.

COLD-CATHODE IONIZATION GAGE

The cold-cathode gage consists of an open anode loop between two cathode surfaces with a high voltage impressed between them (see Figure 1-61). A magnetic field deflects electrons from travelling directly to the anode and causes them to oscillate among the magnetic lines of flux. With the increased mean free path, a significant number of ionizing collisions with gas molecules occur. The charge on the field builds up to an equilibrium, where each ion leaving the field causes an ion to enter. This current is then a measure of molecular density or pressure. The cold-cathode gage accounts for half the ionization gages used. This covers the range 10^{-5} to 10^{-2} torr.

KNUDSEN GAGE

The Knudsen gage consists of a very light vane and mirror supported on a delicate torsion suspension (see Figure 1-62). Adjacent to the suspended vane are two fixed vanes that are heated electrically. A rarefied gas in the regions between the vanes produces a force whereby the suspended vane is repelled against the torsion of its suspension. The resulting deflection is read on a calibrated scale. The Knudsen gage finds little industrial application because of its mechanical

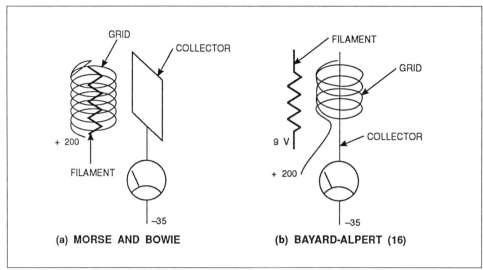

Figure 1-60. Hot-Cathode Ionization Gage

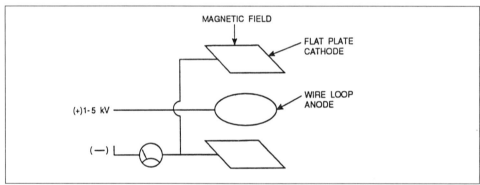

Figure 1-61. Cold-Cathode Ionization Gage

awkwardness and lack of convenient remote indication. It is relatively insensitive to gas compensation except for variation of accommodation coefficients from one gage to another.

Mass spectrometers of the magnetic deflection type are used in the measurement of ultrahigh vacuums. High sensitivity is provided in one model by a nine-stage electron multiplier with magnesium-silver anodes, which gives a gain of 10^5 to 10^6 electrons per ion. This sensitivity permits partial gas pressure measurement to 10^{-12} mm Hg. For even lower pressures, they allow identification of the partial pressures of components in gas mixture.

HIGH-PRESSURE MEASUREMENT

Applications for pressures above 100,000 psi use electrical gages based on the resistance change of manganin or gold-chrome wire with hydrostatic pressure.

The sensing component is a loosely wound coil with one end grounded to the cell body and the other end brought out through a suitable insulator (see Figure 1-63). The coil is enclosed in a flexible, kerosene-filled bellows that transmits the measured pressure to the coil. The change in resistance is linear with change in pressure and is sensed by a conventional Wheatstone bridge. As gold-chrome is less temperature sensitive than manganin, gold-chrome is used. Kerosene experiences a sudden increase in temperature due to a transient and sudden change

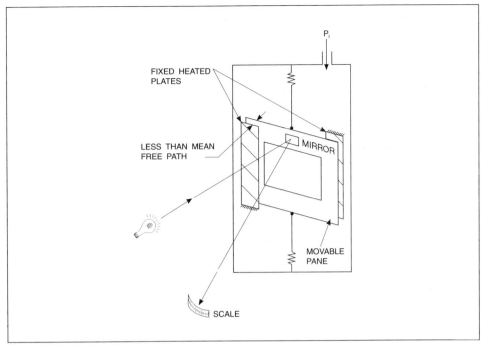

Figure 1-62. Knudsen Gage

in pressure. This change in temperature rise will affect the measurement if the wire sensitivity to temperature variation is high.

Differential Pressure Measurement

Differential pressure measurement is of prime importance in industrial processes, especially to measure flow rate and level but also to measure the pressure drop across a filter element. The application of differential pressure cells was described in the section on level measurement. All the pressure-sensitive elements already described are adaptable to differential pressure measurement. Most

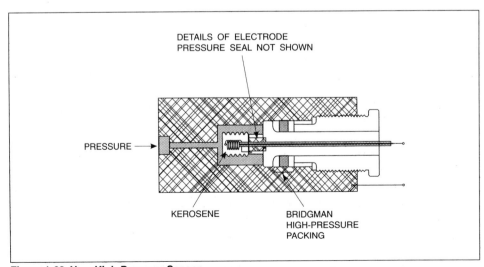

Figure 1-63. Very High-Pressure Sensor

applications of this type involve differential pressure signal transmission. The signal can be either pneumatic or electronic.

Electrical Pressure Sensors

Electrical pressure sensors are normally selected for remote indication purposes. Therefore, they are often used in pressure transmitters. Factors used to select this type of pressure sensors are:

(1) compatibility with electronic control equipment,

(2) pressure range and span,

(3) very long distance signal transmission,

(4) fast response requirement,

(5) availability of instrument air supply, and

(6) high accuracy requirement.

Most electronic pressure instruments incorporate one of the elastic deformation elements in such a way that a measurable electrical quantity produces a proportionately variable electronic signal. Because the energy form is transferred from a mechanical to an electrical nature, these instruments are often classified as electromechanical pressure transducers. The most used transducers for pressure detection are on strain gages; capacitive, inductive, and variable reluctance devices; linear variable differential transformers; and piezoelectric types.

STRAIN GAGES

Strain is defined as a deformation or change in the shape of a material as a consequence of applied forces. A strain gage is simply a fine wire in the form of a grid. When the grid is distorted, the resistance of the wire changes according to the formula:

$$R = \frac{KL}{A}$$

where:

R = resistance in ohms
K = constant for the wire material
L = length of wire
A = cross-sectional area of the wire

As the strain gage is distorted by the elastic deformation element, its length is increased and its cross-sectional area is reduced. Both changes increase the resistance. Little distortion is required to change the resistance of a strain gage through its total range. This type of transducer can be used to detect very small movements and, therefore, very small pressure changes. Figure 1-64 shows a schematic of a strain gage element.

The flexible backing is usually made of epoxy, and it either functions as a diaphragm or is attached to another type of elastic deformation element. There are two general types of strain gages: bonded and unbonded. The unbonded strain gage is a strain-sensitive wire with one end fixed and the other end movable. Because of its properties, the unbonded strain gage is used for tension measurement only. A typical application of a bonded strain gage schematic is shown in Figure 1-65.

Another method of manufacturing a bonded strain gage is to form the grid around a hollow tube (see Figure 1-66). The compression load of the diaphragm causes the tube to both shorten and increase in diameter. Two strain wire wind-

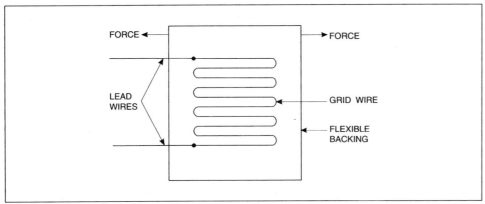

Figure 1-64. Strain Gage Element

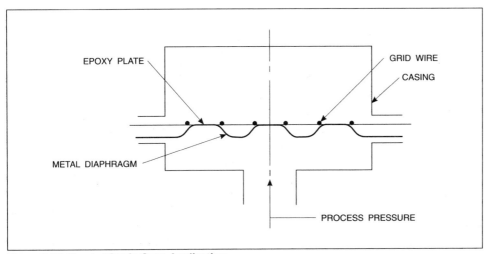

Figure 1-65. Bonded Strain Gage Application

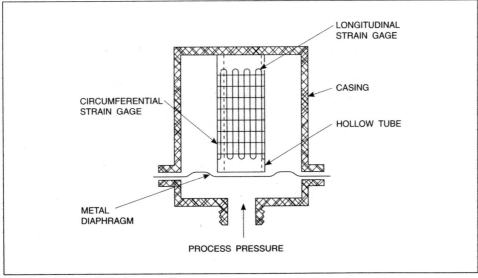

Figure 1-66. Strain Tube Pressure Transducer

ings, one with the effective grid placed axially and the other wound circumferentially, form two active arms in a resistance bridge. This type of strain gage provides important advantages over a flat type: less thickness is required for the sensing diaphragm, and temperature has negligible effects on its flexibility.

Most strain gages are used in a Wheatstone bridge circuit, which is necessary to transmit an electronic signal to a remote indicator or controller. There is another reason why the Wheatstone bridge circuit is used. The measured process fluid may contain heat, which will increase the temperature of the strain element. Therefore, with a bridge circuit it is possible to compensate for the increase of resistance due to high-temperature exposure. A recent development in strain gage transducers is the "thin-film" type. These gages can be made of various resistive materials. The fabrication method is similar to that of integrated circuits. A metal substrate is used to provide the needed mechanical properties, and ceramic film is deposited on the metal to provide electrical insulation. The strain gage is then deposited on the insulator and integrated with a Wheatstone bridge circuit. Obviously, this type of transducer is considerably reduced in size.

Other resistance-type pressure transducers combine a bellows or Bourdon tube with a slide wire resistance or variable resistor.

Another strain-type pressure transducer is the semiconductor strain gage, which consists of a silicon strain element sealed in a silicon fluid and protected from the process fluid by a metal diaphragm.

CAPACITIVE PRESSURE TRANSDUCER

Capacitive pressure transducers consist of two conductive plates separated by a dielectric (see Figure 1-67). The dielectric may be the process fluid, air, or another substance. This type of transducer operates on the basis that a change occurs due to the movement of an elastic deformation element. The movement physically changes the distance between the two capacitor plates. Changes in the process pressure deflect the diaphragm, and the resulting change in capacitance is detected by a bridge circuit. A high-frequency, high-voltage oscillator energizes the sensing element. The capacitance is converted through the bridge circuit to a proportional DC signal. This signal is amplified to a standard milliampere output signal range. The capacitor replaces one of the resistors in a conventional Wheatstone bridge circuit, and by suitable calibration the resistance value re-

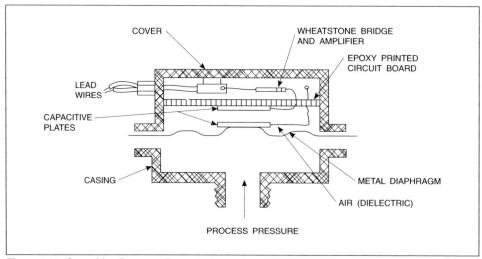

Figure 1-67. Capacitive Pressure Transducer

quired for balance can be correlated with the process pressure that acted upon the diaphragm (hence, one of the capacitive plates).

INDUCTANCE PRESSURE TRANSDUCERS

The linear variable differential transformer (LVDT) is an inductive device that consists of three parts: an elastic deformation element, a coil, and a moveable magnetic core. The core movement, which is a function of the element movement, produces voltage variations that are measured directly. The sensing element is attached to the magnetic core. As the pressure varies, the element causes the core to move within the coil. An alternating current passes through the coil and, as the core moves, the inductance of the coil changes. The current passing through the coil increases as the inductance decreases. This type of pressure transducer is used in a current-sensing circuit. For increased sensitivity, the coil can be divided in two, using a center tap. This construction actually provides two coils. As the core moves inside the coils, the inductance of one coil decreases as the inductance of the other increases. The required core displacement is within a range that provides a high degree of linearity. Many instrument manufacturers utilize an iron core instead of a magnetic core; therefore, an additional coil is required to induce a magnetic field. Figure 1-68 shows a schematic diagram of a linear variable differential transformer pressure transducer that employs an iron core instead of a magnetic one. This pressure transducer provides an AC output, which has to be converted into a standard DC output.

Other types of LVDT pressure transmitters are available. Manufacturers have developed a variety of designs. One operates on the force-balance principle, with the force bar attached to a bellows and range spring at the process inlet side and to a ferrite disk at the transmitter side. The ferrite disk acts as an iron core, and it faces two stacked ferrite coils, which serve as a detector. There is an air gap between the coils and the disk. Any position change of the ferrite disk changes the output of the differential transformer, thus determining the output to an oscillator.

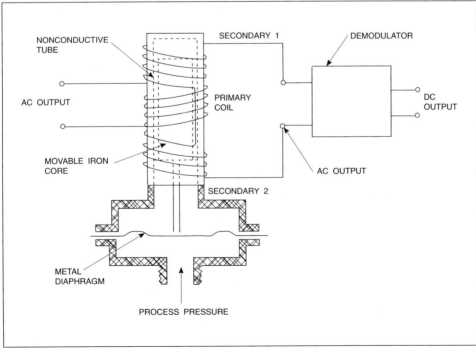

Figure 1-68. LVDT Pressure Transducer

This output is rectified to a DC signal and amplified to a standard DC transmitter output. A feedback device rebalances the movement of the force bar. This system is analogous in its principle to the pneumatic pressure transmitter. See Figure 1-69.

VARIABLE RELUCTANCE PRESSURE TRANSDUCERS

The variable reluctance transducer is an inductance-type transducer. The deflection of a metal diaphragm moves a ferrite armature between two ferrite cup cores, which, in turn, change the inductance ratio in a bridge circuit whose imbalance is detected and amplified to a DC transmitting signal that is proportional to pressure variations in the sensor. Figure 1-70 shows a schematic diagram of the reluctance pressure head showing the measuring diaphragm, the armature, and the transformer coils.

PIEZOELECTRIC PRESSURE TRANSDUCERS

Piezoelectricity is defined as the conversion of mechanical energy into electrical energy or vice versa. Because of this property, crystals such as quartz, Rochelle salt, barium titanate, and lithium sulphate can be employed in a pressure transducer to produce an electric potential due to the application of pressure on the crystal. This effect is reversible. This electromechanical energy conversion principle is used in both directions. Synthetic crystals produce the same electric potential but offer higher sensitivities than natural crystals. However, the only drawback for all crystals is that in a static pressure condition its potential drops off, thereby producing an error. This characteristic limits its use to reaction processes where pressures change rapidly. The piezoelectric is direction-sensitive; that is, tension and compression produce opposite voltage polarity. Figure 1-71 shows a schematic diagram of a piezoelectric pressure transducer.

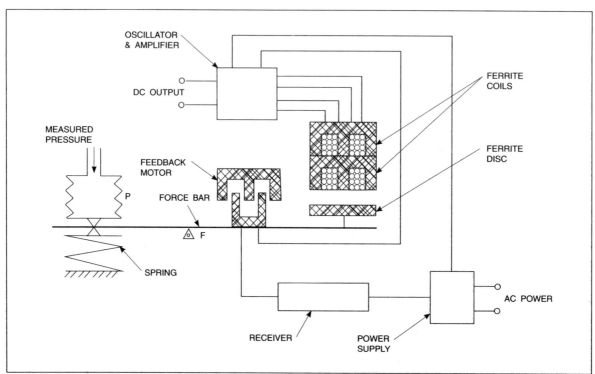

Figure 1-69. LVDT Pressure Transmitter

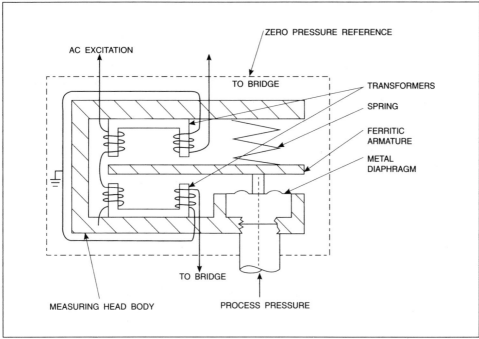

Figure 1-70. Variable Reluctance Pressure Transducer

Pressure Switches

The operating principle of a pressure switch is similar to that of a pressure gage without a dial or indicator. Instead of having a pointer it has a micro or a mercury switch to energize or de-energize an alarm system, solenoid valve, or motor control at a predetermined pressure setting. It is used to sense absolute, compound, gage, and differential pressures. The sensing elements are similar to the ones used in other pressure instruments. The diaphragm type is the most frequently used element. Switch types available are single-pole/single-throw (SPST), single-pole/double-throw (SPDT), double-pole/single-throw (DPST), and double-pole/double-throw (DPDT) arrangements.

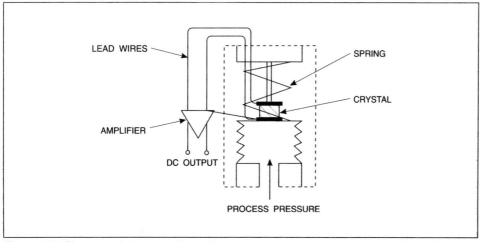

Figure 1-71. Piezoelectric Pressure Transducer

Impulse Lines

The arrangement of the impulse lines that connect the pressure instrument to the process will vary depending upon the specific application. For gas service the line and instrument should be free of liquid. For liquid and vapor service, the lines should be full of liquid. The reason for this stipulation is to keep a constant static pressure on the instrument. If the amount of liquid in a line changes, the resulting change in hydrostatic head on the pressure element will cause a zero shift.

Figure 1-72 shows a typical application of a vertical impulse line connecting a pressure instrument to a pressure process. Air or other gases can be vented at point D as indicated. Entrapped gases in vertical lines can be vented easily or usually will seek a higher elevation and be returned to the process vessel.

When it is necessary to locate a pressure instrument at a point from the process where horizontal lines are required, care must be taken to prevent entrapped air or gases from accumulating in the lines. For this reason the horizontal lines should be run at a gradual and continuous slope of about 1/2 inch to 1 inch per foot. When initially putting the instrument in service, air and noncondensable gases should be vented through a valve provided for this purpose. Valve C can be used in this way.

Differential Pressure Instrument Installation

In differential pressure applications, it is important to note that both the pressure taps or points of measurement should be at the same vertical elevation. If one tap were located above the other, an error would result from the difference in liquid head to only one side of the differential pressure instrument.

Differential pressure instruments for flow application fall into the following categories:

(1) Liquid flow with no seal liquid in vertical and horizontal lines

(2) Steam, liquid, or gas flow with a liquid seal in vertical and horizontal lines

(3) Gas flow with no liquid seal in vertical horizontal lines

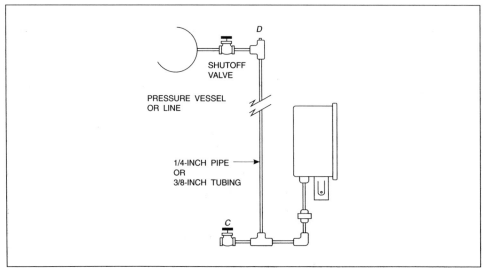

Figure 1-72. Installation of Pressure-Sensing Instrument

In categories (1) and (2) the instrument and connecting lines must be full of liquid, and in category (3) the instrument and lines must be free of liquid. Figure 1-73 shows category (1) service, Figure 1-74 shows category (2) service, and Figure 1-75 shows category (3) service. Note that the instrument is below the line for service where the instrument is liquid-full and above the line where it is liquid-free. It is most important that both sensing lines be kept at the same elevation to negate the effect of liquid legs. Also, in addition to the two service valves in the high- and low-pressure impulse lines, a third valve is provided to equalize the pressure on the dP cell for field zeroing and pressurizing the system. The bypass valves should always be opened before opening either of the line service valves so that both sides of the dP cell chamber receive equal amounts of pressure. This will prevent damage to the pressure element. Figure 1-76 shows a typical dP cell installation with the signal connected to a receiver instrument. The pipe taps in the horizontal gas line are at the top of the pipe so that slugs of liquid will not enter the impulse lines. The taps in the horizontal liquid line are located on the side of the flange to avoid a buildup of scale and debris in the taps.

When process fluids freeze at adverse or normal operating temperatures, the connecting lines and instrument must be protected. Liquid seals can be used for this purpose, unless traces of the seal fluid enter the process line, in which case other winterizing methods, such as heat tracing, must be used.

Heat Tracing

Instead of filling the instrument systems with nonfreezing or low freezing point liquids, the temperatures of the connecting lines and instrument housing can be maintained above the freezing point. This is done by wrapping the exposed surfaces with tubing through which an appropriate amount of steam flows. The entire surface area is then insulated. Electrical heat tape can be used instead of the steam tubing when it is more desirable to do so. The service valves normally

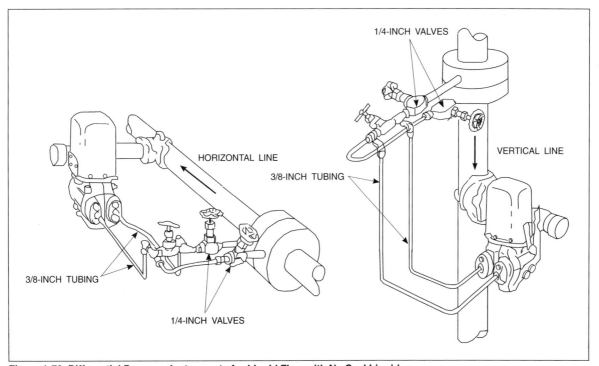

Figure 1-73. Differential Pressure Instruments for Liquid Flow with No Seal Liquid

are left exposed, so field zeroing can be accomplished without disturbing the insulation.

Where liquid sealing or heat tracing is not a viable option, the instruments can be mounted in a housing whose interior is controlled by an appropriate temperature control system.

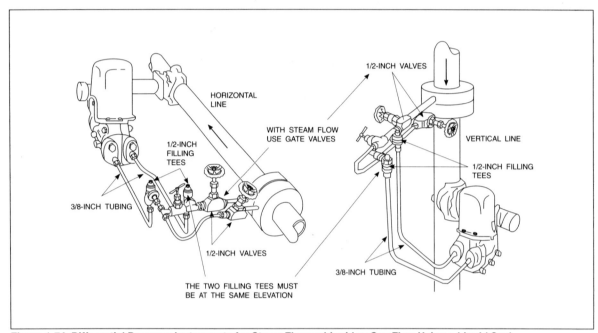

Figure 1-74. Differential Pressure Instruments for Steam Flow or Liquid or Gas Flow Using a Liquid Seal

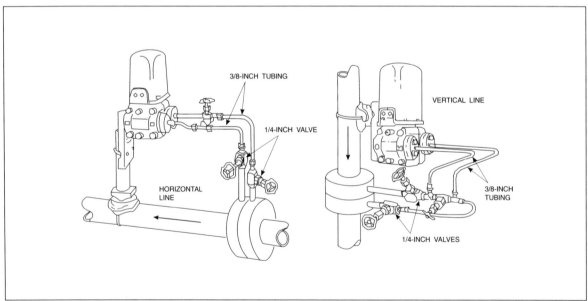

Figure 1-75. Differential Pressure Instruments for Gas Flow with No Liquid Seal

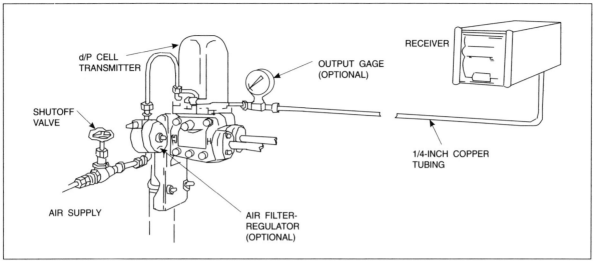

Figure 1-76. Typical dP Cell Connection

Flow Measurement

Fluid Properties

Flow must be accurately measured and controlled to maintain the process conditions required for maximum plant production and product quality. In most process applications, proper regulation of flow rates ensures control of the process reaction and helps regulate other variables such as pressure, level, and temperature. Often, flow measurements are used as indicators of overall process performance.

Flow is one of the most difficult process variables to measure accurately. One simple way to determine a fluid's rate of flow is the weight per unit time method, which assumes a basic premise of fluid mechanics: mass is a conserved quantity. The mass entering a system is equal to the mass leaving the system when both are measured over the same time interval. In practice, using the weight per unit time method to measure flow requires catching the fluid in a container and weighing it over a given interval of time. Often, this is impractical; for example, applications involving gases that cannot ordinarily be condensed into liquids, and closed-loop processes commonly associated with chemical applications.

Consequently, other methods must be used to obtain accurate flow measurements. These methods must take into consideration two basic properties of fluid — density and viscosity — and their effect on the accuracy of flow measurement. The discussion of density and information pertaining to fluids in a static phase are presented first; viscosity is discussed later.

DENSITY

All substances have density. Density is a measure of the closeness to one another of the molecules that make up a substance. In the simplest terms, density is mass per unit of volume. This can be expressed as:

$$\rho = m/v$$

where:

ρ = density
m = mass
v = volume

Often, density is the property that distinguishes one substance from another. For instance, it is density that causes the same volume of different substances to differ in weight. Figure 1-77 illustrates the impact of density on weight. Two substances of equal volume, a cubic foot of iron and a cubic foot of water, are arranged on a balance scale. It is the greater density of the iron that causes it to weigh more than an equal volume of water. The relationship of density to weight applies to all forms of matter: solids, liquids, and gases.

Density is typically expressed in pounds per cubic foot. In some instances, a conversion to grams per cubic centimeter may be useful. The following equation can be used to convert density to pounds per cubic foot or grams per cubic centimeter.

$$1 \text{ lb/ft}^3 = 0.0160262 \text{ g/cm}^3$$

Given a 2.04 cubic foot container, the density of 100 pounds of liquid that occupies 1.53 cubic feet is:

$$\rho = m/v$$
$$= 100 \text{ lb}/1.53 \text{ ft}^3$$
$$= 65.359 \text{ lb/ft}^3$$

To express this value in grams per cubic centimeter, the calculation is extended:

$$(65.359 \text{ lb/ft}^3)(0.0160262 \text{ g/cm}^3) = 1.0475 \text{ g/cm}^3$$

EFFECTS OF TEMPERATURE AND PRESSURE ON DENSITY

The density of a substance is affected by variables such as temperature and pressure. These effects vary according to the physical properties of the substance. For example, the effects of temperature on the density of solids and gases varies widely. Gases are most affected; solids are least affected; liquids generally range somewhere in between. The effects of pressure on the density of solids and liquids are considered negligible because both are relatively noncompressible. However, the effect of pressure on the density of gases is very pronounced. Consequently, systems designed to measure the flow of gases must include compensation for both variables to ensure accurate results.

SPECIFIC GRAVITY

Another term commonly used to express the density of fluids is specific gravity (SG). While density is a stand-alone measurement, specific gravity is a ratio that compares the density of a fluid at a specific temperature to the density

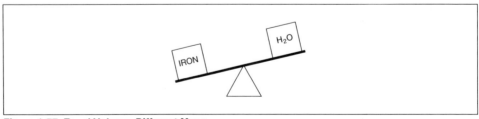

Figure 1-77. Equal Volume, Different Mass

of water or air at the same temperature. The specific gravity of liquids and gases can be represented by the following equations.

Specific gravity = Density of a liquid/Density of water at standard conditions

Specific gravity = Density of a gas/Density of air at standard conditions

Since the units in the equation cancel each other out, the resulting value is a dimensionless number. This makes it possible to use the value of the specific gravity of a fluid in combination with measurements in any units.

Many industries use different standard conditions to calculate specific gravity. This results in a variety of numerical values. For this reason, density is more commonly used to describe gases. However, to avoid error, standard conditions should be used to calculate the specific gravity. Typically, liquid specific gravity is referenced to 60°F and 14.696 psi. Since other standard temperatures may also be used, the following information may be helpful.

$$
\begin{array}{llll}
\text{Density of water at 60°F} & = & 62.33630 \text{ lb/ft}^3 \\
& = & 0.9990121 \text{ g/cm}^3 \\
\text{Density of water at 68°F} & = & 62.31572 \text{ lb/ft}^3 \\
& = & 0.9982019 \text{ g/cm}^3 \\
\text{Density of air at 60°F} & = & 14.696 \text{ psia} \\
& = & 0.0764 \text{ lb/ft}^3 \\
\text{Density of air at 68°F} & = & 14.696 \text{ psia} \\
& = & 0.7528 \text{ lb/ft}^3
\end{array}
$$

Changes in temperature will affect the specific gravity of a substance.

To demonstrate, compare the specific gravities of a liquid at 60°F and at 68°F. A liquid has a density of 1.095 grams per cubic centimeter. Its specific gravity can be calculated as follows:

$$\text{SG} = (1.095 \text{ g/cm}^3)/(0.9990121 \text{ g/cm}^3)$$
$$= 1.096$$

When the temperature is raised to 68°F, the density changes to 1.094 grams per cubic centimeter. This change is reflected in the following calculation:

$$\text{SG} = (1.094 \text{ g/cm}^3)/(0.9990121 \text{ g/cm}^3)$$
$$= 1.095$$

Therefore, the specific gravity of liquids and gases will decrease as temperature increases.

Fluid properties such as density apply to fluid in a static phase. Viscosity is a property that applies to fluids in motion. Viscosity can be described as a measure of how freely a liquid flows.

The force that causes fluids to flow is created by a change or a difference in pressure. While there are many types of pressure, the study of fluid flow and flow measurement is primarily concerned with static pressure, dynamic pressure, and differential pressure.

Static pressure is an important variable in the measurement of fluid flow. It can be defined as the pressure exerted by fluids at rest. Static pressure is independent of the kinetic energy of the fluid. The liquid in the tank exerts static pressure against the walls of the tank.

Dynamic pressure is the increase in pressure above static pressure that results from the transformation of the fluid's kinetic energy into potential energy. In other words, it is the pressure above static pressure caused by the movement of fluids. Dynamic pressure can be produced by gravity (as in the case of an elevated water tank), or mechanically (as in the case of a pump).

Differential pressure is the pressure difference between two related pressures. Differential pressure can be determined by measuring two related pressures and calculating the difference between the two measurements. This difference is the differential pressure value. Differential pressure is frequently used to determine fluid flow rate.

FLOW

While a study of flow measurement requires an understanding of what causes fluid to flow, accurate measurements systems must also take into account the flow profiles of fluids.

Fluids may move in smooth patterns, agitated or turbulent patterns, or in a combination of these patterns. Figure 1-78 shows a comparison of three types of flow: laminar flow, turbulent flow, and transitional flow.

Figure 1-78(a) represents a smooth, layered flow, which is described as laminar flow. In laminar flow, the fluid particles move along parallel paths. If laminar flow could be observed, it would appear as several streams of liquid flowing smoothly alongside each other.

Figure 1-78(b) represents a turbulent flow pattern. In comparison to laminar flow, turbulent flow is agitated and disturbed. Turbulent flow appears to have small, high-frequency fluctuations that travel in all directions, forming eddies.

Transitional flow, which is illustrated in Figure 1-78(c), exhibits characteristics of both laminar and turbulent patterns. In some cases, transitional flow will oscillate between laminar and turbulent flow.

The degree of frictional resistance generated by the three types of flow patterns varies. Laminar flow offers the least amount of frictional resistance to fluid flow, whereas turbulent flow causes a great deal of frictional resistance. The degree to which transitional flow tends toward one pattern or the other is a major factor in determining the amount of friction the flow generates.

VISCOSITY

Viscosity is the property that determines how freely fluids flow. Viscosity can be further described as the property of a fluid that contributes to laminar or turbulent flow characteristics. Fluids have various degrees of viscosity. Such variations result from internal friction between the particles of the substance. If the molecules slide easily over one another, the substance has a relatively low viscosity. A substance with a higher viscosity has a higher resistance to flow. Two substances with different viscosities are water and oil; water pours freely, while oil pours more slowly.

In any situation a fluid's flow profile will depend in part on the viscous forces that resist flow and the forces that act to keep flow moving at a constant rate. For example, fluid traveling through a pipe must overcome the resistance generated by two forces: the internal friction determined by the viscosity of the liquid and the friction determined by the viscosity of the liquid; and the friction that occurs between the liquid surfaces and the walls of the pipe. If the flow is laminar, it can be assumed that there is less frictional resistance at the center of the flow (where the molecules are sliding against each other), than at the pipe walls (where friction occurs between the liquid and pipe surfaces).

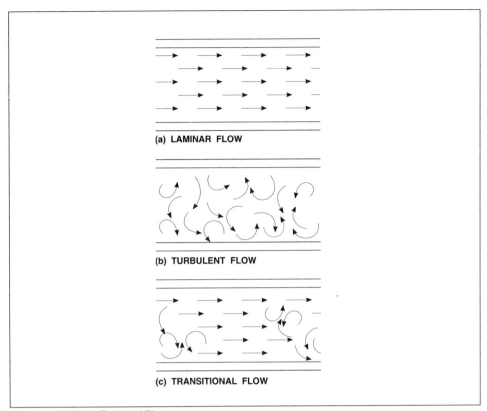

Figure 1-78. Three Types of Flow

If the flow is laminar, the viscous force causes the flow to slow as it approaches the pipe walls. Theoretically, this flow is parabolic, with the central core having a higher velocity and the outermost area having a lower velocity. This is illustrated in Figure 1-79.

Turbulent flow is less affected by viscous forces along the walls of the pipe. Due to relatively low viscous forces, turbulent flow exhibits a more uniform profile than laminar flow. However, the fluid layer next to the wall remains laminar even in the case of fully developed turbulent flow.

EFFECT OF TEMPERATURE ON VISCOSITY

Temperature has a significant effect on the viscosity of a substance. Relatively small changes in temperature may produce significant changes in a fluid's viscosity. Generally, changes in temperature have an inverse effect on viscosity. If the temperature of a fluid decreases, its viscosity will increase. On the other hand, if the temperature of the fluid increases, its viscosity will decrease. For ex-

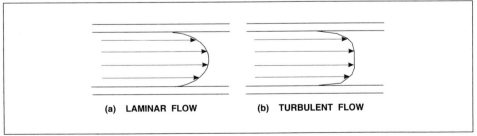

Figure 1-79. Viscosity and Laminar Flow

ample, if molasses stored at room temperature is placed in a refrigerator, its viscosity will increase and it will become harder to pour. However, if the molasses is heated, its viscosity will decrease.

If the viscosity of a liquid at one temperature is known, its viscosity at another temperature may be estimated by using the generalized viscosity curve for liquids (see Figure 1-80). The value is expressed in centipoise (cP), a unit of viscosity measurement. For example, if the viscosity of a liquid at 50°C is 25 cP, the vis-

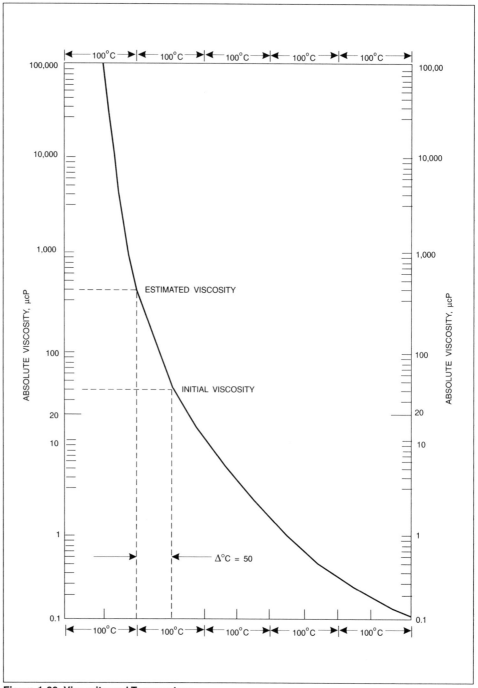

Figure 1-80. Viscosity and Temperature

cosity of the liquid at 0°C can be determined using the generalized viscosity curve for liquids.

The initial viscosity is indicated where the generalized viscosity curve intersects at 25 μcP and 50°C. As each division represents 50°C, the estimated viscosity at 0°C is approximately 400 cP. This example illustrates the magnitude of the effect on a substance's viscosity even with small temperature changes.

REYNOLDS NUMBER

Flow is often measured in terms of velocity. Therefore, when different portions of the flow are moving at different velocities, measurement accuracy will be affected. Further, as stated earlier, the flow profile depends on a combination of factors, including the forces that resist flow and the forces that act to keep flow moving at a constant rate. The relationship between these forces is expressed by the Reynolds number, which is a ratio of inertial to viscous forces specific to flow conditions.

$$R_D = \text{inertial forces/ viscous forces} = \rho \times v \times D/\mu$$

Like specific gravity, the Reynolds number is dimensionless and is not absolute for a given liquid substance but varies with the conditions of flow. The following equations are used to calculate the Reynolds number for liquid and gas flow through a pipe, given that ρ is in pounds per cubic foot and D is in inches.

$$\text{Liquid R} = (3160 \times Q_{\text{gpm}} \times \text{SG})/(\mu\text{cP} \times D)$$

where:

$$
\begin{align}
3160 &= \text{units constant} \\
Q_{\text{gpm}} &= \text{flow in gallons per minute} \\
\text{SG} &= \text{specific gravity} \\
\mu\text{cP} &= \text{viscosity in centipoise} \\
D &= \text{inside pipe diameter}
\end{align}
$$

$$\text{Gas R} = (379 \times Q_{\text{acfm}} \times \rho)/\mu\text{cP} \times D$$

where:

$$
\begin{align}
379 &= \text{units constant} \\
Q_{\text{acfm}} &= \text{flow in cubic feet per minute} \\
\rho &= \text{density}
\end{align}
$$

It is understood that the Reynolds number reflects fluid effects only; it disregards such factors as pipe roughness, pipe obstructions, and pipe bends. However, boundary limitations have been established that serve as estimates for practical applications. The boundaries that follow are not established by controlled laboratory experiments. When the Reynolds number is less than 2000, flow is in the laminar region. When the Reynolds number is greater than 4000, flow is considered to be in the turbulent region. When the Reynolds number is in the range of 2000 to 4000, flow is transitional.

Viscosity is the factor that most affects the value of the Reynolds number. In cases where small changes in temperature cause relatively large changes in viscosity, corresponding changes in the Reynolds number will also occur. These changes affect whether the flow is in the laminar, turbulent, or transitional regions. The Reynolds number is often a significant factor in determining how well a specific fluid-measuring device will perform and which applications are most appropriate to its use.

Measuring the flow of a substance requires a thorough, detailed understanding of the process and the substance being measured. Process applications may involve the flow of gases, liquids, or solids — singly or in combination. These substances may flow through pipes or open channels depending on the nature of the process. Two factors that determine the method of flow measurement of the flow-meter most suited to an application are the quantity of the flow and the type of substance being measured.

The flow rate of a substance can be described using a number of terms, including feet per second, gallons per minute, cubic feet per minute, and tons per hour. The unit chosen to indicate flow rate is an important factor in flow measurement applications and varies according to the indicating requirements specific to the process. There are several methods used to measure flow, including flow rate, volumetric flow rate, and mass flow rate.

FLOW RATE

The measurement unit used to express the rate of flow actually refers to the velocity of the flow or how rapidly the substance moves. A flow rate is a measure of the distance a particle of a substance moves in a given period of time.

Feet per second is a unit commonly used to measure flow rate. An application of this method is illustrated in Figure 1-81. If a molecule of the fluid in the pipe takes one second to move from point A to point B, and the distance between these points is ten feet, the flow rate of the fluid in the pipe is 10 feet per second (10 ft/sec).

VOLUMETRIC FLOW RATE

The method of measurement used to indicate the volume of fluid that passes a point of time is volumetric flow rate. Volumetric flow rate is usually expressed in gallons per minute (gpm) or cubic feet per second (ft^3/sec).

One means of measuring volumetric flow is to transfer a specific volume of fluid from one vessel to another and time the procedure. If two minutes are required to transfer 500 gallons of fluid from one vessel to the other, the volumetric flow rate is 250 gallons per minute or 250 gpm. In general, this method has limited practical applications. A more widely used method of measuring volumetric flow rate is typically adopted when two factors — flow velocity and pipe inside diameter — are known. Tables are available to determine the inside diameters for pipes of different sizes.

Volumetric flow rate can be calculated using the formula:

$$Q = A \times v$$

where:

Q = volumetric flow rate
A = cross-sectional area of the pipe
v = flow velocity

In an application in which a fluid has a flow velocity of 14 feet per second through a pipe with an inside diameter of two and one-half feet, the volumetric flow rate is determined to be 68.7 cubic feet per second, using the following equation:

$$Q = A \times v$$
$$= (\pi\ r^2) \times 14 \text{ ft/sec}$$
$$= 4.90625 \text{ ft}^2 \times 14 \text{ ft/sec}$$
$$= 68.7 \text{ ft}^3/\text{sec}$$

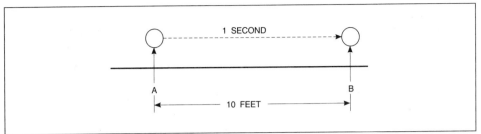

Figure 1-81. Velocity and Flow Rate

MASS FLOW RATE

Another system of flow measurement that can be used to determine the flow rate of a process is mass flow rate. Mass flow rate determines the amount of mass that passes a specific point over a period of time. Mass flow rate applications determine the weight or mass of the substance flowing through the system.

Provided the volumetric flow rate and the fluid density of the process substance are known, mass flow rate can be calculated using the following equation:

$$W = Q \times \rho$$

where:

$$
\begin{aligned}
W &= \text{mass flow rate} \\
Q &= \text{volumetric flow rate} \\
\rho &= \text{density}
\end{aligned}
$$

Therefore, if a process fluid with a density of 65.4 lb/ft^3 has a volumetric flow rate of 59 cubic feet per second, the mass flow rate could be determined as follows:

$$
\begin{aligned}
W &= 59 \text{ ft}^3/\text{sec} \times 65.4 \text{ lb/ft}^3 \\
&= 3858.6 \text{ lb/sec}
\end{aligned}
$$

When mass flow rate is measured, the effect of temperature and pressure on the density of the fluid must be considered. This is especially true in processes that involve gases. In such cases, means must be established to compensate for the changes in density caused by other process variables.

Flow Calculations — General

Correction factors are necessary in making accurate flow calculations. One of these factors is the discharge coefficient C or C_d. For fairly general flow measurements, as in comparing multiphase flow; the following values for C are established:

Orifice plate, C = 0.62 × Area of orifice (d)

Flow nozzle, C = 0.98 × least internal area

Venturi tube, C = 1.0

However, in some applications the Reynolds number allows a calculation of the value of C in the basic flow equation.

EXPANSION FACTOR — Y

When gases or vapors hit an orifice plate, they will be somewhat compressed, and as they pass through the restriction they expand. The net effect would produce a differential pressure around the restriction that is greater than it should be, due to the quantity of fluid alone. A correction called the expansion factor is given for most differential pressure ratios with respect to orifice plate ratios, nozzles, Venturis, air, steam, and diatomic gases (hydrocarbons). Y is a relationship between ΔP and β. These values can be found in tables established by the American Gas Association and several instrument manufacturers.

VELOCITY OF APPROACH FACTOR

When a fluid approaches an orifice plate, its velocity becomes most important at an imaginary plane across the center of the plate. Molecules in the center of the aperture pass through perpendicular to the plane and are discharged in the same fashion as they are received. Other molecules have to pass around the sharp edge of the orifice, and their paths become curved. The interior of the jet may be subjected to excess pressures from these curved paths, and this excess pressure will cause a reduction of the fluid velocity approaching the orifice plate. The larger the ß ratio, the greater the fluid velocity. The equation for the velocity approach factor is:

$$M = 1\sqrt{1 - (\beta)^4}$$

The values for M may also be read from an AGA table on the velocity of approach factor. After all factors have been gathered, it is now possible to make flow calculations employing the $YMCA$ equation, where YMC is equal to the constant k in the basic equation:

$$Q = kA\sqrt{(h/P)} \quad or \quad Q = kA\sqrt{(2gh)}$$

which can be modified to:

$$Q = YMCA\sqrt{(2gh)}$$

where:

Q = flow in ft^3/sec or lbs/sec
Y = expansion factor
M = velocity of approach factor
C = coefficient of discharge
A = area of aperture
g = acceleration due to gravity 32.2
h = differential pressure across orifice

TEMPERATURE CORRECTIONS FOR HEAD FLOWMETERS

An allowance for thermal expansion has to be made especially when the fluid temperature is much higher than the ambient temperature. Tables are available to find the increase in orifice diameters that are exposed to certain temperature differentials with respect to different plate materials. For instance, an austenitic stainless steel plate will expand its bore diameter by 0.5% when subjected to a temperature change of 520°F (271°C). Therefore, corrections must be made to d and D, and the calculation of flow must be based on the two corrected diameters, because an increase of the bore diameter by 0.5% will cause a 1% increase in the flow rate. The temperature change will also cause a viscosity change of the fluid, depending on the state of the fluid.

CRITICAL PRESSURE DROP

The velocity of flow through the orifice bore will become equal to the velocity of sound when the ratio of the absolute downstream pressure P_2 to the absolute upstream pressure P_1 is less than the critical pressure.

The critical pressure ratio is 0.53 air and gases, 0.545 gases, 0.545 for saturated steam, and 0.58 for superheated steam. The critical pressure ratio depends on the ratio of the gas specific heat and constant volume (C_p/C_v) and to the ratio of the orifice bore area and pipe area (d^2/D^2).

The velocity related to critical pressure is called the critical velocity, and it will cause turbulence in the pipe, making corrections for the Reynolds number redundant.

In the *YMCA* equation, the flow rate is proportional to the bore area, critical velocity, and gas density at the bore (aperture). For any gas, the critical velocity is proportional to the square root of the upstream absolute temperature ($\sqrt{T_1}$), and the gas density in the aperture is proportional to P_1T_1.

THE MACH NUMBER

The Mach number is important because its values denote supersonic flow and it indicates the presence of shock waves in the pipe. The shock wave may affect the ratio of throat pressure to upstream pressure, especially where a sharp-edged orifice or Pitot tube is employed. The mach number is dimensionless and is expressed mathematically as:

$$\text{Mach number} = V_{\text{actual}}/V_{\text{sonic}} = V_{\text{actual}}/\sqrt{(g\,k\,\text{Re}\,T)}$$

where:

g = acceleration due to gravity
k = constant 1.4 for air and gas; 1.3 for steam
Re = Reynolds number
T = absolute temperature
V = velocity

Flow Measurement

HEAD FLOWMETERS

The most frequently used flow measuring method is the variable head or differential metering method, in which the rate of flow is determined by measuring the differential pressure across an engineered restriction. The restriction devices are usually orifice plates, flow nozzles, and Venturi tubes. Many formulas for flow calculations have been developed, mainly by empirical methods. The equations used for flow calculations through a restriction in a pipe are based upon Daniel Bernoulli's theorem.

Bernoulli's theorem expresses the law of conservation of energy as it applies in fluid flow. It states that in any fluid flow in a conductor, the sum of the potential energy, kinetic energy, and pressure energy is always constant. Thus, if the potential energy remains constant and the kinetic energy is increased, the pressure energy will decrease. That is, if we increase the velocity (kinetic energy), there will be an accompanying pressure drop.

The basic equations are:

$$V = k\sqrt{(h/\rho)}$$

$$Q = kA\sqrt{(h/\rho)}$$

$$W = kA\sqrt{(h/\rho)}$$

where:

$$V = \text{velocity}$$
$$Q = \text{volume flow rate}$$
$$W = \text{mass flow rate}$$
$$A = \text{cross-sectional area of pipe}$$
$$h = \text{differential pressure across restriction}$$
$$\rho = \text{density of flow fluid}$$
$$k = \text{constant}$$

The constant k contains the following factors: ratio of the pipe cross-sectional area to the cross-sectional area of the restriction, correction factors, units of measurement, pressure and flowing temperature, type of differential device used, the expansion or contraction of the restriction, velocity of approach, compressibility, etc.

Figure 1-82 shows a cross section of a typical orifice plate installation and the variation in pressure. An orifice plate is a thin, circular metal plate in which an engineered restricted opening is drilled and machined. Depending on the application, the hole is drilled either concentric or eccentric.

The maximum pressure is found at point A, and the minimum pressure is at point B. Although the pressure restores at C, it is not equal to the pressure at A because of the pressure loss across the orifice plate. Point B is also called the vena contracta point. At the vena contracta the lowest pressure in the system under consideration occurs simultaneously with the highest velocity. The position of the vena contracta in the flow depends less on the velocity and more on the ratio of the orifice aperture diameter to the inside diameter of the pipe (beta ratio), for which the equation is:

$$\beta = d/D$$

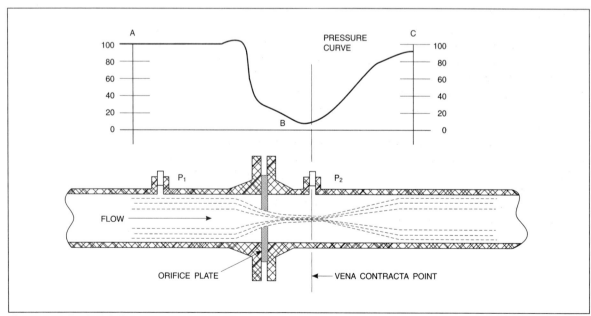

Figure 1-82. Typical Orifice Plate Flow Element

where:

d = orifice diameter

D = pipe inside diameter

ß is also a factor of the constant k in the basic flow equations. Figure 1-83 shows a concentric orifice plate.

The plate thickness W is a function of the pipe size. Edges beveled at 45 degrees are often used to minimize friction resistance to flow. The beta (β) ratio should fall between values 0.15 to 0.75 for liquids and 0.02 to 0.70 for gases in order to maintain measurement accuracy.

Other orifice plate configurations are available for different purposes; for example, eccentric orifices have the aperture (bore) offset from center to minimize service problems involving solids containing fluids. Segmental orifices are also installed for solids-containing fluid flow measurements. The aperture is a segment of a circle, the diameter of which is practically 98% of the pipe nominal diameter.

> Quadrant edge orifices produce a relatively constant coefficient of discharge for services with low Reynolds numbers. It is preferable to specify stainless steel orifice plates.

FLOW NOZZLES

The flow nozzle consists of a restriction with an elliptical or near-elliptical contour approach section that terminates in tangents with a cylindrical throat section (see Figure 1-84). Three standard designs that differ in detail of approach: section, contour, and throat length; the difference in performance is almost negligible.

The approach curve must be proportioned to prevent separation between the flow and the wall, and the parallel section is used to ensure that the flow fills the throat.

The differential pressure measurement taps are usually located one nominal pipe diameter upstream and one half nominal pipe diameter downstream from the inlet faces of the nozzle.

Discharge coefficients of flow nozzles are larger than those for orifices; their values change with a change in Reynolds number. The nozzle type offers all the advantages of a Venturi tube to a slightly lesser extent, but it does take considerably less space.

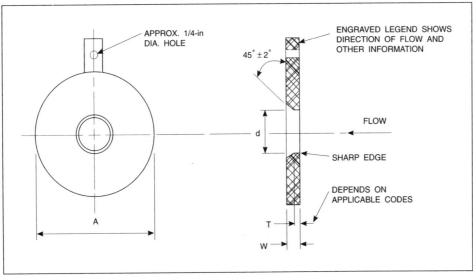

Figure 1-83. Typical Concentric Orifice Plate

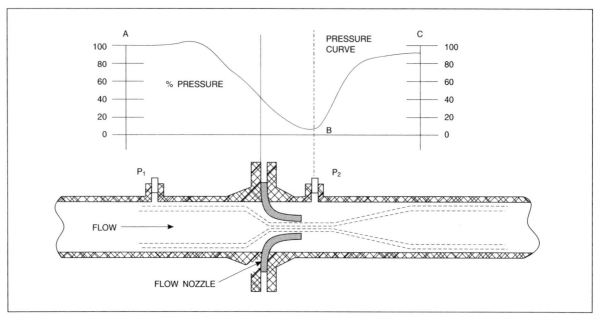

Figure 1-84. Typical Nozzle Flow Element

VENTURI TUBES

The Venturi tube (Figure 1-85) consists of a converging conical inlet section in which the cross section of the stream decreases with consequent increase of velocity head and decrease of pressure head. A cylindrical throat provides a point for measurement of the decreased pressure in an area of constant flow rate, and there is a diverging recovery cone in which velocity decreases and the decreased velocity head is recovered as pressure. Pressure taps are taken at the upstream of the inlet cone at a distance of one-half nominal pipe diameter and at the middle of the throat. The design of Venturi tubes has been standardized by the American Society of Mechanical Engineers (ASME).

A Venturi tube is usually selected to measure the flow of compressed gas (compressor discharge). Discharge coefficients of Venturi tubes are larger than those

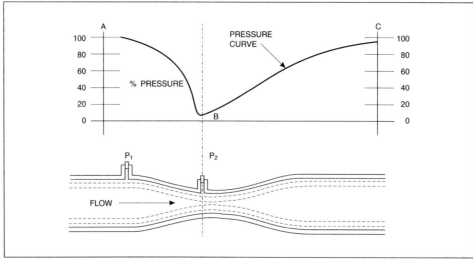

Figure 1-85. Typical Venturi Tube Flow Element

for orifices and nozzles; their values change with a change in Reynolds number. The flow calculations for orifice plates also apply to nozzles and Venturi tubes.

BASIC ORIFICE FACTOR — F_b

This factor F_b (sometimes identified as O_f) is a dimensionless number selected through a basic orifice factor. The factor is derived from the relationship between the orifice diameter vs. the measured pipe I.D. These are empirical values, and it is a result of many years of experimental work.

REYNOLDS NUMBER FACTOR — F_r

The Reynolds number depends on the pipe diameter, flow density, velocity, and viscosity. For any orifice, the coefficient of discharge is equal to the same value of F_r when measuring the fluid through it. The coefficient will have the smallest value when F_r is infinite. AGA tables are required to make correction for the Reynolds number to apply to the F_r equation.

$$F_r = 1 + b \sqrt{(hwP_f)}$$

where:

$$b = \text{a factor found in AGA Table 2}$$
$$hw = \Delta P \text{ in water (inches)}$$
$$P = \text{static pressure}$$
$$f = \text{pressure extension factor found in AGA Table 10}$$

AGA data shows the relationship between the orifice diameter and the pipe diameter. The b values were found by empirical methods.

EXPANSION FACTOR — Y_2

The expansion factor depends upon differential pressure, static pressure, and the gas specific heat ratio C_p/C_v. The adiabatic expansion assumes that the differential pressure in inches of water across an orifice is a head that represents perfect expansion. "Adiabatic" means without heat transfer. However, a perfect adiabatic flow does not exist in practice. Compressible fluids with a ratio of specific heats greater than 1.0 and subjected to adiabatic flow will decrease in temperature in the direction of flow. Therefore, corrections must be made for the inefficiency and resulting inaccuracy of the head in inches of water. The correction factor can be found in the AGA literature as a relationship between the ratio of the average differential head to the average absolute static pressure and the beta ratio.

$$\text{Average differential/Average static} = hw/P_f \text{ and } ß = d/D$$

CORRECTION TO PRESSURE BASE — F_{pb}

This correction factor is just a multiplier that converts one pressure base to another. This factor is either specified by local regulations or mutually agreed on by vendor and purchaser. The value can be found in the AGA literature, and it is based on the equation:

$$F_{pb} = 14.73/\text{base pressure}$$

CORRECTION TO TEMPERATURE BASE — F_{tb}

This correction factor is just a multiplier that converts one temperature base to another. This value can found in the AGA literature and is based on the equation:

$$F_{tb} = T_b/520$$

where a base temperature of 60°F (15.6°C) is established. 520° is a constant that is derived from 460° + 60°F = 520°F, where 460°F is the absolute temperature, because the absolute zero temperature is –460°F (–237.78°C). T_b = 460° + new base temperature.

CORRECTION FACTOR FOR FLOWING TEMPERATURE — F_{tf}

Since the volume of a gas changes when its temperature changes, corrections are made to calculate the gas flow. The standard gas temperature is 60°F (15.6°C). The correction factor at 60°F = 1.000. When the flowing temperature is under 60°F at the orifice, the factor will exceed 1.000, because the gas volume is reduced. The opposite will happen when the flowing temperature exceeds 60°F. The F_{tf} value can be found in the AGA Report 3, Table 6, using the following equation:

$$F_{tf} = \sqrt{(520/T_f)}$$

where:

$$T_f = 460° + \text{flowing temperature}$$
$$520° = \text{constant of base temperature}$$

CORRECTION FACTOR FOR SPECIFIC GRAVITY — F_g

The gas volume depends on the specific gravity of that gas; therefore, adjustments are required in order to achieve an accurate gas volume measurement. The F_g value can be obtained from the AGA literature using the equation:

$$F_g = \sqrt{(1/G)}$$

where:

$$G = \text{specific gravity of the gas}$$

SUPER COMPRESSIBILITY FACTOR — F_{pv}

When a gas is compressed under high pressure, it no longer obeys Boyle's law, and the actual density is generally greater than the theoretical density. Therefore, in Boyle's law for a perfect gas, $P_v = CT$, a deviation factor Z must be included, resulting in the practical equation:

$$P_v = ZCT$$

where Z is a correction factor to compensate for deviations caused by pressure, temperature, gas composition, and other factors agreed on by all parties to the sales contract.

The value of F_{pv} can be obtained from the AGA literature, on the relationship of mol. % nitrogen content versus mol.% carbon dioxide content.

The seven correction factors multiplied will give a corrected orifice constant C, which can be incorporated into the $YMCA$ equation to obtain the orifice size for a gas flow measurement. Bear in mind that the tables referred to in this section are valid only for natural gas flow calculation. To calculate factors for other gases, one must refer to other sources (for example, handbooks issued by companies such as The Foxboro Company, Rockwell Manufacturing, and American Meter).

MULTIPHASE FLOW

In some processes pipes may convey vapor and liquid simultaneously. When this is the case, special calculations must be made. This is a complex subject and, although much work has been done in the field, many questions remain unanswered. In comparing two-phase flow to single-phase flow, the principal difficulty is the variety of flow patterns in a gas-liquid system. The patterns encountered depend on fluid properties, flow rates, and equipment geometry. Unit pressure loss may vary significantly between patterns.

Two of the more widely used two-phase flow correlations are the Beggs and Brill correlation and the Dukler AGA–API correlation (American Gas Association – American Petroleum Institute):

Seven basic flow patterns for horizontal flow were visually observed by Alves — bubble, plug, stratified, wavy, slug, annular, and spray — and five basic flow patterns for vertical flow were visually observed by Davidson — bubble, slug, froth, annular and mist. Two-phase flow can also occur in steam lines and natural gas pipes, especially coming from offshore sources where gas condensate may be transported with the natural gas.

HEAD METER LOCATION

The location of an orifice, flow nozzle, or Venturi tube in a piping system is important to obtain an accurate flow measurement. The flow should approach the head meter in a normal turbulent state, without the influence of swirls, cross currents, eddies, or other disturbances that create helical paths of flow. Nor can there be any disturbance following the head meter restriction that would in any way cause interference with the static pressure measurement at that point. Therefore, it is necessary to have adequate lengths of straight pipe on both sides of the restriction.

Standards and graphs have been established by the American Society of Mechanical Engineers and the American Gas Association, which can be found in the ASME transactions published in July 1945.

LOCATION OF PRESSURE TAPS

There are five common tap locations for taking differential pressure measurements across restrictions (primary elements): flange, vena contracta, pipe, radius, and corner. Figure 1-86 shows the pressure tap location of the flange, vena contracta, and pipe for concentric orifices.

The location of the flange taps is shown in the bottom of the flange. All the taps upstream of the orifice flange are high-pressure taps, and the ones downstream are the low-pressure taps. These taps are to be connected to a differential pressure gage or transmitter.

FLOW TRANSMITTERS — dP CELL TYPE

The dP cell flow transmitter is in fact a force-balance differential pressure transmitter. The mechanism of this type of transmitter was discussed in the previous section on pressure measurement. Both pneumatic and electronic dP cell flow transmitters are available; however, the electronic type is becoming more

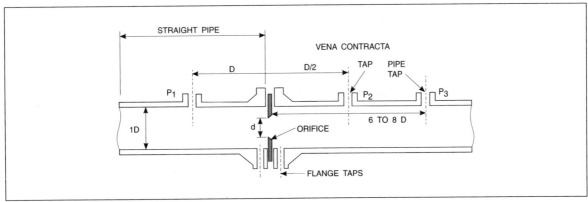

Figure 1-86. Pressure Tap Location Diagram

popular. Figure 1-87 shows the hookup of a flow transmitter on a pair of orifice flanges.

The tap orientation on orifice flanges is very important; this orientation depends on the type of fluid that is to be measured. On the installation detail shown in Figure 1-87 the taps are made on top of the flanges because this is wet gas, which could have condensate in the line, and the tubing is usually made to slope back into the flange. On liquid flow applications the taps are normally made at the side of the flange to prevent trapped gas bubbles or particles from entering the dP cell. On steam lines condensation pots are added between the taps and the dP cell.

As could be seen from the flow calculations, this type of flow measuring device has a square root relationship (flow is proportional to the square root of the differential pressure across the restriction). Therefore, the output of this flow transmitter is not linear. In order to get a linear output, a square root extractor must be connected between the transmitter and the remote flow indicator, recorder, or controller.

VARIABLE AREA FLOWMETERS

The operation of a variable area flowmeter is basically the same as the head flowmeters that employ orifices. However, the head meter has a fixed aperture and the flow rate is a function of the differential pressure, whereas the variable area flowmeter has a variable aperture and the pressure drop is relatively constant. Therefore, the flow rate is a function of the area of the opening through which the flow is passing. The position of a float in the meter depends on that area. The three main groups of variable area flowmeters are:

(1) orifice and tapered plug meters,

(2) piston-type meters, and

(3) rotameters.

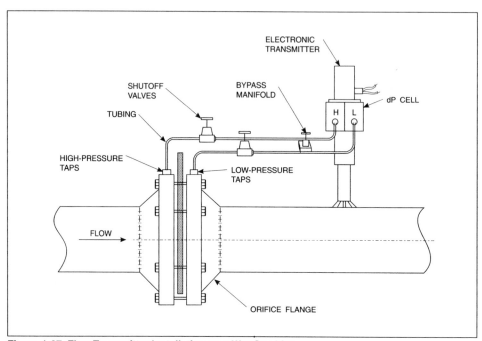

Figure 1-87. Flow Transmitter Installation on a Wet Gas Line

Only the rotameter is discussed here, because it is the most frequently used variable area flowmeter in the process industry. Variable area flowmeters are used mainly to measure low flow rates.

ROTAMETERS

The name rotameter was chosen in the past because at one time the floats were designed with slots to produce a rotational movement for the purpose of centering and stabilizing the float. Figure 1-88 shows a schematic of a rotameter.

The float is free to move up or down within the tapered tube. The tube is tapered to provide a linear relationship between the flow rate and the float position within the tube. The fluid enters at the bottom and passes upward around the float and out at the top. The float rises and falls depending on the flow rate, giving a variable area of the annular passage. The upward hydraulic forces that act on the float are in balance with its weight less the buoyant force; this will maintain the float position in equilibrium. The scale shows flow rate in percent.

The rotameter sizing is also based on Bernoulli's theorem. Rotameters may be used to measure gas or liquids and even some light slurries. Therefore, the sizing gas measurement is based on the flow of air and for liquids on the flow of water. Where very low flows are measured, capacities are usually given in cc/minute of air or water; these rotameters are then called purge meters.

The same fundamental equation used for the head meter also applies to the rotameter:

$$Q = kA \sqrt{(2gh)}$$

In this equation the differential pressure (h) is replaced by the factor that causes it to remain constant. This factor denotes the volume and area of the float, as well as the density of the float and the fluid. The equation will then be:

$$Q = kA_m \sqrt{(2g \, V_f/A_f)} \times (D_f/D - 1)$$

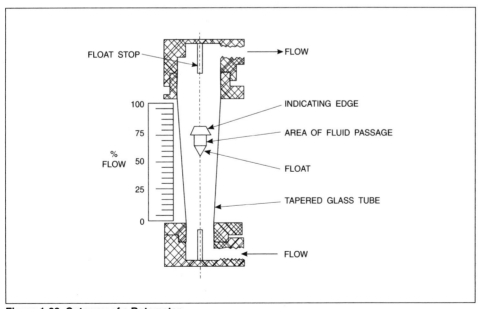

Figure 1-88. Cutaway of a Rotameter

where:

$$Q = \text{flow rate}$$
$$k = \text{taper constant}$$
$$A_m = \text{annular area measured at indicating edge}$$
$$g = \text{gravity}$$
$$V_f = \text{float volume}$$
$$A_f = \text{float area}$$
$$D_f = \text{float density}$$
$$D = \text{fluid density}$$

The rotameter may be subject to error due to changes in the fluid density. Small variations in fluid viscosity will not affect measurements if the float is designed to be insensitive to them. This will make $D_f/D = 2$. Therefore $(D_f/D - 1) = 1$.

Since $\sqrt{1} = 1$, the effect of density is eliminated. The float also must be designed to ignore small changes in the viscosity of the fluid. The rotameter is available with a tapered glass tube or a metal tube. The choice of tube depends on the application. Where toxic or flammable fluids are to be measured, a metal tube is selected for the rotameter. However, in this case the float is not visible, and a different indication technique is required. The linear movement of the float must be converted into a rotating movement for local indication or signal transmission. The metering float is extended with a permanent magnet on top of it. This magnet is imbedded in the extension rod or another float. The converter consists of a magnetic iron helix that is permanently attached in an aluminum cylinder and mounted adjacent to and parallel with the rotameter extension tube. The leading edge of the helix is continuously attracted to the permanent magnet in the float extension. The linear motion of the permanent magnet causes the helix to rotate, and this rotating motion is the indication of the flow rate.

A characterized cam, which is sensed pneumatically or electronically to produce linear transmission signals, is attached to the helix cylinder. In a pneumatic transmitter, a pneumatic cam follower that will convey a signal to a pneumatic relay is employed. This relay amplifies the signal to a suitable transmission pressure.

The electronic transmitter is equipped with a sensing coil underneath the characterized cam. This coil is connected to an amplifier to transmit an output signal. See Figure 1-89 for a pictorial explanation.

OPEN-CHANNEL FLOWMETER

This type of flowmeter is usually selected for the measurement of large volumes of flow, such the waste water or sewage disposal. These types of flow measurements are required for water pollution control. The primary elements used in open-channel applications are weirs, flumes, and open nozzles. When these primary elements are employed, the flow rate is measured using a level sensor placed in a still well adjacent to the flow channel. As the flow through the channel increases, the level rises simultaneously in both the channel and the still well. Hence, the float position or level in the still well reflects the flow rate. The level scale is calibrated in flow units (millions of gallons or cubic meters per day). For remote signal transmissions, a capacitance level probe is selected.

A weir is a flat bulkhead with a specially shaped notch along its upper edge. It is situated across the open fluid stream, forcing the fluid to rise up the notch as the flow rate increases. The notch comes in different shapes: V-notch, rectangular, Cippoletti, or trapezoidal (see Figure 1-90).

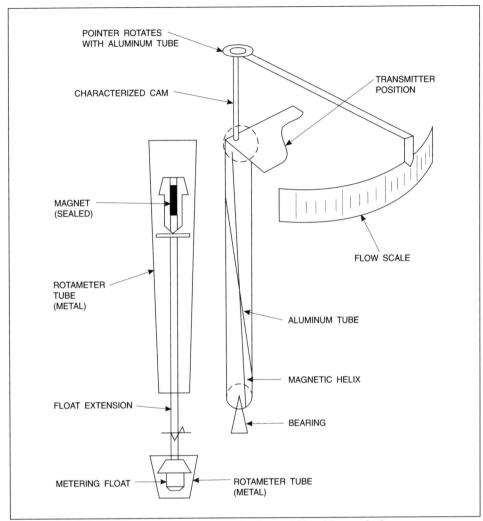

Figure 1-89. Metal Tube Rotameter for Local Indication and Signal Transmission

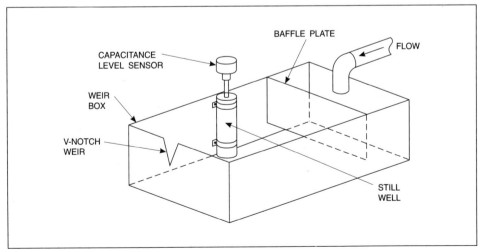

Figure 1-90. V-Notch Weir Box

A flume frequently called a Parshall flume (Figure 1-91) is a formed structure installed in the open fluid stream that forces the fluid level to rise within it as the flow rate increases. The flow measurement and the signal transmission method are similar to those of the weir plate arrangement. The open nozzle, often called a Kennison nozzle, is shaped so that the level of the fluid in the nozzle rises uniformly as the flow rate increases.

OPEN-CHANNEL FLOW CALCULATIONS

For the rectangular weir with side contraction, the theoretical flow rate is:

$$Q = (2b/3) \sqrt{(2gh^3)}$$

where:

b = width of the rectangular notch, ft
h = head over the weir crest, ft
g = gravitational constant, 32.2 ft/sec^2

Actual flow requires use of a flow coefficient, which is 0.644 ±5%:

$$Q = (0.644 \pm 5\%)(2b/3) \sqrt{(2gh^3)}$$

However, the flow through the V-notch is:

$$Q = (0.605 \pm 6\%)(8/15 \tan \pi/2) \sqrt{(2gh^5)}$$

The flow through the Parshall flume is expressed in almost the same fashion as that for the rectangular weir; except for the coefficient, this will give the equation:

$$Q = (0.63 \pm 3\%)(2b/3) \sqrt{(2gh^3)}$$

TARGET FLOWMETERS

The principle of a target meter consists of measuring flow by measuring the force on a disk (target) positioned in the center of the pipe perpendicular to the direction of fluid flow (Figure 1-92). The force on the target caused by the flow is proportional to the square of the flow. Two methods of signal transmission are available:

(1) The force-balance type — pneumatic or electronic

(2) The strain gage type — electronic only

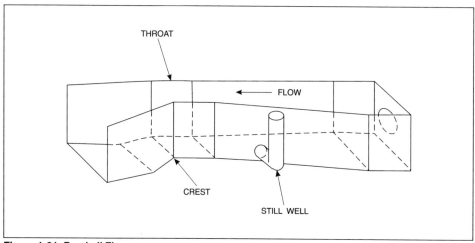

Figure 1-91. Parshall Flume

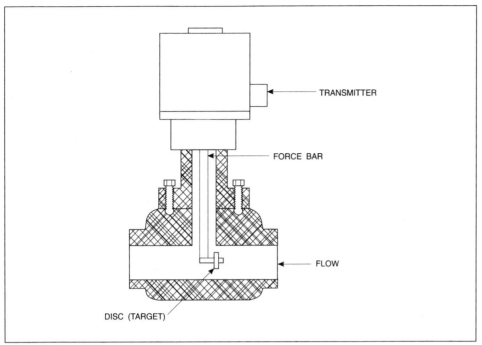

Figure 1-92. Target Meter

In the force-balance type the disk is mounted to a force bar; its mechanism is similar to that of a pressure transmitter that employs the flapper and nozzle arrangement for pneumatic transmission or the detector coil and ferrite armature construction for electronic transmission. The force limit is usually within 2 to 16 pounds. The relationship between flow rate and force is expressed by the equation:

$$Q = k\sqrt{F}$$

where:

k = a given coefficient
F = a force between 2 to 16 lbs
Q = flow rate

The bonded strain gage-type target flowmeter uses the same principle, where the force is expressed as:

$$F = C_d A_\rho (V_2/2g)$$

where:

F = force
C_d = drag coefficient
A = target area
ρ = fluid density
$V_2/2g$ = *velocity head*

This force is converted into an electrical output by the strain gage bridge circuit, which is proportional to the square of the flow. Obviously for this type of transmitter, a square root extractor is required to linearize the signal.

Target meters are well suited to measure viscous and corrosive fluids, even under high pressures and with high temperatures. For example, the force-balance types can handle pressures to 1,500 psig and temperatures to 750°F, and strain gage types — 5,000 psig and 600°F.

TURBINE FLOWMETERS

The turbine meter (Figure 1-93) is actually a pulse generator that consists of a precision turbine wheel mounted on bearings inside a housing, which is connected to a pipe and a magnetic pickup coil. The velocity of the fluid flow imparts a force to the turbine blades, which rotate at a speed that is proportional to the flow rate. As each turbine blade passes the magnetic pickup coil, one AC voltage pulse is induced and each pulse is equal to a definite flow quantity. The electrical pulses can be totaled, subtracted, or manipulated by digital techniques into an analog output signal. This type of flowmeter can handle only clean fluids, although modified types can also handle slurries.

Straight runs of pipe upstream and downstream are recommended for turbine meters as is required for differential head meters.

SWIRL FLOWMETERS

The operation of a swirl meter is based on the principle of vortex precession. The swirl meter has a stationary swirl agitator at one end, which causes the fluid stream to rotate around the centerline of the meter. A de-swirler, located at the opposite end of the meter, restores the fluid stream to its original flow pattern. The internal shape of the meter resemblance a Venturi tube, with an entrance cone, a throat, and an exit cone. When the fluid passes through the expanded portion of the meter, the axis of rotation moves perpendicular to the center of rotation (precesses). The frequency of the precession is proportional to the flow rate. This precession or oscillation of the fluid causes variations in temperature, which are sensed by a thermistor placed in the throat area of the meter. The temperature variations also vary the resistance of the thermistor. These variations are converted into voltage pulses that are amplified, filtered, and transformed into constant amplitude, high-level pulses of square waveform. The pulse frequency is measured by an electronic counter, or it may be transduced into an analog transmission signal. Figure 1-94 shows the swirl vortex flowmeter construction.

VORTEX SHEDDING FLOWMETERS

The principle of operation of the vortex shedding flowmeter is the use of the formation of vortices. A specially shaped "bluff body," which spans the meter area and creates a vortex inside the meter, has a triangular-shaped cross-sectional

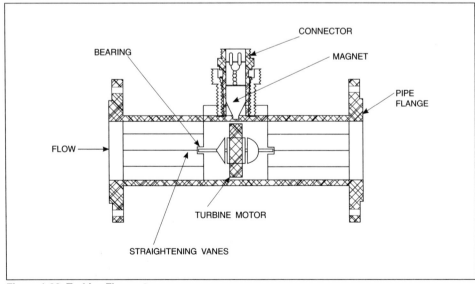

Figure 1-93. Turbine Flowmeter

area and is installed with the base facing upstream. The fluid vortices are shed off its downstream faces in a regularly oscillating pattern. The frequency of oscillation is directly proportional to the volumetric flow rate of gas or liquid. Vortex shedding frequency is sensed by a pair of thermistors embedded in the upstream face or by a piezoelectric crystal attached to the "bluff body." The type of sensor used depends on the instrument manufacturer. Another bluff body and sensor configuration employs a single removable thermistor set in a passage drilled through the flow element for easy replacement. Figure 1-95 shows a schematic diagram of a vortex shedding flowmeter employing two heated thermistors; they are heated to a temperature above that of the fluid flow and sense the cooling effect of the fluid vortices by changing temperature and, therefore, resistance at the vortex shedding frequency, which is proportional to the flow rate.

MASS FLOW METERING

Some process applications require the mass flow as a process variable in order to control the reaction or heat transfer and efficiency determination. Mass flow is

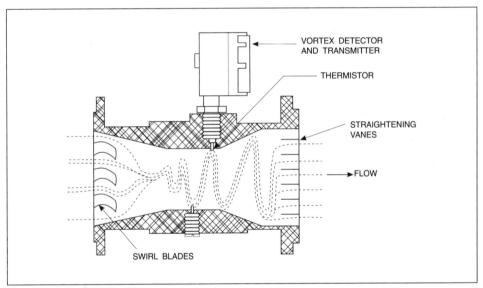

Figure 1-94. Swirl Vortex Flowmeter

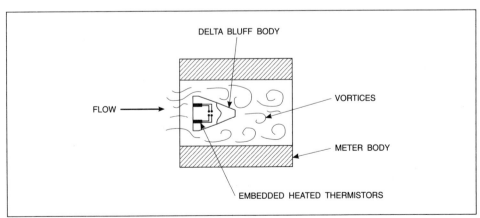

Figure 1-95. Schematic of a Vortex Shedding Flowmeter

also used for accounting procedures in which units of standard volumes are invariably required. The text will discuss the mass flow measurement of solids-bearing slurry.

Although true mass flowmeters are available, they are rarely applied due to their high cost. Therefore, two variables will be measured — fluid velocity and fluid density. These two variables are to be multiplied and their product will be the mass flow.

The mass flow equation is:

$$M = A\rho V$$

where:

M = mass flow rate
A = cross-sectional area of pipe interior density of fluid
ρ = density of fluid
V = velocity of fluid flow

The mass flow rate can be expressed in any desired units of measurement (English or metric).

To measure the fluid flow velocity, either a magnetic flowmeter or a Doppler flowmeter may be used. To measure the fluid density, a nuclear density gage with a built-in solid-state multiplier will be used. The density gage will be discussed in the section on density. Figure 1-96 shows a schematic of a mass flow metering arrangement utilizing a magnetic flowmeter and a nuclear density gage.

Discrete Quantity Flow Measurement

Discrete quantity flowmeters are mainly used for accounting applications or in batch processes. The two categories are positive displacement meters and positive displacement metering pumps.

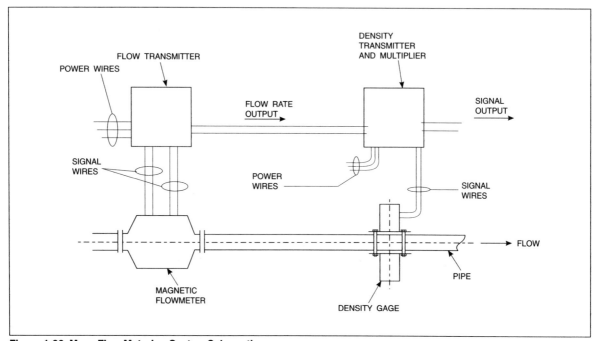

Figure 1-96. Mass Flow Metering System Schematic

Positive displacement meters are measuring devices that separate a flowing stream into individual volumetric increments and count those increments. The internal parts are machined so that each volumetric increment is accurately sensed, and the summation of the increments gives a very accurate measurement of the total volume passed through the meter. Most of these type of meters are mechanical types that are used primarily to measure total quantities of fluids to be transferred. They may be equipped with auxiliary devices for flow indication, totalizing, on–off contacts for remote indication, control, computer input, annunciation, etc. Positive displacement meters may be classified by the movement of the metering element.

THE NUTATING DISC METER

The disc employed in this meter makes a wobbling or nutating motion when the liquid flows through the meter. For each nutation of the disc, a specific volume of liquid passes through the meter. A metal pen on top of the disc causes a shaft to turn; the shaft is coupled to an external counter, register, totalizer, or transmitter.

THE OSCILLATING PISTON METER

The internal design of this meter is similar to the nutating disc type, except that the measuring device is a split ring that oscillates in one plane only. Each oscillation delivers a very precisely measured volume.

THE ROTATING METER

This type of meter is available with different measurement devices such as a lobed impeller, a rotor, or a star rotor. Each rotation of the impeller or rotor gives an accurate measurement of flow due to the precise, close fitting of the operating parts of the meter.

THE RECIPROCATING PISTON METER

This meter is very similar in design to a reciprocating steam engine piston and cylinder. The flow enters the cylinder and forces the piston down; this vertical movement is transferred to a rotary movement by way of a crankshaft. The flow inlet and discharge are regulated by slide valves. Every rotation of the shaft represents a volumetric quantity.

Miscellaneous Flow Measuring Devices

PITOT TUBE FLOWMETERS

This type of flowmeter measures fluid velocity at one point in the pipe. It consists of a tube with two small holes perpendicular of each other, one of which faces the flow. The bottom of the tube is sealed. The differential pressure across these holes is proportional to the velocity of flow. Quantity rate measurement is calculated from the ratio of average velocity to the velocity at the point of measurement. See Figure 1-97 for a detailed drawing of a Pitot tube flowmeter, which requires a sufficiently long straight run of pipe upstream of the tube. Other designs are available in combination with Venturi tubes, especially where a higher differential pressure than that produced by impact pressure alone is required. Usually, signal transmission for Pitot tube flowmeters is accomplished by differential pressure transmitters.

ANNUBAR™ TUBE FLOWMETER

This flow measuring device is a simple but unique design based upon the principle of the fundamental Bernoulli theorem on flow. Its operation is almost identical to that of a Pitot tube flow element, except that there are two tubes instead of one. One probe tube has four sensing ports and it faces the flow. A second probe

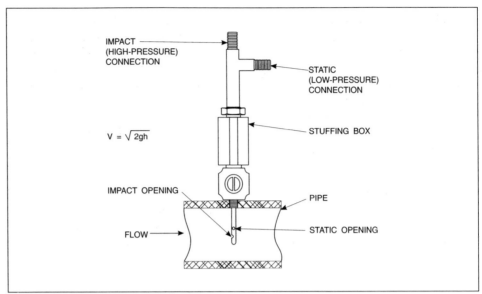

Figure 1-97. Pitot Tube Flow Sensor

behind the first, with its opening facing downstream, senses the static pressure. The impact probe senses each annular segment of the pipe interior, and an internal equalizing tube senses the average of the four pressures. A description of this flow element is shown in Figure 1-98. A differential transmitter is employed to transmit the flow-sensing signals to a remote controller or indicator.

ELBOW TAPS

Another economical approach to flow measurement is the installation of pressure-sensing taps in a pipe elbow. A differential pressure is developed by a centrifugal force as the direction of the fluid is changed in a pipe elbow. Because of one basic law, which states that the force is directly proportional to the magnitude of the acceleration and to the mass so accelerated, any material, such as a fluid flowing in a pipe, that is made to change direction will require that a force

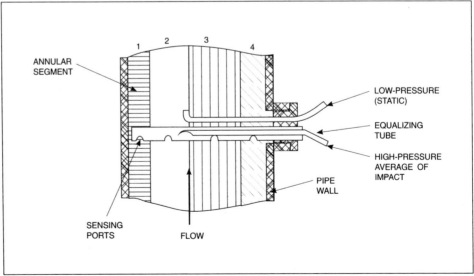

Figure 1-98. Annubar™ Tube Flow Sensor

be applied to it. Without this force, no acceleration can take place (see Figure 1-99).

If an elbow is inserted into a pipe line, then the liquid will be accelerated as it changes direction. However, if the elbow has the same cross-sectional area as the pipe, no change of speed will occur. The accelerating force will be a maximum where the change in velocity is the greatest at the "outer" curve of the elbow. There will be a reduction in pressure at the opposite curve on the inside of the bend. Taps are located either at 45° or a 22-1/2° The differential pressure is proportional to the flow rate, and this pressure is transmitted by a dP cell transmitter.

FLOW SWITCHES

Some parts of the process require a flow or no-flow signal to annunciate, to start and stop equipment, or for status indication. Several types of flow switches are available for particular applications. Most of the flow devices are based on the operating principle of the previously discussed flow-sensing devices, such as target types, plunger types, magnetic vane-actuated types, rotameter types, nuclear gage types, and ultrasonic types. All switches are available with the standard contacts: SPST, SPDT, DPST, and DPDT.

Magnetic, Thermal, and Ultrasonic Flowmeters

MAGNETIC FLOWMETERS

Magnetic flowmeters are widely used to measure the flow rate of conductive liquids in process applications. Recent technological developments have made them relatively easy to install and, often, more economical than other designs. A primary advantage of this type of instrument is that magnetic flowmeters provide virtually unobstructed flow. Because the flow path is relatively unobstructed, they are well suited to use with high-viscosity fluids and process liquids that contain solids. Important applications include flow measurement of sludge in sewage treatment plants, quarries in mining operations, and liquid metals in various industrial processes.

In addition, magnetic flowmeters do not create pressure drop. Velocity is measured directly, and, thus, variations in density do not affect their accuracy. Further, these instruments are free of Reynolds number constraints; they may be used for measuring the velocity of liquids with any flow profile.

In general, magnetic flowmeters are accurate, reliable measurement devices that do not intrude into the system. Often, these instruments can be maintained without shutting down the process. Further, since they produce an electrical output, this type of flowmeter is compatible with electronic control systems.

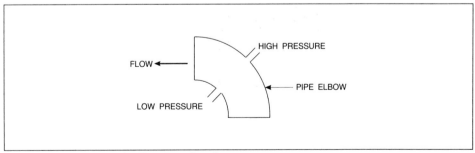

Figure 1-99. Elbow Taps for Flow Rate Sensing

PRINCIPLE OF OPERATION

Magnetic flowmeters operate on the principle of Faraday's law of electromagnetic induction. This law states that the voltage (E) induced in a conductor moving through a magnetic field at right angles to the field is directly proportional to the number of conductors, or, as in this case, the distance between the probes (D), the intensity of the magnetic field (B), and the velocity of the motion of the conductor (v).

$$E = \text{Constant} \times D \times B \times v$$

When these three factors are present, an electrical voltage is induced in a conductor that is moving through a magnetic field. The faster the conductor moves through the magnetic field, the greater the voltage induced in the conductor.

Figure 1-100 demonstrates how a magnetic flowmeter applies this principle. The flowmeter itself consists of a straight length of pipe. This is called the metering section; it is made of nonmagnetic material. In this case, the process fluid serves as the conductor. The direction of the induced voltage will be perpendicular to both the motion of the conductor and the magnetic field. The magnetic field is produced by magnets that are positioned on opposite sides of the short length of the nonmagnetic pipe.

As the fluid passes through the magnetic field, a voltage is generated. Two electrodes that project through the metering tube lining pick up the induced voltage. Wires connected to the electrodes can be attached to a meter that will measure the amount of voltage produced. Since the magnetic field is constant, the induced voltage is directly proportional to the velocity of the liquid and, thus, to

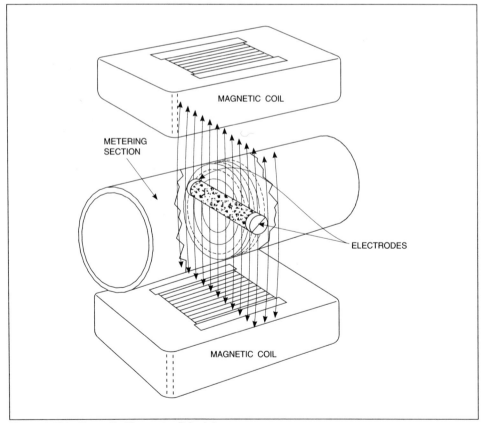

Figure 1-100. Magnetic Flowmeter Principle

the volumetric flow rate of the liquid passing through the meter. This measured voltage can be used to calculate flow rate.

In pipes, the induced voltage can be determined from Faraday's equation:

$$E = \text{Constant} \times B \times D \; v$$
$$E = (\text{Constant} \times B \times 4/\pi \times D) \times Q$$

where Q is determined by the equation $Q = A \times v$ and A is $\pi \times D^2/4$.

All the terms in parentheses are held constant in a magnetic flowmeter. Therefore, the induced voltage output is linearly proportional to the changes in flow rate (Q). This rate of flow can be calculated and indicated as flow rate, volumetric flow rate, or mass flow rate.

AC MAGNETIC FLOWMETERS

Alternating current (AC) magnetic flowmeters excite the electromagnetic field with AC current. A typical AC magnetic flowmeter is shown in Figure 1-101.

Two problems must be considered when using AC flowmeters to measure flow. The first is signal distortion and interference caused by extraneous voltages or noises. Noise may either be induced by or already present in the system. Because voltages induced by the electromagnetic field are relatively small in comparison to the extraneous voltages, measurement accuracy can be seriously affected.

Noise may be produced within the meter or within the process. The liquid itself may bear trace voltages, induced from exposure to a source of electromagnetic interference upstream, or a static charge that may be produced by friction. Stray voltages caused by this noise can be picked up by the electrodes and can introduce substantial inaccuracies in the measurement. Such sources of noise can be partially eliminated with a zero adjustment in the measuring instrument. The zero must be adjusted when the flowmeter is full of process fluid at zero flow.

The second problem that can adversely affect the accuracy of AC flowmeter measurements is that the sensitivity of the electrodes may be reduced if the electrodes become coated with a nonconductive material, either from electrolytic by-products of the process or by clinging process material. Buildup usually occurs gradually, but the effects over time can be significant.

Some magnetic flowmeters are available with removable electrodes. This option allows the electrodes to be inspected periodically and cleaned, if necessary, without dismantling the meter. Ultrasonic cleaning systems are also available to remove accumulated nonconductive material from electrodes without opening the system.

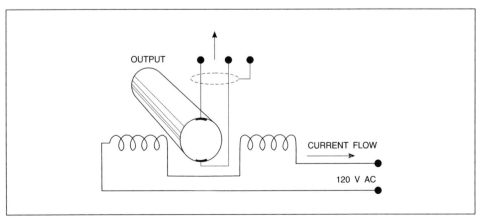

Figure 1-101. AC Magnetic Flowmeter

DC MAGNETIC FLOWMETERS

Direct current (DC) magnetic flowmeters excite the electromagnetic field with a DC current. A typical arrangement of a DC magnetic flowmeter is shown in Figure 1-102. When the liquid flows through the magnetic field, a voltage is induced in the liquid. The voltage is picked up at the electrodes. The voltages measured represent the sum of the flow induced in the moving conductor and the noise. The device is zeroed when there is no process flow and the electrodes detect only extraneous voltages. With flow restored, the output of the meter will indicate only voltages that are induced by the process.

DC magnetic flowmeters are not subject to inaccuracies due to the coating of electrodes. As long as electrode sensitivity remains high enough for a DC flowmeter to operate, its performance is relatively unaffected. Miniature DC magnetic flowmeters are also widely used in miniaturized electronic circuits. These instruments weigh less and have reduced power requirements.

THERMAL FLOWMETERS

In a thermal flowmeter, flow rate is measured either by monitoring the cooling action of the flow on a heated element placed in the flow or by the transfer of heat energy between two points along the flow path. Hot wire anemometers and calorimetric flowmeters are two common types of thermal flowmeters.

Either type can be used to measure mass flow rate, making both types especially suitable for gas applications. However, neither instrument is grouped with mass flowmeters because they do not measure mass directly; rather, mass is inferred from the thermal behavior and properties of the fluid.

HOT WIRE ANEMOMETERS

Hot wire anemometers have probes inserted into the process flow. These probes are usually connected in a typical bridge circuit. In the configuration shown in Figure 1-103, one of two probes is heated to a specific temperature. The second probe measures the temperature of the fluid. As the flow increases, it causes a heat loss in the heated probe. Consequently, more current is required to maintain the probe at the correct temperature. The increase in current flow reflects the energy necessary to compensate for the heat loss from the probe that was caused by the changing fluid flow. This change in current flow can be measured and used to calculate mass flow rate.

Proper operation requires that thermal conductivity (the ability of the heat to be conducted from the probe to the fluid) and heat capacity (the quantity of heat that a given mass requires to raise its temperature a specified amount) are assumed to be constant.

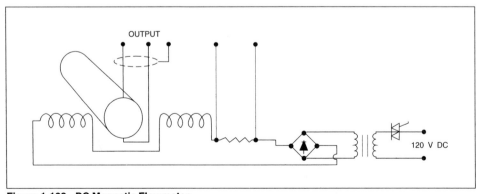

Figure 1-102. DC Magnetic Flowmeter

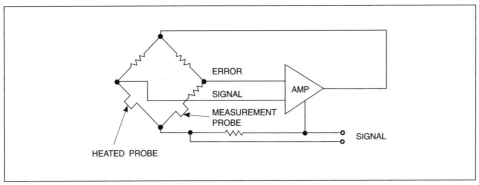

Figure 1-103. Hot Wire Anemometer

CALORIMETRIC FLOWMETERS

Calorimetric flowmeters work on the principle of heat transfer by the flow of fluid. Typically, calorimetric flowmeters are situated along the direction of the flow. Figure 1-104 illustrates this configuration. A heating element is placed in the flow. A sensor is positioned to measure the temperature upstream of the device; a second measuring device reads the temperature of the flow downstream from the heater. The rate of flow is determined by the difference in the two temperatures.

With a constant power input, this difference in temperature is a linear function of the mass flow and the heat capacity. The flowmeter can then be calibrated to indicate directly in mass flow units.

Applications of thermal flowmeters are limited to use with fluids that have known heat capacities. Usually, these are clean gases or clean mixtures of pure gases of known composition where heat capacity is known and is constant during flowmeter operation. Liquid applications are less common because liquids generally contain more impurities than gases.

Thermal flowmeter designs can measure fluid flow at temperatures as high as 450° Celsius, although most have a temperature rating between 100 and 150 degrees Celsius. Pressure ratings are normally limited by the pressure rating of the flange or connection.

Conductive surfaces of a thermal flowmeter can become contaminated and should be routinely cleaned to maintain performance levels. In addition, thermal flowmeters are sensitive to the thermal conductivity between the probes and the fluid, and any change will affect the measurement. Therefore, these devices are

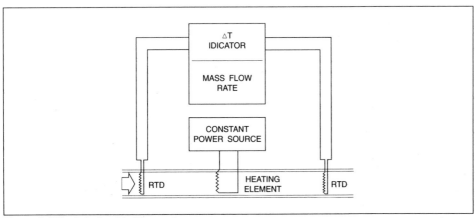

Figure 1-104. Calorimetric Flowmeter

generally not applied to abrasive fluid service. Flowmeters are available with replaceable sensors.

ULTRASONIC FLOWMETERS

Ultrasonic flow instruments measure the velocity of sound as it passes through the fluid flowing in a pipe. Some designs allow measurements to be made external to the pipe, or they may be mounted in a section of pipe that is installed in the system.

Ultrasonic devices use sound waves or vibrations to detect the flow in a pipe. The measurement is based on the time of flight of the sound waves. Pulses are transmitted along and against the fluid flow. The transmitted beam is usually projected at an angle in the pipe. There are usually two transducers, one located upstream of the other. The time of transit of the ultrasonic beam is measured and used to calculate the flow through the pipe.

Two types of transducers are generally used in ultrasonic flowmeter application: clamp-on transducers and inserted transducers. Clamp-on transducers are attached to the exterior of the pipe (see Figure 1-105(a)). Since the transducers are not in contact with the process fluid, the materials of construction are not a consideration. Clamp-on transducers are capable of operating faster than inserted transducers because the sonic echo is away from the receiver in clamp-on applications and not caught up in an "echo change," as is the case with transducers that face each other.

Inserted or wetted ultrasonic transducers (see Figure 1-105(b)) are considered more accurate than the clamp-on type. Direct contact with the liquid results in a superior signal-to-noise ratio. However, variations in sonic velocity due to changes in fluid properties will affect the performance of the flowmeter.

In addition, since the transducer is in contact with the process fluid, the materials of construction must not react with the process. Stainless steel is typically used for ultrasonic transducers. Some designs allow for removal of the transducer while fluid is flowing in the pipe. Others require that the pipe be drained before removing the transducer.

Clamp-on transducers can be designed to operate in the shear or the axial beam mode, both of which are illustrated in Figure 1-106. In the shear mode, the

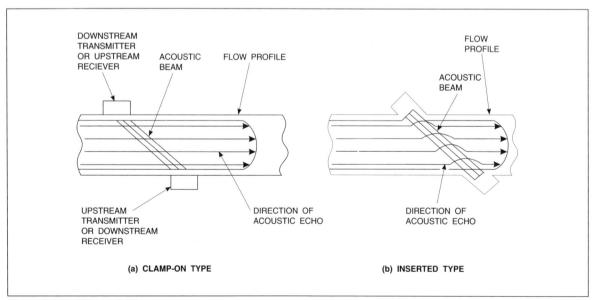

Figure 1-105. Ultrasonic Flowmeter

ultrasonic energy is focused into the liquid so that the signal at the receiver shifts positions. As the flow varies, the signal can miss the receiver and not be sensed.

Axial beam injection (see Figure 1-106) avoids this problem by focusing the energy axially along the length of the pipe, covering the receiver with the signal. Thus, the flowmeter is less sensitive to changes in the sonic properties of the liquid.

Time of flight types of devices can measure flows of clean liquids as well as liquids that contain up to 30 percent solids. Clamp-on designs, however, require that the pipe be sonically conductive, since the ultrasonic energy is transmitted to and received from the gap or interface between the transmitter and the pipe wall or the sensor and the pipe wall.

Thermometers

Thermometers are classified as mechanical temperature-sensing devices because they produce some type of mechanical action or movement in response to temperature changes. There are many types of thermometers, including the familiar liquid-in-glass thermometers; liquid-, gas-, and vapor-filled systems; and bimetallic thermometers.

LIQUID-IN-GLASS THERMOMETERS

Liquid-in-glass thermometers can be read directly and are very accurate and stable when used properly. Consequently, these thermometers are often used in laboratories to monitor baths and to check calibrations of other temperature sensors.

The features of a typical liquid-in-glass thermometer are shown in Figure 1-107. The bulb is usually a thin-walled glass chamber that serves as a reservoir for the liquid. The stem is a glass tube that contains the capillary for the liquid. A capillary is a narrow passage within which the liquid can rise and fall. The scale is a series of markings that is used to read the temperature. In addition, an immer-

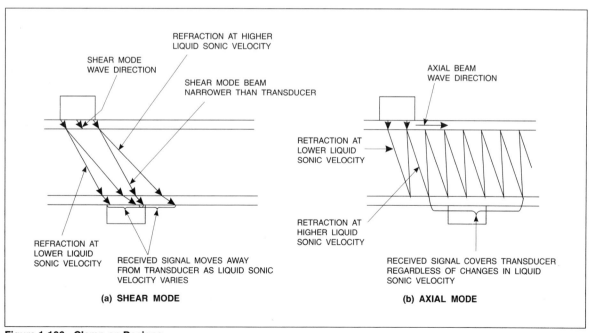

Figure 1-106. Clamp-on Designs

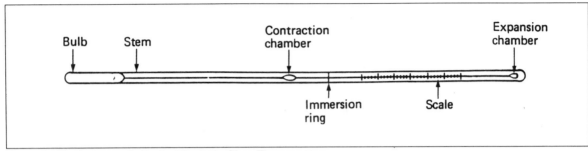

Figure 1-107. Features of a Typical Liquid-in-Glass Thermometer

sion ring may be provided to indicate the proper immersion depth on partial immersion thermometers.

A precision unit also includes contraction and expansion chambers. These chambers are enlargements of the capillary. The contraction chamber is located between the bulb and the scale. It increases the volume of the capillary and prevents total contraction of the fluid into the bulb at low temperatures. The expansion chamber is located beyond the top of the scale in order to contain fluid at high temperatures if it moves past the scale; in this way, the expansion chamber protects the thermometer from rupture at high temperatures. However, neither chamber is effective in cases of extreme high or low temperatures.

The operation of a liquid-in-glass thermometer depends on the difference in thermal expansion of the liquid and the glass. The liquid is usually mercury, which has a volume coefficient of expansion that is about eight times that of glass. For a given temperature change, the change in the length of the liquid column in the capillary will depend on the cross-sectional area of the capillary.

Since mercury freezes at $-38.9°F$, organic fluids, such as alcohol $(-80°F)$, toluene $(-130°F)$, or pentane $(-330°F)$ are used for low temperature measurements. Organic fluids are also used in inexpensive thermometers or in applications in which the release of mercury could not be tolerated in the event of breakage.

There are three basic types of liquid-in-glass thermometers: partial immersion, total immersion, and complete immersion. A partial immersion thermometer is inserted to a fixed point that is indicated by the immersion ring. This is the least accurate liquid-in-glass thermometer because the temperature of the stem and any capillary liquid that is above the immersion ring may differ significantly from the temperature of the immersed portion. Since the glass stem is exposed to different temperatures, this will cause a variation in the diameter of the capillary. It will also affect the column of liquid above the surface. Since the amount of variation will depend on the specific application, there is no way to avoid or compensate for the problem through calibration.

A total immersion thermometer is immersed to the height of the fluid column, not the entire length of the thermometer. Therefore, a total immersion thermometer does not usually have an immersion ring marking.

A complete or full immersion thermometer is totally submerged in the fluid to measured. These thermometers usually bear the inscription "complete immersion." They are often used in applications where the scale can be read through a glass wall, window, or port.

The accuracy of any thermometer depends on its construction, use, calibration, and scale markings. However, the deeper a thermometer is immersed, the more accurate its reading will be. The maximum achievable accuracy for industrial total immersion mercury-in-glass thermometers ranges from $0.01°C$ for lower temperature thermometers $(0-150°C)$ to $1°$ for higher temperature thermometers $(300-500°C)$. For partial immersion industrial thermometers, the maximum achievable accuracy ranges from 0.1°C for lower temperature thermometers (0 –150°C) to 2°C for higher temperature thermometers (300 –500°C).

In some cases, a total immersion thermometer may have to be used in a partial immersion mode. In these situations, accurate measurements require calculation for stem correction, in which case, the following equation is used:

$$dt = Kn(T_b - T)$$

where:

dt = temperature correction
K = temperature correction factor
n = number of degrees on the scale of the fluid being measured and the height of the liquid column in the capillary
T_b = bulb temperature
T = average temperature of the portion of the thermometer between the fluid surface and the end of the liquid column in the capillary

The correction factor depends on the type of liquid and the glass the thermometer is made of. Charts are available for determining the correction factor to be used for a specific thermometer.

The bulb temperature is the best estimate of the fluid temperature and is generally taken as the uncorrected thermometer reading. The value of T may be determined with a small auxiliary thermometer attached to the exposed proportion of the main thermometer. Once the correction factor, bulb temperature, and the value of T have been obtained, the equation for stem correction can be solved.

Suppose a total immersion borosilicate glass thermometer is partially immersed in a fluid. The mercury in the capillary rises to the 100°C mark on the scale, the fluid surface is at the 40°C mark on the thermometer, and a small thermometer attached to the exposed portion of the thermometer reads 25°C. In this case, the stem correction would be calculated as follows:

$$dt = 0.000164(60)\,(100 - 25)$$
$$= 0.74°C$$

FILLED THERMOMETERS

Filled thermometers contain a gas or a volatile liquid and rely on pressure measurements to provide temperature indications. There are several types of vapor- and gas-filled systems. Although each type of system is slightly different, they have similar components and share the same principle of operation.

Figure 1-108 shows the components of a typical filled thermometer: a bulb that is exposed to the fluid being measured, a capillary tube, a pressure element such as a Bourdon tube, and a scale. The bulb, capillary tube, and Bourdon tube are filled with a liquid, vapor, or gas. When the temperature changes, the fluid either expands or contracts, which causes the Bourdon tube to move, thereby moving the position of the needle on the scale.

A liquid-filled system is completely filled with a liquid. This type of system operates on the principle that liquid expands with an increase in temperature. When the liquid expands, it causes the pressure to increase, which causes the Bourdon tube to uncoil and move the needle on the scale.

Various liquids are used in liquid-filled systems. Typically, inert hydrocarbons such as xylene are used because of their low coefficient of expansion. In some

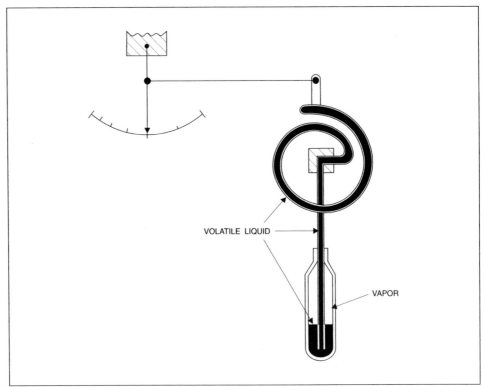

Figure 1-108. Class IIA Vapor System

cases, even water may be used. Another liquid commonly used is mercury. Mercury-filled systems are often considered to be in a separate class from liquid-filled systems because the unique characteristics of mercury offer several advantages over other liquid-filled systems. For example, mercury responds more quickly to temperature changes and has a higher degree of accuracy than other liquids.

A vapor system contains a volatile liquid and vapor. This type of system operates on the principle that pressure in a vessel containing only a liquid and its vapor increases with temperature and is independent of volume. In a vapor system, temperature is measured at the interface between the liquid and the vapor.

For proper operation of a vapor system, the interface must remain in the bulb. Certain conditions can affect the position of the interface. For instance, if the bulb temperature is below that of the capillary and the Bourdon tube, the bulb will be full of vapor and the liquid-vapor interface will not be within the bulb.

To handle different operating conditions, there are four subclasses of vapor systems. The class IIA vapor system (see Figure 1-108) is designed to operate with the measured temperature above the temperature of the rest of the system. In this system, the higher process temperature vaporizes the volatile liquid, which then condenses in the capillary and the Bourdon tube.

The Class IIB vapor system (see Figure 1-109) is designed to operate with the measured temperature below the temperature of the rest of the system. In this case, the capillary and Bourdon tube are filled with vapor.

The Class IIC vapor system (see Figure 1-110) is designed to measure temperatures above and below the temperature of the system. When bulb temperature is lower than the temperature of the rest of the system, the bulb will be mostly filled with liquid and the capillary and Bourdon tube will contain vapor. When bulb temperature is higher than the temperature of the rest of the system, the bulb will

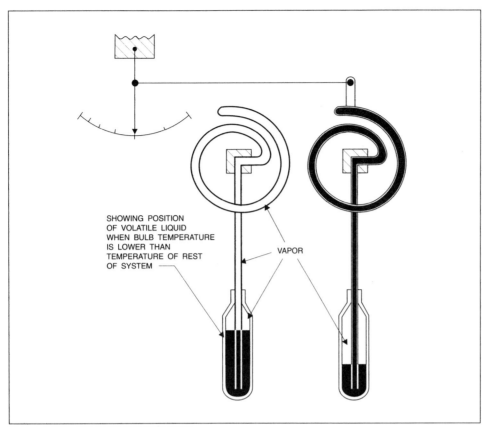

Figure 1-109. Class IIB Vapor System

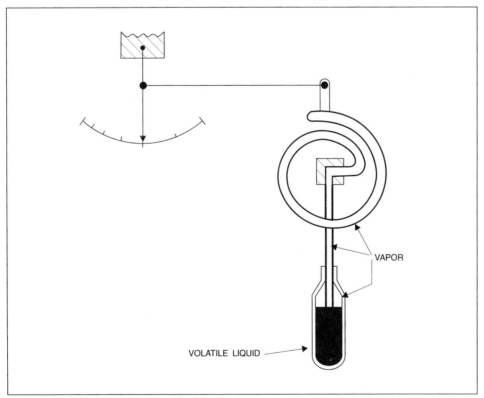

Figure 1-110. Class IIC Vapor System

be mostly filled with vapor, and the capillary and Bourdon tube will be filled with liquid.

With the Class IIC type of vapor system, there is a cross-ambient effect. When the temperature of the measured fluid changes so that the bulb temperature approaches and meets the temperature of the rest of the system, there will be a delay as the vapor and liquid reverse positions. Because of this cross-ambient effect, vapor system thermometers are often used either exclusively below ambient or exclusively above ambient.

The Class IID vapor system (see Figure 1-111) is designed to overcome the cross-ambient limitation by using a second nonvolatile liquid. In this case, the capillary and Bourdon tube remain full of the nonvolatile liquid. This liquid hydraulically transmits pressure changes to the Bourdon tube. These pressure changes result from changes in the interface of the vapor and the volatile liquid in the bulb.

Vapor systems have a relatively narrow temperature measurement span. The minimum and maximum temperatures for a system depend on the particular type of fluid used. Therefore, the type of fluid and class of system selected depend on the temperature range of the specific application.

Gas-filled systems are commonly used for industrial applications and, in some cases, for laboratory measurements. The operation of gas-filled systems is based on the ideal gas law and is an approximation at normally encountered temperatures and pressures. According to this law, the pressure of an ideal gas confined to a constant volume is proportional to absolute temperature. In a typical gas-filled system, the gas (usually nitrogen) is not perfect, so there may be a slight

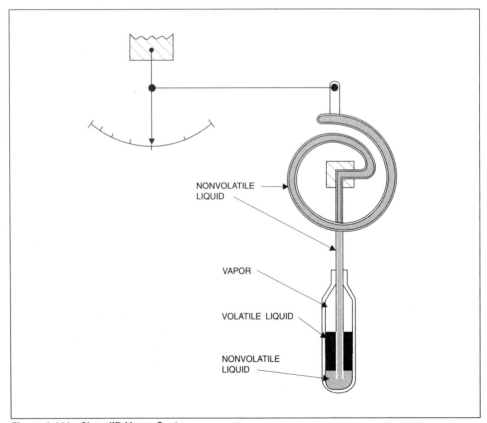

Figure 1-111. Class IID Vapor System

change in volume. However, these differences are minor and do not prevent the use of pressure measurement to indicate temperature.

The ideal gas law is exact at low temperatures and pressures. Therefore, when a gas-filled system is used as an absolute thermometer in a setting such as a laboratory, measurement is performed at low pressure.

BIMETALLIC THERMOMETERS

Bimetallic thermometers use the differences in the thermal expansion properties of metals to provide temperature measurement capability. Strips of metals with different thermal expansion coefficients are bonded together but at the same temperature. When the temperature increases, it causes the assembly to bend. When this happens, the metal strip with the larger temperature coefficient of expansion expands more than the other strip. The angular position versus temperature relation is established by calibration so that the device can be used as a thermometer. Refer to Figure 1-112.

Iron-nickle alloys are commonly used for bimetallic thermometers. One alloy, Invar™ (35% nickel), has a very low temperature coefficient of expansion. Many bimetallic thermometers use Invar as the low expansion coefficient metal and another iron-nickel alloy for the high expansion coefficient metal.

Typical commercial bimetallic thermometers usually use spiral or helical configurations of the bonded metal combination. One end is fixed, and the other is attached to a pointer that indicates temperature on a scale.

Bimetallic thermometers are inexpensive, rugged, and simple to use. They usually provide a visual temperature reading, rather than an electrical signal, that is suitable for automatic monitoring or recording. They are used for on-off control, overload, and cut-out switches.

SELECTION

Selection of a particular type of thermometer depends on the specific application and the degree of accuracy required. For example, liquid-in-glass thermometers can usually be read directly and are quite accurate when used correctly. However, because glass is fragile, liquid-in-glass thermometers must be handled carefully and stored properly in order to obtain stable readings and avoid breakage. Filled and bimetallic thermometers are generally more durable than liquid-in-glass thermometers. However, they are slightly less accurate.

The range of measured temperatures, which depends on the materials of construction and the type of fill fluid used, also varies with the type of thermometer.

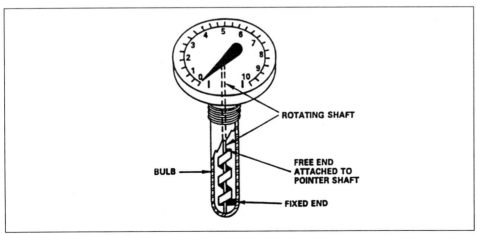

Figure 1-112. Bimetallic Sensor

Thermocouples — Part I

One type of electrical temperature sensor is called a thermocouple. Thermocouples are simple, inexpensive, durable, and relatively accurate sensors that can be used in a wide variety of applications and environmental conditions. They provide a means for sensing temperature in many different types of processes, including power systems, refineries, aeronautical systems, and nuclear reactors. As a group, they have a wide measurement range, covering temperatures from below $-183°C$ to about $2500°C$, with an accuracy of $0.1–1\%$ of absolute temperature.

Most thermocouples are made up of metal wires assembled in a variety of styles to provide direct or differential temperature measurements, ruggedness, durability, and circuit isolation. Thermocouples have a rapid dynamic response to temperature change. However, for protection and ease of calibration and removal, most thermocouples are installed in insulated thermowells, which delays dynamic response considerably.

LAWS OF THERMOCOUPLES

The behavior of thermocouples is described by the Seebeck, Peltier, and Thomson effects.

The Seebeck Effect. The principle of the thermocouple was first described by Seebeck in 1821. Seebeck discovered that a current flowed when wires of two dissimilar metals were joined together to form a circuit of at least two junctions at different temperatures. This phenomenon, called the Seebeck effect, is the basis upon which thermocouples are designed.

The Peltier Effect. If a current flows across the junction of two dissimilar metal conductors that have the same temperature, the heat is either released or absorbed, depending on the direction of current flow. If the current flow is in the same direction as that produced by the Seebeck effect, heat is released at the hot junction and absorbed at the cold junction.

The Thomson Effect. The Thomson effect or the reversible heat effect is the result when an electric current passes through a conductor in which there is a temperature gradient. Heat is absorbed or liberated when current flows in a homogeneous material in which a temperature gradient exists. The effect is reversible with respect to current direction. The Thomson coefficient, watts/ampere/degree, or Thomson emf in volts per degree also depend upon material and temperature.

The laws of thermocouple behavior may be stated as follows:

(1) The thermal emf of a thermocouple with junctions at T_1 and T_2 is totally unaffected by temperature elsewhere in the circuit if the two metals used are each homogeneous.

(2) If a third homogeneous metal C is inserted into either A or B, as long as the two new thermojunctions are at like temperatures, the net emf of the circuit is unchanged irrespective of the temperature of C away from the junctions.

(3) If metal C is inserted between A and B at one of the junctions, the temperature of C at any point away from AC and AB is immaterial. So long as the junctions AC and AB are both at the temperature T_1, the net emf is the same as if C were not there.

(4) If the thermal emf of metals A and C is E_{AC} and that of metals B and C is E_{CB}, the thermal emf of metals A and B is $E_{AC} + E_{CB}$.

(5) If a thermocouple produces emf E_1 when its junctions are at T_1 and T_2, and E_2 when at T_2 and T_3, it will produce $E_1 + E_2$ when the junctions are at T_1 and T_3.

In its simplest form, a thermocouple consists of two wires, each made of a different homogeneous metal or alloy (see Figure 1-113). The wires are joined at one end to form a measuring junction. This measuring junction is exposed to the fluid or medium being measured. The other end of the wires are usually terminated at a measuring instrument, where they form a reference junction (see Figure 1-114). When the two junctions are at different temperatures, current will flow through the circuit. The millivoltage that results from the current flow is measured to determine the temperature of the measuring junction.

The reference junction is held at a constant, or reference, temperature. In many cases, the junction is kept at the temperature of melting ice, which allows temperature to be read directly from an indicator without the need for calculating a correction. In the laboratory setting, the 0°C reference temperature can be maintained using an ice bath (see Figure 1-115). In industrial settings, refrigeration units are often used. These units are usually designed to provide stable termination of groups of thermocouples. In other cases, the reference function may be kept at a temperature above ambient. This is usually accomplished by placing the reference junction in an oil bath or oven that is maintained at a constant temperature.

In many thermocouple installations, the measuring junction is several hundred feet from the voltage-measuring instrument at which the reference junction is located. Since the pure metals and metal alloys used for the thermocouples wires are relatively expensive, extension or lead wires are typically connected to the thermocouple and run to terminals at the measuring instrument. When appropriate extension wires are properly connected to the thermocouple, the reference junction is transferred to the point at which the extension wires are connected to the temperature controlling device.

Even with the use of appropriate extension wires, intermediate junctions of dissimilar metals are created in the circuit. However, the law of intermediate metals states that the algebraic sum of the thermoelectromotive forces in a circuit that is composed of any number of dissimilar materials is zero if the entire circuit is at a uniform temperature. This law emphasizes the point that the driving force in thermoelectricity is a difference of temperature. Therefore, in an actual installation in which there is a difference in temperature among intermediate junctions, the algebraic sum of the millivoltages will not be zero but a value related to the temperature differences at the junctions. For example, if the reference junction is at a point where the extension wires are connected to copper wires at the terminals that are at different (but known) temperatures, the temperature at the measuring junction can be calculated by referring to tables of measured values of junction millivoltages for particular metal combinations. In this way, compensation is provided for these additional junctions.

Typically, reference junction compensation is an integral feature of the instruments that measure the millivolt signals from thermocouples. In these instruments, a temperature-sensitive component is thermally bonded to the reference junction connections. The resistance–temperature curve of the bonded component matches the millivoltage–temperature curve of the thermocouple wires. The voltage change across the component is then equal and opposite to the reference junction thermal voltage over a wide ambient temperature range.

TYPES OF THERMOCOUPLES

About a dozen types of thermocouples are commonly used in industrial applications. Seven of these have been assigned letter designations by the Instrument Society of America (ISA). By convention, a slash mark is used to separate the materials of each thermocouple wire. For example, copper/constantan identifies a thermocouple with one copper wire and one constantan wire. The order in which the wire materials are listed identifies the polarity of the wires. The first

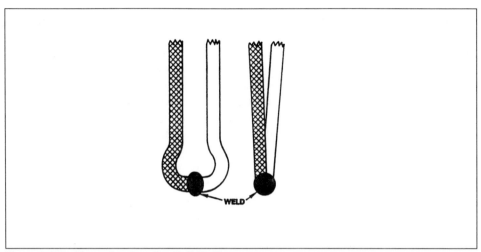

Figure 1-113. Measuring Junction Butt Welds

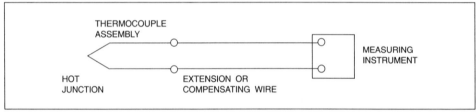

Figure 1-114. Typical Thermocouple Circuit

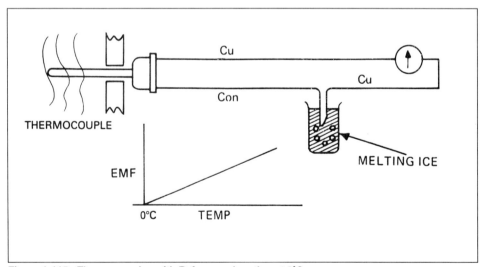

Figure 1-115. Thermocouples with Reference Junction at 0˚C

wire, on the left of the slash, has a positive polarity when the measuring junction is at a higher temperature than the reference junction.

Thermocouples can be divided into three functional classes: base metal, noble metal, and refractory metal. Base metal thermocouples are useful for measuring temperatures under 1000°C. This class includes thermocouples made of iron/constantan (Type J), copper/constantan (Type T), Chromel-Alumel™ (Type K), Chromel™/constantan (Type E), and alloys of copper, nickel, iron, chromium, manganese, aluminum, and other elements.

Noble metal thermocouples are useful to about 2000°C. This class includes tungsten-rhenium alloy thermocouples as well as those made of tantalum, molybdenum, and their alloys.

Each type of thermocouple has individual characteristics that make it desirable for some applications and unsuitable for others. Table 1-5 identifies the wire materials, nominal temperature scale, recommended atmosphere, and features of the more commonly used thermocouples. The ISA wire material designations, particularly for Type K, do not specify a particular alloy combination, only a temperature–voltage relationship.

Table 1-5.   Thermocouple Comparison Chart

Type	Thermocouple Element	Thermocouple Wire Materials*
B	BP BN	Platinum – 30% rhodium Platinum – 6% rhodium
E	EP EN	Nickel Chromium Copper Nickel
J	JP JN	Iron Copper Nickel
K	KP KN	Nickel Chromium Nickel Aluminum Silicon
R	RP RN	Platinum – 13% rhodium Platinum
S	SP SN	Platinum – 10% rhodium Platinum
T	TP TN	Copper Copper Nickel

It is important to note that the copper-nickel alloy used for the negative wire of the Type J thermocouple is not the same material used for the negative wires of Type T and Type E thermocouples. Although both materials are commonly referred to as constantan, they are not thermoelectrically equivalent. Another caution relates to Type S thermocouples. Prior to 1974, the positive wire of Type S thermocouples contained about 9.9% rhodium (nominal 10%). Since then, the National Bureau of Standards (now the National Institute of Standard and Technology) adopted a standard of 10.000.05% rhodium composition for the positive wire. To avoid errors, a "nominal" or "accurate" platinum-rhodium positive wire should be specified.

THERMOCOUPLE REFERENCE TABLES

Many thermocouple circuits contain devices that amplify the millivolt signals from the thermocouple and convert the voltage signals into direct temperature indications. However, some thermocouple circuits do not have converting circuits for direct temperature readout. In these cases, temperature–emf reference tables are available for use in converting voltmeter readings to the equivalent temperature values. Since different types of thermocouples have different thermoelectric relationships, there is a reference table for each type of thermocouple. The values in these tables are based on the International Practical Temperature Scale and U.S. legal electrical units. Tables are available from a variety of sources, including NBS Monograph 125, ANSI Standard MC 96.1, the ASTM Annual Book of Standards, and ISA standards.

A portion of the reference table for a Type J thermocouple is shown in Table 1-6. Thermocouple reference tables equate temperatures (in Celsius) to the voltages (in millivolts) produced by a thermocouple with its reference junction at 0°C and the measuring junction at various temperatures. Therefore, if the reference junction is maintained at 0°C, the temperature that corresponds to the measured voltage can be read directly from the table. For example, if the measured voltage produced by a Type J thermocouple with its reference junction at 0°C is 10.000 mV, the temperature of the measuring junction is 186°C.

> **Suppose the output of a Type J thermocouple is 5.340 mV and the temperature of the reference junction is 20°C. According to the table, the correction factor for 20°C is 1.019 mV. Adding this value to the output of the thermocouple (5.340 mV) produces a corrected output of 6.359 mV. Using this corrected millivolt output value, the temperature of the measuring junction is 120°C.**

If the reference junction of a thermocouple is not maintained at 0°C, the tables can still be used by applying an appropriate correction to compensate for the difference between the reference junction temperature and 0°C. Compensation is required because the millivoltage produced by a thermocouple is decreased. The value of the correction can be obtained from the table. The correction factor is the millivolt value for the reference junction temperature. This value is algebraically added to the millivolt output of the thermocouple. The resulting value is used to determine the temperature of the measuring junction.

Thermocouples — Part II

THERMOCOUPLE DESIGN

The wires used in the fabrication of thermocouples are a critical factor in the accuracy of a thermocouple. Commercially fabricated thermocouples can be purchased, or thermocouples may be assembled on site. Commercially fabricated thermocouples are carefully tested to ensure that they conform to specified calibration limits. When thermocouples are to be assembled on site, thermocouple-grade wire that is low in impurities and has uniform low resistance should be used.

Thermocouples are fabricated using bare, insulated, and sheathed wires. The wire sizes normally used for thermocouples are as follows:

- Types E, J, K — 8, 14, 20, 24, 28, AWG

- Types T — 14, 20, 24, 28 AWG

- Types B, R, S — 24 AWG only

The smaller the diameter of the wire, the quicker the response time of the thermocouple, because heat transfer takes less time in a thermocouple with smaller mass. However, larger wires have greater resistance to corrosion and high temperature, which can cause a shift in calibration. For this reason, as wire size

Table 1-6. Reference Table for Type J Thermocouples

TEMPERATURES IN DEGREES CELSIUS (IPTS-68) REFERENCE JUNCTIONS AT 0°C

THERMOELECTRIC VOLTAGE IN MILLIVOLTS

DEG C	0	1	2	3	4	5	6	7	8	9	10	DEG C
-210	-8.096											-210
-200	-7.890	-7.912	-7.934	-7.955	-7.976	-7.996	-8.017	-8.037	-8.057	-8.076	-8.096	-200
-190	-7.659	-7.683	-7.707	-7.731	-7.755	-7.778	-7.801	-7.824	-7.846	-7.868	-7.890	-190
-180	-7.402	-7.429	-7.455	-7.482	-7.508	-7.533	-7.559	-7.584	-7.609	-7.634	-7.659	-180
-170	-7.122	-7.151	-7.180	-7.209	-7.237	-7.266	-7.293	-7.321	-7.348	-7.375	-7.402	-170
-160	-6.821	-6.852	-6.883	-6.914	-6.944	-6.974	-7.004	-7.034	-7.064	-7.093	-7.122	-160
-150	-6.499	-6.532	-6.565	-6.598	-6.630	-6.663	-6.695	-6.727	-6.758	-6.790	-6.821	-150
-140	-6.159	-6.194	-6.228	-6.263	-6.297	-6.331	-6.365	-6.399	-6.433	-6.466	-6.499	-140
-130	-5.801	-5.837	-5.874	-5.910	-5.946	-5.982	-6.018	-6.053	-6.089	-6.124	-6.159	-130
-120	-5.426	-5.464	-5.502	-5.540	-5.578	-5.615	-5.653	-5.690	-5.727	-5.764	-5.801	-120
-110	-5.036	-5.076	-5.115	-5.155	-5.194	-5.233	-5.272	-5.311	-5.349	-5.388	-5.426	-110
-100	-4.632	-4.673	-4.714	-4.755	-4.795	-4.836	-4.876	-4.916	-4.956	-4.996	-5.036	-100
-90	-4.215	-4.257	-4.299	-4.341	-4.383	-4.425	-4.467	-4.508	-4.550	-4.591	-4.632	-90
-80	-3.785	-3.829	-3.872	-3.915	-3.958	-4.001	-4.044	-4.087	-4.130	-4.172	-4.215	-80
-70	-3.344	-3.389	-3.433	-3.478	-3.522	-3.566	-3.610	-3.654	-3.698	-3.742	-3.785	-70
-60	-2.892	-2.938	-2.984	-3.029	-3.074	-3.120	-3.165	-3.210	-3.255	-3.299	-3.344	-60
-50	-2.431	-2.478	-2.524	-2.570	-2.617	-2.663	-2.709	-2.755	-2.801	-2.847	-2.892	-50
-40	-1.960	-2.008	-2.055	-2.102	-2.150	-2.197	-2.244	-2.291	-2.338	-2.384	-2.431	-40
-30	-1.481	-1.530	-1.578	-1.626	-1.674	-1.722	-1.770	-1.818	-1.865	-1.913	-1.960	-30
-20	-0.995	-1.044	-1.093	-1.141	-1.190	-1.239	-1.288	-1.336	-1.385	-1.433	-1.481	-20
-10	-0.501	-0.550	-0.600	-0.650	-0.699	-0.748	-0.798	-0.847	-0.896	-0.945	-0.995	-10
-0	0.000	-0.050	-0.101	-0.151	-0.201	-0.251	-0.301	-0.351	-0.401	-0.451	-0.501	-0
0	0.000	0.050	0.101	0.151	0.202	0.253	0.303	0.354	0.405	0.456	0.507	0
10	0.507	0.558	0.609	0.660	0.711	0.762	0.813	0.865	0.916	0.967	1.019	10
20	1.019	1.070	1.122	1.174	1.225	1.277	1.329	1.381	1.432	1.484	1.536	20
30	1.536	1.588	1.640	1.693	1.745	1.797	1.849	1.901	1.954	2.006	2.058	30
40	2.058	2.111	2.163	2.216	2.268	2.321	2.374	2.426	2.479	2.532	2.585	40
50	2.585	2.638	2.691	2.743	2.796	2.849	2.902	2.956	3.009	3.062	3.115	50
60	3.115	3.168	3.221	3.275	3.328	3.381	3.435	3.488	3.542	3.595	3.649	60
70	3.649	3.702	3.756	3.809	3.863	3.917	3.971	4.024	4.078	4.132	4.186	70
80	4.186	4.239	4.293	4.347	4.401	4.455	4.509	4.563	4.617	4.671	4.725	80
90	4.725	4.780	4.834	4.888	4.942	4.996	5.050	5.105	5.159	5.213	5.268	90
100	5.268	5.322	5.376	5.431	5.485	5.540	5.594	5.649	5.703	5.758	5.812	100
110	5.812	5.867	5.921	5.976	6.031	6.085	6.140	6.195	6.249	6.304	6.359	110
120	6.359	6.414	6.468	6.523	6.578	6.633	6.688	6.742	6.797	6.852	6.907	120
130	6.907	6.962	7.017	7.072	7.127	7.182	7.237	7.292	7.347	7.402	7.457	130
140	7.457	7.512	7.567	7.622	7.677	7.732	7.787	7.843	7.898	7.953	8.008	140
150	8.008	8.063	8.118	8.174	8.229	8.284	8.339	8.394	8.450	8.505	8.560	150
160	8.560	8.616	8.671	8.726	8.781	8.837	8.892	8.947	9.003	9.058	9.113	160
170	9.113	9.169	9.224	9.279	9.335	9.390	9.446	9.501	9.556	9.612	9.667	170
180	9.667	9.723	9.778	9.834	9.889	9.944	10.000	10.055	10.111	10.166	10.222	180
190	10.222	10.277	10.333	10.388	10.444	10.499	10.555	10.610	10.666	10.721	10.777	190
200	10.777	10.832	10.888	10.943	10.999	11.054	11.110	11.165	11.221	11.276	11.332	200
210	11.332	11.387	11.443	11.498	11.554	11.609	11.665	11.720	11.776	11.831	11.887	210
220	11.887	11.943	11.998	12.054	12.109	12.165	12.220	12.276	12.331	12.387	12.442	220
230	12.442	12.498	12.553	12.609	12.664	12.720	12.776	12.831	12.887	12.942	12.998	230
240	12.998	13.053	13.109	13.164	13.220	13.275	13.331	13.386	13.442	13.497	13.553	240
250	13.553	13.608	13.664	13.719	13.775	13.830	13.886	13.941	13.997	14.052	14.108	250
260	14.108	14.163	14.219	14.274	14.330	14.385	14.441	14.496	14.552	14.607	14.663	260
270	14.663	14.718	14.774	14.829	14.885	14.940	14.995	15.051	15.106	15.162	15.217	270
280	15.217	15.273	15.328	15.383	15.439	15.494	15.550	15.605	15.661	15.716	15.771	280
290	15.771	15.827	15.882	15.938	15.993	16.048	16.104	16.159	16.214	16.270	16.325	290
300	16.325	16.380	16.436	16.491	16.547	16.602	16.657	16.713	16.768	16.823	16.879	300
310	16.879	16.934	16.989	17.044	17.100	17.155	17.210	17.266	17.321	17.376	17.432	310
320	17.432	17.487	17.542	17.597	17.653	17.708	17.763	17.818	17.874	17.929	17.984	320
330	17.984	18.039	18.095	18.150	18.205	18.260	18.316	18.371	18.426	18.481	18.537	330
340	18.537	18.592	18.647	18.702	18.757	18.813	18.868	18.923	18.978	19.033	19.089	340
DEG C	0	1	2	3	4	5	6	7	8	9	10	DEG C

Table 1-6 (continued)

TEMPERATURES IN DEGREES CELSIUS (IPTS-68)   REFERENCE JUNCTIONS AT 0°C

THERMOELECTRIC VOLTAGE IN MILLIVOLTS

DEG C	0	1	2	3	4	5	6	7	8	9	10	DEG C
350	19.089	19.144	19.199	19.254	19.309	19.364	19.420	19.475	19.530	19.585	19.640	350
360	19.640	19.695	19.751	19.806	19.861	19.916	19.971	20.026	20.081	20.137	20.192	360
370	20.192	20.247	20.302	20.357	20.412	20.467	20.523	20.578	20.633	20.688	20.743	370
380	20.743	20.798	20.853	20.909	20.964	21.019	21.074	21.129	21.184	21.239	21.295	380
390	21.295	21.350	21.405	21.460	21.515	21.570	21.625	21.680	21.736	21.791	21.846	390
400	21.846	21.901	21.956	22.011	22.066	22.122	22.177	22.232	22.287	22.342	22.397	400
410	22.397	22.453	22.508	22.563	22.618	22.673	22.728	22.784	22.839	22.894	22.949	410
420	22.949	23.004	23.060	23.115	23.170	23.225	23.280	23.336	23.391	23.446	23.501	420
430	23.501	23.556	23.612	23.667	23.722	23.777	23.833	23.888	23.943	23.999	24.054	430
440	24.054	24.109	24.164	24.220	24.275	24.330	24.386	24.441	24.496	24.552	24.607	440
450	24.607	24.662	24.718	24.773	24.829	24.884	24.939	24.995	25.050	25.106	25.161	450
460	25.161	25.217	25.272	25.327	25.383	25.438	25.494	25.549	25.605	25.661	25.716	460
470	25.716	25.772	25.827	25.883	25.938	25.994	26.050	26.105	26.161	26.216	26.272	470
480	26.272	26.328	26.383	26.439	26.495	26.551	26.606	26.662	26.718	26.774	26.829	480
490	26.829	26.885	26.941	26.997	27.053	27.109	27.165	27.220	27.276	27.332	27.388	490
500	27.388	27.444	27.500	27.556	27.612	27.668	27.724	27.780	27.836	27.893	27.949	500
510	27.949	28.005	28.061	28.117	28.173	28.230	28.286	28.342	28.398	28.455	28.511	510
520	28.511	28.567	28.624	28.680	28.736	28.793	28.849	28.906	28.962	29.019	29.075	520
530	29.075	29.132	29.188	29.245	29.301	29.358	29.415	29.471	29.528	29.585	29.642	530
540	29.642	29.698	29.755	29.812	29.869	29.926	29.983	30.039	30.096	30.153	30.210	540
550	30.210	30.267	30.324	30.381	30.439	30.496	30.553	30.610	30.667	30.724	30.782	550
560	30.782	30.839	30.896	30.954	31.011	31.068	31.126	31.183	31.241	31.298	31.356	560
570	31.356	31.413	31.471	31.528	31.586	31.644	31.702	31.759	31.817	31.875	31.933	570
580	31.933	31.991	32.048	32.106	32.164	32.222	32.280	32.338	32.396	32.455	32.513	580
590	32.513	32.571	32.629	32.687	32.746	32.804	32.862	32.921	32.979	33.038	33.096	590
600	33.096	33.155	33.213	33.272	33.330	33.389	33.448	33.506	33.565	33.624	33.683	600
610	33.683	33.742	33.800	33.859	33.918	33.977	34.036	34.095	34.155	34.214	34.273	610
620	34.273	34.332	34.391	34.451	34.510	34.569	34.629	34.688	34.748	34.807	34.867	620
630	34.867	34.926	34.986	35.046	35.105	35.165	35.225	35.285	35.344	35.404	35.464	630
640	35.464	35.524	35.584	35.644	35.704	35.764	35.825	35.885	35.945	36.005	36.066	640
650	36.066	36.126	36.186	36.247	36.307	36.368	36.428	36.489	36.549	36.610	36.671	650
660	36.671	36.732	36.792	36.853	36.914	36.975	37.036	37.097	37.158	37.219	37.280	660
670	37.280	37.341	37.402	37.463	37.525	37.586	37.647	37.709	37.770	37.831	37.893	670
680	37.893	37.954	38.016	38.078	38.139	38.201	38.262	38.324	38.386	38.448	38.510	680
690	38.510	38.572	38.633	38.695	38.757	38.819	38.882	38.944	39.006	39.068	39.130	690
700	39.130	39.192	39.255	39.317	39.379	39.442	39.504	39.567	39.629	39.692	39.754	700
710	39.754	39.817	39.880	39.942	40.005	40.068	40.131	40.193	40.256	40.319	40.382	710
720	40.382	40.445	40.508	40.571	40.634	40.697	40.760	40.823	40.886	40.950	41.013	720
730	41.013	41.076	41.139	41.203	41.266	41.329	41.393	41.456	41.520	41.583	41.647	730
740	41.647	41.710	41.774	41.837	41.901	41.965	42.028	42.092	42.156	42.219	42.283	740
750	42.283	42.347	42.411	42.475	42.538	42.602	42.666	42.730	42.794	42.858	42.922	750
760	42.922											760
DEG C	0	1	2	3	4	5	6	7	8	9	10	DEG C

decreases, the upper temperature limit at which a thermocouple can be used will also decrease.

Bare wire thermocouples are typically insulated with hard-fired ceramic insulators (see Figure 1-116). Insulators are available in single bore, double bore, or multi-bore as well as in a variety of sizes, shapes, and lengths. For noble metal thermocouples (Types B, R, and S), the use of one-piece, full-length aluminum oxide insulators is recommended for maximum protection (Types E, J, K, T). Insulation of braided glass or other fabric is sometimes used. Thermocouples are also made from insulated thermocouple wire. In this case, the insulation must be suitable for the exposure temperature and must not contaminate the process fluid.

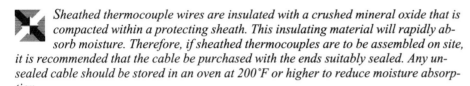 *Sheathed thermocouple wires are insulated with a crushed mineral oxide that is compacted within a protecting sheath. This insulating material will rapidly absorb moisture. Therefore, if sheathed thermocouples are to be assembled on site, it is recommended that the cable be purchased with the ends suitably sealed. Any unsealed cable should be stored in an oven at 200°F or higher to reduce moisture absorption.*

For identification purposes, symbols have been assigned for the positive (P) and negative (N) wires of thermocouples. Table 1-7 shows the wire symbols of different types of thermocouples. For example, BP indicates the positive lead of a Type B thermocouple, while BN indicates the negative lead of a Type B thermocouple.

Table 1-7. Symbols for Types of Thermocouple Wires

Type*	Thermoelements	
	Positive	Negative
B	BP	BN
E	EP	EN
J	JP	JN
K	KP	KN
R	RP	RN
S	SP	SN
T	TP	TN

* Any thermocouple material having temperature–emf relationships within the tolerances for any of the above-mentioned tables shall bear that table's appropriate "type-letter" designation.

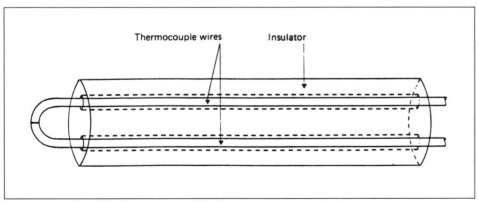

Figure 1-116. Bare Wire Thermocouple Insulation

To ensure accuracy when extension wire is used with a thermocouple, the extension wire must have the same temperature–emf characteristics as the thermocouple. The positive and negative leads of extension wire have also been assigned symbols. These symbols and their materials of construction are identified in Table 1-8.

Table 1-8. **Symbols for Types of Extension Wires**

Type	Combination	Positive	Negative
B	BX**	BPX	BNX
E	EX	EPX	ENX
J	JX	JPX	JNX
K	KX	KPX	KNX
R or S	SX*	SPX	SNX
T	TX	TPX	TNX

*Both Type R or S thermocouples use the same SX compensating extension wire.
** Special compensating extension wires are not required for reference junction temperatures up to 100°C. Generally copper conductors are used. However, proprietary alloys may be obtained for use at higher reference junction temperatures.

The insulation used on extension wires may be divided into four general classifications: waterproof, moisture-resistant, heat-resistant, and radiation-resistant. The materials used for insulating extension wires are selected to perform a variety of functions, including physical protection, bonding, mechanical separation, and electrical insulation. For dry locations, these functions can be performed by nonconducting materials such as cotton, glass, paper tapes, and ceramic beads. Where moisture may be present, enamel coatings, asphalt or wax impregnations, plastics, or rubber or lead sheaths may be required. Where heat resistance is necessary, glass and ceramics may be used. If the extension wire will be exposed to varying degrees of heat and moisture, a combination of materials may provide satisfactory insulation.

To assist in visual identification of wires and to avoid inadvertent cross wiring, many thermocouple wires and extension wires are color coded. Table 1-9 identifies the insulation color for duplex-insulated thermocouple wire. The color codes for single-conductor insulated extension wire and duplex-insulated extension wire are identified in Tables 1-10 and 1-11. Note that the color red is always used to identify the negative lead.

Table 1-9. **Color Code — Duplex-Insulated Thermocouple Wires**

Thermocouple			Color of Insulation		
Type	Positive	Negative	Overall*	Positive*	Negative
E	EP	EN	Brown	Purple	Red
J	JP	JN	Brown	White	Red
K	KP	KN	Brown	Yellow	Red
T	TP	TN	Brown	Blue	Red

* A tracer color of the positive wire code color may be used in the overall braid.

Table 1-10. Color Code — Single-Conductor Insulated Extension Wires

Extension Wire Type			Color of Insulation	
Type	Positive	Negative	Positive	Negative*
B	BPX	BNX	Gray	Red-Gray Trace
E	EPX	ENX	Purple	Red-Purple Trace
J	JPX	JNX	White	Red-White Trace
K	KPX	KNX	Yellow	Red-Yellow Trace
R or S	SPX	SNX	Black	Red-Black Trace
T	TPX	TNX	Blue	Red-Blue Trace

Table 1-11. Color Code — Duplex Insulated Extension Wires

Extension Wire Type			Color of Insulation		
Type	Positive	Negative	Overall	Positive	Negative*
B	BPX	BNX	Gray	Gray	Red
E	EPX	ENX	Purple	Purple	Red
J	JPX	JNX	Black	White	Red
K	KPX	KNX	Yellow	Yellow	Red
R or S	SPX	SNX	Green	Black	Red
T	TPX	TNX	Blue	Blue	Red

* A tracer having the color corresponding to the positive wire code color may be used on the negative wire color code.

THERMOCOUPLE ASSEMBLY COMPONENTS

A thermocouple assembly consists of a thermocouple and one or more associated parts, such as a terminal block, connection head, and protecting tube (see Figure 1-117). The terminal block is a block of insulating material used to support and join the terminations of the thermocouple wires to the extension wires or electrical measuring device. The connection head is a housing that encloses the terminal block.

While some thermocouples may be exposed directly to the process medium, they are typically installed in a protecting tube or a thermowell. Protecting tubes and thermowells prevent contamination of the thermocouple by protecting it from the process. They also provide mechanical protection and support.

Protecting tubes are thin-walled metal or ceramic tubes and are used in low-pressure applications. A protecting tube may have external threads designed for direct attachment to a connection head, or a bushing or flange may be provided for attachment to a vessel (see Figure 1-118). Protecting tubes are not primarily designed for pressure-tight attachment to a vessel.

On the other hand, thermowells include external threads or other means for pressure-tight attachment to a vessel. These devices are designed for high-pressure applications. Thermowells are usually solid cylindrical metal pieces with a hole drilled for insertion of the thermocouple. Figure 1-119 shows two common

A connection head provides easy access for removal and replacement of the thermocouple. It usually has threaded openings for attachment to the protecting tube that contains the thermocouple, and a conduit through which extension wires may be run.

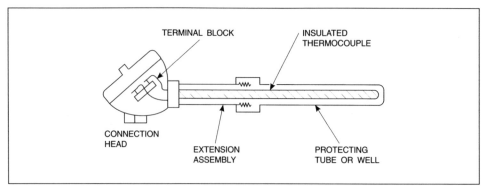

Figure 1-117. Typical Thermocouple Assembly

thermowell designs. One has a lagging extension designed to extend through the insulation of a vessel. The other design has a connection head extension installed between the thermowell and connection head.

Many types of protecting tubes and thermowells are available. Some of the common materials of construction and their maximum operating temperatures are provided in Table 1-12. Depending on the application, the protecting tube or thermowell should have some or all of the following properties:

Table 1-12. Protecting Tube Materials

Protecting Tube Materials	Maximum Operation Temperature	
	Deg. C	Deg. F
Carbon Steel	540	1000
Wrought Iron	700	1300
Cast Iron	700	1300
304 Stainless Steel	870	1600
316 Stainless Steel	870	1600
446 Stainless Steel	980	1800
Nickel	980	1800
75 Nickel – 15 Chromium-Iron	1150	2100
Porcelain	1650	3000*
Silicon Carbide	1650	3000
Alumina-Silica	1650	3000*
Aluminum Oxide	1750	3200*

* Horizontal tubes should receive additional support above 1480°C (2700°F)

(1) Mechanical strength to withstand pressure and resist sagging at high temperatures

(2) Temperature resistance to withstand the temperature being measured

(3) Thermal shock resistance to prevent damage to the tube or well in cases of sudden temperature change

(4) Corrosion resistance to avoid chemical action with the process medium

(5) Erosion resistance

(6) Low porosity at the operating temperature, particularly in furnace applications since furnace gases are generally damaging to thermocouples

A thermowell or protecting tube will increase the radial heat transfer in the gap between the thermocouple and the wall of the well or tube. For this reason, a filler material is often used to improve heat transfer. However, it should be noted that a filler is subject to aging and/or redistribution. Therefore, its effect on the response time of the thermocouple is uncertain. Filler materials used to improve heat transfer should have certain characteristics:

(1) High thermal conductivity

(2) Chemical compatibility with the thermocouple, thermowell, and process medium in case of a leak

(3) Long-term stability of chemical and physical properties at operating temperatures (Some fillers can improve heat transfer when new but reduce heat transfer after aging.)

(4) Adequate fluidity or plasticity (A filler must not compact at the tip of the well because it will prevent complete insertion of the thermocouple. Plasticity is required to prevent the filler from running out if the thermowell is installed with its tip higher than the head.)

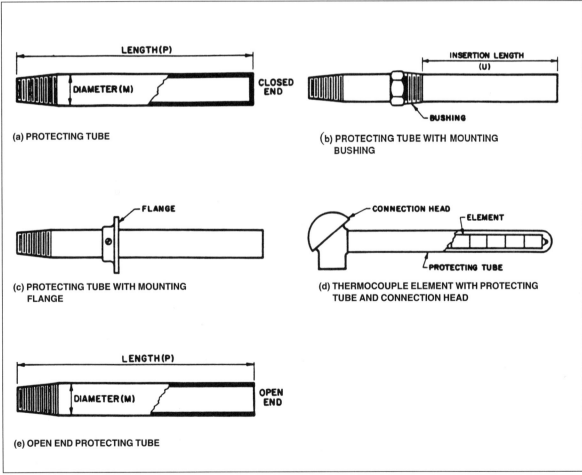

Figure 1-118. Protecting Tubes

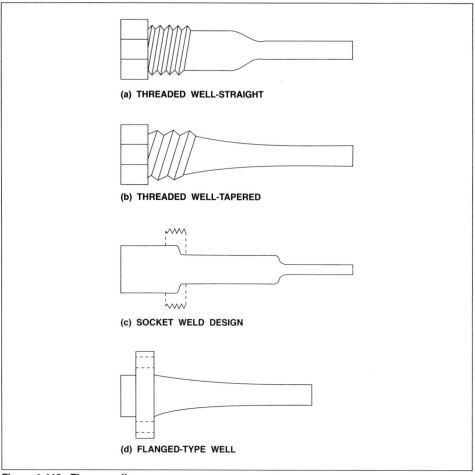

(a) THREADED WELL-STRAIGHT

(b) THREADED WELL-TAPERED

(c) SOCKET WELD DESIGN

(d) FLANGED-TYPE WELL

Figure 1-119. Thermowells

To improve the response time, a thermocouple is usually bottomed in the protecting tube or thermowell. Bottoming ensures that the measuring junction is pressed tightly against the end of the tube or well. This may ground the thermocouple, causing difficulties with some types of installations.

INSTALLATION CONSIDERATIONS

Proper installation of a thermocouple will provide protection of the thermocouple and help ensure the accuracy of the measurement. To determine the appropriate type of thermocouple, assembly components, and extension wire for a particular application, several other factors must be considered when planning a thermocouple installation.

For example, since the input of a thermocouple is based upon the temperature of the measuring junction, it is important to determine the best location for the thermocouple. Since temperature can vary from one point to another within a process, the best location is the one that provides the most representative measurement. One method for determining the best location is to use a portable thermocouple to take measurements at several places.

In addition to location, the depth of immersion is also critical to the accuracy of the measurement. The immersion length of a thermowell or protecting tube is the length from the free end to the point of immersion in the medium being measured. Immersion length should not be confused with insertion length, which is the length from the free end of the well or tube to, but not including, the external threads or other means of attachment to a vessel or connecting head.

Figure 1-120 illustrates these two lengths. A minimum immersion length of 8 to 10 times the tube or well diameter is recommended in order to minimize conduction errors. However, when a thermocouple is used in a high-velocity liquid, it may not need to be immersed as deeply.

The position of the thermocouple is also an important installation consideration. Thermowells are usually installed perpendicular to a pipe or vessel wall, at an angle, or in an elbow. These installations are illustrated in Figure 1-121.

In gas applications, the thermocouple should be located where the mass velocity is as high as possible. This will ensure good heat transfer by convection. However, if the velocity is greater than 300 ft/sec, a specially designed stagnation-type probe should be used. When a thermocouple must be installed in a location where the gas velocity is very low, it may be necessary to induce gas flow past the junction. Several aspirating pyrometers are available for this purpose.

In high-temperature applications such as ovens or furnaces, vertical installation of the thermocouple through the top of the vessel will prevent the thermocouple from bending or sagging. It is also recommended that the protecting tube or thermowell extend beyond the outer surface of the vessel furnace or processing equipment so that the temperature of the connection head is close to the ambient atmospheric temperature. This is particularly important for Types B, R, and S thermocouples with compensating extension wires. The connection head temperature should never exceed the temperature limits for the thermocouple extension wires.

Another factor to consider is the extension wire. Thermocouple extension wire should be installed in such a way that it is protected from excessive heat, moisture, and mechanical damage. Whenever practical, it should be installed in con-

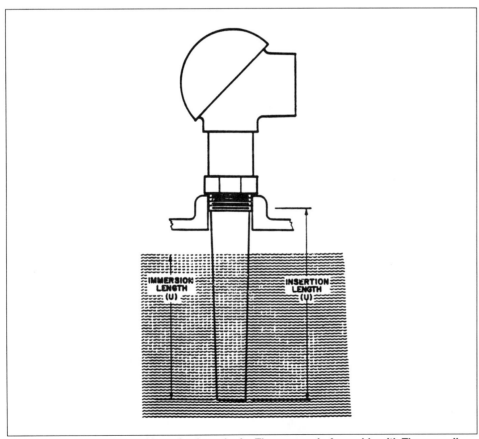

Figure 1-120. Immersion and Insertion Lengths for Thermocouple Assembly with Thermowell

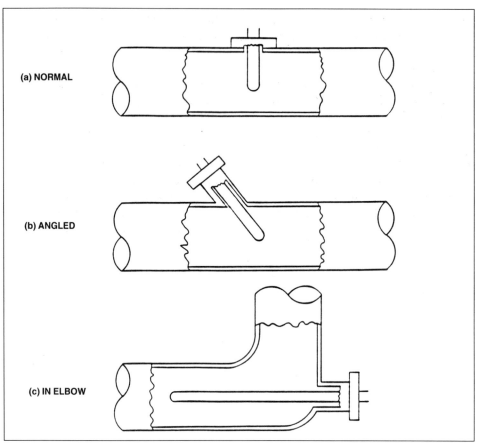

(a) NORMAL

(b) ANGLED

(c) IN ELBOW

Figure 1-121. Thermowell Installations

duit so that it is not subjected to excessive flexing or bending, which can change the thermoelectric properties. Long radius bends should be used in place of elbows, because pulling the wire through elbows introduces additional mechanical stress. Also, to minimize errors that can be introduced by junction boxes, the extension wire should be run from the connection head to the measuring instrument terminal in one continuous length. Finally, to prevent induced extraneous voltages, electrical wires should never be run in the same conduit with extension wires. The compensating cables and insulation are color coded, and connection must be properly made at the terminals of the thermocouples. Polarity checks must be done at the measurement and instrument ends. Grounding at both the ends must be avoided.

> **Since the thermocouple body is grounded to a metallic pipe or the surface, the grounding of leads at the instrument end is not recommended.**

On lines at high-steam pressure pipes or high fluid pressure pipes, it is advisable to weld the thermocouple; the gaskets could give away, causing leakage of steam or fluid, and the plant may require a shutdown. Whenever a new thermocouple must be installed, ensure proper matching of the thread sizes of the thermocouple and the stub. Proper care must be taken to ensure that when the thermocouple is removed for maintenance, the stub should be screwed with a proper sized threaded cap. The porcelain block of the terminal block must be replaced if it is found broken since there is a chance of loose connection and play at the terminal block. Ensure that the connection head is in its position to prevent dust settling on the screwed connections as this may be a source of error in reading millivolts. For thermocouples with a total length exceeding 800 mm, thermocouples with large connections are preferred. Generally, these can be found installed in flow ducts. Thermocouples may also be installed to measure the surface temperatures of a body.

In addition to using one thermocouple to provide single-point measurements, multiple thermocouples can be installed in different configurations to provide other approaches to temperature measurement. For example, thermocouples may be connected in parallel to provide an average temperature measurement. The measurement instrument will provide a readout of the temperatures at the measuring junctions. One disadvantage of this approach is that the average measurement will not show a hot spot.

Thermocouples can also be installed in series to form a thermopile. In this case, the thermocouples are connected so that alternate junctions are at a known temperature, such as the melting point of ice, and the other junctions are exposed to the temperature to be measured. In a thermopile, the voltages of all the thermocouples are added so the output is n times the number of measuring junctions. This type of thermocouple arrangement magnifies the output signal, providing a way to detect small changes in temperature. However, one disadvantage is that if a short circuit occurs, it would be undetected since the temperature of individual junctions is not indicated.

CHECKING THERMOCOUPLE ACCURACY

The properties of thermocouple wires, fabricated thermocouples, and extension wires are carefully controlled and tested during manufacture to ensure that they are within the tolerance limits of their appropriate temperature–emf specifications. The output of a new, unused thermocouple will be determined solely by the temperature of its measuring junction. However, after the thermocouple has been used, the thermocouple material will no longer be homogeneous. In this situation, the output of the used thermocouple will not be determined solely by the temperature of its measuring junction. Thus, there are different procedures for checking the accuracy of unused and used thermocouples.

Various methods for testing new thermocouple materials are described in Publication 300, Volume II, *Precision Measurement and Calibration — Temperature*, published by the National Bureau of Standards, and in E2220-72, *Calibration of Thermocouples by Comparison Techniques*, published by the American Society for Testing and Materials.

One common method for testing the accuracy of a new thermocouple involves placing the reference junction in an ice bath to establish the reference temperature and placing the measuring junction in a variable calibrating temperature bath. The voltage output of the thermocouple is read with a calibrated potentiometer, high-impedance voltmeter, or thermocouple indicator. Then the voltage outputs are tabulated with the corresponding temperatures in the furnace and compared with the appropriate thermocouple reference table to determine whether or not the thermocouple output agrees with the table within the prescribed limits of error throughout its usable range.

Used or installed thermocouples that are exposed to high temperatures in various atmospheres may change characteristics. This thermoelectric nonuniformity results from contamination or deterioration of the thermocouple wires and/or junction. While the reference junction of a used thermocouple will normally be like new, the measuring junction will also be affected to some degree. Therefore, the thermocouple material is no longer homogeneous. To avoid continued use of thermocouples with excessive deviation from their original properties, it is good practice to check thermocouples at regular intervals. In general, thermocouples used in high temperatures or contaminating atmospheres must be checked more frequently.

The purpose of checking an installed thermocouple is to determine the temperature error in actual service, not the temperature–emf characteristics of the thermocouple. Therefore, a used thermocouple should always be checked in its normal, installed location. If a thermocouple were removed from its installed location and placed in a calibrating furnace for checking, it is highly improbable that the temperature gradient in the furnace would match the temperature gradient of the normal installation.

The accuracy of an installed thermocouple is checked by comparing its readings with the readings of a new or checking thermocouple of the same type. When the diameter of the protecting tube is large enough, the checking thermocouple may be inserted beside the service thermocouple. When the protecting tube is not large enough to accommodate another thermocouple, the service thermocouple can be removed and the checking thermocouple temporarily inserted in its place. When this replacement method is used, it is essential that stable temperature conditions be maintained.

If the installed thermocouple is used to measure a wide range of temperatures, it should be checked at more than one temperature within the range of its use. While testing a thermocouple of one temperature provides some information, it is not safe to assume that the changes in the emf of a thermocouple are proportional to the temperatures or to the emf.

In addition to being the same type as the installed thermocouple, the checking thermocouple should be homogeneous and uncontaminated. Any new thermocouple may be used, but it should be checked against a primary standard and tagged with its deviation from the standard curve. The accuracy of a checking thermocouple will become questionable after extended use. Noble metal thermocouples used as checking thermocouples may normally be relied upon for a considerable period of use. However, base metal checking thermocouples should not be used below 480°C if they have been exposed between checks to temperatures above 760°C.

Resistance Temperature Detectors

Another class of electrical temperature-measuring devices, called resistance thermometers, respond to temperature by changing their electrical resistance. Two common types of resistance thermometers are resistance temperature detectors (RTDs), which have metallic sensing elements, and thermistors, which have semiconductor elements.

PRINCIPLE OF OPERATIONS OF RTDS

An RTD consists of a sensing element fabricated of metal wire or metal fiber that responds to temperature change by changing its resistance. The sensor is connected to a readout instrumentation that monitors the resistance to a temperature value.

The principle of operation of an RTD is based on the fact that the electrical resistance of some metals varies directly with temperature changes. The relationship between resistance and temperature is based on the temperature of the metal. The coefficient of resistance is the fractional change in resistance per degree Celsius. Since no two metals have the same coefficient, different metals have different temperature versus resistance curves, none of which are strictly linear. Figure 1-122 shows the relationship between temperature and resistance for some metals that are commonly used for RTDs. On this graph, the resistance axis values represent the ratio of wire resistance at the measured temperature (R_T) to wire resistance at 0°C (R_0). The variation of resistance R_T with temperature T can be represented by an equation of the form $R = R_0 (1 + \alpha_1 T + \alpha_2 T_2 + \alpha_n T^n)$. R_0 is the resistance at temperature $T = 0$. The number of terms depends on the material, accuracy, and temperature range to be covered.

The resistance property of a metal is usually identified by its resistivity (r). The resistivity of a metal is a constant factor and is used in evaluating the effect of the length and the cross-sectional area of a metal on its total resistance. The resis-

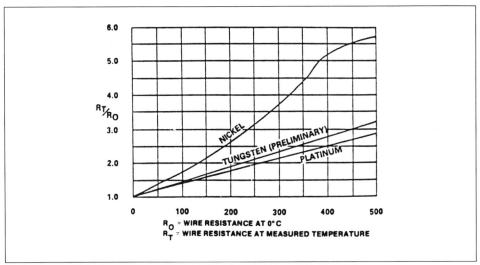

Figure 1-122. Resistance versus Temperature for Common RTD Materials

tance is proportional to length and inversely proportional to cross-sectional area, as shown in the following equation for total resistance.

$$R = (r \times L)/A$$

Analysis of this equation also shows that when two different metals of the same length and cross-sectional area are compared, the metal with higher resistivity will have a higher total resistance.

The metals used in the fabrication of RTD sensing elements must meet several requirements:

(1) High resistivity, so the size of the element will be of practical use

(2) High temperature coefficient of resistance to provide a readily measurable resistance change when temperature changes

(3) Good ductile or tensile strength to permit its being made into wire or film and then wound or coiled to form a sensing element

(4) Chemical inertness with other materials used for structural components, electrical insulation, and packaging of the sensor

Table 1-13 shows some of the properties of metals used for RTDs. Platinum is the most widely used metal in the manufacture of RTD elements. The resistance of platinum is nearly linear with temperature; it has a reasonably high resistivity, a high melting point, and a relatively large temperature coefficient of resistance. Other metals used for RTD elements include copper, nickel, nickel alloys, and tungsten. While copper has the most linear of all known temperature versus resistance relationships, its useful temperature range is narrow. Nickel has a high temperature coefficient of resistance; however, its temperature–resistance relationship becomes quite nonlinear above 300°C. Tungsten is being developed as an RTD material because of its usefulness for high-temperature measurements; however, tungsten cannot be fully annealed and is less stable than platinum.

> Another desirable property for metals used in RTDs is good linearity between resistance and temperature. While this property is not essential, it simplifies the readout instrumentation.

SENSOR DESIGNS

The sensing element of an RTD usually consists of a wire cut to a length that provides a predetermined resistance at 0°C. The wire may be coiled within or wound around an insulating material.

A critical factor in the design of RTD sensors is the manner in which the metal wire or film is supported. This is because strain, as well as temperature, can cause

Table 1-13. Properties of Metals Used for RTDs

	Platinum	Copper	Nickel	Balco[a]	Tungsten
Average temperature coefficient of resistance over 0° to 100°C (Ω/Ω°C	0.00385 to 0.003925	0.0042	0.0067	0.0052	0.0045
Resistivity (Ω cm)	9.81×10^{-6}	1.529×10^{-6}	5.91×10^{-6}	20.0×10^{-6}	4.99×10^{-6}
Linearity of resistance versus temperature	excellent	excellent	poor	poor	poor
Useful temperature range (°C)	−260 to 800	−100 to 150	−100 to 500	−100 to 300	−70 to 2700

[a] An alloy of 70% nickel and 30% iron.

a change in the resistance of the metal. Ideally, the mounting should impose no strain on the metal for the entire range of temperatures for which it would be fragile and impractical for industrial use. For this reason, the various sensor designs have some degree of support for the wire or film.

The bird cage design offers the least amount of support. In this design (see Figure 1-123), the wires hang in a vapor space and are threaded through thin disks that separate the wires. The disks are usually made of mica or ceramic material. With this design, the wires are free to move, so strain is negligible. Due to their fragility and cost, bird cage design RTDs are used most often for laboratory measurements in which a high degree of accuracy is required.

RTDs with partially supported elements are well suited for industrial applications. Several types of designs resist shock and can be used for temperature measurements from −260°C to 800°C. One type of partially supported design is shown in Figure 1-124. With this design, tight coils of wire are inserted into small axial holes in an insulating mandrel. The mandrel material must be pure to prevent contamination of the wires. It is typically made of aluminum oxide or magnesium oxide. An adhesive is injected into the holes and the assembly is fired. This secures part of each turn of the coil to the mandrel, leaving the remainder of each turn free to move.

Another type of partially supported design, a wall-mounted RTD, is shown in Figure 1-125. In this case, insulating cement is used to secure a coil of wire to the interior wall of sheath. Only part of each loop is secured, leaving the remainder free to move.

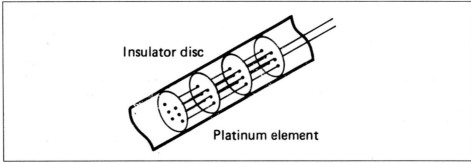

Figure 1-123. Bird Cage RTD

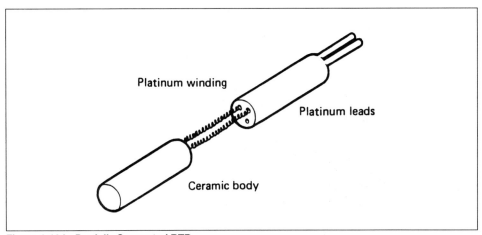

Figure 1-124. Partially Supported RTD

Fully supported element designs offer greater shock resistance and are more rugged than partially supported designs. While fully supported element designs are satisfactory for most industrial applications, they place more strain on the element. One type of fully supported element is shown in Figure 1-126. This type of sensor is manufactured by winding a wire around an insulating mandrel and coating it with an insulator by means of painting, dipping, or flame spraying. Since the wires are totally constrained, any difference in thermal expansion between the wire and the mandrel or insulation will cause strain on the wire. While careful design can minimize this stress, it cannot be totally eliminated.

Another type of fully supported design uses metal film (see Figure 1-127). With this design, a film element sensor is manufactured by depositing a thin film of platinum on the surface of an insulating support piece that is usually made of ceramic material. Unlike wire element sensors, in which resistance can be adjusted by the length of the wire, film elements are designed to allow resistance to be adjusted during manufacture. For example, the film may be manufactured in the form of a grid. Resistance of the element can then be adjusted by severing one or more filaments in the grid with a laser beam. With other element designs, resistance may be adjusted by removing a thin layer with a laser beam.

In some applications, an RTD sensing element can be immersed in the medium being measured, provided the fluid is nonconducting or the sensor is properly insulated and protected. In these applications, a fully supported sensor with a coat-

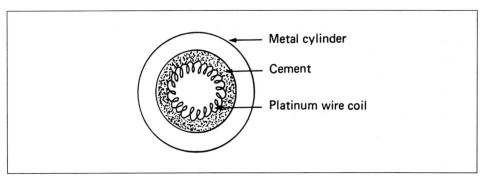

Figure 1-125. Wall-Mounted, Partially Supported RTD

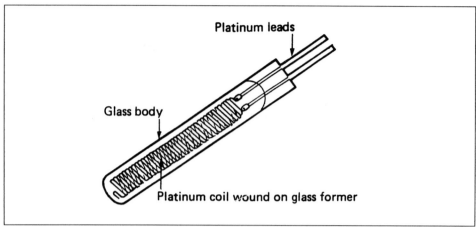

Figure 1-126. Fully Supported RTD

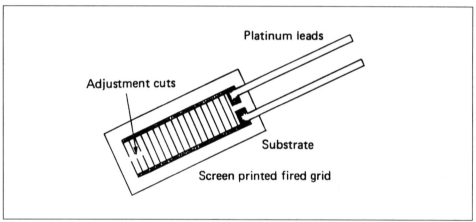

Figure 1-127. Film-Type Platinum RTD

ing of glass, ceramic, organic insulator, or Teflon™ is typically used (see Figure 1-128). However, most applications require packaging of the sensor element. For immersion sensors, the sensing element is usually mounted inside a metal sheath (see Figure 1-129). A filler is used to hold the element in place and insulate it from the sheath. The filler may be a cement or powder, such as aluminum oxide or magnesium oxide. Proper handling of the element and its leads must be assured; otherwise, there is a chance of breakage of the leads at the junctions of the element and lead wires.

RTD sensors are also manufactured for surface temperature measurement. Surface sensors may be clamped, bolted, or cemented to a surface. Figure 1-130 shows two typical surface sensors.

RTD READOUT INSTRUMENTATION

Temperature measurement with an RTD is actually a measurement of the sensor's resistance, using the sensor calibration to convert the measurement into temperature. This is achieved by connecting the sensor to a transducer that has a bridge circuit, typically a Wheatstone bridge or a Mueller bridge. The sensor leads are connected so that the RTD forms one leg of the bridge circuit. Figure 1-131 shows the connection of a two-wire RTD to a Wheatstone bridge.

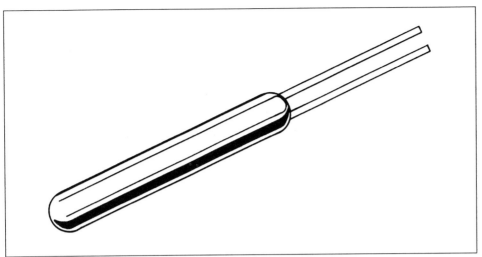

Figure 1-128. Fully Supported RTD Sensor with Protective Coating

Initially, R_S is adjusted to match the resistance value the sensor exhibits at some reference temperature, such as 0°C. The bridge then operates in the non-balanced mode. In this mode, the three circuit resistors are fixed and the sensor (R_T) acts as a variable resistor. A change in the resistance of the sensor will cause a proportional change in the measured voltage drop (E). The voltage output of the circuit (E) is then converted to a temperature that corresponds to the resistance of the sensor.

Sensors are usually located far from the readout instrumentation. Copper leads are commonly used as connecting wires. The lead wires represent a resistance in series with the RTD. Over long distances, the resistance of the copper leads may be significantly greater than the resistance of the RTD sensor, resulting in meas-

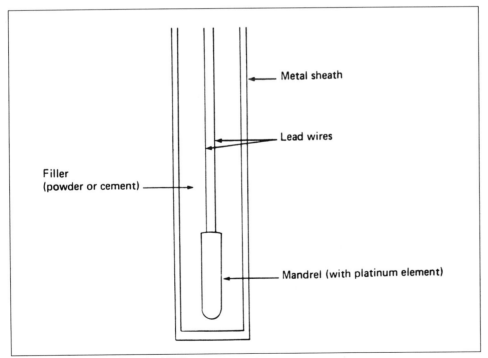

Figure 1-129. Typical Sheathed RTD

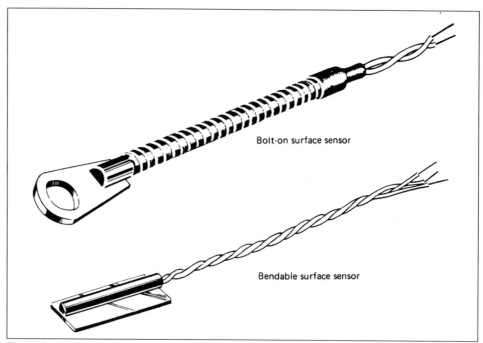

Figure 1-130. Typical Surface Sensors

urement errors. For this reason, RTD sensors are available with two, three, or four wires. A two-wire sensor does not allow for lead compensation. However, when the readout instrument is located near the sensor installation, which allows the use of short leads, a two-wire sensor can provide good accuracy.

Three-wire and four-wire sensors provide compensation for lead resistance. Figure 1-132 shows the connection of three-wire and four-wire sensors to bridge circuits. In the three-wire lead circuit, L_1 and L_2 are in opposite legs of the bridge, which cancels their effect on the bridge. L_3 is connected in series with the input voltage and cannot unbalance the bridge.

The three-wire configuration is most commonly used for industrial applications. The four-wire sensor connection provides greater accuracy than the three-wire connection. However, since this configuration requires frequent rebalancing and lead reversal, it is not commonly used in industrial applications.

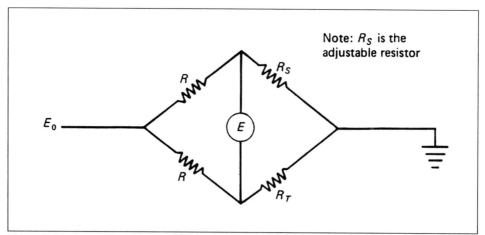

Figure 1-131. Two-Wire RTD and Wheatstone Bridge Circuit

RTD ACCURACY

In addition to lead wire resistance, which can be compensated through the use of a three-wire or four-wire sensor, several other factors influence the accuracy of RTD temperature measurements. One of these is a lack of a single standard for RTDs. Several standards specify different values for the temperature coefficient of resistance for a given metal. In addition, manufacturers observe different tolerances in the manufacture of RTDs. Therefore, accuracy problems can occur when RTDs from different manufacturers are used in the same system, or when an RTD from one manufacturer is replaced with an RTD from another manufacturer.

Self-heating can also affect accuracy. An RTD is a passive element that requires the application of a current flow in order to measure its resistance. Heat generated by the current can affect accuracy if it raises the temperature of the RTD element above the process temperature, resulting in transfer of heat to the process. The self-heating effect depends on the design of the RTD element as well as its environment (i.e., fluid velocity past the element). To minimize self-heating, industrial readout instrumentation usually limits current in the bridge to 0.003 amp.

Other factors that affect the accuracy of RTDs are physical and chemical changes that can cause shifts in the accuracy of RTDs, leading to drift in readings. Drift rates for platinum RTDs are generally approximated by manufacturers

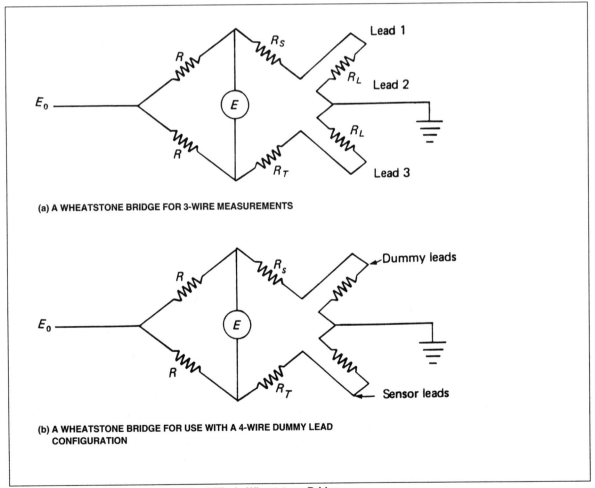

(a) A WHEATSTONE BRIDGE FOR 3-WIRE MEASUREMENTS

(b) A WHEATSTONE BRIDGE FOR USE WITH A 4-WIRE DUMMY LEAD CONFIGURATION

Figure 1-132. Three-Wire and Four-Wire RTDs in Wheatstone Bridge

at 0.05°C per year. Studies of platinum RTDs have found that physical changes occur because of thermal or mechanical factors, such as mechanical shock due to rough handling, constraints that lead to stress during thermal expansion, and vibration. Chemical changes can be caused by chemical reaction of the platinum or impurities dissolved in the platinum. These reactions are often ignored because platinum is a noble metal. However, one effect that has been observed in steel-sheathed RTDs is the migration of iron into the platinum. This migration can have a significant effect on the accuracy of sensors operated at about 500°C.

RTDS FOR SPECIALIZED APPLICATIONS

RTDs are also manufactured in designs that make them suitable for specialized applications. Some of these designs include averaging RTDs, annular element RTDs, and combination RTD–thermocouples. An averaging RTD has long resistance elements (i.e., 15 feet). This design measures the average temperature over the length of the sensor.

In annular element RTDs, the sensors are made with annular elements that provide a tight fit against the inner wall of a thermowell. The tight fit and small heat capacity of the sensor and its attached components cause this RTD to have a quick response to changes in temperature.

Combination RTD–thermocouple designs are available with both an RTD and a thermocouple enclosed in the same sheath. This design allows two simultaneous measurements based on different measurement principles. Another combination design consists of a four-wire platinum RTD with one Chromel™ lead and one Alumel™ lead. This configuration creates two thermocouples in addition to the RTD, permitting three simultaneous temperature measurements at the same location. Combination designs alleviate concerns about degradation that could be common to sensors using the same measurement principle but that are implausible for sensors with different principles.

Thermistors

Thermistors are another type of electrical temperature-measuring device. They are made of solid semiconductor materials that have a high coefficient of resistivity. The relationship between resistance and temperature and the linear current-voltage characteristics are of primary importance. Typical thermistors are suitable for temperatures as low as –200°C and others as high as 600°C. While thermistors and RTDs have many similarities, some important differences affect their use and accuracy.

CHARACTERISTICS OF THERMISTORS

Thermistors are semiconductors formed from complex metal oxides, such as oxides of cobalt, magnesium, manganese, or nickel. They are available with positive temperature coefficients of resistance (PTC thermistors) and with negative temperature coefficients of resistance (NTC thermistors). NTC thermistors are used almost exclusively for temperature measurement.

Because it is a solid device, a thermistor conducts like a transistor. However, the negative coefficient is an important factor in understanding thermistor operation. Unlike an RTD, which increases its resistance, an NTC thermistor will decrease its resistance when temperature increases, causing an increase in current or electron flow through the thermistor. Therefore, the current flow is directly proportional to the temperature. Figure 1-133 shows the resistance versus temperature curve for a typical NTC thermistor. Their resistance/temperature relation is generally of the form:

$$R = R_0 \, e^{B} \, (1/T - 1/T_0)$$

$$
\begin{array}{rcl}
R &=& \text{resistance at temperature } T \\
R_0 &=& \text{resistance at temperature } T_0 \\
B &=& \text{constant characteristics of material, } K \\
T, T_0 &=& \text{absolute temperatures, } K
\end{array}
$$

A thermistor's temperature coefficient of resistance is very large compared to an RTD's coefficient of resistance. This large coefficient make thermistors more sensitive to small changes in temperature, and therefore ideally suited to applications that require precise measurements. A thermistor's change in resistance per degree change in temperature is typically as large as 3 to 5% per °C, compared to 0.4% per °C for most RTDs. As a result, the resistance of a thermistor at operating temperatures can be much larger than the resistance of an RTD. A thermistor's resistance will normally be tens of thousands of ohms, while an RTD's resistance is normally hundreds of ohms.

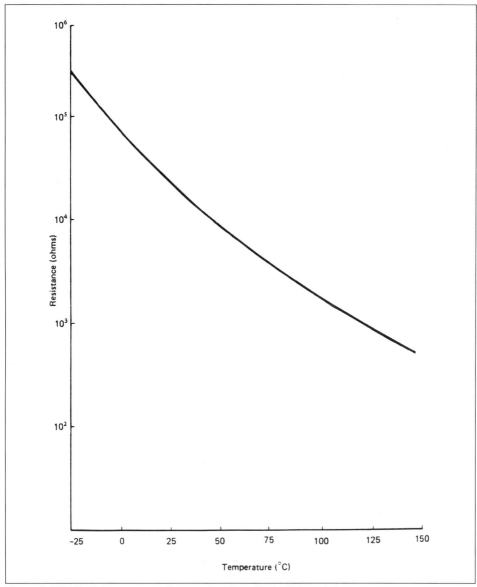

Figure 1-133. Resistance versus Temperature for a Typical Thermistor

On the other hand, thermistors have a very nonlinear relationship between resistance and temperature. For this reason, the range of operating temperatures for which they can be used is much smaller than the temperature ranges RTDs are capable of handling. However, a thermistor will provide stable and repeatable performance if used within its specified temperature range.

Another characteristic of thermistors is that they are made with two lead wires instead of the typical three-wire and four-wire designs used for RTDs. This is because the large resistance values for thermistors virtually eliminates the concern about lead wire resistance.

A variety of circuit designs are used to measure temperature with a thermistor. For example, a thermistor may be connected to a Wheatstone bridge, much like a two-wire RTD circuit, or a series circuit that includes a battery, the sensor, and a microammeter may be used. In this case, as long as voltage is constant, the current flow will be determined only by the resistance of the thermistor.

Despite the nonlinear nature of thermistors, readout instrument circuits have been developed to provide a nearly linear output voltage versus temperature or resistance versus temperature. The simplest circuit for linearizing the voltage versus temperature relationship is a voltage divider (see Figure 1-134). In this circuit, the voltage drop may be measured across the fixed resistor (R) or across the thermistor (R_T).

The voltage versus temperature relationship can also be linearized by using modifications of the voltage divider circuit. For example, multiple thermistors can be added to the voltage divider circuit to achieve better linearization. Figure 1-135 shows a two-thermistor circuit and a three-thermistor circuit. In these circuits, the thermistors are matched so they have identical resistance versus temperature curves. Assemblies with multiple matched thermistors and the required fixed resistors can be obtained.

Linearizing for resistance measurements can be achieved by using a parallel circuit arrangement (see Figure 1-136). As with the circuits for linearizing voltage versus temperature, a number of circuits using multiple matched thermistors have been developed for linearizing resistance versus temperature.

THERMISTOR DESIGNS

Thermistors are manufactured in a variety of sizes and configurations. Some of the more standard designs include beads, discs, washers, and rods.

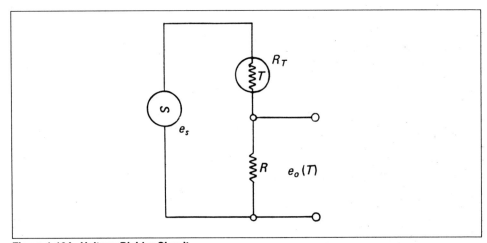

Figure 1-134. Voltage Divider Circuit

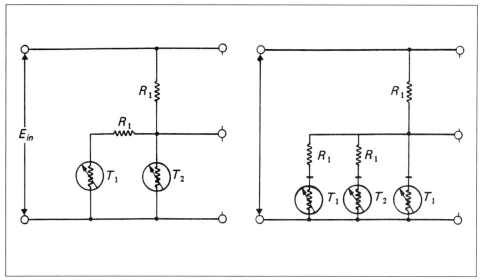

Figure 1-135. Multiple Thermistor Linearizing Circuits

The bead thermistor is made of a small bead of thermistor material to which a pair of leads is attached. The bead is usually enclosed in glass.

A disc thermistor consists of a disc of thermistor material and a pair of leads. The leads may be attached radially or axially to the top and or the bottom of the disc. Some disc thermistors have no leads and are fabricated with metal-plated faces that can be clipped or soldered in the circuit.

A washer thermistor resembles a disc thermistor but has a center hole and metal-plated faces for contact. The center hole enables the thermistors to be held by a mounting bolt or stacked with other washer thermistors and electrical components.

A rod thermistor is basically a stick of thermistor material to which a pair of leads is attached. The leads may be attached axially or radially to each end of the rod.

Thermistors may be used for immersion or surface measurements, and sheathed thermistors are fabricated in a variety of sizes and designs. Surface sen-

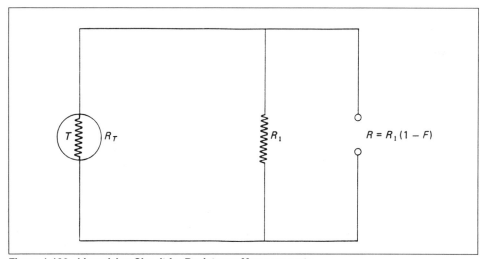

Figure 1-136. Linearizing Circuit for Resistance Measurements

sors with thermistor elements are also available. Matched thermistors that have the same resistance versus temperature curve can be used when interchangeability or equivalence is required for an application. In addition, there are probes that contain the necessary matched thermistors and associated resistors to provide linearized voltage versus temperature or resistance versus temperature.

THERMISTOR ACCURACY

A thermistor will maintain a stable and repeatable resistance versus temperature relationship when it is used within its specified temperature range. The most common problem related to thermistor accuracy is interchangeability. The shape of the resistance versus temperature curve is greatly dependent on the composition of the thermistor material, which is difficult to control. In addition, the small size of thermistors makes it difficult to adjust the resistance by altering the amount of material through cutting or grinding. These factors make it difficult to satisfy tight tolerances. However, through careful manufacture and selection of in-tolerance thermistors from large production batches, thermistors that follow a reference resistance versus temperature curve to within 0.1˚C over limited temperature ranges can be obtained.

Tolerance for thermistors may be expressed as a tolerance on indicated temperature (˚C) or a tolerance on resistance (%). Figure 1-137 shows the tolerances for a typical high-quality thermistor. Analysis of these tolerance curves shows the importance of using thermistors only in their specified operating temperature range. The flattened areas of the curves indicate that stability is achieved only in an approximate range of 0˚C to 75˚C.

When the current is passed through a thermistor in order to measure resistance, the effect of Joule heating or self-heating can result in measurement errors. This effect is greater in thermistors than in RTDs because the larger resistance of a thermistor gives greater Joule heating for any given current. Therefore, the voltage supplied by readout instrumentation to a thermistor must remain small enough to ensure that the self-heating effect is insignificant.

In addition, the larger range in resistance over the useful range of a thermistor may cause some complications in the readout instrumentation. For example, the resistance of a thermistor at the low-temperature end of its operating range can be 10,000 times greater than its resistance at the high-temperature end of its operating range.

Thermistor accuracy can also be affected by several mechanical or chemical actions that change its electrical resistance:

(1) Surface contamination may cause electrical shunting or shorting.

(2) Physical damage can chip the surface of the thermistor. Removing some of the semiconductor material changes the electrical resistance of a thermistor.

(3) Mechanical or thermal stress at the connection between the thermistor and the lead wires may increase the connection resistance.

(4) If extension wires are soldered to the lead wires, metal used in the solder may diffuse along the lead wires to the thermistor's lead wire junction and alter the electrical contact of the junction. This may also create a thermocouple effect that would produce extraneous emf. The use of thermally immune solder can minimize this problem.

DIGITAL THERMOMETER

The following is a brief description of a system that accepts temperature inputs and provides digital signals and/or displays as outputs.

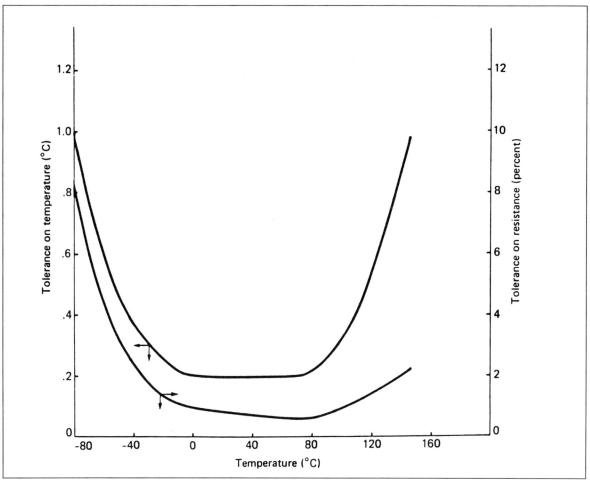

Figure 1-137. Tolerance for a Typical Thermistor

Electronic oscillators with piezoelectric quartz crystals as the resonant element that establishes the frequency of oscillation have been widely used. For the most critical applications it has been necessary to place the crystal in a temperature-controlled oven, since the natural frequency of the crystals varies with temperature, causing drifts in oscillator frequency. This difficulty is turned to good advantage and serves as a temperature-sensing element. Changes in probe temperature cause a frequency change in proportion. By applying the oscillator voltage to an electronic counter for a definite time interval, a direct digital reading of temperature is obtained.

Radiation Pyrometers

Temperature measurement through the use of mechanical thermometers, thermocouples, RTDs, and thermistors requires placing the sensor in physical contact with the medium or object being measured, either directly or within a thermowell. However, in some applications, contact measurement is not suitable. For example, contact measurement is impractical when an object to be monitored is moving or when the medium is corrosive, abrasive, or at an extremely high temperature that could destroy a sensor. In these situations, noncontacting temperature measurement can be achieved through the use of radiation or optical pyrometers. The high-temperature limits of radiation pyrometers exceed the limits of most other temperature sensors. Radiation pyrometers are capable of measuring temperatures to approximately 4000°C without touching the object being measured.

PRINCIPLES OF RADIATION PYROMETRY

Temperature measurement with radiation pyrometers is based on the principle that all objects emit radiant energy. Radiant energy is emitted in the form of electromagnetic waves, considered to be a stream of photons that travel at the speed of light. The wavelengths of radiant energy emitted by a hot object range from the visible light portion (0.35 to 0.75 microns) to the infrared portion (0.75 to 20 microns) of the electromagnetic spectrum. In the visible light portion of the spectrum, radiant energy appears as colors. The expression "red hot" is derived from the fact that a sufficiently hot object will emit visible radiation. Common examples include a piece of red hot steel and a tungsten filament lamp.

Although some of the radiant energy emitted by a hot object is in the visible light portion of the spectrum, much more is emitted in the infrared portion where it is not visible. For example, steel at 850°C emits 100,000 times more infrared radiation than visible light.

Radiation pyrometers measure the temperature of an object by measuring the intensity of the radiation it emits. The intensity and wavelength of the radiation emitted by an object depends on the emittance and the temperature of the object. Emittance is a measure of an object's ability to send out radiant energy. It is inversely related to reflection of the object's surface.

Since emittance will differ from one object to another, a standard, called a blackbody, is used as a reference for calibrating radiation pyrometers and serves as the basis for the laws that define the relationship of the intensity of radiation and wavelength, having a surface that does not reflect or pass radiation. It is considered a perfect emitter because it absorbs all heat to which it is exposed and emits that heat as radiant energy.

The relationship of radiant intensity and wavelengths for a blackbody at different temperatures is shown in Figure 1-138. Total radiant intensity is given as the area under a curve. This illustration shows two important effects:

(1) The intensity of radiant energy increases as temperature increases.

(2) The peak of radiation moves to lower wavelengths as temperature increases. In the visible light portion of the spectrum, this effect can be seen by the change in color of heated metals. They change from red to yellow to white to blue-white as temperature increases.

The dependence of the intensity of radiation on temperature and wavelength for a blackbody is provided by the formula for Planck's radiation law:

$$H(\Gamma, T) = C_1 / \Gamma^5 (e_{C_2} / \Gamma T - 1)$$

where:

$H(\Gamma, T)$ = intensity of radiation emitted by a blackbody at temperature T and wavelength 1
Γ = wavelength (cm)
T = temperature (K)
C_1 = a constant of 3.74×10^{12} Wcm^2
C_2 = a constant of 1.44 cmK

This formula provides the curves shown in Figure 1-138. Other formulas used for radiation pyrometry can be derived from Planck's radiation law. For example, for a given temperature, the total radiation intensity for a blackbody is given by the area under the curve. The Stefan-Boltzmann law provides the following formula for calculating total radiation for a blackbody at a given temperature:

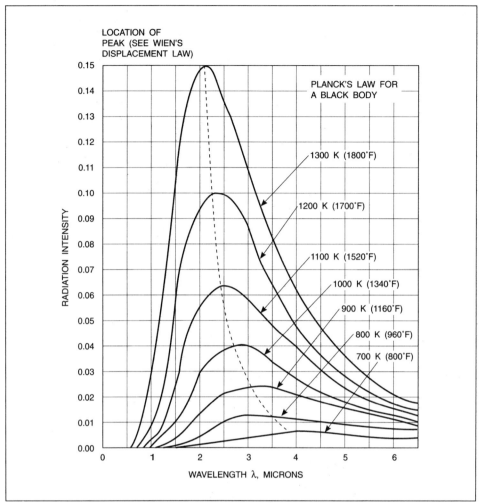

LOCATION OF PEAK (SEE WIEN'S DISPLACEMENT LAW)

PLANCK'S LAW FOR A BLACK BODY

1300 K (1800°F)
1200 K (1700°F)
1100 K (1520°F)
1000 K (1340°F)
900 K (1160°F)
800 K (960°F)
700 K (800°F)

RADIATION INTENSITY

WAVELENGTH λ, MICRONS

Figure 1-138. Radiation Intensity vs. Wavelength and Temperature

$$H(T) = \sigma T^4$$

where:

$H(T)$ = total radiation per unit area
σ = Stefan-Boltzmann constant
  = $5.669 \times 10^{-12} W/cm^2 K^4$
I = temperature (K)

This law illustrates that total emitted energy is proportional to the fourth power of the temperature. Therefore, a small increase in temperature will result in a significant increase in total radiation emitted by a blackbody. For example, increasing the temperature by a factor of 2 will increase the total radiation by a factor of 16, or 2^4.

The theoretical laws that relate temperature to radiation apply for a perfect blackbody. However, few materials have an emittance property that even approximates that of a blackbody. As a result, a nonblackbody emits less radiant energy than a blackbody radiates at the same temperature. Therefore, in practice, modifications must be made to account for the emissivity of the object, or nonblackbody, being monitored for temperature. Emissivity is the ratio of the total radiation emitted by a nonblackbody to the total radiation emitted by a blackbody at the same temperature.

Blackbodies may be obtained commercially or they may be fabricated on site. Typically, a blackbody consists of a hollow metal sphere with a small hole drilled through the wall. The interior wall is coated with a substance that is a good emitter, such as carbon black, platinum black, zinc black, or carborundum. When the exterior of the sphere is heated, the radiant energy emitted through the hole in the sphere closely approximates the radiant energy a perfect blackbody would emit under the same conditions.

The emissivity of a nonblackbody can be determined using a radiation pyrometer to measure the total radiation emitted by a blackbody and the total radiation emitted by a nonblackbody. A value of 1 is assigned for the emissivity of a blackbody. Since a nonblackbody emits less radiant energy than a blackbody at the same temperature, the emissivities of nonblackbodies range between 0 and 1. If the emissivity of a body is constant at all wavelengths, the body is said to be gray. If the emissivity varies with wavelength, the body is referred to as nongray.

To account for the emissivity (ε) of an object when determining temperature, the Stefan-Boltzmann law is amended as follows:

$$\varepsilon H(T) = \varepsilon \, \sigma T^4$$

Table 1-14 shows the emissivities of some common metals at different temperatures. Table 1-15 shows emissivities of some materials at wavelengths of 0.65 microns. These values are representative and should not be viewed as exact because surface conditions vary. The majority uncertainty in most pyrometric measurements is due to uncertainty in the values for emissivity.

TYPES OF PYROMETERS

A radiation pyrometer consists of optical components that collect the radiant energy emitted by the target object, a radiation detector that converts the radiant energy into a electrical signal, and an indicator that provides a readout of the measurement. Several different types of radiation pyrometers are used for industrial temperature measurement. These include the optical pyrometer, the total radiation pyrometer, and the ratio pyrometer.

The optical pyrometer, also known as the brightness pyrometer, requires manual adjustment based on what it views through a sighting window. Because it relies on what can be seen by the human eye, an optical pyrometer is designed to respond to a very narrow band of wavelengths that fall within the visible light portion of the electromagnetic spectrum. A filter is used to obtain the desired range of wavelengths. Typically, a red filter is used because it will transmit radiation with wavelengths above 0.63 micron and will block radiation with smaller wavelengths. The human eye has diminishing sensitivity to wavelengths above 0.63 micron. Therefore, the red filter and the human eye act together as a band-pass filter for the radiation that enters the pyrometer.

In addition to the filter, the pyrometer housing contains an objective lens that focuses the radiant energy emitted by the target object on a lamp filament. The lamp filament, supplied with current from the measuring circuit, also emits radiant energy and serves as a reference.

Figure 1-139 shows the basic configuration of a typical optical pyrometer and its principle of operation. The target object is viewed through a microscopic lens and filter and appears as a background behind the lamp filament. The brightness of the filament light is manually adjusted by changing the lamp current until the filament disappears against the background. This causes the lamp current to be proportional to the energy emitted by the target object. Since the lamp current of the pyrometer is calibrated against the temperature of a blackbody, measurement of the current provides a temperature reading of the target object.

Optical pyrometers are versatile, provide reliable performance, and, when properly calibrated and used, provide a good degree of accuracy.

Table 1-14. Emissivities of Metal (Surface Unoxidized)

Material	Emissivity (of)					
	25°C	100°C	500°C	1000°C	1500°C	2000°C
Aluminum	0.022	0.028	0.060			
Bismuth	0.048	0.061				
Carbon	0.81	0.81	0.79			
Chromium		0.08				
Cobalt			0.13	0.23		
Columbium					0.19	0.24
Copper		0.02		(liquid 0.15)		
Gold		0.02	0.03			
Iron		0.05				
Lead		0.05				
Mercury	0.10	0.12				
Molybdenum				0.13	0.19	0.24
Nickel	0.045	0.06	0.12	0.19		
Platinum	0.037	0.047	0.095	0.152	0.191	
Silver		0.02	0.035			
Tantalum					0.21	0.26
Tin	0.043	0.05				
Tungsten	0.024	0.032	0.071	0.15	0.23	0.28
Zinc	(0.05 at 30°C)					
Brass	0.0350	0.035				
Cast iron		0.21		(liquid 0.29)		
Steel		0.08		(liquid 0.28)		

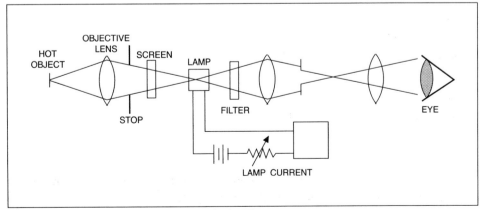

Figure 1-139. Typical Configuration of an Optical Pyrometer

Table 1-15. Emissivities of Materials at 0.65 Micron

Material	Emissivity at 0l.65μ	
	Solid State	Liquid State
Beryllium	0.61	0.61
Carbon	0.08-0.93	
Chromium	0.34	0.39
Cobald	0.36	0.37
Columbium	0.37	0.40
Copper	0.10	0.15
Erbium	0.55	0.38
Gold	0.14	0.22
Iridium	0.30	
Iron	0.35	0.37
Manganese	0.59	0.59
Molybdenum	0.37	0.40
Nickel	0.36	0.37
Palladium	0.33	0.37
Platinum	0.30	0.38
Rhodium	0.24	0.30
Silver	0.07	0.07
Tantalum	0.49	
Thorium	0.36	0.40
Titanium	0.63	0.65
Tungsten	0.43	
Uranium	0.54	0.34
Vanadium	0.35	0.32
Yttrium	0.35	0.35
Zirconium	0.32	0.30
Steel	0.35	0.37
Cast Iron	0.37	0.40
Monel	0.37	
Chromel P (90 Ni–10Cr)	0.35	
80 Ni–20 Cr	0.35	
69 Ni–24 Fe–16 Cr	0.36	
Alumel	0.37	
90 Pt–10 Rh	0.27	

Table 1-16 shows the calibration uncertainty for both primary standard and good commercial optical pyrometers at different temperatures.

Automatic optical pyrometers use solid-state photodetectors, or photomultiplier tubes, to replace the human eye as the detector. In this type of pyrometer, the lamp current is automatically adjusted by a detector that alternately views the target and the lamp filament. Figure 1-140 shows the basic components of a typical automatic optical pyrometer and its principle of operation.

Table1-16. Optical Pyrometer Calibration Uncertainty

Range		Range 1			Range 2			Range 3			Range 4		
Temperature	(˚C)	800	1064	1235	1100	1400	1750	1500	2300	2725	2500	2725	3524
IPTI Uncertainty	(˚C)	0.5	0.12	0.15				0.32	1.5	2.4			
Pyrometer Instability	(˚C	0.5	0.4	0.4				1.3	1.4	1.6			
Transfer Error		0.2	0.2	0.21				0.4	0.5	0.5			
Maximum Error		1.2	0.7	0.8	1.2	1.2	1.6	2.0	3.4	4.5	5.8	5.7	8.7

Light from the target enters the pyrometer through an objective lens and is focused on an optical modulator. The modulator is usually a mirror or disk that oscillates at high speed and alternately passes the radiation from the target and the lamp filament to a filter. The filter passes radiation at the selected wavelength from the target and the lamp filament to a filter. The filter passes radiation at the selected wavelength from the target and the lamp filament to the photodetector. The detector then compares the radiation emitted by the target with the radiation from the filament. Using a null-balance system, the current through the lamp filament is automatically adjusted until the radiation emitted by the filament equals the radiation emitted by the target. Measurement of the adjusted lamp current provides an indication of target temperature. Since lamp current is continuously adjusted by a self-balancing operation, temperature is indicated continuously.

Another type of pyrometer that is commonly used for industrial temperature measurement is the total radiation pyrometer. A total radiation pyrometer responds to wavelengths in both the visible and the infrared portions of the spectrum. Ideally, it would measure all wavelengths within this range. However, the glass window filters out some wavelengths. Any gases or vapors between the target and pyrometer will also attenuate certain wavelengths.

In typical radiation pyrometer designs, radiant energy emitted by the target passes through a lens or is reflected by a mirror and focuses on a heat-sensitive element that serves as the detector. The temperature of the element is converted into an electrical signal that provides a temperature readout. A thermopile (multiple thermocouple) is commonly used as the detector. Other devices used for detection include photocells, RTDs, and thermistors.

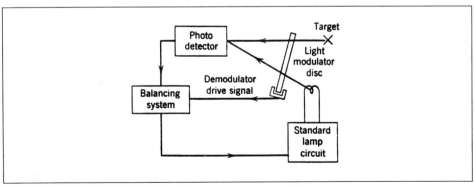

Figure 1-140. Basic Components of a Typical Automatic Optical Pyrometer

While thermal detectors are responsive to a wide range of wavelengths, different optical components are used for different applications. Pyrometers with glass lenses are used for high temperatures, quartz lenses are used for medium temperatures, and mirrors are used for low temperatures (down to 150°C).

Total radiation pyrometers are based on the Stefan-Boltzmann law, which states that the total radiation is proportional to the fourth power of the temperature. These pyrometers are calibrated using a blackbody and, therefore, measure the temperature based on the total radiation a blackbody would emit. As a result, it is necessary to account for the emissivity of the target object to determine the true temperature. The true temperature (T) is related to the indicated temperature (T_1) amending the Stefan-Boltzmann law as follows:

$$T = T_1 (\varepsilon)^{-\frac{1}{4}}$$

The filtering effect of glass windows and any intervening gases or vapors may also affect calibration. Consequently, total radiation may not vary exactly as T^4. One way to overcome the filtering effect in field applications is to use an optical pyrometer to calibrate the total radiation pyrometer under actual operating conditions. This technique is feasible because optical pyrometers can be designed to be quite insensitive to the absorption of selected wavelengths by intervening glass, gas, or vapor.

Another type of pyrometer, which is closely related to the total radiation pyrometer, is the ratio pyrometer, which provides accurate measurement of surface temperature. A ratio pyrometer uses a quantum or photon detector to measure the radiation at two different wavelengths in the visible light spectrum. The ratio of the two measurements is used to determine the temperature. The operation of a ratio pyrometer is based on the principle that energy radiated at one wavelength increases with temperature at a different rate from radiation at another wavelength. Measuring radiation at two wavelengths reduces the effect of the target's emissivity on the measurement. However, the accuracy of the temperature measurements depends on the target's having the same emissivity at the two wavelengths or emissivity that does not change rapidly with wavelengths.

In one type of ratio pyrometer, the two wavelengths are provided by a rotating filter wheel. The filter wheel consists of a rotating disk with red windows that pass radiation only at the red wavelength of the spectrum, and blue windows that pass radiation only at the blue wavelength. As the disk rotates, it provides the detector with alternate radiation pulses at two different wavelengths. The detector senses incident energy in the form of surface electrons that are released as a result of radiation incident at certain frequencies. The pulses from the two wavelengths are amplified, separated, and compared in amplitude. The resulting ratio is a function of temperature.

Another ratio pyrometer design uses two detectors. The two wavelengths from the target are split by a special mirror, which then passes each signal to a separate detector. To prevent measurement errors, both detectors must be maintained at the same temperature or any difference in temperature must be compensated.

The ratio pyrometer provides measurement of the true temperature even though the target is not a blackbody. It is considered to be the most accurate instrument for measuring surface temperature, provided the target's emissivity at the filter wavelengths is constant or changes very slowly.

Each type of radiation pyrometer has distinctive features that influence the selection of a pyrometer for a specific application. Optical pyrometers are simple to use and provide accurate measurement of radiation within the relatively narrow band of visual light. Total radiation pyrometers respond to both visible and infrared radiation and have a high output signal level that provides good sensitivity. Ratio pyrometers provide very accurate measurement of surface temperature, providing the object has a constant of slow changes in emissivity.

PYROMETER ACCURACY

In addition to proper calibration, several techniques can help to ensure accurate measurement with radiation pyrometers. It is important that the radiation observed by the pyrometer is actually emitted by the target object, not radiation reflected from its surface. There is no relationship between the temperature of a surface and the radiation that it reflects. One technique for ensuring that emitted radiation rather than reflected radiation is being observed is to drill a hole in the target object and aim the pyrometer into the hole. It is recommended that the depth of the hole be about five times its diameter. This technique provides a closer approximation to blackbody radiation than surface radiation.

Measurement accuracy can also be affected by the presence of gases or vapors between the target and pyrometer. Gases and vapors can filter out some radiation wavelengths. One technique for resolving this problem is to use fans to disperse any gases or fumes. A film of dirt on the viewing window or lens will also affect measurement accuracy. In some applications, it may be necessary to use a purge to prevent soot or other particles from being deposited on the viewing window or lens.

Uncertainty regarding the emissivity of target objects also affects the accuracy of measurements. One method for overcoming problems due to uncertainties in emissivity is to use a sighting tube to create a condition that approximates a blackbody. A long tube with a sealed end provides a good approximation of a blackbody if the pyrometer is sighted at the closed end. The sighting tube is inserted into the medium being measured. This technique can be used in applications where suitable tube materials, such as refractory metals, are available.

Weight Measurement

Weight is a variable that is required to determine the level of solids in a storage bin or silo, the transfer of solids on a conveyor belt or the discharge rate of a screw conveyor, and, of course, the net weight of the product itself.

Weight is defined as the force exerted on the object by gravity. Gravity determines the weight of an object. Gravity drives a mass at a downward acceleration of about 32.2 ft/sec^2 (9.8 m/sec^2). There is a relationship among weight, force, and mass, which can be expressed as:

$$W = mg$$

where :

W = weight
m = mass
g = acceleration due to gravity

Since gravity is considered constant, weight values are used instead of mass. It may be noted that the terms "weight" and "force" are used interchangeably. Weighing devices are divided in five groups:

(1) Mechanical lever scales

(2) Hydraulic load cells

(3) Pneumatic load cells

(4) Electric load cells

(5) Nuclear weigh scales

Mechanical Lever Scales

The mechanical lever scale is the oldest known weighing device. There are three classes of levers, each class with a different "multiple." The multiple of a lever of a given length varies with its class. For a lever to be in equilibrium, the following equations apply:

$$\text{Load} \times \text{Load arm} \quad = \quad \text{Power} \times \text{Power arm}$$

$$\text{Mechanical advantage} \quad = \quad \text{Load/Power}$$

$$\text{Multiple} \quad = \quad \text{Power arm/Load arm}$$

Load arm is defined as the distance from the fulcrum to the point of application of the load, measured perpendicular to the direction of the load force.

Power arm is defined as the distance from the fulcrum to the point of application of the power (the counterbalancing force in a scale), measured perpendicular to the direction of the power force. Figure 1-141 shows an even arm balance scale.

The scale mechanism is designed in such a way that without any weights or loads on either pan, both arms will be level. An unknown weight (L) is placed on one pan and is balanced using very accurate weights (P) on the other pan. The total of the known weight (P) is equal to the unknown weight (L). Another simple scale is the steel yard. With this type of scale, the length of the power arm is variable and the length of the load arm is constant. The length of the power arm is directly proportional to the weight.

Other mechanical scales are the pendulum scale, the spring-balance scale, and a combination of levers and a spring-balance scale. Most mechanical scales are a combination of simple levers, with the final indication utilizing an indicating dial or a transmitting device.

Hydraulic Load Cells

The operation of hydraulic load cells is based on the principle of a force counterbalance. It consists of a piston and cylinder, whereby the load exerts a force upon the piston, which in turn is supported by a hydraulic fluid confined within a sealed chamber. The pressure exerted in the hydraulic fluid chamber is proportional to the force exerted on the weight supporting piston.

$$P = F/A$$

where:

P = pressure, lbs/in.^{2} or kg/cm^2
F = lbs or kg
A = area, in.^{2} or cm^2

Since the area of the piston is constant, the hydrostatic pressure of the fluid is directly proportional to the force. Figure 1-142 shows a section of a hydraulic load cell. The full-cale deflection of the diaphragm is very small, approximately 0.03 inch. The hydraulic pressure is indicated by a Bourdon tube manometer, which has a calibrated scale in weight units. In practice it is necessary to calibrate a hydraulic load cell weighing system such that a zero is indicated when a load is exerted by the supporting structural members, tanks, load transmitting members, and the weight of the hydraulic fluid, without the load of the product. This calibration method is called zeroing the system or the tare weight compensation.

When several load cells are utilized to obtain the total weight of a structure or tank being supported at several points, the outputs of the cells may be added by using a hydraulic totalizer.

The output of a hydraulic load cell may be connected to a pressure transmitter, which will convert the pressure value to any type of signal for input to a controller, recorder, or computer.

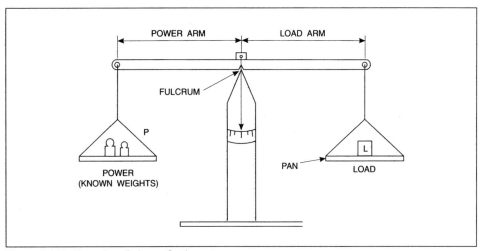

Figure 1-141. Even Arm Balance Scale

Pneumatic Load Cells

Pneumatic load cells are quite similar to hydraulic load cells in that the applied load is balanced by a pressure acting over a resisting area, with pressure becoming a measure of the applied load. Pneumatic load cells commonly employ diaphragms of a flexible material rather than pistons, and they are designed to automatically control the balancing pressure. A typical arrangement is shown in Figure 1-143. Air pressure is supplied to one side of the diaphragm and allowed to escape through a position-controlling bleed valve.

Application of an unknown weight or force (F_i) causes a diaphragm deflection (x) and an increase in pressure (P_o) since the bleed valve is more nearly shut off. The increase in pressure acting on diaphragm area (A) produces an effective force (F_p) that tends to return the diaphragm to its original position. For any constant

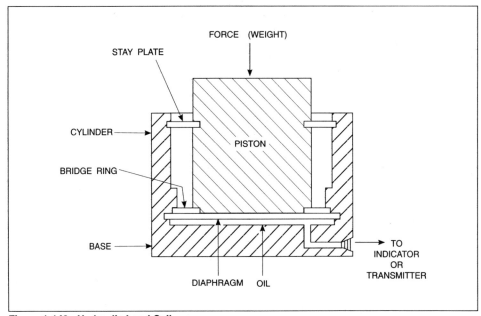

Figure 1-142. Hydraulic Load Cell

input force (F_i) the system will come to equilibrium at a specific bleed valve opening and corresponding pressure (P_o). The static behavior is given as:

$$P_o = (F_i - P_o A) k_d k_n$$

where:

k_d = diaphragm compliance
k_n = bleed valve gain
A = effective area

and k_n may vary somewhat with x.

This type of load cell is applied to weighing systems in which the maximum load per cell does not exceed 9,800 lbs. The same calibration procedure is required for the pneumatic load cell as for the hydraulic load cell with respect to zeroing.

Electric Load Cells

Electric load cells, the most widely used weight sensors today, are of two types: strain gage cells and induction cells. Cantilever load cells using a strain gage is a configuration designed to measure very small loads.

Strain gage load cells are available for either compression, tension, or universal loading of either type.

Tensile Load Cells

Tensile load cells, as illustrated in Figure 1-144, are widely used with lifting machinery. Load is transmitted to the tensile member by two end fittings; to increase its sensitivity, the tensile member is sometimes made with a tubular cross section. An end-load-measuring strain gage bridge is employed, and usually the dummy gages for temperature compensation are bonded to the specimen at right angles to the line of principal strain.

Compressive Load Cells

Compressive load cells are often used in weighing platforms and structures that may be supported by a compressive member, such as the one illustrated in Figure 1-145.

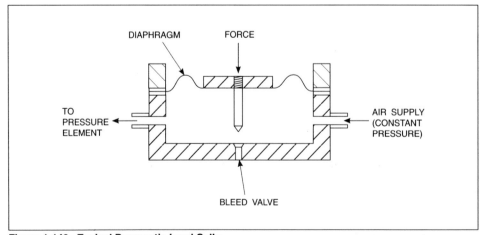

Figure 1-143. Typical Pneumatic Load Cells

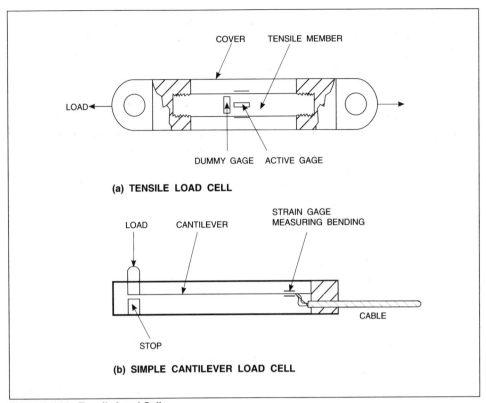

Figure 1-144. Tensile Load Cells

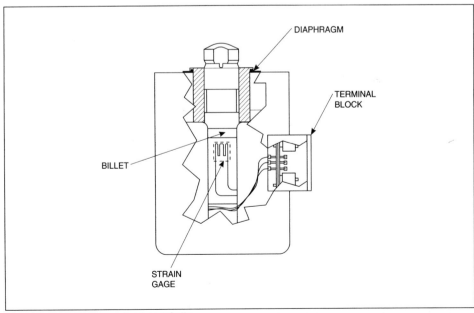

Figure 1-145. Compressive Load Cells

Universal load cells are made in such a way that they can be used for either type of load. Some strain gages have "double bridges" for a dual output signal. Output signals of strain gages are small and are proportional to the excitation. The values range from 1 to 3 millivolts per volt of excitation. Excitation voltage can be AC or DC.

The load cells should be protected from angular or nonaxial loads. Since strain gage load cells are often exposed to temperature changes, they are provided with temperature compensation for zero shift and span. This is accomplished by making the strain wires out of temperature-insensitive alloys and using compensating resistors in the bridge circuit.

The output of several cells may be added electrically to obtain the total weight or force exerted when it becomes necessary to support the load with multiple load cells.

Figure 1-146 shows a schematic wiring diagram of a load cell arrangement using a Wheatstone bridge circuit with strain gages as arms in the bridge.

An induction type of load cell consists of a cell case that is the load bearing member of the unit. One end of the case is the cell dome, which will deform slightly when a load is applied to it. This deformation is detected by ferrite inductors and a movable armature that is attached to the dome. When the inductor coils are excited by a low-voltage, high-frequency electrical input, the output voltage of the cell varies directly with the change in the position of the armature. The output, therefore, is directly proportional to the load when allowances for tare weight are considered. The input voltage may range from 5 to 150 millivolts. The deflection of the dome is 0.003 inch at full load. These cells are free from drift due to temperature changes.

Conveyor Belt Scales

Belt conveyor weighing systems are utilized to measure the flow of solids. A section of the conveyor belt will be weighed and this measurement combined with the belt speed to obtain a total weight measurement. The weight of the belt and the material it contains is transmitted to the load cell. If the belt is running at a continuous speed, the total amount of weight passing over the scale is easily weighed by integration.

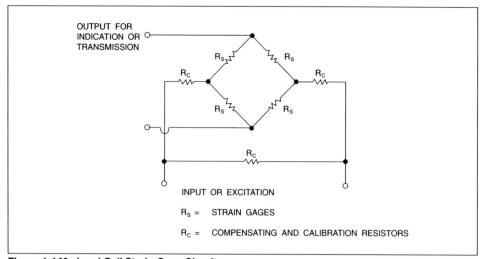

Figure 1-146. Load Cell Strain Gage Circuit

If the belt speed is variable or if close accuracy is required, a belt speed detector and transmitter must be installed. The total weight is obtained by multiplying the two signals. The belt speed is usually mounted under the belt with the roller in contact with the belt. Strain gage load cells are used for this purpose. The load is transmitted via a lever system to the load cell. A tachometer, generator-driven off the idler roll, supplies an excitation voltage to a potentiometer, whose output is the product of the belt speed and the load on the load cell. This signal is converted to a standard electronic signal that represents feed rate.

The Sankyo Impact Line Flowmeter

Another method to measure a solid flow is by way of an impact flowmeter. It measures the amount of solids coming off a conveyor belt or discharged from a screw conveyor (see Figure 1-147).

The flowmeter is designed to function on the principles of momentum and impact. If a particle of mass (M) is dropped from a height (H) onto a plate inclined (θ) degrees from the vertical, it will hit the impact plate with a velocity V_1, and be deflected and bounced off with a velocity V_2. When the particle hits the plate, it will impart an impulse to the plate that is proportional to the change in velocity of the particle. The impulse is sensed by an LVDT transmitter, and transients are smoothed out in a viscous damper.

Since force is proportional to flow rate, movement is proportional to force, and voltage is proportional to movement. Therefore, the voltage out of the linear voltage differential transformer is directly proportional to the flow rate of material over the plate. In some applications a strain gage load cell is used instead of an LVDT transmitter. An amplifier supplies an excitation voltage to the force sensor (load cell or LVDT) and amplifies the output of the force sensor. The output can be used for digital indication, totalizing, or analog signal conversion.

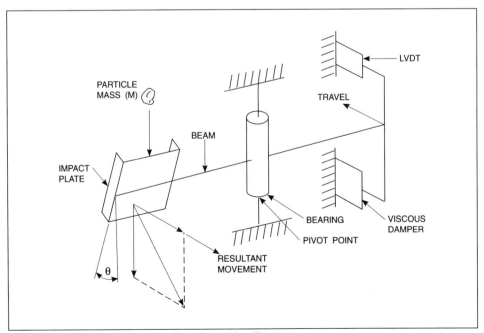

Figure 1-147. Theoretical Model of an Impact Line Flowmeter

Nuclear Gage Conveyor Belt Scale

This type of belt scale offers a noncontact weight measurement. The principle of radiation is similar to that of the nuclear level gage discussed previously. Gamma rays emitted from a source impinge on the target (the material load on the belt), where the rays are absorbed. An increase of material on the belt increases the amount of radiation absorption and decreases the input signal to the detector, which, in turn, produces a low-level current output. This signal and a belt speed signal go to a multiplier that provides a signal that represents the solids flow rate.

In-Line Weighing

In high-speed weighing, manual reading is not possible. Thus, weighers are divided into three categories. The first category includes those weighers with low natural frequencies and a display. These weighers are very accurate, have good repeatability, and are rugged. A second category of weighers has stiff balances with natural frequencies in the hundreds of cycles per second. The third category of weighers combines the principles of the spring balance with electrical output. These weighers have the advantage of comparatively large movements to operate secondary transducers.

A load cell, or weigher, in a dynamic system must not be considered in isolation. It must function as part of a complete feed, weigh, and disposal system. The feed system is made up of conveyors, hoppers, and gates that bring the product to the weigher platform and dispense it as demanded by the weighing cycle. The form the feeder takes depends on the product. It can be a simple roller conveyor that carries slabs of metal over a load cell, or it can be a complex vibrator conveyor. Similarly, the disposal system may simply terminate in a stacking area, or it can be a complex bagging operation.

In-Line Package Weighing

With in-line weighing systems, the arrival of the load on the weighing platform produces a shock to the weigher, followed by an unstable period as the package runs onto the platform. This causes the weigher to oscillate. Following this, a stable period occurs when the package is completely on the platform, and a measurement is possible. The size of the package, its velocity, and the natural frequency of the weigher all determine the rate at which packages may be weighed. A typical roller conveyor and weighing platform is shown in Figure 1-148, and the load cell response to a typical weighing cycle is shown in Figure 1-149.

In-Line Product Weighing

In-line weighing of products presents a different set of problems from those encountered in package weighing. Figure 1-150 is a sketch of a typical weighing system.

The product is brought to the weigher by a conveyor and discharged into the hopper through a lip gate. When the correct amount of product has been collected in the hopper, the lip gate is closed and the discharge gate is opened to fill one of the containers. As a further change is being loaded into the hopper, a conveyor moves the containers forward, so that the next one is in position.

Careful design considerations to avoid spillage after the lip gate has closed must be taken in consideration.

Rate Weighers

A rate weigher operates on the principle that a conveyor system and its contents may be weighed and the speed of the motor adjusted such that a constant feed rate is maintained. The product of feed rate and time provides a measurement of the amount delivered. An alternative system weighs the product in an auxiliary hopper, and the rate at which the hopper loses weight is a measure of the feed rate. The system illustrated in Figure 1-151 employs a conveyor that is weighed by a load cell and its speed is measured by a coil and magnet tachometer, illustrated in Figure 1-152. The tachometer is a pulse counter and converts the valve into a measurement of speed.

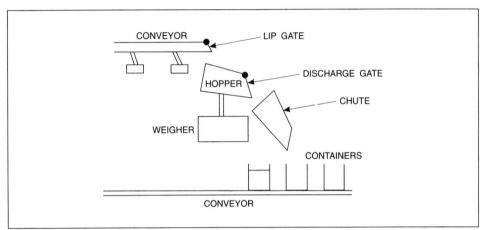

Figure 1-148. Typical Roller Conveyor and Weighing Platform

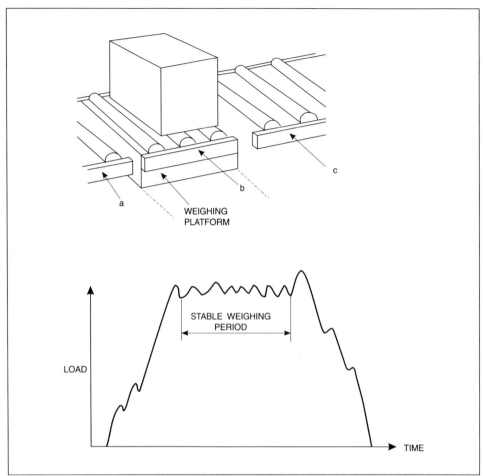

Figure 1-149. Load Cell Response to Typical Weighing Cycle

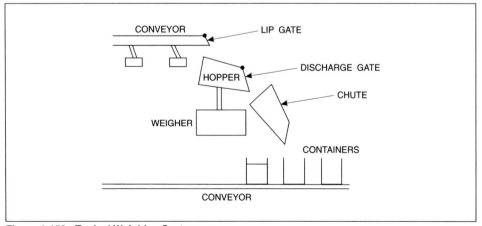

Figure 1-150. Typical Weighing System

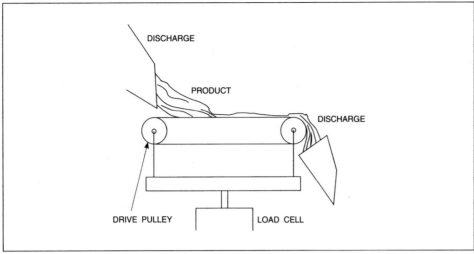

Figure 1-151. Conveyor Weighed by a Load Cell

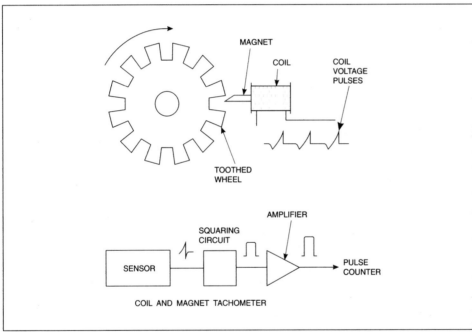

Figure 1-152. Coil and Magnet Tachometer

Bibliography

1. Albert, Charles, *Industrial Process Control and Automation*: McGill University, General Studies Department, 1982.

2. Eckman, D.P., *Automatic Process Control*: John Wiley, 1958.

3. Gupta, S.C., and L. Hasdorff, *Fundamentals of Automatic Control*: John Wiley, 1970.

4. Hougen, J.O., *Measurements and Control Applications*, 2nd ed.: Instrument Society of America, 1979.

5. Hughes, T.A., *Measurement and Control Basics*: Instrument Society of America, 1988.

6. Kallen, Howard P., *Handbook of Instrumentation and Controls*: McGraw-Hill, 1961.

7. Liptak, Bela G., *Instrument Engineer's Handbook*: Vol. I-Process Measurement Control: Chilton Book Co., 1970.

8. Luyben, W.L., *Process Modeling, Simulation, and Control*: McGraw-Hill, 1973.

9. Murrill, P.W., *Automatic Control of Processes*: International Textbook Co., 1967.

10. Shinskey, F.G., *Process Control Systems*: McGraw-Hill, 1979.

11. Smith, C.A., and A.B. Corripio, *Principles and Practice of Automatic Process Control*: John Wiley, 1984.

12. Weber, T.W., *An Introduction to Process Dynamics and Control*: John Wiley, 1973.

13. Weyrick, R.C., *Fundamentals of Automatic Control*: McGraw-Hill, 1975.

About the Authors

Sudhendu N. Banerjee is Deputy Manager of Instrumentation, Ltd., Kota India. He was involved in the development of an expert system package for maintenance forecasting of high voltage power transformers at Institut de recherche d'Hydro-Quebec and McGill University, 1990-1991. He earned degrees at L.D. College of Engineering, the Indian Institute of Technology, and McGill University. His experience includes performance and guarantee testing and acceptance testing for systems after plant commissioning and computer-based system design, engineering, installation, and commissioning of data acquisition systems for thermal power plants.

Zdzislaw Victor Barski earned his degrees in science and engineering at the Technical University of Lody and the Technical University of Warsaw (Poland). The patent holder and inventor of twenty-nine regulation and control devices, Dr. Barski's principal research interests include analysis modeling and simulation of the dynamic and static properties of thermal and industrial processes and the automation of industrial procedures by microcomputers, dynamic systems, and applied automation in the areas of sales energy and energy consumption.

Donald A. Coggan is owner/engineer of his own consulting firm in Quebec, Canada, which is involved in all areas of control, automation, and training including computer-aided training. A graduate of McGill University in Montreal, Mr. Coggan's early interest in the computer as an instrument for control and training led to his developing Computer*Ease, a computer literacy program, and CHINEASE, which is software that teaches writing and pronunciation of the Chinese language. His expertise in HVAC instrumentation and controls resulted in his writing and presenting short courses and training manuals in the field.

Mr. Coggan has written more than 60 technical papers and articles. He is a member of the Order of Engineers of Quebec, the Instrument Society of America, the Association of Professional Engineers of Ontario, and the American Society of Heating, Refrigerating, and Air-conditioning Engineers.

2

Analytical Instrumentation

Analytical instruments (also called analyzers) are one step removed from the measuring devices discussed in Chapter 1, but they still play a role that is similar to most other measuring instruments in process control (see Figure 2-1). The measurement principles involved in analyzers, however, are more sophisticated and generally involve chemicals—either as a variable or as an aid to measurement.

Analyzers have become essential to the process industry, especially when the variable concerned is the composition of a chemical. They are now indispensable instruments for quality control, process control, process troubleshooting, yield improvement, inventory measurements, safety, and waste disposal.

The need for more and better analyzers is the result of increased industry demand for improved productivity, greater competitiveness, reduced environmental pollution, and greater safety. There has been a tremendous increase in the types of analyzers and the applications for which they can be used. It is now no easy task for engineers, managers, and technicians to find the one analyzer that will get the job done at the right price.

A process analyzer may be defined as an unattended instrument that continuously or semicontinuously monitors a process stream for one or more chemical components or physical properties. The cost of an analyzer increases with its complexity, so one must first determine exactly what is wanted in an analysis.

If the process is to be controlled by a computer, the time needed to perform an analysis becomes important—it should be as short as possible, say, under five minutes. To avoid potential problems, a designer will select a continuous analyzer rather than a discontinuous one, wherever possible. To this end, the *ISA Directory of Instrumentation* is an excellent source of names and addresses of analyzer manufacturers.

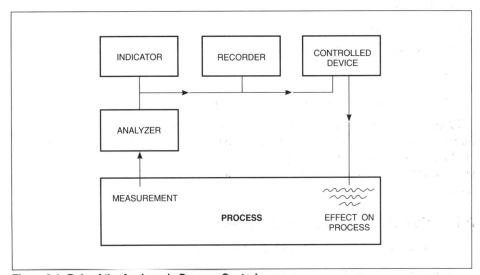

Figure 2-1. Role of the Analyzer in Process Control

Since no single analyzer combines all the needed features, a compromise must be made. The essential features are usually described in a standard specification form that helps the designer obtain quotations from analyzer manufacturers (see Figure 2-2).

Process Chromatography

The 1980s witnessed a significant increase in the use of on-line chromatographs in process control. In a typical process stream, the number of analyses that a process chromatograph performs in a week usually exceeds an entire year's output of traditional laboratory analyses. Chromatography is a versatile analytical technique in that it can make a variety of measures. For this reason, and also for its increasingly important role in process control, on-line chromatography is given a considerable amount of space in this chapter relative to other types of analyzers.

Process chromatography installations are not without their problems. Many early installations failed or required an enormous effort in man-hours to obtain a working system. People were often misled by the simplicity of the early laboratory chromatographs, and they assumed that merely enclosing one in an explosion-proof housing would turn it into an on-line process analyzer. Many difficulties ensued, the most serious one being sample handling. Laboratory samples are generally clean and easily handled; however, a process stream is comparatively dirty, given its usual content of particles, water, condensate, and so on. Nevertheless, sample-handling technology has evolved steadily to its greatly improved present state.

A frequently encountered problem results from trying to handle multiple streams with the same analyzer. It's virtually impossible for an operations department to resist the temptation to connect every stream in a unit to a single chromatograph, particularly where the streams have similar compositions. Be careful, however, because too many variables can lead to complications—and to mistakes. Keep the system simple, analyze for the critical components only, and display the minimum required number of variables.

When selecting a particular instrument, one should consider many factors, particularly:

(1) reliability,

(2) maintenance needed,

(3) downtime due to failure,

(4) overall design, and

(5) ability of the instrument to perform as specified.

Price is also important, but remember that an instrument that initially seemed like a bargain may ultimately cost far more because of changes made to correct "hidden weaknesses." Although process chromatographs are considered an expensive item, the resulting savings can frequently return the investment in a short time.

Principles of Process Chromatography

The basic process chromatograph system consists of:

(1) a means of sampling the gas to be analyzed,

(2) a regulated supply of carrier gas,

A successful installation is the culmination of careful planning and close attention to detail.

ANALYZER SPECIFICATION FORM

Project: .... Date: ....

Specification No.: .... Code: ....

Information Compiled by: ....

A. GENERAL INFORMATION:
1. Plant: .... 2. Unit: ....
3. Process: ....

B. CONDITIONS AT PLANNED ANALYZER LOCATION:
1. Ambient Temperature Range: ...... to ...... Normal ...... °C ☐ °F ☐
2. Protected from Weather: Yes ☐ No ☐
3. Unusual Ambient Conditions ....
   (Corrosive or explosive atmosphere, excessive moisture, dust, etc.)
4. Power Available: .... Volts........ to .... Hertz
   (a) Voltage Variation: .... to .... Volts
   (b) Frequency Variation: .... to .... Hertz
   (c) Grounding Facilities Available: Yes ☐ No ☐
5. Lighting Level: Good Average Poor
   (a) Front of Instrument ....
   (b) Back of Instrument ....
   (c) Direct Sunlight Will Strike Instrument: Yes ☐ No ☐
6. Steam Lines Near Location: ....
7. Instrument Air Available:
   (a) Pressure Range: from .... to .... Normal .... PSIA
   (b) Temperature Range: from .... to .... Normal .... °C ☐ °F ☐
   (c) Contaminants: ....
   (d) Size of Header: .... Volume: .... ft.³/min.
   (Use Separate Page for Each Stream to be Analyzed)

C. SAMPLING INFORMATION
1. Form of Sample: Gas ☐ Liquid ☐ Other ☐
2. Temperature Range: from ...... to ...... Normal ...... °C ☐ °F ☐
3. Pressure Range: from .... to .... Normal .... PSIA
4. Dew Point: .... °C At .... PSIG
5. Quantity Available: .... Per Hour
6. Low-pressure Return Line Available: Yes ☐ No ☐ Back-pressure .... PSIG
7. Specific Gravity ....
8. Contaminants in Sample:
   Oil ☐ Wax ☐ Solids ☐ Particle Size ....
   (Identity and concentration to be included in list of components below)
9. Corrosive Nature: Acid ☐ Basic ☐ Other ☐
10. Other Data (Viscosity, Unusual Surges, etc.): ....
11. Materials of Construction That May Be Used in Contact with
    Sample: ....
12. Distance from Sample Tap to Analyzer Location: .... ft.

13. Size of Tap at Process Line if Any: ....
14. Concentration Ranges of All Components (even if only traces) in
    Stream: (Specify unit of measurement: % by volume, % by weight or
    ppm for each component)

Components to be Analyzed				Other Components				
Max	Min	Normal	Unit	Water (vapor) (liquid)	Max	Min	Normal	Unit

15. The Above Stream Composition Information Is Considered Proprietary:
    Yes ☐ No ☐
16. Method of Lab Analysis Used to Measure Sample: ....
17. Desired Response Time of Analyzer: .... Minutes .... Seconds
18. Accuracy Required: % of full-scale reading ....

D. INSTALLATION REQUIREMENTS:
1. Type of Installation: Permanent ☐ Temporary ☐ Portable ☐
2. Type of Mounting: None ☐ Rack ☐ Panel ☐
   Other
3. Electrical Code: Class .... Group .... Division ....
4. Recorder or Indicator Required:
   (a) To Be Supplied with Analyzer: Yes ☐ No ☐
5. Location of Recorder: At Analyzer ☐ Distance from Analyzer ..... ft.
6. Recorder Mounting: None ☐ Rack ☐ Panel ☐
7. Accessories:
   (a) Alarms: High ☐ Low ☐ (b) Controls: ....
   (c) Others: ....
8. Date Required: ....
9. Sketch of System Indicating Sample Points and Distances to Analyzer:

Figure 2-2. Analyzer Specification Form
(From *Instrument Engineers' Handbook*, Liptak and Venczel, eds., © Bela G. Liptak, courtesy Chilton Book Company)

(3) an analytical column to achieve separation of the components of interest in the sample,

(4) a means of detecting when each component leaves the columns,

(5) a readout device for indicating and/or recording the peak values that result from the presence of the various components, and

(6) increasingly, a controller that integrates the measurements to produce a signal that can be used for closed-loop control.

Process chromatography involves combining the sample with a carrier gas and passing the combination of gases through a packed column, which is made of metal tubing and filled with an absorbent such as charcoal, silica gel, activated alumina, or ceramic balls (molecular sieve). The effect of moving the gases through the stationary bed of the column is the separation of the constituents of the sample gas. The bed substrate has an affinity for each component in the sample, and, since this affinity varies with each component, generally the lighter components are extracted or swept away (eluted) faster than the heavier ones. By selecting the proper column length, the components are separated from each other, and the individual elution time is used in the identification of a particular component. The carrier gas, which forces the sample gas through the column, emerges from the column continuously so that the constituents of the sample gas actually leave the column in combination with the carrier gas. The most common carrier gasses are helium, air, hydrogen, and nitrogen. A typical sample gas might contain constituents such as ethane, propane, butane, pentane, acetylene, and so on.

The immobile part of the column may be a solid that separates gas by adsorption, in which case it is defined as gas-solid chromatography, or it may be a liquid spread thinly over a support of crushed firebrick (packed column) or on the walls of tubing (capillary), in which case it is defined as gas-liquid chromatography.

Chromatographic System

The key feature of the chromatograph is its ability to effect a wide range of separations while still retaining simplicity of design. The basic chromatograph components are a carrier gas, a carrier gas regulator, a sample valve, a column, and a detector (see Figure 2-3).

The carrier gas carries the sample through the column and the detector and vents it to atmosphere. The carrier regulator maintains a constant pressure of carrier gas, which results in a constant carrier flow rate. The purpose of the sample valve is to inject a measured amount of sample, which is then separated into individual components by the column. The detector detects the individual components as they leave the column.

CARRIER GAS

Carrier gas to the column is provided by a cylinder of high-pressure gas equipped with a pressure regulator that ensures a constant flow. Commonly used gases are helium, nitrogen, argon, and hydrogen. The carrier gas should be:

(1) inert to avoid interactions with the sample or liquid phase,

(2) suitable for the detector used in the system,

(3) readily available at high purity,

(4) inexpensive, and

(5) nonhazardous in certain areas.

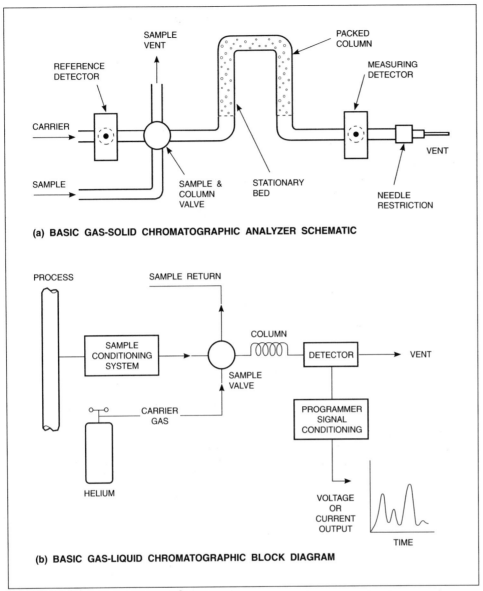

(a) BASIC GAS-SOLID CHROMATOGRAPHIC ANALYZER SCHEMATIC

(b) BASIC GAS-LIQUID CHROMATOGRAPHIC BLOCK DIAGRAM

Figure 2-3. Basic Chromatograph Components

Helium best meets these characteristics and is, therefore, the mostly widely used carrier gas.

VALVES

Valves are used for both gas and liquid samples. To automatically inject a repeatable sample into the column, several different types of valves can be used (rotary, piston, linear slide, and diaphragm, as described in detail in the following paragraphs). The valves that are used for injection are also used for column switching on backflush operations. Valves must be rugged in order to stay on line without failure. A sampling valve actuated every 10 minutes will perform 50,000 operations per year. Some valves have been tested for ruggedness in the laboratory for half a million operations without a failure. Valve reliability is a "must."

Rotary Valves. These valves consist of a flat Teflon™ disc with grooves cut into the face to connect the valve ports. The disc is rotated on a flat metal plate with the two surfaces spring-loaded to stay in contact with one another. Rotary valves are not usually recommended for process applications because of a tendency to leak readily. They are also restricted to low pressures.

Piston Valves. Piston valves are constructed in the form of a cylinder with a series of ports along it. A piston fitted with rubber O-rings fits into the cylinder with the intervals between the O-rings connecting the ports. Movement of the piston back and forth provides the flow operation. These valves are widely used in the laboratory for sampling gases and for backflushing operations. They are restricted to low pressures. The major disadvantage of the valves is the O-rings, which tend to leak eventually. While replacement requires only a few minutes in the laboratory, it is a major job on a process chromatograph.

Linear Slide Valves. These are sometimes referred to as sandwich valves, because a Teflon™ insert is sandwiched between two smooth stainless steel blocks. Grooves are cut in the insert to connect to ports in the steel blocks. The insert slides linearly back and forth to provide a matrix of port connections. The grooves are sized for injecting liquid samples (minimum, one microliter) or connected to ports for larger gas volumes. These valves are widely used, reliable, and capable of high-pressure operation exceeding 2000 kPa. They are pneumatically actuated, with switching done by solenoid valves.

Diaphragm Valves. These valves have a flat, stainless steel surface with six ports located in the form of a hexagon. A flexible diaphragm is fitted over the ports, and a set of six pneumatically actuated pistons is located between each pair of ports. The pistons operate in sets of three. In one mode, three pistons are pressed down on the diaphragm while the other three pistons are raised. To change mode, the piston positions are reversed. These valves have very low internal volumes and are capable of very high operational speeds and high pressure.

SAMPLE VALVE APPLICATION

For most installations, one will normally use either a linear slide valve or a diaphragm valve. Since both can be similarly represented schematically, and since both are pneumatically actuated, the linear slide valve configuration will be used in the figures in this chapter, although it should be noted that most applications can also be carried out with the diaphragm valve. The diaphragm valve would be preferred for use at pressures in excess of 1400 kPa.

Consider the application of a liquid sample valve in the sample inject position or in purge or normal flow position (see Figure 2-4). The volume of the sample is determined by the diameter of the shaded cylindrical part in the slider. The volume can be changed simply by changing the size of this cylindrical hole. Typical values would be 0.5 to 10 microliters. For the injection of liquids, sample pressure must be high enough to prevent the formation of gas bubbles in the liquid when the sample is heated to oven temperature, because the smallest bubble will have an effect on the volume of the sample injected. It is advisable to inject below 900 kPa. Sample valves are rated up to 2000 kPa sample pressure.

Consider now the application of a gas sample valve (see Figure 2-5). In this case, the desired flow pattern is achieved by cutting two slots and two holes in the slider, and the bulk of the sample injected is the volume of an external loop of 1/8- or 1/4-inch outside diameter stainless steel tubing. Gas volumes used would vary between 0.5 and 5.0 milliliters. For highly sensitive detectors and capillary columns, a small gas sample is necessary, and, therefore, a liquid sample valve can be used.

(continued)

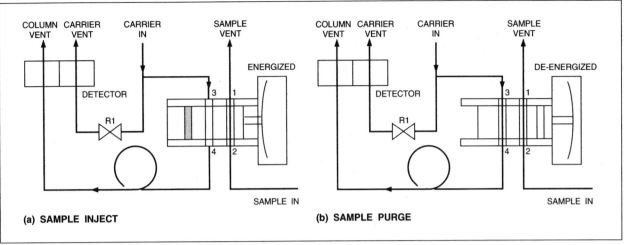

Figure 2-4. Liquid Sample Valve Application
(From *Analyzer Training Course Manual,* Gulf Oil Canada Ltd.)

(continued)

Since gas samples are taken at atmospheric pressure, a sample shutoff valve is recommended in order to stop sample flow prior to injection. In both valves, normal flow with the valve deenergized allows samples to purge through the required sample volume, while the carrier gas flows to the column. When the valve is energized, the volume of sample trapped in the loop or in the cylindrical hole is swept away by the carrier gas into the column. Throughout this period during which the sample is injected, the sample stream flows continuously through a bypass port in the valve and out the sample vent. After sample injection, air pressure on the outer side of the sampling valve diaphragm returns the slider assembly to normal position.

Watch out for high sample pressures. A check valve should be installed on the carier gas entrance of the analyzer to prevent process samples at high pressure from backing up into the helium and thus contaminating all the lines and other analyzers that are on the same manifolded system.

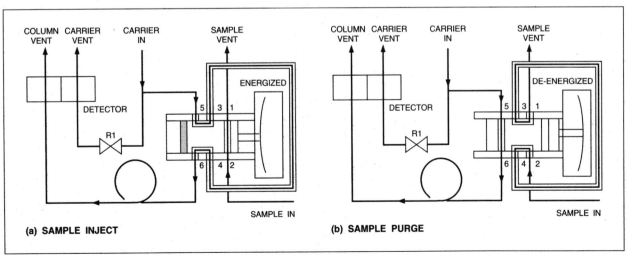

Figure 2-5. Gas Sample Valve Application
(From *Analyzer Training Course Manual,* Gulf Oil Canada Ltd.)

COLUMNS

The chromatographic column is the heart of the entire gas chromatographic process. The column is housed in a temperature-controlled oven maintained at ± 0.1°C of set point. The column tubing can be made from a number of materials, such as stainless steel, copper, aluminum, and glass, and may be formed into a number of shapes to fit into the chromatograph. Stainless steel is commonly used because it is less reactive and more resistant to corrosion than most other materials. Copper is highly reactive to hydrogen sulfide, mercaptan, and acetylene and would find little application around refineries and gas plants. Aluminum is useful in certain applications because it is easily formed into coils and is corrosion resistant under certain conditions. It is not suitable where acidic or basic components are involved. Due to potential breakage, glass is excluded from use in most process applications.

Packed columns vary in length from a few centimeters to 15 meters and are typically two to six meters. The efficiency of a column increases with length. Capillary columns can be considerably longer than the packed type and can range from 22 to 90 meters and, in some exceptional cases, even longer. Column diameters vary from 0.02 to 10 centimeters inside diameter, the smaller diameters having greater efficiency but requiring considerably smaller samples. Larger column diameters are associated with what is called *preparative chromatography*, which employs large samples (up to 150 milliliters) to obtain pure compounds and is frequently used in the commercial production of chemicals.

Gas chromatography is limited almost solely by the availability of columns to achieve the separations required. The columns may be packed with active solids or inert solids covered with a thin film of liquid to suit the requirement of continuous reliable operation demanded by process chromatography. The column materials selected must be able to retain their separating properties over long periods of time and in the face of possible contamination from the process stream. Hence, when active solids are required in the separation of light gases, precautions must be taken to ensure that water, which can have a fouling effect, does not enter the analytical column. When liquids are used as the separating medium, they should be used well below their published maximum temperatures to minimize bleeding of liquid phase off the column. Loss of liquid in this way would lead to loss of separation and changes in retention time. The peaks would seem to disappear because elution (a cleansing in the sense of separation) would not occur at the same time as the appropriate component gate is opened.

Columns are of two types: partition and adsorption.

Partition Columns. A partition column consists of granules of firebrick coated with a liquid phase. The granules of firebrick are called the support, and one brand name is Chromosorb™. The liquid coating that goes on the firebrick can be either nonpolar or polar. Nonpolar coatings are usually some type of oil, such as DC-200™ (Dow Corning 200 silicon oil), squalane (shark liver oil), and OV-101™ (Ohio Valley silicon oil). The liquid phases separate components according to boiling points. If there is a large enough difference in the boiling points of the components for them to be separated, a nonpolar coating will work well. Lower boiling point components have less solubility with the liquid phase than higher boiling point components. Solubility refers to the length of time the component stays in the liquid phase. If the components boil at temperatures that are too near one another, a polar coating should be used. An example of such a coating would be propane or propylene. Polar coatings are types of liquids consisting of adipate, nitriles, and Carbowax™. A polar coating works like a magnet. The more polar the component is, the longer it stays in the column. The less polar the component is, the faster it moves through the column. All components have polar characteristics.

Select a column that gives a good baseline separation between peaks. Where measurement of a trace component is required, select a column that elutes a minor component before a major one. This prevents the tail of the larger peak from swamping the small one.

If a partition column is not separating components, increase the liquid phase or length of the column to increase separation. If a partition column goes bad, flush the column with carrier gas only and increase the temperature of the oven (ensuring all the while that the temperature does not go so high that it destroys the column by losing the liquid phase). This is called reconditioning a column. If this doesn't put the system back into operation, replace the column.

Adsorption Columns. Adsorption columns separate permanent gas and light hydrocarbons according to molecular size—larger molecules go through the column faster than smaller molecules. Adsorption columns are normally used to separate components such as oxygen, nitrogen, carbon dioxide, carbon monoxide, methane, and ethane. The types of column packing materials used include Porapak™ (porous polymer), Chromosorb™ (porous polymer), Porasil™ (porous silica beads), and molecular sieve (zeolites).

COLUMN CONFIGURATIONS

When the analytical requirements of the process are too demanding for a single column, it is often possible to solve the problem by using a column switching valve to alter the configuration of the columns used. For most purposes, satisfactory results are obtained with a diaphragm valve or with a six-port linear slide valve of the type used for gas samples. The main techniques used are: (1) stripper column or backflush to vent, (2) dual column or peak park, (3) backflush to detector, and (4) heart cut.

Stripper Column. This configuration is used when some components in the stream could poison the analysis column. It has a stripper column followed by an analysis column, with provision for backpurging the stripper column to vent (see Figure 2-6). The analysis is started with both columns in series (this is called mode I). As soon as the components of interest are calculated to have entered the analysis column, the setup is switched (to mode II) in such a way that any unwanted component is vented. This can help prevent water from poisoning an active solid column. This configuration is also used when the stream contains components that take a while to elute and when it is not necessary to measure the concentration of these so-called heavy ends. In the analysis of condensate product from a natural gas plant, components up to N-pentane can be measured, while C6 and above are vented.

Dual-Column. This configuration (see Figure 2-7) allows the use of a combination of two columns, each used for the specific portion of the sample analysis for which it is necessary. The use of this configuration can decrease total analysis time and allow an analysis that would otherwise be impossible with a single column. If the stream to be analyzed covers a wide boiling range, the analysis can be arranged so that the light components first go through both columns in mode I.

To establish that a component of interest has entered the analysis column, an ITC (intercolumn) detector arrangement is a nice feature to have on a process gas chromatograph.

> To analyze a mixture of oxygen, nitrogen, methane, carbon dioxide, and nitrous oxide, for example, inject the sample in mode I with both columns in series. After the composite oxygen-nitrogen peak is deduced to have entered column 2 (but before methane has left column 1), change the configuration to mode II, thus parking the composite peak on column 2. Methane, carbon dioxide, and nitrous oxide are then eluted directly from column 1 into the detector. Once nitrous oxide has emerged, switch the configuration back to mode I to allow oxygen and nitrogen to elute from column 2 (see Figure 2-8). To calculate when a composite peak elutes from column 1, run a sample of air in mode II.

Switching to mode II then results in only the heavy components going through one column. The first column may accomplish a boiling point separation, whereas the second column may be polar in nature. If time does not allow the complete elution of components from column 2 prior to components from column 1, then once the required peaks are in column 2 they may be parked there in mode I, thus allowing the peaks to elute from column 2.

Backflush. This configuration (see Figure 2-9) is used to separate a mixture with a wide range of boiling points and obtain an analysis, at least in groups, of

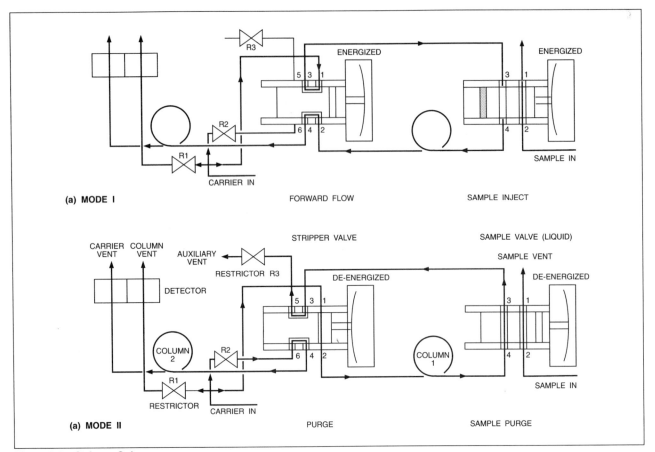

Figure 2-6. Stripper Column

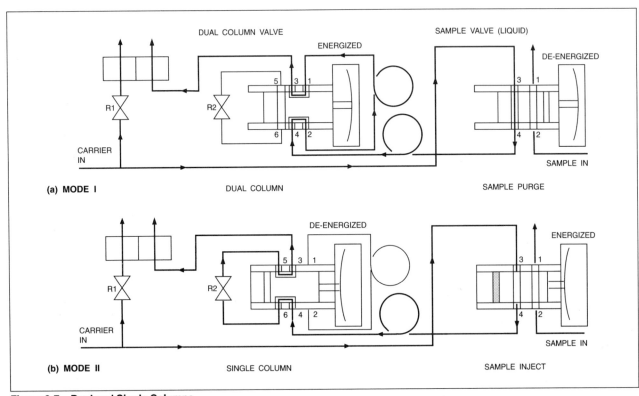

Figure 2-7. Dual and Single Columns
(From *Analyzer Training Course Manual,* Gulf Oil Canada Ltd.)

all components. The sample is injected in mode I. After elution of the light components of interest, the column is switched to mode II and the heavy components are eluted through the detector as one peak. This ensures a much faster analysis time than if the heavies were allowed to pass through the column in one direction.

Heart Cut. This technique is used to measure a trace component of the tail of the major peak in situations where it is not possible to reverse the order of elution. The manual monitor valve is set up so that the entire sample goes through both columns and the heartcut valve (see Figure 2-10). In this way, one can

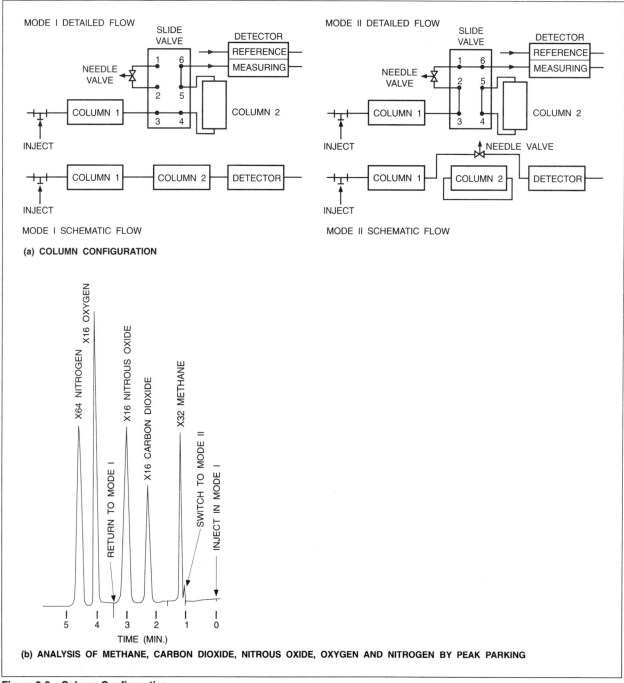

Figure 2-8. Column Configuration
(From *Analyzer Training Course Manual,* Gulf Oil Canada Ltd.)

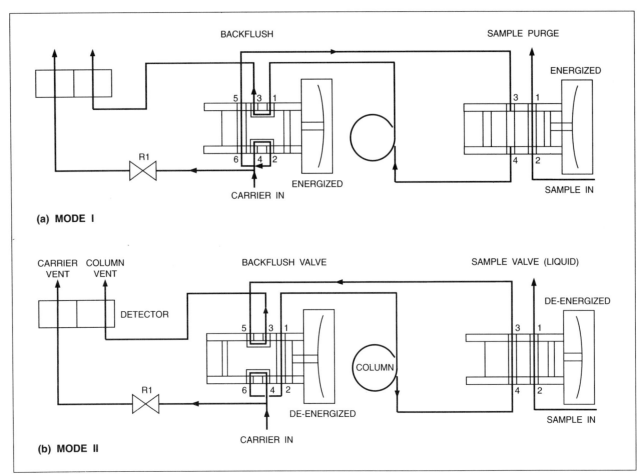

Figure 2-9. Backflush Configuration
(From *Analyzer Training Course Manual,* Gulf Oil Canada Ltd.)

> Assume the analysis of a refinery reformate stream requires separating C1 to C5 hydrocarbons and obtaining a value for the concentration of the heavy ends. Allow components up to C5 to elute in one direction, and then reverse the column to obtain C6s and above as a single regrouped peak. Should regrouping into one peak not be achieved, the addition of a short column between the detector and the valve will usually accomplish this regrouping.

deduce when the tail of the large peak and the minor component emerge from column 1 and enter the valve. In actual operation, the manual valve is set to direct the effluent of column 1 to vent with the column switch valve in mode I. By switching to mode II as the tail of the main peak is leaving column 1, it is possible to inject that portion of the tail containing the minor peak into column 2. The sample is now more easily separated in column 2, since the components are in a ratio more favorable to good separation. This technique is employed in the measurement of acetylene in high purity ethylene.

 All the above column configurations, including both liquid and gas sampling, are arranged to be fail-safe; that is, in case of a power failure, the columns will revert to a harmless configuration.

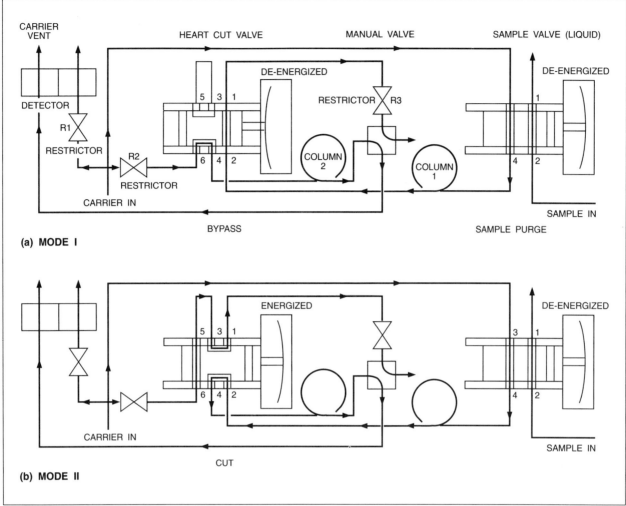

Figure 2-10. Heart Cut Configuration
(From *Analyzer Training Course Manual,* Gulf Oil Canada Ltd.)

DETECTORS

Many detectors are used in process chromatography. It is usually the analyzer vendor who selects the detector to match the specific application. The most commonly used types are thermal conductivity and flame ionization detectors.

Thermal Conductivity Detector (TCD). This type of detector measures the thermal conductivity of the carrier gas with and without a sample in it, and from these measurements the presence of a particular component in the carrier can be inferred. TCDs employ either a filament of heated metal (such as tungsten-rhenium alloy), commonly called a "hot wire," or a thermistor (a semiconductor of fused metal oxides) as the element used to sense changes in thermal conductivity. Hot wire detectors, more commonly used than thermistors, have much lower upper temperature limits, are noisier, and are subject to hydrogen attack (hydrogen reduces the metal oxides). For a complete description, see the later section that covers TCDs in more detail.

Thermal conductivity is not the sole mechanism for heat loss from the connecting filaments; there are three others—convection, end losses (conduction through leads), and radiation. For this reason, a TCD requires calibration data. Tables of calibration data, known as relative response factors, are available. The tables are prepared by determining the specific peak area per mole of unit weight and relating this to the specific area of a pure reference compound.

Flame Ionization Detector (FID). In this type of detector, the column effluent is mixed with hydrogen fuel and burned at the end of a metal jet, in an excess of air. When hydrogen only is burning at the jet, it produces a very low current. A component mixed with hydrogen produces ions on combustion; these ions collect at an electrode and increase the current. An electrometer amplifier converts the signal to a millivolt level for conventional recorder display. The FID is an ultrasensitive detector (1,000 to 10,000 times more sensitive than the TCD) capable of parts-per-billion detection on a routine basis. Another very convenient feature of this detector is that its response to all hydrocarbons, on the basis of weight, is virtually constant. This response characteristic is attributable to the production (on combustion) of carbon ions, which are, in turn, proportional in number to the carbon content of the eluted compound, thus producing a constant weight response for hydrocarbons.

Sample Handling

A major problem in most analyzer installations is obtaining a clean sample and supplying the instrument with this sample as being continuously representative of the process stream. The function of the sample system is to provide this sample. Note that this definition includes those parts (such as the pressure station) of the system that may be located in the sampling line remote from the box that contains the main sample conditioning system. It must be emphasized here that unless the sample system is correctly designed, constructed, and maintained to fulfill this function, the overall system will not work—no matter how efficient the chromatograph is.

REPRESENTATIVE SAMPLING

Since the purpose of sampling is eventual analysis, it follows that the sample should be taken from the process without introducing a significant change in the composition. The usual way to take the sample is by inserting a probe through the process pipe wall. This should be designed to minimize pickup of entrained liquids and solids and to prevent pickup of contaminants from the pipe wall.

Probes can be fairly complicated, with variations such as water sprays and coolers, but, in general, the simplest design that does the job well is the best.

TRANSPORTATION OF SAMPLE

Once withdrawn from the process, the sample must be conditioned and transported to the instrument without affecting the composition. Phase preservation of the gases must be retained as gases, and liquids as liquids. Under certain conditions, liquefied gases may be vaporized in a vaporizing regulator and handled as a vapor. In general, however, phase change is to be avoided in order to eliminate any possibility of fractionation or enrichment of one phase at the expense of another. System pressure and temperature should be compared to dew and bubble points to ensure phase preservation and should include a safety margin.

TIME LAG

The time lag consists of dead time and a time constant. The dead time here is that amount of time necessary for straight replacement purge. The time constant is determined by the extent of mixing, which will be large if dead-ended volumes are in the system. Chromatography is a batch type of operation that introduces its own inherent time lag; it is essential, therefore, to minimize all other time lags to ensure keeping in touch with the process (especially if it's a continuous operation). For this reason, sample lines should be as short as possible and of the lowest possible volume without introducing an unacceptable pressure drop. Sample flow rates must be rapid, and bypass loops should be used whenever possible. Straight replacement purge time, *t,* for vapor is:

$$t = \frac{VL + C}{F_S} \frac{P + 101}{101} \qquad (2\text{-}1)$$

where:

V = tubing volume, cc per meter
L = total tubing length, meters
F_S = flow rate at standard conditions (101 kPa, 15°C), cc per minute
P = sample line pressure, kPa
C = volume of modified components (filters), cc

For liquids, the pressure term would be omitted. The prediction of flow and pressure drop is necessary in order to establish correct transport time and to size sample lines effectively. The effect of changes (such as pressure drop, temperature, specific gravity, and so on) across needle valves can be estimated using traditional valve sizing formulae. For example, for gases,

$$Q_G = \frac{C_v \sqrt{p \times P}}{(460 + T) G} \qquad (2\text{-}2)$$

where:

Q_G = gas flow rate, scfh
p = pressure difference, kPa
P = absolute inlet pressure, kPa
T = fluid temperature, °C
G = specific gravity, referenced to air
C_v = valve sizing coefficient

For liquids,

$$Q_L = \frac{C_v \sqrt{P}}{G} \qquad (2\text{-}3)$$

where:

Q_L = liquid flow rate, liters per minute
P = absolute inlet pressure, kPa
G = specific gravity, referenced to water
C_v = valve sizing coefficient

When needle valves are used, pressure adjustments are necessary either to reduce and control pressure or to raise and control pressure to obtain proper system control. Pressure reduction and pressure control require an automatically adjusted restriction, and these devices are generally used in two different ways. A reducing regulator automatically senses downstream pressure and operates an internal valve to maintain this pressure constant. Usually a diaphragm is used to sense pressure change, with the diaphragm-to-orifice area ratio determining sensitivity. Temperature changes are a source of error and are related to expansion or contraction; the remedy is to house the regulator in an environment in which the temperature is constant or at least variable within a small range.

A backpressure regulator is an automatic device that senses inlet pressure and adjusts an internal valve to maintain this pressure constant. Its operation is opposite that of a reducing regulator: the valve closes rather than opens when pressure falls. Again, the same relationship between diaphragm and orifice area determines sensitivity.

Pumping may be needed to increase the pressure high enough to ensure working above the bubble point of the sample or to ensure enough sample flow to the analyzers. In general, any pump can be used, but selection should be made only after due consideration of such things as materials of construction, possible contamination or leakage, electrical hazard, change of phase of sample going through the pump, and so forth. For repeatable measurements with gas samples, pressure must be kept constant.

SAMPLE SYSTEM DESIGN PITFALLS

The sample system should not be oversimplified. Lack of a filter could put the analyzer sample valve out of commission. All materials in contact with the sample must be corrosion-resistant; this includes pressure regulator diaphragms, the Bourdon tube inside pressure gages, and so on. Corrosive products will ultimately plug the system. Stainless steel should be used almost exclusively. On the other hand, the sample system should not be so complicated that maintenance is costly.

Dead volume should be kept to a minimum. This applies both to the sample line that transports the stream up to the sample system and to the sample system itself. If the sample lines are too long, excessive lag time is introduced, and the analysis is liable to be past history rather than what has just happened.

In a multistream sample system, excess dead volume can mean contamination of one stream with another due to the time required to purge the line after switching streams. Too much dead volume in dead-end branch lines (such as at pressure gages and relief valves) can also cause contamination due to back diffusion. Where possible, these dead ends should be placed on a bypass or vent line, and, failing that, the volume should be minimized.

Calibration and Maintenance

IMPORTANCE OF CALIBRATION

For accurate quantitative meaning, the peak heights obtained from the chromatograph must be compared to standard samples. The standard samples can be commercially obtained, they can be prepared by the user, or they can be previously analyzed samples of the process stream under consideration. These samples must be uniform, representative, and stable.

Samples should be kept under conditions that will not cause changes in composition. The concentrations of the various components in the samples may change because of evaporation, polymerization, stratification, and contamination. Storage conditions should be selected according to the type of calibrant used. Standard samples should be obtained from a reliable source and should be checked for composition in the user's laboratory. This also gives a direct correlation between laboratory and process instruments that is invaluable during setup and troubleshooting procedures.

GAS MIXTURES

For gases that do not condense under storage conditions, the calibration blends can be prepared according to Dalton's law of partial pressures, which gives:

$$\%P_A = \frac{P_A}{P_A + P_B} \times 100 \tag{2-4}$$

where:

$\%P_A$ = component A as percentage of total mixture

P_A = partial pressure of component A (kPa)

P_B = partial pressure of component B (kPa)

The vessel is evacuated and each component is added, via a manifold, up to its required partial pressure, starting with the lower concentration components. If the blend is made at 100 psig or less and care is taken to allow the system to come to equilibrium prior to measuring the pressure, the mixture should be accurate to two percent. For higher pressures, compressibility factors must be taken into account, or, alternatively, the blend can be made up to nominal values and checked by analysis.

 Precautions must be taken in the case of gases or vapors that are liable to condense due to the combined effect of the pressure in the vessel with the lowest possible temperature that could occur in storage.

LIQUID MIXTURES

Liquid-mixture blends can be prepared by weighing each component into an evacuated vessel. In cases where a small concentration of condensable gas is required in a high boiling background, the background liquid can be "spiked" with the gas and the resultant blend analyzed prior to use.

Maintenance

The cost of purchasing and installing process analyzers is high enough and their role important enough to warrant a full preventive maintenance program by qualified analyzer technicians. Such maintenance should include:

(1) regular inspections and calibration at a frequency determined by the complexity of the equipment and the sensitivity of the application,

(2) ongoing training of the technicians carrying out the maintenance,

(3) a combination of in-house service and contracted service work by the equipment manufacturer, and

(4) an adequate inventory of spare parts.

Summary

A process gas chromatograph must:

(1) continuously remove and condition a repetitive sample from the process,

(2) periodically inject a constant volume of conditioned sample into the flowing carrier gas and onto the column,

(3) trap and backflush unwanted components,

(4) perform the desired separations,

(5) detect and produce an electric signal for each component leaving the column,

(6) integrate the detector signal,

(7) produce results in a form that is meaningful for monitoring or process control,

(8) monitor its own operation

(9)   provide an alarm upon equipment failure,

(10) prevent unreliable results from being generated

(11) automatically restart after a power failure, and

(12) transmit results only when it is operating properly.

Thermal Conductivity Detectors

The thermal conductivity detector (TCD) is used to analyze the proportion of binary gases in a mixture by measuring gas conductivity. Various gases differ considerably in their ability to conduct heat, and the overall thermal conductivity of a gas mixture depends upon the nature and concentration of the constituents.

Operation of the TCD is simple because only three components are needed: a detector cell, a Wheatstone bridge with a regulated power supply, and a controlled temperature enclosure (see Figure 2-11). A detector element and matching reference resistance are used together in the bridge circuit. When the heated elements are cooled by the pure carrier gas and come to a definite resistance, the bridge is balanced to give a zero or baseline reading of the recorder. When a mixture of carrier gas and a sample component pass over the detector element, the different thermal conductivity of the mixture changes the rate of heat loss, and the resistance of the element changes. This difference in resistance between reference and detector is a function of the concentration of the component in the carrier gas. The net result is that the out-of-balance signal appears as a peak on the output recorder.

The most common way to measure thermal conductivity is with a hot wire cell (or cells). An electric current passing through the wire maintains the cell at an elevated temperature with respect to its surroundings. Heat loss due to gaseous conduction is maximized as a result of cell design and geometry and by limiting the temperature rise of the heated element. This temperature rise at constant electric power input is inversely proportional to the thermal conductivity of the

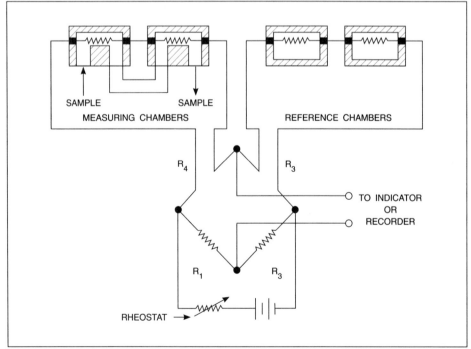

Figure 2-11. Two-Cell Thermal Conductivity Bridge

gas confined within the cell. The relative thermal conductivity of a gas is expressed as:

$$TR = \frac{K_{gas}}{K_{air}} \qquad (2\text{-}5)$$

where:

TR = relative thermal conductivity, no units
K_{gas} = thermal conductivity of gas
K_{air} = thermal conductivity of air

The sample gas flows through the measuring side of the chamber and the reference gas flows through the reference side. Alternatively, the reference ports are capped if the cell utilizes a sealed-in, nonflowing reference gas. Thermal conductivity gas analyzers are used most commonly to determine the amount of hydrogen in air, the amount of hydrogen in nitrogen, the amount of helium in hydrogen, and so on, as well as to determine the combustibility of a gas. Another common use of the TCD is in process chromatography, as described in the previous section.

The thermal conductivity of gases and vapors varies considerably (see Table 2-1 for some typical data). Helium's comparatively large value and its inertness make it a common choice as a carrier gas.

Table 2-1.   Relative Thermal Conductivity (TC) of Gases at 100° C

Gas	TC
Air	1.0
Hydrogen	6.990
Helium	5.840
Argon	0.725
Carbon dioxide	5.3
Methane	1.450
Ethane	0.970
Nitrogen	0.996
Benzene	0.573

 Be careful when applying hot wire detectors. The heated wires will oxidize (burn out) if the carrier gas runs out and the filament current is left on.

Process chromatographs that measure component concentrations greater than 0.1 mole percent use a thermistor rather than a hot wire. A thermistor detector is a semiconductor, in the shape of a small glass bead, whose resistance changes in inverse proportion to temperature. A small electrical current constantly flows through the bead and generates heat. The sensitive bead is located at the end of the column and is exposed to carrier gas flowing from the column. As the separated pure components emerge from the end of the column, they surround the bead and its temperature rises. This happens because the separated components absorb less of the heat generated in the bead than does the carrier gas.

The resistance of the bead decreases as its temperature rises. An identical bead, exposed to flowing carrier gas, is called the reference detector. These two detectors, connected as resistors in a Wheatstone bridge, complete the detector circuit.

The TCD is a universal detector in that everything is detected except the carrier gas itself. The minimum detectable quantity is two to five micrograms or 100 ppm in five ml of gas. The advantages of the thermal conductivity detector are (1) broad application, (2) high repeatability, (3) low cost, and (4) low maintenance.

Combustion Analyzers

To properly monitor the combustion efficiency of a boiler or a reformer, one must analyze the flue gas before it leaves the chimney or the flare stack. Such monitoring is also done for reasons of air pollution control. The products of complete combustion, in addition to heat, are carbon dioxide and water vapor. When sulfur is present in the fuel, sulfur dioxide is also a combustion product. When carbon monoxide is detected in the flue gas, it is an indication that combustion is not complete. Process analyzers for combustion control are able to analyze flue gas for oxygen, carbon monoxide, carbon dioxide, opacity, and hydrocarbons.

Infrared Analyzers

The infrared process analyzer measures the concentration of one component in the mixture. The infrared absorption of a compound is a characteristic of the type and arrangement of the atoms that make up its molecules. Most manufacturers use the Luft-type infrared analyzer technology. The Luft-type detector is highly sensitive but susceptible to temperature variations and external vibration. Two similar Nichrome™ filaments are used as sources of infrared radiation. Beams from these filaments travel through parallel cells. One beam traverses the sample cell; the other beam traverses the comparison cell. The emergent radiation is directed to the single detector cell. As the gas in the detector absorbs radiation, its temperature and pressure increase. An expansion of the detector gas causes the membrane of a condenser microphone to move. This movement, when converted and electrically amplified, produces an output signal. A rotating interrupter alternately blocks the radiation entering the sample cell and the comparison cell. When the beams are equal, an equal amount of radiation enters the detector from each beam. The amplifier is tuned so that only those variations in infrared intensity that occur at the interrupting frequency produce an output signal; therefore, the beams are equal and the output is zero.

When the gas to be analyzed is introduced into the sample cell, it absorbs some infrared energy and thus reduces the amount of radiation that reaches the detector from the sample beam. As a result, the beams become unequal and the radiation entering the detector flickers as the beams are alternated. The detector gas expands and contracts in accordance with the flicker. In this way, the detector directly indicates the difference between the two beams.

The microphone membrane moves in response to the pressure changes of the detector gas. The membrane movement varies the condenser microphone capacity. This variation in capacity generates an electrical signal that is proportional to the difference between the two radiation beams. This signal is amplified and rectified and can be output in current or voltage.

Density Analyzers

Differential Pressure Density Measurement

The most widely used density sensor and transmitter is the bubbler system, which consists of two bubbler tubes in a vessel of fluid at different levels (see Figure 2-12). The difference between the two levels is constant, and the level of the tank is kept constant as well. The pressure required to bubble air (or an inert gas such as nitrogen) into the fluid is a measure of the pressure at the level of the tube outlet. The difference between these two pressures is measured by a differential pressure device and is equal to the weight of a constant-height column of the liquid. The change in differential pressure as density changes is proportional to the fluid density. A density-to-pressure transmitter may be used as the differential device for the purpose of signal transmission.

Note the difference between density and specific gravity. Density is mass per unit volume. The specific gravity of a liquid is the ratio of its density to the density of water at 4°C.

Densitometer Gas Density Measurement

Measuring the density of gases is essential in the petrochemical industries. This measurement can be accomplished using a densitometer with a probe (see Figure 2-13). The probe contains a vane that is symmetrically positioned across the supporting cylinder. The probe is installed in the pipeline in which the gaseous fluid flows. The vane oscillates in a simple, harmonic motion. The frequency of the oscillation varies with the density of the fluid. As the density increases, the oscillation frequency decreases. The relationship is expressed as follows:

$$d = \frac{A}{f^2} - \frac{B}{f + C} \qquad (2\text{-}6)$$

where:

d = measured density
f = frequency of oscillation
A, B, C = constants related to the size of the probe, properties of the fluid, and units of measure

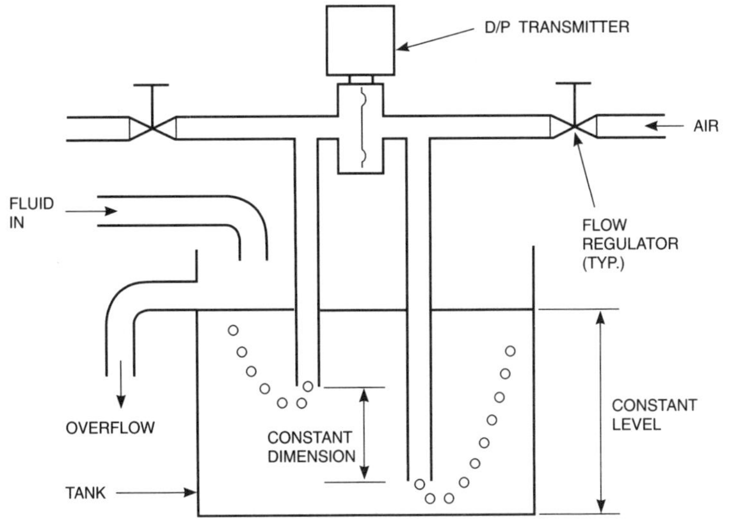

Figure 2-12. Bubbler-Type Liquid Density Measuring System

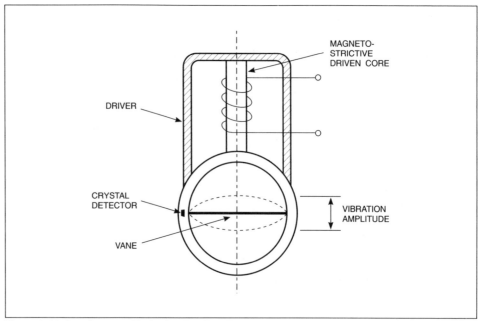

Figure 2-13. Densitometer Probe for Gas Density Measurement

The transmitter amplifies the signal and also energizes a driver with the probe. The driver sustains the oscillation. The transmitter converts the frequency to a standard output current.

Displacement-Type Liquid Density Measurement

Displacement density instruments operate on the principle of buoyancy of a completely submersed body. The operation is similar to that of the displacer used in liquid level measurement. The buoyancy force acting on the balance or torque arm is directly related to the density of the fluid displaced by the float. The liquid must be clean to prevent the settling of solids in the chamber and the buildup of material on the float. The temperature must remain constant.

Nuclear Density Gage

The density of a slurry or a liquid can be inferred using the radiation from a suitable radioactive isotope. The operation of this type of radioactive source and detector is similar to the nuclear level gage. The source and detector are clamped on a pipe in which the process fluid flows (see Figure 2-14). The detector is usually an ionization chamber. Gamma rays from the source pass through the measured material and are absorbed in proportion to material density; that is, the number of ions passing through the fluid is inversely proportional to the density. The gamma rays energize an ionization detector, which then produces a proportional electrical signal. An amplifier-indicator unit can be remotely mounted to present the signal as required. Collimated beam geometry restricts radiation in all directions, except to the detector (collimate means to make straight or parallel).

Vibrating U-Tube Density Detector

The vibrating U-tube operates on the principle that a body may be caused to vibrate by a pulsating force, and the amplitude of this vibration is proportional to its mass. The total mass of the U-tube includes the mass of the flowing fluid, and, therefore, it changes as the fluid density changes. An increase in density increases the total mass of the U-tube, thereby decreasing the vibration amplitude.

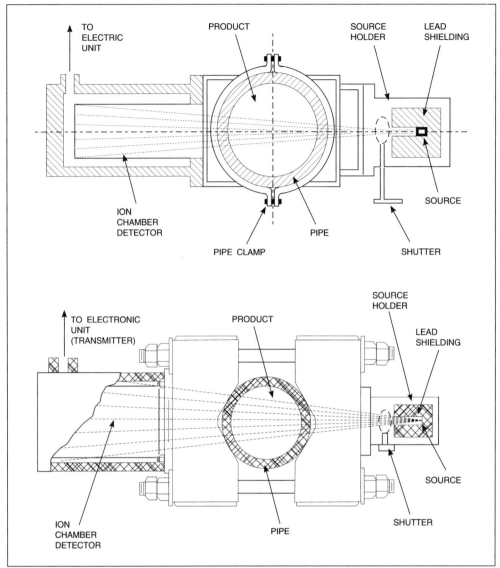

Figure 2-14. Nuclear Density Gages (Ion Chamber Detectors)
(Courtesy of Kay-Ray/Sensall, Inc.)

An armature and coil assembly form a pickup circuit in which an AC voltage is induced by the vibration of the armature. This voltage is proportional to the amplitude of the vibration and, therefore, to the density of the fluid. The AC voltage is converted to a DC millivolt signal as required for indication or control.

Viscosity Analyzers

The viscosity of a fluid is its resistance to deformation under shear. Viscosity measurements are used in the control of the flowability of fluids, in the determination of the molecular weight of fluids during the production process, and in the determination of solids concentration and/or size in slurry streams. The most common unit for expressing absolute viscosity is the centipoise (cP). Kinematic viscosity is the ratio of the absolute viscosity to the mass density. The unit of kinematic viscosity is the centistoke (cSt).

Vibrating Reed Viscometer

The detector of this viscometer consists of a driver section, a vibrating probe in the form of a U-shaped paddle, and pickup circuitry. The drive coil uses 60-Hz power to set up a 120-Hz vibration in the drive armature. The vibration is transmitted through a stationary node point to the probe rod, where the amplitude of probe vibration is proportional to fluid viscosity. The resistance to the shearing action caused by the probe vibration increases with an increase in viscosity and results in a decrease of vibrational amplitude, which is transmitted through a second node to the pickup assembly. A permanent magnet inside the pickup coil induces a 120-Hz signal, the magnitude of which is proportional to the viscosity. The output of the unit is converted to the desired millivolt signal as required. Additional components can be installed for on-off or proportional control.

Miscellaneous Viscosity Meters

The *ultrasonically vibrated probe* coupled with a digital computer measures the product of viscosity and density (cP × gr/cc). Acting as a transducer, the probe shears the liquid under measurement and measures the damping effect of the liquid. The computer translates the damping action of the viscous fluid on the probe into a useful electronic signal that is proportional to viscosity when the density remains constant. The output of the unit, in DC millivolts (0 – 100), can be used for indicating or controlling viscosity of both Newtonian and non-Newtonian fluids.

A *rotational viscometer* measures the torque required to rotate an element in a liquid and relates this torque to viscosity.

The *float viscometer* is similar in construction to the rotameter flowmeter. The float of a rotameter is designed in such a way that the viscous drag area is relatively small, and the float is viscosity insensitive. In the viscometer, the viscous drag area is increased, the flow is held constant, and the float responds to forces due to kinematic viscosity.

Acidity and Alkalinity Analyzers

pH Measurement

Because acids and bases are of vital importance in the chemical industry, food technology, and water and sewage treatment, measurements of the concentration and strength of acids and bases are the most common measurements made on chemical systems. The pH measurement technique employs special selective electrodes that develop an electromotive (emf) force that is proportional to the hydrogen-ion concentration in the solution into which they are immersed. pH meters generally cover a range of 0 to 14 pH units (see Figure 2-15, which shows the relationship among pH, hydrogen-ion concentration, and various common liquids and substances).

In general, the practice is to speak of hydrogen-ion concentration as corresponding to a given pH, when actually what is meant is "effective concentration" or "activity."

pH Electrodes and Meters

Two electrodes must be employed to obtain a measurement. One electrode produces a change in emf (voltage) since the pH of the solution in which it is immersed changes. The other electrode maintains a constant (voltage), which is the reference electrode.

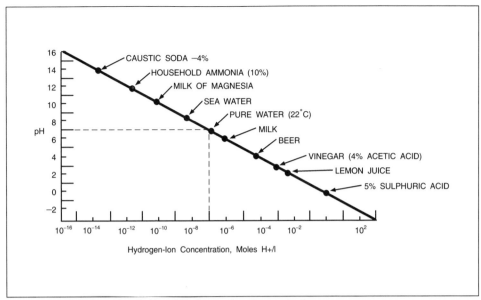

Figure 2-15. Relationship of pH and Hydrogen-Ion Concentration

The glass electrode pH measurement involves high resistances in the measuring loop. A high-impedance amplifier is required to indicate the pH value. The asymmetry potential of a glass electrode can be observed by coupling the glass electrode to the reference electrode and noting any change in the electrode pair output over a prolonged period of time while in a stable buffer solution. The glass electrode's bottom is made of a thin glass membrane; it is filled with a buffer solution. The top consists of an inner and an outer glass tube (see Figure 2-16). The reference electrode must produce a predictable potential that is compatible with the glass electrode, and it must be linear with respect to temperature change.

The most commonly employed reference electrodes are the mercury-mercurous chloride (calomel) type, and more recently, the silver-silver chloride type. The calomel electrode consists of an inner tube packed with a mercury-mercurous chloride mixture. A hole in the bottom of the tube communicates with a saturated potassium chloride solution, and this, in turn, is contained in a larger glass chamber. The glass chamber has an asbestos fiber junction with the measured solution (see Figure 2-16).

Together these electrodes form an electrolytic cell whose output equals the sum of the voltage produced by the two electrodes. This voltage is applied as the input to a null balance millivolt potentiometer, similar to that used with a thermocouple. A temperature-compensating resistor, which is immersed in the solution, is frequently included in the circuit. Its resistance changes with the temperature of the solution so that the pH measurement is correct at the operating temperature (see Figure 2-17).

Continuous pH analyzers can be found in practically every industry that uses water within its processes. Applications range from industrial water and waste treatment to pH control of flotation processes in the mining industry. Many pH applications can be found in the pulp and paper industries, metals and metal-treating fields, and in petroleum refining, synthetic rubber manufacturing, power generation plants, pharmaceuticals, chemical fertilizer production, and a broad spectrum of the chemical industry.

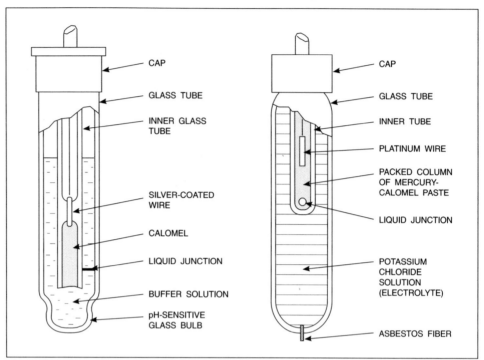

Figure 2-16. Glass Electrode and Reference Electrode

Electrical Conductivity Meters

Electrolytic conductivity is a measure of the ability of a solution to carry an electrical current. The conductivity of a process stream is usually measured to determine stream purity. Any sudden change in the conductance of a process stream is often translated as the appearance of an undesirable contaminant. The unit of electrolytic conductivity is the mho/cm, which is the reciprocal of resistance.

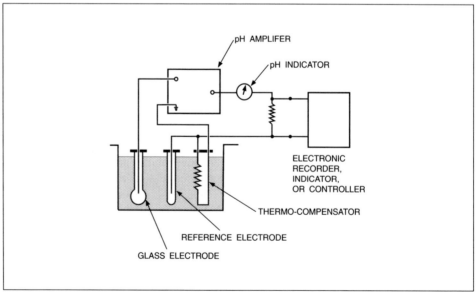

Figure 2-17. Electrometric pH Measurement Circuit

AC current is used more often than DC current, since DC current leads to "polarization" in the electrode area. An AC Wheatstone bridge is utilized for this purpose. Two electrodes of known area are immersed in the solution, this arrangement being called a conductivity cell. A conductivity cell is used with an AC null balance bridge. Similar to the pH measuring system, the circuit uses a temperature-compensating resistor to correct variations in the temperature of the solution. The cells are available in probe form. They contain titanium-palladium or graphite electrodes with plastic insulation. A thermistor is enclosed in the center electrode for temperature compensation. The thermistor is part of a series-parallel network in the measuring circuit.

Electrodeless Conductivity Measurement

In this system, two toroidally wound coils are mounted on nonconductive or nonconductive-lined metallic piping. One coil is connected to a transmitter (oscillator) that supplies a voltage in the high audio frequency range and is stable and free from drift in both frequency and amplitude. This coil will generate a current in the loop of the solution; the current is proportional to the conductivity of the solution. This current generates a current in the second coil, which is the output of the probe (see Figure 2-18 for a schematic diagram of this system).

Miscellaneous Analyzers

Oxygen Process Analyzers

Oxygen process analyzers are essential in the monitoring of combustion, oxidation, and other industrial process applications. The analysis of dissolved oxygen is becoming important in the control of waste or flue gas emissions into the atmosphere. The most commonly used types are deflection paramagnetic, thermal magnetic, catalytic combustion, micro fuel, and galvanic cell oxygen analysis. For dissolved oxygen analysis, polarographic, galvanic, and thallium cells are used.

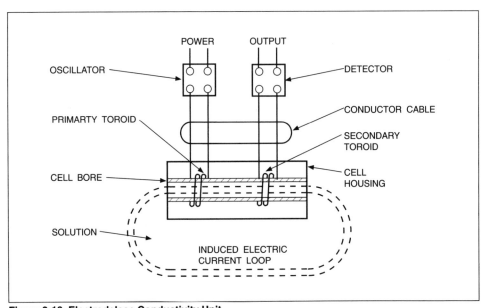

Figure 2-18. Electrodeless Conductivity Unit

X-Ray On-Stream Analyzers

The mining industry utilizes X-ray fluorescent on-stream analyzers in mineral dressing plants to make chemical analysis of specific elements in the flotation pulp. The elements commonly analyzed are chromium, manganese, iron, cobalt, copper, zinc, and lead. Backscatter radiation is used to determine pulp density. As the source of primary radiation, both X-ray tubes and radioisotope sources are presently in use. Many systems are available; therefore, only one system will be described briefly. The Courier 300 Analysis System™ of Outokumpu Oy Instrument Division consists of: (1) a sampling system that includes primary samplers, a pumping system that pumps the slurries to the analyzer site and/or back to the process, secondary sampling, and feeding the sampling streams to the analyzer; (2) an on-stream spectrometer system comprising an on-stream X-ray spectrometer for a maximum of l4 slurries, measuring a maximum of six elements in each slurry; (3) an X-ray generator; and (4) a water-cooling system. Data processing equipment includes: a digital computer, interfaces for assay output to report-type printers, a process computer, recorders or analog controllers, and alarm systems.

A continuous sample flow is taken from each process slurry to be analyzed. The final sample flow is obtained after sampling in two or three stages, depending on the quantity of the process flow. Each slurry sample flows through a separate sample cell in the analyzer. Fluorescence radiation intensities are measured through thin windows in these cells. The X-ray excited fluorescence radiation contains all the data needed to compute the actual element contents. A timing device determines the measurement time for each slurry sample.

On-Stream Particle Size Analyzers

The mining industry monitors ore milling operations by on-stream particle size analyzers. This type of analyzer is installed in the grinding circuit, where it is used as a sensor for the automatic and manual control modes and for circuit diagnosis. The system utilizes an ultrasonic sensor. There are two pairs of ultrasonic transducers, each pair consisting of a transmitter and a receiver (see Figure 2-19).

In the transmitter, high-frequency electrical signals are converted into ultrasonic energy, which travels through the slurry between the transducer and the receiver. In the receiver the ultrasonic energy is converted back into high-frequency signals and fed to the electronics section by means of coaxial cables.

The output format of the electronics section is adjustable for any standard control signal. The amount of ultrasonic energy that gets through the slurry is measured to arrive at independent particle size and percent solids signals. Electronic comparison of these two signals yields an output that varies as slurry particle size only and is independent of percent solids changes. A separate output that indicates percent solids is also provided. The system includes sample-gathering equipment as well.

Bibliography

1. Annino, R., and Villalobos, R., *Process Gas Chromatography: Fundamentals and Applications,* Research Triangle Park, NC: Instrument Society of America, 1992.

2. Clevett, Kenneth J., *Process Analyzer Technology*, New York: John Wiley and Sons, Inc., 1986.

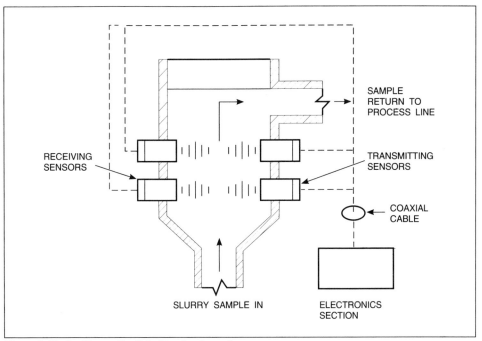

Figure 2-19. On-Stream Particle Size Analyzer Sensor
(Courtesy of Denver Autometrics)

3. Griffin, D. E., and Webb, P. U., "Process Chromatographs and Computers in Optimizing Control Systems," *INTECH*, Research Triangle Park, N.C.: Instrument Society of America, July, 1979.

4. Jutila, M., "Multicomponent On-Stream Analyzers for Process Monitoring and Control," *INTECH*, Research Triangle Park, N.C.: Instrument Society of America, July, 1979.

5. Krigman, A., "On-stream Analyzers: Progress with New Technologies," *INTECH*, Research Triangle Park, N.C.: Instrument Society of America, October, 1983.

6. Liptak, B., and Venczel, K., eds., *Instrument Engineers' Handbook*, New York: Chilton Book Company, 1985.

About the Author

Jean Claude Moisan, a graduate of Eastern Ontario Institute of Technology in chemistry, was selected by Petromont, Inc., for specialized training in the maintenance of process analyzers and in their selection and specific application. As First Process Analyzer Specialist, he is responsible for a group of technicians and specialists. His group adapts the engineering of chromatographs to suit the applications and supervises installation.

3

The Process and Process Control

No discussion of industrial process control would be complete without a few words about the process itself. Countless activities in industry are called processes. This chapter does not attempt to cover every possibility; instead, it discusses general and fundamental principles that the practitioner can extrapolate to a particular situation.

Processes

Definition of a Process

Process is a general word that can mean different things to different people, depending on their area of work, their expertise, or the social context in which the word is used. For industrial or technical people, a process is a method or procedure followed to achieve a result; the type of result expected helps define the type of process to be considered.

For scientists interested in discovering the method of formulating a new detergent or in explaining chemical reactions that occur when wood pulp is bleached, a process will be the set of laws that govern the chemical transformation or combination of substances into new products, or the set of laws that define the physical conversion, mixing, or separation of a mixture of substances under the effects of heat, pressure, or contact with other substances with which they have special affinities. For example, the conversion of wood into bleached fine paper is obtained through a series of chemical processes in which caustic soda, chlorine, and other chemicals under pressure and at a relatively high temperature react with and dissolve out the lignin that bonds wood fibers together, to leave only white fibers with which paper is made. Similarly, certain household detergents are produced by the reaction of liquid heavy organic alcohols with a very explosive gas, ethylene oxide, in large heated kettles. On the other hand, the desalination of seawater in the desert to produce potable water is primarily a physical separation of water vapor from an increasingly concentrated salt solution; the water is boiled off in successively lower pressure vessels using steam as the heating medium. The laws of thermodynamics and the properties of steam, water, and brine solutions govern the change in concentration in the successive evaporator bodies.

For engineers concerned with building a plant to produce chemical products for commercial purposes, a process is either the sequence of steps that lead to the preparation of the product or the collection of equipment, reactors, pipes, tanks, distillation columns, evaporators, etc., in which the chemical or physical phenomena occur that lead to the preparation of the product. Hence, the term "process" refers either to the "recipe" to be followed or to the "utensils" to be used, depending on whether one is interested in developing a new formula or in building a plant to commercialize a product.

Whether a process is relatively simple, such as a boiler heating water to make steam, or very complex, such as a petrochemical complex or a pulp mill, the key functions performed to shape the product to its final specification are a combination or a succession of the following steps:

(1) The transportation of raw materials and/or energy between various processing units

(2) The transfer of heat between different streams

(3) The transfer of mass between different streams

(4) The reaction of a mixture of compounds to yield new materials

(5) The separation of components from a mixture into a number of different streams of varying degrees of purity

(6) The storage of materials between processing steps and at the end of the process prior to commercial distribution

The type of process chosen for a particular final product will depend on its production requirements and quantities. For some products batch processes are suitable, while for others only continuous processes are economically feasible.

Batch Processes

Batch processes are those in which a number of successive and different chemical or physical operations are carried out on the feed chemicals to transform them to the final desired product in a single or a few reactors or tanks. A batch process is operated according to a recipe. The mix of ingredients remains in a vessel or reactor until its characteristics are changed to meet the specifications. By performing all the manipulations in a limited number of containers, the required space and equipment, as well as the plant's capital cost, are minimized. However, because of the nature of the process, the quantities of a product that can be achieved in a reasonable time span are limited.

Batch processes are concerned primarily with any of the following:

(1) Products that are produced in many different formulations but in relatively small quantities, such as pharmaceutical products, perfumes, and synthetic detergents

(2) Expensive products with limited markets or short lifetimes, such as perfumes

(3) Products that require long fermentation or maturing times in undisturbed conditions, such as wine, beer, whiskey, etc.

In addition, initial processing of many chemicals is done in the batch mode until market demands grow and eventually justify the dedication of a whole train of equipment, or processes, to the production of a single product (see Table 3-1).

A batch procedure provides at least two types of directives to make a product: the batch recipe and the batch sequence. A batch recipe consists of a list of parameters such as temperature of reaction, quantities of chemicals, reaction time, reaction pressure, and profile of the operating conditions with time. The batch sequence defines all the necessary states of the process, the transition triggers, and the order in which the states should follow each other. Between active states, the waiting, idle, and holding times can also be programmed.

There are three basic modes of batch process unit operation: manual, semiautomatic, and automatic. In the purely manual mode, stepping through the se-

The main advantage of batch processes is that the same equipment can be used to produce many different products in succession once the vessels are cleaned and readied for the next recipe. The usual drawbacks are the relatively inefficient use of heating and cooling media to perform the operations, the limitation on production quantities, and the need to clean the equipment frequently to prepare it for the next batch.

quence is done by the operator through specific commands issued to the equipment from the operating console, using mainly push buttons to start and stop the various operations. In the semiautomatic mode, every consecutive batch sequence is initiated separately by the operator, but then the process proceeds automatically through the various steps, controlled by programs in programmable logic controllers (PLCs). In the automatic mode, once a sequence is initiated, it can repeat itself a number of times without operator intervention, as long as the operating parameters or "recipes" remain unchanged.

Table 3-1. Common Batch Processes

Alcoholic beverages
Explosives
Food products
Liquid detergents
Metal ore processing
Paints and dyes
Perfumes
Pesticides and soil nutrients
Pharmaceutics

In general, batch process controllers are timers, switches, alarms, and indicators for variables such as flow, level, temperature, and pressure. Few operations require a constant flow or temperature for any length of time; but, because the same few pieces of equipment are used to perform most operations, their filling and emptying times are critical. It is also essential that they reach the desired temperature or pressure as expeditiously as possible. Programmable logic controllers play a very important role in the control of batch processes, and the types of valves most often employed operate in an on-off manner, as opposed to modulating.

> The preparation of a particular household detergent is a typical semi-automatic batch process. The process is illustrated in Figure 3-1. A heavy organic alcohol (such as lauryl alcohol) is stored in an atmospheric tank where the temperature is controlled by adjusting the flow of steam in coils installed in the tank. The tank level is monitored by a level indicator only. Liquid ethylene oxide is stored in a pressurized but unheated tank; its level is monitored by measuring the tank pressure. Nitrogen, used to displace air in the kettle during the reaction, is stored in a pressurized but also unheated tank; its pressure is also used to monitor the remaining quantity. The reactor, in which the ethoxylation reaction takes place, is mounted on load cells that weigh the tank contents. The load indicator is biased to read zero weight when the reactor is empty. The reactor is also known as a "kettle." The reactor contents are heated or cooled by circulating through an external heat exchanger at strategic times. An external jacket around the reactor assists in the heating or cooling. A vacuum pump draws vapors from within the reactor to create a vacuum when necessary; the vapors are passed through a condenser to recover any entrained vapors that are recyclable. A catalyst pot holds a preset
>
> (continued)

(continued)

quantity of caustic soda diluted in the alcohol, which serves as a catalyst to initiate the ethoxylation reaction. The final product is stored in another tank that is also heated by steam to keep the product viscosity sufficiently low for pumping it out to the bottling plant. A centrifugal pump circulates the chemicals from the kettle through the heat exchanger or pumps the product out to the product storage tank.

The desired weight of alcohol is keyed into the load-indicating device (WI-1), and the pump from the alcohol storage tank is started. A control valve (FV-2) on the line feeding the alcohol to the kettle is opened and controls the flow of alcohol into the kettle. When the desired weight is reached, the pump is stopped, and the weight valve is closed. The circulation valves (HV-3 and HV-4) are opened, and the circulation pump is started. Steam is allowed into the jacket and the exchanger by opening the steam inlet and condensate outlet valves (HV-5, HV-5B, HV-6, and HV-6B). The rate of heating of the alcohol is controlled by a valve on the circulation line (TV-7), which modulates the flow through the exchanger.

At the same time, the kettle mixer is started. While the contents are heating up, the vacuum pump is switched on, and the vacuum valve (PV-8) is opened to withdraw from the alcohol all traces of water, which would denature the product. The water inlet and outlet valves to the condenser (HV-9 and HV-10) are opened at the same time. The catalyst inlet valve (HV-11) is opened, the catalyst is sucked into the kettle by the effect of the vacuum, and the valve is closed. Any condensed alcohol is collected in a collection pot and recycled in the next batch; the water evaporated is exhausted by the vacuum pump.

When the contents are hot, the steam flow to the exchanger is stopped by closing valves HV-5 and HV-6, and the pressure indicator on the kettle is watched until it reaches a required vacuum level. A nitrogen purge is then initiated by opening the nitrogen valve (HV-12) to displace traces of air and moisture in the kettle and its contents, while the vacuum pump is still in operation. The pressure gage on the kettle monitors the pressure in the kettle and shuts the nitrogen inflow valve when the reaction pressure is reached.

The weight indicator (WI-1) is zeroed again and the weight of ethylene oxide needed for the reaction is then entered. The ethylene oxide inflow valve (FV-14) is opened, and the flow is monitored very closely because of the hazardous nature of the very exothermic reaction. As soon as the kettle temperature starts rising, the cooling water flow to the exchanger is started when valves HV-15 and HV-16 are opened. The flow of ethylene oxide is controlled (FV-14) to keep the temperature just below some predetermined maximum value, above which the product properties change drastically. Upon reaching the set weight of ethylene oxide in the kettle, valve FV-14 closes.

The product continues circulating in the exchanger and cools down to some intermediate temperature. A second nitrogen purge is effected by reopening the nitrogen feed valve (HV-12) and an exhaust valve (HV-17) for a fixed duration of time. This purge dilutes and evacuates unconsumed ethylene oxide and other harmful gaseous by-products through a tall stack. A timer shuts the exhaust valve at the end of the purge. The kettle is once more pressurized by the nitrogen as it cools down to avoid sucking in air and moisture while the product is still hot and subject to degradation. The kettle pressure indicator turns the nitrogen feed on and off to maintain the pressure within an acceptable range.

When the contents are cooled down sufficiently, product transfer valve HV-18 opens, circulation valve HV-4 closes, and the product is pumped to storage. At that time the nitrogen flow is stopped by closing valve HV-12,

(continued)

(continued)

and the agitator is also stopped. The pump is stopped when the weight indicator (WI-1) indicates a zero kettle weight. The batch process is then complete. The next batch can be started, or the kettle can be washed and readied for the production of another type of product.

This batch process can be run manually or a PLC can be programmed to run it in the semiautomatic or the automatic mode. A switch on the control panel selects the mode of operation. Manual operation is chosen to train new operators and to test new formulations. The semiautomatic mode is chosen for single batch runs, and the fully automatic mode is chosen when many consecutive and identical batches are needed to produce a large quantity of product.

Continuous Processes

In continuous processes a product is manufactured continuously day in and day out without interruption, except for emergency repairs or for periodic annual or semiannual maintenance shutdowns. Raw materials enter the process at a constant rate and proceed through a number of transformation operations in succes-

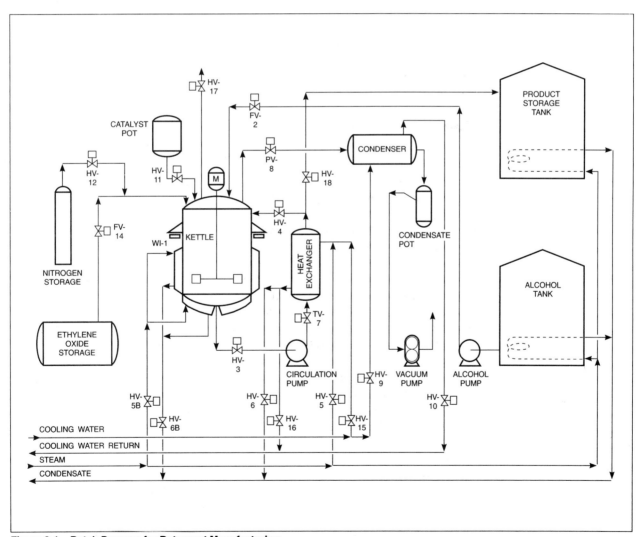

Figure 3-1. Batch Process for Detergent Manufacturing

sive pieces of equipment until the product emerges after the last operation. Materials do not remain stationary in any processing unit, other than in intermediate and final storage tanks. Each piece of equipment is dedicated to a single task; it receives the same feed materials at the same conditions and delivers the same transformed materials at the same conditions in a continuous way. Each unit is designed for a maximum operating capacity, and all operating parameters are designed for one or very few running conditions. Little variation is expected in the operation other than the occasional reduction in production capacity to meet smaller market demands or temporary upsets elsewhere in the process. Because of the large number of processing steps and physical units, transportation equipment (such as pumps, blowers, and conveyer belts) plays a much larger role than in batch processes.

The type of control required is also different from that in batch processes. Controllers have to ensure that the process operates steadily with as few upsets and peaks as possible. The majority of control loops are concerned with keeping pressure, flows, temperatures, compositions, and levels constant. Few operations require timing operations or the sequencing of process steps.

Distributed control systems (DCSs) are becoming more popular in the larger plants. They minimize the space required to display all the controllers and alarm panels; they provide data gathering, logging, and storage for long periods of time; and they permit plant-wide dissemination of information about the overall operations. The DCS can, through connection to larger computers, perform ongoing plant optimization and update the controllers' set points to maximize production despite variable feed conditions of raw materials. The most common type of control valves used are modulating valves operated by elaborate process control models.

Semicontinuous Processes

Semicontinuous processes are less commonly applied to large scale processes. They are usually reserved for specific, repetitive applications. As their name implies, they share characteristics of both batch and continuous processes and require the instrumentation peculiar to both kinds, such as continuous controllers, timers, and programmed controllers.

Certain types of processes (such as gas cleaning, liquid filtering or centrifuging, air drying, water treatment, etc.) are semicontinuous, or cyclic, because their tasks are repetitive in nature. Their function is to separate components from a stream, such as suspended particles from a liquid stream, moisture from instrumentation air, or dissolved chemical ions from water destined for a high-pressure boiler. The separated component accumulates in the processing unit until it fills it or until it opposes any further inflow or treatment of the incoming stream. Periodically, the process is stopped, and the unit is cleaned either by being purged of the accumulated undesirable component or by having the treatment medium chemically regenerated. Then the process is repeated. This semicontinuous operation requires either manual intervention by the operator or automated sequencing of the purge or regeneration steps. While the unit is still unsaturated with the contaminant or waste, the process behaves as a continuous process and proceeds undisturbed. The cleanup cycle is triggered either by a high pressure drop across the unit or a lower purity of the output stream as measured by increased conductivity. When the frequency of interruption is known from experience or anticipated in the design and fixed ahead of time to avoid overloading the unit, a series of timers controls the entire sequence.

High pressure drop or high conductivity switches are used only as emergency backup in case the unit's efficiency drops earlier than expected.

The Development of Process Designs

The development of a process design starts with an idea, a marketing opportunity, or the necessity to dispose of the by-products of another process. The initial step in this development sequence is a literature survey to gather sufficient information about the desired product: its properties, known methods of synthesis, existing patents describing the product, production methods, and so on. With the information obtained, chemists and engineers devise and perform laboratory experiments to test different routes to reach the desired objectives.

When the chemistry leading to the right formula or recipe is established, the industrial processing steps necessary for commercial scale production are developed, first in laboratory-scale process equipment, then in a pilot-scale process plant. A pilot plant is a small-scale replica of the final commercial-size plant. It is used to set the sequence of unit operations, to obtain operating and engineering data for the design of the full-scale process, to iron out process problems and hazards, to optimize the full-scale process, to identify by-products, and to develop means of using or disposing of them. A new process may be devised from the ground up by evolution of the processing steps one at a time and formulation of new ways of performing the required physical and chemical operations. It may also take shape through the assembling of a number of known processing steps to achieve the same results with less research and development.

When the process is fully defined at the pilot plant, market surveys and large-scale economic feasibility studies are conducted to establish the viability of the desired plant. If proven feasible, detail engineering design proceeds, followed by process construction and start-up.

A large multi-disciplinary team of engineers is assembled to conceptualize, design, and build the industrial-size plant. The first task in the design of a new chemical processing plant usually belongs to the process or chemical engineer, who must specify the sequence of operations necessary to produce the final product. Only after the process units and the processing sequence have been selected can any detail engineering by other disciplines start.

Civil engineers will survey the site, blast rock and make the site ready for construction, and run underground lines and services such as water and fuel lines, storm and process sewer lines, and so on. Structural engineers will prepare the foundations and design the buildings that will house the process equipment and control rooms, the offices, the warehouses, the maintenance areas, and other such buildings. Mechanical engineers will design the physical processing units and order them from suppliers or builders, prepare their optimum layout configuration, run pipes between them, perform stress analyses on the larger pipes, and design all the moving equipment and material-handling equipment. The electrical engineer will design the electrical substation that will feed power to the new plant, specify and purchase all electrical motors and switchgear, run cables to the power users, and design the motor control centers. The role of the process engineer initially is the preparation of all the material and energy balances to define the process operating conditions and to size the equipment. The material and energy balances set the desired quantities of raw materials needed as feed to the process, the quantities of steam, water, fuel, and other utilities required to operate the plant, as well as the amounts of all the by-products generated that must be disposed of or reconverted into saleable products. The balances are also used to estimate the peak flows and upset conditions, which will dictate the safety margin to add in to the design of equipment and storage tanks. This information is presented on flowsheets that show the equipment and pipes that make up the process as well as the flow of the various substances in the pipes. Information is also presented in equipment lists that describe the physical parameters of the equipment.

The process engineer is responsible also for developing the start-up, shutdown, and normal operating procedures of the process. These activities involve instrumentation and process control, at least to the point of defining the tasks to be performed by the various control loops and process interlocks and alarms. The process engineer is the person most familiar with the process sequence and hazards. The design control information is presented on a more elaborate version of the process flowsheets, which show not only the major equipment and pipes but also the ancillary equipment, service piping, drains, overflows, and so on. These drawings, called piping (process) and instrumentation diagrams cr P&IDs, show every single piece of equipment, every single pipe, and every single control loop needed to run the plant.

As these drawings evolve, the instrumentation engineer designs the control loops and selects the hardware and software most appropriate to perform the desired control functions. For large processes, a team of process engineers and control engineers will work simultaneously or successively on these activities. When the process and mechanical engineers have completed their design and equipment sizing and layout, the instrumentation specialists are provided with the equipment details and process configuration in order to devise the optimum control strategies and program the programmable logic controllers (PLC) or the distributed control system (DCS) that will run the plant.

By this time, all tank sizes and pipe lengths are established so that all instruments can be located, instrument ranges defined, and process dynamics developed. When the plant is completely built, or as sections of it are readied, the instrumentation and control specialists tune their controllers and calibrate their sensors, test the logic in the PLCs and DCS, and perform all tasks necessary to verify the soundness of their control philosophy.

The Material and Energy Balances

STEADY-STATE AND UNSTEADY-STATE BALANCES

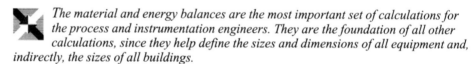 *The material and energy balances are the most important set of calculations for the process and instrumentation engineers. They are the foundation of all other calculations, since they help define the sizes and dimensions of all equipment and, indirectly, the sizes of all buildings.*

The most widely used material and energy balances are listed in Table 3-2. The basis of all balances is the generalized law of conservation of material or energy, which applies to all processes except nuclear reactions (where matter is converted into energy). This law is defined as follows:

Accumulation within the system = Input through system boundaries (3-1)

— Output through system boundaries

+ Generation within the system

— Consumption within the system

The accumulation may be positive or negative and applies with or without chemical reactions in the process. When there is no generation within the system, the equation reduces to:

Accumulation = Input − Output (3-2)

Table 3-2. Most Common Process Balance Calculations

Steady-State Balances	Unsteady-State (Dynamic) Balances (sequential calculations repeated at short time intervals)
Material balances around individual units	Material balances around individual units
Sequential calculations; with no recycles	Energy balances around individual units
Iterative calculations; with recycles	
Energy balances around individual units	
Sequential or iterative calculations	
Reaction kinetics; coupled with heat of reaction calculation	
Vapor-liquid equilibrium balances	
Chemical consumption balances	
Overall process water and steam balances	
Effluent balances	
pH calculations	

Furthermore, when there is no accumulation within the system, the equation becomes:

$$Input = Output \tag{3-3}$$

Inherent in the generalized equation is the concept of "system," which can be assigned to any arbitrary portion of a process or to the whole process. These equations are applied to every component that enters and leaves the system, and they yield all output streams and compositions.

In systems where no accumulation of material takes place, a solution of the balance equations for all components and for all units yields a "steady-state" picture of the process and is used to size most units and pipelines, because the balances represent the conditions at which the process will operate most of the time. The steady-state balance equations are algebraic equations only.

Steady-state material or energy balances usually cannot be uncoupled. The balances around a number of units in a process are done in either a sequential or an iterative method. For systems with no recycle streams, all inputs to each unit in succession are known from the results of the previous unit calculations, and one can calculate the composition of the unit's output stream by applying the generalized equation to each component.

In systems with recycle streams, such as distillation columns, water circuits in pulp mills, mineral washing sequences, and others, the flow and the composition of the recycle streams are usually not known; they must be guessed to start the sequential calculations. Their values will eventually be recalculated in the sequence and used to repeat the sequence, as they will probably be different from the initial guesses. The iterative process is repeated until the values of the guessed streams over two successive iterations do not change significantly.

Iterative hand calculations are very long and laborious. Fortunately, many computer programs have been written to speed up these calculations and present the results directly on flowsheets or P&IDs. These programs contain subroutines that simulate most of the known unit operations; they also have very large data banks of physical, chemical, and thermodynamic properties. Some programs will also size equipment and pipe lines automatically based on the results of the balances. Others simulate the dynamic process balances over defined time spans and include models of all types of instrumentation to control the process behavior. In systems where accumulation does take place, the generalized equation yields a picture of the system at a particular moment in time only, like a snapshot. The process conditions are in a continuous state of change, and the balances are referred to as "unsteady-state" balances or "dynamic" balances. The solution of the dynamic balances around all units in the process must be repeated at successive time intervals over the desired time span to reproduce the process behavior during a state of transition, such as start-up, shutdown, grade change, etc. These calculations differ from the iterative calculations in that they yield successive pictures of the plant at different time periods, as opposed to aiming for a single solution. Unsteady-state balances require the solution of differential equations or their approximation by pseudo steady-state equations over very short time intervals.

REACTION KINETICS

When reactions occur in the vessel and if heat is generated or absorbed, the reaction stoichiometry adds an extra complication, since the rate or extent of the reaction depends on the temperature. The calculations are iterative in this case as well, because information such as the unit's temperature is not known from the beginning.

VAPOR-LIQUID EQUILIBRIUM (VLE) CALCULATIONS

VLE calculations occur when there is a possible transfer of material and/or energy between streams that are in different physical states, or phases, and are in contact for any length of time. This happens in processes such as distillation, condensate stripping, leaching, or washing. In the gaseous state, components mix and intermingle freely without separating into distinct partitions; however, in the liquid or solid state more than one phase can coexist simultaneously, each with a different composition (such as mixtures of oil and water, which separate into two immiscible phases, or different crystalline forms of the same metal that can exist in the same solid mass). In multiphase systems, there is an exchange of constituents between the various phases until every component in one phase is in equilibrium with the same component in the other phase or phases. Equilibrium conditions are established by complex thermodynamic rules and are very temperature-dependent.

CHEMICAL CONSUMPTION BALANCES

These balances establish the storage requirements of raw materials and feed chemicals and help assess the operating costs of running a plant. Of primary importance to the process and instrumentation engineers are the type and size of raw material storage and the conditions of storage. Their specification differs from that of the actual process equipment in the plant, since they must be tied in to delivery methods and schedules, as well as to levels of operations comfort of the plant.

WATER AND STEAM BALANCES

These are a special case of the material balances around a process. In processes that consume large volumes of water or steam in different parts of the plant, such balances are most important to identify the users and attempt to control their consumption.

EFFLUENT BALANCES

Effluent balances identify and quantify the sources and composition of plant effluents to establish their degree of compliance with environmental regulations. They are used to devise treatment systems, such as pH control, and primary (suspended solids removal) and secondary (effluent oxygenation) treatment processes.

pH CALCULATIONS

The neutralization of acidic or basic streams is controlled by measuring the pH of the product stream and adjusting the flow of the neutralizing stream. Calculations of pH are difficult because they depend very heavily on the chemical species involved, their relative concentrations, the presence of buffering agents that retard the pH response to increased dosage rates, and so on. Values of pH are logarithmic calculations of the concentration of the hydrogen ion (H^+) in the solution; a unit change in the pH value reflects a tenfold change in the H^+ concentration due to chemical reactions.

Documents Developed for a Process Design

Throughout a process study or design activity, numerous documents are developed by the process and instrumentation disciplines to establish the process design and convey the information to the disciplines concerned as well as to the client or plant owner. Some documents take the form of written text, whether as descriptions, tables or lists; others are pictorial and come in the form of diagrams or drawings.

TEXT DOCUMENTS

Numerous textual documents and lists evolve through the development of a project. Text documents help describe the scope of the project to the client, transmit information between engineering departments, communicate with vendors, and train the plant operators. Some of the more important and relevant documents for process and instrumentation engineers are described below.

The Scope Document. The scope document is issued at the study phase to describe the process to be designed, its boundary limits, its contents, the operating parameters, and the major pieces of equipment. The document is updated at the end of the project to reflect the final process design; it is not unusual for design changes to occur as a result of evolving technology and a better understanding of the process.

The Design Specifications. Design specifications are process operating conditions specified by the client. They are used to define such conditions as production rate, product specifications and quality, available water and steam qualities, environmental regulations to be adhered to, raw materials properties, and other parameters that the client wishes to define independently of the ensuing material and energy balances and that will affect these balances and the process design.

The Design Criteria. Design criteria are tabular presentations of the results of material and energy balances around every major piece of equipment. They are used by the mechanical engineers to size equipment or to specify to vendors the operating conditions of the equipment to be purchased. The criteria tabulate the running flow rate, the temperature and pressure of the various feed and products around equipment, the equipment important process characteristics such as residence time or operating pressure, and the rate of consumption of raw materials and production of by-products.

The Equipment List. The equipment list enumerates every single piece of independent equipment purchased for the plant construction. It lists its equipment number, its vendor, and some sizing information such as dimensions or horsepower or capacity. The list is used to keep tabs on equipment delivered to the plant, to order spare parts, and to assist in preparing maintenance records.

The Pipe List. A pipe list records every pipe to be run in the field. This list assigns a number to every branch of a pipe line and shows the start point and termination of every pipe, its length, diameter, material of construction, thickness, maximum operating flow rate, temperature and pressure, whether it is insulated or heat traced, and the fluid (commodity) it carries. The pipe list also includes a section on tie points. Tie points are connections to pipes in an existing plant, which are made ahead of time during regular plant maintenance shutdowns, to allow new parts of the plant to be built and connected to the existing plant without interrupting its operation. A pipe list is developed in parallel with a P&ID. It is usually used to define the piping to be installed by each contractor in each area of the plant and for material takeoff for ordering.

The Instrument List. The instrument list is similar in concept to the pipe list and is prepared simultaneously with the loop diagrams (described later). It lists every loop in the plant, assigns it a loop number, and describes its function, its components, their vendors, their characteristics, and their sizes.

The Process Operating Manuals. The team of process, mechanical, electrical, and instrumentation engineers complete the design project by writing a number of operating manuals that summarize the steps to be followed to start the process and the plant, to run it at normal operating conditions, and to shut it down both normally and under emergency conditions. This task is distinct from the task undertaken by the mill operating crew, who rewrite these manuals in a jargon more familiar to the operators as part of their prestart-up training.

GRAPHICAL DOCUMENTS

Block Diagram. The function of the block diagram is to depict the main processing steps in the process. This is usually done through the use of a series of rectangular blocks, each representing a unit operation or a process area. It represents the first document around which the client and the engineering firm doing the design discuss and agree on the main steps of the process. The block diagram differentiates between new and existing equipment and may present the major flows between departments. A single sheet is usually sufficient to represent a process block diagram.

This diagram is developed at the very beginning of a study or design activity and may be updated for reference only as the project develops, but it has little information to convey past the first few weeks of a project.

Process Flowsheet. A flowsheet is a more elaborate form of the block diagram. Each unit is shown pictorially by a symbol that represents more closely the actual shape of the equipment. All the major pieces of equipment are illustrated, and all the principal lines and flows are indicated. The major control instruments are shown to help clarify the mode of operation, but the bulk of the instrumentation used for start-up, shutdown, and minor control is usually omitted. Known tank capacities are also shown, and all major motors are identified.

To prepare the flowsheet, material and energy balances must be performed, processing details must have been gathered, and some equipment vendors contacted to learn about the operation of their particular equipment and the secondary equipment required, such as dust collection or exhaust gas scrubbers on boilers or kilns. The flowsheet of the entire process normally spans many drawings, and the interconnections between drawings are well defined. Flowsheets are prepared during the first few months of a project as the details of the processing steps are elaborated and as the balances are being performed. They contain sufficient information to complete a feasibility study; however, their usefulness stops once detail design for a construction job starts. They are much too primitive to be of assistance to the piping and instrumentation disciplines. They are soon developed into or replaced by the P&IDs.

Piping (or Process) and Instrumentation Diagrams (P&IDs). P&IDs are the ultimate process design drawings and are of great use to all the disciplines involved in the plant design. P&IDs are initially prepared by the process engineers as an extension of their flowsheets, but their development is soon shared with piping designers, mechanical engineers, and instrumentation specialists, all of whom have a substantial input into their completion. They are also used by structural designers who must be aware of all the equipment in the process.

P&IDs show, first of all, all the pieces of equipment in the process whether major or minor, including all motors, agitators, etc. Each unit is identified by an equipment number, a short description, and perhaps a few details about capacity. This task is shared by the process and mechanical engineering disciplines. They also show every single pipeline connecting the units together or connecting the main utility headers to the units, including bypasses around control valves, tank drain lines and overflows, and so on. Flow data is copied from the flowsheets onto the P&IDs; additional flows are calculated and shown for lines not initially on the flowsheets. Lines branching from other lines are indicated in their correct relative order so as to indicate the correct placing and sizing of pipe reducers and fittings and changes in line diameters. Every pipe is assigned a line identifier that indicates the line diameter, its material of construction, the commodity or fluid it transports, and whether it is insulated or heat traced. Each pipe is also assigned a unique line number to allow the piping designers to prepare their line lists for the construction contractors. This task is achieved jointly by the process engineers, who define the lines required in the process, and the piping designers, who must run the lines in the new or the existing plant buildings and hence are in a better position to choose line branching points and show them on the P&IDs. The process engineers may have to redo their balances to define the correct flows throughout different sections of a line after its routing has been established.

P&IDs are finally used to indicate every single control loop as well as every manual valve in the process design. The process and mechanical engineers and instrumentation specialists work together at this task. Instrumentation on P&IDs may be represented in one of two ways, depending on the client's standards or wishes. Normally every loop is shown in its entirety on the P&ID, including the measurement element, the transmitter, the pneumatic/electric signal converter, the control function and location of controller, the control valve or control element, and the actuator and air feed line to the actuator. This loop representation adds a lot of detail to a P&ID and requires the process to be shown on numerous drawings for clarity of presentation. A simpler method of indicating a control loop is sometimes chosen, where only the control element is shown connected to a symbol that identifies the control loop function, with a line connected to the measured stream. In both cases, separate loop schematic diagrams are prepared for each individual loop, where not only are the above details shown but more details about the wiring and cabling of the signals are identified to assist in field running and testing of each loop.

Service Diagrams. Service diagrams are a special type of P&ID dedicated to showing the distribution of certain services in a section of a plant, such as seal water to all pumps and agitators, steam to all unit heaters, etc. They follow the same principles as conventional P&IDs.

Logic Diagrams. Logic diagrams are prepared by the process and instrumentation or control specialists to show, in Boolean notation or some other convention, all the steps required to run a process or to sequence a batch system. They resemble computer programming flowcharts, with branching steps based on decision nodes, comparison of signals for deciding on the next step, and so on. They are usually prepared towards the end of a project, when the process operating procedures are being developed and coded into the control system.

Piping Orthographic Drawings. These drawings show the exact routing in the plant of every pipe two inches and above in diameter, in plan and elevation view (one set of drawings for each type of view). Small diameter pipes are usually field run after the large pipes are installed. The drawings identify the pipes by single lines for small bore pipes or by two parallel lines for large diameter pipes.

They also indicate every pipe fitting required to route the pipes, every manual and control valve, and all tie points and connection points to existing lines or equipment. Every pipe is identified by a tag and is listed in the piping list using this tag as identifier. For multiple-level plants, a pipe routing drawing is produced for every level, depending on the spread of the buildings and the complexity of the piping. The equipment layout drawings are used to identify the beginning and termination of each pipe and to identify the interferences and obstacles around which the pipes must be routed.

Instrument Location Diagrams. These diagrams are prepared by the instrument specialists with the piping designers. Their objective is to indicate to the piping contractors where to install the sensing elements and the control valves in order to respect the many constraints imposed by good instrumentation design principles, particularly with respect to the length of straight pipe ahead of an element. The instrument location drawings use the piping orthographic drawings as background to indicate the exact position of each instrument element on the piping.

Loop Diagrams. Loop diagrams are produced by the instrumentation department. Each loop diagram completely describes a single control loop. It shows a schematic of the components that make up the loop, the wiring between these components, and the algorithm followed in the control philosophy. The level of detail may vary depending on the client or company internal design standards. A typical loop diagram, with the minimum required information items, for a pneumatic control system is illustrated in Figure 3-2. Figure 3-3 illustrates the same loop diagram for an electronic control system.

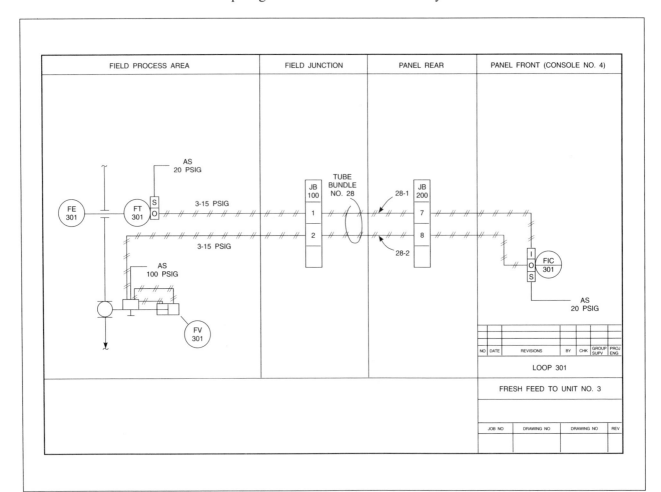

Figure 3-2. Typical Loop Diagram for a Pneumatic Control System [Ref. 1]

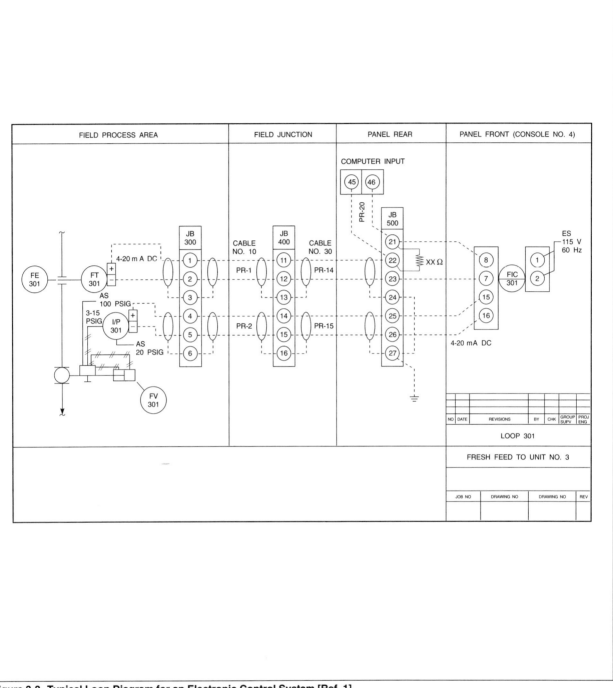

Figure 3-3. Typical Loop Diagram for an Electronic Control System [Ref. 1]

Conventional Process Measurements

Some basic measurement and control loops, in widespread use throughout all processing industries, are listed in Table 3-3. They are used on or around the types of process equipment common to most chemical, petrochemical, pulp and paper, and other processing industries. Table 3-4 presents examples of such applications and indicates whether the variables are controlled or merely displayed on control panels.

Table 3-3. Conventional Process Measurement and Control Parameters

Flow
Temperature
Pressure
Level
Weight
Viscosity
Density
Conductivity
pH
Concentration
Power consumption
Conveyer speed
Motor speed

Table 3-4. Instrumentation around Common Process Units

Process Unit	Controlled Variables	Control Function
Pipes	Flow rate	I or C
	Temperature	I or C
	Pressure	I or C
	Fluid conductivity	I
	Fluid pH	I or C
Pumps, blowers, and compressors	Discharge flow rate	I or C
	Discharge pressure	I or C
	Power requirement, for larger units	I
	Cooling water temperature	I
	Cooling water pressure	I
	Cooling water flow rate	I or C
Conveyors	Conveyor speed (for flow control)	I or C
	Weight of contents	I
	Moisture of contents	I
Storage tanks	Level	I or C
	Temperature	I or C
	Pressure	I
Agitators	Power consumption (as a measure of fluid viscosity or suspended solids concentration)	I
Heat exchangers	Temperature of all streams	I
	Flow of the cooling/heating stream	I and C
	Pressure drop across the exchanger on both hot and cold sides	I
Combustion chambers (boilers, kilns)	Flue gas temperature	I
	Flue gas analysis (for O_2, CO, CO_2, SO_2)	I
	Opacity	I
	Flow of fuel to burner	I and C
	Combustion air damper control	I and C

Table 3-4 (continued)

Process Unit	Controlled Variables	Control Function
Distillation columns	Pressure on various trays	I
	Temperature on various trays	I
	Level on reflux condenser	I and C
	Level of column bottoms	I and C
	Feed flow rate to column	I and C
	Top product flow rate	I and C
	Bottom product flow rate	I and C
	Flow of intermediate output streams	I and C
	Stream flow rate to reboiler	I and C
	Cooling water flow to top condenser	I and C
Reactors	Flow rate of reactants	I or C
	Temperature	I or C
	Pressure	I or C
	Viscosity	I
	pH	I
Filters	Flow rate through filter	I
	Pressure drop	I
	Conductivity	I or C

Note: I = Indication only of a measurement
C = Control of a variable (includes display of measurement, alarm, and interlocks)

Process Information for Instrument Specification

The selection of control equipment and control parameters and the purchase of the various control loop elements are the responsibility of the instrument designer. He or she does need, however, to work closely with the process and mechanical engineers on the design team and with vendors and technicians to obtain the vital information to properly size or specify the control elements. There are no formal methods for communicating the relevant process information to the instrumentation group, but the data is available in a number of different documents and from a number of different sources, which will be described below.

The P&IDs are the first source of process information. They indicate all the control loops in the process and their functions; they also indicate normal and maximum flow rates of most major lines and many secondary lines. Mechanical drawings of vessels, tanks, and so on must be consulted to obtain dimensional information used to design level control loops. General arrangement drawings define the location of equipment and show suction and discharge points of streams; they are used to calculate pressures and pressure drops in lines. Stream properties (such as temperature, pressure, pH, moisture, conductivity, and so on) needed to specify valves for other control purposes may be obtained from either the design criteria document or the process engineers. Data required to fully specify control loops is shown in Table 3-5. When all the information is available, control valve spec sheets, such as shown in Figure 3-4, are filled in by the instrumentation designer and checked by the process engineer. These sheets are then used to size the control valves.

At later stages of the project, the control algorithms are chosen by the instrumentation group and discussed with the process group. Both disciplines collaborate to develop the control logic and interlocks. Since these decisions affect mainly the programming of the control equipment, the data gathering and logic design schedules are less tight. It also requires the process to be fully developed and the operating manuals for start-up, shutdown, and normal operations to be well in progress.

The most efficient way initially to exchange information and agree on process values and control variables is through face-to-face meetings, where information is discussed, control philosophies argued, and specifications selected. When the basic plant control scheme is agreed upon by all concerned, then design data is sought.

PROJECT_____ DATA SH_____ OF_____
UNIT_____ SPEC. _____
P.O. _____ TAG _____
ITEM_____ DWG _____
CONTRACT_____ SERVICE _____
MFR SRL NO_____ _____

	UNITS	MAX FLOW	NRM FLOW	MIN FLOW	SHUTOFF
SERVICE CONDITIONS					
FLOW RATE					
SOURCE PRESSURE, ABSOLUTE					
PUMP, BLOWER, ETC					
ELEVATION CORRECTION					
LINE LOSS					
EQUIPMENT LOSS, 1					
EQUIPMENT LOSS, 2					
VALVE INLET PRESSURE, ABS					
RECEIVER PRESSURE, ABS					
ELEVATION CORRECTION					
LINE LOSS					
EQUIPMENT LOSS, 1					
EQUIPMENT LOSS, 2					
VALVE OUTLET PRESSURE, ABS					
INLET TEMPERATURE					
SPEC WT / SPEC GRAV / MOL WT					
VISCOSITY / SPEC HTS RATIO					
VAPOR PRESSURE. ABSOLUTE					
CRITICAL PRESSURE, ABS					
REQUIRED C_v					
RATED C_v					
PERCENT TRAVEL					

Row groups at left: SOURCE TO VALVE · RECEIVER TO VALVE · AT VALVE INLET

NOTES, PROCESS DATA SCHEMATIC, PIPING CONFIGURATION, ETC.

Figure 3-4. Process Data Worksheet
(From ISA-RP75.21-1989, Process Data Presentation for Control Valves)

Table 3-5. Process Information for Controller Specification

Fluid state (gas, liquid, slurry, multi-phase)
Fluid composition
Minimum, normal, and maximum flows
Minimum, normal, and maximum temperatures
Minimum, normal, and maximum pressures
Fluid pH
Fluid viscosity
Fluid density
Fluid solids suspension
Acceptable pressure drop across control valve
Total system pressure drop

Basic Concepts of Process Control

The Role of Process Control

A process requires a number of events to take place in a particular sequence in the most efficient and economical method. These events take place efficiently only if the results of a preceding event are always predictable and repetitive, thereby requiring the least effort to run the subsequent step in an optimum way.

In an ideal process, all inputs are constant in quality (for example, composition, temperature, and pressure) and in quantity (flow rate, for example). Every process step has fixed and constant parameters, such as constant heating rates, constant flow-splitting characteristics, and so on. The process runs smoothly and produces even quality and quantity products. In reality, processes are never ideal and do not run in a steady state for any appreciable length of time. There are all sorts of upsets in the process operating conditions—feed stream compositions or flow rates, climatic conditions, and so on—that subject it to dynamic variations, which cause the final product to fluctuate from its desired properties more or less widely.

Variations can occur in a stepwise fashion (for example, when a raw material of a different composition is substituted suddenly in a process), or in a ramped fashion if a new raw material is bled in slowly until it replaces the previous raw material. Some variations are cyclic in nature, such as the temperature of the ambient air over a day or over a year. Others are instantaneous and occur only for very short durations like spikes. Yet others are completely random in nature, shape, or occurrence.

Variability in the product is caused not only by external forces but also by the deterioration with time of the machinery that manufactures the product. Pump impellers wear out and push less fluid at the same operating conditions, tanks leak, heat exchangers foul up, and so on. Sufficient production overcapacity is usually designed into the process to overcome these variations; nevertheless, if a process is left to run by itself with no external supervision and with no manipulation of variables that can be affected, it is bound to run itself into the ground and stop or break down.

Industrial processes today are not capable of functioning without the use of automatic control systems that track the symptoms through which variations manifest themselves and react so as to change the variables that are available to be controlled. Processes are controlled automatically for a number of reasons:

(1) To eliminate or reduce human error while operating a plant, therefore providing greater safety for the operators and for the environment

(2) To reduce the amount of labor and labor costs, which tend to drive up the price of a product

(3) To minimize the consumption of energy

(4) To improve the product quality and product consistency

(5) To reduce the size of the plant and of intermediate storages and inventories, which results in better control of the processing steps because the magnitude of disturbances is reduced or limited

(6) To better control plant emissions to meet more stringent pollution control regulations

Automatic control systems relieve the human operator of the tedious and repetitive tasks of monitoring and adjusting the process parameters manually and allow more time for inspecting the operating condition of the equipment, performing maintenance tasks, and assisting in the optimization of the plant operation.

Steady-State Balanced Condition

The design of a process always starts with the assumption that the process will operate under balanced conditions and that it will run continuously at the design conditions. A process or a unit operation is said to be at steady state (or in balanced condition) when the sum of all flows entering it equals the sum of all flows leaving it. The same applies for the energy input and output. Any change in the condition of one of the inputs, either from a material or from an energy point of view, that is not accompanied by an appropriate change in one of the outputs unbalances the process and sets it to an unsteady-state condition until it regains its equilibrium at the same or at a different equilibrium state.

Self-Regulation

A process unbalance will manifest itself in a number of ways. There may be a change in level in a tank, in a product of different composition, in a hotter stream, and so on. Some processes have an inherent capacity to limit the effect that external variations in some input condition may have on the value of a variable that needs to be controlled, be it the product stream or some intermediate stream. For example, in a tank that has a single input and a single output, both completely uncontrolled, the liquid in the tank reaches a level at which the static head of the liquid above the discharge orifice pushes out just as much liquid as is coming in. The flow of the liquid through the discharge orifice creates a pressure drop equal to the static head above the orifice. When the input flow rate increases, the level rises in the tank, raising the static head in the vessel and increasing the output flow until it equals the new input flow rate, at which the level stabilizes. The opposite is true when the inflow drops.

This inherent ability of certain processes to balance their mass or energy input and output is termed "self-regulation."

 Safety valves on pressure vessels and tank overflows are a different kind of crude self-regulation but do not constitute control mechanisms other than in extreme conditions to prevent major upsets or accidents.

Automatic Controls

The majority of processes are neither perfectly immune to external disturbances nor are they self-regulating. Even the best designed and newest processes cannot maintain the desired operating conditions indefinitely. It is therefore necessary to ensure that the balance of material and energy between inputs and outputs remains in control under most, if not all, running conditions. Automatic (external) control can be defined as a mechanism that measures the value of a process variable and operates to limit the deviation of this variable from a predefined set point. Thus, the primary function of automatic process control is to manipulate the material or energy input-to-output relationships in order to keep process variables within desired limits.

Numerous parameters are involved in the description and characterization of automatic control systems and controllers. Some of the more important ones are presented below.

TIME CONSTANT

A time constant is defined as the time required to complete 63.2 percent of the total rise or decay of a process parameter after being subjected to a step change in one of the external forces affecting it. This precise percentage value is derived from the solution of mathematical balances around simple processes.

Figure 3-5 illustrates a closed, continuously stirred tank of defined volume V (in units of volume, such as cubic meters) into which a flow F comes (in the same units of volume per chosen unit of time, such as hours). Let T_i be the temperature of the input flow and T_o be the temperature of the output flow. Let C_p be the heat capacity of the fluid (in units of heat per unit of mass per unit of temperature, such as cal/kg/°C) and D be its density, assumed constant over the temperature range under consideration (in units of mass per unit of volume, such as kg/m^3). In this closed vessel, the volumetric output flow rate is equal to the volumetric input flow at all times. At steady state, T_o equals T_i. If at some time, defined as $t = 0$, the temperature of the input flow is suddenly raised from T_i to a new temperature T_n, the temperature in the reactor will change with time until it eventually reaches the new input value T_n. The rate of change, q, of the total heat content of the reactor is defined by the difference between the heat that comes in and the heat that goes out of the reactor. Equations (3-4) to (3-8) represent this change:

Heat change, in cal/h, between the input and output streams is

$$q_{io} = FDC_p (T_n - T_o) \qquad (3\text{-}4)$$

Heat change, in cal/h, in the tank contents is

$$q_{TC} = VDC_p \frac{d}{dt} T_o \qquad (3\text{-}5)$$

Since there must be conservation of energy, both changes in heat are identical. Hence, by equating both terms, rearranging this equation, and combining terms, the equation becomes:

$$\frac{F}{V} \int dt = \frac{1}{T_n - T_o} \int dT_o \qquad (3\text{-}6)$$

which, when integrated, yields:

(continued)

(continued)

$$T_o = T_n - (T_n - T_i)\, e^{[-t/(V/F)]} \qquad (3\text{-}7)$$

The term V/F has units of time only (hours in this example) and is referred to as the time constant of this reactor. One time constant unit after the change in input concentration, the change in the reactor concentration is given by:

$$T = T_n - 0.3678\,(T_n - T_i) \qquad (3\text{-}8)$$

If $T_i = 0°C$ and $T_n = 100°C$, then T (one time constant) = 63.2°C, which represents 63.2 percent of the difference between the original temperature T_i and the final temperature T_n. The change of the output stream temperature with time is shown in Figure 3-6.

CAPACITY AND CAPACITANCE

The capacity of a piece of equipment is a measure of the total amount of material or energy it can absorb and store and with which it is able to buffer the effect of variations in the incoming stream in order to produce an outgoing stream that exhibits less variability. In thermal systems, heat is stored by increasing the temperature of solids until it is released again. In electrical systems, capacitors store a certain amount of electrical charge. In mechanical systems, inertia is the measure of energy capacity stored in a moving object.

Industrial process control systems deal primarily with the first two types of capacity systems mentioned above, namely, the material and thermal systems. For instance, a well-stirred tank full of water has a large capacity to absorb and dissolve crystal table salt. The capacity of the tank is given by the product of

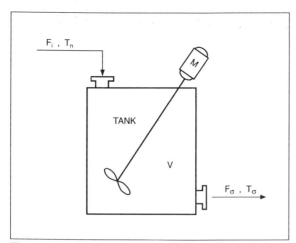

Figure 3-5. Closed, Continuously Stirred Tank

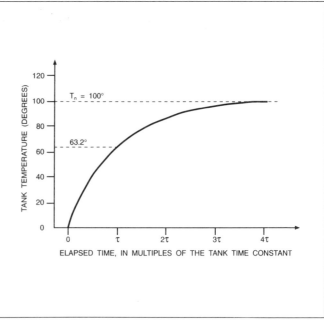

Figure 3-6. Temperature Rise in Tank, Over Time

the volume of the water contained in the tank and the volumetric saturation concentration of the water:

$$\text{Capacity (kg)} = \text{Volume (m}^3) \times \text{Saturation concentration (kg salt/m}^3) \qquad (3\text{-}9)$$

Salt crystals metered into the tank will be absorbed and dissolved until the saturation concentration is reached. Thereafter, excess salt drops to the bottom of the tank, the tank dissolving capacity having been reached. The same applies to an electric capacitor that leaks excess electricity once it is saturated, or to a thermal system that no longer absorbs heat once it reaches the same temperature as the medium supplying heat to it.

Capacitance is defined as the change in material or energy required to produce a unit change in some measure of material or energy content in the vessel or unit. If one kilogram of salt is added to a tank of volume V (m^3) full of water, the change in the tank salt concentration due to one unit of mass (kg) is given by:

$$\text{Capacitance (concentration kg/m}^3) = 1 \text{ (kg)}/V \text{ (m}^3) \qquad (3\text{-}10)$$

In the example described above, the time constant F/V is a function of the process capacity V.

RESISTANCE

All processes offer opposition to the flow of energy or material either through the effect of retarding or resisting the imposed driving force, such as friction in pipes retarding the flow of water, or through imposing barriers between two fluids that are exchanging heat. The ratio of the change in driving force to the change in flow is called "resistance." In electrical terms, the change of voltage across a conductor divided by the electrical current through the conducting element is called resistance. In thermodynamics, resistance to the flow of heat is the change in heat content of a vessel in a unit of time divided by the difference in temperature between the vessel and the surroundings. Were it not for thermal resistance, both media would equilibrate instantaneously at a common temperature. Resistance is a function of time; capacitance is not. The product of resistance and capacitance is the time constant of the system under consideration. Table 3-6 gives a broader view of the analogy between the various physical systems.

Types of Process Responses to Disturbances

In order to control a process, one has to know more about the type of reaction that a control variable will demonstrate to changes in the input variables when no control is applied.

DEAD TIME (OR ZERO-ORDER) SYSTEMS

Zero-order systems exhibit a variation in the output variable identical to that in the input variable, with a time shift or time lag. Pipelines are typical dead time systems. These systems have neither capacitance nor resistance to dampen the effect of the incoming variations. Occurrences of pure dead time systems are rare. Most processes exhibit some capacity to store material or energy, with or without a certain amount of dead time.

FIRST-ORDER SYSTEMS

The previous example for time constants represents a continuous first-order system, which are systems that have single capacitances. They exhibit first-order lag due to the capacitance of the process rather than to process dead time. They are characterized by two parameters: a time constant and a gain. They have a

Table 3-6. Analogy between Different Physical Systems [Ref. 5]

Variable	Electrical	Liquid Level	Thermal	Pressure
Quantity	Coulomb	Cubic foot	Btu	Cubic foot
Potential or force	Volt	Foot	Degree	$\dfrac{\text{Pounds}}{\text{Square inch}}$
Flow rate	$\dfrac{\text{Coulombs}}{\text{Second}}$ = Amperes	$\dfrac{\text{Cubic feet}}{\text{Minute}}$	$\dfrac{\text{Btu}}{\text{Minute}}$	$\dfrac{\text{Cubic feet}}{\text{Minute}}$
Resistance	$\dfrac{\text{Volts}}{\text{Coulombs per second}}$ = ohms	$\dfrac{\text{Feet}}{\text{Cubic feet per minute}}$	$\dfrac{\text{Degrees}}{\text{Btu per minute}}$	$\dfrac{\text{Pounds per square inch}}{\text{Cubic feet per minute}}$
Capacitance	$\dfrac{\text{Coulombs}}{\text{Volts}}$ = farads	$\dfrac{\text{Cubic feet}}{\text{Foot}}$	$\dfrac{\text{Btu}}{\text{Degree}}$	$\dfrac{\text{Cubic feet}}{\text{Pounds per square inch}}$
Time	Seconds	Minutes	Minutes	Minutes

maximum rate of change immediately following the application of a disturbance and always reach 63.2 percent of the change of state after one time constant, because they follow the same rise rate described by Equation (3-7) and illustrated in Figure 3-6.

SECOND- AND HIGHER-ORDER SYSTEMS

In continuous processes, higher-order systems can be the result of several different situations:

(1) Several first-order processes may be encountered in series.

(2) The installed feedback controller may introduce a characteristic differential equation, which, when considered in series with the other system components, makes the overall description of the system higher ordered.

(3) Mechanical or fluid components of the system may be subject to accelerations, i.e., to inertial effects (usually this is a minor possibility).

(4) The process may be a distributed process that gives a response curve that can be described only by a higher-ordered differential equation or by partial differential equations.

The time response of a second-order system to a step change in the input variable depends on the inertia or "damping ratio E" of the system. Figure 3-7 illustrates the complexity of the response, compared to a simple first-order system, for different values of the damping ratio E.

In batch processes, second-order characteristics are inherent when two operations occur more or less simultaneously, one operation depending on the other and both having distinct capacitances that delay the response of one variable to changes in the other. Heat transfer in batch reactors exhibits typical second-order response.

The S-shaped curve, as in Figure 3-9, is typical of such processes. Initially, there is a dead time during which no liquid heating occurs while the jacket and reactor wall are heating up. This resistance to heat transfer between two capacitances separated by a resistance causes a transfer lag. This process block is known as a second-order system, and the S-shaped curve is a second-order lag.

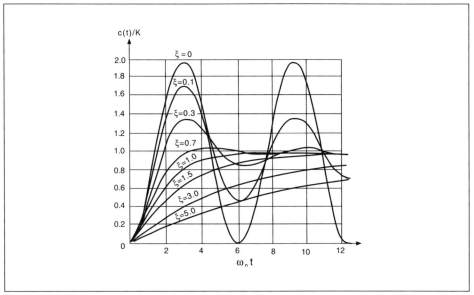

Figure 3-7. Second-Order Lag Step Response
(From B. Liptak, *Instrument Engineers' Handbook*, Chilton Book Co., 1985.)

Figure 3-8 illustrates a jacketed, well-stirred reactor that contains a liquid to be heated to a predetermined set point by introducing steam into its external steam jacket. A control valve starts and stops the steam flow to the jacket. Initially, the liquid is cold and no steam is fed to the jacket. When the valve opens, steam flows into the jacket, pressurizes it, and starts heating the jacket wall and the reactor wall; only then does the liquid start heating up. The heat transfer to the liquid is proportional to the difference in temperature between the steam temperature and the liquid temperature. As it heats up, the rate of heat transfer diminishes until it reaches the set point, at which time the steam is stopped. Figure 3-9 illustrates the liquid temperature rise with time.

Frequency Response Analysis and Characteristics

Load changes and disturbances to inputs of processes are not neat step changes as described above; they tend to fluctuate in magnitude and frequency. Disturbances that are of very short duration, compared to the time constant of the process, may repeat rapidly and frequently and change in magnitude at a predictable frequency around some average condition, like a sine wave.

It is possible to learn more about a process characteristic and its response to a control system by deliberately subjecting it to a sine wave type of variation in one of its inputs, then varying the wave frequency and amplitude and plotting the output response as a function of time and the input variable. The variation of the output of a linear first-order type of process will respond very similarly to its input variation and will follow a sine wave curve; however, very few systems are linear. If both input and output waves are sinusoidal, both the system gain and its phase lag may be calculated.

BIAS AND AMPLITUDE

Bias is the average level of the sine wave. The amplitude of the sine wave is the value from the bias or zero line to the peak of the wave. Bias is also called the instrumental or systematic error, meaning the difference between the average of a

In practice, rather than measuring amplitude as the distance from the bias line to the peak, the peak to the bottom of the wave is measured and divided by two.

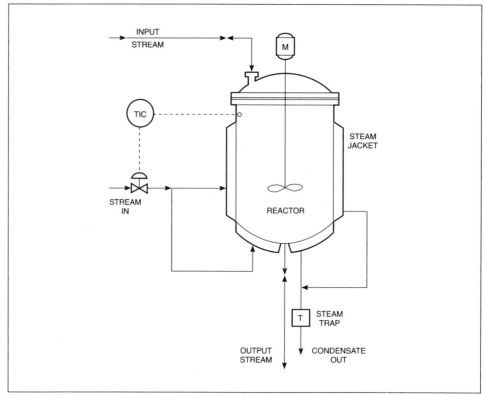

Figure 3-8. Jacketed Well-Stirred Reactor

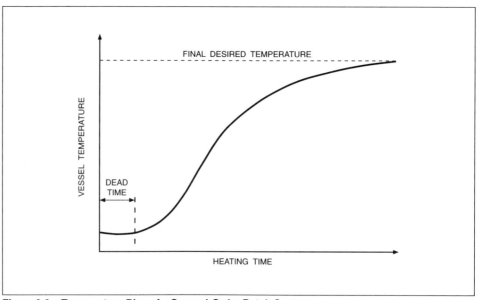

Figure 3-9. Temperature Rise of a Second-Order Batch System

series of up and down values of an instrument output and the corresponding true values of input.

> In the Figure 3-5 example, assume that the tank has a volume of 1,000 liters and the input and output streams flow at 100 liters per minute (the time constant is 10 minutes). The tank is initially full of warm water at 50°C, which is the same temperature as the input stream. At some time, the temperature of the input stream starts to vary sinusoidally around the same temperature as that of the warm water in the tank, with a maximum temperature of 60°C and a minimum of 40°C. Hence, the input stream bias is 50°C and its amplitude is 10°C (see Figure 3-10).

In a well-stirred tank, the output stream has the same temperature as the tank contents. If the change in the input steam temperature is slow, the tank temperature change will follow rapidly and may reach a value close to the input stream's maximum temperature before dropping again, closely following a sine wave similar to the input. If the frequency of change in the input stream accelerates, the tank temperature change will not follow as quickly; it will lag the input stream variation, and its maximum value will not have time to reach the same maximum value before dropping again, still sinusoidally.

Figure 3-10 shows the change in output stream temperature for an input stream varying slowly in magnitude at a frequency of one cycle per hundred minutes. Figure 3-11 shows the same for a faster changing input stream, namely, one cycle every twenty minutes. Both these responses would occur only after sufficient time has elapsed for the initial transient response to have died away, typically after three or four time constants in elapsed time. Notice the significant dif-

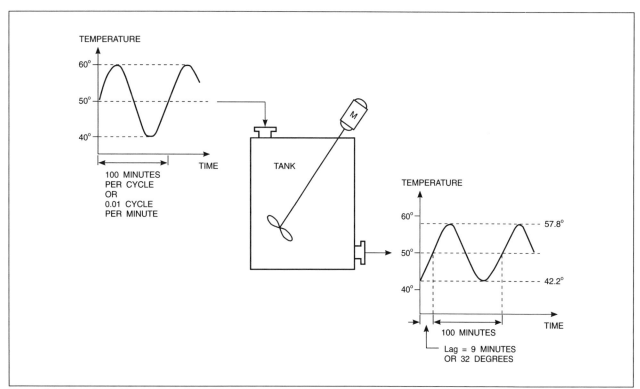

Figure 3-10. Amplitude Response at Low Frequency

ference in amplitude of the output and the lag time, in degrees, at these two frequencies.

GAIN AND GAIN EQUIVALENT IN DECIBELS

The ratio of the maximum change in the tank temperature to the maximum change in the input stream temperature is called the gain:

$$\text{Gain} = \frac{\text{Maximum output amplitude change}}{\text{Maximum input amplitude change}} \tag{3-11}$$

In electrical control of processes, it is more common to refer to gain in decibel (dB) units. The decibel is a logarithmic unit equivalent to twenty times the logarithm (to the base 10) of the magnitude ratio of the amplitude of the output signal to the amplitude of the input signal, provided that the input and output impedances are equal:

$$dB = 20 \log \frac{(\text{Output amplitude change})}{(\text{Input amplitude change})} \tag{3-12}$$

or

$$dB = 20 \log (\text{Amplitude ratio}) \tag{3-13}$$

or

$$dB = 20 \log (\text{Gain}) \tag{3-14}$$

Thus, an amplitude ratio or gain of unity corresponds to zero decibels.

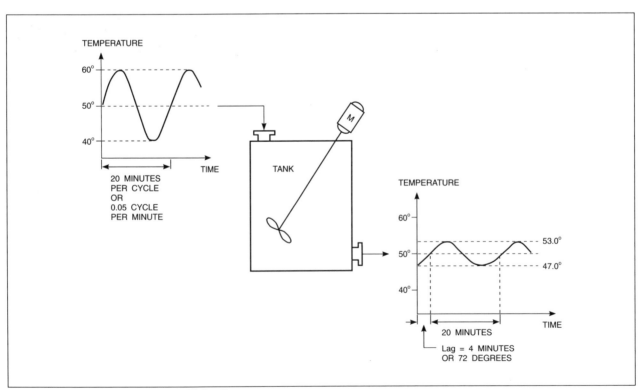

Figure 3-11. Amplitude Response at Higher Frequency

PHASE ANGLE

When a system is subjected to a sinusoidal variation in one of its inputs, the difference in time between when the input stream reaches its maximum value and the output reaches its own maximum temperature, converted to angles of the input sine wave, is referred to as the "phase lag" or "phase angle."

As the frequency of variation of the input stream increases, the phase lag increases and the gain drops. The procedure of running a number of sine wave variations of different frequencies through a process, with the resulting decay in gain accompanied by an increasing phase angle lag, may be shown graphically on a Bode plot or Bode diagram.

The Bode diagram for the previous example is shown in Figure 3-12. This diagram consists of two sections: an amplitude ratio curve and a phase angle curve, which are both plotted against the product of the frequency of change of the signal and the time constant of the process. Both axes are made dimensionless for convenience. The phase angle is shown as negative when the output signal lags the input signal; it is positive when the output signal leads the input signal.

Traditional Control Strategies

Process Control Loops

Figure 3-13 illustrates a hot water tank that receives water at some uncontrolled temperature that may vary with the season of the year or because of some other factor. The tank outputs hot water at a fixed temperature, which is maintained constant by steam heating coils in the tank. The amount of steam injected depends on the difference between the inlet and outlet temperatures. A thermocouple embedded in the tank relays the heated water temperature to some control unit, which increases or decreases the flow of steam to the coils by varying the opening of the control valve on the steam feed line. The collection of elements that contribute to the automatic control of the output temperature is known as the control loop, as illustrated in Figure 3-13 in conventional pneumatic instrumentation symbology. The combined functions of temperature (or process variable) measurement, transmission of the measurement, comparison of the measurement to a desired value or set point, decision on the corrective action to

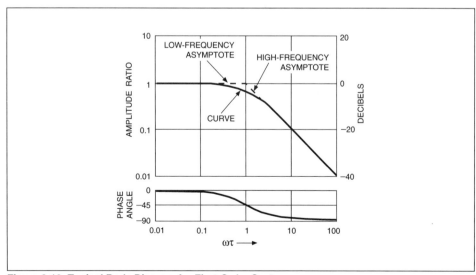

Figure 3-12. Typical Bode Diagram for First-Order System

take on other variables, and actual implementation of this corrective action through a control valve are known as the "control loop."

Pictorial Representation of Control Loops

The block diagram is a convenient pictorial method of depicting these elements for computational purposes, as shown in Figure 3-14. The block diagram helps in visualizing the mathematical relationships between the various signals. In the plant, the individual schematic blocks of a control loop are actual physical hardware and/or software elements that perform each of the above mentioned tasks.

The Control Elements

Every control loop consists of at least four fundamental elements, which work together to achieve some measure of control on one or more desired variables. These elements are:

(1) a sensor, or primary element,

(2) a transmitting element,

(3) a controller element, and

(4) a final control element.

The sensor is the unit that detects a change in the property that is measured. The thermocouple in the example is a sensor. A ball float resting on the surface of the liquid in a tank is a level-sensing element. It rises and drops as the level changes. The tank level could also be measured by a capacitance level probe or by a pneumatic tube bubbling air through the liquid. An orifice plate on a pipe is a flow sensor. A sensor may or may not transmit the measurement onwards. The ball float or the orifice plate are elements used only to sense a property, such as the position of the fluid level or the pressure drop across the plate.

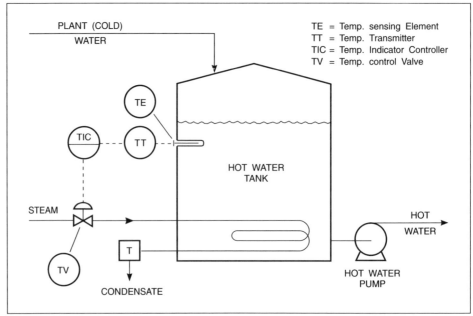

Figure 3-13. Hot Water Tank and Temperature Control Loop

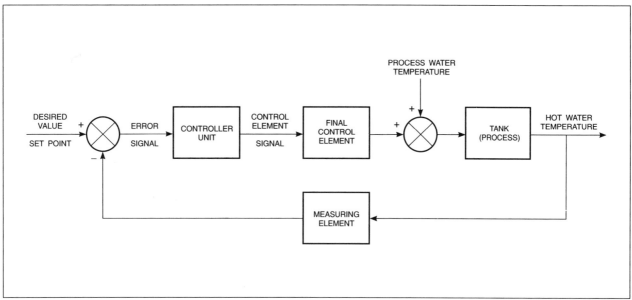

Figure 3-14. Block Diagram of Control Loop for Hot Water Tank

A transmitting element receives the signal from the sensor, converts it into a different type of signal, and transmits it to a remote area such as the control room or control rack room. The rope attached to the ball float, the electric signal from the capacitance probe, or the pressure drop across the orifice plate sends a signal to the transmitting element, which converts it into a value or a relative change in value of the measured property.

The measured value is converted into a pneumatic signal or an electrical signal and is transmitted through tubes or over a cable to a unit that compares the signal to a desired set point. This unit performs some arithmetical function to assess the change required in the stream that is affecting the measured variable to bring it back to its desired value. Then it sends out either a pneumatic or an electrical signal to the final control element to effect this change.

The final control element may be a modulating unit (such as a control valve, a damper, a pump, a conveyor belt, a variable speed drive, and so on) whose position changes with the magnitude of the signal it receives, or it may be a non-dynamic element, such as a recorder, an alarm, a counter, a computing device, or a shutoff device. The signal and the control function of a process control loop end at the final control element.

Open (Feedforward) and Closed (Feedback) Loops

Two concepts are used in most automatic control strategies: (1) feedforward or open-loop control, and (2) feedback or closed-loop control.

Feedforward is based on measuring the change in a process input, anticipating the effect that the change will have on the process, and automatically taking corrective action on another input variable to counteract the change. Feedforward control is usually specified where changes in process inputs occur so frequently that the feedback controller cannot keep up or if the disturbances are so large that the controlled variable cannot be kept within tolerable limits.

Feedforward control is very powerful and fast but does not yield accurate results by itself because of a number of major drawbacks:

(1) The models prepared to represent the controlled process and those used to devise the control schemes must be exact.

> It is better to solve a problem at the source than to correct the problem from some after-the-fact symptoms.

(2) All instruments must be perfectly calibrated.

(3) Disturbances other than those controlled by the feedforward loops are not controlled.

(4) There is no way to backcheck and correct the control signal.

Figure 3-15 shows a simple process heat exchanger that uses steam to heat a liquid flowing through the exchanger. The steam flow is manipulated by a controller that measures the liquid input flow and temperature. The controller calculates the flow of steam from the steady-state energy balance around the exchanger:

$$W \times D \times C_p \times (T_{sp} - T_i) = F \times H \qquad (3\text{-}15)$$

where:

$$
\begin{aligned}
W &= \text{the liquid flow (liters/minute)} \\
D &= \text{the liquid density (kg/liter)} \\
C_p &= \text{liquid heat capacity (calories/}^\circ\text{C/kg)} \\
T_i &= \text{liquid inlet temperature (}^\circ\text{C)} \\
T_{sp} &= \text{liquid set point temperature (}^\circ\text{C)} \\
F &= \text{steam flow (kg/hr)} \\
H &= \text{heat released by steam (calories/kg)}
\end{aligned}
$$

Equation (3-15) is solved for the steam flow F:

$$F = WDC_p \frac{(T_{sp} - T_i)}{H} \qquad (3\text{-}16)$$

Most of the measurements and constants used in Equation (3-16) are temperature-dependent, but the modelling equation does not account for this dependency. Similarly, two measurements (W and T_i) affect each other; if both are off by five percent, the steam flow could be controlled at up to 10 percent over the required flow, and the liquid output temperature would be nowhere close to the desired set point. With no "check" on the actual value of the liquid output temperature, this control mode can cause serious problems.

Feedforward is normally combined with a feedback loop, which changes the feedforward loop set point to make up for unforseen disturbances or drifts in instrument calibration. Together, coupled feedforward and feedback loops are powerful control tools.

Feedback is designed to achieve and maintain the desired process output conditions by measuring the output variable and checking its value against a preset target point (the set point), then modifying the controlled input variable to reach this set point. Feedback control is, therefore, more successful in reaching the target set point than feedforward control, but because of the process capacitance, it responds more slowly since it reacts only to an output upset that is detected only a certain length of time after a process input has changed and caused the upset. Feedback is potentially less stable if the corrective action taken is too big and causes an output variable swing in the opposite direction and of a larger magnitude—hence, the need to choose the appropriate control mode.

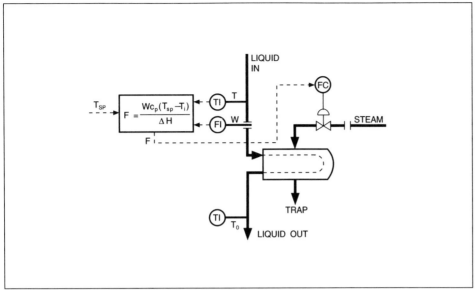

Figure 3-15. Feedforward Control of a Heat Exchanger [Ref. 5]

Control Modes

The hardware inside a control element may be constructed or configured in one of many ways to give several different types of relationships between the response of the controller and the detected error in the process. The relationships between error and controller outputs are called "actions" or "modes"; thus, a controller mode is the particular response of a controller block to an error signal.

Modern industrial controllers are designed to produce one or a combination of the following control modes:

(1)   On-off (two-position) control

(2)   Floating control

(3)   Proportional control

(4)   Integral (reset) control

(5)   Derivative (rate) control

Certain combinations of these, such as proportional-plus-integral, or proportional-plus-integral-plus-derivative modes, are widely applied in industry.

A control mode recognizes the existence, size, direction, and speed of an error signal and acts to determine when to change the controller output and by what amount to correct for the error. It also determines what the output signal will be when the error no longer exists.

Traditional pneumatic control boxes were built differently for each type of control mode. Their hardware was different, and they had to be purchased for the desired function. Modern electronic controllers supplied with distributed control systems (DCSs) can perform all modes of control. The appropriate mode is selected by software configuration of each individual controller.

ON-OFF CONTROL

The on-off or two-position control mode is the simplest and cheapest of closed-loop control modes. The on-off controller compares the measurement signal to a given set point and moves the final control element from one preassigned fixed position to another preassigned fixed position. The positions of the control element, such as a valve on a steam line to a tank heater, are either wide open or completely closed. Examples of on-off controllers are motor push buttons, household thermostats, and refrigerators. An on-off controller cannot make exact corrections, and no stable balanced conditions are achievable. In practice, however, the mechanical nature of the controller provides some inertia and introduces a narrow band, called a deadband, around the zero error position that the measurement must first overcome before causing the final control element to change positions. This inertia helps dampen the otherwise excessive cycling of the measurement.

Figure 3-16 shows the response of a simple on-off controller to a sinusoidal variation in the measured variable and the effect of the deadband around the zero-error position. Usually, a process has sufficient capacity that, once the controlled variable has shot slightly past the deadband in either direction, it will take some time before it changes again and moves past the opposite side of the deadband. For instance, a heated room does not cool down instantly once the heat is turned off.

TWO-POSITION DIFFERENTIAL GAP CONTROL

The two-position differential gap controller is a special case of the on-off controller. The final control element moves from one position to the other when the controlled variable reaches a preset value on either side of the set point. Hence, the deadband around the set point is built into the controller, as opposed to being a random mechanical feature of the controller. The response is identical to that in Figure 3-16, except that the differential gap is wider and is preset by the control specialist. A typical example of a two-position differential gap controller is a household hot water tank.

FLOATING CONTROL

In the floating control mode a predetermined relation exists between the deviation of a measured variable and the rate of travel of the final control element (the control valve or other element). This element moves relatively slowly toward either one or the other of its extreme positions, depending on whether the

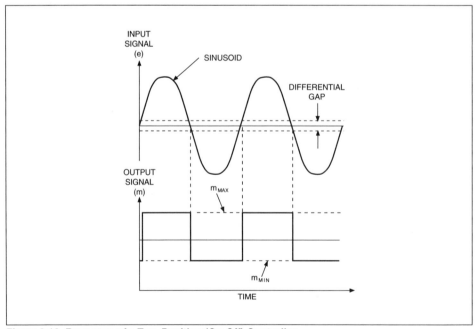

Figure 3-16. Response of a Two-Position (On-Off) Controller
(From B. Liptak, *Instrument Engineers' Handbook*, Chilton Book Co., 1985.)

measured variable is above or below its set point. The rate of change is dependent on the error signal. Different types of floating control action are:

(1) single-speed floating control,

(2) multispeed floating control, and

(3) proportional-speed floating control.

In the single-speed floating mode (see Figure 3-17), the rate of output change is constant regardless of the magnitude of the error. The multispeed floating mode allows the final control element to move at two or more different rates, depending upon the amount of the deviation from the set point. In proportional-speed floating mode, the rate increases with increasing error signal.

Floating mode controllers are used in systems that have little or no capacitance and, hence, little or no lag in process response to the change in the position of the final control element.

PROPORTIONAL CONTROL MODE

The simplest modulating control mode is proportional control. The controller output (or corrective action) signal is equal to a constant multiple of the deviation between the measured value of the controlled variable and the desired set point, to which is added a constant value. The equation representing the proportional control mode is:

$$\text{Output} = K_c e + p_o \qquad (3\text{-}17)$$

where:

K_c = the constant of proportionality or proportional gain

e = the error or difference between the set point and the measured variable value

P_o = the controller output when there is no deviation from the set point

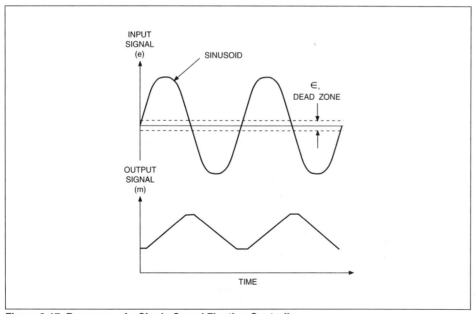

Figure 3-17. Response of a Single-Speed Floating Controller
(From B. Liptak, *Instrument Engineers' Handbook*, Chilton Book Co., 1985.)

There is a range of measured values of the controlled variable within which the controller output is proportional to the deviation. The boundaries of this range correspond to the zero and maximum values of the controller output. The range of measured values, expressed as a percentage of the instrument scale, is called the "proportional band" (PB). Most controllers have internal adjustments to alter the proportional band. Many manufacturers calibrate their adjustment in "gain" rather than proportional band. Gain is defined as the inverse of the proportional band multiplied by 100. It is also referred to as the proportional sensitivity of the controller.

$$\text{Gain} = \frac{100}{\text{PB}}$$ (3-18)

Figure 3-18 shows the relationship between gain and proportional band. The wider the proportional band, the less sensitive will be the response to upsets. It takes a large deviation from the set point to move the controller.

Offset

Proportional control has a serious deficiency. It hardly ever keeps the controlled variable at the desired set point, particularly if there are frequent disturbances to the process. There will always be a difference between the process variable measurement and the set point. After a change in the measured variable due to external sources, there will always be a sustained error that would last forever if other conditions remained constant. This error is called the "offset." Offset can be minimized by increasing the proportional gain (K_c), but a high gain will cause an unstable or cyclic operation.

Figure 3-19 compares the response of an output variable to the change in a free input variable under uncontrolled and controlled conditions using proportional, in-

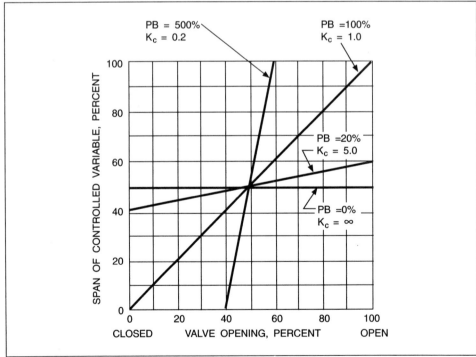

Figure 3-18. Gain (K_c) and Proportional Band (PB) and Their Effect on Valve Opening [Ref. 5]

tegral, and proportional-integral controllers. Notice the offset position reached with proportional-only control. Figure 3-20 compares the output variable response to a set point change. Notice here the difference in response time with each mode.

INTEGRAL CONTROL MODE

The action of the integral control mode is to affect the controller output in proportion to the cumulative deviation, over time, of the measured variable with respect to its set point. As long as the deviation is positive, the controller output increases. When the deviation decreases and becomes negative, the value of the time integral of the deviation reduces and the controller output decreases. Hence, the integral action recognizes the magnitude, the direction, and the time duration of the deviation. The controller output is governed by the following equation:

$$\text{Output} = K \int e \, dt \qquad (3\text{-}19)$$

Integral (also called reset) action alone can eliminate the error, but very slowly, since it depends solely on the lengthy time integral action to average out the error. It is limited usually to processes of small capacitance that respond fast to large changes in primary input, since a large capacitance would make its action too sluggish.

DERIVATIVE CONTROL MODE

The action of the derivative control mode is to provide an output from the controller that is proportional to the rate of change of the deviation from the set point. The controller output is governed by the equation:

$$\text{Output} = K \frac{d}{dt} e \qquad (3\text{-}20)$$

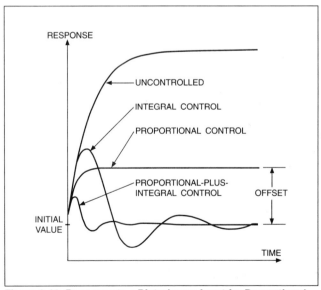

Figure 3-19. Response to a Disturbance Input for Proportional, Integral, and Proportional-plus-Integral Controllers
(From B. Liptak, *Instrument Engineers' Handbook*, Chilton Book Co., 1985.)

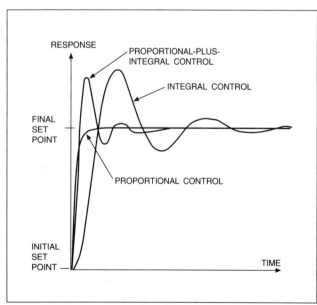

Figure 3-20. Response to a Step Change in Set Point with Proportional, Integral, and Proportional-plus-Integral Controllers
(From B. Liptak, *Instrument Engineers' Handbook*, Chilton Book Co., 1985.)

Derivative action, by itself, is incapable of controlling a process because it does not recognize a constant deviation from the set point ($de/dt = 0$) and since a sudden change in controlled variable value would send an "infinite" signal to the controller and cause the control valve to go fully open or fully closed.

Derivative control is selected for systems that have long lags or high capacitances, where it can give a large amount of correction to a rapidly changing error signal while the error is still small. It is almost as if the controller were looking into the future to correct a situation that could worsen with time because of its large capacitance.

Essentially, derivative action is a high gain mode. Where high gain with excessive cycling is possible but obviously undesirable, a type of derivative action known as "inverse derivative action" reduces the gain inversely to the deviation from the set point.

Comparison of All Combinations of Control Modes

The individual control modes are not used separately, except for occasional uses of the proportional mode, since they each have individual properties that are more useful in combination.

PROPORTIONAL-ONLY (P) CONTROL

This type of control, described earlier, is the most basic control response mode of automatic controllers. It has the fastest response time to changes in set point and responds immediately to upset conditions. However, it is adequate only for systems that have small capacitance and where load changes are small and where close control or offset is not critical.

PROPORTIONAL-PLUS-INTEGRAL (PI) CONTROL

Integral action is rarely used alone because of its slow response to upsets. It is normally combined with proportional action with which its transient response is much faster. The controller output is governed by the equation:

$$\text{Output} = K_1 e + K_2 \int e\,dt + p_o \qquad (3\text{-}21)$$

The combined effect of the proportional control mode and integral action ensures that when an upset occurs, the proportional mode reacts immediately to change the controller output, since there is not yet a time integral of the deviation. Then, as the deviation decreases, the time integral of the deviation will not necessarily have reached zero, and the controller output will remain changed from its previous steady-state value, and the proportional action offset is eliminated. The term "automatic reset" is applied when an integral action controller is used in combination with a proportional-only controller. Figure 3-21 shows the effect of each action on the PI controller as well as their combined action; Figure 3-19 illustrates the controlled variable response to a step change under PI control.

PROPORTIONAL-PLUS-DERIVATIVE (PD) CONTROL

Derivative or rate action increases the response speed of a closed loop. On difficult-to-control processes such as multicapacity systems, the addition of derivative action to the proportional mode is often preferable to adding reset. It enhances both the speed and stability of control responses, particularly on slow responding systems. Its action is the reverse of integral; it leads rather than lags the proportional action. This type of mode combination is not desirable, however, for systems that are subject to noise problems (such as turbulent flow systems), for its derivative action amplifies the noisy error signals and produces instability. In addition, there is no resetting action as with integral control, nor is the offset of

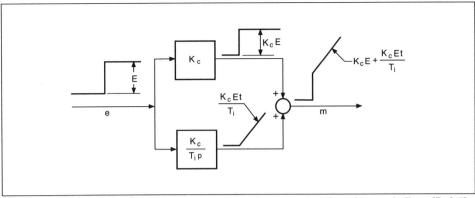

Figure 3-21. Proportional-plus-Integral Controller's Response to a Step Change in Error [Ref. 5]

the proportional controller eliminated. Figure 3-22 illustrates the effect of each of the two control actions and their combined effect.

PROPORTIONAL-PLUS-INTEGRAL-PLUS-DERIVATIVE CONTROL (PID)

Full three-mode control is achieved by combining all the modes simultaneously. Mathematically, the controller output is governed by the equation:

$$p = K_1 e + K_2 \int e\, dt + K_3 \frac{d}{dt} e + p_o \qquad (3\text{-}22)$$

where K_1, K_2, K_3, and p_o are system-dependent constants and e is the error.

The three-mode controller is the most complex to apply in industry because it has the greatest number of parameters to specify (or tune) correctly. It is recommended for large process time constants, large load changes, and fast load changes (except for flow control).

Derivative response will cause overcorrection of the final control element for process lags. To accommodate process load changes without offset, integral action is added.

The three-mode controller offers an ideal control known as the "quarter-amplitude decay" and its main characteristic is that the amplitude of the second peak is one-quarter of the amplitude of the first peak.

Figure 3-23 summarizes the typical response of an output variable controlled by the three control modes, independently and in the combinations described above, to different types of disturbances in an input variable. Table 3-7 summarizes the particular features of each combination.

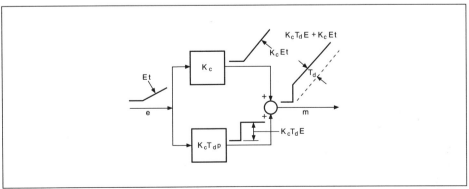

Figure 3-22. The Output of a Proportional-plus-Derivative Controller for a Ramp Input [Ref. 5]
(T_d is the time derivative.)

Table 3-7. Characteristics of Conventional Controller Modes

Two-position controller
    Inexpensive
    Extremely simple

Proportional (P)
    Simple
    Inherently stable when properly tuned
    Easy to tune
    Experiences offset at steady state

Proportional-plus-integral (PI)
    No offset
    Better dynamic response that integral alone
    Possible instability due to response lag

Proportional-plus-derivative (PD)
    Stable
    Less offset than proportional-only
    Reduces lag and responds more rapidly

Proportional-plus-integral-plus-derivative (PID)
    Most complex
    Most expensive
    Rapid response
    No offset
    Difficult to tune
    Best control when properly tuned

Other Control Techniques

CASCADE CONTROL

In a conventional control loop a process output variable (such as a reactor temperature) is measured, and an external input variable (such as the flow of a heating fluid) is controlled to keep the reactor temperature steady. When the controlled variable varies in inherent properties, it may be difficult to control accurately. For instance, if the fluid were steam and the inlet pressure of the steam varied uncontrollably, a control loop using the temperature measurement to vary the steam flow rate would provide poor temperature control because the heat content of the incoming steam would change with its pressure and temperature, hence, changing its mass.

To ensure a more reliable temperature control, a better scheme would be to use the difference between the temperature measurement and its set point as a signal to a second or "cascade" controller to control the flow rate of the steam. Then, if the steam pressure varied, the flow controller would compensate automatically for this fluctuation and provide more accurate control. Figure 3-24 illustrates this control mode.

RATIO CONTROL

In certain processes, it is often necessary to maintain the relative proportions of two variables constant. Typical examples would be the ratio of air to fuel in a combustion chamber or the ratio of two reactants entering a reactor. When the flow of one of the two reactants changes (the "independent" or "free" variable), either voluntarily to increase production or randomly due to external constraints, the flow of the second variable should change correspondingly. If both flows were controlled independently, the constant ratio of both flows could be maintained only if the operator varied the flow control set point manually and frequently to maintain the same proportion.

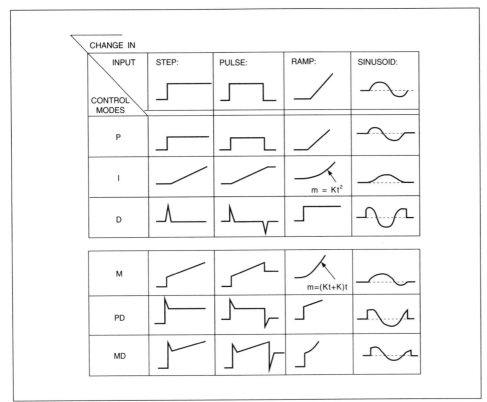

Figure 3-23. Response of Combination of Controller Modes [Ref. 5]
(Shapes will vary with actual values of K_c, T_i, and T_d.)

In practice, it is easier to measure both flows and send the flow measurement signal of the independent flow to the second flow controller, where it will be multiplied by the proportionality ratio and used as a set point. Then, the variations in both flows will be synchronized and the proportionality will be always be maintained. Figure 3-25 illustrates the ratio control loop configuration.

OVERRIDE CONTROL

It is sometimes necessary to limit the value a variable can take to maintain a safe operation or to protect process equipment. If this variable is a function of the system's primary controlled variable, the two variables can be interlocked in an override system. The primary variable maintains control as long as the secondary variable does not exceed a safe limit, at which point the second variable assumes control. This mode of control is also called "protective control."

SPLIT-RANGE CONTROL

Split-range control is a system that operates two or three control valves from a single controller. This mode applies to a process in which different actions have to be initiated if the controlled variable exceeds or drops below the desired set point.

To achieve the split-range mode of control, the signal from the controller goes to both valves simultaneously. One of the two valves is configured to start from its fully open position when the error is large and positive and closes gradually to its fully shut position when the error reaches zero or the controlled variable is at the set point. The valve then remains shut when the error becomes negative. The second valve is closed for any positive value of the error signal, then opens gradually to its fully open position when the error signal drops below zero and reaches some large negative value. In some cases it is desirable to provide a small deadband between the closing of one valve and the opening of the other valve. Other applications may require an overlap in the action of both valves or may

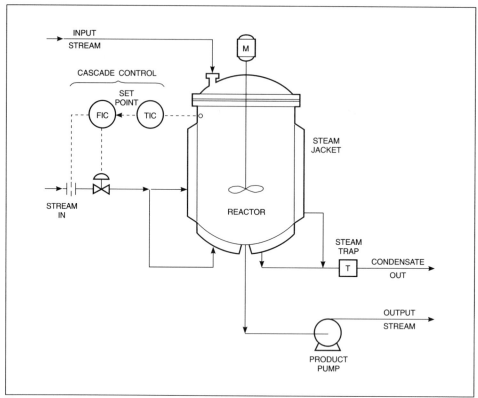

Figure 3-24. Cascade Control on Reactor Steam Heating

even require the control action to be split among three separate control valves instead of two, each operating over a defined range of the error signal. Many possibilities are achieved by setting the controller signal required to fully open or fully close each valve independently at some point other than the mid-value point.

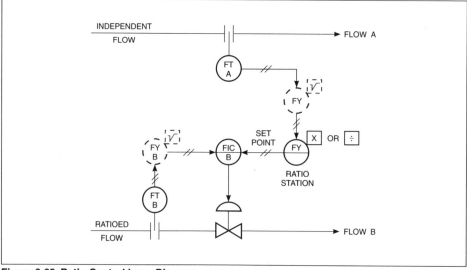

Figure 3-25. Ratio Control Loop Diagram
(From B. Liptak, *Instrument Engineers' Handbook*, Chilton Book Co., 1985.)

Consider, for example, the nitrogen-padded pressurized reactor shown in Figure 3-26. The pressure in the reactor is controlled by a split-range controller. The set point is situated in the middle of the pressure range expected in the reactor. At this set point, both valves are closed. If the pressure drops below the set point, the valve regulating the incoming flow of nitrogen from a high-pressure storage tank (valve A) opens to allow nitrogen to flow in and raise the pressure while the depressurization valve (valve B) remains closed. Should the pressure rise above the set point, perhaps because of an increase in reactor temperature or raw material inflow, the depressurization valve B opens to the atmosphere, and its opening is modulated to bleed off excess nitrogen until the pressure returns back to its set point, while the nitrogen feed valve A remains closed.

TIME-CYCLE AND PROGRAM CONTROL

Time-cycle control applies to semi-batch processes and to processes that are cyclic in nature, such as air drying systems, molecular sieves, moisture removal systems, water treatment plants, and so on. Time control involves one or more circuits, electrical or pneumatic, which activate on-off valves and other control devices to perform repetitive operations in a process.

In a process such as a water demineralization system, a steady-state operation in the demineralizers is maintained by a conventional control system that controls the water flow to each unit, until the resin in the demineralizer is saturated and needs to be regenerated. This point in the sequence is sensed either through a conductivity meter on the demineralized water or by a preset timer. The demineralization process is interrupted by stopping the water flow to the unit, and the regeneration process is initiated. This latter process requires the flow of the regenerant to enter and exit the unit through different valves and pipes and to be controlled by a different control sequence. Upon completion of the regeneration step, the demineralization procedure is repeated. To achieve this change in control modes and in time- or property-dependent cyclic operation, special control sequences are programmed into programmable logic controllers or are hardwired in local control panels near the units.

END-POINT CONTROL

End-point or predictive control techniques are sometimes used in batch processes for target or end-point control. These control techniques are helpful when the controlled variable is difficult to measure directly. Such applications use process models along with process measurements to predict the value of certain variables. When the predicted variable value reaches the set point, the process operation is stopped. For example, in batch pulp digesters, the pulp kappa number, which is a measure of the degree of lignin separation from the wood fibers, is predicted by calculating the area under the digester temperature-versus-time curve. When the prediction equals the desired final value, the digester is emptied.

Predictive control is sometimes combined with adaptive control so that the differences between the actual variable measurements after processing and the predicted values are used to update the process model on which predictions are made. Predictive control can substantially improve the efficiency of processes.

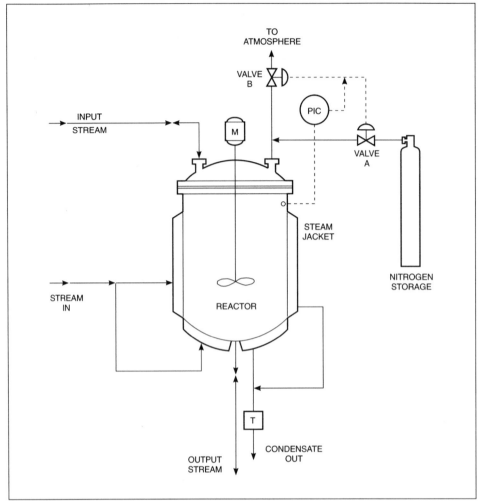

Figure 3-26. Split-Range Control on Pressurized Reactor

References and Bibliography

1. ANSI/ISA-S5.4-1981, *Instrument Loop Diagrams*, Research Triangle Park, N.C.: Instrument Society of America, 1981.

2. Himmelblau, D. M., *Basic Principles and Calculations in Chemical Engineering*, New York: Prentice-Hall, Inc., 1967.

3. Lavigne, J. R., *An Introduction to Paper Industry Instrumentation*, New York: Miller Freeman Publications, 1977.

4. Liptak, B., and Venczel, K., eds., *Instrument Engineers' Handbook*, New York: Chilton Book Company, 1985.

5. Murrill, P. W., *Fundamentals of Process Control Theory*, 2nd ed., Research Triangle Park, N.C.: ISA, 1991.

6. Murrill, P. W., *Application Concepts of Process Control*, Research Triangle Park, N.C.: ISA, 1988.

7. Pollard, A., *Process Control*, New York: American Elsevier Publishing Company, Inc, 1971.

8. Sandler, H. J., and Luckiewicz, E. T., *Practical Process Engineering*, New York: McGraw-Hill Book Company, 1987.

9. Gunkler, A. A., and Bernard, J. W., *Computer Control Strategies for the Fluid Process Industries*, Research Triangle Park, NC: ISA, 1990.

About the Author

Norman Peters is a 1971 Chemical Engineering graduate of McGill University, Montreal, and has received his Ph.D. in process simulation from the University of Aston, Birmingham, U.K. He has worked in software development for process simulation as well as in the design and operation of processes for the production of organic and inorganic chemicals and of pulp and paper products.

4

Final Control Elements

The final control element is that portion of the loop which directly changes the value of the manipulated process variable. Any final control element, such as a valve, positioner, damper, or metering pump, is merely another block in a process control loop. The main function of any block in a loop is the transfer of energy with respect to time. The energy transfer block representing a control valve has a pronounced characteristic of gain and phase lag and is as important as any other energy transfer block, with the exception, perhaps, of the controller. Final control elements include valves, dampers, louvers, governors, pumps, feeders, and variable resistors.

Control Valves and Actuators—An Introduction

A control valve is the most widely used final control element in a process plant. It is defined as a final controlling element through which a fluid passes that adjusts the size of the flow passage as directed by a signal from a controller to modify the rate of flow of the fluid.

Two distinct parts make up a control valve:

(1) The valve body, which is installed in a pipe, its end connections being screwed, flanged, or welded.

(2) The valve actuator, which supplies the necessary force required to drive the valve.

There are endless applications, and the number of actuator-valve combinations is enormous. For this reason, no attempt is made here to cover all the possible applications and combinations of control valves. There are many sources of information on control valves, including ISA and the valve manufacturers.

Control Valve Functions

The functional properties are those that have to do with the type of control that will result from the multiplication of the energy transfer functions that make up a loop. Since pneumatic control valves account for the majority of final control elements in the process industry, this chapter focuses on the fundamentals of pneumatic control valves—these fundamentals applying, for the most part, to electric and hydraulic valves. In any case, the energy transfer function may be considered to be independent of whether a control block is electronically or pneumatically actuated. Functional valve properties include dead band, speed of response, flow characteristic, and rangeability.

DEAD BAND

This is the range through which an input can be varied without initiating an observable response. For example, in a diaphragm-actuated control valve, dead band is the amount the diaphragm pressure can be changed without causing valve stem movement. It is usually expressed as a percentage of diaphragm pressure span.

SPEED OF RESPONSE

This is a function of the volume of air required to produce any desired pressure on the diaphragm of a valve actuator. To a lesser extent, it depends on the friction between the valve stem and packing.

FLOW CHARACTERISTIC

This is the relationship between flow through the valve and the position of the valve stem as it varies from 0 to 100 percent of travel. This is an important term and should always be designated as either inherent flow characteristic or installed flow characteristic. Inherent flow characteristic is the flow characteristic obtained with a constant pressure drop across the valve. Installed flow characteristic is the flow characteristic obtained when pressure drop across the valve varies as dictated by flow and related conditions in the system in which the valve is installed. Most control valves used in the process industry have one of three flow characteristics: linear, equal percentage or quick-opening (see Figure 4-1).

Linear describes an inherent flow characteristic that can be represented ideally by a straight line on a rectangular plot of flow versus percent of rated valve stem travel. A linear valve gives approximately equal increments of flow per increment of valve stem travel at a constant pressure drop.

Equal Percentage describes an inherent flow characteristic that for equal increments of rated travel will ideally give equal percentage changes of the existing flow.

Quick-opening describes an inherent flow characteristic in which there is maximum flow with minimum travel. A quick-opening valve has an approximately straight-line characteristic near its seat (from zero to about 60 percent flow at 30 percent travel). Beyond this point, flow increases too rapidly with valve opening for the quick-opening characteristic to be useful in a flow proportioning application.

 These characteristics are theoretical and assume that the pressure drop across the valve is constant (or that the total loop pressure is across the valve), which is not the normal case. Usually the pressure drop decreases with increasing flow, making the installed characteristic quite different from the theoretical. Equal percentage

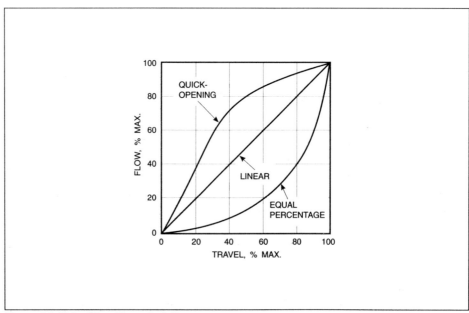

Figure 4-1. Valve Flow Characteristics
(Courtesy of Fisher Controls International, Inc.)

valves tend to become linear, linear valves lean toward quick-opening, and quick-opening valves become useless except for on-off service.

RANGEABILITY

This is the ratio of maximum controllable flow to minimum controllable flow. It may also be defined as the ratio of maximum to minimum usable sizing coefficient. Sometimes it is referred to as inherent rangeability. Rangeability of globe valves varies generally from 30:1 to 50:1, V-notched ball valves in excess of 100:1, butterfly valves between 10:1 and 20:1, and Saunders (patented) and pinch valves as low as 5:1. The operating or installed rangeability may be defined as the relationship between rangeability and pressure drop. It is mathematically expressed by:

$$R = \frac{q_1}{q_2} \frac{\sqrt{\Delta P_2}}{\sqrt{\Delta P_1}} \qquad (4\text{-}1)$$

where:

R = rangeability, dimensionless
q_1 = maximum flow, gpm
q_2 = minimum flow, gpm
ΔP_1 = pressure drop, psi, at maximum flow
ΔP_2 = pressure drop, psi, at minimum flow

Control Valve Selection

Valve selection is based on many considerations in addition to controllability. Materials of construction, tight shutoff, cost, and special features such as split-body or sanitary construction also play a role in valve selection.

Process conditions, fluid type, and the effect of piping and control system design are other criteria that also need to be considered in valve selection. Refer to Figure 4-2 for an example of a valve specification form filled in with typical information.

When considering valve selection from a control point of view, apply the following rules of thumb:

- For simple processes, valve characteristics are relatively unimportant. The less expensive quick-opening valve should be considered.

- If the required flow range is 3:1 or less, there is little difference between linear and equal percentage valves. If the flow range is 8:1 or more, linear valves are preferred.

- If the pressure drop across the valve varies more than about 2 or 3 to 1 with the valve opening, equal percentage is probably the better choice, even if linear is the desired theoretical characteristic.

- If the valve is oversized, the equal percentage characteristic will allow somewhat better control because an oversized linear valve will require a lower controller gain at the operating point.

- For flow control where the primary element is an orifice plate or other differential pressure device, use a linear valve if the pressure drop across the valve decreases with the valve opening.

A double-seated valve will have greater capacity than the same size single-seated valve, but tight shutoff may be sacrificed.

©ISA S20

Specification Forms for Process Measurement and Control
Instruments, Primary Elements and Control Valves

			CONTROL VALVES				SHEET ____ OF ____		
							SPEC. NO.		REV.
			NO	BY	DATE	REVISION			
							CONTRACT		DATE
							REQ.	P.O.	
							BY	CHK'D	APPR.

GENERAL	1.	Tag No.	710-TV-10305					
	2.	Service	GLYCOL SUPPLY UNIT TEMPERATURE					
	3.	Line No./Vessel No.						
	4.	Line Size/Sched. No.						
BODY	5.	Type of Body	GLOBE					
	6.	Body Size | Port Size	2"					
	7.	Guiding | No. of Ports	—					
	8.	End Conn. & Rating	150 # R.F.					
	9.	Body Material						
	10.	Packing Material						
	11.	Lubricator | Isolating Valve	— | —					
	12.	Bonnet Type						
	13.	Trim Form						
	14.	Trim Material Seat/Plug	STAINLESS STEEL					
		Shaft Mtl.						
	15.	Required Seat Tightness	CLASS IV					
	16.	Max. Allow., Sound Level dBA	85 dBA					
ACTUATOR	17.	Model No. & Size	MASONEILAN 38					
	18.	Type of Actuator	DIAPHRAGM					
	19.	Close at | Open at						
	20.	Flow Action to	OPEN					
	21.	Fail Position	CLOSE					
	22.	Handwheel & Location	—					
POSIT.	23.	MFR. & Model No.	MASONEILAN					
	24.	Filt. Reg. | Gages | Bypass	✓ | ✓ | —					
	25.	Input Signal	3-15 PSIG					
	26.	Output Signal						
	27.	Air Supply Pressure	20 PSIG					
TRANSDUCER	28.	Make & Model No.						
	29.	Input Signal						
	30.	Output Signal						
OPTIONS	31.							
	32.							
	33.							
SERVICE	34.	FLOW UNITS kg/h | LIQUID | STEAM | GAS	755					
	35.	Fluid	SATURATED STEAM					
	36.	Quant. Max. | Cv	46					
	37.	Quant. Oper. | Cv	31					
	38.	Valve Cv | Valve FL	46 | 0.9					
	39.	Norm. Inlet Press kPa(g) ΔP kPa	780 | 105 | 35←MIN					
	40.	Max. Inlet Press.	↑MAX					
	41.	Max. Shut Off ΔP kPa	875					
	42.	Temp. Max. | Operating						
	43.	Oper. sp. gr. | Mol. Wt.						
	44.	Oper. Visc. | % Flash						
	45.	% Superheat | % Solids						
	46.	Vapor Press. | Crit. Press.						
	47.	Predicted Sound Level dBA	< 80 dBA					
	48.	Manufacturer	MASONEILAN					
	49.	Model No.	38-21115					

Notes:

ISA FORM S20.50

Figure 4-2. Completed Valve Specification Form for Typical Application

Valve Body Designs

A valve body is a housing for internal parts that have inlet and outlet flow connections. It includes a bonnet assembly, a bottom flange (if used), and trim elements. The trim includes the valve plug, which opens, closes, or partially obstructs one or more ports. Many styles of control valve bodies have been developed through the years. Some have found wide application, while others have been designed for meeting specific service conditions and their usage is less frequent. The following summary describes some of the more popular control valve body designs in use today.

GLOBE VALVES

Globe valves are the most common type in use today. They are divided into several groups such as single-port, double-port and three-way. Split-body and angle valves are special types of globe valves. A globe valve is a valve design with a linear-motion, flow-controlling member with one or more ports normally distinguished by a globular-shaped cavity around the port region.

SINGLE-PORT VALVES

Single-port valves are simple in construction and the most commonly used body style. They are frequently selected for sizes two inches and smaller; however, they may also be used in four-to eight-inch sizes with high-thrust actuators. Figure 4-3 shows a typical single-port globe valve.

The simplicity of this valve can be deceiving. Because the flow usually enters beneath the seat and tends to open the plug from the seat, a well designed actuator must be selected that will close the valve against any line pressure. These unbalanced forces acting on the plug make the globe valve less suitable than others for some applications.

DOUBLE-PORT VALVES

Double-port valves have balanced forces acting on the plugs (one force upward and one downward). Generally they have higher flow capacities and require less stem force to operate than do single-port valves of the same size. Most of the problems encountered with the double-port valves result from misapplication or use in a dirty flow medium. They are frequently specified for sizes larger than two inches. Many double-ported bodies are reversible, so the valve plug can be installed as either "push down to open" or "push down to close." Figure 4-4 illustrates a double-ported valve.

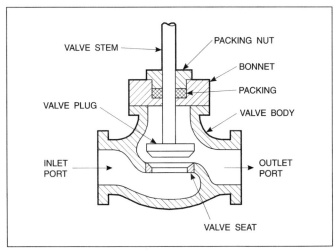

Figure 4-3. Typical Single-Port Globe Valve

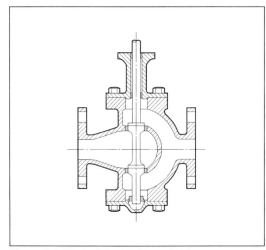

Figure 4-4. Typical Double-Ported Valve

THREE-WAY VALVES

A three-way valve has three openings and is designed to blend (mix) or to divert (split) flow. It can replace two straight-through valves in many applications. In blending service, there are two inlet ports and one outlet, whereas in diverting service there are two outlets and one inlet. Total flow is proportioned between the two inlets, or the two outlets, so that flow is constant through the single common port in either service. Most three-way valves require powerful actuators because of unbalanced forces that act on the plug. A schematic diagram of their operation is shown in Figure 4-5.

SPLIT-BODY VALVES

The split-body design is essentially a streamlined version of the single-port valve, except that the body is split in two halves and bolted together. Its construction minimizes erosion effects, allows parts to be replaced easily, and is relatively inexpensive. Figure 4-6 depicts a split-body valve.

ANGLE VALVES

Angle valves are often used in pressure and level control systems. Angle valves are usually single-ported and are often selected where space is at a premium. They are also suitable for applications that require high pressure drops or where the effects of turbulence, cavitation, or impingement present problems. Several designs of angle valves are available, all with good control characteristics, high rangeabilities, and high pressure and temperature ratings. They can be removed from the line with ease and can handle sludges and erosive materials. Figure 4-7 shows an angle valve.

NEEDLE VALVES

A valve design that belongs to the same category of globe valves is the needle valve. Its application is for high pressure requiring small flows and high rangeabilities. The design of this valve is useful for pilot plant facilities, for control of liquid catalyst or additive flows to various processes, for pressure letdown services to analytical instruments, and in cryogenic gas plants as a Joule-Thomson valve (to create a Joule-Kelvin effect).

BALL VALVES

A true ball valve is seldom used as a control valve because of its poor throttling capability. Other variations of ball-type designs are selected instead. An example is the partial ball (V-notched) body design. This construction is similar to a

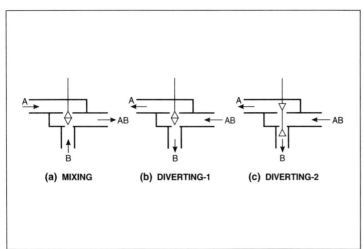

Figure 4-5. Three-Way Mixing and Diverting Valve Operations

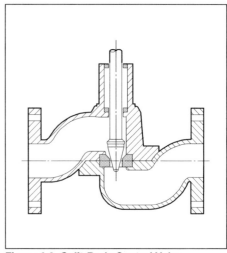

Figure 4-6. Split-Body Control Valve

conventional ball valve, but with a patented, contoured V-notch in the ball. The V-notch produces an equal-percentage flow characteristic. These control valves have good rangeability, control, and shutoff capability. They are widely used in the paper industry (thermo-mechanical pulp process), chemical plants, sewage treatment plants, the power industry, and petroleum refineries. See Figure 4-8 for details.

ECCENTRIC DISC VALVES

This valve belongs to the category of ball valves. It is also called an eccentric rotary stem valve by some manufacturers. A spherically faced plug segment, eccentrically mounted, rotates 50 degrees to the in-line seat ring. It may directly replace conventional globe valve designs. It can also be considered a cheaper alternative to the globe valve where process applications warrant it. This type of control valve can handle fluid with temperatures to 540°C (1000°F). A splined-shaft, actuator-lever connection prevents lost motion and improper positioning of the plug. Because of its high pressure recovery factor, the eccentric disc valve can be used as a possible solution for cavitation. It should be noted that this type of valve requires that the process fluids be relatively clean because of the plug construction. It would, therefore, be unsuitable for applications such as stock services in the paper industry. Figure 4-9 shows a section of this valve design.

BUTTERFLY VALVES

The butterfly valve may be of the wafer type, in which the disc rotates to the open position in the adjacent pipe, or it may be of the bulkier, double-flanged cylinder type. A butterfly valve consists of a shaft-supported vane or disc that is capable of rotating within a cylindrical body. Butterfly valves have large capacity, which is why they are sometimes substituted for globe valves in large diameter pipes. They may be selected for slurry and entrained-liquid applications. Generally, they have little tendency to cause significant pressure drop in a line. This has a negative aspect in that it can lead to poor control.

As good as they seem, butterfly valves are not the answer to all final control element problems. Rangeability is poor. A pneumatic actuator usually does not have the power to hold the vane exactly in the desired position. As the valve is opened, forces that are caused by the flow will decrease, whereas force required by the actuator to rotate the shaft will increase since the force is not always at right angles to the shaft arm.

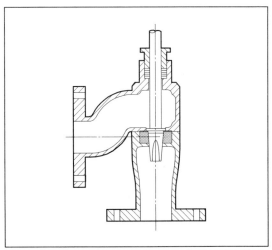

Figure 4-7. Angle Valve with Split-Body Construction

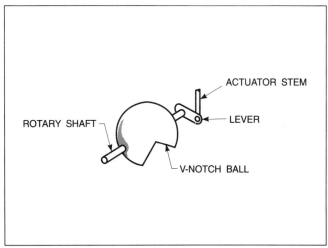

Figure 4-8. Ball Valve with V-Notch
(Courtesy of Fisher Controls International, Inc.)

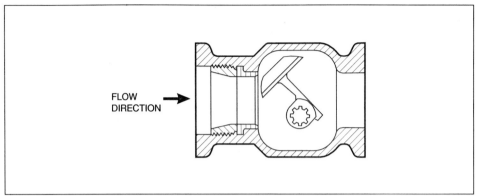

Figure 4-9. Eccentric Disc Valve
(Courtesy of Masoneilan/Dresser)

The basic design has been improved upon by a valve manufacturer who has added a "fish tail" to the trailing edge of the disc. This retards the flow at low opening angles and controls the flow at high angles where the normal disc edge has been shadowed by the hub. Figure 4-10 shows a schematic of a typical wafer-type butterfly valve.

SAUNDERS DIAPHRAGM VALVE

This valve consists of a valve body, bonnet, and flexible diaphragm. Figure 4-11 shows the Saunders patent valve, in which a heavy fabric-reinforced diaphragm serves as both a seal and a closure member when forced down against a weir in the body. This type of valve is excellent for slurry and viscous fluid applications. Although it has a high capacity, it controls poorly and has a low rangeability because its inherent characteristic is essentially quick-opening. It requires a positioner for any intermediate position between the extremes. Every Saunders valve requires a somewhat different force to open than to close its diaphragm as both upstream and downstream pressures try to open it.

PINCH VALVES

Another useful design is the mechanical pinch valve, which is simply a flexible tube compressed mechanically, pneumatically, or hydraulically. It will handle very coarse and extremely viscous fluids such as those found in the mining industry. Pinch valves have high capacities but poor control characteristics and low rangeabilities.

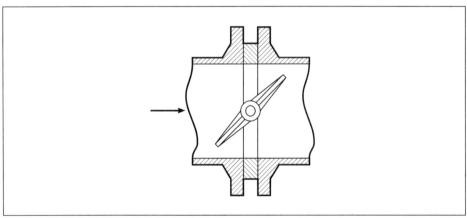

Figure 4-10. Typical Wafer-Type Butterfly Valve

GATE VALVES

Knife gate valves are used for both on-off and throttling applications. They consist of a cast body and a gate that is a beveled, knifelike component that slides up and down and pushes aside or cuts through solids in the flow stream. This feature makes them suitable for many applications in the pulp and paper industry. The seat may be either metal or resilient, the metal seat being more suitable if tight shutoff is required. A variation of the gate, a V-port design, is especially advantageous for reliable throttling control of thick slurries such as paper stock. The V-port orifice is maintained from open to closed to prevent bridging or plugging and to assure maximum control accuracy. Figure 4-12 shows a typical knife gate valve.

Pressure Differential

The design pressure drop across a valve is often referred to as a percentage of the pressure drop through the process exclusive of the valve. Many designers adopt as a rule that 50% of this friction drop should be available as drop across the valve. Thus, one third of the total system drop, including all heat exchangers, mixing nozzles, piping, and other high resistance blocks, is assumed to take place in the control valve.

In most systems, the design should be changed if the pressure drop across the valves is less than one third of the total system pressure drop. The amount of pressure differential needed for good control is a function of the pressure differential across the valve with respect to the drop across the rest of the system. If variations in flow are small, this ratio may be reduced.

As the proportion of the system drop across the valve is reduced, it loses its ability to rapidly increase or decrease flow. If the flow resistance is too small a fraction of the total system resistance, large movements of the valve will cause very small changes in the flow. The resulting control will be sloppy and slow responding. If the control valve is too large a portion of the total resistance in a control system, small movements in the valve will cause large, sudden changes in flow.

It is sound practice to keep control valve pressure drop as low as possible consistent with the fact that a valve energy transfer block regulates by absorbing and giving up pressure drop to the system.

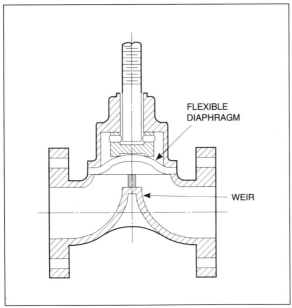

Figure 4-11. Saunders Patent Valve

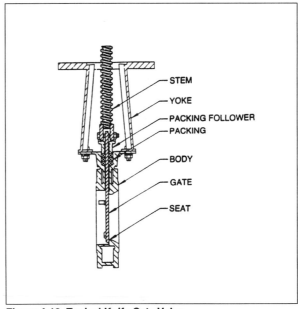

Figure 4-12. Typical Knife Gate Valve

Friction Losses — Calculating P_1 and P_2

In flow systems, the pipe is usually one of the major sources of flow resistance. This flow resistance, which is due to viscous friction, varies as a function of pipe size, pipe material, and flow rate. Actual flow resistance values have been determined experimentally and recorded in pump or piping handbooks. Resistance is given as a pressure loss per unit length of pipe as a function of the flow rate.

The inlet pressure (P_1) to a control valve is determined by finding the pressure at the source and subtracting all the static and friction losses up to the valve. The outlet pressure (P_2) is calculated by finding the required pressure at the outlet of the system and adding to it all the friction and static losses between the valve outlet and the system outlet. Pipe fittings, measurement elements, and manual valves are also sources of flow resistance. In many handbooks, these resistances are expressed as being equivalent to a certain length of pipe.

In many flow systems, there are other sources of flow resistance. Heat exchangers, filters, and other process equipment all restrict flow and must be accounted for when computing pressure drops in a system. The static pressure losses or gains must also be considered in flow systems. This static pressure is a result of changes in potential energy due to differences in elevation. The static pressure can be calculated using Equation (4-2).

> If the source is at a higher elevation than the receiver, the static pressure differential is a gain rather than a loss.

$$\Delta P_{static} = \frac{\gamma h}{144} \tag{4-2}$$

where:

$$\Delta P_{static} = \text{static pressure differential, psi}$$
$$\gamma = \text{density, lb/ft}^3$$
$$h = \text{difference in elevation, ft}$$
$$144 = 144 \text{ in.}^2/\text{ft}^2$$

The following example shows how P_1 and P_2 are determined.

Saturated water (i.e., pressure = vapor pressure) at 35 psia is pumped to a boiler (see Figure 4-13). The pressure at the inlet of the boiler is 500 psia. The maximum flow is 600 gpm. The specific gravity of the water is 0.939 (density is therefore $62.4 \times 0.939 = 58.6$ lb/ft^3). The system uses six-inch piping for which the friction loss is 1.04 psi per 100 feet of pipe. The pressure loss through the heater is seven psi at full flow. Preliminary data from the process engineers says that the pump discharge pressure is 532 psig (546.7 psia).

Calculating P_1:

Up to the valve, the static loss is 10 feet of water; therefore, substituting in Equation (4-2),

$$\Delta P_{static} = (58.6 \times 10)/144 = 4.07 \text{ psi}$$

The pressure loss due to piping and fittings is calculated by summing the equivalent pipe lengths for the fittings and adding the result to the upstream pipe length of 79 feet.

(continued)

(continued)

Item	Qty.	Equivalent Pipe Length	Total
Check valve	1	63	63
90° elbow	1	8.9	8.9
Tee (thru-flow)	1	3.8	3.8
Gate valve	2	3.2	6.4
Total fittings			82.1 feet
Piping			79.0 feet
Total			161.1 feet

Therefore, total friction loss due to piping and fittings is

$$(161.1 \times 1.04) / 100 = 1.67 \text{ psi}$$

and,

$$P_1 = 546.7 - 1.67 - 4.07 = 540.96 \text{ psia}$$

Calculating P_2:
The static loss after the valve is 70 feet; therefore, substituting again in Equation (4-2),

$$\Delta P_{static} = (58.6 \times 70) / 144 = 28.49 \text{ psi}$$

The pressure loss due to the heater is given as

$$\text{delta } P_{heater} = 7.0 \text{ psi}$$

The pressure loss due to piping and fittings is calculated by summing the equivalent pipe lengths for the fittings and adding the result to the downstream pipe length of 99 feet.

Item	Qty.	Equivalent Pipe Length	Total
90° elbow	2	8.9	17.8
Tee (thru-flow)	1	3.8	3.8
Gate valve	2	3.2	6.4
Total fittings			28.0 feet
Piping			99.0 feet
Total			127.0 feet

Therefore, total friction loss due to piping and fittings is

$$127.0 \times 1.04 / 100 = 1.32 \text{psi}$$

and,

$$P_2 = 500.0 + 1.32 + 28.49 + 7.0 = 536.8 \text{ psia}$$

It's usually necessary to calculate friction losses for several flow rates in order to predict noise and cavitation conditions.

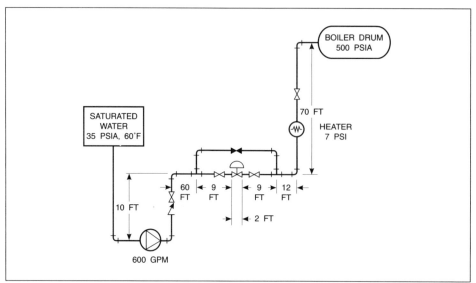

Figure 4-13. Calculating P₁ and P₂

Control Valve Sizing

C_V Factor

Before selecting a control valve type, one must calculate the correct flow coefficient, or C_v factor. Other data such as pressure drop across the valve, fluid type (liquid, gas, steam, or multiphasic), flow capacities, specific gravity, viscosity, and potential flashing or cavitation conditions may be required in whole or in part for the calculations. In short, one needs to have a total knowledge of the process to determine the proper valve size and type. The correct valve size selection is made primarily from the C_v factor, an indispensable piece of information found in every valve manufacturer's catalog.

Valve capacity is the rate of flow through a valve under stated conditions. Its measurement is the nondimensional flow coefficient C_v expressed by:

Without reducers:

$$C_v = \frac{Q}{N_1} \frac{\sqrt{G_f}}{\sqrt{\Delta P}} \tag{4-3a}$$

With reducers:

$$C_v = \frac{Q}{N_1 F_p} \frac{\sqrt{G_f}}{\sqrt{\Delta P}} \tag{4-3b}$$

where:

Q = volumetric flow rate through the valve, gpm
ΔP = pressure differential across the valve, psi
G_f = specific gravity of the fluid
N = numerical constants based on units used
F_p = piping geometry factor

In other words, C_v is the number of gpm of 60°F water that will flow through a valve with one psi pressure drop. Rated C_v (also referred to as the manufacturer's C_v) is the value of C_v at the rated full-open position of the valve.

 Viscosity becomes of great importance in valve flow C_v calculations when handling highly viscous liquids. For liquids below 100 ssu (Saybolt seconds universal), no corrections are required. The correction factor R of the viscosity effect can be found in nomographs that are supplied by most valve manufacturers.

Critical Flow

Every valve or flow orifice exhibits some degree of pressure recovery downstream (at the vena contracta) of the principle restriction. Whenever the static pressure at the vena contracta is lowered to the vapor pressure of a liquid medium or the critical pressure of a gas causes sonic velocity, no further flow will be obtained as the valve outlet pressure decreases. In other words, a further increase in pressure drop (with the inlet pressure remaining constant) produces no further increase in flow. This can otherwise be defined as choked flow, which is a condition of critical flow.

Cavitation

For liquid service, the vapor pressure may be reached at the vena contracta, causing localized flashing (some liquid remains in the vapor phase and the bubbles formed in this region will collapse or implode downstream as the pressure rapidly increases). This causes cavitation that results in physical damage to valve trim, body, or downstream. The collapsing bubbles can result in localized pressures of up to 100,000 psi! Figure 4-14 best illustrates these phenomena for incompressible and compressible fluids.

Cavitation will be at its worst at minimum flow with maximum supply pressure and minimum outlet pressures.

The critical flow factor, F_L, is a dimensionless expression of the ratio of the C_v obtained under critical conditions such as liquid vaporization or gas sonic velocity at the vena contracta, and the C_v measured under normal pressure recovery conditions. For maximum accuracy, F_L factors should be obtained from the valve manufacturer.

Cavitation exists when the actual pressure drop is greater than the allowable pressure drop. This condition can be verified by using the equation:

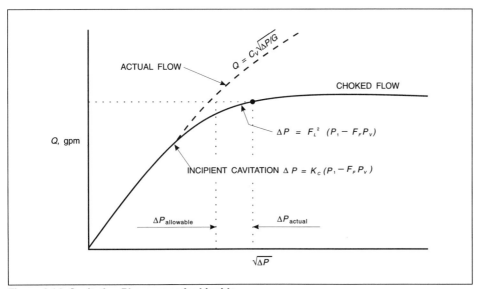

Figure 4-14. Cavitation Phenomena for Liquids
(Courtesy of Fisher Controls International, Inc.)

$$\Delta P_{\text{allowable}} = F_L^2 (P_1 - F_F P_v) = \Delta P_{\text{actual}} \tag{4-4}$$

where:

$\Delta P_{\text{allowable}}$ = allowable differential pressure

F_L = valve recovery factor (from manufacturer's literature)

P_v = vapor pressure of liquid at body inlet temperature (from manufacturer's table)

$F_F = 0.96 - 0.28 \dfrac{\sqrt{P_v}}{\sqrt{P_c}}$ = liquid critical pressure ratio factor (determined from manufacturer's tables) or by using the formula where P_c = thermodynamic critical pressure of a fluid

Choked flow is reached when the allowable pressure drop equals the actual pressure drop, which, in turn, is equal to $F_L^2(P_1 - F_F P_v)$. Further increase in pressure drop will not produce increased flow through the valve.

 Nomenclature for the F_L factor varies from one manufacturer to another. For example:

$F_L^2 = K_m$ = *Fisher Control Valve Recovery Coefficient*

$F_L = C_f$ = *Masoneilan Control Valve Recovery Coefficient*

The ISA standard F_L factor is used by most other valve manufacturers.

Flashing exists when P_2 (valve outlet pressure) is less than or equal to P_v (fluid vapor pressure).

EFFECT OF PIPE REDUCERS

Control valves are often smaller than the pipe size for many applications, so that the size of the pipe entering and leaving the valve must be reduced. The use of pipe reducers diminishes valve capacity due to the additional pressure drop they create.

For noncritical liquid sizing when inlet and outlet reducers are used, the piping correction factor F_p is applied in Equation (4-5):

$$F_p = \sqrt{1 - 1.5 \left(1 - \frac{d^2}{D^2}\right)^2 \left(\frac{C_v}{30d^2}\right)^2} \tag{4-5}$$

where:

d = valve diameter

D = line diameter

and when only outlet reducers are used, F_p is applied in Equation (4-6):

$$F_p = \sqrt{1 - \left(1 - \frac{d^2}{D^2}\right)^2 \left(\frac{C_v}{30d^2}\right)^2} \tag{4-6}$$

Sidebar:

Two ways to eliminate or compensate for cavitation are: (1) relocating the valve to a lower elevation to increase the actual pressure differential, and (2) considering the use of a different type of valve (one whose valve recovery coefficient F_L is higher). A ball valve will have a higher F_L factor than a butterfly valve, for example.

For critical liquid, gas, or steam sizing, the piping correction factor F_p is replaced by F_{Lp}, the valve reducer correction factor, and is expressed in the Equation (4-7) as:

$$F_{Lp} = \left[\frac{1}{F_L^2} + \left(\frac{C_v}{30d^2}\right)^2 \left(1 - \frac{d^4}{D^4}\right)\right]^{-\frac{1}{2}} \qquad (4-7)$$

where:

F_L = valve critical flow factor = $\sqrt{P_1 - P_2} / \sqrt{P_1 - P_{vc}}$

F_{Lp} = combined pressure recovery and piping geometry factor

C_v = valve flow coefficient

d = valve size, inches

D = line size, inches

P_{vc} = pressure at the vena contracta

The F_{Lp} factor can best be defined in terms of the critical (allowable) pressure drop, which is the pressure drop in a valve at which cavitation occurs. The relationship is described by Equation (4-8) as follows:

$$\Delta P_{critical} = \left(\frac{F_{Lp}}{F_p}\right)^2 (P_1 - P_v) \qquad (4-8)$$

where:

$\Delta P_{critical}$ = pressure drop in a valve at which cavitation exists

F_{Lp} = valve reducer correction factor

F_p = piping correction factor

P_1 = valve inlet pressure

P_v = vapor pressure

EFFECT OF VISCOSITY

When flow velocity is very low or when viscosity is high (for liquid flow), laminar rather than turbulent flow may result (Reynolds number 100). Laminar flow should be assumed for very low flow rates or very viscous fluids, in which case, a correction must be made to the C_v calculation. The case of turbulent flow is described by Equation (4-9), which shows that flow is directly proportional to the pressure drop:

$$Q \; \alpha \; \sqrt{\Delta P} \qquad (4-9)$$

Equation (4-10) describes a quite different situation for laminar flow:

$$C_v = 0.072 \left(\frac{\mu Q}{\Delta P}\right)^{2/3} \qquad (4-10)$$

where:

Q = flow rate, gpm

ΔP = pressure drop, psi

μ = viscosity, centipoise

For some valves, especially those in larger sizes, the effect of reducers can be ignored if the valve is one standard pipe size smaller than the pipe. The effect of pipe reducers depends upon the valve design, so manufacturer's literature should be consulted. Note that many valve manufacturers supply values for F_{Lp}, F_L, and F_p factors. Therefore, an alternative method for calculation is to consult their tables for the corresponding values.

LIQUID SIZING CORRECTION FOR CONSISTENCY

Flow resistance through pipes, fittings, and valves may be increased due to the consistency level of substances such as pulp stock. A pulp stock correction factor F_C is used in liquid sizing equations to compensate for these frictional losses. Although the F_C factor is used primarily in pulp and paper suspensions of wood fiber and water, it may also be used to correct for flow calculations on other fibrous slurries such as sewage sludge (see Table 4-1).

Table 4-1. Pulp Stock Correction Factor F_C for Different % Consistencies

Consistency, %	Chemical Stock	Mechanical Groundwood Stock
1	1.0	1.0
2	0.97	0.99
3	0.90	0.95
4	0.84	0.92
5	0.80	0.90

To determine the effective corrected C_v of an existing valve, multiply the standard published C_v by the F_C factor. [5]

$$C_v = \frac{Q_F \dfrac{\sqrt{G_f}}{\sqrt{\Delta P}}}{F_C} \qquad (4\text{-}11)$$

where:

C_v = valve flow coefficient

Q_F = volume rate of flow, gpm

G_f = specific gravity

(relative to water, std. temp.)

ΔP = pressure drop, psi

F_C = pulp stock correction factor

Liquid Sizing

The sizing equations shown here are applicable for most valves described in this chapter. The basic liquid sizing equation is as shown in Equation (4-3). The first step is to calculate the required C_v. The pressure drop used in the equation must be the actual valve pressure drop or allowable differential pressure drop, whichever is smaller. When solving for critical flow (cavitation or flashing), Equation (4-3) is modified to become:

Without reducers:

$$C_v = \frac{Q}{N_1} \frac{\sqrt{G_f}}{\sqrt{\Delta P_{allowable}}} \qquad (4\text{-}12a)$$

$$= \frac{Q}{N_1} \frac{\sqrt{G_f}}{\sqrt{F_L{}^2 (P_1 - F_F P_v)}}$$

$$= \frac{Q}{N_1 F_L} \frac{\sqrt{G_f}}{\sqrt{P_1 - F_F P_v}}$$

where $\Delta P_{\text{allowable}}$ is as defined in Equation (4-4).

With reducers:

$$C_v = \frac{Q}{N_1 F_{LP}} \frac{\sqrt{G_f}}{\sqrt{P_1 - F_F P_v}} \qquad (4\text{-}12b)$$

The second step is to consult the manufacturer's catalog and select a valve with a C_v in the range of the calculated C_v.

 Accurate valve sizing for liquids requires the use of the dual coefficients of C_v and F_L. A single coefficient is insufficient to describe both the capacity and the recovery characteristics of the valve.

. The use of the additional cavitation index factor K_C (coefficient of incipient cavitation), is appropriate in sizing high recovery valves that may develop damaging cavitation at pressure drops well below the level of choked flow.

The pressure differential for incipient cavitation is calculated using the following equation:

$$\Delta P = K_C (P_1 - F_F P_v) \qquad (4\text{-}13)$$

Gas Sizing

A sizing procedure for gases can be established, based on adaptations of the basic liquid sizing equation, by introducing conversion factors to change flow units from liters/minute to cubic meters/hour and to relate specific gravity in meaningful terms of pressure. Equation (4-14) applies to gas service:

   An example can illustrate typical valve sizing and selection for liquid services.

   Given Information:

Application:	**Condensate recirculation**
Fluid:	**Water**
Piping	**8 inch Schedule 40 carbon steel**

   Service Conditions:

Valve inlet pressure:	$P_1 = 470$ **psig**
Valve pressure drop:	$\Delta P = 200$ **psi**
Fluid specific gravity:	$G_f = 1$
Inlet temperature:	$T = 83°$**F and** $P_v = 0.56$ **psia**
Valve $F_L{}^2$	$F_L{}^2 = 0.3$
Flow rate:	$Q = 5000$ **gpm**

   (continued)

(continued)

From Equation (4-3), calculate the C_v to determine the valve size:

$$C_v = \frac{Q}{N_1} \frac{\sqrt{G_f}}{\sqrt{\Delta P}}$$

$$= \frac{5000}{1} \frac{\sqrt{1}}{\sqrt{200}}$$

$$= 354$$

Preliminarily, a 6-inch ball valve is chosen with a manufacturer's $C_v =$ 1040.

At 70% opening, the manufacturer's $C_v = 450$.

The next step is to check for cavitation using Equation (4-4). Taking F_F 0.94 and $P_v \approx 0.56$ psia at 83°F gives

$$\Delta P_{allowable} = F_L{}^2 (P_1 - F_F P_v)$$

$$= 0.3[485 - (0.94)(0.56)]$$

$$= 145 \text{ psi}$$

Because $\Delta P_{actual} = 200$ psi $> \Delta P_{allowable}$, there is cavitation.

Having found that a ball valve cavitates at the given service conditions, consider now a globe valve, which has a higher pressure recovery factor. Since the process fluid is clean, using a globe valve is not a problem. For a 6-inch globe valve, $F_L{}^2 = 0.5$. Substituting this value in Equation (4-)4 gives a $\Delta P_{allowable} = 242$ psi $> \Delta P_{actual}$. From this, it can be deduced that cavitation no longer is evident.

$$C_v = \frac{Q}{N_7 P_1 Y} \frac{\sqrt{G_g T_1 Z}}{\sqrt{X}} \qquad (4\text{-}14a)$$

or

$$C_v = \frac{Q}{N_9 P_1 Y} \frac{\sqrt{M T_1 Z}}{\sqrt{X}} \qquad (4\text{-}14b)$$

where:

Q = gas flow rate at 14.7 psia and 60°F

T_1 = absolute inlet temperature

Z = gas compressibility factor

P_1 = upstream pressure (psia)

N = numerical constant based on units

X = pressure drop ratio, $\dfrac{\Delta P}{P_1}$

M = gas molecular weight

G_g = gas specific gravity; ratio of gas density at standard conditions

Y = gas expansion factor, where $Y = 1 - \dfrac{X}{3F_K X_T}$ and

F_K = gas specific heat ratio factor (air = 1.0) = $\dfrac{K}{1.4}$ and

X_T = pressure drop ratio factor (refer to manufacturer's table)

Figure 4-15 illustrates a typical curve of flow of gas versus the square root of the pressure differential, assuming gas flow at a constant P_1.

At critical flow, $P_1 - P_2 > P_2$, sonic velocity is reached. Pressure impulses cannot move upstream against the sonic barrier, and, therefore, further changes in downstream pressure cannot affect the upstream flow. When this occurs, mass flow rate becomes completely independent of outlet pressure, and only a change in inlet pressure will affect flow rates. When P_2 is less than or equal to 0.5 P_1, any additional decrease in P_2 will provide no increase in flow.

Steam Sizing

The equation used in sizing valves for steam service is derived from the gas sizing Equation (4-14). The formulae for saturated and superheated steam vary slightly. Equation (4-15) is normally used for saturated steam:

Without reducers:

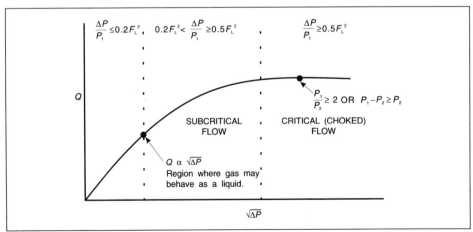

Figure 4-15. Critical Flow for Gases
(Courtesy of Masoneilan/Dresser)

$$C_v = \frac{w}{N_6 Y \sqrt{X P_1 \gamma_1}} \qquad \text{(4-15a)}$$

where:

w = flow, weight basis
N = numerical constant based on units
γ_1 = specific weight, upstream conditions
X = pressure drop ratio, $\Delta P / P_1$
P_1 = upstream pressure, psia

Y = gas expansion factor = $1 - \dfrac{X}{3 F_K X_T}$ and

F_K = gas specific heat ratio factor = $K / 1.4$ (air = 1) and
X_T = pressure drop ratio factor
(refer to manufacturer's tables for F_K and X_T).

With reducers:

$$C_v = \frac{w}{N_6 F_P Y \sqrt{X P_1 \gamma_1}} \qquad \text{(4-15b)}$$

For superheated steam (below 1000 psig),

Without reducers:

$$C_v = \frac{w(1 + 0.0007 T_{SH})}{N_6 Y \sqrt{X P_1 \gamma_1}} \qquad \text{(4-16a)}$$

where:

T_{SH} = steam superheat temperature, °F or °C

With reducers:

An example will illustrate typical valve sizing and selection for steam service.

Given information:

Application:	**Temperature control—steam to HVAC unit**
Fluid:	**Saturated steam**
Piping:	**3-inch**
Service conditions:	
Valve inlet pressure:	P_1 = 780 kPa(g) = 880 $kPa(a)$
Valve pressure drop:	ΔP = 105 kPa
Flow Rate:	w = 955 kg/h

(continued)

(continued)

At 780 kPa(g), the temperature of steam is 346°F (174°C), and the specific volume = 3.507 ft³/lb, which is equivalent to a density of 0.0285 lb/ft³ or 4.6 kg/m³ (0.0285 × 16.0184).

Let us assume a preliminary selection of a globe valve for steam service, since the steam is saturated and without any particles.

Using Equation (4-15), calculate the C_v to determine the valve size:

$$C_v = \frac{w}{N_6 Y \sqrt{X P_1 \gamma_1}}$$

From Table 4-1,

$$N_6 = 2.73$$

$$X = \frac{\Delta P}{P_1}$$

$$= \frac{105}{880}$$

$$= 0.1$$

Using values of F_K and X_T from manufacturer's literature,

$$Y = 1 - \frac{X}{3 F_K X_T}$$

$$= 1 - \frac{0.1}{3(0.94)(0.68)}$$

$$= 0.84$$

Therefore,

$$C_v = \frac{955}{(2.73)(0.84)\sqrt{(0.1)(880)(4.6)}}$$

$$= 21$$

A 2-inch globe valve with manufacturer's C_v of 46 is selected. The body should be carbon steel because of steam service. The valve actuator should be fail-closed, because of steam service; therefore, the actuator's action is indirect, air-to-open, fail-closed.

$$C_v = \frac{w(1 + 0.0007 T_{SH})}{N_6 Y F_P \sqrt{X P_1 \gamma_1}} \qquad (4\text{-}16b)$$

Valve Noise Calculations and Reductions

CONTROL VALVE NOISE

Fluid velocity in a process line can be a major source of noise, causing both vibration and mechanical damage to valve and piping. Noise problems can be attributed to: (1) high pressure at the valve inlet, (2) elevated liquid flow, and (3) high pressure differential across the valve. Specific sources of control valve noise are: (1) mechanical vibrations, (2) hydrodynamic noise caused by flashing or cavitation, and (3) aerodynamic noise.

The unit of measurement for noise levels is the decibel (dB), which is defined as the measure of noise intensity or its force. The decibel is calculated on a logarithmic scale as the ratio between two numerical quantities. This means that a measurement of 100 dB is 10 times more intense than 90 dB and 100 times more intense than 80 dB

 Noise levels greater than 85 dBA can damage one's hearing if exposure to it is for a prolonged period.

For a sound level range of 90 – 100 dBA, corrective measures in the process piping give the best results for the 10 – 15 dBA noise reductions required to bring the noise level down to an acceptable level.

Noise problems can be treated at the source or in the process piping (path). Treatment of noise at the source requires selection of special valves or pressure differential devices installed directly at the valve outlet. These accessories reduce the total pressure differential, resulting in a total pressure that is reduced in steps. This method is applicable for valves where the noise level exceeds 100 dBA. Treatment of noise in the process piping can be divided into the four categories of distance, transmission loss, dispersion/dissipation, and velocity.

Distance. Locate the valve as far away as possible from plant personnel. The sound level may be checked by verifying the sound pressure level (*SPL*) and is determined using Equation (4-17):

$$SPL_{distance} = SPL_{source} - 10 \log \left(\frac{distance\ (ft)}{3\ ft} \right) \qquad (4\text{-}17)$$

Transmission loss. Consider increasing the thickness of the piping or adding acoustical insulation. Acoustical insulation may reduce noise by up to five dBA for an inch of insulation, but this value decreases proportionally as the insulation thickness increases. In other words, four inches of insulation will reduce the noise only by approximately seven dBA. For acoustical insulation to be effective, it must be used throughout the downstream system.

Dispersion/Dissipation. This is done by installing a silencer in the line directly downstream of the valve. Note that a silencer may not be used as a pressure-reducing device.

Velocity. Keep the velocity at a minimum value in order to reduce the noise level. Line diffusers at the valve outlet can be used. The pipe should be sized in such a way as not to exceed a velocity of 300 ft/s.

Types of Noise and Recommended Solutions

MECHANICAL VIBRATIONS

Mechanical vibrations are caused by turbulence, pressure oscillations, or certain process variables due to fluid velocities and/or large flows. These vibrations may be eliminated by changing the valve plug type or using a cage-guided valve for greater stability in the line.

HYDRODYNAMIC NOISE

Hydrodynamic noise is caused by cavitation or flashing and may be reduced by careful selection of acoustical insulation, a special valve configuration, or special pressure-reducing devices. One could consider using valves with multistage

The decibel is generally used to express a sound power level relative to a chosen reference. A sound level in decibels A-scale (dBA) is a sound pressure level that has been adjusted according to the frequency response of the A-weighting filter network. With reference to valve noise, sound level figures are generally referred to standard conditions such as laid out in ISA-S75.07-1987, Laboratory Measurement of Aerodynamic Noise Generated by Control Valves.

trims that are available from a number of manufacturers. Valve bodies are compatible with those of standard valves, but the inner design reduces noise and vibrations. However, certain piping accessories are still required to reduce the noise level.

Equation (4-18), along with the necessary tables and figures, can be used to predict hydrodynamic noise (see Tables 4-2 to 4-4 and Figures 4-16 to 4-20):

$$SL = 10 \log C_v + 20 \log \Delta P - 30 \log (t) + 5 \qquad (4\text{-}18)$$

where:

SL = A–weighted sound level (1 m downstream and 1 m from pipe surface)
C_v = actual required flow coefficient
ΔP = pressure drop, psi (bar)
t = pipe wall thickness, mm (in.)

AERODYNAMIC NOISE

Aerodynamic noise is the direct result of the conversion of the mechanical energy of the flow into acoustic energy as the fluid passes through the valve

Table 4-2. Numerical Constants for ISA Liquid Flow Equations

Constant		Units Used in Equations				
N	w	Q	P, ΔP	d,D	γ_1	
N_1	0.0865	—	m³	kPa	—	—
	0.865	—	m³	bar	—	—
	1.00	—	gpm	psia	—	—
N_4	76000	—	m³/h	—	mm	—
	17300	—	gpm	—	in.	—
N_6	2.73	kg/h	—	kPa	—	kg/m³
	27.3	kg/h	—	bar	—	kg/m³
	63.3	lb/h	—	psia	—	lb/ft³

Table 4-3. Numerical Constants for Gas and Vapor Flow Equations

Constant		Units Used in Equations				
N	w	Q*	P, ΔP	γ_1	T_1	
N_6	2.73	kg/h	—	kPa	kg/m³	—
	27.3	kg/h	—	bar	kg/m³	—
	63.3	lb/h	—	psia	lb/h³	—
N_7	4.17	—	m³/h	kPa	—	K
	417	—	m³/h	bar	—	K
	1360	—	scfh	psia	—	R
N_8	0.948	kg/h	—	kPa	—	K
	94.8	kg/h	—	bar	—	K
	19.3	lb/h	—	psia	—	R
N_9	224	—	m³/h	kPa	—	K
	2240	—	m³/h	bar	—	K
	7320	—	scfh	psia	—	R

* Q is in cubic feet per hour measured at 14.73 psia and 60°F, or cubic meters per hours measured at 101.3 kPa and 15.6°C.

Table 4-4. Attenuation Factor *SL* (Not Equal to *TL*)

Pipe Size, in.	Pipe Schedule										
	5S	10S	20	40	80	100	120	160	STD	XS	XXS
1.0	35.5	29.0	—	26.5	22.5	—	—	18.0	26.5	22.5	13.5
1.5	39.0	32.5	—	28.5	24.5	—	—	20.0	28.5	24.5	15.5
2.0	41.5	35.0	—	30.5	26.0	—	—	20.0	30.5	26.0	17.0
3.0	42.0	37.0	—	29.5	25.0	—	—	20.5	29.5	25.0	16.0
4.0	44.5	39.5	—	31.0	26.0	—	23.0	20.5	31.0	26.0	17.0
6.0	44.5	41.5	—	32.0	26.5	—	23.0	20.0	32.0	26.5	17.5
8.0	47.0	43.0	36.0	33.0	27.0	25.0	22.5	19.5	33.0	27.0	20.0
10.0	46.0	43.5	38.0	33.0	27.0	24.5	22.0	18.5	33.0	29.0	20.0
12.0	46.0	44.0	39.5	33.5	26.5	24.0	21.5	18.0	34.5	30.5	21.5
14.0	47.0	44.5	38.0	33.5	26.5	24.0	22.0	18.5	35.5	32.0	—
16.0	47.5	46.0	39.5	33.0	26.5	23.5	21.5	18.0	37.0	33.0	—
18.0	48.5	47.0	40.5	32.5	26.0	23.0	21.0	17.5	38.0	34.0	—
20.0	48.0	46.0	39.0	33.0	25.5	23.0	20.5	17.0	39.0	35.0	—
24.0	40.5	45.5	40.5	32.5	25.0	22.0	20.0	16.5	40.5	36.5	—
30.0	38.5	45.0	38.5	—	—	—	—	—	42.5	38.5	—
36.0	—	—	40.0	35.0	—	—	—	—	44.0	40.0	—

Note: For other schedules compare wall thickness to nearest schedule shown.

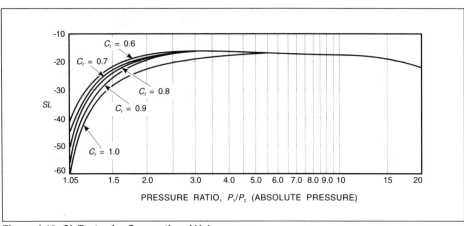

Figure 4-16. *SL* Factor for Conventional Valves
(Courtesy of Masoneilan/Dresser)

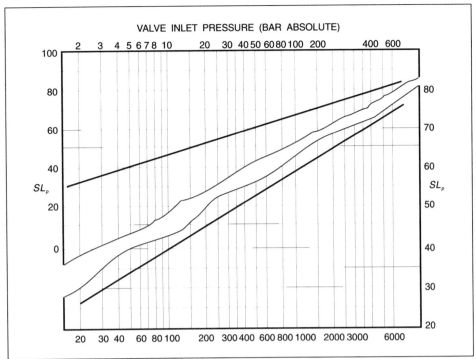

Figure 4-17. SL_P **Factor for Valve Inlet Pressure**
(Courtesy of Masoneilan/Dresser)

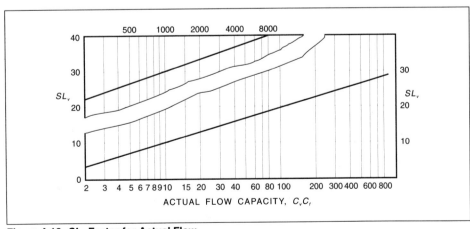

Figure 4-18. SL_T **Factor for Actual Flow**
(Courtesy of Masoneilan/Dresser)

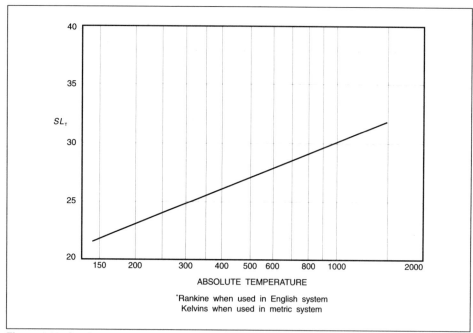

Figure 4-19. SL_T Factor for Absolute Temperature
(Courtesy of Masoneilan/Dresser)

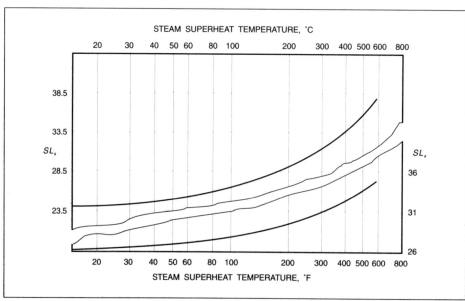

Figure 4-20. SL_S Factor for Steam Superheat
(Courtesy of Masoneilan/Dresser)

restriction. Equations (4-19) and (4-20) can be used to predict aerodynamic noise.

For all gases except for steam,

$$SL = SL_e + SL_P + SL_T + SL_v + SL_g + SL_a \qquad (4\text{-}19)$$

where:
SL = A–weighted sound level 1 m downstream
and 1 m from the pipe surface, dBA
SL_e = accoustical efficiency factor, dBA
SL_P = valve inlet pressure factor, dBA
SL_T = fluid temperature factor, dBA
SL_v = flow capacity factor, dBA
SL_g = gas property factor, dBA
SL_a = pipe attenuation factor, dBA
SL_s = steam temperature factor, dBA

For steam,

$$SL = SL_e + SL_P + SL_v + SL_a + SL_s \qquad (4\text{-}20)$$

Aerodynamic noise may be reduced by trim components that are specially designed for noise abatement and are similar to the multistage trims used for hydrodynamic noise reduction.

 Multistage trims should be considered only for clean processes. Saturated steam with wood fibers, for example, would clog the pores of the trim, thereby eventually reducing valve capacity and efficiency.

The following example will help illustrate aerodynamic noise prediction using the graphical method.

Given information:

Fluid:	**Saturated Steam**
Piping:	**4-inch Schedule 160**

Service Conditions:

Valve Inlet Pressure:	P_1 = 2500 psia
Valve Outlet Pressure:	P_2 = 400 psia
Flow Rate:	w = 955 to 62,000 lb/hr
Temperature:	T = 668°F

F_L = 1; valve has low noise trim

C_v = 14

Using Equation (4-20), calculate the total sound pressure level (SPL):
$$SPL = SL_e + SL_P + SL_v + SL_a + SL_s$$

(continued)

(continued)

From Figure 4-15,

$$SL_e = -18, \; P_1/P_2 = 6.25, \; \text{and} \; F_L = 1$$

From Figure 4-16,

$$SL_P = 68, \; P_1 = 2500 \; \text{psia}$$

From Figure 4-18,

$$SL_v = 11.5, \; F_L \times C_v = 14$$

From Table 4-4,

$$SL_a = 20.5, \; \text{pipe} = 4 \; \text{inches}, \; \text{schedule} \; 160$$

From Figure 4-19,

$$SL_s = 35, \; T = 668°\text{F}$$

Substituting the above values into Equation (4-20) results in

$$SPL = 117 \; \text{dBA}$$

Using a low noise trim recommended by the supplier increases SL_e to -33, thus reducing the noise to 102 dBA. Further reduction in noise can be achieved by proper pipe insulation upstream and downstream of the valve.

Acoustical insulation may also be considered in order to reduce aerodynamic noise.

Trim Design

Trim is defined as the internal parts of a valve that are in flowing contact with the controlled fluid.

The shape of plugs and seats to obtain the desired flow characteristic is a function of trim design. For example, in a cage-guided globe valve, trim would typically include valve plug, seat ring, cage, stem, and stem pin. A seat is that portion of the seat ring which a valve plug contacts for closure. The seat ring is a separate piece inserted in a valve body to form a port. The valve plug is a movable part that provides a variable restriction in a port. Valve flow characteristics are determined primarily by the valve plug shapes or patterns.

Stainless steel, 304 or 316, is frequently used as material for seats, plugs, guides, bushings, and other trim parts. Carbon steel or bronze are usually selected for city water, air, and steam services. Monel™, Hastelloy™, and other alloys are used for corrosive applications.

Actuators

Control valves may be actuated pneumatically, electrically, hydraulically, and/or manually. The pneumatic actuator is the most widely used. The actuator overcomes forces that are unbalanced due to friction, the weight of moving parts, stem unbalance, and pressure drop across the valve.

Pneumatic Actuators

TYPES

Pneumatic actuators may be the spring-and-diaphragm type or the piston (springless) type. The spring-and-diaphragm is the most frequently used (see Figure 4-21).

Diaphragm actuators may be direct or reverse acting. A direct-acting actuator is designed so that increasing air pressure (20-100 kPa, or 3-15 psig) on top of the diaphragm moves the stem downward, thus closing a direct-acting valve. This actuator type is defined as air-to-close or fail-open; loss of air pressure allows the spring to open the valve.

With a reverse-acting diaphragm actuator, air pressure below the diaphragm moves the stem up, opposing the spring action. This action is called air-to-open or fail-closed. Some designs allow the diaphragm to reverse to obtain the desired action. Figure 4-22 shows several actuator actions.

The diaphragm is usually made of rubber or neoprene with a fabric insert, most of the area being firmly clamped in a diaphragm plate. The spring is usually high temper alloy steel with parallel ground and squared ends designed for linear changes in stress throughout its working range. The spring has two functions: diaphragm return and diaphragm position. The diaphragm plate against which the spring butts is rigidly fastened to the stem.

The springless or piston actuator is for use in high-pressure and high-pressure-differential conditions that exceed those in which the spring-and-diaphragm actuator gives accurate service. One side of the diaphragm is "cushioned" with constant air pressure from an air pressure regulator. This, in effect, takes the place of the spring. The type of valve body and the desired action will determine which side of the diaphragm is to be cushioned. Pressure applied to the controller output side of the diaphragm will depend on the energy transfer function of the controller.

A reversing pneumatic relay may be substituted for the constant air pressure regulator. A change in controller output to one side of the diaphragm is transmitted to the relay, which alters the air pressure on the other side of the diaphragm so that the sum for the two pressures is constant.

Pneumatically operated piston actuators provide integral positioner capability and high stem force output for demanding service conditions. Adaptations of

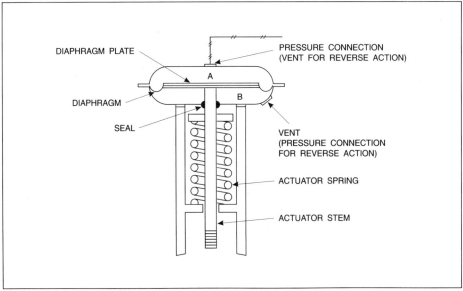

Figure 4-21. Spring and Diaphragm Actuator

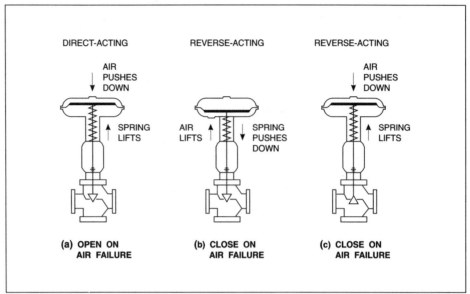

Figure 4-22. Direct- and Reverse-Acting Actuators

both spring-and-diaphragm and piston actuators are available for direct installation on rotary-shaft control valves such as butterfly and V-notch ball valves.

FAIL-SAFE POSITION

 An actuator's fail-safe position should always be based upon process safety considerations in the event of control valve air or electrical failure. The three fail-safe positions are fail-open, fail-closed, and fail-as-is.

Fail-Open. This action can limit overpressure in a process system that may be caused by an air or electrical failure. It can permit discharge of gas (steam) or liquid to atmosphere in case of air or electrical failure.

Fail-Closed. This action can limit sudden pressure drops in case of air or electrical failure. It permits the process to be shut down in the case of a leak or a break in the process line and can prevent toxic gases or liquids from being released to the atmosphere.

Fail-As-Is. This action allows the actuator to stay in the position it was in at the moment of air or electrical failure. This action can be used where an air or electrical failure won't cause undue disturbances in the process. The fail-as-is position for valves over eight inches in diameter installed with piston actuators can also be considered where process conditions are considered less critical, thus reducing the need for expensive accessories such as volume tanks, which ensure that the valve fails in a certain position.

Certain piston actuators that do not use a spring for fail-safe positioning utilize stored air capacity to meet the valve fail-safe requirements. A lockup valve and check valve are placed between the air supply line, a storage bottle, and the actuator or positioner. In case of air supply failure, the check valve will trap air in the storage bottle, which can then be used to close or open the valve as desired by selectively loading the actuator piston. This approach to fail-safe design is not as positive as a spring, but it is sufficient in less critical installations.

ACTUATOR SIZING

The use of too large an actuator adds unnecessary expense and increased response time to a control valve, while the use of an undersized actuator might make it impossible to open the valve or close it completely. However, selection of an optimally sized actuator for a given control valve application is a subject of greater scope than can be completely covered here. In general, the actuator must provide sufficient force to stroke the valve plug to the fully closed position with sufficient seat loading to meet the required leak class criteria. With spring return actuators, the spring selected must be sized to properly oppose the force provided by the air supply pressure. To put it simply, sizing an actuator involves solving a problem in statics. The forces, and the direction in which each force acts, depend upon actuator design and flow direction through the valve. The free body diagram in Figure 4-23 illustrates the forces involved in achieving static equilibrium.

The figure depicts a direct-acting (push-down-to-close) valve body with the flow tending to open the valve plug. The actuator is a reverse-acting, spring-and-diaphragm construction that is air-to-open (fail-closed). The actuator force available is the product of the air supply pressure and the area against which that pressure is applied (diaphragm area). Packing friction varies with the stem size, packing material, and packing design. The unbalance force is the product of the force of the flowing medium and the area against which that force is applied (total port area):

$$\text{Unbalance force} = \text{Shutoff differential pressure} \times \text{Unbalance area} \quad (4\text{-}21)$$

Seat load is usually expressed in pounds per linear inch of port circumference. The seat load is the product of the port circumference and the pounds-per-linear-inch force recommended by the valve manufacturer. The actuator force available must be greater than the sum of the forces that the actuator force must oppose to achieve static equilibrium.

Electric Actuators

Electrically operated control valves are usually selected where no instrument air supply is available, such as remote areas of a plant or storage and loading facilities. The type of actuator must be compatible with the electric controller output, as shown in Table 4-5. Several variations of electric actuators are available.

The switch of the single-relay output section may be mechanically or electrically operated in an on-off controller. It may be double-throw with or without dead

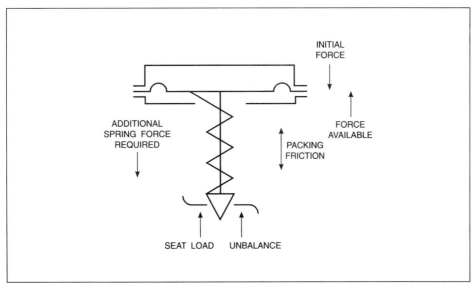

Figure 4-23. Forces Involved in Static Equilibrium

Table 4-5. Electric Controller Output Compatibility

Output Section	Actuator
Single-relay output section	Relay, solenoid, or two-position actuator
Current output section and feedback	Variable-speed motor internal controlled drive, electropneumatic converter, SCR
Dual-relay output section including provisions for direct position feedback from the drive	Reversible electric motor

band between the open and closed circuits. The switch may be operated by a proportional-plus-integral action controller varying the ratio of the open to closed time of a continuously pulsed relay. Because loss of controller power deenergizes a relay, an open relay contact must correspond to a safe operating condition.

The current output section provides a direct current that is proportional to the required output. This current is derived from a standard PID controller. On loss of controller power, the valve must return to a safe operating condition.

In the dual-relay output section, the controller switches are interlocked to prevent simultaneously energizing both motor fields. They may be mechanically or electrically operated in an on-off controller for two-position control. A time-pulse switch is in the common lead to the motor for floating control. The switches may be in a controller with proportional action or PID. On loss of controller power, the motor will remain in its last position.

In some situations, an infinite position actuator is operated in a two-position manner. Typical electric controller output connections are presented in Figure 4-24.

Valve Positioners and Accessories

Positioners

FUNCTION

The function of a valve positioner is to sense both the instrument signal (controller output) and the valve stem position, and from these measurements to ensure that the valve position is always directly proportional (or related in a known way) to its controller output signal. A positioner may be considered as a closed-loop controller that has the instrument signal as input, output to the actuator, and feedback from the valve stem position. The positioner must be mounted somewhere on the control valve to be able to measure the stem position.

A positioner can be used to: (1) provide a split-range valve operation, (2) reverse the signal to a valve, (3) overcome forces within a valve caused by friction or high pressure across the valve, and (4) help bring about fast, accurate control. It overcomes errors caused by the imbalance of forces on the valve plug, and the hysteresis effects of the diaphragm and spring. Applications include temperature control, liquid level control, gas flow control, and mixing and blending. In certain fast systems, such as for liquid pressure control or liquid flow control, a volume or ratio booster is more advantageous than a positioner.

From the standpoint of system dynamics, the positioner adds another functional block to the control loop, and the effect of this change can be to make the system more sensitive. It may also increase the overall response time of the loop. While this may be desirable in some processes, it could lead to instability.

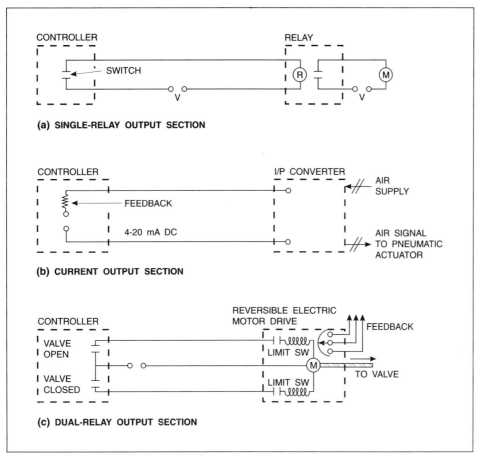

(a) SINGLE-RELAY OUTPUT SECTION

(b) CURRENT OUTPUT SECTION

(c) DUAL-RELAY OUTPUT SECTION

Figure 4-24. Typical Electric Controller Outputs

SPLIT-RANGE OPERATION

Split-range signals (when one common controller signal commands two or more control valves) are typically used in temperature control applications. For example, the temperature in a product storage room might be maintained by a temperature controller that operates both heating and cooling valves. The valves would typically have nonoverlapping operating ranges of 3 – 9 psig and 9 – 15 psig, respectively.

POSITIONER TYPES

The basic categories of positioners are pneumatic, electropneumatic, and electrical. Input signal ranges may be 3 – 9, 3 – 15, 6 – 30, or 9 –15 psig for pneumatic positioners, and 1 – 5, 4 – 20, and 10 – 50 mA for electronic devices. Supply pressures may be up to 100 psig. Most positioners installed today are the pneumatic and electropneumatic types, even though the control system may be electronic or direct digital control. An electropneumatic transducer is used separately or is incorporated in the positioner to convert the electronic signal into a pneumatic signal. Transducers are usually mounted on the actuator, although for reasons of space and economy many transducers are now rack-mounted in field panels.

BASIC POSITIONER DESIGNS

There are two basic designs for positioners: motion-balance and force-balance. In the motion-balance type, the stem motion is compared directly with a similar motion produced by a bellows expanded by the air signal. Force-balance means that the feedback derived from the valve position provides a force to balance the

In many control systems, a properly sized spring-and-diaphragm actuator will do an excellent job without the use of either a positioner or a booster.

controller signal that is acting on an input diaphragm or bellows. See Figure 4-25, which shows schematics of both designs.

In a motion-balance pneumatic positioner, the input signal from a controller or a transducer goes to the bellows. A beam is fixed to the bellows at one end and, through linkage, to the valve stem at the other end. A relay nozzle forms a flapper-nozzle arrangement with the beam. As the bellows moves in response to a changed control signal, the flapper-nozzle arrangement moves, either admitting air to, or bleeding air from, the diaphragm until the valve stem position corresponds to the input air signal, and the positioner is once again in equilibrium.

With a force-balance pneumatic positioner, the controller signal acts on a diaphragm, creating a signal force that is opposed by a feedback spring. A temporary offset in the diaphragm position moves a spool valve, which in turn allows supply air to flow to the diaphragm of the valve actuator. The resultant stem motion is sensed by a lever that rotates a cam. This cam displacement is then converted by a suitable lever arrangement into compression of the feedback spring, which in turn produces an equivalent force to match the signal level.

Use of a cam to characterize the feedback motion has gained increasing importance. One advantage is that certain rotary valves that have an unsuitable inherent characteristic (butterfly valves and ball valves, for example) can be modified so that the characteristic matches the requirements of the system. Note that cam feedback is truly effective only when the system process loop is slower than the positioner-valve combination.

An electropneumatic positioner combines an electropneumatic transducer and a pneumatic positioner. See Figure 4-26.

The operation is similar to the pneumatic positioner shown in Figure 4-24, except that the bellows is replaced with a magnetic motor unit. An electrical rather than a pneumatic input signal is received.

Handwheels

One of the principle accessories on automatic control valves is the handwheel, which is an arrangement to override the pneumatic actuator manually in case of air failure or during certain maintenance operations. Two types are available: top-mounted, and side-mounted. The top-mounted handwheel is a simple device that **permits manual repositioning of the actuator stem.**

Side-mounted handwheels are usually found on large control valves where required height would make it inconvenient for the operator to actuate the valve manually from the topmost part of the valve. This design is more expensive, but one of its advantages is the ease of maintenance of the actuator itself. It is usually possible to service and replace the diaphragm while the valve itself is held in position by the handwheel. The handwheel can also provide the function of a limit stop in either direction of travel.

Limit Switches

The limit switch is attached to the actuator yoke by a suitable bracket, which in turn senses the motion of the valve stem through a takeoff arm. Limit switches can be either single- or multi-throw, and they are used to signal that the valve stem has reached a predetermined position. Such information on valve opening can be used to actuate safety or other interlocks.

Limit Stops

When use of the handwheel as a limit stop is not practical, other limit stops can be provided as part of the body assembly. A typical limit stop may be mounted on the bottom of a globe valve that is guided top and bottom. This stop consists of an adjustable spindle sealed by a cap. This stop can be adjusted while the valve is in operation to meet the exact required minimum or maximum settings.

Stem-Position Indicators

Stem-position indicators show the exact valve position to operating personnel at a remote location. Such valves may be located in unmanned pumping stations or in a hazardous area closed off to operating personnel (near an atomic reactor, for example). Remote position indicators can be electrical, with a linear variable resistor suitably connected to the valve stem. The electrical signal is then shown on a calibrated panel meter. Another way to indicate stem position remotely is with a pneumatic signal. This is a desirable alternative in areas where there is a risk of explosion. Most pneumatic positioners can be inverted to work as position transmitters. When properly modified, they will transmit a pneumatic signal as a function of valve stroke, which is indicated on a calibrated receiver gage.

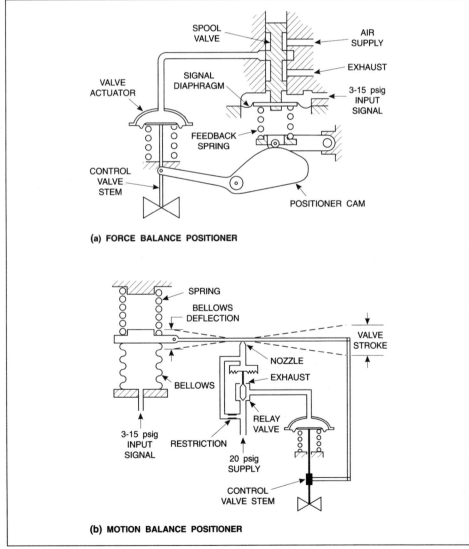

Figure 4-25. Valve Positioners

Airsets

The most common of all accessories is the airset. This is a compact self-contained air pressure regulator with an integral filter and drip valve and a maximum flow capacity around 20 scfm of air. This air filter regulator is used to supply pressurized air to either the positioner or a yoke-mounted controller. Its main advantage is that it provides a way to set the individual pressure supply to a positioner. (Pressures from 20 to 80 psig may be needed to meet the power requirement for a particular valve.)

 Piping the usual plant air supply of perhaps 80 psig directly to the valves could overstress smaller valve stems or damage receiver bellows in positioners or controllers. Also, instrument air must be clean and moisture-free.

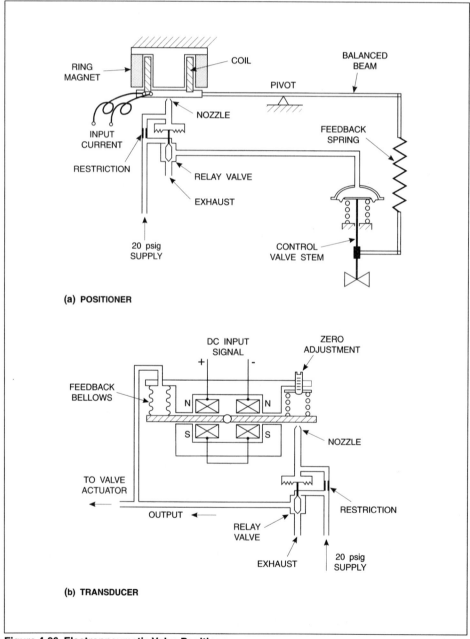

Figure 4-26. Electropneumatic Valve Positioners

Boosters

Booster relays are essentially air-loaded, self-contained pressure regulators. They can be classified as: (1) volume boosters, (2) ratio relays, and (3) reversing relays.

Volume boosters multiply the available volume of the air signal. The air regulator sends its output signal to the volume booster instead of directly to the valve. Volume relays can be used to increase the frequency response of a control valve. This is sometimes preferable to the use of positioners on fast control loops. Refer to Figure 4-27.

Ratio relays multiply or divide the pressure of an input signal. Ratio relays help in split-ranging applications. For example, a 1:2 ratio relay could change a 3 – 9 psig controller signal to a 3 – 15 psig output signal, and another 1:2 ratio relay could change a 9 – 15 psig controller signal to a 3 –15 psig output signal. Thus one controller signal of 3 – 15 psig could operate two valves of 3 – 15 psig without overlapping operation of the valves.

Reversing relays produce a decreasing output signal for an increasing input signal. Reversing relays are employed when two control valves, one air-to-open and the other air-to-close, are operated from the same controller. They might also be used to reverse part of the output pressure from a single-acting positioner to a double-acting piston actuator.

Reversible Electric Motor Drives

This type of actuator or servomotor is suitable for operating valves, dampers, and other lever-operated process regulators. Models are available with torques ranging from 5 to 75 lb-ft and with full-stroke speeds ranging 10 to 60 seconds. Operating voltage is usually 120 V AC, single-phase, 50 – 60 Hz with maximum running current of one ampere. A handwheel is provided for local manual operation. Internal worm gears prevent back drive caused by unbalanced loads. The linkage may be arranged for characterizing the process regulator. Extra limit switches and feedback transmitters have to be added for complex control applications.

Actuators with higher torques are also available. Units with torques ranging from 150 to 4,000 lb-ft are designed to operate large valves and dampers. Their full stroke speeds range from 30 to 300 seconds. Operating voltage is 220 V AC, 3-phase, 50 – 60 Hz with a power consumption at full running load ranging from 0.5 to 5.0 kVA.

Stepping Motors

Stepping motors are also used as electrical actuators for small size valves. A stepping motor is an electromechanical device that rotates a discrete step angle when energized electrically. The step angle usually is fixed for a particular motor and thus provides a means for accurately positioning in a repeatable, uniform manner. Typical step angles vary from 0.72 degrees to 90 degrees. Several means for electrically energizing stepping motors include DC pulses, square waves, fixed-logic sequences, and multiple-phase square waves.

Basic design types for stepping motors are solenoid-operated ratchet, permanent magnet, and variable reluctance. Variations of these basic types may be combined with gears or hydraulic amplifiers to provide increased-output-torque stepping motors.

 Although most stepping motors can be driven from switches or relays, most drive circuits incorporate solid-state devices that permit high-powered, fast operation.

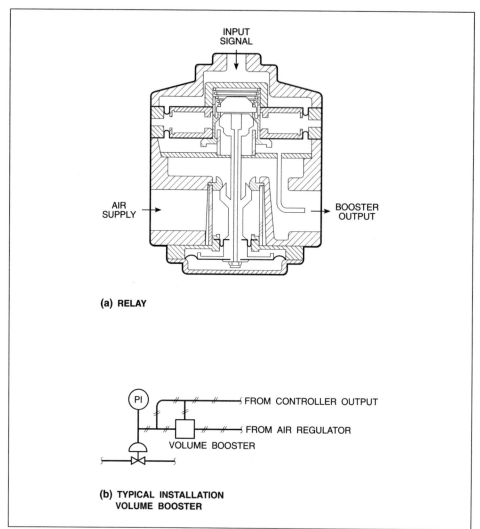

(a) RELAY

(b) TYPICAL INSTALLATION VOLUME BOOSTER

Figure 4-27. Volume Booster Design and Installation

Simple drives convert low-level pulses into power pulses or correctly phased power. More complex drives adjust the power levels to allow running at very high rates.

At very low rates the step movement resembles the classic damped oscillation curve. Mechanical or electrical damping can be added to modify the curve to provide critical damping. Since stepping motors move in discrete steps, they do not have the problems related to stability and feedback inherent in most servo devices. Direct digital control in an open-loop mode is possible. Closed-loop mode may be provided by coupling a pulse feedback to the motor.

Solenoid Valves

Function

A solenoid valve consists of an electromagnetic coil and a valve. The electromagnetic coil actuates an armature or a valve stem in a magnetic field to control fluid flow. Solenoid valves provide an on-off switching option in the system and are actuated by electric signals from remote locations. They accomplish in a pneumatic or liquid system what an electrical relay accomplishes in an electrical system. They are frequently used in conjunction with control valves to open or close the valve at predetermined conditions or limits.

When electrical power is supplied to the electromagnet, a magnetic field is created that causes the plunger to be positioned in the solenoid coil. The plunger is connected to a valve disc that opens or closes the orifice depending on the valve action, that is, whether the valve is energized-to-open or energized-to-close. See Figure 4-28.

Types

The four basic types of solenoid valves are: (1) two-way, (2) three-way, (3) four-way, and (4) pilot-operated.

Two-way solenoid valves have two ports and provide a simple on-off switching action.

Three-way solenoid valves have three pipe connections. A typical application is for two of the ports to be used to load or unload cylinders or diaphragm actuators.

Four-way solenoid valves are used principally for controlling double-acting cylinders.

Pilot-operated solenoid valves apply pressure to a diaphragm or piston, or they may release pressure, allowing higher upstream pressure to open the valve. A widely employed device is a small solenoid pilot valve to supply pressure to a diaphragm or piston for a wide range of output forces. Solenoid valves are used either directly to supply signal air to the actuator or indirectly by blocking the air supply to the positioner.

Solenoid valves are the least expensive form of electrical actuation. They come in a variety of styles and flow capacities and with explosion-proof housings. It is usually specified that the valve supplier mount them directly to the diaphragm case by means of a suitable nipple, although gang-mounting of large quantities of solenoid valves in local panels is also often done.

Electric Motor Drive Control

Motor controls are usually selected and specified by the electrical engineer. However, the instrumentation or control engineer and designer should have some knowledge of this technology, since some of the process controls are interfaced with the motor control center (for example, the use of electric motors for variable

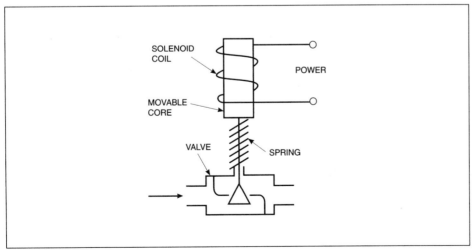

Figure 4-28. Schematic of a Solenoid Valve

speed pumps, metering pumps, screw conveyers, belt feeders, and so forth). Electric motor drives can be classified as constant-speed and variable-speed.

Constant-Speed Drives

Constant-speed applications have employed induction motors, wound-rotor motors, and synchronous motors mainly because alternating current is the more readily available source of electrical power. These AC motors can be started by full-voltage starters or by reduced-inrush starters. The latter method reduces the initial drain on the power system. The wound-rotor motor with a secondary control is also suitable for starting and accelerating a high-inertia load when low inrush currents are a requirement. Figure 4-29 shows diagrams of a full-voltage and a reduced-inrush starter.

These starters may be interlocked with process control devices such as programmable controllers for sequencing control (a start/stop loop, for example), limit switches, level, pressure, temperature, flow switches, and safety switches. The interlock contacts are usually in series with the push button contacts and are activated by relays.

Variable-Speed Drives

Variable-speed drives can use both AC and DC motors. Three major categories for DC drive systems are: (1) constant-potential with motor-field control, (2) rotating motor-generator set, and (3) static converter.

Most variable-speed DC motor drives are regulated by a silicon controlled rectifier (SCR), except for the rotating motor-generator type. A DC motor can be wired to large, constant-potential DC sources and regulated over a limited speed range by motor-field control. Very wide speed ranges can be achieved with rotating motor-generator sets and static (SCR) conversion technologies capable of providing adjustable armature voltage. Figure 4-30 shows a schematic diagram of a variable-speed DC drive control using an SCR.

The firing control is the open and close switching activity of the SCR when the AC voltage passes through zero. One limiting factor in DC motor applications is that energy is supplied to the rotor of the motor through the commutator. Large amounts of power must pass from the DC line to the rotor through stationary carbon brushes rubbing against the rotating commutator bars. This energy transfer reduces the practical top speed and upper voltage rating of large-horsepower DC drives. Such a method of energy transfer also precludes the application of DC motor drives in erosive and corrosive environments and in areas where explosive gas is present. The ideal drive system for such conditions would be a variable-speed AC drive.

Many combinations of variable-speed AC drives are available, especially with the latest solid-state technology, such as power transistors, gate turn-off thyristors, and microprocessors. The following classification of variable-speed AC drives shows how many possibilities exist:

(1) Adjustable voltage with wound-rotor motor.

(2) Adjustable voltage with high-slip induction motor.

(3) Adjustable frequency by:

- Rotating DC/AC motor-generator set with induction motor.

- Static converter/inverter with induction motor.

- Static cyclo-converter with induction motor.

(4) Synchronous motor adjustable-frequency drive.

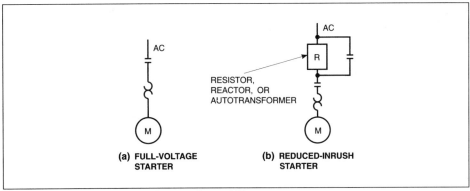

Figure 4-29. Constant-Speed AC Motor Control Diagram

(5)   Induction motor with eddy current coupling.

(6)   Induction motor with mechanical transmission.

Figure 4-31 shows a schematic diagram of a static SCR converter/inverter, power-conversion type, adjustable-frequency AC drive system.

In this system, constant-frequency AC power first is converted to constant-voltage DC power by means of a static rectifier. This DC power, in turn, is inverted into an adjustable-frequency power supply by means of static logic circuitry (microprocessor).

Metering Pumps

Metering pumps, also known as controlled-volume pumps or proportioning pumps, are utilized in process control as a final control element. Combining the functions of a pump, a measuring instrument, and a control valve, they control the rate at which a volume of fluid is injected into a process. They have inherently high steady-state accuracy and can be adjusted while in operation.

All three liquid-end designs rely on positive or swept-volume displacement to meter a wide range of substances. In each case, a metered pulse of fluid results from the combined motion of plunger or diaphragm and the one-way check valves located at the inlet and discharge ends of the pump. In all three designs, the check valves operate 180 degrees out of phase to permit filling of the displacement chamber during the suction stroke and prevent backflow during the discharge stroke.

The powered drive unit may be either an AC or DC motor. The speed of the metering pump can be automatically varied, through an SCR unit, in proportion to changes in process line flow so as to maintain a fixed ratio of chemical addi-

> The key to constant, precise liquid delivery is the liquid-end of a metering pump. There are three basic types of liquid-ends: (1) packed plunger, (2) disc diaphragm, and (3) tubular diaphragm.

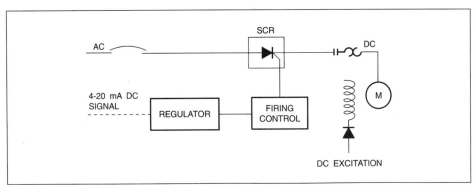

Figure 4-30. SCR Power-Conversion Variable-Speed DC Drive System

tive to the process line flow. A controller may also transmit a signal to automatically reset the stroke length of the pump by way of a stroke adjustment actuator. Figure 4-32 shows an application of a metering pump.

In this example, both flow and pH values are measured, but only the pH value is controlled. The motor speed depends upon the flow rate, and the stroke adjustment is a function of the pH value. This is an example of the additional applications where a variable-speed motor drive may be used as final control element.

Regulators, Relief Valves, and Other Control Elements

Regulators

It is a self-contained device that performs all control functions necessary to maintain a constant, reduced downstream pressure, flow, level, or temperature. All the energy required to operate it is derived from the controlled system.

In general, regulators are simpler than the alternative of a control valve with its external power sources and transmitting and controlling instruments.

One of the simplest of automatic process controllers is the regulator valve.

REGULATOR TYPES

All regulators, whether they are being used for pressure, level, or flow control, are either direct-operated or pilot-operated. Direct-operated regulators are adequate for narrow range control and where the allowable change in outlet pressure can be 10 to 20 percent of the outlet pressure setting. Pilot-operated regulators are preferred for broad-range control or where the allowable change in outlet pressure is required to be less than 10 percent of the outlet pressure setting. They

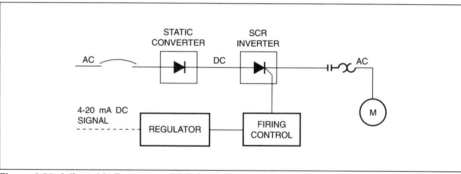

Figure 4-31. Adjustable Frequency AC Drive System

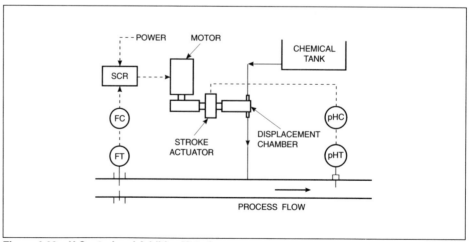

Figure 4-32. pH Control and Additive Metering

are also commonly used when remote set point adjustment is required for a regulator application. Figure 4-33 gives a comparison of regulators and control valve characteristics.

LEVEL

Level regulators are used to maintain liquid level within a tank. A complete level regulator consists of a float as a sensing element, an actuator, and a valve body. A level regulator may be an integral unit assembled from a ball float sensor, a level actuator, and a valve. Figure 4-34 shows different types of level regulators.

PRESSURE

A pressure-reducing regulator maintains a desired reduced outlet pressure while providing the required fluid flow to satisfy a variable downstream demand. The value at which the reduced pressure is maintained is the outlet pressure setting of the regulator.

A direct-operated pressure-reducing regulator, as shown in Figure 4-35 senses the downstream pressure through either an internal pressure tap or an external control line. This downstream pressure opposes a spring that moves the diaphragm and the valve plug to change the size of the flow path through the regulator. The addition of a pilot to a regulator provides a two-path control system. The main valve diaphragm responds quickly to downstream pressure changes, causing an immediate correction in the main valve plug position. The pilot diaphragm responds simultaneously, diverting some of the reduced inlet pressure to the other side of the main valve diaphragm to control the final positioning of the main valve plug. See Figure 4-35.

FLOW

A self-contained flow regulator or differential pressure-reducing regulator maintains a pressure difference between two locations in the pressure system. The value at which the pressure difference is maintained is the differential pressure setting of the regulator.

As shown in Figure 4-36, a differential pressure-reducing regulator has two pressure taps. Output pressure from a remote-mounted instrument or a pressure loader is applied through an external pressure tap to the top of the main diaphragm. The outlet or control pressure is applied to the bottom side of the diaphragm through an external pressure tap.

In some differential pressure-reducing regulators, this control pressure is applied to the bottom side of the diaphragm through an internal pressure tap. The differential pressure is applied to a spring-and-diaphragm mechanism that moves the valve plug to change the size of the flow path through the regulator.

SELECTED REGULATOR CHARACTERISTICS	SELECTED CONTROL VALVE/INSTRUMENT CHARACTERISTICS
Purchase price, installation, and maintenance costs are normally lower.	Wide variety of construction materials and accessories available.
Requires no additional power sources for basic operation.	Transmitting and controlling instruments are separate and may be remote mounted.
Less complex, and often lighter and more compact.	Specific construction has broad application flexibility.
Controller, which provides fixed-band proportional control only, is built in.	Separate controller allows for adjustable-band proportional control with reset and/or rate options for excellent control response.

Figure 4-33. Regulator vs. Control Valve
(Courtesy of Fisher Controls International, Inc.)

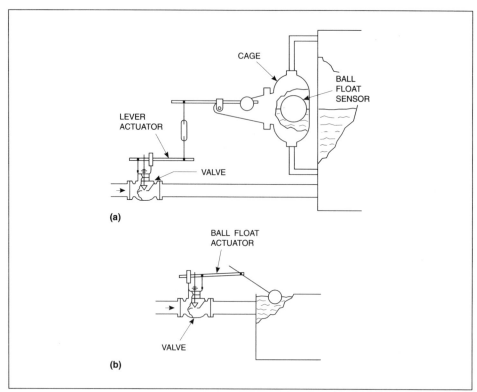

Figure 4-34. Level Regulators
(Courtesy of Fisher Controls International, Inc.)

TEMPERATURE

The temperature regulator is a self-contained control device consisting of a primary detection element or bulb, a measuring element or actuator, a reference input adjustment, and a final control element or valve. As with pressure regulators, there are direct-operated and pilot-operated devices.

With the direct-operated temperature regulator, the components of the actuator (bellows, diaphragm) are connected directly to the valve plug, thus developing the force and travel necessary to open and close the valve. Direct-operated temperature regulators are generally of a more simplified construction and operation and are less expensive than the pilot-operated type.

In the pilot-operated type, the actuator moves a pilot valve (internal or external). The pilot controls the amount of pressure from the fluid through the valve to a piston or diaphragm, which in turn develops power and thrust to position the main valve plug. Pilot-operated temperature regulators have smaller bulbs, faster response, and higher proportional gain. They can also handle higher pressures through the valve.

Temperature regulators may be either self-contained or remote-sensing. Self-contained regulators contain the entire thermal actuator within the valve body, the actuator being part of the primary detecting element. They can sense only the temperature of the fluid flowing through the valve. The regulator regulates the fluid temperature by regulating the fluid's flow.

In remote-sensing regulators, the bulb is connected to the thermal actuator by flexible capillary tubing. This construction allows them to sense and regulate the temperature of a fluid distinct from that of the fluid flowing through the valve. This type of regulator, although frequently lower in cost, is limited in application to such uses as regulating the temperature of water or some other type of coolant.

Valve action (direct or reverse) is selected as a function of the process. Direct action is used for heating control, the direct-acting valve reducing the flow of the

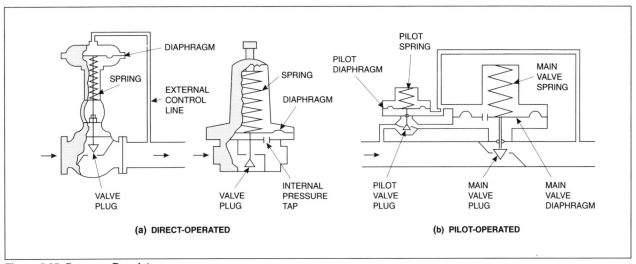

Figure 4-35. Pressure Regulators
(Courtesy of Fisher Controls International, Inc.)

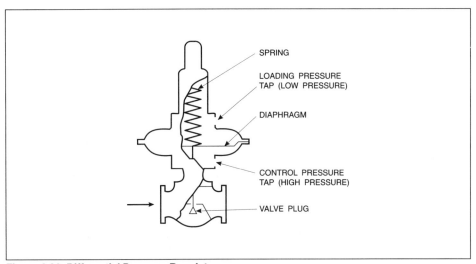

Figure 4-36. Differential Pressure Regulator
(Courtesy of Fisher Controls International, Inc.)

heating medium on temperature rise. The reverse-acting regulator is generally used for cooling control to increase the flow of coolant on rising temperature.

The mixing of two media at different supply temperatures to control the mixed temperature is accomplished with three-way valves.

Bronze and cast iron are standard body construction materials. Composition discs are used for tight shutoff on low pressure or temperature applications.

Dampers

APPLICATION

Dampers may be used to control the flow of gases and vapors as well as solids, or to throttle the capacity of fans and compressors. Dampers are suitable for control of large flows at low pressure where high control accuracy is not a requirement. They are usually larger in size compared to control valves and, therefore, are restricted to lower operating and shutoff pressures. Typical applications include air conditioning systems and furnace draft control.

TYPES OF DAMPERS

Two qualities of dampers are available: (1) commercial, which is used primarily for HVAC (heating, ventilating, and air conditioning) applications, and (2) industrial, which is used in process control to handle higher pressures, temperatures, and corrosive vapors.

Dampers are classified as to type of construction as louvered, guillotine, butterfly, and iris.

Louvered, or multiblade, dampers consist of two or more rectangular vanes mounted on shafts one above the other and interconnected so as to rotate together. The vanes are operated by an external lever that can be positioned manually, pneumatically, or electrically. They may be parallel-blade or opposed-blade. See Figure 4-37.

Guillotine or slide-gate dampers are similar in principle to knife-gate valves.

Butterfly dampers are similar in principle to butterfly valves.

Radial-vane dampers are used on blowers and fans. The damper consists of a number of radial vanes arranged to rotate about their radial axis. This type of damper has a high leakage rate. See Figure 4-38.

Iris dampers are also known as variable-orifice valves. The closure element moves within an annular ring in the valve body and produces a circular flow orifice of variable diameter. Tight shutoff is not possible, and the maximum allowable differential pressure is 15 psi. Dual valve units are available with a common discharge port for blending two streams. When iris dampers are used for throttling on solids service, they must be installed in a vertical line. See Figure 4-39.

DAMPER ACTUATORS

Damper actuators may be manual, electric, hydraulic, or pneumatic. Standard spring ranges include spans of 3 – 7, 5 – 10, or 8 – 13 psig. Actuators may be provided with positioners, which assure more accurate throttling.

Limit switches can be installed to detect blade angles when remote indication of damper status is desired.

DAMPER CHARACTERISTICS

Just as with valves, dampers have inherent and installed characteristics. For some reason, this fact is nearly ignored compared to the attention it is given with respect to valves. As a result, dampers tend to be oversized, making control more difficult.

Figure 4-40 shows the installed characteristics of parallel-blade and opposed-blade dampers for different values of, where is the ratio of the duct system pressure drop (excluding the damper) to the pressure drop across the damper in the full-flow, wide-open position. Oversizing a damper generally leads to high values of which, in turn, makes the loop more difficult to control.

According to one control manufacturer, research in damper characteristics has established that, to achieve linear performance between measured variable and air flow, the proper static split between a wide open damper and the rest of the system is five to 10 percent for opposed-blade dampers and 20 to 30 percent for parallel-blade dampers. These important numbers are supported by the intuitive picture presented by Figure 4-41.

Safety Relief Valves

One of the important responsibilities of an instrumentation designer is the safe operation of a process plant. This includes the protection of equipment against failure due to overpressure. Overpressure may develop because of one or more of the following factors:

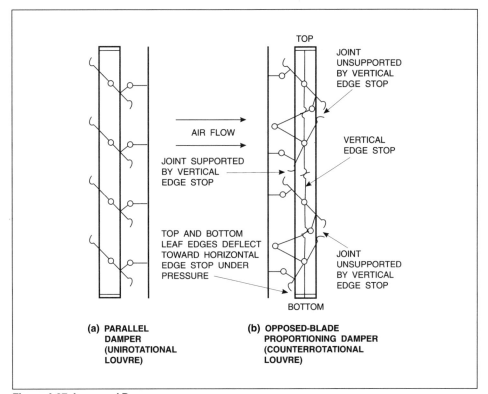

Figure 4-37. Louvered Dampers
(From Liptak, B., *Instrument Engineer's Handbook*, Chilton Book Co., 1985.)

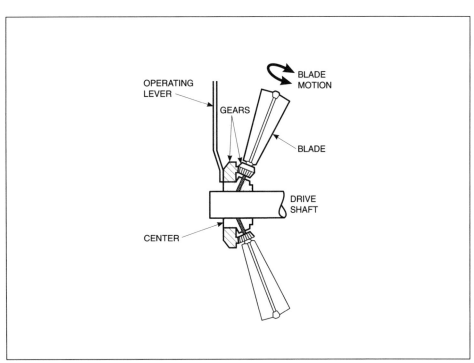

Figure 4-38. Radial-Vane Damper
(From Liptak, B., *Instrument Engineer's Handbook*, Chilton Book Co., 1985.)

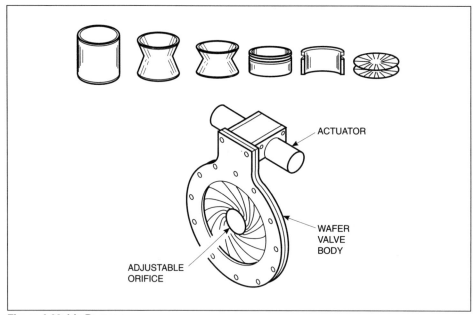

Figure 4-39. Iris Damper
(From Liptak, B., *Instrument Engineer's Handbook*, Chilton Book Co., 1985.)

(1) Operating failures and errors

(2) Thermal expansion

(3) Chemical reactions

(4) Explosions

(5) External fires

(6) Material fatigue

(7) Power failure

(8) Automatic control failure

(9) Cooling failure

Provisions must also be made for the safe disposal of the material released by the operation of the relieving devices.

Codes and recommended practices pertaining to pressure-relieving devices have been prepared by the American Society of Mechanical Engineers (ASME) and by the American Petroleum Institute (API) (refer to Section VIII of the ASME Boiler and Pressure Vessel Code and API RP250 and 521).

Pressure-relieving devices are installed for the following reasons:

(1) To provide safety for operating personnel.

(2) To protect the environment.

(3) To prevent the destruction of capital investment.

(4) To prevent material loss.

(5) To prevent downtime that might result from overpressure.

SAFETY VALVES — DESIGN

A safety valve is an automatic pressure-relieving device actuated by the static pressure upstream of the valve and characterized by rapid full opening (a pop action). It is used for gas or vapor service and also for steam and air. This action is

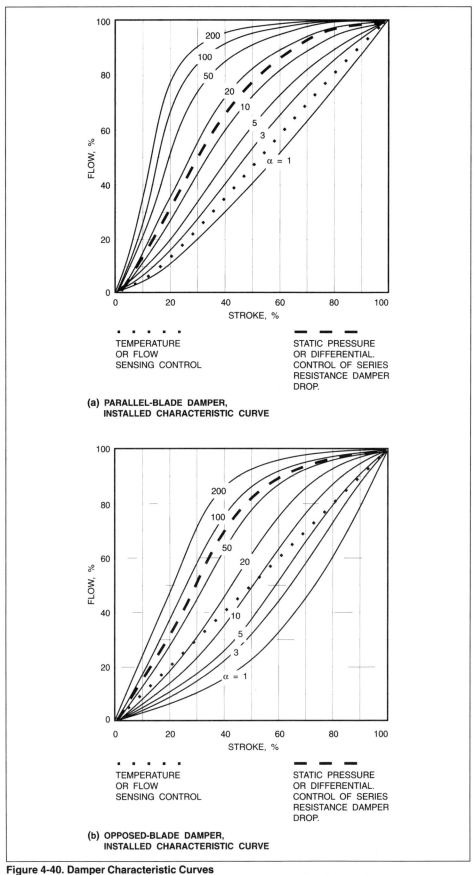

Figure 4-40. Damper Characteristic Curves
(Courtesy of American Society of Heating, Refrigerating, and Air-Conditioning Engineers, Inc.)

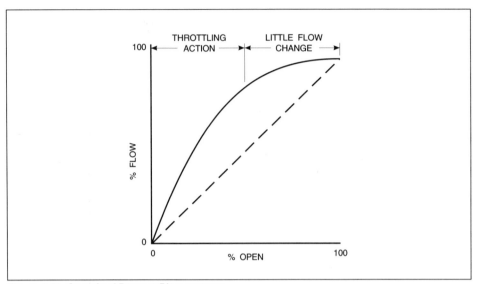

Figure 4-41. Oversized Damper Phenomenon
(Courtesy of Honeywell, Inc.)

accomplished by a force-balance system acting on the closure of the relieving area (see Figure 4-42). The orifice area of the pressure relief valve is selected to pass the required flow at specific conditions. This area is closed by a disc until the set pressure is reached. The contained system pressure acts on one side of the disc and is opposed by a spring force on the opposite side.

SAFETY VALVES — SIZING

Formulae are available to size safety valves for liquid, gas, and steam services. Consider as an example the sizing of a safety valve for steam service. This formula is extracted from API RP 520, Appendix D:

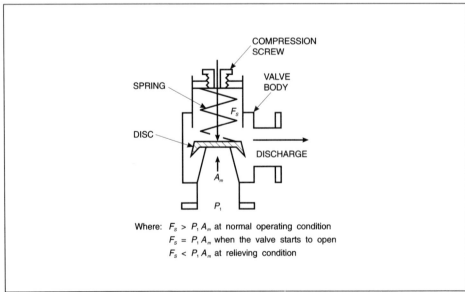

Where: $F_s > P_1 A_m$ at normal operating condition

$F_s = P_1 A_m$ when the valve starts to open

$F_s < P_1 A_m$ at relieving condition

Figure 4-42. Pressure Balance in a Relief Valve

$$W = 50\, A P_1\, K_{SH} \qquad\qquad (4\text{-}22)$$

or

$$A = \frac{W}{P_1\, K_{SH}}$$

where:

W = flow rate (lbs/hr)

A = effective discharge area (sq. inches)

P_1 = upstream pressure (psia). This is

the set pressure multiplied by 1.03 or 1.10
plus the atmospheric pressure.

K_{SH} = correction factor due to the amount of superheat.

For saturated steam at any pressure, the factor = 1.0.

50 = constant used in API equations;
however, Section VIII of the ASME Power Boiler Code
uses a value of 51.5.

Terms used synonymously with safety valve are "safety relief valve" and "pressure relief valves."

RUPTURE DISCS

A rupture disc consists of a thin metal diaphragm held between flanges. Its purpose is to fail at a predetermined pressure, serving essentially the same purpose as a pressure relief valve. Rupture discs are fabricated from carefully selected pieces of metal. They have defined limitations that are basic to their ultimate tensile or compressive strength and to creep, fatigue, or corrosion resistance.

Rupture disc types are categorized as solid metal, composite, reverse-buckling, and shear.

The construction materials for rupture discs are usually stainless steel, copper, nickel, aluminum, Monel™, Inconel™, and sometimes titanium and tantalum. Metal foils and sheets in soft annealed conditions are required.

Summary

Proper control valve selection is critical to satisfactory process control and should be undertaken with great care.

Nevertheless, Table 4-6 is included here to provide a summary of global characteristics that can be used by the designer as a guide to matching a valve type to an application.

Bibliography

1. Albert, Charles A., *Industrial Process Control and Automation*, Chapter 3 — "Final Control Elements," Quebec, Canada: St. Lambert , 1982.

2. *ASHRAE 1991 Applications Handbook*, Atlanta, GA: American Society of Heating, Refrigerating, and Air-Conditioning Engineers, 1991.

3. Dezurik, "Gate Valves," CVS-13 9-75 Bulletin 33.00-1, Sartell, MN: Dezurik Co., a Unit of General Signal.

4. Fisher Controls, *Control Valve Handbook*, 2nd edition, 4th printing, Marshalltown, Iowa: Fisher Controls International, Inc. , 1977.

Table 4-6. Quick Reference Valve Selection

Key Features	Reciprocating							Rotary			
	Single Seat	Double Seat	Cage Trim	Angle	Y	Three Way	Split Body	Camflex General Purpose	Ball	Butterfly Valve	HPBV
Capacity	1	1.1	1.2	1 to 2	1.5	0.7	1	1.3	3	3.2	2
Shutoff % Rated C_V	0.01 Class IV	0.5 Class II	0.01 Class IV or III	0.01	0.01	—	0.01	0.01	0.01 Class IV	1.0 Class I	Drop Tight
C_V Ratio Rangeability	50	50	50	50	50	50	50	100	100	25	100
Cavitation	S	G	G	S	P	S	P	S	S	S	S
Noise	S	G	G	S	P	S	S	S	P	P	P
High Pressure High ΔP	S	G	E	S	S	P	S	S	P	P	S
High Temperature Low Temperature	S	S	E	S	S	S	S	G	P	P	S
Erosion/Slurry	S	S	P	G	S	S	G	G	S	P	P
Corrosion	S	S	P	S	S	S	G	G	S	S	S
Maintenance	G	S	S	S	S	S	E	G	G	G	G
Cost	1.0	1.06	1.12	1.2	1.5	1.8	0.97	0.83	0.73	0.4	0.5

Note: Single seat is the base.
E = excellent, G = good, S = satisfactory, P = poor

5. Fisher Controls, *General Catalog 501*, Section 7, 5th ed., Marshalltown, Iowa, U.S.A.: Fisher Controls International, Inc. , 1989.

6. Fisher Controls, *Noise Control Manual*, Catalog 10: 3-1 - 3-51, Marshalltown, Iowa, U.S.A.: Fisher Controls International, Inc. , 1984.

7. *Honeywell Damper Sizing Manual*, Form No. 77-0078, Rev. 11-79, Minneapolis, MN: Honeywell, 1977.

8. Hutchison, J. W., and Merwick, A. R., eds., *ISA Handbook of Control Valves*, Research Triangle Park, NC: Instrument Society of America, 1976.

9. ISA-S51.1-1979, Process Instrumentation Terminology, Research Triangle Park, NC: Instrument Society of America, 1979.

10. Liptak, Bela G., *Instrument Engineers' Handbook*, Revised Edition, Radner, PA: The Chilton Book Company, 1985.

11. Masoneilan/Dresser, *Masoneilan Handbook for Control Valve Sizing*, Bulletin OZ 1000E, 7th ed., Houston, Texas: Dresser Industries, Inc., 1987.

12. Masoneilan/Dresser, *Masoneilan Noise Control Manual*, Bulletin OZ 3000E, 3rd ed., Houston, Texas: Dresser Industries, Inc., 1984.

13. Tabachinik, R. L., and Kwasik, L., SNC Cellulose Inc., *Control Valve Sizing and Selection Manual*, Montreal, Quebec, Canada: Cellulose, Inc., 1981.

About the Author

Helen Beecroft is experienced in the development and design of instrumentation for application in the pulp and paper and a range of other industries. As Instrumentation Design Technologist at Sandwell, Inc., Ms. Beecroft was responsible for the configuration of a factory-wide integrated graphics and control system from specification and procurement through design of standards, databases, and graphics, as well as start-up, system analysis, and operator training. A graduate of Concordia University in Montreal, she earned the degree of Bachelor of Science in Physics.

5

Computer Technology

Computers, in a very general sense, are devices that process information to solve problems. This definition includes the abacus, the adding machine, the hand calculator, and the library card catalog, among many other tools. So what is special about computers as we normally speak of them?

As implied by our everyday use of the word "computation," computers deal with quantitative information. Non-numerical information such as words of text must be converted into a numerical representation (such as ASCII code) before computers can deal with it.

Beyond this fact, computers tend to be defined by how they store and process information and how they receive and carry out instructions from their users.

The Digital Computer

How to Represent Information

Early efforts to represent calculations by mechanical analogs (basically large systems of gears) ran into accuracy problems due to the difficulty of machining the many gears accurately and consistently enough. Reliability was also a problem due to the many contact points that offered opportunities for mechanical breakdown. Mechanical computation, therefore, never grew much beyond the scope of the desktop calculating machine.

More recently, electrical analog principles were used, with voltages or currents proportional to the quantities to be represented. Analog computers served well in many applications with a structure such that voltage or current operations worked well as a physical model. Programming an analog computer, however, never ceased to be a mysterious art, and the computer itself required a fair amount of somewhat delicate "fine tuning" to work effectively. A simpler system was required before computing could move out of the scientific or engineering laboratory and into the everyday working world.

Digital computers use a numerical code to represent information. In practice, this is always a binary code (one for on, zero for off) because such a code is easily implemented by switchable electrical and electronic devices.

Information written in binary code can also be manipulated by binary (base 2) arithmetic, which is well defined and understood. Because of the simplicity of storing and manipulating information in the digital environment, digital computing has become the predominant form of computing today.

Digital Operations

Operations that can be performed on a string of binary digits (bits) include arithmetic and logical operations along with various others that do not fit these categories. Arithmetic operations are the familiar foursome of addition, subtraction, multiplication, and division, plus exponentiation and root taking. All of these operations return a numerical result.

Logical AND and OR operations compare two binary strings bit by bit, returning a true or false value. If the corresponding bit in each of the two patterns is a 1, the AND function will return a value of "true." If either or both is a 1, the OR function will return "true." The logical NOT function, applied to a single string, simply reverses the bits: all the zeros become ones, and all the ones zeros.

A logical operation can be summarized by a truth table that tells you, for each combination of zeros and ones input to the operation, whether a zero (false) or one (true) will be output. Figure 5-1 shows several common logical operations and their associated truth tables.

Other operations can be applied to bit strings. Shift operations move the whole string of bits one position to the left or right in a storage area (either dumping one bit off the end or recirculating it into the other end of the string). Mask operations compare the string to a superimposed pattern and keep only those bits for which an AND of the string and the pattern returns "true."

The Stored Program Concept

The stored program concept, developed by the mathematician John von Neumann in the early 1950s, was one of those revolutionary ideas that enabled a whole new technology to grow. It involved nothing less than a fundamental change in how a computing machine was told what to do.

When you operate a standard typewriter, one without the new quasi-computer features like buffers and memory that typewriters now come with, you give it instructions in a very direct mechanical way. You press keys, levers move, type heads strike paper, and out comes typed text. The "program," if we can dignify it with that name, goes straight from your head to your finger to the typewritten page in a very direct manner, with minimal encoding or storage.

As computers evolved and were used to tackle more complex problems than balancing a checkbook or typing a business letter, ways were developed to set up or "program" a computer beforehand to follow a certain sequence of calculations automatically. Analog computers often included patch panels (like an old-fashioned telephone switchboard) to allow the pattern of interconnections between their physical elements to be changed just by plugging in jacks in different locations. More recently, separate programming units are used to "burn" programs into read-only memory chips or to change the program in some programmable logic controllers (PLCs) in cases where for security reasons it is desired to keep the programming function entirely separate from the controller itself.

The important point here is that, although real programs were (and are) definitely entered and run by these means, the programming function and the programming technology remained separate from the operation of the computer and the technology used in it. With the stored program concept, however, all this was about to change.

The essence of the stored program concept is that a computer program is just another kind of information, which can be stored in memory in the same way as the data it will operate on. This means that a computer can have one kind of memory to store both programs and data (although in practice they may be placed in dif-

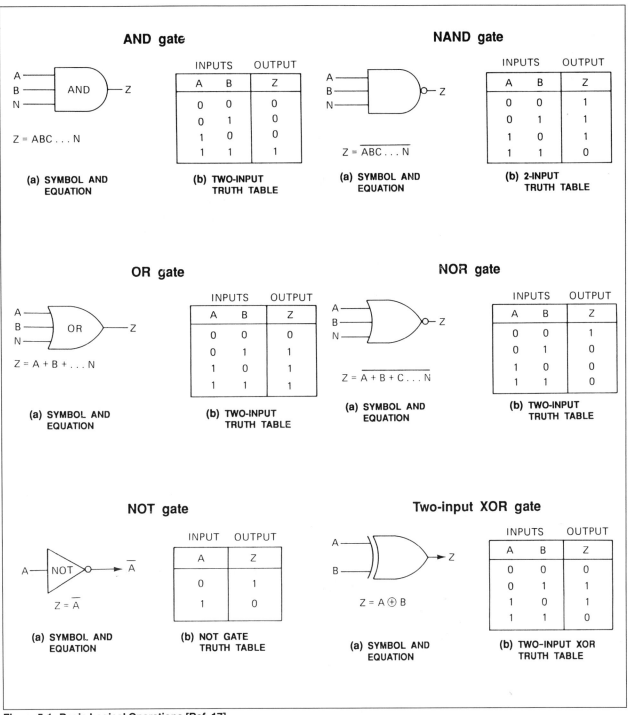

Figure 5-1. Basic Logical Operations [Ref. 17]

ferent areas) and can use one information processing technology both for decoding and executing instructions and for reading and writing data.

The impact of this was that computer architecture (how the physical "building blocks" of a computer are structured and connected) could become very much simpler. Development effort could also focus on the digital technology already being used to store data, instead of dividing itself between data storage technology and a separate programming technology.

Programming also became much easier. Instead of working with hard-wired electrical connections, programmers could manipulate ones and zeros. This meant that programming became accessible to people who were not electrical engineers, once they learned basic programming skills. Now that we are well along in this evolution, children can write programs in Basic.

Why Use Computers?

The previous discussion leaves unanswered the question of why a machine that is good at information processing should be so interesting to us. Why do we need computers to the point that now we wonder how we lived without them?

One major reason is that the amount of information available to us is increasing at an exponential rate. Over half the scientists that have ever lived are now alive—and they are all madly writing papers to keep their jobs! Closer to home, the quantity of information that can be produced and archived by a modern distributed control system (DCS) in a single day is far greater than any human being can read and absorb in that same day. We need help just to cope with the flood of information battering at our doors.

Flexibility is another asset that computers give us, and one that we badly need. Gone are the days when we can "hard-wire" a process and have it make money for us until the plant falls down of old age. The changing and expanding world market, upheavals in society and culture, and the drive of affluent and educated people to have exactly what they want right now, all ensure that what we make and how we make it has to change on an ongoing basis. To make the connection clear, our manufacturing processes (as well as much of the rest of our lives) have to be reprogrammable—over and over again—and therefore must make use of a technology that makes reprogramming easy: the modern digital computer.

Another advantage of digital computers is their intrinsic repeatability and reliability. Once threshold voltage values are set for when a bit will be considered "on" and when "off," the same inputs to the computer will produce the same internal bit patterns and the same outputs—essentially always. (Analog computers used to give us problems in this area as their components aged and changed their characteristics). As for component reliability, although computers do fail, their reliability in terms of the number of operations they perform is impressive. Mean time between failures (MTBF) in the billions of operations is standard for many computer components. How many other devices do you own that work a billion times before breaking down?

 Computers do not replace us; they extend us.

A computer cannot recognize the face of a loved one, or enjoy beautiful music, or laugh, or dream. But it can calculate pi to thousands of decimal places and send a spaceship on an orbit to the planet Neptune; it can remember more than we can read in a lifetime, store it in a database, and pull out on demand a set of references on medieval water wheels; it can pay attention to a process long after we have fallen asleep or gone crazy from boredom; and it can find among thousands of process alarms the three we need to address to prevent the plant from blowing up. The partnership of human beings with computers is already well established in the industrial world, and with good reason.

The Central Processing Unit (CPU)

The heart of a computer is the central processing unit (CPU), which applies instructions to data to generate results. This includes fetching from memory the data the instructions need and returning the results to the appropriate memory locations.

Some Basic Vocabulary

The most basic logical or physical element of a digital computer, the entity that takes on a one or zero value, is called a bit. Eight bits make up a byte, and from one to four bytes form a word, (see Figure 5-2).

Components and Structure

"Where the action is" in the CPU is the arithmetic logic unit (ALU), which contains the electronic components and logic to implement mathematical operations (as its name implies) and also logical and other operations.

In the ALU and elsewhere in the CPU we find registers, which function as temporary storage areas for data moving around in the CPU. Registers in the CPU are normally specialized by the kind of information they process (data, addresses, or instructions).

A special kind of register is the program counter, which always contains the address in memory of the next instruction to be executed.

The timing and control unit contains another special register, the instruction register, which holds the instruction currently being executed. Controlling and sequencing logic also reside here. In addition, the timing and control unit accepts the clock signal and sends it on to other parts of the CPU that need it.

The internal bus moves information around between different parts of the CPU. Note that the internal bus is distinct from the system bus, which serves as a communications highway among the components of a computer.

The CPU communicates with its "outside world" (the system bus and memory) through data and address lines. Additional lines exist to transmit the clock pulse into the CPU and provide power and ground connections.

Figure 5-3 illustrates the major components of a CPU and how they are connected.

The Instruction Cycle

The CPU executes a program one instruction at a time and needs to perform essentially the same operations for each instruction. These operations constitute the instruction cycle (also known as the load-operate-store cycle, since this is what it does). Figure 5-4 illustrates one pass through the instruction cycle.

The first step is to read the program counter, which contains the address of the next instruction to be executed. The CPU can then fetch the actual instruction from memory and, upon decoding it, can obtain the addresses of the required data (operands), which the CPU also fetches from memory. With both instruction and data in hand, the CPU can then execute the instruction. Finally, the results are

> Note that 1K (one kilobyte) is not exactly one thousand bytes, as you might think, but 1024 bytes, which is 2 to the tenth power. Therefore, of course, 1 MB (one megabyte) is not exactly one million bytes, either, but 2 to the twentieth power.

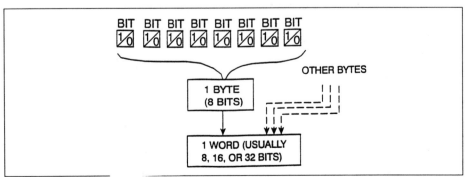

Figure 5-2. Bits, Bytes, and Words

written to memory and the program counter is advanced to point to the next instruction.

The System Clock

Most computer programs require some coordination between data fetches and stores and instruction executions, if only because later instructions need the results of earlier ones. Imagine the chaos if instruction B grabbed its input data from memory before instruction A had finished updating it, or sent its results out on a set of data lines at the same time instruction C was trying to load its inputs from memory using the same data lines!

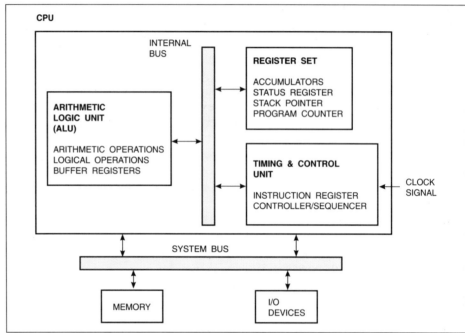

Figure 5-3. Major Components of a Central Processing Unit (CPU)

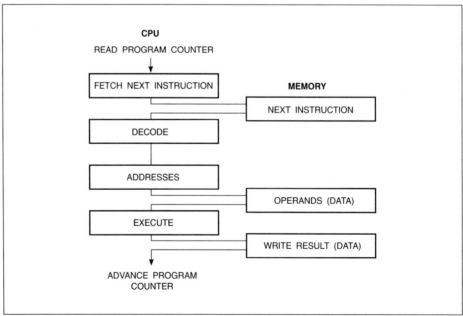

Figure 5-4. Instruction Cycle

Another consideration is that some electronic components respond more slowly than others. A common situation is that the microprocessor (CPU on a chip) is faster than the data memory and somehow the CPU must be told to wait until the memory is ready for the next operation.

The system clock is actually a vibrating crystal that emits x pulses per second as more or less square waves (see Figure 5-5). The faster the vibrational frequency of the clock crystal (usually measured in MHz (megahertz)), the shorter the computer's basic unit of time will be, the more clock pulses the computer will cram into each second, and (other components permitting) the faster the computer will be. Advertising claims referring to x MHz crystals are talking about the vibrating frequency of the clock crystal.

Some of the electronic and other devices in a computer need the clock signal to work properly, while others do not. Synchronous devices wait to detect the leading edge of the clock pulse one or more times before taking action, thus ensuring that they are coordinated (synchronized) with the clock—sort of like old-time railway conductors setting their watches before the train left the station. Asynchronous devices don't care about the clock signal because they have other means of making sure they do not get "out of step" with the devices they interact with. For example, they may send "handshaking" pulses that must be acknowledged by the other device to confirm that a data transmission has been received.

The Microprocessor

Clearly the logical functions of a CPU could be implemented in various configurations of electronic devices. Early computers used imposing arrays of amplifiers, gates, resistors, and other paraphernalia to build a CPU. This mass of electronic components was space-consuming, failure-prone, and hot, although with tender loving care it could be made to work.

As electronics technology advanced in the 1960s and 1970s, the same functions could be implemented in less and less space with more and more processing power per component. Some components began to be packaged as chips with input-output pins that could be inserted into sockets, eliminating much soldering of connections.

Finally, it became possible to put an entire CPU on a chip. The microprocessor made it possible to put a standard CPU any place you could plug in a chip. The promise of high volume meant that developers could afford to put a lot more effort into building and testing a microprocessor with fast cycle time and a capable

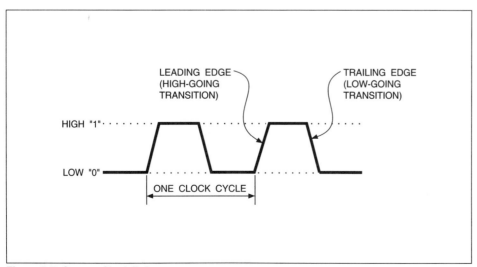

Figure 5-5. System Clock Pulses

and efficient instruction set. Virtually all computers nowadays contain a microprocessor, as do numerous other "smart" devices from sensors to microwave ovens.

Computer Architecture: Components and Structure

The term "computer architecture" refers to how a computer is built: what its components are and how they are connected together to form an overall structure. Figure 5-6 shows a somewhat simplified version of a typical computer architecture.

Memory

In the preceding section, we discussed the central processing unit (CPU) and its functions. Of course, the CPU cannot function in a vacuum; it needs to obtain information and to store results. The various kinds of memory store programs and data until the CPU needs them.

Thanks to the stored program concept, we do not necessarily need different kinds of memory to store programs and data. In practice, however, different regions of memory are set aside for different purposes, since this makes the work of system programmers easier and minimizes the chance that a program will be inadvertently overwritten or used as data (which, of course, is possible, thanks to von Neumann). In addition, the operating system needs some memory of its own to work in, and blocks of memory may be set aside to serve the needs of physical devices such as screens or printers.

THE MEMORY CHIP

In the past, devices ranging from mechanical gear trains to magnetic cores (hence the term "core memory") were used to store information. Now, virtually all memory is composed of memory chips, which are similar to microprocessor chips except smaller and less complex (see Figure 5-7). To simplify information processing, normally each memory chip handles one bit of a memory word, which means that a bank of X memory chips is required to process an x-bit word.

Although modern memory chips are very reliable, errors in reading and writing can and do occur. To detect these, an additional bit is often attached to each word, requiring an additional memory chip in the bank. The parity bit is set to the result of some operation on the contents of the original x bits, for example, counting the number of these bits that are set to 1 and setting the parity bit to 1 if this

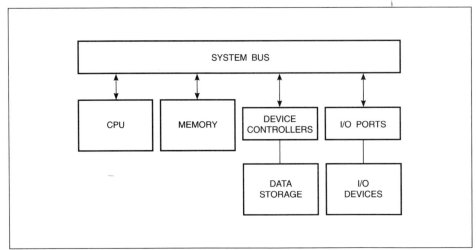

Figure 5-6. Typical Computer Architecture

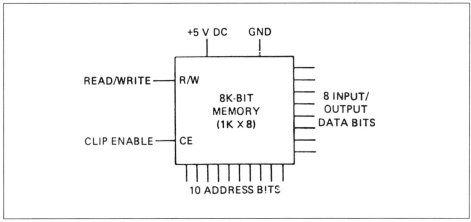

Figure 5-7. Typical 8K-Bit Memory Chip [Ref. 17]

number is odd and 0 if it is even. This operation is performed periodically on the *x* data bits, and an error is signaled if the result does not match the parity bit. Of course, there is a small but nonzero chance that TWO errors would occur in the same word, in which case the parity bit would fail to detect them; but remember that memory is intrinsically very reliable and the probability of two errors in the same word is, therefore, vanishingly small.

Figure 5-8 illustrates how a single 8-bit word of data, with its parity bit, would be stored in memory.

TYPES OF MEMORY: ROM

In preceding sections, we have discussed memory as if it always could be either read or written. This is not, in fact, the case. Sometimes it is unnecessary or undesirable to write to a particular area of memory; for example, you would not want to overwrite a critical control program. On the other hand, you would seldom find it necessary to read the data already on its way to a printer.

To protect important information from being accidentally modified, various types of read-only memory (ROM) have been developed. The simplest and oldest form has the information hard-coded into the physical connections on the chip. Once the chip is manufactured, it can never be modified. Users of this kind of

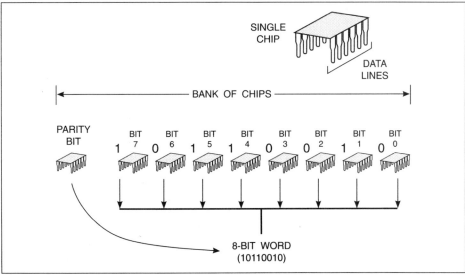

Figure 5-8. Use of Memory Chip Bank to Represent an 8-Bit Word

ROM had better be sure that their program works and that they need enough copies of it to justify the development time to get the program working perfectly! Straight ROM of this kind is often used for the "kernel" (fundamental portion) of an operating system or for control programs where absolute security is essential.

A slightly more adaptable version of read-only memory is the PROM (programmable read-only memory). The chip is made with fusible elements for each bit and comes with all bits set to 1. The chip is programmed by passing a high current through the bits that are to become zero, melting the fusible link. This process is clearly not reversible, which means that a PROM can be programmed once and only once, and any error still means scrapping the PROM. However, the manufacturer can make a "generic" PROM that each user can program for a specific application.

Somewhat more flexibility is provided by the EPROM (erasable programmable read-only memory), which has largely supplanted PROM for obvious reasons. An EPROM chip has a window over the storage area that allows the chip to be erased by exposure to ultraviolet (UV) light. The exact mechanism is that a charge is placed at each location in the memory that is to contain a 1. Since no discharge path is provided, the charges stay there until the chip is reprogrammed. Evidently there is a limit to how many times this can be done, but few practical applications encounter it.

EPROM is not quite as certifiably permanent as ROM or PROM. In particular, the chip can be inadvertently erased by exposure to an ultraviolet light source such as a welding torch or a fluorescent light. To minimize this risk, most EPROMs come with a sticker that is used to cover the window after programming. In principle, rare events such as a cosmic ray impact could change a bit in an EPROM, but this seldom seems to happen.

A variant on EPROM is EEPROM (electrically erasable programmable read-only memory), which is erased electrically instead of by light.

TYPES OF MEMORY: RAM

Memory that can be both read from and written to is normally referred to as RAM (random access memory). This is, in fact, a misnomer since virtually all kinds of memory in use today can be accessed randomly. Random access means that you can go directly to any bit on the chip without having to deal with the bits before or after.

So far as we know, there is no limit to how many times RAM may be modified. RAM is therefore used for data memory that is changing all the time. It does, however, have some disadvantages, which are described in the next section.

PERMANENT VS. VOLATILE MEMORY

One of the advantages cited earlier for the various kinds of ROM was its permanence. In fact, much data in computer memory is at risk not only from accidental overwrites but also from simple loss of power. The data in RAM is maintained by the continued presence of an electrical voltage. If power is lost, the data disappears. Of course, as long as the power is on, changing the data is quick and easy, which is the main advantage of RAM.

STATIC VS. DYNAMIC MEMORY

The basic unit of static RAM (SRAM) consists of six transistors: two for storage of a digit and its complement plus two load transistors and two enable transistors. This kind of RAM is called "static" because, once a digit is stored in it, the digit stays there until it is explicitly changed.

Since using six transistors to store one digit is more complicated and space-consuming than some people would like, efforts were soon made to simplify the basic static RAM cell. It was soon found that by using the inherent capacitance of

---

In addition to using some version of ROM for information whose permanence you are worried about, another approach is to provide battery backup for the RAM. Since the power requirements of RAM are minimal, a small battery can last several years. One must, however, be careful to change the battery as it nears the end of its life (easy to forget if the last time was three years ago), and never, ever to separate the RAM from its battery. If the battery resides on a power supply card and the RAM on a memory card, then clearly, removing either card from the computer will zero out the RAM.

a MOS transistor for storage, the number of transistors could be reduced from six to three. Further simplifications led to a one-transistor RAM cell that consisted of a capacitive storage element and an enable gate.

But this new, simpler memory cell is no longer static. With time the charge in the capacitor will leak away, and the contents of the memory will be lost unless they are refreshed every few milliseconds or so. This kind of memory is therefore called dynamic RAM (DRAM).

Dynamic RAM is popular for high-end computer hardware because it is fast (a benefit of simplicity) and because more DRAM than SRAM can be packed onto a given surface area. However, dynamic RAM is more expensive than SRAM because of the additional circuitry and logic required for refreshing the memory (see Figure 5-9).

WAIT STATES

Memory, especially static RAM, tends to be slower than the CPU. However, memory access is only one of the CPU's tasks, and a computer designer may want to obtain the other benefits of a fast CPU while still ensuring reliable

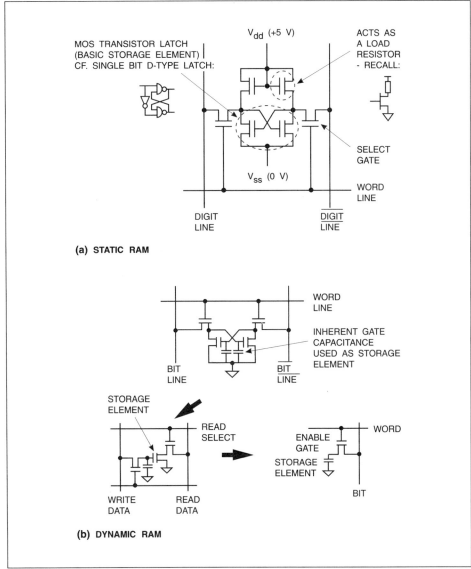

(a) STATIC RAM

(b) DYNAMIC RAM

Figure 5-9. Static vs. Dynamic RAM [Ref. 1]

memory access. The usual solution is to have the CPU, after each memory access, wait a certain number of clock cycles (usually 1 or 2) before accessing memory again. These cycles where the CPU just waits for the memory are called "wait states." In a zero wait state computer, the memory is fast enough to keep up with the CPU and the CPU does not need to wait for it.

ADDRESSING

Every location in memory that can hold a data word (whatever size that is for the computer in question) has an address that is just a unique number that the CPU uses to find that location. The address may be divided into a part that specifies an area of memory (like a chip or bank of chips) and a part that specifies a location within that area. The simplest way for an instruction to specify the address of the data it wants is to state this number—this is called direct addressing (see Figure 5-10). Direct addressing has the great advantage of simplicity, but a number of disadvantages, chief of which is that the data to be referenced by a program must ALWAYS be stored in exactly the same place in memory. This allows the operating system no latitude at all to change the allocation of memory, which it may well want to do in the interests of efficiency as various numbers of different-sized programs start and stop running. Moreover, certain useful programming constructs such as loops cannot be done efficiently with direct addressing.

One solution to this problem is indirect addressing, in which the instruction points not to the memory location itself but to a second location that contains the address of the memory location. The contents of the second location can be repeatedly changed without having to modify the instruction itself.

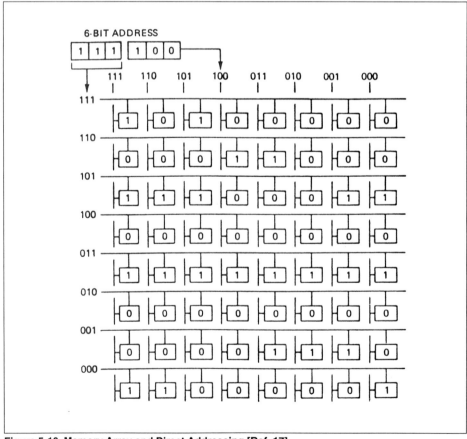

Figure 5-10. Memory Array and Direct Addressing [Ref. 17]

Another approach is indexed addressing, which calculates the desired address by adding together the contents of an index register and a second, variable number. Indexed addressing is frequently used for data tables or arrays. In this case, the index register points to the beginning of the table, and the elements in various rows and columns are accessed by changing the second number of the address calculation.

A variation on indexed addressing is base addressing. Strictly speaking, a base register is not expected to be modified whereas an index register may be, but the difference is unimportant in most practical cases. Base addressing is often used to specify a program's data area at run time, so that the operating system can assign whatever region of memory is convenient at the moment.

Base and indexed addressing are often used together with displacements. In this case, the base address is the start of the program's data area, various displacements are added to the base address to give the starting point of data subsections like arrays and lists, and then various index values are added to these starting points to give the actual data addresses. Figure 5-11 illustrates how these various addressing schemes work.

All of these addressing schemes are ultimately limited by the number of bits on the computer's address bus. For example, with a 16-bit address, one can access 64K (65,536 bytes, since 1K = 1024 bytes) of memory. Memory segmentation was developed to enable the expansion of addressable memory without having to change the bus. In a typical scheme (see Figure 5-12), a 16-bit segment register is logically ANDed with a 16-bit *effective memory address*, but with the two offset by four bits to generate a 20-bit address with which 1 megabyte of memory can be accessed. This 20-bit address can be calculated when and where it is needed from its two 16-bit components, which can be sent over the address bus.

Another approach to expanding addressable memory is memory paging. Memory paging looks at the current chunk (or page) of data in memory as only one of many possible pages that might occupy memory. The other pages are stored on disk, and if a reference is made to a memory location that is not in the

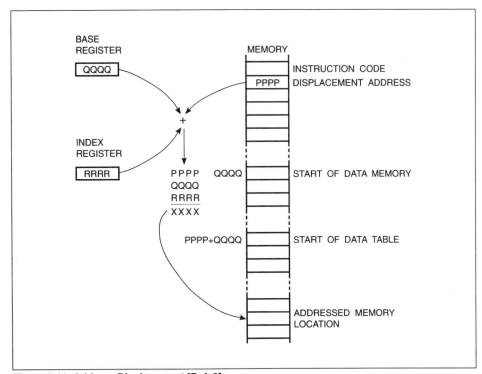

Figure 5-11. Address Displacement [Ref. 2]

page of data currently in memory, that page is swapped out to disk and the page containing the desired location is brought in (see Figure 5-13).

The advantage of paged memory over segmented memory is that you do not have to buy and install a lot more RAM. (Paged memory is also called "virtual memory" because the additional pages are "shadows" or alternative versions of memory and do not exist as banks of real RAM chips.) Of course, if you need to swap pages often, the time required for page swapping starts slowing down your computer, and you need the space to store currently unused pages on a fast disk. Addresses in a paged memory scheme are calculated in a manner similar to segmented addresses.

Paging and other data swapping schemes may make use of a cache to gain speed. A cache is a high-speed buffer that works on the assumption that it is often possible to predict what data will next be required from memory. For example, in the execution of some operation on the ith column of a data table, it is very likely that the $(i + 1)$ column will be needed soon. If the $(i + 1)$ column can be loaded from memory into the cache while the CPU is processing the ith column, the whole table will be processed faster than if the CPU had to wait for each column to be accessed from memory.

The System Bus

The system bus provides a communications channel among the various components of a computer: the CPU, memory, data storage, and input-output devices. Physically the system bus consists of a number of parallel conductors on a printed circuit board (PCB). These lines can be divided into four groups (also individually called buses) according to function (see Figure 5-14).

Address lines transmit memory and other addresses from one part of the computer to another. There must be one address line for every bit of an address, so the width of the address bus will correspond to the length in bits of an address in the computer in question. Data lines do the same for data. There must be one data

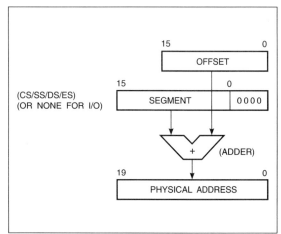

Figure 5-12. Memory Segmentation [Ref. 1]

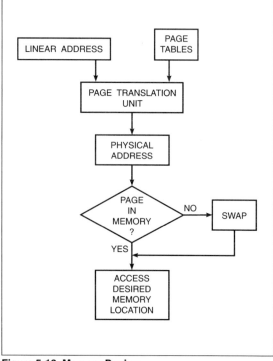

Figure 5-13. Memory Paging

line for each bit of a data word, so again the width of the data bus corresponds to the length in bits of a data word in the computer in question.

Control lines transmit control signal inputs and outputs by the various devices connected to the system bus. There must be one control line for each specific control signal to be expected. Examples of control signals are read, write, interrupt request, and reset.

Clock, power, and ground lines transmit the clock signal and provide power and ground connections.

In addition to the actual lines, a system bus includes a number of auxiliary devices that interface between the bus and the devices connected to it. Address decoders read addresses off the bus and check whether they match the address of their own device. Bidirectional bus transceivers sort out signals coming into a device from the bus with signals going out from the device to the bus. Bus drivers boost signal power to overcome the intrinsic capacitance of the bus.

Buses may be synchronous (which means they use the system clock pulse to coordinate their activities) or asynchronous (relying rather on handshaking and acknowledgment signals to make sure their activities are properly sequenced).

Because more than one device may want to use the bus at the same time, some form of bus arbitration is required. This may be on the basis of "first come, first served" or, more likely, of assigning priorities to devices. For example, a request for data from the CPU will take precedence over a request from the printer.

> Some computers use a system bus that is narrower than their data word size, requiring each word to be transmitted in two pieces (and two clock cycles). For example, a computer may use a 16-bit word but have an 8-bit bus. Obviously, this slows down the computer!

External I/O

The information processed by a computer must be obtained through input devices (keyboard, mouse, etc.) and results eventually transmitted to the outside world via output devices (screen, printer, etc.). The techniques used to move information from the CPU and memory to and from these devices are referred to as external input-output (I/O).

PERIPHERAL SUPPORT CHIPS

A basic design decision in interfacing to an external device is how much processing will be done by the general-purpose operating system and how much by device-specific hardware. At one extreme, interfacing can be handled entirely by the CPU, with very little hardware support; at the other extreme, a dedicated peripheral controller may handle most I/O tasks with minimal intervention from

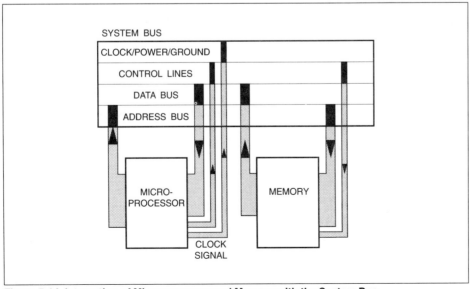

Figure 5-14. Interaction of Microprocessor and Memory with the System Bus

the CPU. The usual solution is somewhere in the middle: a peripheral support chip with a certain number of built-in functions and control registers that are written to by the CPU whenever access to the device is needed. In recent years, some of these peripheral support chips have acquired so much intelligence that they have become, in effect, I/O coprocessors working with the CPU in a quasi-distributed computing system.

The hardware and software that interface an external device to the system bus and the CPU are often collectively referred to as an I/O port. The CPU sees the I/O port as just another address and refers to it using a special subset of I/O instructions.

EXTERNAL DEVICE ACCESS TECHNIQUES

Since the CPU cannot predict when an external device will supply data to it, it needs some way to synchronize itself with the external device so that it is ready to accept data when the device is ready to provide it. There are two basic ways of doing this: polling and interrupts.

Polling involves continually asking an external device, or a series of devices, if it needs servicing. This technique is robust and uncomplicated and works well if I/O servicing is a major part of the computer's workload (many PLCs, for example, use polling quite successfully). If, however, there are many devices to check or the computer has other work to do, the amount of processing time spent polling devices can become burdensome. The scan time to check a large number of devices can also become a problem if fast response is needed. In the worst case, if a device needs servicing right after it has been polled, it will have to wait for the entire scan time.

Instead of having the CPU ask a device whether it has data to send, another approach is to have the device tell the CPU when it has data. The CPU can go ahead with other tasks and attend to data transfer only when a device requests it. The request is sent by the I/O port as an interrupt (see the section on operating systems for more detail on how an interrupt works). Advantages of this technique include speed of response, the ability to use the existing interrupt priority mechanism to get certain devices serviced faster than others, and the elimination of the operating system overhead of continually asking devices if they need servicing. The main disadvantage is the increase in software complexity for device identification and interrupt handling.

A third approach, offering even greater speed, is to move the device servicing task out of software altogether and into a dedicated controller with its own built-in instructions. This controller is dedicated to high-speed data transfer between the external device and memory, bypassing the CPU. This approach, is therefore, called direct memory access (DMA) or memory-mapped I/O.

The direct memory access controller (DMAC) interacts with the CPU via an interrupt like any other ported I/O device. Once it has been recognized, it takes over the data bus from the CPU and performs data transfer to a dedicated region of memory without further CPU intervention. Variants on this scheme allow the DMAC and CPU to use alternative bus cycles or to provide two buses, one for the CPU and one for the DMAC.

DEVICE DRIVERS

The software that manages the interface between the operating system and a particular external device is often referred to as a device driver. The existence of device drivers makes it possible for a general-purpose operating system to access a wide variety of input-output devices and to be easily expanded to accommodate new devices.

The heart of a device driver is the data control block (DCB), which contains the information necessary to deliver data from the device to the system bus and

The inventory of available device drivers for most software is continually being updated. If you are missing a driver for your favorite device, especially a new one, contact your software supplier and one may be on the way.

vice versa. Its capabilities include queuing and buffer management, protocol translation (for example, between the parallel system bus and a serial port), interrupt dispatching, and local error handling.

Data Storage

The various kinds of memory (ROM and RAM) hold programs and data in addressable locations where they can be quickly retrieved by the CPU. Most computer systems, however, need to have access to much more information than can be stored in their memories. Data storage units keep large quantities of information in a form that can be loaded into memory when needed.

In contrast to memories that operate on electrical and electronic principles, most data storage devices use magnetic or optical methods that are described below. They are always nonvolatile, which means the information remains even when the power is turned off. They may, however, be subject to physical damage as well as some long-term degradation.

MAGNETIC STORAGE

Magnetic data storage is based on the imposition of local clockwise or counterclockwise magnetic fields to a magnetic medium to represent zeros or ones.

The data are in principle nonvolatile, since once the medium is magnetized, the magnetic field remains until another magnetic field is imposed, either deliberately (by an erase or rewrite operation) or accidentally (by exposure to an electric motor or other magnetic source). Accidental erasure is not as common as some people believe, since a metal paper clip or ballpoint pen does not normally carry much of a magnetic field, but it is certainly worth being aware of the possibility and keeping magnetic storage media away from strong magnetic fields.

The lifetime of information stored on magnetic media cannot be considered infinite. The magnetic fields intrinsically weaken with time, to the point that after ten or twenty years the field strength may have fallen below the threshold value at which a zero or one is recognized. The substrate or carrier on which the magnetic medium is deposited may also deteriorate (especially with tapes), or the medium may crack or debond from the substrate. In the case of some tapes of satellite data from the 1960s, scientists now find they have only one chance to read the tape because pieces of the magnetic medium go flying off as the tape goes through the reader!

MAGNETIC STORAGE MEDIA AND DEVICES: DISKETTES

The best-known form of magnetic storage is probably the floppy disk or diskette, which comes in various formats but is essentially a plastic disk covered with magnetic material inside a protective casing. The disk is inserted into a drive to be read or written. Floppy disks provide relatively small amounts of data storage in an inexpensive and eminently portable form. The mounds of such disks that accumulate around most personal computers attest to their convenience and popularity.

Floppy disks vary in size (the 3.5- and 5.25-inch formats are the most popular; the 8-inch is an older size that is going out of use), recording density (how many bytes of information can be recorded on each disk), and whether data can be recorded on one or both sides of the disk. Most diskettes sold now are "double sided" (meaning data can be recorded on both sides). Common recording capacities for 3.5-inch diskettes are 720 KB , referred to as "double density" and 1.44 MB, or "high density." The higher density diskettes use a higher quality of magnetic material in order to be able to record information reliably. For 5.25-inch diskettes, double density means 360 KB, and high density is 1.2 MB. Occasionally, one encounters single density diskettes that are 360 KB in the 3.5-inch format and 180 KB in the 5.25-inch format.

The 3.5- and 5.25-inch diskettes differ in more than size: 5.25 inch diskettes have a relatively soft outer casing that can be bent (although bending is not recommended), while 3.5 inch diskettes, a more recent development, have a stiff plastic outer shell (thereby really not deserving the name "floppy" at all) and a sliding metal door that closes automatically over the read-write area when the diskette is removed from the drive (see Figure 5-15).

The 3.5-inch disks include a small plastic slider that can be moved into the "write protect" position (making the diskette read only) or the "write enable" position (allowing both reading and writing). Write protecting a 5.25-inch disk requires placing a separate write protect tab over the notch in the side of the disk.

Normally, diskettes must be formatted before use. The formatting operation sets up the track and sector structure as a template into which data will be written. Users impatient with the thirty seconds to two minutes required to format a diskette can buy preformatted diskettes (for a slight extra charge, of course). Some older types of diskettes are "hard sectored," which means that the sector structure is built into the disk when it is manufactured; these disks do not require formatting.

 Formatting a diskette at a higher information density than that for which it was designed is not recommended. Data sectors of marginal quality may initially accept data recording, but the data will "drop out" after a while and become unreadable.

A typical diskette drive consists of a slot into which the diskette is inserted, a spindle which fits into the center hole (5.25-inch format only), and a read-write head that moves in and out over the disk to access the information (see Figure 5-16). A disk controller takes care of translating read-write requests from the CPU (via the system bus) to movements of the read-write head to the appropriate track and sector for each item of data.

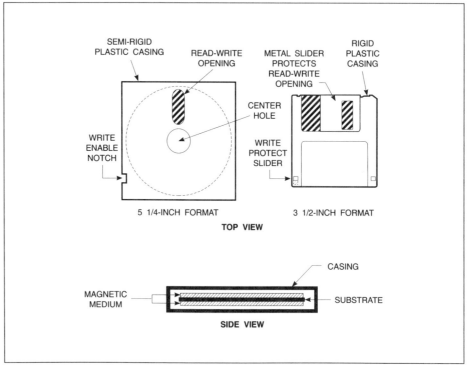

Figure 5-15. Floppy Diskettes

MAGNETIC STORAGE MEDIA AND DEVICES: THE HARD DISK

The hard disk offers larger storage capacity than the diskette along with the easy erasure and rewriting capabilities offered by magnetic technology. Typical capacities range from 40 to several hundred MB as of this writing (late 1990), but are subject to constant revision. Only five years ago a 10-MB hard disk was a prized possession.

A hard disk unit is in some ways like a number of diskettes mounted horizontally on a metal shaft, except that the substrate holding the magnetic medium is usually metal instead of plastic. Reading and writing take place on both sides of each disk, except that the top surface of the top disk and the bottom surface of the bottom disk are often not used. Instead of the one or two read-write heads in the floppy disk drive, a whole bank of heads is used, one for each surface of each disk. The set of heads and their mounting arms look somewhat like a comb (see Figure 5-17).

Hard disks are normally not removable from the computer in which they reside. Some units have been made with removable disk cartridges, though these have experienced reliability problems, probably because of the physical shock of repeatedly being pulled out and reinserted. With portable computers containing 40-MB hard disks the size of a pack of cards, there is probably no reason for most people to want to remove a hard disk.

Access time is a measure of how many milliseconds on average are required to access a file. Typical values are 15-20 ms for a fast disk, 30 ms or so for an average one—but again, these numbers are changing fast. Older hard disks may be much slower (80-120 ms).

When a hard disk is in operation, the read-write heads travel a few thousandths of an inch from the data recording surface. Any physical shock while the disk is in operation can send the head plowing into the magnetic medium, destroying data (this is what is called a "head crash"). The likelihood of serious damage from a head crash has been reduced in recent years with the development of more impact-resistant recording surfaces, but it is still worthwhile to try to avoid bumping or banging the hard disk while it is reading or writing data. Another preventive measure is to park the heads in an area not used for data storage when the disk is not actually in use. Some disks do this automatically after, say, ten or twenty seconds without any disk access requests; others have a head park program that is supposed to run before the computer is shut down.

The data on a hard disk can also be lost through corruption of the control information needed to access it. This topic will be dealt with in the section on operating systems and file structures.

For maximum peace of mind, keep two or three backup copies of the data on your disk, rotating them by reusing the oldest set for the newest backup. Store at least one in another location to guard against fire or other catastrophic accident that renders both the computer and the backups unusable.

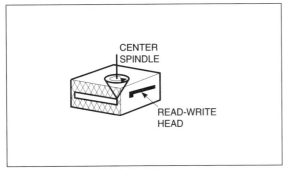

Figure 5-16. Typical Diskette Drive

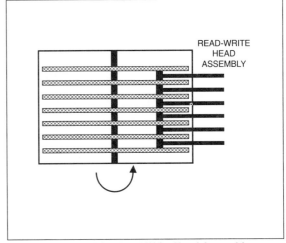

Figure 5-17. Hard Disk Read-Write Head Assembly

Because of the large quantity of data on a hard disk and the fact that it is vulnerable to physical or logical damage, it is important to make frequent backup copies of the contents on some other medium such as diskettes or tape. Note that a second copy of the files on the same hard disk gives minimal protection!

MAGNETIC STORAGE MEDIA AND DEVICES: TAPE

Tape, the oldest magnetic storage medium used on computers, is rapidly falling into disuse except for specialized purposes such as hard disk backup. Its greatest disadvantage is the sequential nature of data access: getting information at the end of the tape means reading the entire tape to get to it. Diskettes or hard disks, by contrast, allow direct access to the desired data by using its track and sector address.

Tape is available in various formats: reels about 8 inches in diameter, which are used extensively on minicomputers and mainframes; cartridges the size of a large paperback book; and cassettes similar to audio cassette tapes.

OPTICAL STORAGE

Optical storage technologies use laser light to burn pits in a disk to encode zeros and ones. The procedure is similar to that used to record music on compact disks (CDs).

Optical storage is seemingly permanent; not only has no degradation of information been observed, but there is no physical reason why it ever should deteriorate. The information is also invulnerable to magnetic fields and other emi (electromagnetic interference). These characteristics make optical storage ideal for archiving information that for legal or other reasons must be kept for long periods of time.

The earliest optical storage units were write-once (non-erasable) since no practical way had been invented to erase the disk. However, capacities were so large (often over 1 gigabyte, or one billion bytes!) that revised information could simply be written elsewhere on the disk. Now some optical storage units can be erased and rewritten a limited number of times. These are actually magneto-optical devices, with the laser used to heat up a small area of the disk before a magnetic field is imposed. Cooling fixes the orientation of the magnetic field and makes it much more resistant to erasure than conventional magnetic media, since a substantial amount of heat must be applied before the field can be changed.

 Because of its extremely large storage capacities, optical storage must be used with retrieval software that is capable of finding stored information in a reasonable time.

Typical uses for optical storage technology include large information banks such as catalogs, government regulations, and videodisc training programs that store realistic sequences of pictures for teaching purposes.

Peripherals

The term "peripherals" refers to a large class of auxiliary devices that may be connected to a computer. Most peripherals are used either for data storage or for input or output. Although some are virtually essential (there are few computers without keyboards and screens), many peripherals are optional adjuncts to the operation of a computer.

INPUT DEVICES: KEYBOARD

Computers originally dealt only with letters and numbers as input, so the keyboard is the oldest and most widespread input device. In addition to the basic typewriter arrangement of letters and numbers plus shift key, tab, and so forth, computer keyboards typically contain additional specialized keys. Control (Ctrl) and alternate (Alt) keys are used in conjunction with letter or number keys to issue commands from the keyboard to programs that are running on the computer. For example, Ctrl-C is a key combination often used to "break" or interrupt a program run and return control to the operating system. Function keys (typically ten or twelve) may have frequently used operations assigned to them by various programs one might run. For example, F1 (function key 1) might be used to get on-screen help, F3 to undo the last operation, and F10 to save the file. Cursor keys are used to move the cursor (a blinking line or rectangle indicating the currently active location on the screen) under keyboard control. Additional keys for insert, delete, page up, page down, and other operations help one move around in files during editing. The numeric keypad found on many keyboards is a convenience if many numbers must be entered, because the hand can stay in one location (typically at the right-hand side of the keyboard) rather than having to move back and forth on the top row. A typical computer keyboard is shown in Figure 5-18.

For specialized applications, such as operator stations, keyboards that do not have the letter and number keys of the typewriter keyboard may be used. Instead, they have keys that correspond directly to tasks the operator will want to perform, such as "alarm acknowledge" or "display chart."

The keyboard communicates with the operating system by issuing an interrupt for every key press. It normally has its own connector to the system unit and so does not occupy an I/O port.

INPUT DEVICES: MOUSE

The mouse, developed by Doug Englebart of Xerox in the 1960s, is perfectly suited for operations like pointing, drawing, and selecting. Programs that involve picking objects on the screen, choosing entries from menus, or manipulating graphic objects often make extensive use of the mouse.

A typical mouse translates the motion of its underside on a flat surface into cursor movement on the screen. The principle may be mechanical (a roller or ball whose rotation translates into cursor movement) or optical (involving reference to a special gridded pad on which the mouse must be used).

A mouse may have one, two, or three buttons. A simple press and release (called a "click") normally selects an item; a click and hold operation is used for dragging an object around the screen or sometimes for moving down to the next level of a menu structure. For a mouse with fewer buttons, double clicks or clicks

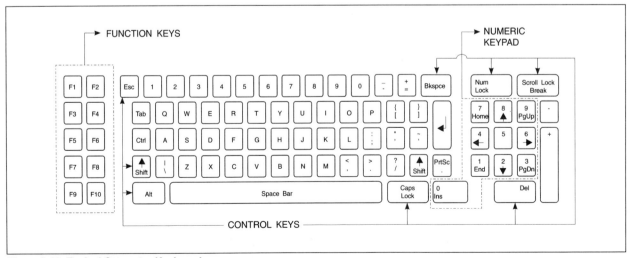

Figure 5-18. Typical Computer Keyboard

combined with keystrokes are often used to extend the range of operations that can be performed.

A mouse may either be connected to the rest of the computer through a serial port (serial mouse) or else have its own interface card and connector, which hooks directly onto the system bus (bus mouse). It often requires a device driver that must be loaded at system start-up.

The tradeoff between an external (serial) or internal (bus) mouse is the same as for other devices such as modems for which these two options are offered. The external device occupies one of the serial ports, may require an external power supply, and contributes to the rat's nest of wires and cables that seems to form around any computer. Fortunately, it can be easily moved from one computer to another, replaced with a newer model, or removed for servicing. The internal device is a board that occupies a slot in the computer chassis and cannot be removed without opening up the computer case.

The mouse is vulnerable to clogging with dirt and grime and so is less often found on the plant floor. It is, however, an essential part of computer-aided drawing (CAD) workstations, and an increasing number of general-purpose programs are now able to recognize mouse input.

Other pointing devices such as light pens (used directly on a computer screen to pick locations using a beam of light) or trackballs (a ball mounted in a holder that can be manually spun to move the cursor) are still used, but none has achieved the widespread popularity of the mouse.

INPUT DEVICES: BIT PAD

The bit pad is a sort of programmable slate that sits flat on a tabletop and is used with a pointing device. The bit pad is divided into regions, each associated via software with a certain computing action (run a program, display a graphic, etc.). When a particular region is selected with the pointing device, its associated action is executed.

The real usefulness of a bit pad lies in its programmability, which means that the same physical device can be used for many applications simply by changing the program that assigns regions to actions. Bit pads often come with plastic overlays that serve as visual maps of different possible assignment patterns. For example, the same bit pad can be used for electrical or mechanical drawing, with a different repertoire of symbols.

The major use of bit pads so far has been in computer-aided design, where they make the required symbol repertoires readily available.

INPUT DEVICES: TOUCH SCREEN

The touch screen is, in a sense, the ultimate in simplicity for the computer user: touch the screen with your finger and something happens. As with the bit pad, different regions of the screen are assigned to different actions and are visually indicated by shape, color, and text. The visitor information systems at EPCOT Center at Disney World use touch screens.

The principle of the touch screen is electrical; the screen is fabricated with conductive and insulating layers in such a way that when a human finger presses on the screen (which is slightly deformable), an electrical connection is made. Making the screen sensitive to pressure rather than simply to touch improves its robustness under dirty conditions, whether industrial grime or chocolate sauce on children's hands.

One problem with the touch screen is that the human finger is a rather imprecise pointing device compared to a light pen or mouse. This limits the number of choices that can be displayed on a normal-sized computer screen at one time and forces the programmer to deal with questions concerning what happens if

Although the light pen can select screen locations with precision, operator fatigue has limited its acceptance. Holding the pen up so it can be used on a vertical screen has been found to be tiring.

someone presses partly inside and partly outside a region. These problems have limited touch screen applications in industry.

INPUT DEVICES: DIGITIZER

The digitizer is a specialized device for getting graphical information such as maps into a computer. The information is represented as a series of (x,y) coordinate pairs. Typically, the map or other graphic is laid out on the digitizer table, and a stylus or other pointing device is used to trace its contours and enter points one by one.

INPUT DEVICES: OPTICAL SCANNER

A more recent and more widely applicable method of entering graphical information into computers is the optical scanner. A scanner takes a black and white paper original and represents it as a set of ones and zeros that correspond to little black and white squares. This kind of image is often called bit-mapped, since the information in the drawing has been "mapped" into a sequence of black and white squares or binary bits. A difficulty with scanning technology is that the resulting image takes up a great deal of disk space if high resolution is desired. Text scanning is even more difficult, since the computer not only has to read in the image but must also decide whether the shape it is seeing is an A, a B, or whatever. Different type fonts or poor quality printing sharply reduce the performance of text scanners, but the technology is developing.

OUTPUT DEVICES: MONITORS (SCREENS)

The monitor or screen is the computer's usual means of communicating with the user. The term "monitor" probably comes from the fact that the user monitors, or keeps track of, what the computer is doing by watching the display on the screen. Screens vary in size, resolution, color and graphics capabilities, and the technology used to produce the screen image.

DISPLAY TECHNOLOGIES: THE CATHODE RAY TUBE (CRT)

The CRT is the oldest and most commonly used display technology in use in the computer world. The principle of operation is similar to a television set in that a tube similar to a picture tube projects dots of colored light onto different parts of the screen to form letters and other images. However, the process by which information is transferred from computer memory to a screen image is different from the reconstruction of a television picture from a broadcast signal, as will be explained shortly.

Monochrome (one color) monitors display either one color (most commonly yellow, green, or white) on a black background, or else black on a white background. The color is determined by the chemical makeup of the phosphor, a substance that coats the inner surface of the CRT and generates the display by glowing when excited by light.

Color monitors have three phosphor dots for each point (or pixel, for a picture cell) on the screen that can be activated by the scanning beam. Each of these dots will glow red, blue, or green when excited by light. Since these are the three light primaries, other colors can be produced by exciting more than one of the dots for a given pixel. More shades of color can be generated if the intensity of the dots can be varied.

As in a television set, the phosphor dots are activated by a light beam that scans the screen in horizontal lines, moving from top to bottom (see Figure 5-19). Scan time or refresh rate is a measure of how long the beam takes to make a complete pass over the screen. Scan time is important because the excited phosphor dots stay bright for only a limited time once the beam moves past them (this time is called the "persistence" of the phosphor). Display quality therefore involves matching the scan time with the characteristics of the phosphor. If the scan is too slow for the phosphor, the display will flicker because some phosphor dots will have a chance to grow visibly dim before the beam comes back to recharge them. A high-persistence phosphor, however, will leave a "ghost" image on the screen for a noticeable time after the light stimulus is removed, which is annoying to some users. The faster scanning needed with a low-persistence phosphor is more expensive because the electronics and control circuitry for the beam must have a faster response time.

Resolution refers to the number of display points on the screen and is usually described by two numbers (e.g., 640 by 320) giving the number of points in the horizontal and vertical directions. The two numbers are normally different because a typical CRT screen is not square. On a low-resolution display, you can see the individual dots, and text and graphics look jagged and clumsy, like a newspaper halftone photograph. A high-resolution screen will give you a more readable, lifelike, and visually pleasing image, more like a good photographic print.

The "standard" CRT screen is 14 inches from corner to corner, though smaller sizes are seen on some portable computers and larger (usually 19 inch) ones on workstations for CAD and similar applications.

DISPLAY TECHNOLOGIES: THE LIQUID CRYSTAL DISPLAY (LCD)

When a computer must be small or portable, CRT technology becomes difficult to implement. The tube cannot be compressed front to back beyond a certain point without compromising image quality (as the scanning beam must move through a wider angle, the dots near the edge become elliptical instead of circular). Moreover, CRTs are highly subject to impact damage or breakage. Therefore, many portable computers use LCD technology instead.

LCDs are widely used in calculators, digital watches, and instruments as well as in computers. The screen consists of a sheet of crystalline material sandwiched between two sheets of glass. The crystals can assume two shapes: one when excited by an energy input, the other when unexcited. In their unexcited state, the crystals reflect most of the incident light and appear pale gray; when excited, they absorb light and appear black (see Figure 5-20).

One advantage of LCDs is that the amount of energy required to make the crystals change shape is much less than that needed to turn on a dot on a CRT phosphor display. The lesser energy input, however, plus the fact that the LCD is merely reflecting or absorbing the ambient light rather than itself emitting light, means that LCDs can suffer from poor contrast, especially in marginal lighting conditions or when viewed at an angle. More recent designs use improved crystals or backlighting to obtain more contrast.

Some portable computers with LCD displays include a port through which a second, better-quality screen can be connected to the computer when it is being used in a fixed location such as in an office.

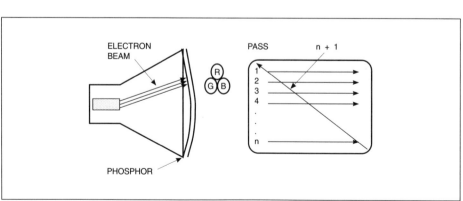

Figure 5-19. CRT Technology

DISPLAY ADAPTERS

The process of translating the binary information in computer memory into letters, numbers, and graphic entities on a screen is taken care of by a display adapter card. This card has two main tasks: putting the information into a form that can be displayed and controlling the actual painting of the pixels onto the face of the screen.

The first task is normally accomplished by means of a virtual display: an electronic image of the screen in the card's onboard memory. If the screen is to display the letter "a" at a particular position on the screen, the appropriate pixels are turned on in the virtual display to form that letter. The adapter then takes the content of the virtual display and passes it through hardware and software that put it onto the screen in the right locations (see Figure 5-21).

The screen and display adapter in a computer must be compatible in certain important ways. The CRT must be capable of responding to the commands issued by the adapter and displaying information at the same resolution as the virtual display the adapter has constructed. For this reason, these two items are often purchased together. The more recently developed "multisync" monitors can decode and implement the signals from several types of display adapter cards.

> Some graphics software will fail to work with a display adapter card that does not include a certain minimum amount of onboard memory. Try adding more memory to a card if problems arise.

DISPLAY STANDARDS

To simplify the job of third-party equipment suppliers and application programmers, a number of display standards have evolved in recent years, particularly for color. The color graphics adapter (CGA) standard was the color option on the original PC in the early 1980s, but its poor resolution (320 by 200 with four colors, 640 by 200 with two) made it unacceptable for many uses. The extended graphics adapter (EGA) standard soon offered 640 by 320 resolution, which was a big improvement for text but was still inadequate for many graphics applications. The video graphics adapter (VGA) standard offered somewhat higher resolution and a larger color palette and has largely superseded the CGA and EGA. Unfortunately, VGA has become less of a standard as different manufac-

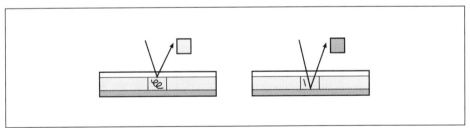

Figure 5-20. LCD Technology

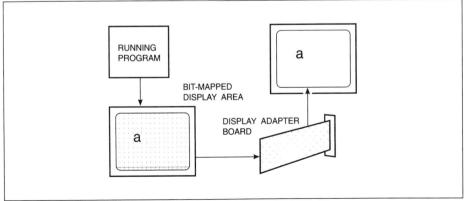

Figure 5-21. Screen-Computer Interface

turers have chosen different ways of enhancing and modifying it; therefore, terms such as "enhanced VGA" and "super VGA" change their meaning depending on context.

Special high-resolution displays are common for applications that rely on detailed graphic displays such as CAD (computer-aided design) or operator stations for process control.

HARDCOPY OUTPUT DEVICES

The two principal means of getting hardcopy (something on paper) out of a computer are printers and plotters. Printers are traditionally for text and plotters for graphics, although laser printers produce excellent graphics also.

The most accepted printing technologies today are dot matrix and laser printing.

DOT MATRIX PRINTING

Dot matrix printing is an impact printing technology in the tradition of line printers long used with mainframe computers. The print head consists of an assembly of movable fine wires perpendicular to the paper. By making different combinations of wires strike an inked ribbon, characters are printed on paper. The resulting characters are readable but obviously made up of dots, especially with print heads with fewer wires (7×7 or 7×9). Better quality can be obtained (at slower speed) by making more than one pass (doublestrike) over the same line of text with the print head slightly offset each time, or by specifying boldface printing. The newer 24-pin dot matrix printers (with a 24×24 wire print head) produce much better quality output. Some drawings can also be printed on a dot matrix printer, but the quality is often marginal and the process very slow because the wires must be activated one by one, whereas to print a letter or number all the requisite wires are moved in parallel.

Typical speeds are 150-200 characters per second for draft (low) quality output and 50-75 characters per second for better quality.

Dot matrix printers are cheap and reliable and are the usual choice for text output where quality is not particularly important.

LASER PRINTING

Laser printing is similar to photocopying, but the image comes from computer memory instead of from a paper original. It produces high-quality text and graphics output at typically 6 to 12 pages per minute. However, laser printers are substantially more expensive (three times the price or more) than dot matrix printers, and they require a continuing supply of print cartridges, ink, and toner. They tend, therefore, to be used where output quality is important, for example, for business correspondence or presentation graphics.

To use anything but basic printer features, the usual software may require access to a printer driver: another small program that translates commands as issued by the software into terms the printer can understand. More and more software packages are now being sold with a large inventory of printer drivers.

PLOTTERS

Plotters produce output by drawing it with pen on paper. They are well adapted to drawings that contain many lines and shapes but not so well adapted to text or to filling in large areas with color or patterns. They are not particularly fast but produce high-quality output and can handle large sizes of output such as engineering drawings, which most printers cannot.

With a carousel of pens, plotters easily produce color output, which is rarely possible with existing printer technology (color laser printers are only now becoming widely available). With appropriately formulated inks, they can also draw directly onto film or transparencies.

Useful features for a dot-matrix printer include the ability to handle both tractor feed (the "computer paper" with the holes down the side) and friction feed (for single sheet paper such as business letterhead) and easy front-panel selection of features such as boldface or compressed printing. With older printers, control codes may have to be sent from software to accomplish anything but standard printing.

Physical Assembly of a Computer

The electronic components of a computer are normally mounted on printed circuit boards, which are then mounted in a metal chassis. Devices such as disk and tape drives may also be mounted in the main chassis (internal devices) or in their own case sitting outside the computer; they may be connected to the rest of the computer by a cable (external devices).

Chips

The components of a computer that perform logical, arithmetic, or input-output operations (microprocessors, specialized processors, or device controllers) or short-term data storage (memory) are physically implemented as chips. These are preconfigured assemblies of microelectronic elements, such as registers, switches, and gates, in a small plastic case with two rows of conductive metal pins extending out the bottom. To mount the chip on a board, the pins are gently pressed into a matching socket.

Specialized Processors

In addition to its general-purpose microprocessor, a computer may contain one or more additional processors that are designed to perform certain tasks more quickly and efficiently than the general-purpose CPU. *The term "coprocessor" is often used for a processor that runs in tandem with the CPU and takes over certain tasks whenever they arise.* A math coprocessor, for example, goes into action whenever a program performs floating-point arithmetic. The calculation is sent to the math coprocessor, which has an instruction set optimized for fast calculations, instead of to the general-purpose CPU. String coprocessors (for character string operations such as searching and splitting) and graphics coprocessors (to handle graphic entities such as lines and geometric shapes) also exist for specialized applications.

Other specialized processors work less intimately with the CPU but also take over particular kinds of work that they are designed to do well. Array processors excel at matrix computations, usually by processing the rows or columns of the matrix in parallel; signal processors perform filtering, averaging, enhancement, and other operations on incoming signals from other devices; video processors refresh and modify screen displays efficiently.

Boards

A printed circuit board consists of a rigid, nonconductive board (usually some form of plastic or fiberglass) onto which a pattern of conductive traces has been imprinted. Other devices such as resistors and batteries may be soldered onto the board as needed. Chips may also be soldered in place but increasingly are provided with sockets to allow for easy removal and replacement if necessary. Edge connectors and cable connectors allow interconnection of the board with other boards or devices.

The term "motherboard" is often used for the board on which the microprocessor (CPU) resides. Typically, the system clock, a certain amount of memory, and possibly some device controllers are also found on this board (see Figure 5-22). Additional boards (or cards) are used to hold devices that will not fit on the motherboard or to provide specialized capabilities for particular applications. Frequently the board will come with associated software that resides either in memory on the board or in the main computer memory.

Memory expansion boards are used to augment the available memory in the computer beyond that which can be installed on the motherboard. Controller or

If your collection of boards exceeds the space available in your computer, an expansion chassis can provide additional slots. Normally, the expansion chassis communicates with the rest of the computer through a parallel connection, either to an existing parallel port or to an associated board and parallel connector installed in the last remaining slot of the main computer.

adapter boards contain the controller for a drive, a display, or some other device, together with its associated circuitry and electronics. Network adapter cards contain the hardware needed to interface to a network. Data acquisition boards contain hardware and software to collect and process data from a number of process input points.

Boards can also be used to substitute for certain external devices or to eliminate the need of these devices for existing serial or parallel ports. For example, an internal modem card can replace an existing modem running off a standard serial port.

Boards are normally installed side by side or one above another in slots in the computer chassis (see Figure 5-23).

Cautions Regarding the Physical Computer Assembly

Because many computer components are low-voltage, static electricity can be lethal to them. Boards and chips are often shipped in special antistatic packaging to protect them from static electricity. When installing or removing chips or boards, take care to remain grounded and wear clothing and use tools that are unlikely to generate or conduct static electricity, especially in dry environments such as heated offices. Special clips, mats, and similar equipment are available if needed.

The pins on electronic chips are small and delicate and are likely to bend or break if forced into or out of their sockets. Specially designed insertion and extraction tools minimize the chance of chip damage.

Boards are not as vulnerable to physical damage as chips, but foil or solder connections can break if the board is bent or dropped.

Hierarchy of Computers

Although all computers share many of the same components, the scope of the computing work they can undertake varies considerably. The computer needed to calculate and correct a space vehicle orbit in real time is different from that needed to monitor and control temperature in a small mixing vat.

Moreover, the capabilities of a given type of computer change over time with changes in available technology and the selection of other computers on the market.

A mainframe was the original "big computer," once the only kind of computer in existence. They have become smaller but more powerful over the years and now typically serve the data processing needs of large organizations.

The supercomputer has been developed recently as a special-purpose "big computer" focused on high-speed scientific and engineering calculation needs. Using the latest in high-performance hardware and software, these computers push back the frontiers of what is possible in computation, running, for example, gigantic atmospheric models for weather forecasting or wind tunnel simulations for aircraft design.

Minicomputers were developed in the mid-1970s as a cheaper, more accessible alternative to the mainframe. They were initially used by midsized businesses and for purposes such as university research, which welcomed access to computing power free from competition with other users' demands on the central mainframe computer. The first attempts at computerized process control were also made using minicomputers, although their speed, storage capacity, and operating systems generally proved inadequate to this task.

The workstation started out as a computer platform for CAD (computer-aided drafting). It provides the extensive data storage, high-speed computation facilities, and high-quality graphic display needed for drafting and design applica-

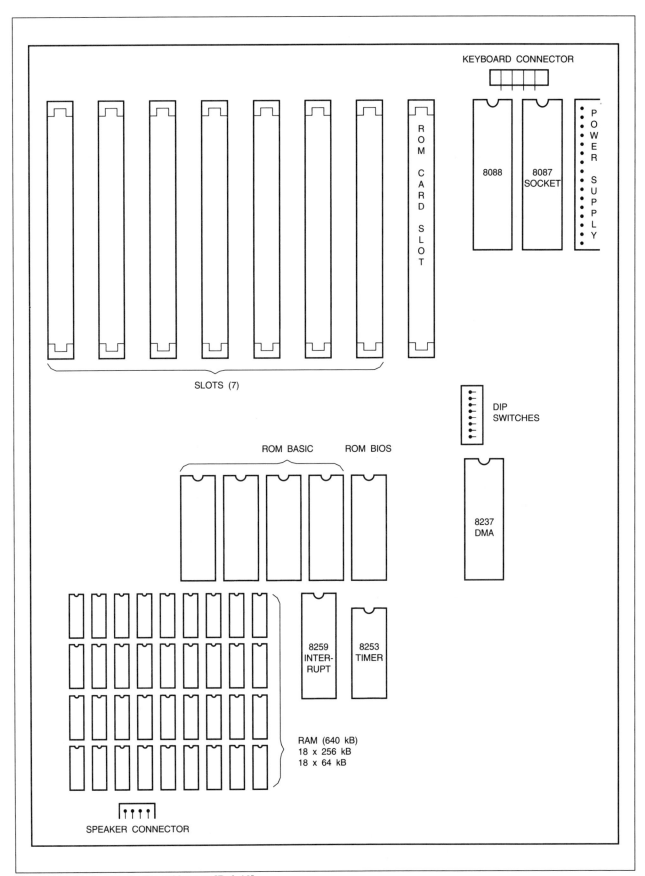

Figure 5-22. Typical Motherboard Layout [Ref. 16]

tions. These endowments have also made workstations useful as high-end computers for a variety of scientific and engineering applications. Workstations are also often used as network servers or for heavy-duty data processing tasks such as managing a large database.

The microcomputer or personal computer (PC) has become the most familiar computer for many people today. The emphasis in its design was to provide low-cost, single-user computing capability and to make it easy to use for people with limited computer experience. With widespread use and acceptance of personal computers, a vast quantity of add-on hardware and software was soon developed, so that now PCs are being used for thousands of different applications. However, large applications or those with specialized needs (e.g., fast real-time control) may find that the microcomputer cannot accommodate them.

The portable computer, an outgrowth of the personal computer, was designed to meet the need for moving a computer around: between home and office, on the road, or on the job site. "Portable" can be a relative term, ranging from suitcase models that are luggable with difficulty, through the "lunch box" size, to notebook-sized laptop models. To achieve portability, compromises have been made in display size and quality, disk storage space, and customizability (optional boards cannot normally be installed). Many portables include a modem as standard equipment so data can be electronically transferred directly to and from a larger computer or over a network.

Many of the above-mentioned types of computers have been adapted for industrial use. Since most computers are not designed to stand up to the rigors of the industrial environment, they must be either "industrially hardened" by redesign or use of special components (e.g., waterproof keyboards and heavy-duty fans), or enclosed in an environmentally controlled room. Additional input-output and communications capabilities and a real-time operating system also may need to be added. Special types of computers have also been designed for industrial use, of which the best known is probably the programmable logic controller (PLC) discussed in Chapter 9.

Computer Operating Environment

The electronic components of digital computers are designed to be operated by electrical current that meets certain specifications and under certain conditions of temperature, humidity, and cleanliness. Failure to respect these conditions may cause components to fail or function erratically.

General-purpose computers are intended for use in homes and offices, where the environment offers few threats to electronic components. Their design, therefore, provides minimal protection against adverse conditions. One way to increase the level of protection is to change the design of the computer to incor-

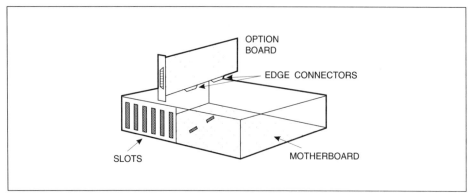

Figure 5-23. Physical Assembly of Computer

porate more rugged components. An extreme example of this is the "battle hardening" of computers in military equipment. Water- and dust-tight cases, shock-absorbent mountings, high-temperature electronics, and other such measures can improve the odds of the computer's survival in a hostile environment.

This approach, however, entails a substantial increase in cost, especially since solving some environmental problems may create others. For example, sealing up the outer case of the computer to exclude water, dirt, and harmful vapors will also cut off air circulation, resulting in excessively high temperatures inside unless large heat sinks and cooling systems are added.

For this reason, many industrial computer installations use the approach of enclosing unmodified general-purpose computers in a protective room where an environment suitable for the computers can be maintained (see Figure 5-24). However, the latest distributed control systems are designed for use in most industrial environments.

It should be noted that for programmable logic controllers (PLCs) the situation is somewhat different. PLCs were designed from the outset to function on the plant floor, and their environmental requirements are, therefore, less stringent. This does not mean, however, that PLCs do not need any protection. They are vulnerable to the same environmental threats as general-purpose computers, only somewhat less so.

The rest of this section discusses specific environmental threats to computers and how to protect against them.

Dust and Dirt

Dust and dirt will quickly form a blanket over all accessible surfaces in a computer. The layer of dirt may prevent physical contacts from closing (as in keyboards) or may clog ventilation and other openings. If the dust is conductive, it may cause short circuits; if insulating, it may lead to heat buildup and component failures.

The obvious preventive for dust infiltration is to minimize the number of openings through which dust can enter. This approach is taken in many PLCs and industrially hardened computers, although it may create heat dissipation problems. The other approach is to filter the air, either at the inlet to the computer room or as it goes in through particular openings in the computer case. This method works well if filter media are appropriate to the contaminants to be guarded against and are changed regularly. Another simple though possibly unpopular measure is to ban smoking in the vicinity of computers.

Chemical Vapors

Chemical vapors, if corrosive, can eat away at contacts and conductive traces inside a computer, eventually breaking their electrical continuity. They can also damage data recording surfaces on disks.

The usual defense against chemical vapors is either to make the computer or the control room tight against the vapors in question or to install appropriate filters. It may be necessary to set up the computer room at a distance from sources of chemical vapors to ensure they remain at a sufficiently low concentration.

Water

Water can get into a computer in two ways: by direct splash (installation near process equipment using water or accidental spillage) and by condensation. Obviously, water can cause catastrophic short-circuiting inside a computer, and a chronic water problem can also lead to corrosion and failure of components. The operating specifications of most computers include a humidity range within which the computer is designed to function.

Designing an entire computer to be watertight is difficult because of heat dissipation requirements, but watertight keyboards or key pads are frequently used on the plant floor and are connected to a computer nearby in a protected room. Keeping coffee cups and other sources of non-process liquids at a safe distance from computers is a sensible preventive measure.

Condensation is usually the result of low temperature or extremely high humidity in the ambient air. If the problem is local, a heater may be used to raise the temperature, which lowers the relative humidity for a given air moisture content per unit volume. In an environmentally controlled room, a humidifier or dehumidifier may also be used. An air conditioning system will control both temperature and humidity.

Vibration

The vibration that often accompanies operating process equipment can break contacts and solder joints inside computers and cause malfunctions and data loss in disk drives. The simplest remedy is to isolate the computer from the vibration source, either by moving it away from the vibrating equipment or by mounting it on a shock-absorbent base.

Electrical Disturbances

Electronic computer components require a relatively unvarying supply of low-voltage, low-current electrical power. Standard distribution voltages (110-120 volts or greater) must be stepped down to supply these components, usually by a

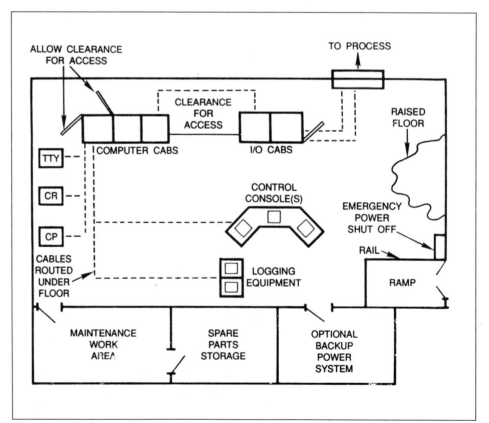

Figure 5-24. Typical Control Computer Installation [Ref. 28]

stepdown transformer (often called the computer power supply), which outputs the voltage used on the system bus (typically 5 volts). Further voltage reductions may be required for specific components.

VOLTAGE AND CURRENT TRANSIENTS

An overvoltage or overcurrent condition, if severe or prolonged, will simply burn out electronic components. Less severe transients may be interpreted as a spurious change of state, causing erratic program function or corruption of data. With too low an initial voltage, the computer will simply fail to function, because the voltage will be insufficient to raise electronic devices above their threshold values. A severe voltage drop while the computer is in operation may trigger an unplanned restart (warm boot). The most serious condition is fluctuating low voltage, which may cause repeated partial restarts, leaving memory and data in unpredictable states and possibly hanging up the computer altogether. The effects of a complete power loss are actually less drastic, except that the contents of volatile RAM are lost and the contents of a file may be corrupted if a write operation is in progress when the power loss occurs.

> One common source of voltage and current transients is a "dirty" plant electrical supply. Isolation transformers and other power conditioning devices will prevent most power line instabilities from reaching your computer.

INTERFERENCE

Noise in the form of radio frequency interference (rfi) or electromagnetic interference (emi) can be harder to track down and correct. Radio frequency interference arises most often from lengths of cable acting as antennae or from mobile communications sources in the plant such as walkie-talkies and lift truck radios. Electromagnetic interference can come from any piece of equipment that generates an electrical or magnetic field (most commonly, motors). Do not neglect to check possible temporary sources of interference; arc welding is a notorious culprit. Note also that computer components such as cables, long communication lines, or disk drive motors may themselves generate interference. Normally, however, computers are required to meet government specifications for rfi and emi emissions so that they do not interfere with the operation of other electronic equipment.

The effect of electrical interference is similar to that of fast transients from the power line: the generation of spurious signals that the computer mistakes for data. These may cause erratic operation, or in extreme cases, total computer failure.

The fundamental remedy for interference is to move the computer away from the source. This may not be easy, especially if the source is the machine the computer is controlling or if the source is difficult to identify. If either is the case, careful shielding and grounding of cables and cabinetry is the best defense. Pay special attention to cable penetrations where ambient electrical noise can leak in. Use of noise-resistant technology such as fiber optic cables can also make for a more robust computer system.

POWER FAILURE

If the continued operation of a computer is critically important, an uninterruptible power supply (ups) should be added to the incoming power line. During normal operation, the line power charges a battery, which, in turn, feeds the power supply and the computer. In case of a blackout, the power stored in the battery can maintain the computer in operation for a limited time, typically one to several hours. If all that is needed is time to copy the contents of memory to disk and shut down the computer in an orderly fashion, twenty minutes to half an hour should be enough; otherwise the cost must be balanced against the importance of the process and the normal length of power outages. The wattage rating of the ups should be sufficient to support the memory, display, and disk drives (typically the biggest power consumer).

STATIC ELECTRICITY

 When working on electrical or electronic components of a computer, take care to avoid connecting devices to inappropriate power sources. A serious but not as obvious threat is static electricity.

A static discharge is typically low current but can reach extremely high voltages that can damage computer equipment. Components such as memory chips and integrated circuit boards are typically shipped in static-proof bags to protect them from accidental discharges during shipment. Once these components are installed, they are protected by the computer's grounding subsystem. During installation, however, they are vulnerable, and personnel should take appropriate precautions, such as wearing ground straps and static-free clothing and using antistatic mats. Maintaining adequate humidity levels helps to control static electricity.

Heat

Electronic components are designed to work within an operating temperature range that is stated in their specifications. Outside this range, their characteristics change beyond design limits, their behavior becomes erratic, and their life is shortened. In extreme cases, a fire hazard may be created.

The usual method of cooling computer components is by convection to the surrounding air, often assisted by a fan that draws air over the components. Other devices such as monitors (screens) have no fan and cool themselves only by convection. For this method to work, the surrounding air has to be substantially cooler than the components, which is one reason computer rooms are often air conditioned. Computers have vents to draw cooling air in and out, which must be left unblocked even if they seem to provide a nice shelf for manuals, drawings, and bag lunches.

If convection alone cannot maintain the components within their specified operating range, components with a higher temperature tolerance must be substituted, or a refrigeration system must be added to the computer. Both of these solutions entail substantial extra cost.

Operating Systems

 The computer components discussed in the preceding sections (collectively referred to often as "hardware") would be a pile of useless electronic junk unless told what to do by programs (collectively known as "software"). The most important piece of software running on any computer is called the operating system, the program that coordinates the computer's activities.

Historical Development

The earliest computers had no operating system, since instructions were fed to them one bit at a time by flipping switches on a panel. This soon became tedious, so short sequences of instructions for frequently performed activities, such as recognizing a keystroke or accessing a disk drive, began to be stored in read-only memory (ROM). General-purpose read-write memory (RAM) was still in very short supply (maximum of 64K), so computer designers could not afford the luxury of using up RAM by reading an operating system off disk. Operating systems were generally customized for each computer in order to make most efficient use of scarce memory. This meant that other programs also had to be customized for each machine, and, therefore, it was difficult to distribute software widely.

The desire to be able to move software from one computer to another was a powerful motive force behind the development of operating systems that would run on a variety of computers. However, many operating systems require that the computer use a particular microprocessor, a situation that persists even today. CP/M, the first truly portable operating system for 8-bit computers, required an Intel 8080™ or 8085™ or Zilog Z80™ microprocessor. The various versions of PC-DOS™ expect to use one of the Intel family of microprocessors (8086, 80286, 80386, et al.), and Apple® operating systems require a microprocessor belonging to the Motorola 68000™ family.

As memory became cheaper and more plentiful, operating systems were divided into a small ROM-resident component (often called the BIOS or "basic input-output system") and a larger component that was read from a storage device (usually diskette or hard disk) into RAM. Typically, the system ROM today contains only enough of the operating system to enable the computer to find the disk files that contain the rest of the operating system. When the operating system does this, it is, in a sense, "pulling itself up by its own bootstraps"; hence, starting a computer is often referred to as "booting up."

Some computers still keep their entire operating system in ROM for particular reasons, such as protection from accidental overwriting (important in many real-time applications) or ease of field upgrading (a simple chip replacement instead of a possibly complex software installation).

Functions of an Operating System

In a typical computing environment, programs and devices compete for access to system resources such as the CPU, memory, system bus, and I/O ports. Allocation of resources is, therefore, one of the operating system's most important tasks. The operating system must continually make decisions about which program or device gets to use a resource and for how much time.

The second major task of the operating system is managing the file system. Since all programs and data are stored as files, files are constantly being created, deleted, modified, stored, and retrieved during normal computer operation. The operating system must make sure that pointers to each file are properly set so the information in that file can be accessed, that enough space for each file is set aside on a storage device, and that any errors in file access operations (such as asking for a nonexistent file) are properly handled.

The operating system must also recognize and deal with a wide variety of unpredictable events that may affect the functioning of a computer. Some of these events are exceptional and undesirable (for example, a sector on a disk becomes unreadable due to damage to the magnetic medium) while others are normal and ordinary (a key is pressed or a printer needs more characters to print).

Some of the operating system's interactions with an application program are diagrammed in Figure 5-25.

ALLOCATION OF RESOURCES
The complexity of this task depends, among other things, on whether the computer has only one program running at any given time (single-tasking) or normally has two or more programs running at once (multitasking). In a single-tasking computer, the only allocation decisions to be made are between the currently running program and any devices that may require the attention of the CPU. The operating system normally attends to the needs of the currently running program. Whenever a device requests the attention of the operating system, the currently running program is paused and the operating system deals with the needs of the device, returning afterwards to continue execution of the program.

In a multitasking computer, the operating system must include a scheduler, which determines which program gets to execute next. Access may be "first come, first served" but is more often granted according to the importance of the various tasks, with the proviso that no task should be kept waiting too long. Normally, each program or device is assigned a priority value that indicates the extent to which that program or device should be able to grab resources away from other tasks. When interactive programs are involved, not keeping the human user waiting too long will also be a consideration.

Some programs whose usefulness depends on instant availability (such as a pop-up calculator or phone book) pre-empt the scheduling function of the operating system, in effect ensuring they always have priority. Programs typically do this by writing values into certain areas of memory that the operating system normally uses for its own scheduling activities. Since these write operations are not controlled by the operating system, conflicts can occur with other programs and devices that expect to use these memory areas, possibly paralyzing the computer's operation. On PC compatible computers, these programs are called TSRs (for "terminate and stay resident") because they copy themselves into memory at the beginning of a computing session and remain there whether or not they are currently running. More recent operating system versions tend to include functions formerly provided by TSRs in order to bring these activities under the control of the operating system and allow it to manage them in a consistent manner.

FILE SYSTEM

The file system is that part of the operating system that takes care of accessing files and making use of the information they contain. A file is simply a block of information stored on disk with a name and address that enables the operating system to retrieve and use it. The form of the file name may indicate what type of file it refers to.

There are two main types of files: program files and data files. Program files (executables) contain the actual machine instructions that the computer fetches and executes one after the other in order to run a program. They are not usually stored in a form that is readable by human beings; trying to display the contents

> Normally, the commands that load memory-resident programs are stored in a start-up file (AUTOEXEC.BAT on PC-compatible computers). Some programs are particular about where in the sequence they are loaded; check the documentation and try changing the order if you have problems. Another debugging strategy is to remove the memory-resident programs one by one from the start-up file until the problem disappears.

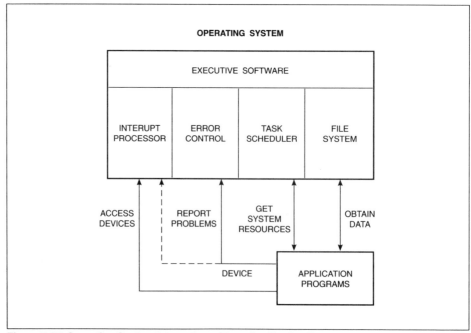

Figure 5-25. Operating System Interaction with an Application Program

of an executable file on the screen usually results in a mass of gibberish. An executable file may also contain some header information that indicates, for example, how much memory it will require to run or whether special resources such as a math coprocessor are needed.

In some cases, programs will have source and object files. Source files contain the programs as they were written by a programmer in a human-readable programming language such as Basic, Fortran, Pascal, or C. Programs in this form are also referred to as source code. These human-readable program statements are, however, not understandable to the computer and must be either interpreted or compiled to be useful. An interpreter (yet another program) goes through the source program and converts it one statement at a time to machine-readable instructions, which are executed immediately. A compiler (another program) converts the entire block of source code to a block of machine-readable instructions, which are stored as an executable file. Compiled code generally runs faster than interpreted code but allows less immediate interaction with the program while it is running.

Larger compiled programs that call subroutines and predefined functions usually go through an intermediate step in their conversion to machine code. The compiler in this case produces a file of object code (similar to machine language instructions except for references to memory addresses that must be resolved among the main program, subroutines, and functions so they all expect to find the same information in the same place). A second program (called a linker) sorts all this out and produces a single executable file from the multiple object code files produced by compiling programs and subroutines.

Data files are the files from which running programs will obtain stored data (of course programs may also obtain information from the keyboard or other external devices). These files may or may not be directly readable by human beings. Data files written using ASCII code (a set of character codes commonly used for data transmission) are used by many programs and are human-readable. Many programs (such as spreadsheet and database programs) have their own data file formats, which contain not only the data itself but information about how that data is to be manipulated or displayed. Such files, not directly human-readable, can be read only by means of the program with which they were intended to be used.

The allocation of physical disk space to files is done by means of sector and track addresses. On a given disk platter, data is stored in concentric circles around the disk hub. The track address tells the computer which circle contains the file wanted. To allow access to smaller storage units than an entire circle, the disk platter is divided into sectors, like pie slices. The sector address indicates which sector of the specified track contains the beginning of the file. With some operating systems, the sector size is fixed; with others the user can choose the sector size, within certain limits, when initially setting up (formatting) the disk.

If a file takes up more than one sector on a track, the rest of the file is written, if possible, into the immediately following sectors on that track. If one or more or those sectors is already in use and the whole file cannot be written in the same place, the rest of the file is written elsewhere and a pointer is set to establish the connection between the two pieces of the file. If the rest of the file will not fit into the second piece of storage space, this procedure is repeated, so a large file can end up being stored in several physical locations on the disk, with pointers from one to the next.

The file allocation table (FAT) maintains a record of where each file is on the disk. To read or write a file, the disk controller obtains its track and sector address from the FAT and moves the read-write heads to the appropriate location (see Figure 5-26). If the file is split among two or more physical locations, the read-write heads must move to retrieve each piece of the file.

A smaller sector size requires more bits for the sector address (this is usually the limiting factor on how small a sector can be) but allows more files to be stored with less wasted space. A file consisting of a single byte will still require one sector of storage space on disk, because the disk controller cannot find a unit of space smaller than one sector.

Disk maintenance programs, either included with the operating system or purchased separately, can be used to clean up any misallocations of disk space. It is important to do so regularly because a disk backup made from a disk containing cross-linked files may not be usable later to restore the disk. These same programs can often detect bad sectors, move the data in them (if any) to a safe area, and mark them as bad to prevent future reuse.

If many files are split into pieces, average disk access time becomes longer because of the extra head movement needed to retrieve portions of files. The disk is then said to be fragmented. Commercially available programs can be used to rearrange the files on the disk so they are stored as much as possible in contiguous locations, restoring disk access times to normal values.

Physical damage to the disk can cause problems, needless to say, in accessing files in the damaged area. The FAT itself is written on a special track on the disk, and the worst problems result if this track is damaged. In the most serious case, the entire disk may become unreadable. If the damage occurs instead in the area occupied by one or more files, those files may become unreadable.

File access problems can also result if the information in the FAT becomes corrupted. Sometimes the same area of the disk ends up being assigned to two files; in this case, part of one of the files may be overwritten or become inaccessible (the files are then said to be "cross linked"). Pieces of files that have become inaccessible due to corruption of the FAT information needed to access them are referred to as "orphan clusters." They are unlikely to cause problems accessing the remaining files, but they take up space needlessly on the disk.

To salvage a disk that has for one reason or another become unreadable, reformat the disk and then restore the files from a recent backup (which one should always have, of course). In this case, the only work lost is that done since the last backup.

INTERRUPTS

As mentioned earlier, one of the tasks of the operating system is to deal with unpredictable events. The occurrence of these events is signaled to the operating system by means of interrupts. These are codes sent by various devices on dedicated interrupt lines in the system bus, which tell the operating system to suspend what it is doing and deal with some situation that has arisen. Different values of the code correspond to different events.

Not all interrupts are generated by abnormal or unusual events. An interrupt is generated every time a key is pressed, an input device sends data, or a communications buffer is full. Most interrupts seen by the operating system are, in fact, of this mundane variety. Of course, interrupts are also produced by more serious problems such as an unreadable disk sector or an attempt to read from or write to a nonexistent device. The various interrupt codes are assigned different

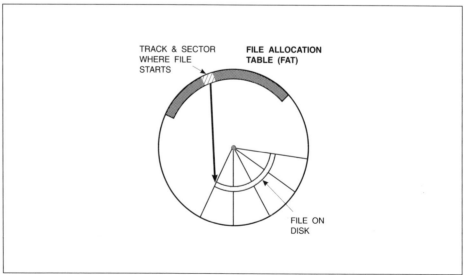

Figure 5-26. Accessing a File on a Disk

priorities so that, for example, inability to read a disk is dealt with before the next character is accepted from the keyboard.

To illustrate how an interrupt works (see Figure 5-27), consider how a key press is handled by the operating system. When a key is pressed, an interrupt is generated that tells the operating system that a piece of input data needs to be processed. Upon receiving the interrupt code, the operating system completes execution of the current instruction, then stores the current state of the CPU, and suspends program execution. Because input data can come from a number of devices, the operating system then sends a request for the keyboard to identify itself. The keyboard does so and the operating system responds with an "OK-go ahead" message to the keyboard. The keyboard then sends the character corresponding to the key press. The operating system processes the character, then restores the stored CPU state, and continues running the program from the point where the interrupt was received.

Special Needs of Real-Time Operation

The tasks just described (allocation of resources, management of the file system, and processing of interrupts) are common to all operating systems. Real-time applications (in particular, process control) impose additional requirements on the operating system. To meet these needs, special real-time operating systems (RTOS) have been developed.

An industrial process must remain under control at all times. This means that the process control computer and its operating system must be highly robust, continuing to function in situations where other computers would "hang up" and have to be rebooted. It would be unacceptable to have to bring up a plant from a cold start because of computer problems! All contingencies must be provided for, including multiple device failures, conflicting or missing data, and communica-

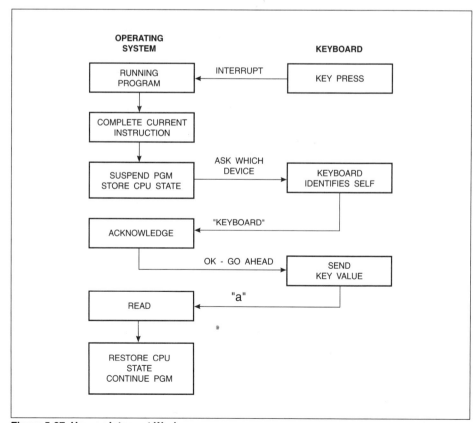

Figure 5-27. How an Interrupt Works

tions breakdowns. Moreover, if a process control computer does go "down," it must fail gracefully, leaving the process in a stable and safe default condition and permitting a smooth transition to manual operation. The state of the system must be periodically saved and documented so that the state of the process at any given time can be reproduced and the development of a problem condition can be traced.

Computers that deal with industrial processes (especially for control) must be fast enough to keep up with the process in order to perform their assigned tasks as fast as the process gives them work to do. For a simple sampling or monitoring job, this may not be too demanding; but, if complex data manipulations must be done, control calculations performed, or elaborate displays or reports generated, time can be all too short! Speed has a number of components, not only the cycle time of the processor itself but also data transfer rates, disk access times, and so forth. A high enough sampling rate must be maintained to provide an accurate picture of process dynamics and feed good enough data to any mathematical or statistical models that are used. Speed is, of course, a consideration for any computer system, but on most personal computers the limiting factor is simply the patience of the user.

Most computer programs run solely with reference to the microprocessor instruction cycle, without taking any notice of external "real-world" time. In real-time applications, on the other hand, time is a critical variable that must be explicitly represented and tracked. Every piece of information has an expiry time and must, therefore, be time tagged; every operation must check the recency of its inputs before using them and get new values if the existing ones are too old. Many operations make more explicit use of time: an action needs to be taken every so often, or only after another action has had time to take effect, or only if nothing has happened in a specified time interval.

Multitasking, the ability to keep several jobs in progress at the same time, is an essential feature of an RTOS. Those of us who like to do several things at once probably already use multitasking in the form of memory-resident utilities (electronic note pads, phone books, and so forth) and multiwindow electronic desktop programs. In a real-time industrial context, however, not only are many tasks running at the same time but their number varies, they may start and stop unpredictably, and their relative priorities may change due to emergencies, deadlines, or trigger values observed during routine process monitoring. Of course, none of the tasks can be simply abandoned but must be continued if at all possible or brought to a safe and stable default condition otherwise. Figure 5-28 illustrates the allocation of processing time among several tasks with different priorities.

Since an industrial computer is often running without human intervention for extended periods of time, many decisions about priorities, scheduling, and job interruption or termination must be made automatically by the operating system. An RTOS must be able to interrupt a job in progress at any time and pick it up later without losing its "train of thought" or any data it has collected and processed—and without disturbing any other task. In particular, it must be able to recognize, accept, and process unscheduled inputs, doing other work in the meantime—an ability that most general-purpose computers lack. The operating system must also avoid getting stuck on one particular task while failing to pay attention to the rest of the process. Built-in "time outs" and well-designed task scheduling algorithms help prevent this from happening.

Since the bread and butter of an industrial computer application is process data, it is not surprising that communications form an important part of an industrial operating system's workload. As a result, software for input and output scanning, data transmission and acknowledgment, status checking for devices

Real-time computing requirements vary depending on whether the process dynamics are fast or slow. For example, a program may be able to obtain data on slowly changing variables by having the process database program write them periodically to a file that the program can read. If the variables change quickly, the program may need to interface directly with the process database, which is probably more difficult to achieve.

and channels, and diagnosis and correction of communications errors often forms an integral part of an RTOS, as opposed to the personal computer situation where communications software is usually an add-on that is purchased separately from the operating system. More sophisticated capabilities may include dynamic checking of the communications parameters of two or more devices and automatic determination of a set of parameters that are satisfactory to all or reconciliation of contradictory information provided by different devices.

The centrality of process data to the operation of an RTOS also means that the software for building and maintaining a process database and reporting the contents in various forms is frequently included with the operating system.

Parallel Processing

As fast as today's processors are and tomorrow's will be, the speed requirements of real time and other applications always seem to be straining their capabilities. One way out of this dilemma is to use more than one processor and have them working on different parts of the problem at the same time (or in parallel). Parallel processing promises orders of magnitude speed improvement for many computations—if the right conditions obtain.

Parallel processing was invented because of what computer scientists call the von Neumann bottleneck, for the mathematician John von Neumann, who first drew our attention to it. Basically, any processor operates on a cycle: fetch information from memory, perform some operations on it, store the results back into memory. The amount of actual computation done is limited by the fact that the processor cannot be calculating while it is fetching and storing. For a job heavy on calculations, this can, indeed, be a bottleneck!

Then someone had the bright idea that if more than one processor worked on a problem, while one processor were fetching or storing, the others could still be doing useful work, and overall throughput could be increased. Such a setup is called a parallel computer since the processors are working in parallel. Systems have been built with as many as 132,000 processors!

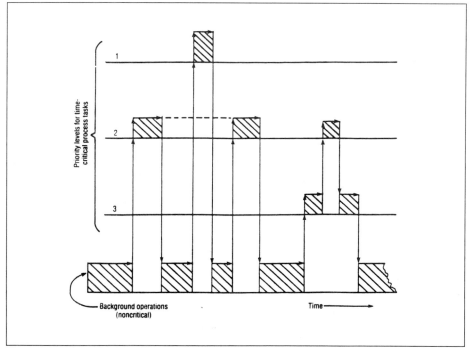

Figure 5-28. Time-Sharing of Tasks of Different Priority [Ref. 28]

Note that parallel processing is different from multitasking, which is another way to get a computer to do several things at once, or at least to appear to. In multitasking, each running program gets a piece—a slice—of a single processor's attention. The computer switches from one job to another so fast that it looks as if all the jobs are running at once—but they aren't. A parallel processing computer, on the other hand, "farms out" a piece of the action to each of a number of processors. Here, the pieces actually are running at the same time.

Like most apparent miracles, this one has a catch. To take full advantage of multiple processors, a program must consist of a number of pieces that can run independently (i.e., not needing each other's results). Most existing programs are not written this way and, therefore, must be restructured from the ground up to make effective use of a parallel computer. Think of the mass of programs that have already been written and the work that would be required to rebuild their logic. Fortunately, compilers are under development that promise to look at an existing, nonparallel program and find and exploit the opportunities for parallel computation it contains.

On the other hand, many real-world situations have an inherently parallel structure that almost cries out for a corresponding computer structure. In a sense, a distributed control system is performing parallel processing. The overall task of making the plant operate smoothly and profitably has been parceled out to control computers and programmable controllers connected by a data highway. The difference here is that the parallelism sort of happened from the bottom up as existing controllers were integrated into a network, whereas most research in parallel processing starts with one large problem and works from the top down to structure a parallel solution.

For many other problems, the reason for using a parallel approach is simply to do the job much faster. For example, large databases or text files can be searched simply by cutting them up into arbitrary blocks and handing each block to a processor for a parallel search.

One way to put together a small-scale parallel architecture, if speed is not critical, is to link several microcomputers together in a network and write a supervisory program to distribute computations among them and coordinate the results. This approach has been used experimentally for process simulation and other engineering calculations. This could be a relatively painless way to experiment with parallel processing without having to buy a Cray supercomputer.

When two or more semi-intelligent entities are put together, disagreements are bound to ensue, and parallel processors are no exception. The arguments arise when two processors try to gain access to the same memory location at the same time. The operating system must, therefore, use a contention resolution algorithm (a "referee"), which makes one of the processors wait while the other accesses memory. Of course, if any number of arguments arise, processors are waiting for memory before they can do work, a situation that parallel processing was supposed to avoid in the first place. The operating system is also spending much of its time "refereeing" instead of coordinating the processors' work. Result: the system slows down. To get around this problem, each processor can be provided with some memory for its own exclusive use. However, if too much memory is local to the processors, speed can again be lost, because processors wait for data to be moved from one processor memory area to another.

How many processors are needed? This depends in part on how many independent pieces the various tasks naturally divide themselves into; there is no point in having more processors than tasks. Remember also that adding more processors increases the overhead for communications and synchronization. For many purposes, the personal supercomputer (yes, they exist) with two or three very fast processors gets more work done than could otherwise be done in the time that is likely to be available on a giant central computer with many more processors.

Methods for determining how much benefit a particular program is actually gaining from a parallel computer are still in their infancy. It is, of course, possible to find out what percentage of the time each processor is working, but finding out what it's working on and what portions of the program are not using parallelism to full advantage is much more difficult. Performance debuggers are under development that provide a visual image on the screen of how a particular chunk of computation is actually being divided up and executed on the various processors.

Communications

A single computer can hold only a limited amount of information in its memory and data storage. Moreover, in a constantly changing environment, much of this information will rapidly become obsolete and must be constantly updated. Human beings can enter a certain amount of new information manually, but the volume of data needed in many applications is too great to be handled in this way, and much of it already resides in electronic form elsewhere. Therefore, computers need to be able to communicate directly with other computers and electronic or electromechanical devices.

This section discusses the hardware and software requirements for communication between a computer and its externally attached devices, or intermittent communication not involving a network. The transfer of information among the different parts of a single computer has already been covered. Communication over a network (a permanent communications link between two or more computers and possibly other devices) is substantially more complex and will be discussed in the following section. Many of the physical and logical elements described in this section also apply, however, to networks.

Range of Complexity of Communications Tasks

In a sense, even the transfer of information between the CPU, memory, and data storage via the external system bus is a communications problem. However, within the restricted environment of a single computer, the communications task is relatively easy. The system bus contains lines that are dedicated to specific types of information (data, addresses, clock signals, etc.), and the synchronization of transmissions with the system clock minimizes the probability that messages will interfere with each other. The transmission errors that can occur are mainly single-bit faults, so simple error-detection procedures like parity bits are sufficient.

Communicating with external devices attached to the computer can lead to complications. The physical connection now typically involves an input-output port with multipin connector and cable and possibly additional hardware (such as a buffer or interface support chip) or software (such as a device driver). The device will supply or demand data at arbitrary times that bear no relation to the microprocessor clock cycle; therefore, communications with an external device must be asynchronous (without reference to the system clock). Handshaking procedures are now required to make sure the communications channel is clear and both the device and computer are ready before transmission is attempted. Acknowledgment procedures are also needed so the sender of a transmission can be notified that it has arrived intact or can be informed of any problems. Since a wider variety of errors can now occur (device not ready, pin assignment conflicts, buffer overflow), more sophisticated schemes for error detection and recovery are needed.

Communicating with a device not permanently connected to the computer requires additional procedures for starting and ending each communications session. Communications hardware and software are likely to be required. The communications path may involve using a network, (with its own additional hardware and software), in which case the message must be translated to and from the form used by the network operating system, and possible conflicts in the use of network resources must be resolved.

Basic Communications System Components

Any communications system can be viewed as consisting of a transmitter (TX) that generates a message, a medium that carries the message, and a receiver (RX) that accepts and processes the message (see Figure 5-29). Between a given transmitter and receiver, more than one communications channel may exist (see Figure 5-30), allowing different types of information to be sent simultaneously to the receiver. The use of different channels can also allow the transmitter and receiver to switch roles so the receiver can, for example, send back a message confirming that the message has arrived intact.

Interfaces

A computer's information gateways to the outside world are referred to as interfaces. The description of an interface tells what procedure the computer will use to talk with an outside device, including what form the signals will take, how they will be sent, received, and acknowledged, how the start and end of a message will be indicated, and how errors will be detected and corrected.

For the purposes of electronic communications, there are two basic types of interface: serial (see Figures 5-31 and 5-32) and parallel (see Figure 5-33). A serial interface sends information one bit at a time, which is cheap because it needs only one wire (plus a ground) connecting the computer to the other device. Serial communication is slow but robust and can work over long distances. Modems, mice, and some printers and plotters are normally connected to computers with a serial interface.

A parallel interface sends a number of bits of information side by side (usually over the familiar "ribbon cable"), which is faster than serial communication but restricted to shorter distances (less than 15-30 meters). Over longer distances, the bits tend to get "out of step" with each other (see Figure 5-34) due to slight differences in impedance between the cable wires, and the message is misinterpreted at the receiving end because some of the bits are missing (the technical term for this problem is bit synchronization delay). Mass storage devices such as hard disks, most printers, and some plotters use a parallel interface.

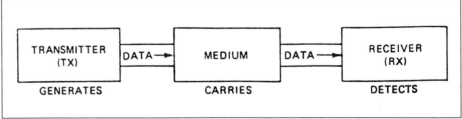

Figure 5-29. Basic Communication System Components [Ref. 17]

Interface Standards

For effective communication to take place, two or more devices must agree on how an interface will work, which means that interface standards are essential.

The most commonly used serial communication standard is RS-232C, the most recent (1969) version of the RS-232 standard from the Electronics Industries Association (EIA). The RS-232C standard specifies the function of each pin in a 25-pin D-type connector, as well as the electrical characteristics of the transmission. In practice, various manufacturers interpret some of the pin functions in slightly different ways, resulting in variants of the RS-232C standard. The V.24 standard closely resembles RS-232C.

Since RS-232C is an older standard, it has difficulty meeting some modern requirements for longer links, higher transmission rates, and greater noise immunity. EIA: RS-422 permits data rates up to 100 Kbaud and transmission distances up to 1.2 km but is still sensitive to noise. RS-423 achieves high noise im-

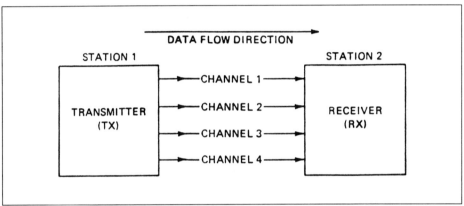

Figure 5-30. Communication Channels [Ref. 17]

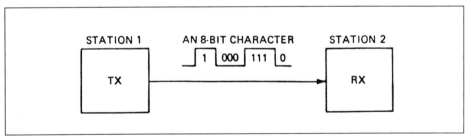

Figure 5-31. Serial 8-Bit Character [Ref. 17]

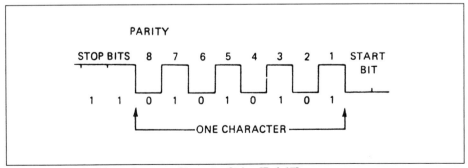

Figure 5-32. Typical Asynchronous Character String [Ref. 17]

munity by using two data signal lines, with the difference between the two signals carrying the information. Ambient noise will likely affect both lines and the noise component will be canceled out. The functional component (pin assignments, plugs, and sockets) of RS-422 has been separated out into a separate standard (RS-449).

The most widely used parallel interface standard is IEEE-488, also called HPIB because it was developed by Hewlett-Packard, or GPIB, which stands for "general-purpose interface bus." IEEE-488 is used in many straight instrumentation and laboratory applications as well as for interfacing devices to computers. IEEE-488 uses 16 parallel lines: eight for data, three for handshaking, and five for general interface management. The Centronics® interface is an older parallel

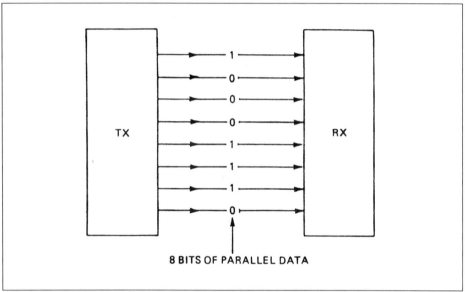

Figure 5-33. Parallel Communications [Ref. 17]

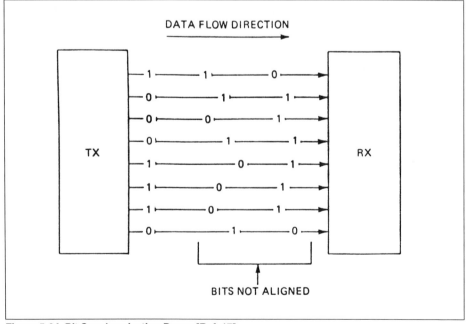

Figure 5-34. Bit Synchronization Decay [Ref. 17]

interface used primarily for printers. The SCSI (ANSI X3T9.2) interface, using nine data lines and nine handshake lines, was developed for connecting hard disk and streaming tape units to computers but has since found wider application, largely due to its ability to support up to eight peripheral devices.

Communications Parameters

Two computers or devices, even when they are using the same interface standard, may send data in slightly different ways that are described by communications parameters. Speed, perhaps the most basic parameter, is measured in baud, which is essentially equivalent to bits per second.

The format of the message itself may also vary. Each character is typically sent as data bits (7 or 8) preceded by a start bit that says, "Here comes another character," and possibly a stop bit that says, "End of character." A parity bit may be used for primitive error checking (more about this later); common settings are none, even, or odd. A more sophisticated error checking protocol may also be selected. At higher transmission rates, data compression codes may be used to achieve higher throughput. Several characters may also be combined into a single message, in which case the message will include header and trailer information specifying, for example, the length of the message and whether any transmission errors were detected.

To keep a human user informed about what is happening during a communications session and to aid in error checking, each character sent is often echoed back and displayed on the sender's computer screen. This requires a full duplex communications link where transmission and reception can take place simultaneously in both directions. A half duplex communications link allows transmission in one direction at a time. Simplex transmission, which is seldom used, allows communication in only one direction (see Figure 5-35).

Communications parameters are usually set by communications software, which is described later. A typical interchange of information between transmitter and receiver to achieve a message transmission is shown in Figure 5-36. The term "line discipline sequence" refers to how the transmitter and receiver will alternate in their use of the communications line and what each will send at what stage of the transmission.

Starting a communications session and then failing to receive any feedback display on screen probably indicates a need to change the communications setup to full duplex. Getting two of everything indicates a need to change from full to half duplex.

Communications Hardware

A serial or parallel device interface is physically implemented as an I/O port, with interface support chip, connector, and cable linking the computer with the device. When the device needs to provide input or receive output, it generates an interrupt that signals the operating system that the I/O port needs servicing.

If the other device or computer is located within 10-15 meters of the computer, a connector known as a null modem cable can be used to establish a direct link between the serial connectors on the two devices. To communicate over longer distances, the usual method is to use a phone line and a modem (short for "modulator/demodulator"). The modem is needed because the phone system is analog rather than digital, operates in a frequency range different from computers, and contains devices like filters and echo suppressors, which would play havoc with digital signals. Modulation is the process of converting digital information into a waveform that can be transmitted through the phone system and then converted back to digital by the other computer (see Figure 5-37).

Most modems use the RS-232C serial interface and the Hayes command set, which provides a means for the computer to tell the modem to do things like dial a number or pause a certain number of seconds. Speeds range from 300 bits per second up; common values are 1200, 2400, and 9600. The faster modems have to use data compression schemes because it is technically impossible to push the raw data that fast across the phone lines (they do things like replacing long runs of 0s or 1s with a more succinct code). If the line quality is poor and the modem makes too many errors transmitting at the high speed, it will usually drop down to a lower speed and try again.

Modems can be external (a little box connected to your serial port and your phone jack) or internal (a card inside your computer, similarly connected). An increasing number of portable and laptop computers have built-in modems.

Communications Software

Using the modem command set for your communications needs would be like doing all programming in assembler. Communications software allows a user to dial up and use another computer and upload and download files with ease. Some programs, however, offer more ease of operation than others, which can be vital since communications seems to be the most trouble-prone of all computing tasks.

A typical communications session using a modem and phone line will involve dialing the number of the other modem and establishing a connection. Once linked to another computer, one can browse through on-line databases, bulletin boards, or an electronic mail system, transfer files, and even run programs

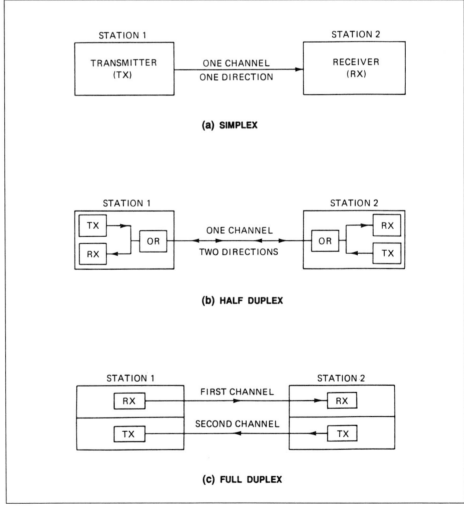

Figure 5-35. Simplex, Half Duplex, and Full Duplex Communications [Ref. 17]

remotely. At the end of the session, commands are issued to hang up the phone and exit the communications program.

Almost all programs let one dial a phone number from within the program. The better ones offer a dialing directory that contains commonly used numbers, together with long distance codes (which can be longer than the phone number when dialing out from some business phone systems) and the appropriate communications parameters. A number is selected from the directory via mouse, cursor keys, or a one- or two-digit code, and the program dials automatically. The number can also be keyed in manually, but the directory is faster, easier, and avoids many errors.

On-screen indications of call progress (dialing, ringing, busy, no carrier (no connection)) help determine whether a call attempt is proceeding normally. A noisy office or plant may overwhelm a modem speaker, and with an internal modem there may be nothing to hear.

Once connected, one can generally log onto and use the other (host) computer using the same procedures as for a terminal. Certain control characters, such as backspace, may not work, which may, in turn, preclude correcting typing mistakes or using certain functions of a program run on the host computer. To get these control characters, use a communications package with terminal emulation, which endows your computer with the command vocabulary of a common terminal. Reading electronic mail or downloading a few files probably does not require terminal emulation, but, for running a program or using an editor remotely, it can be annoying not to have a full set of terminal commands.

One of the most popular uses for communications programs is uploading and downloading files from other computers. Uploading means sending a file from your computer; downloading means bringing a file onto your computer. A "quick

> Programs that do terminal emulation often have two modes: a terminal mode, which uses a blank or almost-blank screen, and another mode displaying, for example, a communications option menu. Look for a program that always has an indication on-screen of how to get to the other mode. Even people with lots of computer experience have moments of panic when unexpectedly confronted with a blank screen.

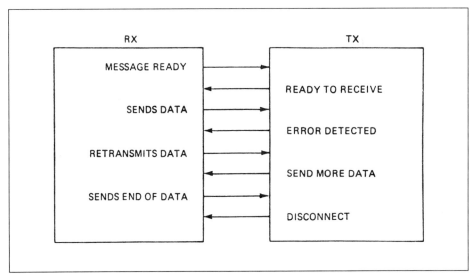

Figure 5-36. Typical Line Discipline Sequence [Ref. 17]

Figure 5-37. How Modems Process a Signal

and dirty" download can be done with a screen capture, which takes everything that appears on the screen and dumps it to a file. All one needs to know is how to display a file to the screen on the host computer. Turning on a session log function can accomplish the same thing. Of course command lines and other "garbage" have to be edited out of the file afterwards.

Uploading or downloading files by name normally requires that the same communications program be running on both computers. Some bulletin board systems (BBS), however, have their own commands that let one upload or download files regardless of the communications program used. A received file first comes into a communications buffer and then is written to disk in chunks, because the host computer will send data continuously regardless of whether the remote computer is ready to receive it. Computers that recognize the XON/XOFF protocol have a little more sensitivity; XON is a signal that says, "I'm ready for data," and XOFF says, "Hold it—I'm busy."

At the end of a session, some packages will hang up the phone automatically; others require a separate hang-up command.

One final desirable feature deserves mention: the capability to diagnose and recover from problems during a session. It seems more can go wrong with communications than with any other aspect of computing: bad connections, incompatible parameters or formats, interrupted transmissions, and countless other problems. It makes a big difference whether you can just press a key to get out of a jam or have to go into a complex and delicate recovery procedure. For example, some programs offer no easy way to hang up on one number and try another if, for example, the first phone tried is busy. Software that can help diagnose problems or recover from them automatically (for example, by trying a slower transmission rate) is even more helpful.

Error Checking

All kinds of important information, from the operating temperature of a reactor to a bank balance, are sent over communications lines. Error checking is vital to ensure that the information sent is what was actually received. The most basic form of error checking is the parity bit, which may, for example, be set to 1 if the character just sent contains an odd number of binary ones and 0 if an even number. This technique catches most errors in data transmission within a computer's memory (this is why nine banks of memory chips are needed to handle an 8-bit data item—the ninth is for the parity bit). However, the parity bit technique is not powerful enough for communications, so many communications programs ignore the parity bit, preferring to use it as an extra data bit.

More advanced techniques such as the cyclic redundancy check (CRC) continuously calculate a mathematical expression based on the contents of the transmission (sort of like a moving average) and use the results to detect errors. Incorrect characters or messages are simply retransmitted.

Minimal error checking is needed to transfer small, human-readable files, more for executable code (which cannot be proofread), large data files, or critically important data.

File Transfer Protocols

File transfer protocols combine error checking with features like XON/XOFF and message acknowledgment to specify a procedure for transferring files. They usually transfer fixed-size blocks of data and recognize a specific ASCII control character such as End of Transmission Block (ETB) or Acknowledge (ACK) as the end of a block. Unfortunately, this is one area of computing where standards were never implemented effectively.

The most widely supported protocols are public-domain ones such as XMODEM and Kermit, but even these exist in versions and do not offer an ironclad standard. Another popular protocol, ZMODEM, offers a "checkpoint restart" feature that allows the resumption of a transmission that was interrupted because of line noise or other problems. Communications packages may include their own proprietary protocols, which can be used only if the other computer uses the same protocol.

Models of the Communications Environment

As evident from the preceding discussion, communications is a complex subject, involving both hardware and software in a complex and dynamic environment. Efforts have been made to model the communications environment to make discussion and standards development easier and to clarify the responsibilities of hardware and software developers. The problem seems to lend itself naturally to a multilayer model, since some aspects of communication are low-level (narrowly focused or of limited scope, such as pin assignments) and others high-level (of broader diversity and scope, such as file transfer protocols).

The International Standards Organization (ISO) has defined a seven-layer Open Systems Interconnection (OSI) model (see Figure 5-38), which is widely used as a conceptual framework for talking about communications systems. The first (lower) four layers are collectively called the transfer service because they deal with moving information from one point to another. The three highest layers, called user layers, deal with providing data access for users over a network. As of this writing, standards have been developed only for the lowermost three layers.

The following are the seven layers and their functions.

Level 1 (Physical Layer) defines the interface in electrical, mechanical, and functional terms to enable it to exchange ones and zeros. Serial and parallel interface standards apply at this level.

Level 2 (Data Link Layer) describes the transmission of data frames (messages) through the interface. Data formats and error detection and recovery are defined at this level. Data link protocols such as HDLC™ (Hewlett-Packard), DDCMP™ (Digital), and BISYNC™ (IBM) belong to this level.

Level 3 (Network Layer) governs the transmission of data frames between stations on a network, establishing an end-to-end transfer procedure for transparent data delivery. Switching and routing of messages on the network are described at this level. Network protocols (described later) belong to this level.

Level 4 (Transport Layer) provides a network-independent interface and ensures a reliable connection between network devices.

Level 5 (Session Layer) deals with the structure and logic of message exchange between points on a network and, in particular, with the tracking and maintenance of multiple simultaneous communications on a network.

Level 6 (Presentation Layer) handles the transformation of messages between various computer, data terminal, and database formats. It also provides the Application Layer (Level 7) with a set of data management, display, and control services.

Level 7 (Application Layer) deals directly with end users and real applications, accepting their data for entry into the network and meeting their data needs from the network.

At this point, the OSI model is far from complete and serves more as a conceptual structure than a set of specifications for an actual communications system.

If you regularly log onto bulletin boards or databases and go through the same sequence of commands over and over, the facility to store these in a macro or script file and replay them can save you a lot of time and effort.

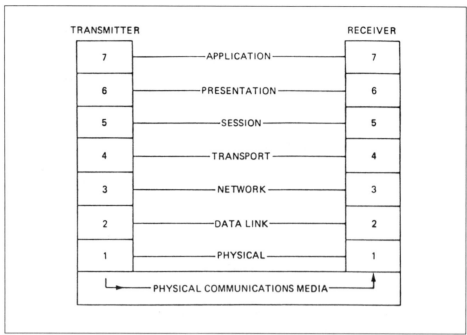

Figure 5-38. ISO/OSI Seven-Layer Model [Ref. 17]

Networks

If communications needs consist of an occasional file transfer, moving the information on floppy disk (the famous "sneaker net") or using a null-modem cable or modem link may be sufficient. If, however, a group of computers and other devices need to exchange information on a regular basis, a local area network (LAN) is probably called for.

Why Networks?

Networks allow two or more computers in the same general area (such as an office or plant) to share resources (printers, disk drives, modems) and/or information (databases, drawing files). Many people must like this idea, since approximately one-fifth of all PCs are connected to a network and almost two out of three offices have a network (*PC* magazine, May 29, 1990, p. 97). In an industrial environment, a network provides access to on-line process data for monitoring and control purposes and allows process models or report generation programs to use the latest information available. A typical industrial plant local area network is illustrated in Figure 5-39.

Of course, none of this comes free. Networks generate extra computing work in the form of data transfer, checking, coordination, and conflict resolution. This work is done by a server (a computer dedicated to network tasks) or is parceled out among the computers on the network. In any case, each computer will need some extra equipment, typically a network card (adapter) and associated software. Various kinds of cables link the components together.

Non-Network Device-Sharing Options

If the primary need is to share access to one or more devices, a simple piece of hardware (possibly with some associated software) may do. It would take care of one particular task such as managing a printer queue or controlling access to a modem line, for example.

> For users anywhere but in the same room with the equipment to be shared, look for equipment that offers remote control via software commands; it's a pain running down the hall to flip a switch to use the printer!

Peer-to-Peer Networks

Peer-to-peer or zero-slot LANs achieve simplicity and low cost by eliminating the server and the network card (hence, the name "zero-slot"). The idea is that every computer on the network has equal access to the resources of any other computer through its serial or parallel port.

One advantage of the peer-to-peer LAN is that the change in individual work habits is minimal; the resources on the other computers are just addressed as additional drives or device ports. This may make the peer-to-peer LAN the solution of choice for less sophisticated (or less frequent) computer users. Peer-to-peer LANs are functioning effectively in many situations where the resources to install and maintain a larger LAN are not available and the demands on the network are relatively light and uncomplicated.

As more computers are added and the workload grows, however, the disadvantages of the peer-to-peer LAN start to become apparent. There is no file security to speak of; everyone can access and change (and possibly screw up) anything on any computer. If some computers are turned off or otherwise unavailable, the resources they control are lost to all users. Keeping track of which files are where can pose a major problem. The memory requirements for the network software on each computer may start to be burdensome, or the fact that the network ties up a serial or parallel port. Finally, the performance of all computers on the network may suffer a serious decline as each computer tries to handle its portion of the network-related computing load in addition to its own work.

Dedicated-Server Networks

A dedicated-server network contains one computer (the server) that does no computing work of its own, but only serves the needs of the network. In principle, any computer can be a server, but for performance reasons the choice is more often a high-end PC (386 or 486) or a small minicomputer or workstation.

Note that the participants in the network are no longer equal. The server runs the show, and if another computer wants to get a file from a hard disk elsewhere on the network, or use a printer or modem not its own, it must ask the server for access. This means that users may have to learn some new commands and proce-

Be careful not to undersize the server with respect to both size and speed. Think of future as well as present needs.

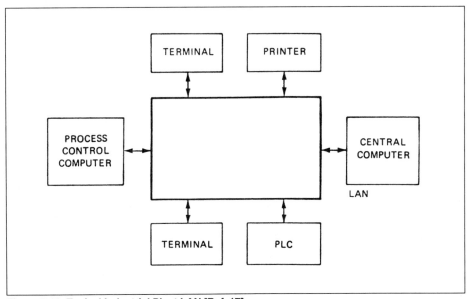

Figure 5-39. Typical Industrial Plant LAN [Ref. 17]

dures to access resources through the network and also that the reliability and speed of the server become of primary importance.

In fact, a network may have different servers for different tasks. File servers manage access to large hard disks or databases. Print servers look after printers or plotters being used over the network. Communications servers may oversee a bank of high-speed modems, a fax link, or a gateway to a mainframe or another network.

CONNECTING TO THE DEDICATED-SERVER NETWORK

Dedicated-server networks require that each computer have a network adapter card installed and run certain software associated with the network. Obviously, this takes up a slot on each machine plus a certain amount of memory, since the software must be memory resident. The memory requirements of the network software are usually not too large, since much of the work can be done either at the server or on the card, but they are worth keeping an eye on if there are other RAM-hungry applications such as CAD or large spreadsheets. Some network software is smart enough to locate part or all of itself in otherwise unused high memory, minimizing the inroads into precious 640K (but beware of new versions of DOS or other clever programs that may try to use that same chunk of high memory).

THE REDIRECTOR

The key to getting a computer to talk to the network is the redirector, which intercepts requests from the operating system or application programs for disk or device access. If the disk or device is attached to a local computer, the redirector just forwards it to the operating system; otherwise, it translates it into terms the network server will understand and sends it out as a message on the network (see Figure 5-40).

Network Hardware

The hardware component of a network normally includes a network adapter card and connector for each computer, some kind of cable to connect the computers, and repeaters to clean up and strengthen the signals as they travel around the network.

THE NETWORK ADAPTER

The network adapter is a card that serves as an interface between a single-user computer and the network. Normally it requires one slot in this computer. Since it includes its own connector, it does not tie up an existing serial or parallel port. However, the network adapter may fail to coexist happily with other boards or hardware. Common areas of conflict are clock speeds, hardware or software interrupts, and memory addresses.

CABLING

Once a message leaves a computer, it will be traveling on some form of cable. Twisted pair cable is used by telephone systems and, therefore, is already in your walls, but it cannot carry data reliably over more than a few hundred feet, has limited capacity, is vulnerable to electromagnetic interference (emi), and provides virtually no data security (not only is it easy to tap, but it tends to function as an antenna).

Coaxial cable can carry more data over longer distances and offers better emi immunity than twisted pair but at a greater cost. It is frequently used in industrial network applications where long communication runs (over 300 meters) are involved and where electrical interference conditions are not extreme.

Fiber optic cables offer much higher data rates, immunity to emi, and high security but at a higher price. Moreover, installing fiber optic cable tends to be costly because special skills are needed to splice, connect, and terminate the cables, and the cable itself is susceptible to physical damage. In many newer buildings, coaxial or fiber optic cables are installed at construction time in the expectation that the occupants will want to set up computer networks.

The various kinds of cables available for networks are illustrated in Figure 5-41.

Since vendors usually include cable with their products, a buyer's choices may be limited, unless different cabling options are offered for the same package.

DISTANCE RESTRICTIONS AND REPEATERS

LANs are not long-distance communication vehicles, and too great a physical separation between any two stations (computers) can lead to problems. These may simply be due to degradation of the signal as it passes down the cable. Since different frequencies tend to travel at different rates, the crisp square waves that represent digital ones and zeros gradually sag and spread out and may eventually overlap.

A repeater can fix this problem by reading and interpreting the corrupted signal and sending it out again as square waves (see Figure 5-42), but eventually another problem arises: most networks allow a certain maximum time between the sending of a signal and the response to that signal, after which they conclude that something is wrong. Obviously, this sets a hard maximum on the distance between any two computers on the LAN. Check with the supplier; this varies from one network to another.

BASEBAND VS. BROADBAND NETWORKS

These terms refer to whether the cable can carry only one digital signal at once (baseband) or many different signals at once (broadband). Broadband networks use a wide bandwidth (up to 450 MHz) and frequency division multiplexing to divide it into a number of smaller channels that can be used simultaneously for different types of communication (for example, to handle data and voice at the same time). They are more expensive because of the extra circuitry and logic

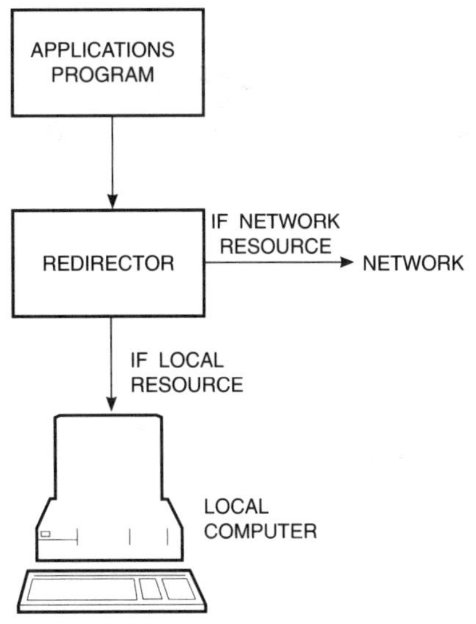

Cable installation can be expensive depending on how much needs to be torn up to provide a physical path for the cable. Remember to check building and fire codes and any restrictions imposed by leases before getting into cable installation.

Figure 5-40. The Redirector

needed to sort out the various messages but provide much more data transmission capability and may be a necessity for high-data-volume applications such as voice mail or CAD.

DATA RATE

Data rate refers to how many megabits per second of information can be moved over the network. Like miles per gallon ratings for cars, data rates quoted by vendors are usually measured under ideal conditions. Heavy network loading, transmission problems, a slow device serving as a bottleneck, or just the composition or traffic pattern of a message stream can reduce the actual throughput.

NETWORK CONFIGURATION

The pattern in which the server and its client computers are cabled together is called the "network configuration" or "topology," which comes in three basic flavors: bus, ring, and star (see Figures 5-43 to 5-45).

The bus configuration essentially broadcasts all transmissions to all stations, avoiding the workload of message routing. The drawback to this is that the transmission medium (cable) must have a large message capacity.

Most ring networks use bidirectional transmission so that the most efficient message routing between any two stations can be used. This allows a faulty station to be bypassed and makes the ring configuration robust.

The star network obviously depends heavily on the reliability of the central server and also on its speed, since only one station can be in communication with the server at a given moment. However, the star configuration does minimize the

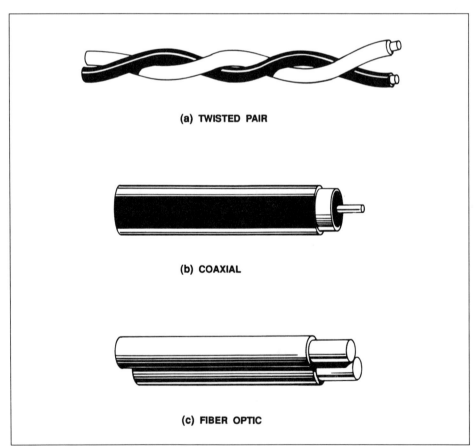

(a) TWISTED PAIR

(b) COAXIAL

(c) FIBER OPTIC

Figure 5-41. Network Wiring Types [Ref. 17]

average number of stations a message must be passed through to reach its destination.

The choice of network configuration is frequently imposed by other design choices made for the network, for example, by the selection of transmission protocol or network operating system software.

MAXIMUM NUMBER OF NODES

A node is a computer or other electronic device connected to a network. A theoretical maximum on the number of nodes is set by the number of bits in the node address. The practical maximum is usually much lower and is determined by factors such as how often the nodes use the network, how much data they send, and how fast the data is transmitted.

Information Transfer over the Network

A computer sending information to another computer or device over a network must put that information in a form that the other computer or device can recognize. Moreover, computers and devices must have a way of knowing when they can send and whether their messages have been properly received.

PACKETS

To make life easier for the network, all messages are sent as one or more packets of a standard format: usually some header information including destination and origin addresses, then a certain number of message bits, and then some end-of-message information such as checksum bits for error detection. It's a bit like putting mail in standard business-size envelopes with the stamp, return address, and postal code in specified locations so the post office can process them more easily.

Character-oriented message transmission relies on a series of control characters within each frame to ensure the accuracy of data transmission. This slows down the network because the accuracy of each frame must be verified and acknowledged before the next is sent. It also makes it difficult to decipher the message packets in case of problems. Bit-oriented transmission dispenses with this complexity and simply identifies each data frame with two or three control char-

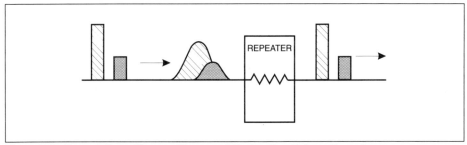

Figure 5-42. Degradation of Square Waves and Restoration Using a Repeater

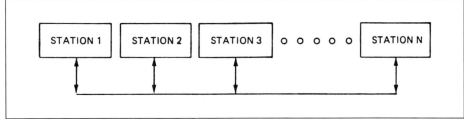

Figure 5-43. Bus Network [Ref. 17]

acters. Some commonly used message formats of both types are illustrated in Figure 5-46.

Every time a packet is sent, it must be assumed that it may not arrive. Interference or collisions between packets may damage the message; the receiving computer may be busy, or its buffer may be full, so the message cannot reach its destination. Therefore, most networks include extensive provisions for detecting transmission failures and re-sending messages.

PROTOCOLS

A protocol is simply a set of procedures and conventions that everyone agrees on so they know what to expect and can do their work without petty arguments. Protocols let us seat the diplomats at a table so they can get on with the negotiations. In a network, they describe how messages will be sent, how contention for network resources will be resolved, and how different types of files or devices will be recognized.

One major message-transfer protocol used on networks is token passing. The token is a special pattern of electrical signals that circulates continuously around

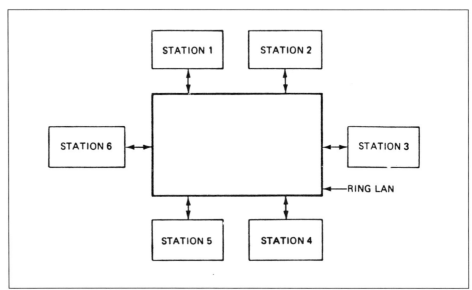

Figure 5-44. Ring LAN [Ref. 17]

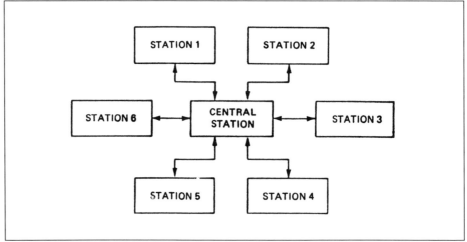

Figure 5-45. Star LAN [Ref. 17]

the network like an empty envelope. A computer wanting to send a message grabs the envelope, puts the message inside, "addresses" the envelope, and returns it to the network. Subsequent computers check the address and pass the token on if the message is not for them. The recipient computer eventually reads the message (but does not clear it from the token). When the token gets back to the originating computer, that computer can reread the message to check whether it arrived intact. Token passing networks are not the fastest (you have to wait for the token before you can send) but are generally very robust and reliable.

The other commonly used message-passing protocol is CSMA/CD, which stands for "carrier sense multiple access with collision detection." The commonly used Ethernet protocol is of this type. The first part of this rather long name, carrier sense, means that a computer wanting to transmit will listen to the network and send its message if it doesn't hear anything else being sent. Of course, somebody else may be doing this at the same time (multiple access), in which case the two messages may bump into and corrupt each other (a collision). Fortunately, the two sending computers can detect when this happens, in which case they both stop and wait a specified amount of time before trying again. Actually, the waiting time has to be varied over a range so it will be different for the two computers—they shouldn't both wait fifty microseconds and do the same thing again! A CSMA/CD network can be very fast and efficient, but performance can deteriorate badly if the network is heavily loaded and there are many collisions or if the collision resolution procedure is inefficient.

Other higher-level protocols help dissimilar computers or networks talk to each other. TCP/IP (Transmission Control Program and Internet Program) was developed by the Department of Defense (DoD) in the United States, which may have the world's largest and most diverse collection of computer equipment to

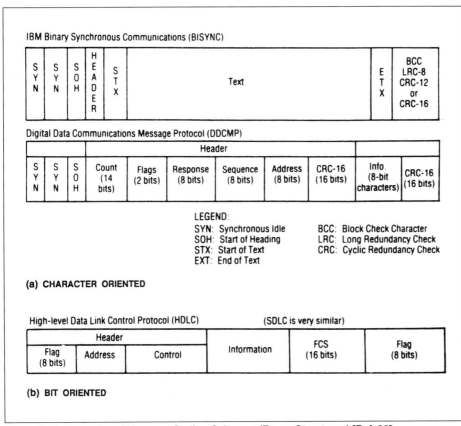

Figure 5-46. Examples of Message Coding Schemes (Frame Structures) [Ref. 28]

link together. The DoD continues to certify software as compatible with TCP/IP and is certainly likely to be around for a long time to support its product. TCP/IP is, therefore, an important standard for networks with different kinds of computers in them.

Note, however, that a file can be sent successfully from one computer to another, yet still be unusable at the other end because the file format is incompatible. Moving a file from, say, an Apple to an IBM-PC so it can be used on the PC involves more than just physically transferring the file.

NetBIOS was designed as an programmable network interface. Application programs generate calls to NetBIOS instead of directly to the hardware or software of the specific network. A number of network operating systems offer NetBIOS emulators.

An example of the variety of interfaces involved in an industrial computer system is shown in Figure 5-47.

Network Operating Systems

A network operating system must perform the same functions as any other operating system—file management, resource allocation, synchronization, interrupt handling—but it must do so in an environment that is much more complex than a stand-alone personal computer.

Since a network is intrinsically multiuser and multitasking, a plain, unadorned single-user operating system such as MS-DOS is incapable of managing it, yet any network containing single-user computers must be capable of accepting and processing the requests their operating systems generate. Some companies have chosen to build their network operating systems on the foundation of existing single-user operating systems in order to minimize the difficulty of interaction with their component computers and to take advantage of widespread familiarity with these operating systems. Added functions include intercepting and processing multiple requests and allocating space and time on processors and devices. This approach is most often used for peer-to-peer LANs.

The other approach is to base the network on an intrinsically multiuser, multitasking operating system such as Unix. These systems sacrifice single-user operating system compatibility and are, therefore, found only in dedicated-server LANs. They must, however, still understand and process requests from single-user systems, which entails an ongoing effort to keep up with the evolution of their operating systems. Additional capabilities such as electronic mail or mainframe gateways are easier to implement in an operating system whose structure was designed from the bottom up to accommodate them.

If a network is operating in a real-time process environment, the requirements for a real-time operating system (time stamping of information, flexible task priority assignment, etc.) are, of course, added to the existing requirements for a network operating system.

Beyond the usual price and performance criteria, selection of a network operating system may depend on the availability of certain special features:

(1) Remote station access (the ability to dial into the network from outside via modem) is indispensable if people off-site will want to use the network.

(2) Ability to run other programs on the server, especially during slack periods (compiling program source code, for example). This, of course, entails a penalty in reduced server processing speed but may be useful if the server is underutilized or computing power in the organization is in short supply.

(3) Disk caching (which tries to make access to data faster on average by storing in RAM the data most likely to be accessed next). Caching is based on the fact that when you read part of a file, it is highly probable that you will

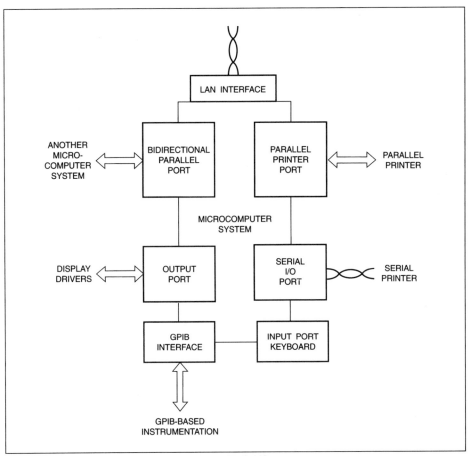

Figure 5-47. Interfaces in a Computer System [Ref. 16]

want to read the rest. The cache software therefore "jumps the gun" and loads the next piece of the file into RAM, where it can be more quickly accessed. Some cache software keeps track of which files are most often accessed and moves them into RAM also.

(4) Print spooling utilities (which provide control over the lineup of jobs waiting for a printer or plotter). These help assign priorities to print jobs, insert jobs into the queue, delete jobs from the queue, and possibly assign characteristics such as two-sided printing to individual jobs.

(5) Network administration and diagnostic utilities, which help the network manager keep track of who is using the system and how, as well as tracking down and solving problems.

Application Software for Networks

Installing a LAN requires investing in network versions of whatever application software (word processors, database programs, etc.) one intends to use over the network. This is partly because LANs are multiuser, and a program designed for a single-user PC has no way to handle several people calling it up at the same time and even trying to work on the same file. There may also be legal problems; multiuser access over a network would violate the terms of most single-computer software licenses.

There may be some flexibility in deciding what portions of a program or its data files are used over the network and what portions are loaded into a user's local RAM at the start of a session. Factors in this decision include expected network loading and speed and memory capacity of both the server and the local computer.

Once software is installed on the server, easy access to it must be provided, via menus or batch files if needed, so that users will not become frustrated trying to type in long, complex commands.

Security on the Network

As soon as two or more people can use the same file or device, the possibility arises that they may try to use it at the same time. In the case of disk files, a common solution is to "lock out" one user until the other has finished. Locking may be done at the volume, file, or record level. With volume locking, no other user can get at any file on the same disk volume until the first user has finished. File locking prevents other users from getting at the file the first user is accessing, while record locking prevents access only to the record the first user is working on. Clearly, the smaller the unit that is locked, the less the users' free access to data is compromised, but managing the locking and unlocking becomes more complicated.

With several users, it becomes more important to control which files each can use and how. It may not be desirable for everyone to have access to payroll or personnel files. On the other hand, it may be desirable for them to be able to look at, say, process data histories or download them to their spreadsheets without being able to change the master copy. Most operating systems include some capability to control file access (setting a read-only bit in the operating system file description, for example). More sophisticated systems provide for restricted access by individuals, groups, or job functions and decide whether each user may look at the file, make a copy of it, write to it, or delete it.

Special Needs of Industrial Networks

The primary function of an industrial local area network (ILAN) is to serve as a data highway connecting sensors, controllers, computers, and other electronic devices in an industrial plant. They also allow human users to query and change the operation of these devices.

High-level ILANs are broadband networks because of the data transfer rates required and must provide a high degree of reliability and equipment redundancy.

Most industrial plants contain a variety of electronic devices from different manufacturers, with many incompatible communications protocols. The effort to interconnect these devices in a network can entail great expense—possibly half of total project cost [Ref. 22]. Moreover, the resulting network can be difficult to change or expand.

In an attempt to solve this problem, General Motors initiated the development of a Manufacturing Automation Protocol (MAP) in the early 1980s. The goal was to achieve a standard for large-scale multivendor communications in industrial plants, allowing products from different vendors to be connected directly to a network without modification. The goal of a single specification has seemed an impossible task, but the project has resulted in some worthwhile achievements along the way.

To achieve good noise immunity and high data rates, most MAP implementations do not use broadband transmission throughout the plant but do use it for a "backbone" network to which are connected smaller "carrier band" subnetworks, typically using coaxial cable. A typical industrial network involving MAP is illustrated in Figure 5-48.

A decision-making guide for industrial network selection, containing some but not all of the factors involved, is diagrammed in Figure 5-49.

Network Operating Problems

Like all large complex systems, LANs have their share of problems, and it is vital to be able to solve them quickly. Once you have a network, you will quickly become dependent on it, with much of your software and data plus services such as printers and electronic mail accessible only via the network. When an industrial process is involved, the efficient operation of the process or the accuracy of process information may be compromised if the network goes down. Clear installation and troubleshooting instructions and good vendor support should be high priorities.

Simple overloading is a prime suspect in frequent network breakdowns or poor performance. Storage gets crowded and fragmented, buffers overflow, or frequent collisions and mistransmissions clog the network. The preventative or cure for this is to get plenty of capacity for present needs and to leave an easy expansion path open for future needs.

Diagnosing network problems is often difficult since there are so many interacting components. The error recovery capabilities of networks actually make diagnosis harder, because in most cases transmissions still get through, although more slowly, so there is no clean break to locate and fix. In fact, today's networks can be so fast that the slowdown caused by a chronic problem may go unnoticed, especially when the network is lightly loaded.

The simplest LAN diagnostic tools simply measure the bits per second transmitted, or the bits per destination. Most network problems, however, cannot be solved without knowing what is being transmitted and why. To do this, you must

Every network needs a person with primary responsibility for keeping the network up and running (a network administrator), even if that responsibility is less than full-time for a smaller network. Needless to say, a sufficient portion of this individual's available working time must be allocated to network-related duties. Networks are sufficiently demanding of attention that no one should be expected to "fit them in" around and above other job demands.

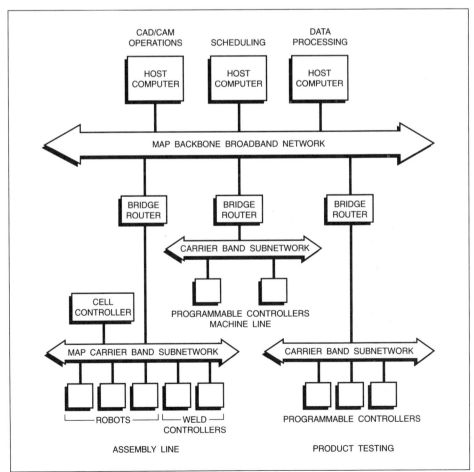

Figure 5-48. MAP Configuration Example [Ref. 22]

look at and decode the actual message packets, which is no simple task, especially since messages often refer to other messages, which must themselves then be decoded.

Network analyzer programs (included with large systems or else available separately) help detect unusual behavior patterns and trace their source, but a fair bit of ingenuity and persistence is still often needed. These may reside on their own separate computer or may require an additional card or other hardware as well as the software.

There is some advantage in a network analysis capability that is physically separate from the network—it won't be unavailable when the network is. However, the required hardware makes this approach expensive and probably justifiable only for large or critically important networks.

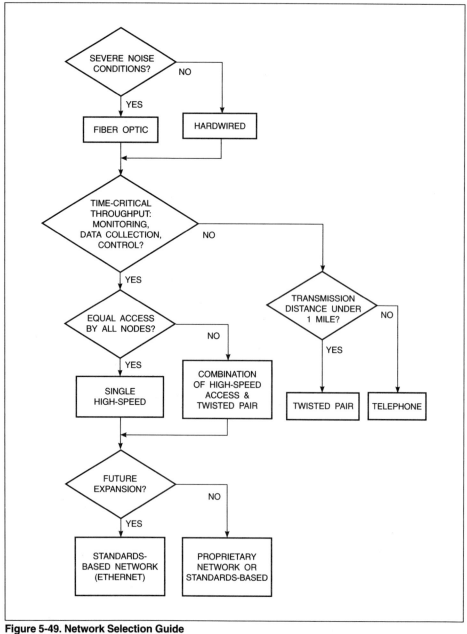

Figure 5-49. Network Selection Guide
(From Carter, Stephen M., "MAP vs. Ethernet in Networking," *INTECH*, April, 1990.)

> Loading a word processor over the network seemed to be taking an exceptionally long time. Inspection of the network message stream revealed that each access to the word processor was preceded by a series of "File not found" messages. It turned out that when the network was installed, the PC user's PATH statement was simply modified to search the network drive instead of the local disk drive. The trouble was, the word processor directory was near the end of the path, and every failure to find the word processor elsewhere on the network drive sent a "File not found" message over the network to the requesting PC. The time needed to process these messages was slowing down access to the word processor. One remedy among several possible ones: call the word processor via a batch file that gives the full correct path name on the network drive.

Applications

What are computers used for in industry? Some applications are directly related to process monitoring and control: obtaining data from sensors and controllers, converting that data into a form usable by other programs, performing analyses and producing reports based on the data, and archiving process data for possible later reference. In a closed-loop system the computer will also send signals to actuators to implement control actions.

Other general-purpose applications, while not specifically related to control, have such widespread usefulness that anyone using a computer for any purpose is likely to encounter them. Databases, spreadsheets, and word processors are an established triad of application programs that almost everyone uses. Technical people in industry will also have occasion to use CADD (computer-aided drafting and design) packages and project management and scheduling software.

Still other computer tools may be used from time to time by technical people involved in particular projects or working in plants with special requirements. One person may need to write customized software using a programming language, another to perform statistical analysis of process data, still another to set up and run computer simulations of a process operation.

Artificial intelligence (AI) is still a new field, and building AI applications is still the province of computer specialists with advanced training, rather than instrumentation and control personnel in industry. Nevertheless, AI deserves mention because its impact over the next five or ten years on what we can do with computers and how we will interact with computers is expected to be profound and far-reaching.

Process Monitoring and Control Applications

Process-related computer applications include the collection of process data from sensors and transducers, the conversion of this data from analog to digital form, the processing of data to render it more useful to programs requiring it as input, and the implementation of closed-loop computer control.

DATA ACQUISITION

The first concern in processing raw sensor or transducer data is to eliminate errors or noise. Such problems may arise from unbalance between the resistance of the signal leads, common-mode voltages, electromagnetic radiation, conductance, or capacitance. Techniques known collectively as signal conditioning can reduce the effect of these electrical problems to less than 1.0% of the signal value.

Electrical or digital filtering may also be used. Electrical filtering reduces or removes signals of particular frequencies; in process control the usual technique is to use a low-pass filter to remove high frequencies, leaving the low frequencies more likely associated with actual process variation. Digital filtering is a computational technique that operates on the digital signal in the computer to accomplish the same purpose.

Once the analog signal is "clean," the next step is to convert it into digital form so that it can be processed by a computer. Analog signals of a continuous process are themselves continuous, so discrete values must first be obtained via process variable sampling. For a few inputs, the sampling can be performed by activating a series of flying-capacitor samplers one by one. For a larger number of inputs, care must be taken to avoid noise or crosstalk between adjacent signals, so two banks of relays or switches are normally used.

The sampled analog signals are then conducted to the analog-to-digital (A/D) converter to be converted to digital signals intelligible by the computer. A typical conversion sequence from raw process data to digital information is illustrated in Figure 5-50.

To obtain values truly representative of the process, averaging may be used to minimize the effect of exceptional readings.

DATA ANALYSIS

Once process data are converted to digital form and stored in the computer, they may be subject to various analytical techniques to obtain information about the current state of the process or the way in which it is changing. Basic statistical techniques will give the mean and standard deviation of individual process variables. Charting techniques such as pie and bar charts allow immediate but subjective visualization of process variation.

To understand the relationship between values observed at different locations or different times, autocorrelation or power spectrum techniques (collectively known as variability analysis) can be used. These can tell you, for example, that there is a strong relationship between the value of one variable observed at the beginning of your process and another variable observed at the end five hours later.

Consider a non-stationary random disturbance (Figure 5-51) representing a typical process output, to which PI control (Figure 5-52) and minimum variance (MV) control (Figure 5-53) are applied. From simple inspection of the observed variation under control, it is impossible to determine which control strategy is performing better.

However, let us now apply an autocovariance function to the PI and MV output. The autocovariance at lag k measures how much of the variation in a variable can be associated with variations in that same variable k time units later.

If the autocovariance drops rapidly to near zero as the lag between samples increases, this means that there is little association between values of a given variable that are more than a few time units apart, or, in other words, that most of the long-term variation or drift in the variable has been stabilized, leaving only short-term fluctuations. Figure 554 shows us that the minimum variance (MV) controller has indeed been more successful than the PI controller in removing the long-term components of process variability.

Another perspective is obtained by looking at the power spectra of the PI and MV controlled output (Figure 5-55). It is mathematically possible to represent these output curves as a weighted sum of sine waves of different frequencies. The power spectrum indicates how much of the "energy" or power of the curve representing the process variation can be accounted for by the sine wave at each frequency. Clearly, the output of the PI controller contains more low-frequency energy than that from the MV controller, indicating again that the latter is more successful at removing low-frequency variation from the process.

(continued)

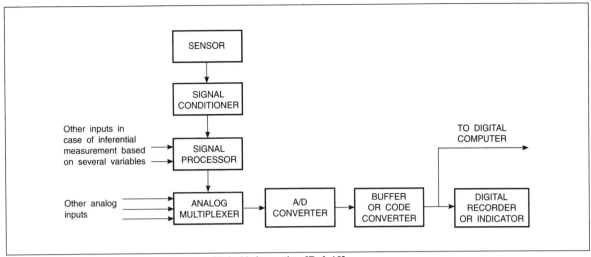

Figure 5-50. Conversion of Sensor Data to Digital Information [Ref. 16]

(continued)

It has now become possible to evaluate the relative effectiveness of the two controllers, which was not possible using simple inspection of the raw process output. [Note: This example was graciously provided by Michel Perrier of the Pulp and Paper Research Institute of Canada for use in this book.]

To summarize relationships between two or more variables, various forms of least squares regression analysis are commonly used. The term "least squares" refers to the fact that these techniques seek to fit to the data that line (or curve) that minimizes the sum of the squared distances of the data points from the line or curve (Figure 5-56). The resulting line or curve serves as a means of predicting furture values of the dependent variable (the one we are trying to estimate) from future observations on the independent variables (the ones we can observe). Care must be taken to ensure that enough observations are taken so that the type of curve in question can be fitted with acceptable confidence; for guidelines, refer to any standard statistical text.

DATA REPORTING

Most process information systems allow the definition of a variety of report formats to extract and present information in whatever form is best known and appreciated by the various departments of a company. Standard formats may also be available to satisfy the needs of, for example, government regulatory agencies or established accounting standards.

DATA ARCHIVING

The volume of data generated by repeated sampling of large numbers of data points can quickly exceed the capacity of all but the largest data storage devices. However, it may be desirable to keep old process data, at least for a while, for reasons ranging from process model validation to regulatory or legal necessity. Many control system users, therefore, write their oldest process data periodically to archival storage, which is no longer on-line but is accessible if needed. Since fast access is unimportant but large capacity is vital, magnetic tapes or optical disks are the preferred media for archival storage. The media should be kept in a secure location such as a data vault because often they constitute the only remaining copy of the information.

Basic General-Purpose Applications

Certain types of programs, are widely used on computers because they fill needs that are almost universal. Database programs allow storage of large quantities of data in structured form, update selected items as they change, and retrieve all or part of the data according to various criteria. Spreadsheet programs enable the creation of tables of formulae and data that can be used as a framework for lengthy, repetitive calculations. Word processing programs facilitate the creation, revision, and formatting of text documents.

In addition, CADD programs provide for engineering drawings the same ease of creation and revision that word processors do for text. They also include special features such as layers to express intrinsic drawing structure and allow suppression of extraneous detail and the assignment of descriptive attributes to drawings or drawing entities to aid in documentation and revision control.

Users' groups have formed for many popular software packages to give people an opportunity to share their experiences with the software. These can be an invaluable source of help when you run into difficulties. Many software vendors also offer an 800 number for telephone support or a bulletin board related to their product.

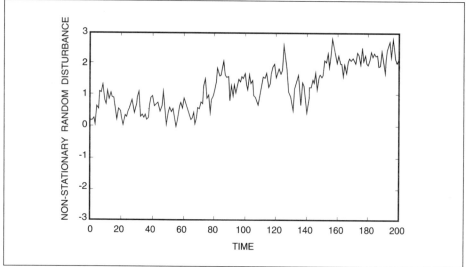

Figure 5-51. Non-Stationary Random Disturbances

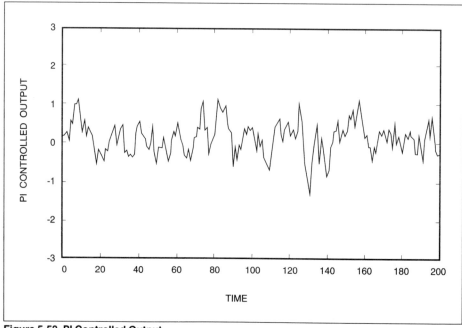

Figure 5-52. PI Controlled Output

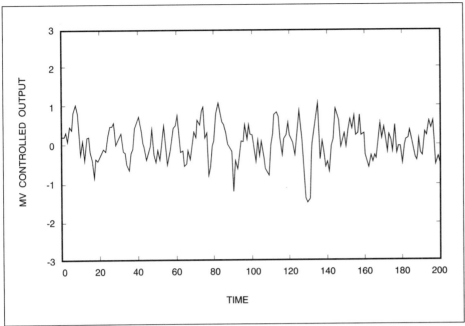

Figure 5-53. Minimum Variance MV Controlled Output

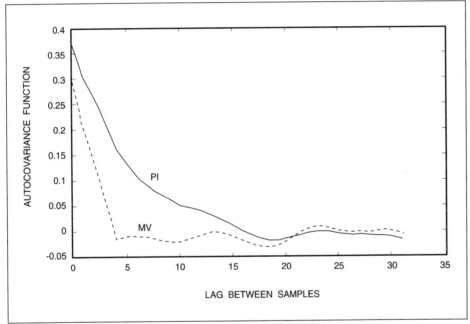

Figure 5-54. Autocovariance Function of PI and MV Controlled Output

Project management packages help in visualizing and understanding the complex interplay of tasks, resources, and deadlines involved in an engineering project. The impact of changes can be readily evaluated and various possible scenarios investigated.

DATABASE PROGRAMS

A database is essentially a structured way of storing and retrieving information—a computerized equivalent of the filing cabinet, but much more flexible. Whereas the filing cabinet restricts us to a single way of organizing files, the computerized database allows us to sort, select, and retrieve information in multiple ways.

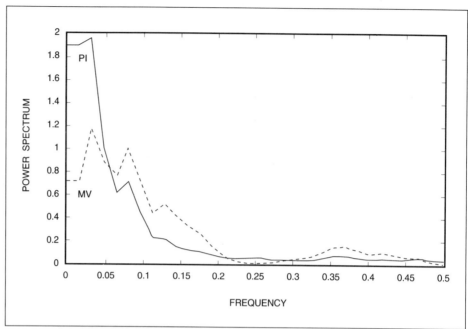

Figure 5-55. Power Spectra of PI and MV Controlled Output

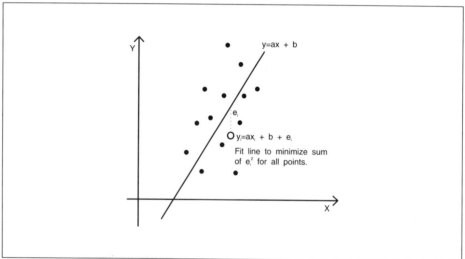

Figure 5-56. Linear Regression by Least Squares

Databases are organized by records and fields. A record corresponds more or less to a file folder containing multiple pieces of information on a particular topic. Each piece of information within the record resides in a field. In more complex structures, the fields can themselves have subfields. A unique record key must be included as one of the fields to serve as an unambiguous record identifier and to provide a link to other databases, as illustrated in Figure 5-57. Normally, all records in a given database have the same field structure, though in some cases the length of certain fields may vary from one record to another.

Imagine for a moment a set of files relating to equipment maintenance in a plant. There would probably be a file folder for each piece of equipment, containing equipment and installation specifications plus records of any checks and servicing performed.

The main problem with this system is that once it has been set up with one organizational scheme (by equipment), it is difficult to extract information any other way than by equipment. Getting a list of all equipment that has been serviced this year, or all equipment from a given manufacturer, requires a painstaking dig through all the file folders.

On the other hand, if the contents of the files were entered into a computer database, the above lists could be generated quickly and easily by selecting all records where a given field had a certain value (the current year, or a manufacturer's name).

Databases can also be designed to check input data against validity specifications for each field to prevent bad data from being entered. If two or more databases need some of the same data items (for example, if a purchasing database needed some of the same equipment information as your maintenance database), many database programs can store the shared information once and have both databases refer to it in the same location. This eliminates the risk of failing to modify all instances of a piece of information (since there is only one instance) and ending up with incompatible information in the database.

The ability to store information once and refer to it from two or more databases is one of the characteristics of relational databases. What makes a database relational is, in fact, much more complex than this. The basic requirement is that data must be stored in tables that have a certain structure and are accessed and manipulated in certain specific ways (and only in those ways). The purpose of these restrictions is to define a set of characteristics that users can assume the database has. The users can then work with the database via a language such as SQL (Structured Query Language) without having to worry about how the underlying structures are implemented.

Unfortunately, many so-called relational databases adhere only partially to the strict relational standard and require the use of complex programming languages to make full use of their power and flexibility.

A database can also be updated by having a process information system feed new values into it, in which case it is called an on-line or real-time process database.

SPREADSHEET PROGRAMS

Spreadsheet programs were originally developed to provide an electronic version of the accountant's paper spreadsheet to make repetitive financial calculations easier and less error-prone. They have, however, been adopted as a general-purpose calculation tool for everything from budget projections to laboratory data analysis.

An electronic spreadsheet is basically a table with rows and columns intersecting to form cells. By convention the columns are referred to by letters and the rows by numbers, hence cell "D24" is found in the twenty-fourth row of the fourth column. The real power of the electronic spreadsheet is that a cell can contain not only data (numbers or text) but a formula referring to data in other cells. This allows the spreadsheet to be used as a simple programming tool. Figure 5-58 shows how a spreadsheet program might handle a compound interest calculation.

The computing power of a spreadsheet can be further extended via macros, which store a sequence of cell operations under a single name by which the macro can be recalled and executed. A spreadsheet with formulas and constants defined, but without variable data values, can be stored as a template and repeatedly copied, filled in, and used for repetitive calculations.

Most spreadsheet programs include a variety of report and graphics generation options and allow results to be exported to other programs, all of which facilitates

Setting up a database should be preceded by careful planning to ensure that the database will be easily maintainable and expandable and that information will be readily available in the form you want it. If several groups of people will be using the database, be sure to consult all of them to make sure everyone's needs will be met.

If you make extensive use of a database, learning its associated programming language (if any) can be a good investment of your time, enabling you to extract and use information much more effectively.

Many databases are accessible via modem and can offer you every kind of information from current stock prices to a list of references on some obscure technical topic. Before getting heavily involved with exploring these databases, however, be sure to check the costs involved. Many of these information services have both a monthly or yearly charge and a second charge based on the actual time logged on.

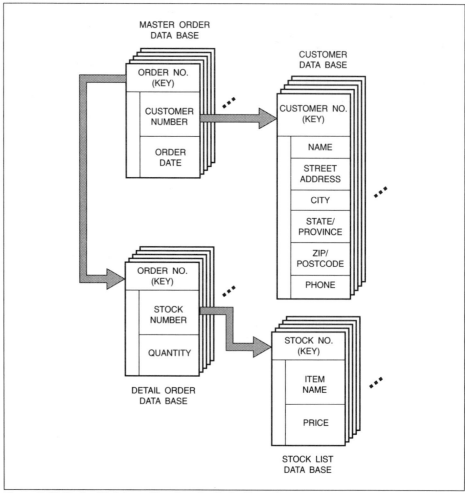

Figure 5-57. Typical Database Structure for Order Entry System

the preparation of reports and presentations using the results of the spreadsheet computations.

For example, a company's standard expense account form could be set up as a spreadsheet template. An employee would then make a new copy for each trip and fill in only the dollar amounts for the expenses for that trip along with factors such as currency conversion rates. The daily expenses and required totals would then be calculated with a single key press.

WORD PROCESSING PROGRAMS

To the dismay of many engineers and technicians who hoped to avoid it forever, writing persists as a necessary part of their daily activities. Specifications, documentation, and reports to various levels of management persist in coming back with copious changes and additions in blue pencil and must be revised again and again. Word processing programs offer a wide selection of revision and reformatting tools to aid in this task. The destructive backspace alone is a sanity-saving antidote to many an engineer's sloppy typing. Block move, copy, and delete capabilities and file import and export allow the reuse and recombination of previously written immortal prose. Spelling checkers save the chronic poor speller from repeated embarrassment, and an on-line thesaurus can provide a fresh choice of synonyms and antonyms when creativity starts to run dry. Some packages even include a style checker to tell you when you are splitting your infinitives, running off at the mouth (figuratively speaking), or spouting jargon.

CADD (COMPUTER-AIDED DRAFTING AND DESIGN)

What word processing has done for text, CADD has done for the engineering drawing. Laborious manual redrafting has become a thing of the past, replaced by the deletion and replacement of electronically defined graphic objects. A typical CADD package allows you to create lines, curves, geometric shapes, and text blocks and combine them into objects that you can then move, copy, resize, store, retrieve, and manipulate in other ways. Color, line thickness, line type, and other characteristics can also be associated with objects. Pan and zoom features let you move around a large drawing, magnify an area of fine detail to work on it, then step back to see the larger picture. Many packages have a layer structure that allows you to separate out different aspects of a drawing (such as layout, mechanical, and electrical). Special engineering features may include automatic dimensioning, interference checking, revision control, and bill of materials takeoff.

Graphics (drawings) tend to require more storage space and processing power than text data, so computers used for CADD tend to have large hard disks, lots of memory, and fast microprocessors. They often have large, high-resolution screens and plotters to produce high-quality graphic output on large sheets of paper or film.

The use of CADD may be extended into CAM (computer-aided manufacturing), in which the computer-generated drawing of, say, a machine part is forwarded to a numerically controlled (NC) machine tool that will actually make the part directly from the electronic drawing.

> In an office that does a lot of CADD, defining standard objects and layer structures helps make everyone's drawings understandable to all.

PROJECT MANAGEMENT SOFTWARE

Project planning is another activity in which technical people often get involved and in which frequent changes make a nightmare of updating manually prepared charts and schedules. Project management software packages allow making changes electronically and then checking out their possible effects in various "what-if" scenarios. Basically, all these programs address the problem of allocating tasks, resources, and time (given certain constraints) and of revising these allocations "on the fly" as circumstances change.

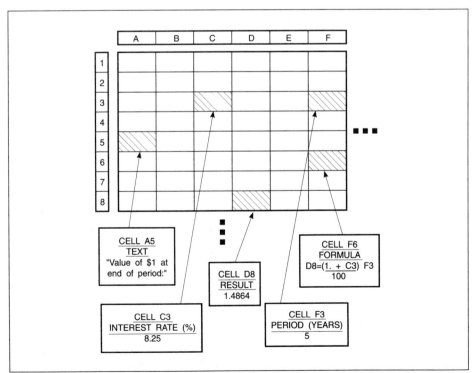

Figure 5-58. Typical Spreadsheet Calculation

Advanced features include resource leveling (adjusting project completion date to fit available resources and time), the ability to plan at various time scales (hours, days, or weeks), and the use of different time units for different tasks or resources. Most packages provide a variety of options, both tabular and graphic, for summarizing their conclusions, including histograms, Gantt charts (a series of horizontal bars representing tasks on a time scale), CPM (critical path method) charts highlighting the tasks where delays will immediately affect the completion date of the project, and PERT charts focusing on the relationships between tasks. Figure 5-59 illustrates the critical path for preparation of a magazine advertisement.

Specialized Applications

In contrast to those discussed in the preceding section, these are applications you may use from time to time depending on your own particular responsibilities.

PROGRAMMING LANGUAGES

If you want direct, "hands-on" access to your computer to make it do exactly what you want, turn to one of the hundreds of computer programming languages on the market today. Languages vary first of all in how close they are to the ones-and-zeros style of machine language. Languages that deal with low-level details such as shift registers and binary-to-decimal conversions, with each statement corresponding to few machine operations, are called low-level languages. The prime example is assembler language, which is microprocessor-specific and amounts to a shorthand for binary code. On the other hand, languages where each statement accomplishes many machine operations ("sort data on this field" or "perform the following series of operations 100 times") are referred to as high-level languages.

The earliest high-level languages (late 1950s) were designed for the leading applications of the day: FORTRAN for scientific computation, COBOL for business data processing. More recently developed languages such as Basic, Pascal, and C were designed for wider applicability and are strong in some areas (graphics display and character string manipulation) that were not even in the picture when FORTRAN and COBOL were developed. However, the older languages have evolved into new versions to keep pace with the times and still do a credible job in the areas in which they were designed to perform well.

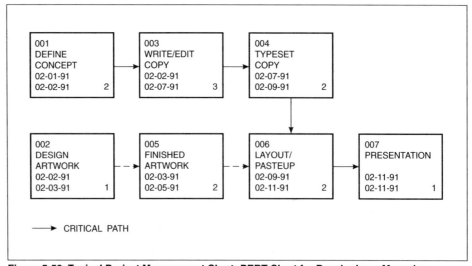

Figure 5-59. Typical Project Management Chart: PERT Chart for Developing a Magazine Advertisement

High-level languages differ in how they treat the source code one writes and how they interact with the code writer. Interpreters simply translate code one line at a time into machine language and execute it. The advantage of this is that if the program hangs up or starts behaving strangely, you know EXACTLY where it happened. The disadvantage is that interpreted languages generally run slowly because the computer is "figuring out" the program as it goes.

To attain greater execution speed, compiled languages translate the source code into machine language first, producing a separate object file. With programs that call subroutines and functions, another step (linking) may be necessary to reconcile the address references in the various program components. The output of the linker is an executable file, which you can then run. Compiled languages tend to produce fast-running executable code, but if anything goes wrong, you may have no indication of where in the source code the problem is.

The decision between assembler and a higher-level language used to be made on the basis that greater speed and compactness could be achieved with assembler, albeit with more programming effort. With modern optimizing compilers this is no longer true, and the effort of programming in assembler can usually be justified only when very high speed or special functions unavailable in a high-level language are needed.

STATISTICAL ANALYSIS

A number of programs exist to perform standard statistical analysis (mean and standard deviation, correlation, regression, analysis of variance (ANOVA)) on one's data and produce common types of graphs such as histograms and bar charts. These methods help separate random fluctuations in information from the underlying patterns and trends.

Beyond this, if a plant is involved in statistical process control (SPC) or statistical quality control (SQC), it can obtain software to generate the charts and numbers commonly used with these techniques. A typical SPC analysis is illustrated in Figure 5-60.

Process Simulation

Simulation involves building a mathematical model of the process, with equations representing the transformations of material and energy that occur. This model, in the form of a computer program, is then repeatedly run with various combinations of inputs. The results indicate how the plant would respond to various process changes.

The great advantage of simulation is that experiments can be performed on the process at minimal cost and without disruption or risk in an operating plant. The mathematical model also serves as a form of process documentation and points out areas where the quality of available process information needs to be upgraded.

SIMULATION SOFTWARE

Although it is possible to build a small simulation model using a programming language or even a spreadsheet program, commercially available simulation software offers a variety of tools to make model building easier, from interactive graphics displays to comprehensive error checking of data files. These package programs provide parts of the simulation model already written and tested (for example, programs to control execution sequence of the modules and test for mass and energy balance, or a set of standard process modules for common operations such as mixing, splitting, and dilution) and thus allow one to concentrate on the aspects of the model that are specific to an application.

THE MODULAR SEQUENTIAL APPROACH

Available simulation packages use two main approaches to process flowsheet calculations. The modular sequential approach represents each piece of equipment or process operation as a block (module) of equations that accepts input values, performs calculations, and produces output values. If the process includes recycle streams, initial assumptions must be made about these streams and the flowsheet recalculated multiple times until a steady state is reached (which can be very time-consuming). This approach does, however, have the advantage that the modules are relatively simple and easy to modify or rearrange.

Figure 5-61 illustrates how a pulp screen room might be represented using modular simulation software. This simple screen room contains primary and secondary screens, with the secondary screen accepts returned to the primary screen feed, and secondary screen rejects being thickened, refined, and then returned to the secondary screen feed.

How the screen room is described for simulation purposes will depend on what information is to be obtained from the simulation. In this case, let us assume that the model builder is mainly interested in the selective removal of coarse or fine fibers by screening and refining. The screen modules will, therefore, contain equations describing how the percentage of each fiber fraction removed by the screen varies as a function, say, of pulp reject rate and the consistencies into and out of the screen. Similarly, the refiner module equations will describe the reduction of the fiber fractions as a function of energy input. General-purpose mixing

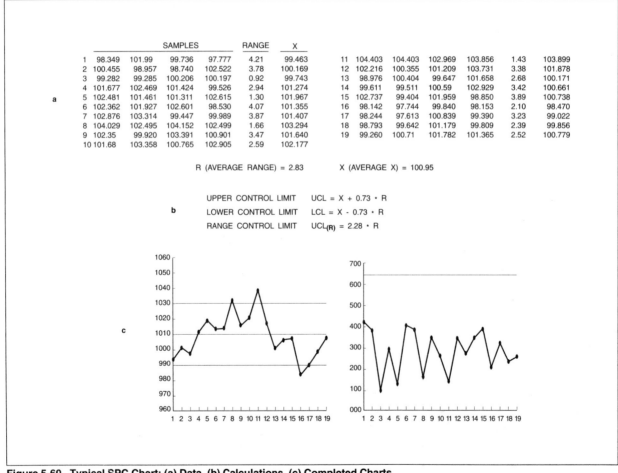

Figure 5-60. Typical SPC Chart: (a) Data, (b) Calculations, (c) Completed Charts
(From Shaw, John, "SPC for the Processing Industries," *INTECH*, December, 1989.)

and dilution modules represent the mixing of pulp streams before the screens and their dilution to the desired inlet consistency.

A modular sequential simulator does not itself optimize the process in any way. Each module simply accepts certain flow characteristics as given and calculates certain others. However, by performing a series of simulation runs under different assumptions, the model user can select the most advantageous from a set of options.

Imposing constraints on a simulation to represent capacity or availability limitations or desired operating ranges can lead to nonlinear situations with a modular simulation. This means that the simulation results must be carefully checked to make sure they are realistic.

MATRIX APPROACH (SIMULTANEOUS EQUATIONS)

The matrix approach represents the entire process as a set of simultaneous equations and applies matrix algebra techniques to solve them. This approach allows us to do some things that are difficult to impossible with modular sequential techniques, such as imposing constraints on process variables or optimizing on the value of an objective function.

Matrix techniques are generally faster than modular, finding a solution in a single iteration with a linear system and usually a few iterations with a nonlinear system. Under certain conditions, however, matrix techniques may not even find a solution, let alone an optimal one. They can also be difficult to debug, since,

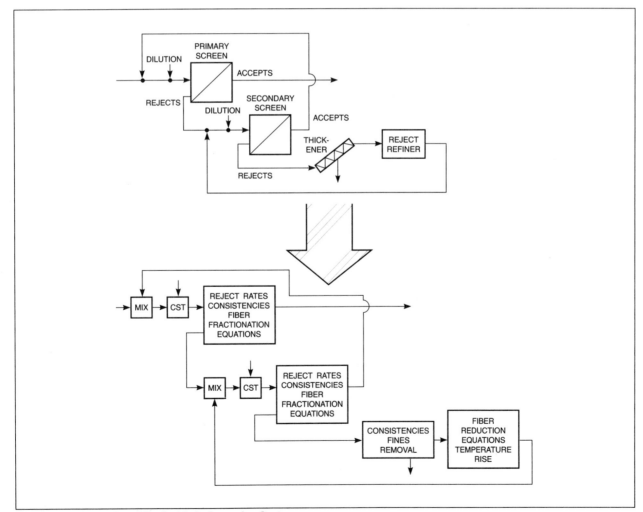

Figure 5-61. Simulation of Typical Pulp Screening System

even if one can locate the mathematical problem that is preventing the solution, it may have nothing obvious to do with anything in the process description.

For some techniques, such as linear programming, all the relationships between the variables must be linear. It may be possible to incorporate nonlinear relationships indirectly by using them to calculate the coefficients in a linear equation. For the many nonlinear systems found in process control, however, this can be an important limitation.

Many matrix-based packages can perform optimization, which involves selecting, among all the solutions that are possible given the conditions of the model, that which maximizes or minimizes some criterion such as profit, cost, resource utilization, or environmental emissions. This selection is performed automatically in the course of a single program run, unlike optimization using the modular approach, which requires repeated runs under different conditions.

In a linear programming context, for example, the function to be maximized or minimized is called the objective function and is of the form

$$a_1x_1 + a_2x_2 + a_3x_3 + ... + a_nx_n = \text{max (or min)}$$

where the x_i are process flow rates and the a_i weights (such as unit costs). The mass and energy balance requirements of the process flowsheet are described using linear flow equations, and the constraints on various process flows are represented by more linear equations of the form

$$x_i + s_i = \text{constraint value}$$

where s_i is a "slack variable" representing how much "room" there is between the current value of a process flow and its maximum or minimum value (see Figure 5-62).

Some simulation packages now use a hybrid approach that allows the user to work with a modular "front end" but translates the resulting specifications into matrix form before performing the model computations.

STEADY-STATE VS. DYNAMIC SIMULATION

For many purposes, it is not necessary to simulate the response of a process to changing conditions over time. As an example, to determine the effect of a new piece of equipment with improved performance characteristics on daily mill production, it is enough to know the new steady state the process would reach after the change. What happened while the process was adjusting to the change would be irrelevant.

A steady-state simulation calculates a mass and energy balance for a flowsheet under the assumption that flows into and out of every unit in that flowsheet balance (see Figure 5-63). Time does not appear as a variable in any equations, and time-related phenomena such as transport lags, tank level variations, and controller response delays are ignored. This approach clearly simplifies the flowsheet calculations.

Some problems, however, cannot be adequately investigated with a steady-state simulation. When simulating the effect of various controllers on a process, for example, the interaction of each controller with the time-response characteristics of process equipment and the intrinsic time delays involved in level, temperature, and concentration changes is an essential part of what we are trying to find out. A steady-state simulation simply cannot answer the sort of questions we are likely to ask about controllers.

A dynamic simulation includes time as an explicit variable. Process equations must include time where appropriate, for example when describing tank levels. Usually these extra equations are differential equations that must be solved using calculus or numerical approximation methods. Because the inputs to a unit may be "absorbed" by a change in the unit's characteristics (such as level, concentration, or temperature), mass and energy flows no longer necessarily balance around every unit at every point in time (see Figure 5-64).

The simplest approach to dynamic simulation is to view time as a sequence of intervals and assume that process variables undergo step changes over these intervals. Depending on the application, a new overall mass and energy balance may be obtained for each interval, or else local imbalances may be allowed to exist temporarily under the assumption that the ongoing process dynamics will redress them in later intervals. This approach avoids most of the differential equations but may not be accurate enough to represent many phenomena of interest.

The more correct but more demanding approach is to represent the process dynamics by differential equations and solve the resulting system of equations. The major difficulty with this approach is that a great deal of information about the process is needed to write the equations. Many equipment manufacturers can supply dynamic response curves for their particular devices, and a certain number of equations have been developed from first principles, but the larger-scale response of a particular process must often be characterized on site.

Another difficulty is that the system of equations may be difficult or time consuming to solve. So-called stiff systems of equations, which cover a wide range of fast and slow dynamics, may prove to be insoluble by all but the most sophisticated techniques, or the results may be unreliable. Many newer dynamic simulation programs offer a variety of numerical techniques from which is picked the one that seems to work best on the problem.

ADVANCED SIMULATION FEATURES

Simulations that use actual process data must cope with the fact that these data are often inaccurate. While often the effort to simulate ends up providing a power-

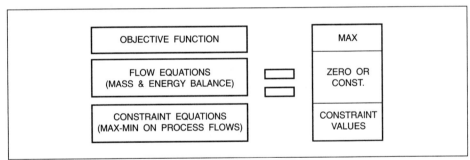

Figure 5-62. Linear Programming Structure

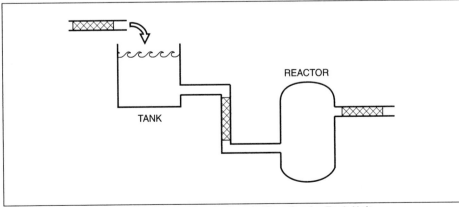

Figure 5-63. Steady-State Simulation: Flow In Equals Flow Out for Each Unit

ful impetus to improve the quality of the data, in many cases this is not possible, and problems like measurement and sampling error will always be with us. Data reconciliation takes a set of observed data values and finds that set of "real" values that provides a best fit to it, usually by least squares techniques similar to regression. It can thus reduce the impact of minor data errors on the simulation results, as well as the manual effort that must otherwise be made to resolve discrepancies between simulated and actual values.

Although most simulation projects start out with a flowsheet drawing on paper, most simulation programs still require that we convert it into a numerical data file for input and view the results in tabular form before transferring them manually onto a drawing. This is an awkward way to work, given the familiarity of most engineers and technicians with drawings, and some simulation programs now allow us to interact directly with an on-screen flowsheet diagram, for example, to set device characteristics, change flow rates, or impose constraints. Many problems, such as over- or under-specification of a piece of equipment, are much easier to identify and correct on a drawing than in a table of numbers. These programs can also display simulation results graphically, either with their own software or via a commercially available CADD package.

WHY DO SIMULATION?

One of the greatest benefits of simulation is often realized before the model is even up and running: a greater awareness of the process, how it works, why problems occur, and how much we may need to learn about it. A simulation project often provides an opportunity to sample or instrument neglected areas of the process, check and update the latest set of mill drawings, and ferret out information about how the plant is actually being operated day to day. Once people begin to know what the trouble areas are, they may begin to change set points or operating practices and bring about improvements even before they see any simulation results!

Once a simulation is working and has been thoroughly enough tested to establish confidence in the results, one can perform an essentially unlimited series of "what if" exercises quickly and at minimal cost. Setting up the simulation initially may have been more work than doing the calculations by hand, but the second and subsequent simulations are much faster, whereas a manual calculation must always start from zero. When the process involves extensive nested recycle loops or dynamic effects, manual calculation becomes close to impossible and highly error-prone. The computer calculations are known to be accurate, and although they may take a while in complex cases, they can be left to run without human intervention.

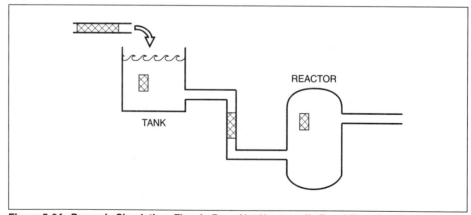

Figure 5-64. Dynamic Simulation: Flow In Does Not Necessarily Equal Flow Out Due to Level Changes and Time Delays

A simulation model will, of course, need to be changed as the process changes. Think of it as another piece of process documentation like drawings and equipment manuals, which also need periodic updating. This maintenance effort can be well worth while to have the power of simulation available to check out the next batch of proposed process modifications.

SOME WORDS OF CAUTION

Running a simulation can be so easy that some people forget the "garbage in, garbage out" principle. If you feed inaccurate data to a simulation model, it will obediently give you inaccurate results—with no warning that they are wrong. Another way to get nonsense output without knowing it is to apply a process equation outside its range of validity. Many of the equations used for simulation were developed empirically in the field from observations in the normal operating range. If conditions move outside this range, the equation may no longer hold; for example, a relationship that is linear within the normal range of a variable may become nonlinear outside, perhaps because by definition the variable must remain within certain bounds (see Figure 5-65).

IMPORTANT USES OF SIMULATION

Simulation is more and more widely used for design, both of individual pieces of equipment and entire processes. Although it is not possible to calibrate a design simulation to an existing process, the characteristics of many equipment and process components are well enough known so the simulation can at least weed out early those design choices that are clearly unsuitable. Simulation also serves as a vehicle for checking out a wide range of design possibilities, including "long shot" or innovative solutions that would never get investigated manually because of the work involved. The final simulation model may be provided to the client along with other documentation.

Simulation of an operating process is increasingly used to fine tune operating practices and to look into possible process changes such as equipment upgrading. It also permits operating staff to gain insight into process behavior under various operating conditions, such as start-ups, shutdowns, upsets, and different production regimes, and to experiment with improved control.

Training simulators are becoming a vital adjunct to today's highly automated processes. Operators can start learning months or years before a new plant starts up, using a realistic operator workstation interfaced to a dynamic simulation model. Usually an instructor supervises the training session "behind the scenes," although in future some of this person's functions may be taken over by artificially intelligent trainers. Normal, abnormal, and emergency conditions can be repeatedly presented until the required actions are thoroughly learned. Areas where the operator has difficulty can be covered again, perhaps with a more gradual progression of cases or more careful explanations. Once the plant is in operation, the training simulator continues to be useful for training new operators or giving refresher courses to those already there.

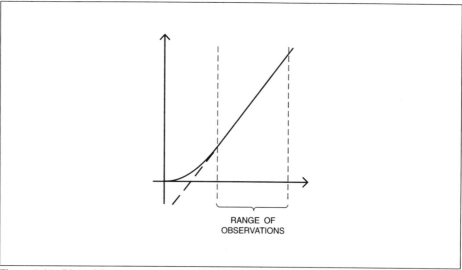

RANGE OF
OBSERVATIONS

Figure 5-65. Risk of Extrapolating an Empirical Relationship

To illustrate the sort of questions that can be investigated using process simulation, let us consider the problem of determining the most economical peroxide bleaching system for a mechanical pulp application.

Pulp can be produced from wood by chemical or mechanical processes. Mechanical pulping offers high pulp yield at low cost, but the pulp still contains much of the lignin (the substance that binds fibers together into wood) and is, therefore, more difficult to bleach than chemical pulp. In recent years, chemical treatment of wood chips prior to mechanical pulping has produced a superior quality mechanical pulp that is suitable for many uses previously reserved for chemical pulp. For these new uses, however, effective and economical bleaching has assumed a vital importance.

Hydrogen peroxide is the most effect known bleaching agent for achieving high brightness in chemically treated mechanical pulps. However, because of the high cost of this chemical, the bleach plant must be designed to make maximum use of it through strategic recycling. Therefore, a simulation study was carried out to investigate the economics and bleaching effectivess of one- and two-stage bleaching systems at various pulp consistencies.

In addition to the usual mass and energy balance equations, a mathematical description of peroxide bleaching was required. The model predicts brightness, residual peroxide, and alkali from temperature, chemical charge, residence time in the bleaching tower, and pulp consistency during bleaching. Some of the relationships incorporated in the model are illustrated in Figures 5-66 and 5-67. By adjusting model parameters, the bleach response of pulps made from different wood species can be simulated.

As a first step, the peroxide bleaching model was turned to both laboratory and mill data. A series of simulation runs was then undertaken to identify the process configuration and operating parameters that would minimize the peroxide cost to reach a given target brightness. The runs represented various combinations of the following variables: number of bleaching stages (one or two), and in each stage, the pulp consistency (medium or high), residence time, and peroxide charge. Figure 5-68 illustrates a one-stage peroxide bleaching process, and Figure 5-69 shows a two-stage process.

The simulation predicted that when bleaching to high brightness (80% ISO), a two-stage process could save $6.50-$7.00 per oven dry metric tonne of pulp in chemical costs over the best one-stage process. The best combination proved to be high consistency in the second stage and medium

(continued)

(continued)

consistency in the first stage, with all peroxide charged to the second bleach tower and then recycled to the first tower.With a lower brightness target (70-75% ISO), a one-stage process might well be sufficient. At lower peroxide charge, peroxide residuals will also be less, and the benefits of recycling are reduced.

[Note: This example is described in greater detail in Strand et al., "Optimization of peroxide bleaching systems," *Tappi Journal*, July 1988, pp 130-134. The authors have graciously given their consent for use of their work as an example in this text.]

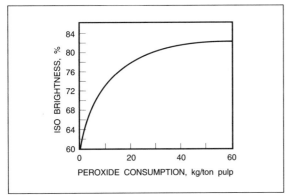

Figure 5-66. Typical Relationship between Brightness and Peroxide Consumption

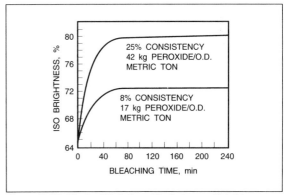

Figure 5-67. Typical Relationship between Brightness and Bleaching Time, Consistency, and Chemical Change

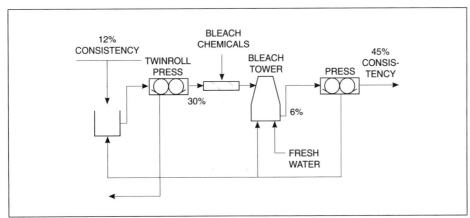

Figure 5-68. One-Stage Bleaching Process

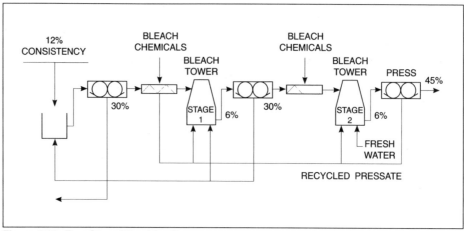

Figure 5-69. Two-Stage Bleaching Process

Artificial Intelligence and Expert Systems

The day will certainly arrive when we expect industrial equipment to possess some measure of intelligence, just as we now expect a kitchen to offer us a toaster, a coffee maker, and a microwave oven. How fast that day will come is, however, difficult to predict. Too many enthusiastic projections have failed to become reality, largely because we have underestimated again and again the complexities of real human intelligence.

What is certain, however, is that whatever intelligence we can build into our processes and equipment will be of immediate practical usefulness. It will provide us with more and better information that will help us evaluate situations and make decisions. It will also feed this information into the increasingly integrated computer systems that are linking our plants together. The total mass of information in the plant will not only get steadily larger (some would say it's too big already), but, thanks to more sophisticated ways of managing and using that information, it will become better organized and more accessible. It is hoped that all this will help us, with our own natural intelligence and whatever artificial intelligences we build to help us, run our plants more efficiently and profitably.

The extremely broad field of artificial intelligence attempts to understand how human mental faculties work and to find ways for computers to replicate some of them. Its diverse spectrum of activities includes expert systems, robotics, machine vision, and speech recognition, among others. These efforts are steadily expanding the boundaries of what computers can do and can be expected to have a major impact over the next ten or twenty years on how computers and other machines work.

Why Artificial Intelligence is Needed

The amount of information in our world is increasing exponentially as science and technology expand to penetrate every aspect of our lives. Our grandparents had minimal contact with advanced technology and needed only a basic stock of general knowledge to get by. Now, even making breakfast may involve dealing with a microwave oven and a programmable coffee maker.

In particular, the industrial environment is becoming more complex and more demanding. Sensors and instruments are everywhere, providing masses of information that need to be interpreted and acted upon. Processes must now be operated to tight tolerances to satisfy quality, cost, and environmental requirements. Hundreds of new products are entering the marketplace, each with new standards to meet and new production challenges to overcome. As business operations become more and more computerized and interconnected, their appetite for information grows, as does that of government and regulatory agencies.

Human beings have coped successfully with the complexities of their environment for tens of thousands of years because they are good at certain things, such as learning, adapting, pattern recognition, and generalization. They are less good at processing large volumes of information or carrying out precise computations. In recent years, computers, which are good at computation and information processing but lack many more "human" abilities such as learning, have taken on many of these tasks for us.

In many areas, however, we need even more help to cope with information overload and demands for sophisticated decision-making. But computers cannot help us beyond what they are doing already unless they can be endowed with some of the abilities that up till now have been the exclusive province of human beings. The development of artificial intelligence involves finding new ways to program computers so that they will have these abilities.

The field of artificial intelligence has many branches, each addressing a different human ability. Expert systems try to capture in the form of a computer program the extensive knowledge bank and sophisticated decision-making strategies that a human expert uses to solve problems. Neural networks try to replicate the pattern recognition abilities of the human brain using electronic neurons. Robotics aims to build intelligent machines that can manipulate objects as well as a skilled human being, and speech recognition and vision systems attempt to give computers a humanlike ability to process sensory inputs.

Expert Systems

The expert system is probably the area in which artificial intelligence concepts have had the greatest practical impact to date. Over 1500 expert systems are in use in industry today, and many of these claim cost:benefit ratios of 10:1 or better.

An expert system contains a significant part (though not all) of the knowledge an expert uses to solve problems in a particular field or domain. Once this knowledge is embedded in a computer program, a non-expert can run the program and gain access to the knowledge to solve problems. This has the effect of making expertise more widely available and raising the performance of average or marginal workers closer to the level of the best.

EXPERT SYSTEM STRUCTURE

Expert systems contain a knowledge base (a set of facts, or declarative knowledge, representing what the expert knows) and a rule base (the IF-THEN or procedural reasoning by which the expert makes use of the knowledge). The rule base may also include control knowledge that embodies the expert's overall problem-solving approach and imposes some high-level direction on the sequence of rule execution.

A particular problem is presented for solution in the form of additional facts that describe the situation at hand. The "executive program" of the expert system, called the inference engine, then takes the facts and the rules and tries to find rules whose IF-condition is matched by the existing facts. A matched rule is then fired (executed), generating more facts, which fire more rules, and so on until conclusions are produced (see Figure 5-70).

SEPARATION OF KNOWLEDGE AND CONTROL

An expert system is a special kind of computer program and works in many ways like a standard data processing application (see Figure 5-71), which processes inputs and historic data from a database to produce outputs. One important difference, however, relates to how the sequence of execution of the program is controlled.

In conventional computer programs such as the data processing application, the flow of control depends on conditionals (IF-statements) within the program code itself. The form of these conditional statements depends on what kind of information is in the database. For example, if five different kinds of records can occur in the database, conditionals that specifically refer to each kind of record will occur in the program. If a new case is encountered (say, a new type of record has been added to the database), not only the database but the program itself must be modified to handle it, at the risk of introducing "bugs."

In an expert system, on the other hand, the sequence of rule firings is controlled by the inference engine, which is completely separate from the knowledge and rule bases (see Figure 5-72). To include a new case in an expert system, only the knowledge base and possibly the rule base will have to be changed. The basic control mechanism remains untouched, making it easier to solve any problems that occur when the new case is introduced.

The control approach taken by the data processing application referred to above is often called procedural, because the sequence of execution is explicitly determined by precoded procedures.

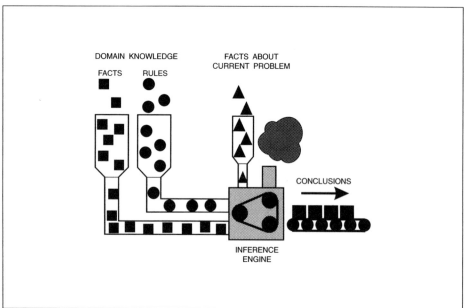

Figure 5-70. How Expert Systems Work

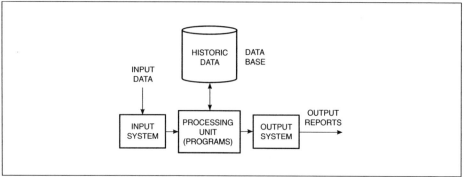

Figure 5-71. Typical Data Processing Application [Ref. 55]

In contrast, the control approach used in expert systems is referred to as data driven, because which rules fire is determined by what facts are present at the moment. There is no fixed sequence of rule firings; rules become active if and only if facts exist to match their IF-conditions. If several rules become active at the same time, some form of conflict resolution involving priorities or control logic determines which rule actually gets fired next. This data-driven style of execution is very flexible and approaches what human experts do.

FORWARD AND BACKWARD CHAINING

If a set of IF-THEN rules were presented with a set of facts as described above, the rules would fire in what is called forward chaining mode (see Figure 5-73). The inference engine would attempt to match the whole set of rules using the whole set of facts. Some rules would be matched and would fire, asserting new facts that would match and fire other rules, until conclusions were reached. This chain of reasoning proceeds from data to conclusions in what we would normally think of as a "forward" direction, from inputs to outputs. Forward chaining is well suited for problems such as classification, where a large mass of data must be resolved into relatively few conclusions (categories).

For certain other problems, however, forward chaining can involve much wasted effort. If there is relatively little data from which to draw many conclusions, as in a medical diagnosis problem, for example, many paths of reasoning are likely to be started that never lead to a conclusion. In following these paths, the system does a great deal of work for nothing.

Backward chaining (see Figure 5-74) involves starting with conclusions (goals) and trying to find chains of reasoning that support them. It involves a backward pass (to identify the chain of reasoning), followed by a forward pass, which fires the rules that constitute this chain of reasoning (as in a standard forward-chaining system) to assert the required intermediate facts and eventually the goal itself.

EXPERT SYSTEM SOFTWARE

The earliest work with expert systems was done either with conventional programming languages or with special artificial intelligence (AI) languages with features well-suited to building these systems: pattern matching capabilities, list and character string handling, and recursion (the ability of a function to call itself

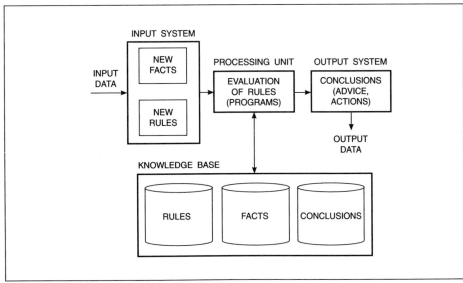

Figure 5-72. Typical Expert System Application [Ref. 5]

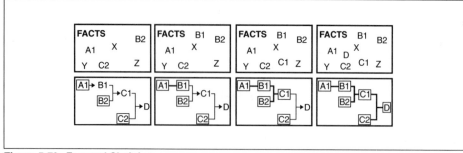

Figure 5-73. Forward Chaining

repeatedly, like the definition of *n* factorial as *n* times $(n - 1)$ factorial). The two best-known and most widely used artificial intelligence languages are Lisp and Prolog.

Most expert system builders today use expert system shell programs, which offer a prewritten inference engine together with a programmer interface, debugging facilities, and hooks for external functions. These shell programs are usually written either in an AI language or a general-purpose language such as C, but competence in the underlying language is not a necessity for using the shell. The idea of a shell is to take care of the "housekeeping details" and let the user concentrate on building a knowledge base and a rule base.

The large number of available shells vary greatly in the features they offer: forward or backward chaining or both; more or less sophisticated pattern matching; frames that work like a database record structure to help organize large knowledge bases; cross-referencing and debugging tools; tool kits for building end-user interfaces. Hardware requirements can range from a standard PC-compatible computer, on which systems of up to 100-150 rules can easily be built, to high-end workstations or mainframes capable of running systems of thousands of rules.

AREAS OF APPLICATION FOR EXPERT SYSTEMS

Any area of activity in which expert knowledge can provide a performance edge over the average person is a candidate for an expert system application. In

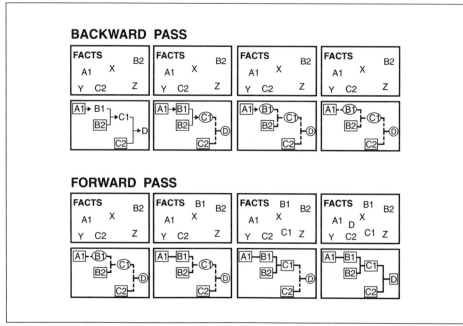

Figure 5-74. Backward Chaining

fact, the most successful applications usually are of medium size (small ones tend to be trivial and large ones, impossible), involve knowledge that can be verbally presented (such knowledge is easier to encode in facts and rules), and can be clearly bounded so a useful application can be developed in reasonable time.

Expert system applications already up and running in various industrial environments include process control, alarm management, diagnosis of operating problems, planning and scheduling, equipment design and configuration, and operator training. A representative list is provided in Table 5-1.

Table 5-1. Sample of Typical Expert Systems Since 1964 [Ref. 55]

Year	Name	Purpose
1964	DENDRAL	Identification of molecular structures
1968	MACSYMA	Aid for mathematicians, scientists and engineers in tackling mathematical problems
1970	META DENDRAL	Identification of molecular structures
1976	CON GEN	Identification of molecular structures
1978	XCON	Configuring of VAX-11/780
1980	VM	Intensive-care monitor
1983	FALOSY	Fault-location system for program debugging
1983	TAX ADVISOR	Tax-planning recommendations
1983	XSEL	Help for salepersons to develop system orders
1984	CRIB	Computer Retrieval Incidence Bank for computing fault diagnosis
1984	MYCIN & INTERNIST	Medical diagnostics
1984	PROSPECTOR	Geological prospecting
1985	TART	Tactical air targeting
1985	SCHOLAR	Geographical tutor

DEVELOPING AN EXPERT SYSTEM APPLICATION

Expert systems is a new technology that differs substantially from most of the other technologies that engineers and technicians are accustomed to using. Even experienced programmers may have difficulty making the transition from procedural to heuristic programming. Therefore, the first step in developing an application often is learning about the technology. If the intent is to build a series of small applications, someone with a general scientific or technical education can usually learn enough in a few months to get started. For larger systems, it is usually necessary to involve AI specialists who have specific training in building expert systems.

Building an expert system requires some knowledge that will constitute the facts and rules. Knowledge acquisition (or knowledge extraction) is the process of getting the knowledge from the expert. The process is nontrivial, since most experts have great difficulty describing how they solve problems. Since computers know nothing to start with, we will need to represent not only the expert's reasoning strategy but the concepts and background knowledge he uses, including things that may seem ridiculously obvious, such as the fact that a washer performs washing.

A common knowledge acquisition technique is to interview the expert, with someone taking notes or recording the interview on audio tape. Care must be taken not to "put words in the expert's mouth" so that what one gets is really the

It is not usually advisable to have the expert build the system. Having at least one other person involved provides another perspective on the expertise, which generally leads to a more accurate representation.

expert's ideas rather than the interviewer's concept of what they should be. As the project advances, diagrams of the rule network or knowledge base structure can serve to focus the discussions on areas in which more information is needed.

Information obtained in this manner tends to be highly verbose and unstructured. Knowledge engineering is the process of taking the raw information, finding the important constructs in it, and representing them in such a way that they will function efficiently as a computer program. For systems of any size, this must be done by a specially trained knowledge engineer.

Many expert system projects involve the building of a prototype system at some early stage. The prototype is a small version of the final system, but with all its important features. It serves as a proof of concept, testbed, and demonstration vehicle. Depending on experience with, and feedback about, its performance, it may be expanded incrementally to become the final system, or it may be scrapped at some point and rebuilt with the benefit of the understanding obtained to that point.

The rest of the project will consist of an iterative expansion and testing of the prototype system (see Figure 5-75). Note that if certain problems occur, it may be necessary to move quite far back in the development cycle and redo earlier work. Depending on the size and complexity of the problem and the completeness of the desired solution, building an expert system can mean one person spending six months on a personal computer with a low-cost shell program, or a team of artificial intelligence specialists spending several years on high-end workstations with state-of-the-art software.

Neural Networks

The preceding section has provided an indication of how difficult it can be to describe the knowledge and rule structures used by human experts in their reasoning. Yet we do not actually refer to such descriptions when we reason. Much of our everyday thinking about the world is, in fact, simply forming associations between phenomena we observe. We learn to abstract out the significant features of what we see so we can categorize, generalize, form analogies, and make an educated guess as to what will happen next.

In physical fact, these operations are performed by special brain cells called neurons, linked to each other by connectors called synapses. One approach to developing artificial intelligence has been, therefore, to replicate these physical elements, rather than the reasoning they support. Such a network of simulated neurons is referred to as a *neural network*.

Structure of a Neural Network

The essence of a neural network is a large number of simple, usually identical, highly interconnected processing elements. There is no separate "memory" where information is stored and no "program" that tells the processing elements what to do. Rather, each processing element has a set of weights and responds to external inputs by modifying these weights. It then uses the new weights to calculate its own stored transfer function. This generates a single output that goes to one or more other processing elements. The resulting pattern of weights in the network "stores" information about the stimuli the network has experienced and the responses that were produced. This is the closest a neural network gets to the stored program and data of a conventional computer.

Because a neural network does not have a program in the usual sense, it is hard to talk about "running" it or getting "results." People tend to use terms more appropriate to brains, such as training, learning, forgetting, feedback, and self-organization.

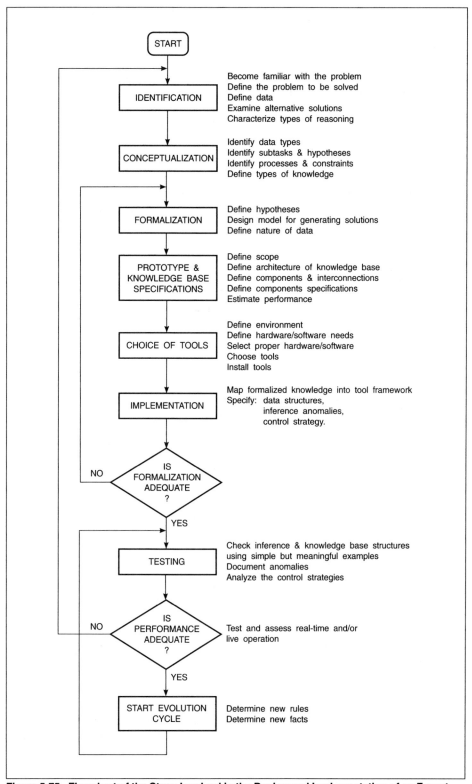

Figure 5-75. Flowchart of the Steps Involved in the Design and Implementation of an Expert System [Ref. 55]

Chemical reactions either absorb heat (endothermic) or release heat (exothermic). Even in exothermic reactions, heat often must be added to a reaction mass to initiate the reaction. This can be a risky business, however, because reaction rates increase exponentially with temperature (Arrhenius' law), which means that they approximately double for every ten degrees Celsius the temperature is raised.

If the temperature in a reactor gets too high, the excess heat makes the reaction proceed even faster, which raises the temperature further—a condition called runaway reaction. This positive feedback situation can require the batch to be dumped or, at worst, lead to fire or explosion.

To help prevent this, some cooling mechanism is normally provided, such as a liquid-filled jacket or a heat exchanger. Alternatively, the process liquid may be allowed to boil and recondense, removing the heat of vaporization. Additional chemicals such as solvents may also be added to correct the reaction conditions.

Traditional control systems respond only when the reaction temperature is already too high. To get early warning, some systems use a rate of temperature rise alarm. Still, there could be other reasons why the temperature is rising faster than a target profile. A knowledgeable operator or engineer (if available) may be called upon to decide whether there is a problem, but often the decision is made on the side of safety and the batch is dumped, leading to substantial economic losses.

An expert system that could evaluate the available data and recommend whether or not to dump the batch would both protect safety and prevent needless waste of resources.

Several types of knowledge would be necessary to build such an expert system. Text book knowledge is data about the physical plant, plus the basic physics and chemistry involved. Operating data or experience consists of a set of records describing how the reaction system has performed under a variety of conditions. Beyond this, the problem extends into the realm of heuristics where reasoning and judgment are required. An experienced operator may realize, for example, that even with a limited heat transfer surface, the reaction can be controlled if cooling water below a certain temperature can be used. At this point a large amount of additional information may be brought into play: production schedules, status of other reactors, maintenance schedules, performance history of interacting parts of the plant, and ambient conditions.

Note that such an expert system may make use of standard computer programs where suitable: process control, simulation, conventional engineering calculations, and so forth. The expert system calls on these sources of information much as a human expert might pull out his calculator to evaluate an engineering formula. One way in which an expert system might be integrated with various other sources of process information is shown in Figure 5-76.

One of the first challenges in building an expert system is often to define the scope of the problem in such a way that the system can actually be built in a reasonable time at acceptable cost. Figure 5-77 diagrams the interaction of the reactor with other components of the plant environment. Clearly it would be impossible to begin by trying to represent all these   interconnections! The preferable approach is a modular one, choosing the first modules to attempt using the following criteria:

- Will produce meaningful results

- Textbook knowledge and operating history available

- On-line data available and easy to represent in knowledge base

(continued)

(continued)

- Operator and engineer expertise available

- Provides a basis for future expansion

The modular decomposition of the problem domain is shown diagrammatically in Figure 5-78. Recipe verification was selected as the first portion of the system to be implemented.

With the domain defined, the next step is to design and build the textbook knowledge base. This consists of physical definitions of the reactor, description of the process in terms of first principles, and details of the recipes and procedures.

The identification of real-time process data and the historical data needed is the next task. This is an iterative process. After the first pass of identifying the data needed, the relationships between these data and the textbook knowledge base are evaluated. Based on this evaluation, the need for additional data may be found. When these iterations are completed, another iterative process begins: gathering the experience and expert knowledge of the operators and engineers that will be used to write the judgmental rules. Figure 5-79 illustrates these two iterative cycles.

The rules as developed may need to be ordered by the knowledge engineer according to some overall reasoning strategy such as the following:

- Evaluate quality of data.

- Evaluate correctness of charge.

- Detect potential problems if reaction proceeds.

- Detect other conditions that might support runaway.

- Detect imminent runaway.

- Determine possible/probable causes.

- Find possible/probable solutions.

A typical set of conclusions drawn by the system (as expressed in messages to the operator) might be:

POTENTIAL PROBLEM DETECTED WITH 90% CERTAINTY

- Solvent charge appears to be deficient.

- When reaction is 50% complete, high viscosity of the system will cause heat transfer rate to fall.

- Available volume in reactor for additional solvent appears satisfactory.

RECOMMENDATION: Add XX pounds of solvent. [Note: this example is condensed from chapter 21 of the *Artificial Intelligence Handbook*, by A.E. Nisenfeld, © ISA, 1989.]

Computer Technology

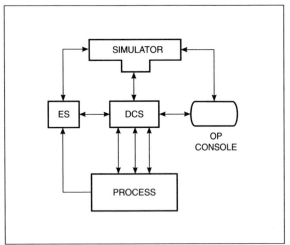

Figure 5-76. Block Diagram of an Expert System with an Integral Dynamic Simulator [Ref. 54]

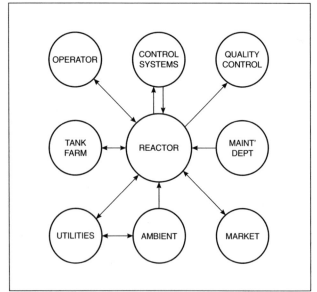

Figure 5-77. Diagram of the Reactor Environment [Ref. 54]

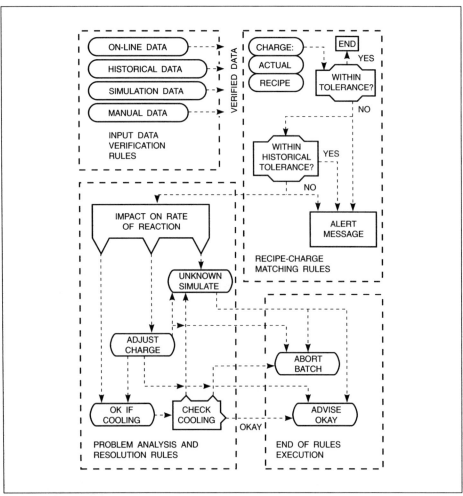

Figure 5-78. Modular Decomposition of the Problem Domain [Ref. 54]

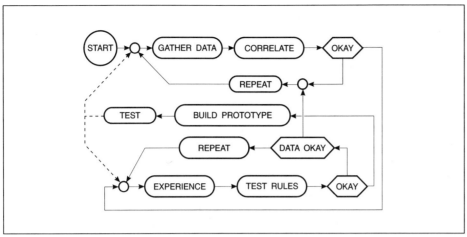

Figure 5-79. Flowchart of an Expert System Project [Ref. 54]

> *It is important to note that each processing element has its own weights and transfer function and operates on its inputs without any knowledge of what other processing elements are doing. In a sense, a neural network is a massively parallel processor. Whatever "intelligence" a neural network has resides in the overall pattern of weights, not in any particular neuron or set of neurons.*

To provide for inputs, outputs, and feedback, the processing elements are normally organized into *layers*. A typical neural network has an input layer, one or more middle layers, and an output layer. Middle layers are also called hidden layers because what they are doing while the network is operating is often hard to see or understand.

As a matter of fact, it can also be hard to understand what a neural network has done afterwards, since all that can be printed out and looked at is a table of weights, which is just a mass of numbers. This fact has some bearing on the selection of problems for neural network solutions. If you need to know how conclusions were reached (for example, an engineer might want to know what equations were used to obtain a result), the neural network approach would be unsuitable. The difficulty of assigning meaning to the weights and transfer functions can also make a neural network difficult to debug, if, indeed, one can talk about "debugging" such a construct.

Training a Neural Network

Since a neural network has no program, it does not initially know what it's supposed to do or how a "good" response is different from a "bad" one. Neural networks must, therefore, be taught or "trained." In *supervised training*, the most common form, the network is presented with an output that is known to be correct for the inputs it was given (e.g., the ASCII code for "A" if it had been given a printed letter "A") and allowed to compare this correct output with the one it actually produced. The difference between the actual and correct output is fed back into the network to modify weights so as to reduce the error next time. The concept is much like fitting a least-squares regression line to a cloud of data points. With time and many examples, the network will "learn" to produce correct outputs more and more of the time (see Figure 5-80).

Unsupervised training involves techniques that strengthen or inhibit the response of certain network elements in ways other than presenting the network with known correct outputs. One method of doing this is to find the processor whose weights are closest to each input and allow the weights of *only* that processor *and* its "neighbors" (defined in various ways) to be modified. It can readily be seen that a network manipulated in this way could develop biases over time and thus generate certain outputs more often than others. Unsupervised training is used in situations where examples of correct outputs may not be readily available. It corresponds very closely with some kinds of human learning, such as recognizing faces.

One problem with feeding error information backwards into the network is that you cannot determine after the fact which processing element's calculations were "responsible" for the error and need to be fixed. A technique called *backpropagation* is therefore used to apportion the error arbitrarily among the processing nodes; essentially, it runs the transfer function calculations in reverse. A neural network using this technique operates in an alternating mode with a forward pass (activation) followed by a reverse pass (backpropagation). All this is time-consuming, which means that this approach is not generally usable for training in real time.

Another important question is how big a change to make to the weights each time—how far to move from the actual toward the ideal output. This problem can be expressed in the form of a *learning factor b* that can take on values between zero and one. Why not jump right to the correct answer, that is, set $b = 1$? Because real systems contain noise (that is, training occurs not with a single perfect output but with a cluster of almost perfect outputs), and networks with $b = 1$ have been found to be excessively sensitive to noise. Too small a value for b will, on the other hand, make the time to train the network very long. Values of 0.2-0.3 are most commonly used for b, but this will vary depending on circumstances.

Once the network has "learned" your task, it may be advantageous to freeze the weight values, therefore speeding up processing of further problems.

How many processing elements to include is also a matter for much debate and experimentation. With a large number of processing elements, each element tends to "memorize" a single, highly specific input pattern rather than developing a sensitivity to important dimensions or features. Elements operating this way are called "grandmother cells" because they work like a person recognizing his or her grandmother. On the other hand, if there are too few processors, each is "struggling to learn" more complex features of the input, and training takes longer.

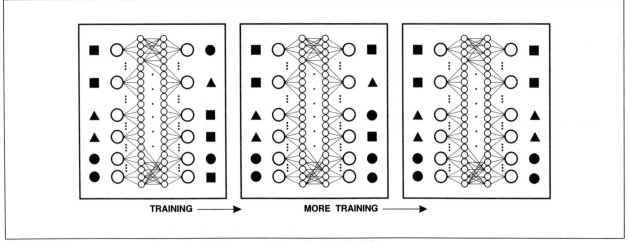

Figure 5-80. Neural Network Recognition Performance Improving with Training

Applications of Neural Networks

Tasks that neural networks do well have three characteristics: human beings know how to do them; there is no formula solution, but lots of examples; and the examples involve associating objects in one set with objects in another set. Typical applications include machine recognition of printed text or handwriting, detection of enemy ships or planes for the military, and inspection of airline baggage for explosives or drugs.

Neural networks are very unlikely to replace conventional computers or programs or other artificial intelligence techniques, since they handle some tasks poorly where other methods perform well. They have no way of expressing rules or strategies, for instance, or physical or logical laws. If a neural network were presented with the statements $a = b$ and $b = c$, it would conclude only that values of a were associated with very similar values of c.

For this reason, no one has built, or is likely to build, a stand-alone computer that operates entirely on neural network principles. The neural network would do a poor job of handling input and output or controlling the disk drive. Neural network applications use either of two approaches: simulate the neural network structure and behavior on a conventional computer, or use a neural network coprocessor board that resides inside an ordinary computer and is treated by it as another peripheral device.

Some of the more successful neural network applications are embedded in hybrid systems, in which each component (conventional program, expert system, neural network, or whatever) takes care of those aspects of the problem for which it is best suited. For example, a neural network might reconcile noisy sensor readings and present the results to an expert system that would perform fault diagnosis.

Robotics

Human capabilities encompass not only thinking but doing. Robots represent an attempt to create machines that can function in the physical world and manipulate objects with the same finesse and flexibility as human beings do. The "mechanical men" of science fiction may not make their appearance for many decades, if ever, but robotic arms and hands are doing useful work in factories today.

Robots occupy an ill-defined middle ground in a spectrum of automation that extends from dedicated automation (replacement of hand work by single-purpose machines, as in the early days of the Industrial Revolution) through flexible automation (multipurpose, reconfigurable machines) to the continuing use of human beings to provide those special capabilities of hand, eye, and brain we have not been able to define or replicate. There is, in fact, considerable disagreement about what is and what is not a robot. One commonly accepted definition put forward by the Robot Institute of America in 1979 [Ref. 67].

> A robot is a reprogrammable multifunctional manipulator
> designed to move material, parts, tools, or specialized devices
> through variable programmed motions for the performance of a
> variety of tasks.

This definition includes the emphasis on re-programmability (implying some sort of controller), multifunctionality (excluding single-purpose machines, no matter how sophisticated), and manipulation of objects (excluding forms of automation that have other purposes). To some, however, a "pick and place" device that simply moves an object from one defined location to another is flexible enough to qualify as a robot, while to others this device would not be sufficiently reprogrammable and would simply be a somewhat flexible automatic device.

Robotics and Artificial Intelligence

To many, robotics is a separate technical discipline with only a loose connection to artificial intelligence. It is true that the early development of robotics was mainly an exercise in mechanical design and automatic control. Now, however, the further advancement of robotics depends largely on providing robots with greater intelligence through machine vision and other sensory abilities plus the ability to process and make inferences from the inputs they receive.

Motivations for Using Robots

The word "robot," first used in a Czech science-fiction film, simply means "worker," and in many cases robots do replace human workers, since, as is often remarked, they do not take coffee breaks or vacations, ask for raises, or gripe about the boss. Just as often, however, robots do work that human beings are either unable or unwilling to do. If working conditions are unpleasant or needed skills are in short supply, human workers may simply not be available. In some environments, such as deep sea, outer space, or the interior of a nuclear reactor, robots can work without the extensive protection human beings would require, roaming the corridors of the sunken Titanic or the surface of Mars. Robots can be endowed with special attributes such as physical strength or heat resistance far beyond those of human beings and can deliver highly consistent performance on a variety of tasks. Moreover, they can be readily interfaced to computers and other process machinery to transfer information about their activities.

Why Robots Are not More Widespread

A brief search through science fiction and futurist literature would yield many predictions that robots should have taken over our working environment already. This has not happened for several reasons. Many applications once envisioned for robots did not, in fact, require their manipulative abilities and are today being performed by non-robotic process control systems. Why get a robot to turn a valve when you can send an electrical signal to accomplish the same thing? Some of the technical problems have also proved more difficult than originally foreseen; for example, the minimum three- to five-second cycle time of most of today's robots is simply too slow for many processes but is difficult to reduce for mechanical reasons. And, understandably, the opposition of labor to the introduction of robots has been intense and vocal in many quarters.

Components of a Robot: Manipulator

Most robots today are not complete "tin men" but consist of a single arm and wrist, to which may be attached a variety of *end effectors* (tools). The entire assembly sits on a stand, which often rotates. Robots vary in how the arm and wrist assemblies move with respect to the stand, leading to the use of different *coordinate systems* to describe the motion of the arm (see Figure 5-81). The design of the arm and the coordinate system used, in turn, determine the *work envelope*, the volume of space within which the arm and wrist can move. By convention this is defined for the arm and wrist alone, without the end effector, since the attachment of a smaller or larger tool can change how far the robot can actually reach.

The relative flexibility of a robot's motions can be described in terms of *degrees of freedom*. A robot on a fixed stand has a maximum of six degrees of freedom: three that describe the movement of the arm and wrist with respect to the stand and three that describe the movement of the end effector and wrist with respect to the arm (see Figure 5-82). For certain tasks, fewer degrees of freedom are adequate. For example, if the end effector is a rotary tool such as a drill, there is no need to rotate the wrist, and a five degrees, of freedom robot would suffice.

An astounding variety of end effectors have been developed for different applications, ranging from specialized grippers and probes to welding tools and hot-metal manipulators.

Components of a Robot: Controller

The multiprogrammable nature of the robot requires some sort of controller. Most early robots had a kind of mechanical controller (cams, valves, pins, etc.), but now almost all robots have electronic memories and control logic just like computers and PLCs. The type and sophistication of control required depends on the task to be performed. For simple "pick and place" tasks, control means simply moving the end effector from a starting point to an end point. This is often done by pointing the arm in the right direction and pushing it outwards with a single hydraulic or pneumatic impulse until it bumps against a preset end stop.

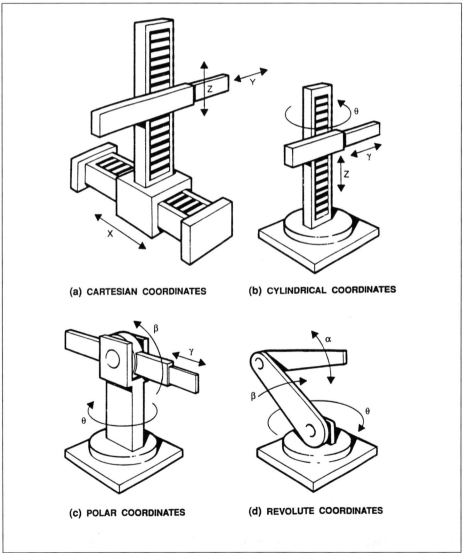

(a) CARTESIAN COORDINATES **(b) CYLINDRICAL COORDINATES**

(c) POLAR COORDINATES **(d) REVOLUTE COORDINATES**

Figure 5-81. Common Coordinate Systems Used for Robots [Ref. 66]

Robots that work this way are often called "bang-bang" robots from their characteristic sound effects.

Point-to-point robots traverse a path defined as a sequence of points connected by straight-line segments. They allow more control over how the end effector moves through the work envelope (useful if there are obstacles or dynamic constraints due, for example, to the weight of objects being moved). However, if the path must be defined with great precision, as for spray painting or welding, it becomes tedious to describe it point by point. Continuous-path robots refer to a description of the curve they are to follow, which has been provided to them either by example (walkthrough) or via mathematical functions.

In both these last cases, a servo mechanism is required so that the robot can determine whether its end effector is actually where it should be. Servo control involves continually comparing actual to desired position and attempting to reduce the difference to zero. A typical robot controller will determine its current end effector position from the angles of its various joints or by sensing distance from a fixed point or plane, then compare this position with what it should be and instruct a motor or other actuator to make any corrections.

Other than manual setting of mechanical controllers, there are two basic methods of programming a robot: teaching it by example, and issuing it preprogrammed instructions. Teaching methods involve moving the end effector through a typical work cycle and having the robot "memorize" what it does. In lead-through methods, a teaching pendant is temporarily attached to the robot and used to issue appropriate commands; in walk-through methods, the robot arm itself or a lighter "teaching arm" is physically moved through the desired work

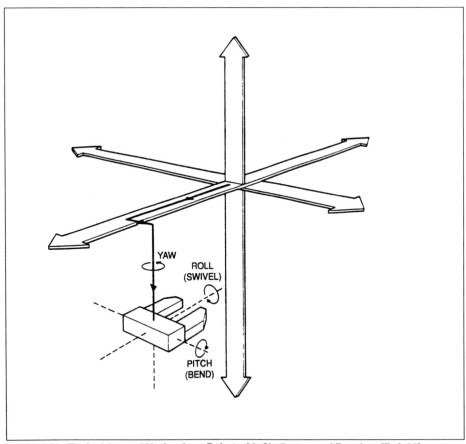

Figure 5-82. Typical Axes of Motion for a Robot with Six Degrees of Freedom [Ref. 68]
(Used with permission of the author and publisher.)

cycle. These methods are most useful where the task is straightforward to do but hard to describe, and where speed is not of the essence (it can be hard for the instructor to "lead" the robot both fast and accurately).

If greater generality or flexibility are desired, an off-line approach can be used, in which the robot is programmed like a computer, but using special languages with the instructions needed to control robot behavior. These are either extensions to standard languages like Pascal and C, or else special languages written specifically for robot control (Unimation's VAL™ and IBM's AML™ are the two most widespread). The difficulty with off-line programming is that it can be challenging to generate an adequate procedural or mathematical description of what the robot is supposed to do.

Components of a Robot: Power Supply

All the above requires some sort of motive power to do useful work. The most common power source is probably electrical, especially since electric motors can be protected from dirty or dangerous environments and run relatively quietly. Pneumatic power may be used in light-duty applications where speed is required; hydraulic power may be used where strength and precision are needed.

Challenges for Further Development

Most of today's robots have severely limited sensory capabilities. Robotics is a prime application area for improvements in machine vision. Improved tactile sensitivity for grippers is also highly sought after, as it would allow robots to work with fragile or easily damaged items (few robots today could pick up an egg intact) and acquire additional information by touch. Mobile robots have a particular need for improved sensory capabilities to allow them to navigate successfully.

Locomotion is another area where many improvements remain to be made. Some robots can now move about slowly and carefully, often with the aid of a guide wire or track, but their skill in avoiding obstacles or dealing with uneven surfaces is limited at best. Robots with legs instead of wheels could, in principle, cope better with rough terrain, but the control of multiple legs has proven to be a formidable challenge—another case in which simply understanding and modeling what human beings (to say nothing of horses, cats, and mice) do every day is still beyond us.

Dealing with a highly unstructured world is difficult for robots. Unfortunately, the extra work needed to structure the robot's world, for example, by presenting parts in a predetermined orientation, often negates the economic benefits of the robots.

For the vast majority of industrial robots, it has been suggested that the equivalent human would be someone who was blind, deaf, dumb, had only one arm with a stub on the end, and had both legs tied together and set in concrete! Despite these incredible handicaps the robot arm has already made outstanding contributions to the manufacturing environment. Yet this is largely only because the environment in which it works has up to now been specially 'structured' for it, and is not identical to the environment that existed when a human did the 'same' job [Ref. 67].

Some Typical Industrial Applications

Many robots today work around other industrial machinery such as forges, presses, and machine tools, preparing and presenting work to the machine, removing it after processing, and either performing finishing work or passing the piece on to another workstation. The advantage here is often consistency and throughput, plus removing human beings from an often dangerous work environment around heavy machinery.

Welding (spot or arc) and spray painting are other jobs robots often do. Here, the work is more skilled and people with the required skills may be in short supply. Heat or fumes in the work environment can make these jobs less pleasant for human beings.

Palletizing (stacking and wrapping objects on pallets for shipping) and warehousing are repetitive and boring jobs, often involving heavy lifting, which robots have in many cases taken over. An additional advantage can be that the robot does not need environmental comforts such as light and heat to the same extent that people do.

Assembly is a challenging job for robots since it involves ongoing assessment of the state of the object being assembled, a lot of careful positioning of components, and recognition and correction of problems as they occur. But since assembly is a major part of the work that goes on in a factory, it is nevertheless a prime target for robotic automation. Some recent progress has been made with robots assembling small electric motors and simple appliances like lawn mowers.

Some recent work has been done [Ref. 68] on using robots to debone smoked bacon backs. At present, skilled workers cut the bones away from the bacon backs using sharp knives before the bacon is sliced. The work is tedious and unpleasant, and recruiting workers is increasingly difficult. An experimental robot senses the location of the bones by pressing a sensor with multiple needles against the meat. Wherever the needles do not sink in, bone is assumed to be present. Once the bone layout is known, a cutting loop is used to lift each bone away from the meat.

A robot reconnaissance vehicle [Waterbury, Robert C., "R-2/D-2 Cousin on Patrol," *INTECH*, August 1990, p. 74] is being developed to patrol remote sites for intruders. A combination of microphones, infrared sensors, radar, and a video camera gathers data on any suspicious entities and transmits it back to a distant control console.

Robots are also being developed [Ref. 68] to perform services for the disabled, relieving the burden on family, friends, and paid caretakers. A teachable robot arm could handle eating utensils, a telephone, a doorknob, a computer keyboard, or whatever came "to hand."

Machine Vision

For most of us, vision is essential to many of the activities we carry out every day. Imagine for a moment how many things you would have to do differently, or could not do at all, if you were blind. There is little wonder, therefore, that many efforts have been made to add vision capabilities to robots and other production machinery.

Until recently, these efforts were held back by the lack of computing power to process the enormous amounts of data needed to represent a visual image. An image with 256 by 256 pixels (or "picture cells," the smallest unit of visual information in an image) and sixteen shades of gray requires over a million bytes of storage capacity to represent it.

Although we now have more computing power, we are still far from understanding how human beings process visual images. After all, the human brain uses biological processing elements ten times more slowly than digital chips, yet processes visual data in real time. We have evidence that the human brain makes some use of parallel processing and a great deal of use of hierarchical abstraction, but we still cannot explain how the brain can process visual data so fast, let alone build a computer to match its performance.

It is important to emphasize here that the problem of building a machine vision system is larger and more complex than that of building an artificial eye. The human vision system consists of the eye (the vision sensor) working integrally with the brain (the vision interpreter). Both must be, to some extent, understood and replicated to provide machines with a vision capability. Fortunately, most industrial applications do not require the full range or sensitivity of human vision. Merely identifying the shape of an object, or the presence or absence of gross manufacturing defects, may be sufficient.

From Image to Action

The machine vision process consists of four main steps:

(1) *Image formation*: a camera senses an illuminated image of, say, a manufactured part and converts it into a series of voltage signals.

(2) *Image preprocessing*: at this stage, the analog camera image is converted to digital form. Additional processing such as windowing or image restoration may be applied to make the resulting digital image easier to process.

(3) *Image analysis*: important features of the image such as object position and geometric attributes are identified.

(4) *Image interpretation*: the image description is matched against stored image models; an identification is made and control or classification decisions on the part are carried out.

Figure 5-83 associates each stage of the machine vision process with the hardware needed to implement it.

The process starts with the sensing of an image using a camera and a light source. The nature of the illumination used is very important and depends on the nature of the task. For simple shape detection, backlighting is usually used; if surface features must be identified, front or side illumination is more appropriate.

Early vision systems, and less expensive ones today, use a vidicon camera similar to that in a home video recorder. The output of a vidicon camera is a series of horizontal scan lines, each represented as a continuous electrical signal, with variations in amplitude corresponding to variations in light intensity. More recent systems use solid-state cameras with an array of vision sensors (typically 128×128 or 256×256). Each sensor converts the light falling on it into an analog electrical signal that is proportional to light intensity (averaged over the pixel, of course). The images provided by these cameras are of higher quality and are less vulnerable to distortion than those provided by vidicon cameras.

Each camera image (typically once each 16 milliseconds) is then presented to an image processor that transforms each analog voltage value into a corresponding digital value. Depending on the requirements of the system, the image may be interpreted in "black and white" (pixels with voltages higher than a cutoff value are white, all others are black) or a gray scale (typically, sixteen value), which provides greater refinement and reduced sensitivity to imperfect lighting at the cost of greatly increased storage requirements.

Since a production vision system is normally required to operate in real time, it is crucial to reduce data processing for each image to a minimum. One approach frequently taken is windowing, which involves placing an electronic mask over the scanned image so that only the region of greatest interest is intensively analyzed. Image restoration techniques may also be used to make subsequent processing easier by compensating for deficiencies in the camera image such as blurred lines and poor contrast. For example, image stretching increases the rela-

The resolution of the camera image is determined by the number of pixels in the image array and the size of the field of view. For a given array, resolution can be improved by using a camera lens with a higher magnification, at the cost of a smaller field of view.

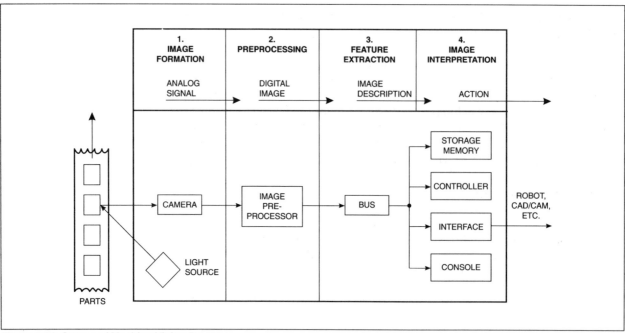

Figure 5-83. Stages of Machine Vision Process

tive contrast between high and low intensity pixels. A more sophisticated technique, Fourier domain processing, analyzes the changes in brightness in the image into a series of sine and cosine waves that can be selectively enhanced or suppressed to provide, for example, better edge definition or a reduction in background noise. Techniques of this nature have been used to make features such as small moons and meteor craters visible on pictures sent back from interplanetary space probes.

Once a good quality image of the region of interest has been obtained, important features of that image, such as position, distance, and orientation, must then be extracted. Position is usually no problem; distance measuring techniques include stadimetry (using the apparent size of an object in the camera's field of view), triangulation, and binocular vision (using two cameras). Orientation (in two or three dimensions, as required) can be determined by geometric techniques (fitting an ellipse to the image or noting the relative positions of three noncollinear points), interferometry, or light intensity distribution. A trickier problem, image segmentation, involves identifying the component regions of a complex object. Machine vision systems find true region boundaries difficult to distinguish from shadow edges or image imperfections.

Image interpretation consists of matching the processed image against a set of stored images to make an identification. Template matching involves superposing the processed image over the stored image and measuring, for example, the percentage of pixels that do not correspond. Feature matching, a more sophisticated approach, involves calculating a weighted function of a number of features of the processed image and comparing it with the same function calculated for the stored image. Once the image has been identified or classified, further action may be taken, whether updating a database or directing a robot arm to move.

Performance of Machine Vision Systems

The performance of a machine vision system depends on several factors: resolution (described above), processing speed (not only the bit-level image processing speed but how fast each item can be examined by the system), discrimination (how many gray levels, and how easily the system can detect edges), and accuracy (the probability of correct interpretation of images). A typical system in use today could process 2-5 simple parts per second with 90% accuracy. Higher speeds will be achieved in future by faster processors, better recognition algorithms, and the use of parallel processing.

> Lumber is classified or graded on the basis of characteristics that are related to appearance and soundness, mainly the number and size of knots per board and the presence or absence of flaws such as resin inclusions and rotten spots. The cut boards are laid sideways on a moving belt, where, typically, human inspectors evaluate each board visually and either mark it or push it off onto the appropriate belt or bin. The work is boring, and lapses of attention can lead to mistakes in grading and customer complaints. A machine vision system can scan the board and analyze the image to determine the number and size of dark spots that indicate defects.

Current machine vision systems can be successful only in highly structured and controlled environments. A system that works in the laboratory to recognize clean parts may fail in the plant where the parts are oily and dirty. Many vision tasks that human beings perform easily, such as picking jumbled parts out of a bin, are far beyond the capabilities of today's systems. Unfortunately, the economic justification of many such systems depends on reducing the need to structure the environment, for example, by eliminating jigs and fixtures or the precise placement of parts on a conveyer belt.

Speech Recognition and Synthesis

Speech is the almost universal mode of human communication, far more widespread than writing or typing. The use of spoken language has, in fact, been proposed as one of the characteristics that defines us as human. It is therefore natural to expect that a computer system with artificial intelligence worthy of the name would be able to deal with the spoken word. This intelligent system would need to perform two tasks: speech recognition, or understanding what human beings said to it; and speech synthesis, or generating an understandable and natural-sounding spoken response.

In an industrial environment, voice input-output is particularly interesting where workers' hands and eyes may be occupied by other tasks. For example, it would be much more convenient for an instrument technician performing field adjustments to speak the results to a voice recognition system rather than fumbling with pencil and clipboard. A good voice recognition system can also achieve error rates lower than keyboard input at a speed easily twice as fast as typing. Data input can also be performed under conditions such as poor lighting that would impede writing or typing.

Recording of Speech Patterns

A computer starts out knowing nothing about speech and must be provided with a set of patterns or templates representing the words or phrases it is to identify. Recorded speech is, of course, analog and must be converted to digital form for storage and later reference. This preprocessing step, known as feature extraction, also samples the signal to reduce the data volume to manageable proportions and identifies a set of characteristic frequencies (known as formant frequencies), each with an energy level typical of that sound. These waveforms are typically averaged over 10 to 20 milliseconds, then sampled 50 to 100 times a second [Ref. 72]. A typical technique for representing the speech signal, called linear predictive coding (LPC), describes the signal by the parameters of a "best-fit" filter.

Comparison of Speech Input to Stored Patterns

Once a set of reference patterns has been stored, actual speech can be input and compared with the stored patterns. This requires time-aligning or synchronizing input with stored patterns, since a word or sound may be spoken more quickly or slowly on different occasions. The beginning and end point of the word or phrase is identified (usually by energy criteria) and used as a basis for the time alignment.

The actual comparison is performed by calculating a measure of similarity between the input item (assume for now it is a word) and each stored word. The stored word differing least from the input word on this measure is assumed to have been recognized. A rejection criterion may be used to eliminate words that are outside the system's vocabulary as well as nonspeech input (sneezes, passing trucks, electric drills, and so forth). Normally, a tradeoff must be made between substitution errors (recognizing the wrong word) and rejection errors (failing to recognize a word when one was, in fact, input).

Figure 5-84 illustrates the operation of a dynamic programming algorithm on the input phrase "Tell me." Dynamic programming works by going through the input word feature by feature. At each stage it calculates, for each possible identification of the next feature, the conditional probability value.

P (next feature is this one)

GIVEN (all features previously identified)

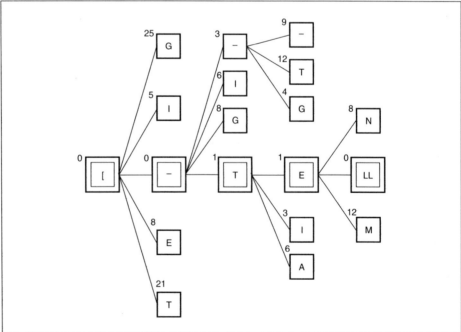

Figure 5-84. Recognition of Spoken Word

In the diagram, the number beside each box is the negative logarithm of this probability—thus the correct recognition path is identified by a sequence of low numbers.

Difficulties in Speech Recognition

It should be clear by now that the longer the unit of speech being analyzed, the more difficult and more computationally laborious the process just described becomes. For this reason, discrete speech recognition (where the speaker must pause between words) is much further advanced than continuous speech recognition (where the speaker "runs words together" as in normal conversation).

The recognition process can also be "thrown off" by the intonation, accent, and other speech habits of speakers other than those who provided the reference pattern—or even if one of the latter is tired or has a cold. The best recognition performance is achieved with a limited group of speakers and a small vocabulary, but even this does not guarantee success: systems are still regularly confusing pairs of words like "go" and "no" or "start" and "stop."

Background noise in itself does not pose a problem for speech recognizers, as long as the system was trained with a similar level and type of noise. Voice recognition systems exist today that could be used in the cockpit of a jet fighter airplane—but if the engine stopped working, so would the system!

Speech Synthesis

Speech synthesis, the "other side" of speech recognition, involves taking stored digitized speech waveforms and reconstituting them into speech output. First attempts produced a "computer voice" familiar to us through science fiction movies and toy robots. More recent systems produce a voice that is much more natural sounding but still far from a perfect imitation of a human voice (for a "real live example," call up the local directory assistance, which will probably use a synthesized voice to give out the phone number). One problem is data storage: current systems may use up to 10 kilobits of storage per second of speech, but compare this to long-distance telephone conversations, which are typically sampled at 64 kilobits per second. Another is that we still do not understand completely what makes a voice sound natural to us. Even today's less-than-perfect computer-generated voices readily grab our attention and can give us information even if our eyes are otherwise occupied (with a computer display, for example?).

Processing Requirements for Speech

Not only storage but processing requirements are highly demanding, both for speech recognition and speech synthesis. Voice I/O must be handled in real time where industrial processes are involved, and, in any case, people quickly become impatient with slow and hesitant speech. One approach often used is to move much of the voice processing out of a general-purpose CPU to specialized (and therefore faster) boards or chips. These "voice processors" will often convert their output into the form expected by other general-purpose programs, which may not even "know" their data is coming from a voice recognition system. Voice recognition and synthesis is also becoming a fruitful application area for parallel processing techniques.

Voice Technology Applications

Defining the requirements for a voice technology application involves looking both at the users and the task. The system design and implementation will vary depending on the number and type of users, how often and for how long they will interact with the system, and how mobile they are (must the terminal be portable?). As for the task, look at the length and complexity of the transaction, the size, nature, and flexibility of the required vocabulary, the working environment, and security considerations.

Voice technology has been slower to gain acceptance than some people would have thought, probably because acceptable recognition rates are a fairly recent achievement in most areas. Some business and industrial applications are parcel routing by zip code, voice annotation during procedures such as equipment repair, remote computer access (no modem required), voice mail, log entries, simple control of process equipment (e.g., start, stop, reverse, slow, fast).

References and Bibliography

CPU and Computer Architecture

1. Fulcher, John, *An Introduction to Microcomputer Systems: Architecture and Interfacing*, Reading, MA: Addison-Wesley, 1989.

2. Osborne, Adam, *An Introduction to Microcomputers: Volume 1, Basic Concepts*, 2nd edition, Berkeley, CA: Osborne/McGraw-Hill, 1980.

3. Robinson, Phillip, "Overview of Programmable Hardware," *Byte*, 12(1): 197-201, January 1987.

4. White, George, "A Bus Tour," *Byte*, 14(9): 296-302, September 1989.

Mass Storage and Peripherals

5. Laub, Leonard, "The Evolution of Mass Storage," *Byte*, 11(5): 161-172, May 1986.

6. Liebson, Steve, "The Input/Output Primer," *Byte*, 7(2-7), February-July 1982.

7. Weston, G. F., and Bittleson, R., *Alphanumeric Displays*, NY: McGraw-Hill, 1983.

Installation and Computer Environment

8. Ciarcia, Steve, "Electromagnetic Interference," *Byte*, 6(1): 48-68, January 1981.

9. Dvorak, Paul J., "Packaging Computers to Survive in the Real World," *Machine Design*, 60(11): 70-76, May 26, 1988.

10. Lawrie, Robert J. (ed.), *Design and Installation of Computer Electrical Systems*, NY: McGraw-Hill, 1981.

11. Smith, R. S., and Meyerhofer, W. L., "Toughening Up PCs for Industrial Use," *Instrumentation and Control Systems*, 58(9): 91-100, September 1985.

Operating Systems

12. Evanczuk, Stephen, "Real-Time OS," *Electronics*, 56(6): 105-111, March 24, 1983.

13. Glass, Robert L., *Real-Time Software*, Englewood Cliffs, NJ: Prentice-Hall, 1983.

14. Paterson, Tim, "An Inside Look at MS-DOS," *Byte*, 8(5): 230-252, June 1983.

15. Tucker, D. M., "Understanding Operating Systems," *PC World*, 2(5): 192-199, May 1984.

Communications

16. Gupta, Sanjay, and Gupta, Jai P., (eds.), *PC Interfacing for Laboratory Data Acquisition and Process Control*, Research Triangle Park, NC: Instrument Society of America, 1989.

17. Hughes, Thomas A., *Programmable Controllers*, Research Triangle Park, NC: Instrument Society of America, 1989.

18. Jordan, Larry, and Churchill, Bruce, *Communications and Networking for the IBM PC*, Bowie, MD: Robert J. Brady Company, 1983.

19. Pouzin, Louis, and Zimmermann, Hubert, "A Tutorial on Protocols," *Proceedings of the IEEE*, 66(11): 1346-1370, November 1978.

20. Schwaderer, W. David, "Communications Concepts," *PC World*, 2(4): 184-194, April 1984.

Networks

21. Derfler, Frank J., Jr., "A Field Guide to LAN Operating Systems," *PC Magazine*, 7(11): 117-128, 135, June 14, 1988.

22. Hughes, Kevin, "Trends in LANs for Plant Automation," *INTECH*, 35(3): 35.40, March 1988.

23. Krumrey, Art, and Kolman, John, "LAN Hardware Standards," *PC Tech Journal*, 7(6): 54-68, June 1987.

24. Sachs, Jonathan, "Local Area Networks," *PC World*, 2(7): 69-75, July 1984.

Computers in Process Control

25. Clune, Thomas R., "Interfacing for Data Acquisition,"*Byte*, 10(2): 269-282, February 1985.

26. Cordova, G. J.; Hertanu, H. I.; and Doyle, G. T., "Microprocessor-Based Distributed Control Systems," *Chemical Engineering*, 92(2): 86-94, January 21, 1985.

27. Englemann, Bill and Abraham, Mark, "Personal Computer Signal Processing," *Byte*, 9(4): 94-110, April 1984.

28. Williams, Theodore J., *The Use of Digital Computers in Process Control*, Research Triangle Park, NC: Instrument Society of America, 1984.

Database

29. Date, C. J., *An Introduction to Database Systems*, 5th edition, Reading, MA: Addison-Wesley, 1987.

30. Neely, Joel, and Stewart, Steve, "Fundamentals of Relational Data Organization," *Byte*, 6(11): 48-60, November 1981.

31. Robbins, Judd, and Braly, Ken, "Database by Design," *PC World*, 4(4): 203-211, April 1986.

Spreadsheet

32. Miller, Harry, "Introduction to Spreadsheets," *PC World*, 3(5): 249-255, May 1985.

33. Urschel, William, "Worksheets by Design," *PC World*, 5(9): 151-157, September 1987.

Word Processing

34. Datz, Terry Tinsley, "Word Processing Tips," *PC World*, 3(5): 242-246, May 1985.

35. Martin, Janette, "New Dimensions in Word Processing," *PC World*, 3(1): 42-51, January 1985.

CAD/CAM

36. Bowman, D. J., and Bowman, A. C., *Understanding CAD/CAM*, Indianapolis, IN: Sams, 1987.

37. Jadrnicek, Rik, "Computer-Aided Design," *Byte*, 9(1): 172-206, January 1984.

38. Meilach, Allen E., "Getting the Picture with CAD," *PC Magazine*, 3(14): 111-127, July 24, 1984.

Project Management

39. Dauphinais, Bill, and Darnell, Leonard, "Project Management: One Step at a Time," *PC World*, 2(9): 241-250, September 1984.

40. Levine, Harvey A., "Seeing the Project Through," *PC World*, 4(4): 145-153, April 1986.

Languages

41. Elfring, Gary, "Choosing a Programming Language," *Byte*, 10(6): 235-240, June 1985.

42. McCoy, Earl, "Strongly Typed Languages," *Byte* 8(6): 418-422, May 1983.

43. Williams, Gregg, "Structured Programming and Structured Flowcharts," *Byte*, 6(3): 20-34, March 1981.

Statistical Process Control

44. Contino, Anthony V., "Improve Plant Performance via Statistical Process Control," *Chemical Engineering*, 94(10): 95-102, July 20, 1987.

45. Grant, E. R., and Leavenworth, R. S., *Statistical Quality Control*, NY: McGraw-Hill, 1980.

46. Levinson, William, "Understand the Basics of Statistical Process Control," *Chemical Engineering Progress*, 86(11): 28-37, November 1990.

47. MacGregor, John F., "On-line Statistical Process Control," *Chemical Engineering Progress*, 84(10): 21-31, October 1988.

Process Simulation

48. Biegler, Lorenz T., "Chemical Process Simulation," *Chemical Engineering Progress*, 85(10): 50-61, October 1989.

49. Bronson, Richard, "Computer Simulation: What It Is and How It Is Done," *Byte*, 9(3): 95-102, March 1984.

50. Roberts, N.; Andersen, D., et al., *Introduction to Computer Simulation*, Reading, MA: Addison-Wesley, 1983.

51. Strand, E.; Moldenius, S.; Koponen, R.; Viljakainen, E.; and Edwards, L., "Optimization of Peroxide Bleaching Systems," *Tappi Journal*, 71(7): 130-134, July 1988.

52. Sugarman, Robert, and Wallich, Paul, "The Limits to Simulation," *IEEE Spectrum*, 20(4): 36-41, April 1983.

General Artificial Intelligence

53. Anderson, Howard, "Why Artificial Intelligence Isn't (Yet)," *AI Expert*, 2(7): 36-44, July 1987.

54. Nisenfeld, A. Eli, and Davis, James R., (eds.), *Artificial Intelligence Handbook* (2 vols.), Research Triangle Park, NC: Instrument Society of America, 1989.

55. Marin, Miguel A., *A Tutorial on Expert Systems, with Some Power Industry Aspects,* IREQ, Varennes, Quebec, 1986.

Expert Systems

56. Butz, Brian P., "Expert Systems," *Abacus*, 5(1): 30-44, Fall 1987.

57. Casey, John, "Picking the Right Expert System Application," *AI Expert*, 4(9): 44-47, September 1989.

58. Hayes-Roth, Frederick; Waterman, Donald A.; and Lenat, Douglas B., (eds), *Building Expert Systems*, Reading, MA: Addison-Wesley, 1983.

59. Myers, Ware, "Introduction to Expert Systems," *IEEE Expert*, 1(1): 100-109, Spring 1986.

60. Rowan, Duncan A., "On-Line Expert Systems in Process Industries," *AI Expert*, 4(8): 30-38, August 1989.

61. Waterman, Donald A., *A Guide to Expert Systems*, Reading, MA: Addison-Wesley, 1986.

Neural Networks

62. Bhagat, Phiroz, "An Introduction to Neural Nets," *Chemical Engineering Progress*, 86(8): 55-60, August 1990.

63. Obermeier, Klaus K., and Barron, Janet J., "Time to Get Fired Up," *Byte*, 14(8): 237-224, August 1989.

64. Wasserman, P. D., *Neural Computing: Theory and Practice*, NY: Van Nostrand Reinhold, 1989.

Robotics

65. Callahan, J. Michael, "The State of Industrial Robotics,"*Byte*, 7(10): 128-142, October 1982.

66. Engelberger, Joseph F., *Robotics in Practice*, American Management Associations, 1980.

67. Milner, D. A., and Vasiliou, V. C., *Computer-Aided Engineering for Manufacture*, New York: McGraw-Hill, 1987.

68. Scott, Peter B., *The Robotics Revolution: The Complete Guide for Managers and Engineers*, New York: Basil Blackwell, 1984.

Machine Vision

69. Dodd, George, D., and Rossol, Lothar, (eds.), *Computer Vision and Sensor-Based Robots*, New York: Plenum Press, 1979.

70. Dunbar, Phil, "Machine Vision," *Byte*, 11(1): 161-173, January 1986.

Speech Recognition

71. Dixon, N., Martin, Rex and Thomas B., (eds.), *Automatic Speech and Speaker Recognition*, New York: IEEE Press, 1979.

72. Doddington, George R., and Schalk, Thomas B., "Speech Recognition: Turning Theory to Practice," *IEEE Spectrum*, 18(9): 26-32, September 1981.

73. Rosch, Winn L., "Voice Recognition: Understanding the Master's Voice," *PC Magazine* 6(18): 261-308, October 27, 1987.

About the Author

Diana Churchill Bouchard is Associate Scientist at the Pulp and Paper Research Institute of Canada. At McGill University, where she earned her B.A., M.A., and M.Sc. degrees, she received the McConnell Graduate Fellowship and the Gulf Oil Graduate Fellowship. She has lectured at McGill and also at Concordia University and Université du Quebec á Trois-Riviéres. Mrs. Bouchard's current research interests include the use of simulation as a process development tool, possible uses of artificial intelligence techniques for self-debugging, and the development of expert systems to aid technical personnel in pulp and paper mills.

6

Control System Theory

In order to design a process control system one must become familiar with many industrial devices. A knowledge of the operation and application of these devices and the behavior of a process is necessary for the instrumentation engineer or designer. The ability to choose between two or more similar devices is critical as well. This chapter has been included to facilitate these tasks by way of instruction on the theory behind control systems. The theory will be restricted to a minimum and emphasis placed on practical application where possible. The rudiments presented will serve to give the reader a "feel" for the subject and the ability to collaborate with suppliers and other control engineering professionals.

A control system may be interpreted as the means by which energy is manipulated and regulated. Although industrial process systems are being considered here, control system theory is, in fact, interdisciplinary and may be applied in many diverse areas. One of the less apparent areas, for example, is the science of economics. Wherever and whenever something may be classified as a system, the theory applies.

The Transfer Function

In instrumentation one encounters many types of systems: mechanical, pneumatic, chemical, electrical, and so on. As a direct result of the interdisciplinary nature of system response, a unified analytical approach has been developed. The main feature of this approach is the *transfer function*, which is represented as a "black box."

In its simplest form an input (or excitation) is applied to a transfer function or "black box," and an output (or response) results. This is graphically depicted in Figure 6-1.

One may consider that the development of the transfer function and its interaction or interconnection is the essence of control system theory.

Open and Closed Loops

There are two categories of control systems: open loop and closed loop. An open-loop system is one in which control action is independent of the response.

As an example (see Figure 6-2), consider a heating system for a room. Adjusting the potentiometer (controller) to a suitable level (set point), supplies current to the element (process element). It is obvious that the ambient temperature has no effect on the current supply. This system is represented graphically in Figure 6-3, where R is the set point or input, C is the response or output (heat), G_1 is the controller and G_2 is the process.

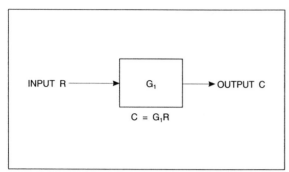

Figure 6-1. The Transfer Function as Black Box

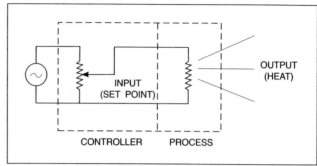

Figure 6-2. Open-Loop Heating System

A closed-loop system is one in which the control action is dependent upon the response. To illustrate this, the previous example can be modified to include feedback. Feedback is the property of a closed loop that permits the output to be compared to the input so that the appropriate control action may be taken. In the revised example, Figure 6-4, the thermostat (comparator) is adjusted to the desired temperature (set point). A thermocouple (feedback element) supplies the status of the present ambient temperature (measured variable) to the thermostat, which sends a signal based on the difference between the desired and ambient temperature (error signal) to the controller. The servomotor in the controller sets the potentiometer to the value required to produce enough heat in the heater element to bring the ambient temperature to the desired set point. When the ambient temperature is correct, the error signal is zero and the controller maintains its output. Thus, one needs only to set the thermostat's set point and the ambient temperature is controlled automatically. This system is represented graphically in Figure 6-5, where R is the set point or input, C is the response or output (heat), G_1 is the controller, G_2 is the process, B is the measured variable, and H_1 is the feedback element.

A second example of a closed-loop system is a water reservoir level control loop (see Figure 6-6) in which the opening of a valve is adjusted to maintain a desired level.

Block Diagrams

A pattern can be seen emerging from the previous examples. Here it will be shown that the G_1 and G_2 of these previous examples may be replaced by a single G. The most general form of a control system can now be examined. In a mathematical sense, the open loop (see Figure 6-7) may be described as:

$$C = GR \qquad (6-1)$$

where the output C is a function G of the input R.

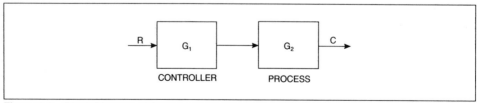

Figure 6-3. Block Diagram of Open-Loop Heating System

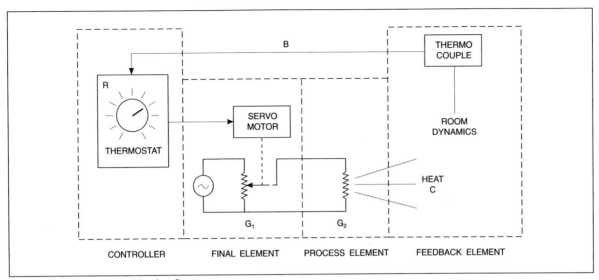

Figure 6-4. Closed-Loop Heating System

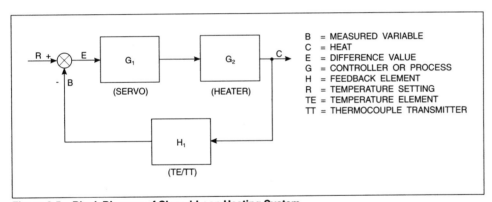

Figure 6-5. Block Diagram of Closed-Loop Heating System

B = MEASURED VARIABLE
C = HEAT
E = DIFFERENCE VALUE
G = CONTROLLER OR PROCESS
H = FEEDBACK ELEMENT
R = TEMPERATURE SETTING
TE = TEMPERATURE ELEMENT
TT = THERMOCOUPLE TRANSMITTER

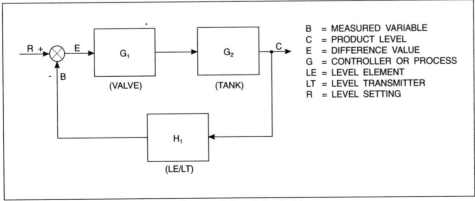

Figure 6-6. Block Diagram of Water Reservoir Level Control Loop

B = MEASURED VARIABLE
C = PRODUCT LEVEL
E = DIFFERENCE VALUE
G = CONTROLLER OR PROCESS
LE = LEVEL ELEMENT
LT = LEVEL TRANSMITTER
R = LEVEL SETTING

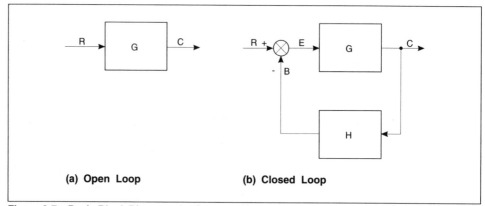

Figure 6-7. Basic Block Diagrams for Open and Closed Loops

The closed loop is not as obvious but may be derived from Figure 6-7 as follows:

$$E = R - B$$
$$B = HC \ or \ E = R - HC$$
$$C = GE \ or \ C = G(R - HC)$$

Then

$$C = GR - GHC$$
$$C + GHC = GR$$
$$C(1 + GH) = GR$$

and

$$C = \frac{G}{1 + GH} R \qquad (6\text{-}2)$$

Equation (6-2) is the basic negative feedback function. It is an interaction of the transfer functions G and H, which together form the closed-loop transfer function. One should notice the negative sign on the summer for the variable B in Figure 6-7. It is a result of the fact that the difference value E is needed to actuate the controller. In Equation (6-2), this fact manifests itself as a positive sign in the denominator. If the feedback were positive, the two signs would be reversed.

In simple block diagram configurations there is one input (the set point R) one output C, and a single, unique loop consisting of G and H. It would be misleading, however, to say that this is always the case. Although larger transfer functions will be discussed later in this chapter, it should be mentioned here that a set point is not necessarily the only type of input to a control system. For example, the chemical reactor temperature control system shown in Figure 6-8 has an additional input called a "disturbance."

In this system, only thermal control is being considered; level control is not included. The fluid entering the vessel is much cooler than the required set point value. Thus, even though the contents of the vessel may have reached the set point, each time new cool fluid enters, a temperature drop occurs. This disturbance is accounted for in the design as shown in Figure 6-9 in which J represents the disturbance.

The concept of "superposition" applies to the analysis of this type of system. It amounts to assuming that the disturbance is zero and then finding the transfer function T_1 for the reference. (How the transfer function is developed will be seen later in this chapter.)

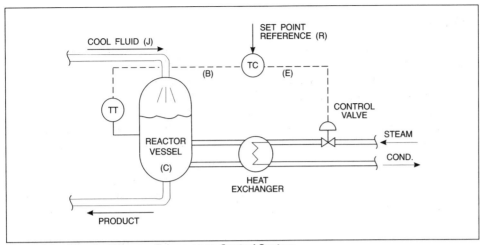

Figure 6-8. Chemical Reactor Temperature Control System

$$C_1 = T_1 R \qquad (6\text{-}3)$$

Next, assume the reference to be zero and find the transfer function T_2 for the disturbance.

$$C_2 + T_2 J \qquad (6\text{-}4)$$

Finally, the total system transfer function is found by applying superposition; that is, C, the combined effect, is the sum of the individual effects of the open- and closed-loop transfer functions.

$$C = C_1 + C_2 \qquad (6\text{-}5)$$
$$C = T_1 R + T_2 J$$

Last to be examined is the feedback element H. Note, in this example, that the output is heat and the input is temperature. In order to compare the two, heat must be converted to temperature. This is accomplished by the function H_1. In some instances the input and output are directly comparable, and the function H reduces to unity (see Figure 6-10).

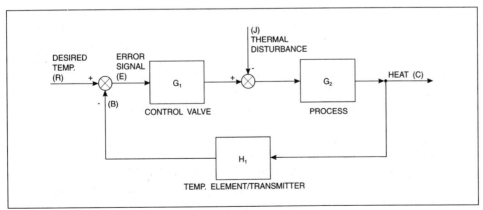

Figure 6-9. Block Diagram of Chemical Reactor System

Modeling

This section deals with what the transfer functions G and H are and how they come about. In a feedback system, one encounters different control modes (relationships between input and output). To understand what a control mode is, examine how it defines the relationship of a "through variable" (T.V.) to an "across variable" (A.V.). These concepts will be examined as they apply to different systems (such as electrical, mechanical, fluid flow, and thermal) for which control devices can be represented directly or in combination. Furthermore, as a result of this, a mechanical system can be modeled by an electrical one, and combinations of systems (for example, electromechanical) are easily handled.

A through variable can be interpreted as a quantity passing a point. Mechanically, this would be a force transmitted through a member; electrically, it would be current flowing through a wire.

An across variable is the difference of a quantity between two points. Velocity may be considered as the across variable of a mechanical system, while voltage or potential difference (between two points) is the equivalent electrical across variable. The word "difference" is important. If there are two drive shafts, one rotating at 510 rpm, the other at 500 rpm, the "difference" for coupling is clearly 10 rpm. When speaking of a 120 V AC source, it is with respect to the ground (0 V AC) or datum. This must be between two points to make sense. In fluid mechanics, the across variable is pressure or, more precisely, pressure difference (between two points). Table 6-1 compares through and across variables of different types of systems.

Table 6-1.   Comparison of Variables for Different Systems

Variable/Type	Through	Across
Electrical	Current (amperes, A)	Voltage (volt, V)
Mechanical	Force (newtons, N)	Velocity (meters per second, m/s)
Rotational	Torque (newton-meters, N-m)	Angular velocity (revolutions per minute, rpm)
Fluid	Volume Flow (cubic meters per second, m^3/s)	Pressure (kilopascals, kPa)
Thermal	Heat Flow (watts, W)	Temperature (degrees Celsius, °C)

A more common way to represent the action of a controller is to describe the relationship between the output signal and the input signal (error). Control modes commonly encountered in feedback control systems are proportional (P), integral (I), and derivative (D). In a proportional relationship the result is directly proportional to the magnitude of the error. In an integral relationship (also called automatic reset) the result follows the sum of the error integrated over time. In a derivative relationship (also called rate) the result follows the time rate of change of the error.

Mathematically, the relationships between an output m and the error e, for the P, I, and D control modes are described in Equations (6-6), (6-7), and (6-8). Table 6-2 compares the constants of proportionality in the different systems. Applying Tables 6-1 and 6-2 to determine the proportional relationship for an electrical system results in Ohm's Law.

$$m = K_1 e \qquad\qquad (6\text{-}6)$$

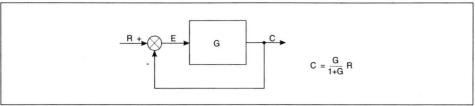

Figure 6-10. Unity Feedback

where K_1 is the proportional constant and, in fact, the gain of the controller.

$$m = K_2 \int_{t_1}^{t_2} e \, dt \qquad (6\text{-}7)$$

where K_2 is the integral constant.

$$m = K_3 \frac{d}{dt} e \qquad (6\text{-}8)$$

where K_3 is the derivative constant.

Table 6-2. Control Modes for Different Systems

Type	K_1 (P)	K_2 (I)	K_3 (D)	Legend
Electrical	R (ohms)	$\frac{1}{C}$ (F)	L (H)	R = resistance C = capacitance L = inductance M = mass J = inertia K = spring constants B = damping coefficients
Mechanical	$\frac{1}{B}\left(\frac{Ns}{m}\right)$	$\frac{1}{M}$ (kg)	$\frac{1}{K}\left(\frac{N}{m}\right)$	
Rotational	$\frac{1}{B}$ (Nsm)	$\frac{1}{J}$ (kgm^2)	$\frac{1}{K}$ (Nm)	
Fluid	$R\left(\frac{Ns}{m^5}\right)$	$\frac{1}{C}\left(\frac{m^5}{N}\right)$	$L\left(\frac{Ns^2}{m^5}\right)$	
Thermal	$R\left(\frac{^{\circ}C}{W}\right)$	$\frac{I}{C}\left(\frac{Ws}{^{\circ}C}\right)$	None	

In a later section on mathematics the use of Laplace transforms in analysis is discussed. For now, it is sufficient to note that the Laplace operators that act on derivatives and integrals are usually in the form of s and $1/s$, respectively. This is the notation used in Laplace transform calculus.

One should also note that some engineers prefer to use displacement (or angular displacement) as opposed to velocity (or angular velocity) for the across variable in mechanical (rotational) systems. These relate simply as follows:

$$v(t) = \frac{d}{dt} x(t) \quad or \quad v(s) = sx(s) \qquad (6\text{-}9)$$

$$\omega(t) = \frac{d}{dt} \theta(t) \quad or \quad \omega(s) = s\,\theta(s) \qquad (6\text{-}10)$$

where:

v = velocity
x = displacement
ω = angular velocity
θ = angular displacement
t = time
s = Laplace frequency domain variable

The following paragraphs present some examples of real devices and their transfer functions. Note that the input and output units are chosen for convenience, and one may mix electrical inputs with mechanical outputs, and so on. These functions are well documented in texts on modeling and are included here for illustration only. They were generated from the above-mentioned concepts as well as from the nature of the devices themselves.

In a DC motor, (see Figure 6-11), the individual components may be examined one by one. The starting point is the electrical area where field current (output) relates to field source voltage (input).

$$V_f(t) = I_f(t)R_f + L_f\frac{d}{dt}I_f(t) \quad \text{(traditional form)} \tag{6-11}$$

$$V_f(s) = I_f(s)R_f + L_f s I_f(s) \quad \text{(Laplace form)} \tag{6-12}$$

$$\frac{I_f(s)}{V_f(s)} = \frac{1}{R_f + sL_f} \quad or \quad V_f(s) \rightarrow \left[\frac{1}{R_f + sL_f}\right] \rightarrow I_f(s) \tag{6-13}$$

The relationship between the field current and the torque of the motor is derived from experimental results and theory beyond the scope of this text; however, it may be summed up as follows:

$$T(s) = I_f(s)K_m \tag{6-14}$$

$$\frac{T(s)}{I_f(s)} = K_m \quad or \quad I_f(s) \rightarrow \left[K_m\right] \rightarrow T(s) \tag{6-15}$$

Torque may also be related with angular velocity.

$$T(t) = \omega(t)B + J\frac{d}{dt}\omega(t) \tag{6-16}$$

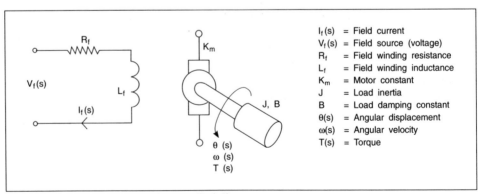

Figure 6-11. Field-Controlled DC Motor (Ref. 3)

$$T(s) = \omega(s)B + Js\,\omega(s) \qquad (6\text{-}17)$$

$$\frac{\omega(s)}{T(s)} = \frac{1}{Js+B} \quad or \quad T(s) \rightarrow \left[\frac{1}{Js+B}\right] \rightarrow \omega(s) \qquad (6\text{-}18)$$

Using Equation (6-10) (since the interest here is in the angular displacement, not the angular velocity), one notes how the units between stages are taken for convenience and how they relate as illustrated in Figure 6-12. These equations may be reduced, as indicated, giving the transfer function of the motor.

Another functional device is the hydraulic actuator (see Figure 6-13). The transfer function will not be derived. The control valve action is to actuate a piston that moves a load.

The *G* functions have been described above. The *H* functions may be found in similar fashion. The *H* functions include feedback elements such as tachometers, level sensors, and so on, as well as transmitters; thus, they are combinations of electrical and pneumatic types of systems. In combination they may be open- or closed-loop.

The next section deals with block diagram reduction techniques, or how to work with loop equations to obtain a system function. From this section, one should have some understanding of how systems can be modeled as well as how the models relate to each other. An example best illustrates this (see Figure 6-14).

The figure shows a vibration stabilizer. The system includes a mass, a dashpot, and a spring. The electrical equivalent components are: a capacitor, a resistor, and an inductor, respectively. Initial conditions are taken to be zero, for simplicity. The mass has a force applied to it (input). The resultant output is the position of the mass. The electrical system equivalent has an input from a current source and its output is voltage. Figure 6-14 also shows the derivation of the equations for the mechanical and electrical models.

As can be seen, one system may be modeled by another system by using the equivalent elements.

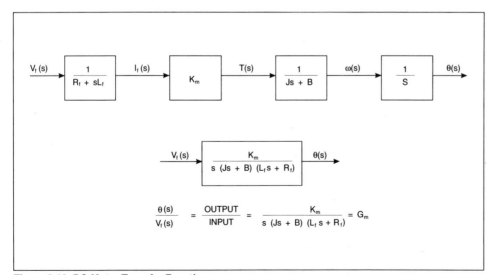

Figure 6-12. DC-Motor Transfer Function

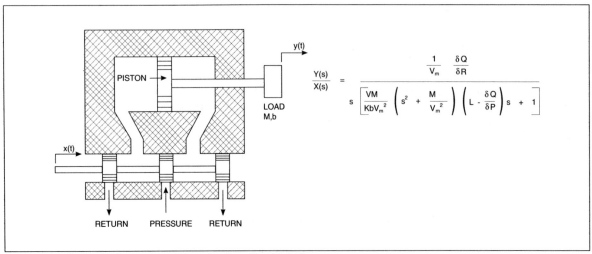

Figure 6-13. Hydraulic Actuator and Transfer Function (Ref. 6)

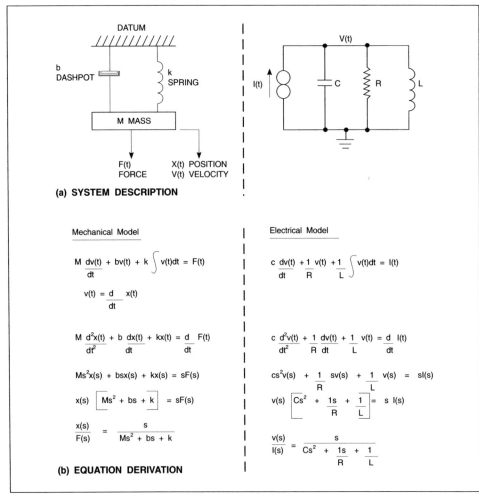

Figure 6-14. Suspension System and Electrical Equivalent

Block Diagram Reduction Techniques

Only simple configurations have been treated so far. In practice a higher degree of complexity is encountered, but no matter how complex a loop may be, it can be reduced to a simple transfer function. From this simple transfer function the response, stability, and analysis of the system are quickly found. These aspects will be discussed later, but for now, it is the method of obtaining the simplest transfer function that is important. Two common methods are in use: block diagram reduction and Mason gain reduction.

A block diagram for a relatively complex system is shown in Figure 6-15. Here, one input is processed through four system elements. Four feedback elements govern the process and act at various stages. The interconnection is fairly complex in this loop. The key to reducing it is the repeated application of several simple replacements, which are summarized in Table 6-3. Note that the problem is represented in nominal form where each simple transfer function is a G or an H. Once the system is reduced, these can be replaced with the corresponding Laplace form. This aspect will be left for later.

Table 6-3. Block Reduction for Different Transformations

Transformation	Original Form	Equivalent Form
1. Combining blocks in cascade	R → G_1 → G_2 → C	R → $G_1\ G_2$ → C
2. Moving a summer behind a block	R →(+) summer (± from X) → G → C	R → G →(+) summer (± from G ← X) → C
3. Moving a pickoff ahead of a block	R → G → C, with pickoff C	R → G → C, with C ← G ←
4. Moving a pickoff behind a block	R → G → C, with pickoff R	R → G → C, with R ← $\frac{1}{G}$ ←
5. Moving a summer ahead of a block	R → G →(+) summer (± from X) → R	R →(+) summer (± from $\frac{1}{G}$ ← X) → G → C
6. Eliminating a feedback loop	R →(+) summer (± from H) → G → C, H feedback	R → $\dfrac{G}{1 \pm GH}$ → C

Clearly there are many ways to apply the rules to reduce this system, as can be seen from the previous example: In Figure 6-16, Rule 4 and Rule 1 are applied to G_4, and the summing point from H_2 and H_1 is expanded. In Figure 6-17, Rule 6 is applied to H_4, Rule 1 is applied to G_2 and Rule 2 is applied twice to G_1. In Figures 6-17 and 6-18, Rule 6 and Rule 1 are applied. The result is the reduced transfer function.

Signal Flow Graphs

A second method for reducing a system employs a signal flow graph. This can be generated from the block diagram directly because there is a one-to-one correspondence between them. Recall that on a block diagram, devices (Gs and Hs) represent functions that act on variables, converting one to another, (i.e, rpm to current in a tachometer or, air pressure to steam flow in a pneumatically actuated valve).

The variables themselves are represented as converted, oriented (with arrows) lines. A summing point deals with one type of variable only. In a signal flow graph, variables (and summing points) are represented by nodes. Simple functions are represented by connecting lines with the process letter written on top. The direction of the "flow" is indicated by an arrow on the line. The sign (as it is at the summing point) of the device (+ or −) is assigned to the letter. To see this, look at the previous example (see Figures 6-15 to 6-20) transposed into signal flow form.

In this form the reduction technique is called Mason's Gain Formula.

Before discussing this, some definitions are in order:

(1) A "path" is a line or series of lines connecting any two nodes where the orientation of each line is in the same direction from start to finish.

(2) A "loop" is a closed path, a path that is connected to itself. Again, the orientation is similar along the loop.

(3) Two (or more) loops are said to be "touching" if they share at least one common node.

(4) Similarly, two (or more) loops are said to be "nontouching" if there are no common nodes.

(5) Two (or more) paths are considered "independent" if there is at least one branch (line) that is not common to each.

(6) A "gain product" is the product (multiplication) of all the gains in a loop or path. A gain of unity (1) may occur if a signal is passed from one node to another without processing, as in the case of unity feedback.

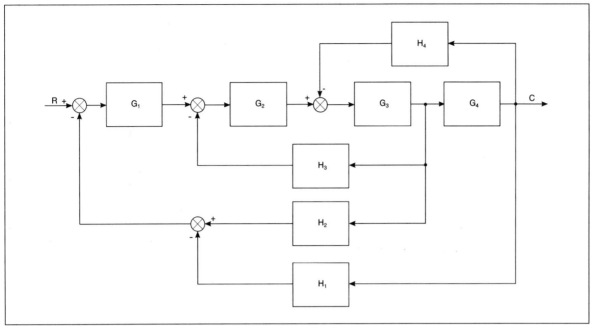

Figure 6-15. An Original Block Diagram

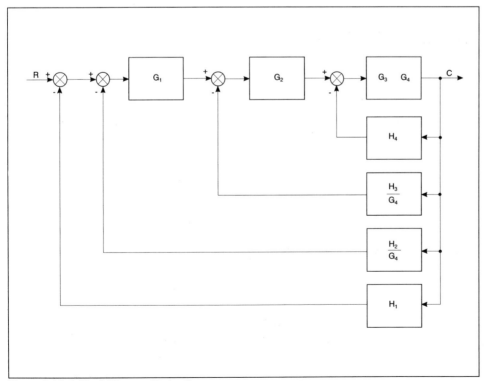

Figure 6-16. A Block Diagram Reduction

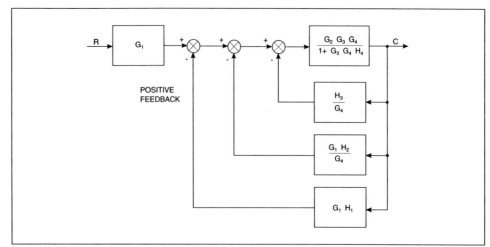

Figure 6-17. A Block Diagram Reduction

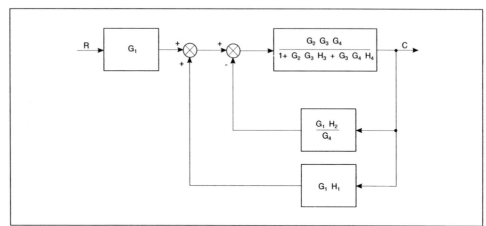

Figure 6-18. A Block Diagram Reduction

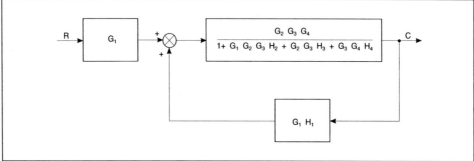

Figure 6-19. A Block Diagram Reduction

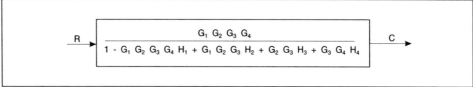

Figure 6-20. A Block Diagram Reduction

Mason's theorem is:

$$\frac{C}{R} = \sum_{k=1}^{N} \frac{P_k \delta_k}{\delta}$$

(6-19)

where:

C = output
R = input
N = number of forward paths
P_k = kth forward path gain product
δ = loop determinant
δ_k = δ, excluding loops touching P_k

and

$$\delta = 1 - \Sigma L_1 - \Sigma L_2 + \Sigma L_3 + \Sigma L_4 + \dots$$

(6-20)

where:

L_1 = loop gain of an individual loop
L_2 = product of loop gains of any two nontouching loops
L_3 = product of loop gains of any three nontouching loops
L_4 = product of loop gains of any four nontouching loops

As can be seen from Figure 6-21, the result is identical to that obtained from block reduction. In this example there are no nontouching loops and only one forward path. For clarity, one more system in which some loops are nontouching

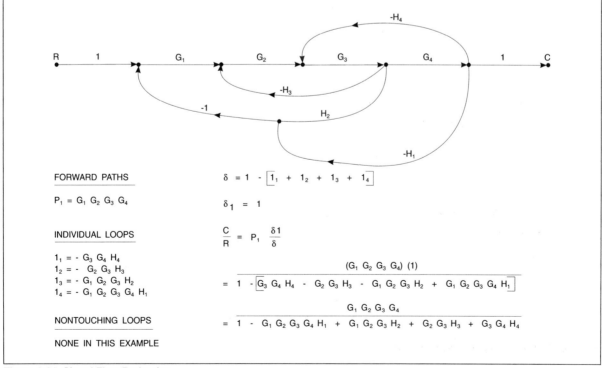

Figure 6-21. Signal Flow Reduction

will be examined. Refer to Figure 6-22. The block diagram reduction of this example will be left to the reader; the results should be the same.

Differential Equations

The mathematics involved in control system analysis and design must now be addressed. For readers not already familiar with Laplace transforms, this part of the theory usually invokes horror at first glance. There is great irony in this, as the use of Laplace mathematics reduces manipulation of differential and integral equations to algebra. The Laplace transform techniques are simple to use and relatively easy to understand. Although the theory here will be limited, the material will be covered in a systematic fashion, with the topic divided into functions, inputs, and responses.

It should be obvious by now that a control function in the traditional sense may be represented by a differential equation of first or second order (see Figure 6-14). It is also known that manipulation of differential equations can be very cumbersome, to say the least. To compound the problem, real systems are usually not in the most convenient state at that point in time when an analysis is started; that is, there are initial conditions. For example, at time $t = 0$, when a system is activated, a capacitor may already have a charge, a tank may have a product in it, and so on.

Traditionally, incorporating these initial states involves adding another set of calculations to an already long analysis. The system is time-dependent and integrals and derivatives may be involved. Thus, the mathematics are complicated.

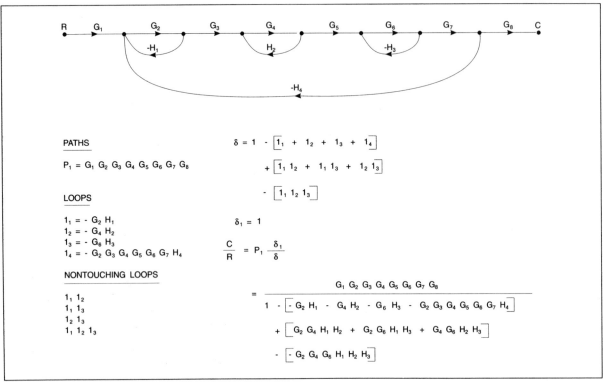

Figure 6-22. Another Signal Flow Reduction

Transform Calculus Using the Laplace Transform

The essence of Laplace transform calculus is as follows:

(1) An equation is first transformed using the Laplace operator. This moves the mathematics to another level. At this level, determining a solution becomes a simple exercise in algebra. Time is compressed such that it disappears. Initial conditions are incorporated directly.

(2) The result is then transformed back to the original form using a reverse transformation process.

(3) Since control functions are dynamic in nature (changing in time), the time domain is traditionally used to describe the function. Symbolically, one writes $f(t)$ and, graphically, one depicts responses as shown in Figure 6-23.

(4) All associated functions (input, transfer function, output) are functions of time.

(5) In Laplace form, which is referred to as the frequency domain, the "s-plane" is used. Recall that a dynamic process (for example, the action of a vibration-reducing suspension) may work well at one frequency but not at another.

(6) Symbolically, one writes $F(s)$ and the response may be in several graphical forms, the most popular of which is the Bode plot. The transfer function, however, is plotted on the s-plane. The s-plane has a real axis and an imaginary one ($j\omega$). The complex frequency s-plane plot graphically protrays the character of the natural transient response of the system.

(7) The transformation itself is accomplished by applying the Laplace integral and subsequently the inverse Laplace integral. These are Equations (6-21) and (6-22), respectively. It should be noted that these equations are rarely used. A set of "transform pairs," which are derived from the equations, is found in most mathematical table references (see Table 6-4). Simply look up and substitute each part of a function.

(8) Several properties assist in this task (see Table 6-5). Note that the initial conditions are incorporated directly; they are values, not variables. Furthermore, these values are set at $t = 0$, when the system is first being considered. The value $y(o)$ could be the initial level in a tank. The value d/dt $y(o)$ would then be the rate of change of the level in the tank at that time. If

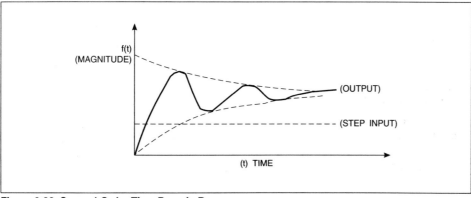

Figure 6-23. Second-Order Time Domain Response

Control System Theory

Table 6-4. Laplace Transform Pairs

Function	$f(t)$	$F(s)$
Unit Impulse	$u_0(t)$	1
Unit Step	$u_{-1}(t)$	$\dfrac{1}{s}$
Ramp	t	$\dfrac{1}{s^2}$
Parabola	t^2	$\dfrac{2}{s^3}$
n^{th} Order Ramp	t^n	$\dfrac{n!}{s^{n+1}}$
Exponential Decay	et^{-at}	$\dfrac{1}{s+a}$
Sine	$\sin(\omega t)$	$\dfrac{\omega}{s^2+\omega^2}$
Cosine	$\cos(\omega t)$	$\dfrac{s}{s^2+\omega^2}$
Exponential Decay Sine Function	$e^{-at}\sin(\omega t)$	$\dfrac{\omega}{(s+a)^2+\omega^2}$

Table 6-5. Properties of Laplace Transforms

Addition and Subtraction	$\mathcal{L}\{f_1(t)+f_2(t)\} = F_1(s)+F_2(s)$
Multiplication by a Constant	$\mathcal{L}\{kf(t)\} = kF(s)$
Derivatives	$\mathcal{L}\left\{\dfrac{d}{dt}f(t)\right\} + sF(s) - f(o)$
	$\mathcal{L}\left\{\dfrac{d^2}{dt^2}f(t)\right\} = s^2F(s) - sf(o) - \dfrac{d}{dt}f(o)$
Integrals	$\mathcal{L}\left\{\displaystyle\int_{-\infty}^{t} f(t)\,dt\right\} = \dfrac{F(s)}{s+1} + \dfrac{1}{s}\displaystyle\int_{-\infty}^{o} f(t)\,dt$
Shifting Theorem (Time Delay T ≥ 0)	$\mathcal{L}\{f(t-T)\ \mu(t-T)\} = e^{-st}F(s)$
Initial Value Theorem	$\lim_{t\to o} f(t) = \lim_{s\to\infty} sF(s)$
Final Value Theorem	$\lim_{t\to\infty} f(t) = \lim_{s\to o} sF(s)$

the tank is initially empty, then $y(o) = 0$. If nothing is going in or out of the tank at that time, then $d/dt\, y(o) = 0$.

$$F(s) = \int_o^\infty f(t)\, e^{-st}\, dt = \mathcal{L}\{f(t)\} \tag{6-21}$$

$$f(t) = \frac{1}{2\pi j} \int_{\delta - j\infty}^{\delta + j\infty} F(s)\, e^{+st} ds = \mathcal{L}^{-1}\{F(s)\} \tag{6-22}$$

There was a transformation in Figure 6-14 (the suspension system). For simplicity at that time, it was stated that the initial conditions were zero. The displacement at the beginning (from the datum) would be $x(0)$ and the initial velocity of the mass would be $d/dt\, x(0)$. A more complete example is now in order.

The governing equation, the driving function r(t), is given by the equation:

$$r(t) = m\, \frac{d^2 y(t)}{dt^2} + B\, \frac{dy(t)}{dt} + ky(t)$$

and the initial conditions with units of measure are:

$r(t)$	=	6 N
m	=	1 kg
$y(0)$	=	2 m
$dy(0)/dt$	=	2 ms
B	=	5 Ns/m
k	=	6 N/m

To find the function $y(t)$, which relates position to time, the method is as follows:

Step 1: Obtain the Laplace transform of the governing equation from Tables 6-4 and 6-5.

$$R(s) = m\left[s^2 Y(s) + sy(0) - \frac{dy(0)}{dt}\right]$$

$$+ B\,[sY(s) - y(0)] + kY(s)$$

Step 2: Solve for Y(s)

$$R(s) = Y(s)\,[ms^2 + Bs + k]$$

$$- [\,my(0)\,s + m\,\frac{d}{dt}y(0) + By(0)\,]$$

$$Y(s) = \frac{R(s) + \left[my(0)\,s + m\,\dfrac{d}{dt}y(0) + By(0)\right]}{\left[ms^2 + Bs + k\right]}$$

(continued)

(continued)

Step 3: Replace the variables with their values and simplify the equation.

$$Y(s) = \frac{\frac{6}{s} + [2s + 2 + 10]}{[s^2 + 5s + 6]}$$

$$= \frac{2s^2 + 12s + 6}{s(s^2 + 5s + 6)}$$

$$Y(s) = \frac{2s^2 + 12s + 6}{s(s+3)(s+2)}$$

Step 4: Do a partial fraction expansion of the equation.

$$Y(s) + \frac{1}{s} - \frac{4}{s+3} + \frac{5}{s+2}$$

Step 5: Now take the inverse Laplace transform of the equation using Tables 6-4 and 6-5.

$$Y(t) = 1 - 4e^{-3t} + 5e^{-2t}$$

This is the response of the system in the time domain.

The next section shows that the frequency response (magnitude and phase) is obtained from Step 3 directly. Note that in Step 4 the transfer function was manipulated algebraically.

System Response and Bode Diagrams

System Response

The time domain response (obtained in Step 5 of the example) may be divided into two basic components: the first component relates to the "steady-state" response; the second component relates to the "transient" response. From studies of differential equations, recall that the possibility of dividing the equation into two components is in the nature of the mathematics.

The "steady-state" component is not a function of time, and its value is dominant when the system has reached equilibrium some time after the period of adjustment. (It is easy to change the set point on a temperature controller in a step increment; however, it takes time for the temperature in the vessel to reach the desired value).

How the system reacts to get to the desired value is functionally found in the "transient" component, which is a function of time.

Recapitulating, in the example we have taken a system and determined its governing equation, driving function ($r(t)$), and initial conditions. To find the time domain response of this system from these equations, one takes the following five steps:

Step 1. Transform into a Laplace, using Tables 6-4 and 6-5.

Step 2. Solve for $Y(s)$.

Step 3. Replace values, simplify.

Step 4. Do a partial fraction expansion.

Step 5. Do an inverse transform.

Before examining the s-plane plot (plot of the Laplace transform), of the function, singularities must be discussed. A singularity is a point in the s-plane where the function, or its derivative, does not exist. The most important singularity is the "pole." At a pole the function will "blow up"; that is, its value becomes infinity.

The most common form of a frequency domain function is given by Equation (6-23):

$$F(s) = \frac{C(s)}{R(s)} = \frac{K(s+Z_1)(s+Z_2) \ \ldots \ (s+Z_n)}{(s+P_1)(s+P_2) \ \ldots \ (s+P_m)} \qquad (6\text{-}23)$$

The poles of Equation (6-23) are the roots of the denominator ($P_1, P_2, \ldots P_m$). When s equals any of these values, $F(s)$ becomes infinite. In the example, the poles are 0, -3, -2 (as per Step 3). These are simple, readily evident poles.

Two poles may appear as "complex conjugate pairs," $(s - a + j\omega)$ and $(s - a - j\omega)$, if the roots are complex. Poles play an important part in stability analysis, as will be seen later. Finally, consider the concept of a "zero." The zeros of a function, $(Z_1, Z_2, \ldots Z_n)$, are the values of s that make $F(s) =$ zero. When a function is mapped onto the s-plane, the poles and zeros are mapped. Remember, this is not the response, only the transfer function. Consider the example of Figure 6-24.

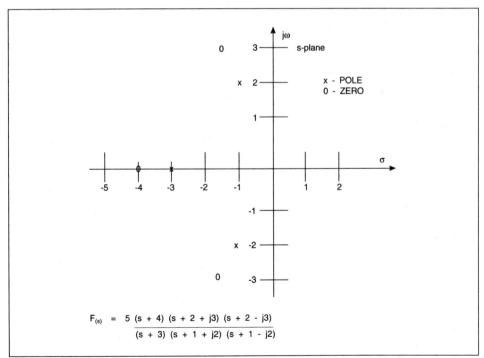

Figure 6-24. s-Plane Plot of a Transfer Function

Input function $R(s)$ may be a simple step, a ramp, a sinusoid, or a combination of these. In control system analysis, the step function is favored because it is the most severe input likely to occur. Table 6-4 gives the Laplace transform pairs for most of the functions used.

Bode Diagrams

As has been seen, any given system has two types of responses: the steady-state response and the transient response.

For the steady-state response, the most useful for analysis is a plot of the response amplitude vs. the frequency. This is called a Bode plot.

A periodic waveform, such as a sinusoid, is chosen as input. The resultant response will be a sinusoid with the same frequency as the input, with amplitude varying, dependent on the characteristics of the system. The resulting plot is representative of the dynamic behavior of the system.

For analysis of the transient response of a system, an amplitude vs. time plot is found to be more useful. Here, a unit step input is employed. The step response plot is an excellent method for evaluating some basic process parameters, such as dead time and steady-state gain, and for estimating controller settings.

In this section, first- and second-order systems will be presented. (The order of the system relates to the order of the differential equation that best describes the system.) For a more detailed understanding of the mathematics, refer to a text on advanced calculus.

A first-order system is not capable of oscillation on its own. This is because a first-order system contains only a dissipative element, an energy storage element, and an energy source; no spring or inductive element is involved.

Examination of Table 6-2 shows that proportional control devices are dissipative elements and that integral and/or derivative control devices are storage elements.

This concept is well illustrated by the example of an RC circuit excited by a step input (see Figure 6-25).

Using Equations (6-1) and (6-2) and Table 6-2, one can develop a response vs. time plot as shown in Figure 6-26. The analysis assumes that closing the switch applies a step input to the system, and that the capacitor is initially in a discharged state.

The appearance of the response of a first-order system in the time domain is that of an exponential curve. The curve could be a decay curve, as in the example, or a growth curve, as would happen if the capacitor were being charged. The quantity RC here is found to be the "time constant." Its units are seconds.

In a decay curve, the time constant corresponds to the amount of time required for the output to reach 36.8% of its initial value [(initial value)/(e)]. After time has reached a value equal to five of these "time constants," the response value reaches to within 0.5% of its final value.

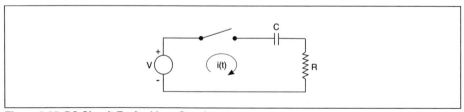

Figure 6-25. RC Circuit Excited by a Step Input

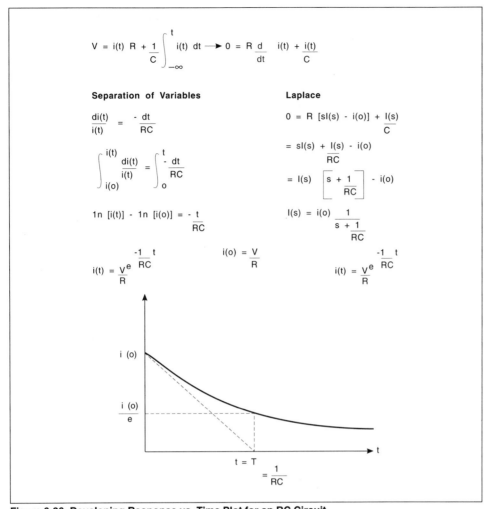

$$V = i(t)\ R + \frac{1}{C} \int_{-\infty}^{t} i(t)\ dt \longrightarrow 0 = R\ \frac{d}{dt}\ i(t) + \frac{i(t)}{C}$$

Separation of Variables

$$\frac{di(t)}{i(t)} = \frac{-\ dt}{RC}$$

$$\int_{i(o)}^{i(t)} \frac{di(t)}{i(t)} = \int_{0}^{t} \frac{-\ dt}{RC}$$

$$1n\ [i(t)] - 1n\ [i(o)] = \frac{-\ t}{RC}$$

$$i(t) = \frac{V}{R} e^{\frac{-1}{RC} t}$$

Laplace

$$0 = R\ [sI(s) - i(o)] + \frac{I(s)}{C}$$

$$= sI(s) + \frac{I(s) - i(o)}{RC}$$

$$= I(s) \left[s + \frac{1}{RC} \right] - i(o)$$

$$I(s) = i(o)\ \frac{1}{s + \frac{1}{RC}}$$

$$i(o) = \frac{V}{R}$$

$$i(t) = \frac{V}{R} e^{\frac{-1}{RC} t}$$

$$i(o)$$

$$\frac{i(o)}{e}$$

$$t = T = \frac{1}{RC}$$

Figure 6-26. Developing Response vs. Time Plot for an RC Circuit

In a growth curve, the time constant is similarly equal to *RC* and corresponds to the time required for the output response to reach 63.2% of the final value.

For an inductive system, the time constant is given as *L/R*.

The order of a system may be seen directly as the largest power of *s* occurring in the denominator of the transfer function. In the example, there was a single s, which indicates a first-order system. The example in Figure 6-24 can be seen to be third-order when the denominator is expanded. Although all systems may be better represented by third-order or higher differential equations, discussion here is restricted to first- and second-order equations. The analysis of third-order and higher differential equations is quite complex, and the results obtained from first- and second-order equation analysis are usually good enough for most purposes. This is because most systems are first- or second-order dominant. That is, the relative values of the first- or second-order response terms are much greater than the responses due to third-order or higher terms.

A system is said to be second-order dominant if the effects of the further roots of the denominator can be considered to be negligible.

A second-order system has at least one pair of storage elements of a different type (integral and derivative). Energy may be stored in one element and transferred to another element (e.g., potential energy in a mass transferred to potential energy in a spring.). Energy can thus oscillate back and forth between the two before it has a chance to dissipate through a proportional element (e.g., heat loss due to viscous friction in a dashpot.). How long a system will oscillate depends on the "damping factor." Differential equations show that there are four possible cases. See Figure 6-27 and Table 6-6.

Table 6-6.   Four Cases of Damping

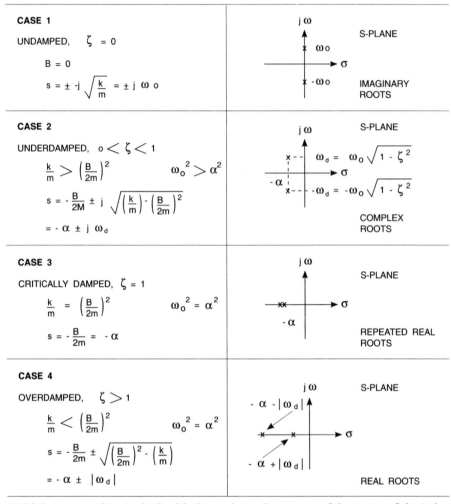

CASE 1 UNDAMPED, $\zeta = 0$ $B = 0$ $s = \pm -j \sqrt{\dfrac{k}{m}} = \pm j\,\omega_o$	$j\omega$ S-PLANE ω_o σ $-\omega_o$ IMAGINARY ROOTS
CASE 2 UNDERDAMPED, $o < \zeta < 1$ $\dfrac{k}{m} > \left(\dfrac{B}{2m}\right)^2$ $\omega_o^2 > \alpha^2$ $s = -\dfrac{B}{2M} \pm j \sqrt{\left(\dfrac{k}{m}\right) - \left(\dfrac{B}{2m}\right)^2}$ $= -\alpha \pm j\,\omega_d$	S-PLANE $j\omega$ $\omega_d = \omega_o\sqrt{1-\zeta^2}$ $-\alpha$ σ $-\omega_d = -\omega_o\sqrt{1-\zeta^2}$ COMPLEX ROOTS
CASE 3 CRITICALLY DAMPED, $\zeta = 1$ $\dfrac{k}{m} = \left(\dfrac{B}{2m}\right)^2$ $\omega_o^2 = \alpha^2$ $s = -\dfrac{B}{2m} = -\alpha$	$j\omega$ S-PLANE σ $-\alpha$ REPEATED REAL ROOTS
CASE 4 OVERDAMPED, $\zeta > 1$ $\dfrac{k}{m} < \left(\dfrac{B}{2m}\right)^2$ $\omega_o^2 = \alpha^2$ $s = -\dfrac{B}{2m} \pm \sqrt{\left(\dfrac{B}{2m}\right)^2 - \left(\dfrac{k}{m}\right)}$ $= -\alpha \pm |\omega_d|$	$j\omega$ S-PLANE $-\alpha - |\omega_d|$ σ $-\alpha + |\omega_d|$ REAL ROOTS

Which case one has to deal with depends on the nature of the roots of the "characteristic equation" (recall from differential calculus). Basically, if the roots have only a real part, expect an exponential decay. If the roots have only an imaginary part, expect some oscillation. If the roots are complex (real and imaginary), expect both exponential decay and oscillations. This may be seen in Figures 6-27 and 6-28. A standard form is used, along with specific nomenclature: the roots are poles; they relate to the *s*-plane as per Table 6-1. The following derivation will clarify the point and introduce the standard form.

The description of the function in the time domain is:

$$m\frac{d^2}{dt^2}y(t) + B\frac{d}{dt}y(t) + ky(t) = \delta(t) \qquad (6\text{-}24)$$

For $y(0) = 0$, and $d/dt\, y(0) = 0$, the transformed function is:

$$Y(s) = \frac{1}{ms^2 + Bs + k} = \frac{\frac{1}{m}}{s^2 + \frac{B}{m}s + \frac{k}{m}} \qquad (6\text{-}25)$$

Given $\alpha = B/2m = \xi\omega_o$ = the damping coefficient, and $\omega_o = (k/m)^{1/2}$ = the undamped natural frequency, the characteristic equation is:

$$s^2 + \frac{B}{m}s + \frac{k}{m} = 0 \qquad (6\text{-}26)$$

and the standard form is:

$$S^2 + 2\alpha s + \omega_o^2 = 0 \qquad (6\text{-}27)$$

Substituting $\alpha = \xi\omega_d$:

$$S^2 + 2\zeta\omega_d s + \omega_d^2 \qquad (6\text{-}28)$$

which is the standard form expressed in terms of $\xi = \alpha/\omega_d$ = the damping ratio, and ω_d = the damped natural frequency. The roots of the equation are:

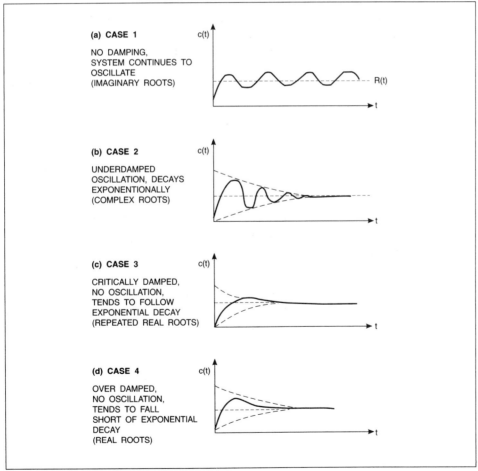

(a) CASE 1

NO DAMPING,
SYSTEM CONTINUES TO
OSCILLATE
(IMAGINARY ROOTS)

(b) CASE 2

UNDERDAMPED
OSCILLATION, DECAYS
EXPONENTIONALLY
(COMPLEX ROOTS)

(c) CASE 3

CRITICALLY DAMPED,
NO OSCILLATION,
TENDS TO FOLLOW
EXPONENTIAL DECAY
(REPEATED REAL ROOTS)

(d) CASE 4

OVER DAMPED,
NO OSCILLATION,
TENDS TO FALL
SHORT OF EXPONENTIAL
DECAY
(REAL ROOTS)

Figure 6-27. Four Possible Cases of Damping

$$s = -\frac{B}{2\,m} \pm \left[\left(\frac{B}{2m}\right)^2 - \left(\frac{k}{m}\right) \right]^{1/2} = \left[\omega_d^2 - \alpha^2 \right]^{1/2} \tag{6-29}$$

$$= -\frac{B}{2\,m} \pm j\left[\left(\frac{k}{m}\right) - \left(\frac{B}{2\,m}\right)^2 \right]^{1/2} = \omega_d [1 - \delta^2]^{1/2}$$

$$s = -\alpha \pm j\,\omega_d$$

The contour formed by the roots as the damping ratio is increased from zero to infinity is called a "root locus" and plays a part in stability analysis. As may be seen from Figure 6-28, a pole in the positive half plane is not acceptable since this means that the system is unstable. It is also undesirable to have this pole on the axis since the system will be prone to oscillation.

The output from a second-order system will now be examined. It is possible to determine whether a system is second-order dominant or not, if, upon examination of the pole placement in the s-plane, it is found that the pole following the principal pole is at least ten times farther away from the origin than this principal pole (i.e., a function with poles $(-1 + 2j)$, $(-1 - 2j)$, $(10 + 0j)$ is second-order dominant). When referring to these outputs, the term "performance specifications" is often used. See Figure 6-29 for a graphical representation of "performance specifications."

The figure shows the underdamped rise time T_r (the time required for the output to go from 0% to 100% of final value) and overdamped rise time T_{r1} (the time required for the output to go from 10% to 90% of final value).

$$Tr = \frac{\pi - \cos^{-1}\delta}{\omega\sqrt{1 - \delta^2}} \tag{6-30}$$

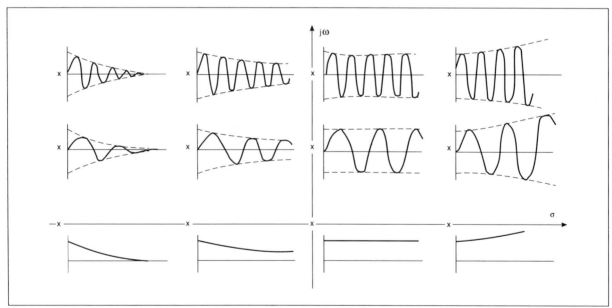

Figure 6-28. Pole Location and Time Domain Output for Impulse Input (Ref. 3)

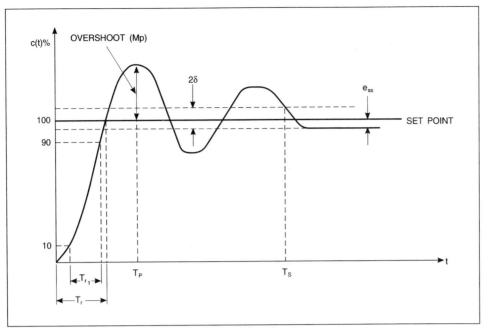

Figure 6-29. Typical Response of Second-Order System to Step Input

Overshoot is generally expressed as percent overshoot (P.O.). It is the ratio, expressed as a percentage, of the difference of the maximum value and the final value, to this final value.

$$P.O. = \frac{M_p - C(\infty)}{C(\infty)} 100 = 100e^{\frac{-\delta\pi}{[1-\delta^2]^{1/2}}} \qquad (6\text{-}31)$$

The time constant (τ) is the same time constant as was used in the first-order response.

$$\tau = \frac{1}{\alpha} \qquad (6\text{-}32)$$

The settling time (T_s) is the time required for the system to settle to a value within a certain percentage of the final response value given as δ. The usual value of δ is 2%.

$$Ts = \frac{-In\ \delta}{\alpha} = \frac{4}{\alpha} \quad (\text{for } \delta = 0.02) \qquad (6\text{-}33)$$

The peak time (T_p) is the time required for the output to reach the maximum value.

$$Tp = \frac{\pi}{\omega_d} \qquad (6\text{-}34)$$

The steady state error (e_{ss}) is the difference between the set point and the final output value. The value of steady-state error depends on the input and on the number of integrators in the open-loop transfer function (G).

$$E(s) = \frac{R}{1 + GH} = \frac{R}{1 + G} \quad \text{(for } H = 1\text{)} \tag{6-35}$$

$$e_{ss} = \lim_{t \to \infty} e(t) = \lim_{s \to o} sE = \lim_{s \to o} \frac{sR}{1 + G} \quad \text{(final value theorem)}$$

Control systems are often described in terms of their "type number" and "error constants." The type number is defined as the number of integrators, $(1/s)$, in the function. The error constants are constant values applying to a system as follows:

$$K_p = \lim_{s \to o} G = \text{position error constant} \tag{6-36}$$

$$K_v = \lim_{s \to o} sG = \text{velocity error constant} \tag{6-37}$$

$$K_a = \lim_{s \to o} s^2 G = \text{acceleration error constant} \tag{6-38}$$

The error constants are: K_p, K_v, and K_a. A summary of steady-state errors with respect to different inputs is listed in Table 6-7. Clearly, the smaller the steady-state error, the more closely the system follows the error signal and the better the system responds. A large error constant implies a better system.

Table 6-7. Summary of Steady-State Errors

Input Type	Step (R(S) = A/S)	Ramp R(s) = A/S^2	Parabola R(s) = A/S^3
0	$e_{ss} = \dfrac{A}{1 + K_p}$	∞	∞
1	$e_{ss} = o$	$\dfrac{A}{K_v}$	∞
2	$e_{ss} = 0$	0	$\dfrac{A}{K_p}$

The frequency response plots (or Bode diagrams) are constructed from the transfer function in either of two ways:

(1) The point-by-point method (more exact)

(2) The asymptotic approximation method (much quicker)

The Bode diagram is obtained by adding the plots for each pole and for each zero. How each of these affects the diagram will now be examined.

Bode diagrams represent both the amplitude and phase responses of a system, usually with frequency as the base. The plots are drawn one below the other, so it is easy to visualize both amplitude and phase responses simultaneously.

The first step in the preparation of a Bode diagram is to put the equation into factored form and replace all the (s) by $(j\omega)$. The next step is to apply the magnitude equation, (see Equation (6-39)), and produce the plot of amplitude vs. frequency. (It is customary to use semilog paper.)

The final step is to apply the phase equation (see Equation (6-40)) and to produce the plot of phase vs. frequency.

$$M(\omega) = 20 \log |\text{ transfer function in } (j\omega)| \tag{6-39}$$

$$\theta\,(\omega) \;=\; \text{phase angle of the transfer function in } (j\omega) \qquad (6\text{-}40)$$

The usual form of a transfer function (T. F.) includes several factors that may be poles or zeros (poles are terms of the denominator, zeros are terms of the numerator). These may be constants (k), origin poles or zeros (s_n), simple poles or zeros $(s + a)$, and complex conjugate poles or zeros $(s^2 + 2Z\omega_0 s + \omega)$.

The typical form of a transfer function, containing one of each of the different types of poles and zeros is given in Equation (6-41):

$$T.F.\,(s) \;=\; k\,\frac{s + a}{s^n \left(s^2 + 2\,\zeta\omega_0\, s + \omega_0^2\right)} \qquad (6\text{-}41)$$

or

$$T.F.\,(s) \;=\; \frac{ak}{\omega_0^2}\; \frac{1 + \dfrac{s}{a}}{S^n \left[1 + \dfrac{2\,\zeta\, s}{\omega_0} + \left(\dfrac{s}{\omega_0}\right)^2\right]} \qquad (6\text{-}42)$$

Factoring, replacing the (s) by $(j\omega)$ yields:

$$T.F.\,(j\omega) \;=\; \frac{ak}{\omega_0^2}\; \frac{1 + j\dfrac{\omega}{a}}{(j\omega)^n \left[\dfrac{\omega_0^2 - \omega^2}{\omega_0^2}\right] + j\left[\dfrac{2\zeta\omega}{\omega_0}\right]} \qquad (6\text{-}43)$$

Equation (6-43) may now be separated into the terms relating to magnitude and to phase, which gives Equations (6-44) and (6-45). Notice that a pole is subtracted and that a zero is added to the sum by this action. Different values of (ω) are then used in the equation to obtain values for the diagram.

$$M\,(\omega) \;=\; 20 \log \left|\, T.F.\,(j\omega) \,\right| \quad \text{(in dB)} \qquad (6\text{-}44)$$

$$=\; 20 \log \left|\, \frac{ak}{\omega_0^2} \,\right| \;+\; 20 \log \left(1 + \frac{\omega^2}{a^2}\right)^{1/2}$$

$$M(\omega) \;=\; -20N \log(\omega) - 20 \log \left\{\left[\left(\frac{\omega_0^2 - \omega^2}{\omega_0^2}\right)^2 + \left(\frac{2\zeta\omega}{\omega_0}\right)^2\right]\right\}^{1/2}$$

$$\theta(\omega) \;=\; \text{Phase angle of the T.F. } (j\omega) \text{ (in radians or degrees)}$$

$$\theta\,(\omega) \;=\; arg\left(\frac{ak}{\omega_0^2}\right) + \tan^{-1}\left(\frac{\omega}{a}\right) - n\,90° - \tan^{-1}\left(\frac{2\zeta\omega_0\omega}{\omega_0^2 - \omega^2}\right) \qquad (6\text{-}45)$$

$$0° \text{ if } k > 0$$
$$-180° \text{ if } k < 0$$

(ω) is angular frequency (measured in radians per second) and is related to frequency (f), (measured in cycles per second), by the equation $\omega = 2\,\pi f$, $(\pi = 3.1416)$. Each factor in the transfer function has a distinctive Bode diagram. In the asymptotic method the factors to generate each individual curve are used, then their values are "added" graphically. This is not an exact method but is quickly and easily accomplished and gives a good indication of the resulting plots. Table 6-8 summarizes the effects of each type of pole (or zero).

It is found that these approximations closely resemble the exact functions with one small exception, the complex singularity. This special case is strongly dependent on the damping ratio, ξ (The effect of varying this parameter is illustrated in Figure 6-30 for completeness). An example derived by each method is shown in Figures 6-31 through 6-33.

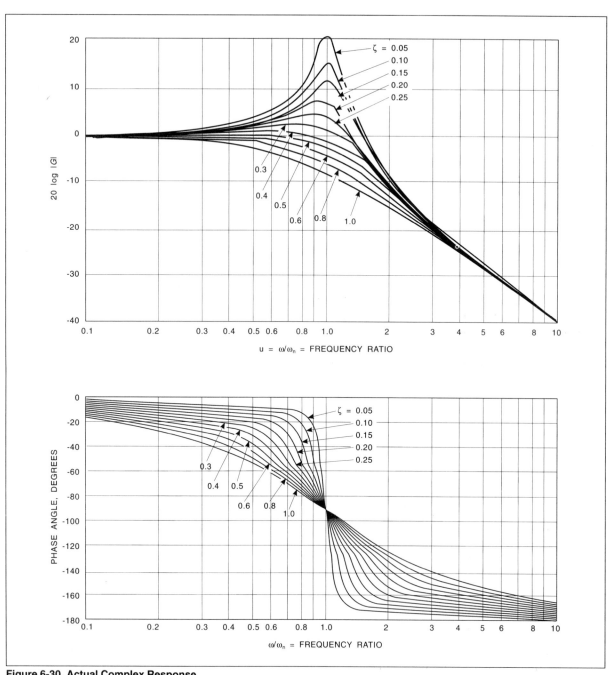

Figure 6-30. Actual Complex Response

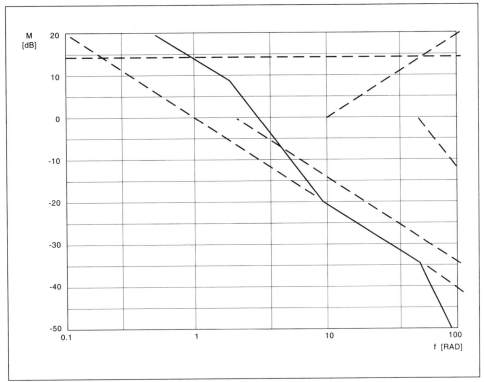

Figure 6-31. Bode Magnitude Approximation

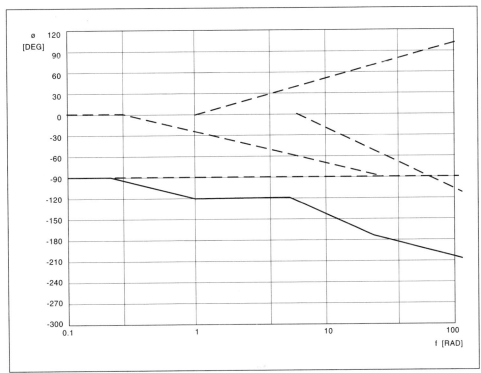

Figure 6-32. Bode Phase Approximation

Table 6-8. **Asymptotic Approximation of Poles and Zeros**

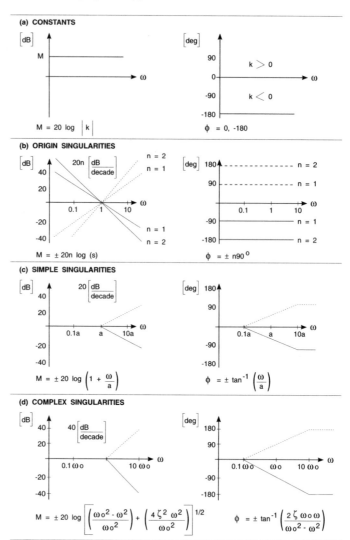

Consider the following transfer function:

$$GH(s) = 2500 \frac{s + 10}{s(s+2)(s^2 + 30s + 2500)} \tag{6-46}$$

Factoring yields:

$$GH(s) = \frac{5\left(1 + \dfrac{s}{10}\right)}{s\left[1 + \dfrac{s}{2}\right]\left[1 + \dfrac{0.6s}{50} + \left(\dfrac{s}{50}\right)^2\right]} \tag{6-47}$$

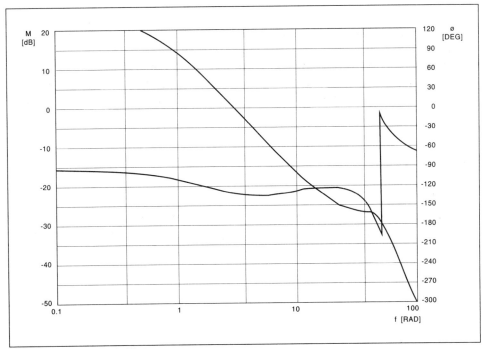

Figure 6-33. Exact Bode Plot

Substituting $j\omega$ yields:

$$GH(j\omega) = \frac{5\left(1 + j\dfrac{\omega}{10}\right)}{j\omega\left[1 + j\dfrac{\omega}{2}\right]\left[\dfrac{50^2 - \omega^2}{50^2} + \dfrac{j0.6\omega}{50}\right]} \tag{6-48}$$

Therefore:

$$M(\omega) = 20\log|5| + 20\log\left(1 + \frac{\omega^2}{10^2}\right)^{1/2} - 20\log(\omega) \tag{6-49}$$

$$M(\omega) = -20\log\left(1 + \frac{\omega^2}{2^2}\right)^{1/2} - 20\log\left[\left(\frac{50^2 - \omega^2}{50^2}\right)^2 + \left(\frac{0.6\,\omega}{50}\right)^2\right]^{1/2}$$

and

$$\theta(\omega) = 0^o + \tan^{-1}\left(\frac{\omega}{10}\right) - 90° - \tan^{-1}\left(\frac{\omega}{2}\right) - \tan^{-1}\left(\frac{0.6(50)\omega}{50^2 - \omega^2}\right) \tag{6-50}$$

This section is concluded with a note on the Bode plot's frequency axis. As mentioned, angular frequency has been used in radians per unit of time. It is often presented in hertz (cycles/sec). The difference is only a constant of multiplication; thus, it is easy to go from one to the other.

To make the asymptotic graphs, the form of Equation (6-49) is used, while for the exact graphs the values of (ω) must be inserted into Equations (6-49) and (6-50). Software is available that will create Bode diagrams. The methods used by computers are similar to those shown here.

A small difference between the "approximate" and the "exact" plots is a result of the complex pole as seen in Figure 6-30 (the effects of the damping ratio are excluded when the approximation method is used).

Although the Bode diagram analysis method is preferred in practice (phase and gain may be associated directly with frequency in this method), several other representations are in use today. Two of the more prominent methods used are polar diagrams and the Nichols Chart.

Stability

Definitions

The "sensitivity" of a system is defined as the ratio of the percentage change in the system transfer function to the percentage change in a particular parameter, x. Mathematically, it is denoted as follows:

$$S = \frac{\dfrac{\delta \text{ T.F.}}{\text{T.F.}}}{\dfrac{\delta \chi}{\chi}} \tag{6-51}$$

Recall that, due to its inherent feedback, a closed-loop system can use hardware that is less precise and still offer good quality of control, compared to an open-loop system. Equation (6-51) gives the mathematical reason for this. Consider the following example (see Figure 6-34).

> As can be seen, if the product of *GH* is large, the sensitivity of the closed loop to variations in *G* is less than for the open loop. If, for example, *G* were a motor, feedback control would allow the use of a less precise motor. This is an economic consideration: precision motors are more expensive and require more field calibration. Note, however, that the sensitivity of the system to variations of *H* is not the same as that for *G*. To visualize this, apply Equation (6-51), using *H* as a variable parameter.

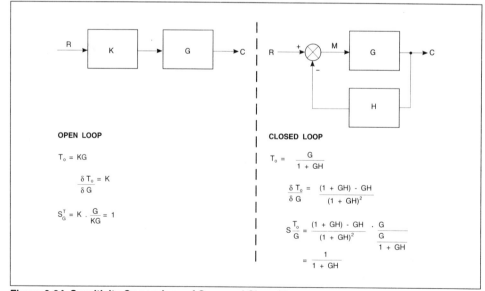

Figure 6-34. Sensitivity Comparison of Open and Closed Loops

The topic of stability is a course on its own. This discussion will be limited to definitions. The stability of a control system is an important consideration when dealing with feedback because, as discussed earlier, closed loops are prone to oscillation. A system must remain stable (that is, the system response must not show unnecessary oscillation and must remain within bounds), when it is subjected to input commands, disturbance inputs, and extraneous factors such as power supply variations and elemental parameter changes (due to temperature, etc.).

Generally, if for a bounded input the output is also bounded, a control system is said to be stable.

Various techniques can be employed to determine if a system is stable. In addition to the Bode diagram, which may be used to evaluate a control loop for stability, the following techniques are briefly explained:

(1) The Bode Stability Criterion

(2) Root Locus Analysis

(3) The Routh-Hurwitz Criterion

(4) The Nyquist Criterion

All stability analysis deals with the location of the poles of the transfer function and the manipulation (compensation) of these poles. Recall the effect of a pole in the positive half of the s-plane (see Figure 6-28). This is not permitted for stability reasons.

There are other considerations, such as how much oscillation will occur for 5% overshoot. Is this a stable condition?

The Bode Stability Criterion (Ref. 2)

The Bode stability criterion was a key result of the work of Bode in the study and design of communications and control systems in the 1920s and can be stated as follows:

A closed-loop system is unstable if the frequency response of the open-loop transfer function $G_{OL} = G_c G_v G_p G_m$ has an amplitude ratio greater than one at the critical frequency. Otherwise, the closed-loop system is stable. The critical frequency is defined to be the frequency at which the open-loop phase angle is -180).

The Bode stability criterion allows the stability of closed-loop systems to be calculated from the open-loop transfer function G_{OL}. Because the criterion can be applied directly to systems that contain time delays, this method is preferred to other stability criteria based on the characteristic equation. The Bode stability criterion is applicable only to open-loop stable systems with phase angle curves that exhibit a single critical frequency. This situation occurs in most process control problems. One exception is the so-called *conditional stability* case, where multiple values of the critical frequency can occur.

When a closed-loop system is at the stability limit (that is, when the open-loop amplitude ratio is 1 at the critical frequency represented by an open-loop phase angle of $-180°$, the feedback control system produces a sustained oscillation in the controlled variable. To understand how such sustained oscillations can occur, consider the analogy of pushing a child on a swing. The child will continue to swing in the same arc if a person pushes the child at the right time (in phase) and with the right amount of force. If the timing or the amount of force is incorrect, the cyclic movement of the swing changes, and the arc will be reduced or increased. A similar phenomenon occurs when a person bounces a ball.

Suppose that the feedback control system in Figure 6-35 is subjected to a sinusoidally varying set point $R(t) = A \sin \omega_c t$ for a period of time $0 < t < t_f$ and

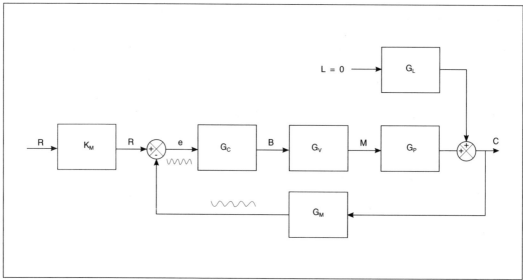

Figure 6-35. Feedback Control System

the comparator is disconnected during this time period. Assume that no load change occurs for $t < 0$ ($L(t) = 0$). The signal $R(t)$ oscillates at the critical frequency to ω_c; after an initial transient period, this causes B to oscillate at the same frequency. At $t = t_f$, the set point signal R is set to zero and the comparator is connected again. If the control system is marginally stable, the controlled variable C will exhibit a sustained sinusoidal oscillation (neither growing nor attenuating) with a frequency $\omega = \omega_c$. To understand why this special type of oscillation occurs only when $\omega = \omega_c$, note that the sinusoidal signal e passes through transfer functions G_c, G_v, G_p, and G_m before returning to the comparator. To maintain the oscillation, signal B must have the same amplitude as signal e and it also must have a -180 phase shift. After signal B passes through the comparator, it is identical to e and the oscillation continues indefinitely. At these conditions the open-loop transfer function has an amplitude ratio equal to unity at the critical frequency ω_c, (which is also called the resonant frequency or the phase crossover frequency). The period of oscillation will be $P_u = 2\pi/\omega c$. On the other hand, if the open-loop transfer function has an amplitude ratio greater than one at ω_c, then the amplitude of B is larger than the amplitude of e and the oscillation will grow with time. This behavior implies that the closed-loop system is unstable.

Root Locus Analysis (Ref. 5)

Root locus analysis is a graphical method of evaluating the stability of a control loop.

The roots and poles of the charateristic equation are found and their values plotted on a complex s-plane graph (x-axis = σ, yaxis = $j\omega$), while a parameter is varied from zero to an infinitely large positive value. The parameter may be any parameter of the system that is being analyzed for stability, such as the total gain of the system or the controller gain. K will be used to represent this parameter.

When $K = 0$, the roots of the characteristic equation are the poles of $P(s)$, where $P(s)$ is the characteristic equation with gain K factored out.

When K approaches infinity, the roots of the characteristic equation are the zeros of $P(s)$.

The locus of the roots of the characteristic equation ($1 + KP(s) = 0$) begins at the poles of $P(s)$ and ends at the zeros of $P(s)$ as K increases from zero to infinity.

Using these basic rules, the locus (path of the roots while K varies) is plotted. The exact values of the roots need not be calculated throughout the complete path since the last rule sets the general area of the values and their breakaway points and directions. This is all that needs to be known for the analysis.

Once the root locus is plotted, the graph is analyzed for stability:

The root locus on the real axis always lies in a section of the real axis to the left of an odd number of poles and zeros.

(1) The system is stable for those values of the root loci that are on the left-hand side of the graph. Therefore, all values of K that gives roots on the left-hand side will be stable.

(2) The system is then designed with values of the controller parameters that will have loci on the left-hand plane.

The Routh-Hurwitz Criterion

The Routh-Hurwitz criterion for the stability of control systems is a mechanical operation in mathematics that may be used to calculate the values of variable parameters of a system for which the system will be stable.

The characteristic equation of the system, in the Laplace variable form, is written as (Ref. 3):

$$\delta(s) = q(s) = a_n s^n + a_{n-1} s^{n-1} + \ldots + a_1 s + a_0 = 0 \qquad (6\text{-}52)$$

It is necessary to determine if any of the roots of $q(s)$ lie in the right half of the s-plane. The equation is rewritten in factored form:

$$a_n (s - Y_1)(s - Y_2) \ldots (s - Y_n) = 0 \qquad (6\text{-}53)$$

where $r_1 r_2 \ldots r_n$ is the nth root of the characteristic equation. Multiplying the factors together:

$$\begin{aligned} q(s) = {} & a_n s^n - a_n (r_1 + r_2 + \ldots + r_n) s^{n-1} \qquad (6\text{-}54) \\ & + a_n (r_1 r_2 + r_2 r_3 + r_1 r_3 + \ldots) s^{n-2} \\ & - a_n (r_1 r_2 r_3 + r_1 r_2 r_4 + \ldots) s^{n-3} + \ldots \\ & + a_n (-1)^n r_1 r_2 r_3 \ldots r_n = 0 \end{aligned}$$

Examine this equation, and note that all the coefficients of the polynomial must have the same sign if all the roots are in the left-hand plane. Also, it is necessary for a stable system that all the coefficients be nonzero. However, while these requirements are necessary for stability, they are not sufficient; that is, if they are not satisfied, the system is unstable. However, if they are satisfied, one must proceed to ascertain the stability of the system (Ref. 3).

The Routh-Hurwitz criterion is a necessary and sufficient criterion for the stability of linear systems. Here are the steps that form the criterion:

Order the coefficients of the characteristic equation:

$$a_n s^n + a_{n-1} s^{n-1} + \ldots + a_1 s + a_0 = 0 \qquad (6\text{-}55)$$

into an array such that:

(6-56)

$$
\begin{array}{c|cccc}
s^n & a_n & a_{n-2} & a_{n-4} & \cdots \\
s^{n-1} & a_{n-1} & a_{n-3} & a_{n-5} & \cdots \\
s^{n-2} & b_{n-1} & b_{n-3} & b_{n-5} & \cdots \\
\cdot & \cdots & \cdots & \cdots & \cdots \\
\cdot & \cdots & \cdots & \cdots & \cdots \\
s^0 & h_{n-1} & & &
\end{array}
$$

where:

$$
b_{n-1} = \frac{a_{n-1}\,a_{n-2} - a_n\,a_{n-3}}{a_{n-1}} = \frac{-1}{a_{n-1}} \begin{vmatrix} a_n & a_{n-2} \\ a_{n-1} & a_{n-3} \end{vmatrix}
\tag{6-57}
$$

and:

$$
b_{n-3} = \frac{-1}{a_{n-1}} \begin{vmatrix} a_n & a_{n-4} \\ a_{n-1} & a_{n-5} \end{vmatrix}
\tag{6-58}
$$

and:

$$
c_{n-1} = \frac{-1}{b_{n-1}} \begin{vmatrix} a_{-1} & a_{n-3} \\ b_{n-1} & b_{n-3} \end{vmatrix}
\tag{6-59}
$$

and so on.

The Routh-Hurwitz criterion states that the number of roots of $q(s)$ with positive real parts is equal to the number of changes in sign of the first column of the array. It requires that there be no changes in sign in the first column for a stable system. This requirement is both necessary and sufficient (Ref. 3). From this, one may calculate the values of the variable parameter that will guarantee that the criterion is met. Three cases must be treated differently:

Case 1: No element in the first column is zero. This case is simple, since one needs only to count the number of sign changes in the first column to know how many of the roots lie in the right-hand plane and thus how many stable roots there are.

Case 2: Zeros in the first column while some other elements of the row contain a zero in the first column are nonzero. The zero may be replaced by a small positive number, such as epsilon, and the analysis is the same as for all nonzero first column, except for the consideration that epsilon will cause instability.

Case 3: Zeros in the first column and the other elements of the row containing the zero are also zero. The polynomial contains singularities that are symmetrically located about the origin of the s-plane. The order of the auxiliary equation is always even and indicates the number of symmetrical root pairs.

The Nyquist Criterion (Ref. 2)

The Nyquist stability criterion is similar to the Bode criterion in that it can be used to determine closed-loop stability from open-loop frequency response characteristics. The Nyquist plot is a polar plot of the frequency response characteristics; consequently, it conveys the same information as the Bode plot. Unlike the Bode criterion, the Nyquist criterion is applicable to open-loop unstable systems and to systems with more than one critical frequency. It thus provides a more general approach.

> **Nyquist Stability Criterion: If N is the number of times that the Nyquist plot encircles the point $(-1, 0)$ in the complex plane in the clockwise direction, and P is the number of open-loop poles of $G_{OL}(s)$ that lie in the right-half plane, then $Z = N + P$ is the number of unstable roots of the closed-loop characteristic equation (those roots lying in the right-half plane).**

Several comments regarding the Nyquist criterion can be made:

(1) The reason that the $(-1, 0)$ point is so important follows from the characteristic equation, $1 + G_{OL}(s) = 0$, which can also be written as $G_{OL}(s) = -1$. This condition corresponds to a complex transfer function with an amplitude ratio of $+1$ and a phase angle of -180.

(2) Typically, the open-loop system is stable and thus has no poles in the right-half plane (i.e., $P = 0$). For this situation, $Z = N$, and the closed-loop system is unstable if the Nyquist plot encircles the $(-1,0)$ point one or more times (see Figure 6-36).

(3) A negative value of N indicates tha the encirclements of the $(-1,0)$ point occur in the opposite direction (i.e., counterclockwise).

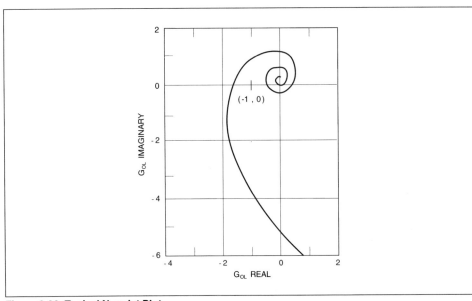

Figure 6-36. Typical Nyquist Plot

Control System Theory

(4) The Nyquist diagram and associated theory can be used to determine if an open-loop unstable process can be stabilized by feedback control. While a polar plot of $G_{OL}(j\omega)$ is meaningful in this case, the earlier physical interpretation of sinusoidal forcing is not, since the longtime response will be unbounded rather than sinusoidal.

(5) The Nyquist stability criterion can be applied when multiple critical frequencies occur, which is the so-called conditional stability case. The phase angle for G_{OL}, ϕ_{OL}, may cross $-180°$ at more than one point. The system can be closed-loop stable for two different ranges of controller gain, which is quite different from the normal result. In other words, increasing the absolute value of K_c can actually improve the stability of the closed-loop system for some ranges of K_c.

Performance Indices

The performance of a control system may be described in several ways. One is an analysis of the steady-state error (real time). This subject was covered earlier (see Figure 6-29).

In the optimization of a control system, one attempts to eliminate as much steady-state error as possible. Evaluation of the degree of optimization is done with performance indices. There are several. Each performance index gives a measure of how much error is accumulating over some period of time. Table 6-9 lists the popular indices in use; the formulas are fairly self-explanatory.

Table 6-9.   Performance Indices

$ISE = \int_0^t e^2(t)\, dt$	I = integral $\quad S$ = square $\quad A$ = absolute $\quad E$ = error $\quad T$ = time $\quad e$ = steady-state error $\quad t$ = time
$IAE = \int_0^t \|e(t)\|\, dt$	
$ITAE = \int_0^t t\|e(t)\|\, dt$	
$ITSE = \int_0^t t\,e^2(t)\, dt$	
$ISTAE = \int_0^t t^2\|e(t)\|\, dt$	
$ISTAE = \int_0^t t^2\,e^2(t)\, dt$	

Compensation

Compensation is the addition of a compensation network (otherwise called a *controller*) to a system. This topic has already been discussed in Chapter 3, but the theoretical aspects must be covered in more detail.

A system may be considered for compensation if the following points are observed:

(1) The system is stable.

(2) The system responds to command input signals in an acceptable fashion.

(3) The system is insensitive to component parameter changes.

(4) The response of the system to a step change in input results in a minimum steady-state error.

(5) Adding compensation to the system eliminates the effect of disturbances.

With the exception of (1), optimization always involves compromise. Variation of system parameters often cannot satisfy the demanding performance requirements placed upon a system. In these cases, the possibility of modifying (compensating) the system becomes the usual approach. The procedure involves the insertion of a "compensator" into the system. The "compensator" may be electronic (PID), mechanical, hydraulic, and so on. Figure 6-37 illustrates common insertion points, where G_c is the compensating network.

Selection of the compensation network depends upon consideration of the system parameters. The compensation of a system is concerned with the alteration of

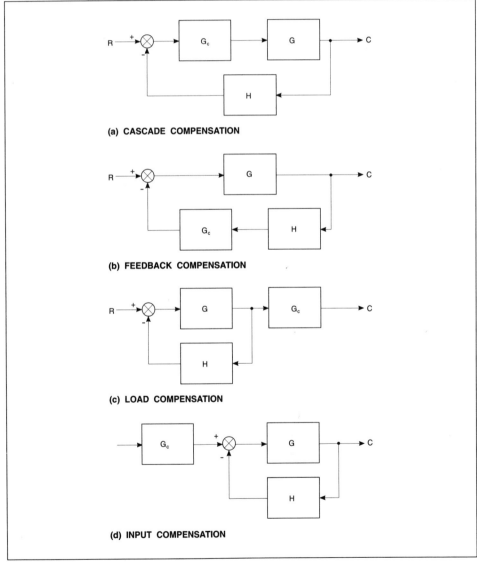

(a) CASCADE COMPENSATION

(b) FEEDBACK COMPENSATION

(c) LOAD COMPENSATION

(d) INPUT COMPENSATION

Figure 6-37. Compensation Network Placement

the frequency response (or the location of the roots) of the system in order to obtain suitable system performance.

If the specifications are time-domain in nature (such as peak time, maximum overshoot, settling time, and, of course, maximum allowable steady-state error), then these quantities may be defined in terms of desirable locations of the poles and zeros (s-plane) of the closed-loop system transfer function. The root locus method of analysis may be used for the theoretical design of the compensation network. (Note that this is not the same as the field calibration of an already designed system.)

Alternatively, there may be frequency performance criteria to be adhered to (peak response, resonant frequency, bandwidth, phase margin, etc).

On Bode diagrams, additional pole and zero effects are additive. This is the preferred method for the calculation of frequency response compensation. Note that a polar plot or a Nichols chart would do the job as well.

For a more in-depth discussion on the topic of compensation, please refer to a text on the subject. This discussion, as previously stated, has been written as an introduction only, sufficient for the understanding of process control system behavior.

The Z-Transform

Today's controller (compensator) systems are computer, or microprocessor-based. Traditional controllers worked with analog or continuously varying signals. Treatment of the signals within these controllers was also of the analog or continuously varying type. Signal treatment within today's microprocessors is of a binary nature; that is, only discrete quantities with fixed steps can be treated. Controllers must, therefore, have analog-to-digital (or digital-to-analog) converters at the input and output. This implies that the signal must be *sampled*, averaged, and converted into discrete values for treatment within the controller. The signal thus takes the appearance of discontinuous steps. Systems with controllers utilizing data treated in this manner are called *sampled data systems*.

The Laplace transform, designed for continuous time-based mathematical analysis, cannot be used to analyze the discontinuous steplike signals of a sampled data system with a reasonable degree of reliability. For this purpose, the Z-transform was developed as an offshoot of the Laplace transform. In fact, the Z-transform is defined by the following:

$$z = e^{ts} \tag{6-60}$$

Solving for s:

$$s = \frac{1}{t} \ln z \tag{6-61}$$

where t is the sampling period in seconds and z is a complex variable whose real and imaginary parts are related to those of s through:

$$z = e^{t_o} \cos \omega t \quad \text{(real part)} \tag{6-62}$$

and:

$$z = e^{t_o} \sin \omega t \quad \text{(imaginary part)} \tag{6-63}$$

with:

$$s = 0 + j\omega \tag{6-64}$$

The relation between s and z is defined as the Z-transformation. If the output of a sampled data system is designated as: $f*(t)$, then:

$$L[f*(t)] = F*(s) = \int_{k=0}^{\infty} f(kt)\, e^{-kts}\, dt \qquad (6\text{-}65)$$

where:

$$f*(t) = f(kt)\, d\,(t-kt) = f(t)\, dt_t\,(t) \qquad (6\text{-}66)$$

where $f(t)$ is the input of the sampled data system. Sampling is assumed to begin at $t = 0$.

Taking the Laplace transform of both sides gives:

$$F*(s) = \int_{k=o}^{\infty} f(kt)\, e^{-kts}\, dt \qquad (6\text{-}67)$$

All this math is necessary for someone who wishes to derive from first principles the Z-transforms of complex transfer functions. Remember that the Z-transform is handled in much the same way as the Laplace transform. The real-time transfer function is converted to a Z-transform transfer function. In this form the math is simpler, and the transfer function is simplified. The inverse Z-transform is then applied and evaluation is then possible.

The method of taking the Z-transform involves three steps as follows:

(1) $f(t)$ is sampled by an ideal sampler to give $f*(t)$.

(2) The Laplace transform of $f*(t)$ is taken to give:

$$F*(S) = L[f*(t)] = \int_{k=0}^{\infty} f(kt)\, e^{-kts}\, dt \qquad (6\text{-}68)$$

(3) Replace e^{ts} by z in $F*(s)$ to get $F(z)$.

$$F(z) = \int_{k=0}^{\infty} f(kt)\, z^{-k} \qquad (6\text{-}69)$$

Unfortunately, the last equation is an infinite series; additional effort is required to obtain the equivalent closed-form function. See Reference 4, pg. 82, example 3.1.

Some of the problems associated with sampled data systems result from the time delay induced in the control loop due to the sampling period. This time delay may cause the loop to be unstable at values of control parameters (proportional, integral, derivative) for which an analog controller would be stable.

Ringing, for example, is the oscillation of a system caused by the inability of the digital sampled data to read the exact value of the feedback and thus obtain a true zero error. This may be due either to the size of the increments of discernibility of the digital conversion or to the size of the time period between each sample, both of which cannot be the optimum desired value: zero.

State-Space Approach to Digital Control Systems

Frequency analysis and root locus techniques of control system analysis and design are being replaced by a more abstract control system theory that allows solving more complex problems. The method, state-space analysis, is more easily adapted to computer-aided numerical solution.

State-space analysis is based on the theory that a control system may be fully described at any instant of time by the values of its state variables. Thus, if the values of the state variables of a process are known at one instant of time and the equations that describe how these state variables change (due to known input variable changes) are also known, the state of the process at any other instant of time later than the first instant can be calculated.

State-space analysis will give excellent results for continuous process control systems, but the method is equally well suited for discrete value sampled systems (traditionally designed for continuous process systems and with many limitations when used with discrete data sampled control systems). Since most applications today utilize computer control systems, which are inherently discrete data types, state-space analysis is proving to be more flexible and more useful.

State-space analysis is a name for a series of different methods that are already in use for solutions of other problems, such as the theory of stability and methods for solving ordinary differential equations.

The description of systems in state-space analysis is based on representation that uses matrices and vectors (one column matrices). The advantages of state-space analysis are:

(1) uniform formulation of problems,

(2) easy solution of control system problems with multivariable inputs and outputs,

(3) solution of control system problems where asynchronous sampling rates or nonperiodic sampling rates are used, and

(4) solution to time variant and nonlinear control systems problems.

The worst known disadvantage is that the state variables are sometimes so difficult to identify that connection with physical reality is lost.

As can be seen from the above discussion, three sets of variables are identifiable in state-space analysis:

(1) Inputs

(2) Outputs

(3) State variables

Vectors (or one column matrices) are used to express the three; for example:

$$\bar{u} = \begin{bmatrix} u_1 \\ u_2 \\ \cdot \\ \cdot \\ \cdot \\ u_m \end{bmatrix} \tag{6-70}$$

is a vector of m inputs;
and:

$$\bar{x} = \begin{bmatrix} x_1 \\ x_2 \\ \cdot \\ \cdot \\ \cdot \\ x_n \end{bmatrix} \tag{6-71}$$

is a vector of n state variables;
and:

$$\frac{d}{dt}\bar{x}(t) = \begin{bmatrix} \dfrac{d}{dt}x_1(t) \\ \dfrac{d}{dt}x_2(t) \\ . \\ . \\ . \\ \dfrac{d}{dt}x_n(t) \end{bmatrix} \qquad (6\text{-}72)$$

is also a vector of n state variables.

In fact, the choice of the vector representing the state variables is unlimited, provided that the state of the system is fully described by the state variables chosen.

Outputs are also vectors and are unique functions of the state vectors. The output vectors may even be identical to the state vector. Here is some nomenclature:

Space of inputs **u** — The set of all possible system inputs u.

State space **x** — The set of all possible system states x.

Space of outputs **y** — The set of all possible system outputs y.

As stated previously, the difficulty of the method lies in the proper formulation of the state variables.

Choose vectors of state variables that are simple to calculate or to measure.

For example, consider a spring-mass-damper system as shown in Figure 6-38.

The equation that represents the typical spring-mass-damper of the figure is:

$$m\frac{d^2}{dt^2}x(t) + f\frac{d}{dt}x(t) + kx(t) = P(t) \qquad (6\text{-}73)$$

Choosing the state vector to be:

$$\bar{x} = \begin{bmatrix} x(t) \\ \dfrac{d}{dt}x(t) \end{bmatrix} \qquad (6\text{-}74)$$

gives the following state equation:

$$\begin{bmatrix} x(t) \\ \dfrac{d}{dt}x(t) \end{bmatrix} = \begin{bmatrix} x(t) \\ v(t) \end{bmatrix} \qquad (6\text{-}75)$$

Equation (6-73) then becomes:

$$M\frac{dv}{dt} + fv(t) + kx = P \qquad (6\text{-}76)$$

(continued)

(continued)

or:

$$\frac{dv}{dt} = \frac{P}{m} - \frac{fv}{m} - \frac{kx}{m} \qquad (6\text{-}77)$$

and from the state equation:

$$\frac{dx}{dt} = v \qquad (6\text{-}78)$$

Therefore, in state-space:

$$\begin{bmatrix} \dfrac{dx}{dv} \\ \dfrac{}{dt} \end{bmatrix} = \begin{bmatrix} 0 & 1 \\ -\dfrac{k}{m} & -\dfrac{f}{m} \end{bmatrix} \begin{bmatrix} x \\ v \end{bmatrix} + \begin{bmatrix} 0 \\ \dfrac{1}{m} \end{bmatrix} P \qquad (6\text{-}79)$$

Generally:

$$[x'] = [F] \quad [x] + [g]u \qquad (6\text{-}80)$$

Because the output is "position" x, the relation

$$y(t) = x(t) \qquad (6\text{-}81)$$

can be established, and the output equation:

$$y(t) = h^T x(t) \qquad (6\text{-}82)$$

implies:

$$h^T = [1 \ 0] \qquad (6\text{-}83)$$

because:

$$[1 \ 0] \begin{bmatrix} x(t) \\ v(t) \end{bmatrix} = x(t) \qquad (6\text{-}84)$$

This is all that is required to input into a computer to calculate the state of the mass-spring-damper system for any instant of time given the original state. Note also how easily these equations can be programmed into a computer.

Although this example is quite elementary, it serves as an excellent example of the method and problems associated with state-space analysis of control systems.

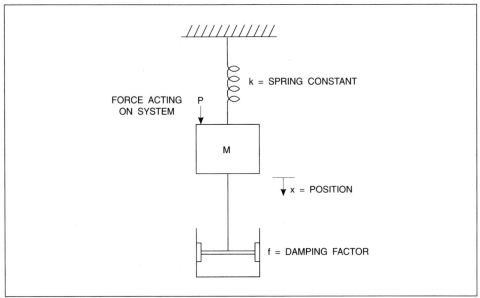

Figure 6-38. Spring-Mass-Damper System for State-Space Analysis

Predictive Control (Ref. 2)

Predictive control techniques employ advanced controllers of the digital type that utilize a (usually computerized) model of the system, including the process, to predict the reaction of the system to variations of the load variable or of the set point. The result is used to calculate the optimum output of the compensator (controller) to the manipulated variable.

Controller design in model predictive control is based on the predicted behavior of the process over the predicted horizon. Values of the manipulated variables are computed to ensure that the predicted response has certain desirable characteristics. One sampling period after the application of the current control action, the predicted response is compared with the actual response. Using corrective feedback action for any errors between actual and predicted responses, the entire sequence of calculations is then repeated at each sampling instant. The control objective is to have the corrected predictions approach the set point as closely as possible.

Predictive control methods are useful, for example, in the design of controllers for multiple input, multiple output (MIMO) systems, where standard design methods for standard controllers cannot be used efficiently. An example would be when the process exhibits unusual dynamic characteristics or when it is crucial to meet constraints on the manipulated and/or controlled variables.

Adaptive Control (Ref. 2)

Process control problems inevitably require on-line tuning of the controller settings in order to achieve a satisfactory degree of control. If the process operating conditions or the environment change significantly, the controller may have to be retuned. If these changes occur frequently, adaptive control techniques should be considered. An adaptive control system is one in which the controller parameters are adjusted automatically to compensate for changing process conditions. Many different adaptive control techniques have been proposed for situations in which the process changes are largely unknown or for the easier class of problems in which the changes are known or can be anticipated. This section principally concerns automatic adjustment of feedback controller settings.

Examples of processes in which changes in the dynamics of the process are not directly measurable are:

(1) catalytic processes;

(2) heat exchanger fouling;

(3) unusual operational status, such as failures, start-up and shutdown, or batch operations;

(4) large, frequent disturbances such as feed composition, fuel quality, etc.;

(5) ambient variations such as rain storms, daily cycles, etc.;

(6) changes in product specifications, grade changes, and changes in flow rates; and

(7) inherent nonlinear behavior (for example, the dependence of chemical reaction rates on temperature).

It is convenient to distinguish between two general categories of adaptive control applications. The first category consists of situations in which the process changes can be anticipated or measured directly. If the process is reasonably well understood, it may be feasible to adjust the controller settings in a systematic fashion (called programmed adaptation) as process conditions change or as disturbances enter the system. The second category consists of situations in which the process changes cannot be measured or predicted. In this more difficult situation the adaptive control strategy must be implemented in a feedback manner, since there is little opportunity for a feedforward type of strategy such as programmed adaptation. Many such controllers are referred to as self-tuning controllers. They are generally implemented via digital computer control.

Programmed Adaptation

If a process is operated over a range of conditions, improved control can be achieved by using a different set of controller settings for each operating condition. Alternatively, a relation can be developed between the controller settings and the process variables that characterize the process conditions. These strategies are examples of programmed adaptation. Programmed adaption is limited to applications in which the process dynamics depend on known, measurable variables and the necessary controller adjustments are not too complicated. Usually the adaptation is simple enough in structure that it can be implemented with some analog and with all digital controllers. The most popular type of programmed adaptation is gain scheduling, where the controller gain is adjusted so that the open-loop gain $K_{OL} = K_c K_v K_p K_m$ remains constant.

Consider a once-through boiler. Here, feedwater passes through a series of heated tube sections before emerging as superheated steam, the temperature of which must be accurately controlled.

The feedwater flow rate has a significant effect on both the steady-state and the dynamic behavior of the boiler. For example, Figure 6-39 shows typical open-loop responses to a step change in flow rate for two different feedwater flow rates, 50% and 100% of the maximum flow. Suppose an empirical first-order plus delay model is chosen to approximate the process. The steady-state gain, time delay, and dominant time constant are all twice as large at 50% flow as the corresponding values at 100% flow. The proposed solution to this control problem is to make the PID controller settings vary with the fraction of full-scale flow ($0 \le \omega \le 1$) in the following manner:

$$K_c = \omega K_c \; ; \; \tau_I = \frac{\tau_I}{\omega} \; ; \; \tau_D = \frac{\tau_D}{\omega} \qquad \text{(6-85)}$$

where K_c, τ_I, τ_D are the controller settings for 100% flow. Note that this recommendation for programmed adaptation assumes that the effect of flow changes is proportionally related to flow rate over the full range of operation.

In this example, step responses were available to categorize the process behavior for two different conditions. In other problems, dynamic response data are not available, but there is some knowledge of process nonlinearities. For pH control problems that involve a strong acid and/or a strong base, the pH curve can be very nonlinear, with gain variations over several orders of magnitude. Consequently, special nonlinear controllers, both adaptive and nonadaptive, have been developed for pH control problems. In this case the process gain changes dramatically with the operating conditions, necessitating the use of gain scheduling (where $K_c K_p$ = constant) to maintain consistent stability margins.

For some types of adaptive control problems, the changes in steady-state and dynamic response characteristics can be related to the value of the controlled variable. For example, in a temperature control loop where the process gain varies with temperature, the controller gain could be made a function of the controlled variable, temperature. Commercially available feedback controllers allow the user to vary K_c as a piecewise linear function of the error signal e, as shown in Figure 6-40. If the process gain K_p varies in a known manner, K_c should also be varied so that the product $K_c K_p$ is constant. This strategy would tend to keep the open-loop gain constant and thus maintain a specified margin of stability (assuming the process dynamics do not change also).

Self-Tuning Control (Ref. 2)

Programmed adaptation cannot be used where the changes in the process cannot be anticipated and/or measured. An alternative approach would be to utilize a model of the process within the controller and to update the parameters of this model as new data are available (possibly on-line). Calculations to define the action of the controller would then be based on the updated model. Such a controller is called a self-tuning or self-adaptive controller. The block diagram for a typical self-tuning controller is shown as Figure 6-41.

Three sets of computations are employed within the "self-tuning" controller:

(1) Estimation of process model parameters

(2) Calculation of the controller settings

(3) Implementation of the settings in a standard feedback control loop

In order to calculate the required control parameters, most real-time calculation techniques require that the process be "jogged" by an external forcing signal at regular intervals, after which the effect of the "jog" can be examined and the calculations performed. Such an input can be introduced deliberately through the set point or added to the controller output.

The first type of self-tuning controller, called the *self-tuning regulator*, was proposed by Aström and Wittenmark in 1973. Subsequent modifications, resulting in the *self-tuning controller* and the *generalized predictive controller*, have also been used to control industrial processes. These are digital models of the process based on a difference equation, and a "minimum variance" criterion is used to reduce the error in the controlled variable.

The self-tuning controller is directed towards applications where the process disturbances are of a random (or stochastic) nature, as compared to those of deterministic nature.

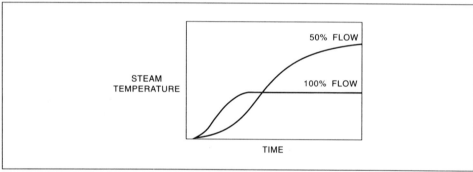

Figure 6-39. Open-Loop Responses to Step Change in Flow Rate

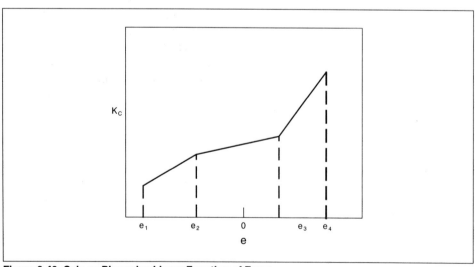

Figure 6-40. Gain as Piecewise Linear Function of Error

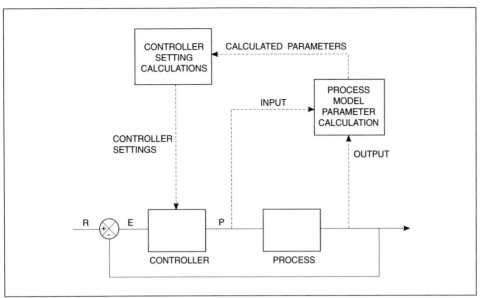

Figure 6-41. Block Diagram of Self-Tuning Controller

Several self-tuning controllers that are commercialized and in use at the present time are based on:

(1) minimum variance stochastic control algorithms as mentioned above;

(2) making step changes in the set point;

(3) an "expert system" approach; and

(4) placing the process in a controlled oscillation of very small amplitude with Ziegler-Nichols rules used to calculate the control parameters (gain, integral, derivative).

Statistical Process Control (Ref. 2)

Statistical process control involves the application of statistics or statistical calculation methods to determine whether a system is operating satisfactorily. These statistical concepts are over 50 years old, but their application to controller design is recent and caused by the necessity to increase productivity (more production, more quality). If a process is operating satisfactorily, the variation of product quality will fall within specific, acceptable bounds. Figure 6-42 shows the "normal," or Gaussian, variation curve for a controlled variable c.

The RMS deviation, which is also called the *standard deviation*, is a measure of the spread of observations about the *mean*. A large value of σ indicates that wide variations in c occur. The probability that the controlled variable lies between two arbitrary values, c_1 and c_2, is given by the area under the histogram between c_1 and c_2. If the histogram follows a normal probability distribution (that is the name given to the shape of the curve in Figure 6-42), then 99.7% of all observations lie within 3 σ of the "mean." These upper and lower control limits are used to determine whether the process is operating as expected.

Statistical process control is a diagnostic tool. It is an indicator of problems with the quality of the product. It does not identify the source of the problem nor the action to be taken.

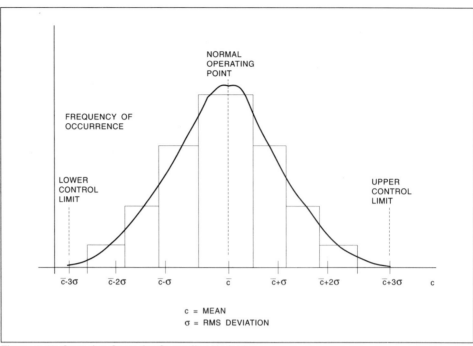

Figure 6-42. Gaussian Curve for Controlled Variable

In continuous processes where automatic feedback control has been implemented, the feedback mechanism theoretically ensures that product quality is at or near the set point, regardless of process disturbances. This requires that an appropriate manipulated variable be identified for adjusting the product quality. However, even under feedback control, daily variations of product quality may result from disturbances or equipment or instrument malfunctions. These occurrences can be analyzed using the concepts of statistical process control.

Expert Systems (Ref. 2)

In a manufacturing facility, the goal of process control is to maintain product quality under safe operating conditions. When operating conditions vary outside of acceptable limits due to external causes, equipment malfunctions, or human error, product quality deteriorates, energy consumption becomes suboptimal, and unsafe conditions can occur. Such process excursions may require plant shutdown or lead to catastrophic events such as explosions, fires, or the discharge of toxic chemicals. In most existing control systems, abnormal measurements trigger alarms that alert the process operator. The operator then must take remedial action to either return the plant to normal levels of operation or shut the facility down.

The success of the manual strategy for handling abnormal conditions relies heavily on the operator being able to respond correctly to process alarms. However, the operator's response depends on many factors: the number of alarms and the frequency of occurrence of abnormal conditions, how information is presented to the operator, the complexity of the plant, and the operator's intelligence, training, experience, and reaction to stress. Because of the many factors involved in determining the appropriate response to an alarm situation, computational aids for the operator are crucial to the success of operating complex manufacturing plants. Such computer-based assistance can be developed as software systems. These so-called *expert systems* are based on emulating the actions of a human expert who is acknowledged to perform the required tasks at a high level of proficiency. The use of expert systems is a branch of artificial intelligence (AI). AI is popularly defined as the science of enabling computer systems to learn, reason, and make judgments. Most expert systems utilize a set of procedures that simplify the application of inductive and deductive reasoning to the database ("knowledge") of the system.

Figure 6-43 shows the architecture of an expert system. The expert system is usually written as part of a *shell*, which is a general software package designed to facilitate implementation. The shell contains the following components:

(1) A knowledge base that consists of data and rules. Data can be entered at system start-up or via the knowledge acquisition system in real time. Rules are structured with "if-then" statements.

(2) An inference engine. This software provides the means of scanning the available rules to draw conclusions or select the appropriate action to be taken.

(3) A user interface. This component displays information, asks questions of the user, and so on.

A simple example that illustrates the need for expert system fault detection and diagnosis is a situation in which the actual flow in a process is significantly higher than the set point for a long period of time. Possible faults that could cause this event include the following:

(1) The sensor is malfunctioning.

(2) The controller has failed in a saturated mode.

(3) The valve is failed open.

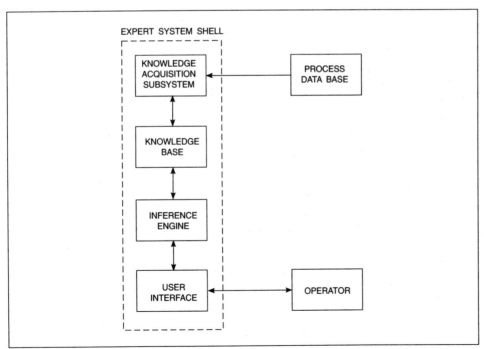

Figure 6-43. Architecture of an Expert System

The identification of the actual fault or failure in the system could be obtained by checking other measurements, performing material balances, or having the operator take certain actions and then observe the results. Interaction with the operator is employed to reach the correct decision.

References

1. Albert, Charles L., *Industrial Process Control and Automation*, Montreal, QC: Instrument Society of America, Montreal Section, 1982.

2. Seborg, Dale E.; Edgar, Thomas F.; and Mellichamp, Duncan A., *Process Dynamics and Control*, New York, NY: John Wiley & Sons, 1989.

3. Dorf, Richard C., *Modern Control Systems, 3rd edition*, Reading, MA: Addison-Wesley Publishing Company, 1981.

4. Kuo, Benjamin C., *Automatic Control Systems, 3rd edition*, Englewood Cliffs, New Jersey: Prentice-Hall, Inc., 1975.

5. Houpis, Constantine H., and Lamont, Gary B., *Digital Control Systems, Theory, Hardware, Software*, New York,NY: McGraw-Hill Book Company, 1985.

6. Shinners, Stanley M., *Control System Design*, New York, NY: John Wiley & Sons, 1964.

7. Aström, K. J., and Hagglund, T., *Automatic Tuning of PID Controllers*, Research Triangle Park, NC: Instrument Society of America, 1988.

About the Author

Gilles J.P. Bouchard has a Master of Mechanical Engineering degree (control systems option) from Concordia University in Montreal. He is presently completing a Ph.D. in mechanical engineering, specializing in controls systems. Mr. Bouchard is employed as a control systems specialist with Sandwell Inc., consultants in the pulp and paper industry. His publications include a paper on the modeling of hydraulic accumulators presented at the ASME winter conference, 1978, and one on boiler controls presented at the International Conference on Controls in Montreal, June 1992. He is also active as a lecturer in measurement and controls for the ISA Montreal Section and in digital controls at the Université of Québec.

7

Analog and Digital Control Devices

Automatic Controllers

Automatic controllers are classified into two main groups: pneumatic and electric/electronic devices. A third category is the hydraulic controller, which is used mainly in heavy industry. Another group is fluidic control devices. This chapter will emphasize electronic and pneumatic controllers.

All controllers have similarities yet are distinct from each other. With the large number of controllers on the market, it is impossible to be familiar with all of them.

Pneumatic Controllers

Two types of pneumatic controllers are available: motion-balance and force-balance. The only difference between the two types is that in a motion-balance controller the nozzle moves away from the flapper to restore a balance, while in a force-balance system the nozzle is static and the flapper is forced away from the nozzle to restore balance.

The basic functions of both pneumatic controller types are the same (see Figure 7-1). They include:

(1) receiving a measurement or process variable signal (PV),

(2) comparing that value to a desired value or set point (SP),

(3) determining the magnitude and direction (positive or negative) of any deviation or error (E), and

(4) providing an output (O) as some function of the deviation.

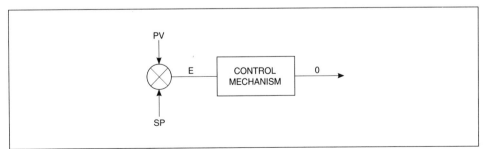

Figure 7-1. Block Diagram of a Basic Controller

As a matter of fact, these functions also apply to electronic or hydraulic controllers, the only difference being the pneumatic medium with which pneumatic devices operate.

On-Off Controller

An on-off controller recognizes and bases its output on only the sign of the error (positive or negative). The controller has no feedback. The nozzle detector is a high-gain mechanism, which means that a flapper need change position only slightly to make the controller output vary through a wide range.

Figure 7-2 illustrates an on-off controller. Half the time the value of the measured or process variable (PV) will hold the clearance between the flapper and the nozzle either at zero or at full open clearance.

At full clearance, the rate of air bleed through the nozzle equals the supply entering through the restriction. When nozzle pressure is very low, amplifier (relay) output drops to zero value (3 psi). For the output pressure to change from minimum to maximum, the bleed through the nozzle must be closed off by the flapper. When this occurs, full pneumatic pressure (15 psi) passes from the amplifier to the final control element.

The value of the measured variable at which the flapper closes the nozzle differs from the value of the measured variable at which the flapper opens the nozzle. That means that the "on" action takes place at a different value than the "off" action. And this difference in value is the neutral zone or differential gap. This gap can be varied by adjusting the pivot point. Therefore, this controller can also be used as a differential gap controller.

Proportional Controller

In proportional action, the output of the controller must have some proportional relationship to the value of the error signal. A proportional relationship eliminates on-off cycling. An adjustable proportional controller is shown in Figure 7-3.

The output of the controller is applied to the feedback bellows (C) to reduce the change in flapper position caused by the initial change in input. Negative feedback has a canceling action. For every value of the controlled variable, which is equivalent to the movement of flapper or baffle, there will be only one clearance

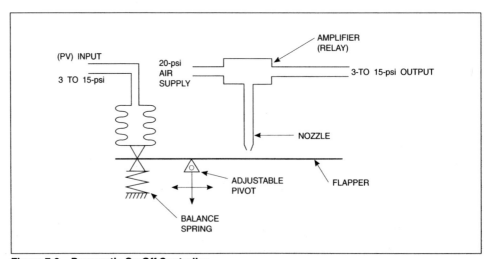

Figure 7-2. Pneumatic On-Off Controller

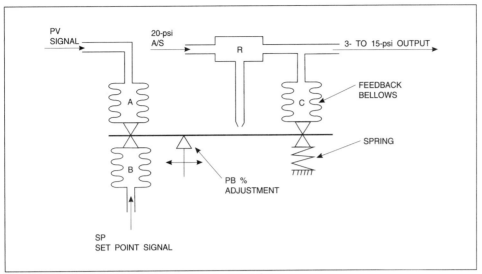

Figure 7-3. Proportional Controller with Adjustable Gain

between the nozzle and flapper, and the output to the final control element will always be proportional to this clearance.

The error detection takes place where the measured process variable (PV) bellows counteracts the set point (SP) bellows or vice versa. The proportional band (PB) or gain can be adjusted by moving the pivot point to the left or to the right.

The controller PB setting depicted in the diagram of Figure 7-3 is 100% or gain = 1, since the distance between the pivot and nozzle is equal to the distance between the pivot and the centerline of bellows A and B. If the pivot is moved to the left the PB will be less than 100% (gain > 1), and when it is moved to the right the PB will increase above 100% (gain < 1). Most controllers have a dial for the PB or gain settings.

Proportional-plus-Derivative Controller

To facilitate the understanding of controller hardware, it will be discussed before the proportional-plus-integral controller.

To add derivative (rate) action to the controller of Figure 7-3, it is necessary only to insert a restriction in the air line to the feedback bellows, as shown in Figure 7-4. The restriction causes a time delay in expansion or contraction of the feedback bellows (C), which allows for an immediate and large change in output for a change in input. Momentarily, the controller becomes a simple on-off device until the full output pressure seeps through the restriction and causes repositioning of the flapper. This action will establish some kind of equilibrium. The derivative action adjustable restriction has a calibrated dial in units of time.

Proportional-plus-Integral Controller

Integral action (reset) gradually raises the gain of the controller from that determined by the proportional band setting to a higher value if the input error persists for a time. The addition of a reset bellows (D) in opposition to the feedback bellows (C) and the addition of an adjustable pneumatic restrictor calibrated in time units provide integral or reset response. The motion of the bellows is the negative feedback motion and that of bellows (D), the positive feedback motion. Figure 7-5 shows the operation of the proportional-plus-integral controller.

As the input (PV) changes, the output pressure change is applied to the needle valve (adjustable restriction) as well as to the feedback bellows (C). The pressure

Often, a capacity tank is built in the line that connects the output line with the positive feedback bellows, especially if the line is of fine bore.

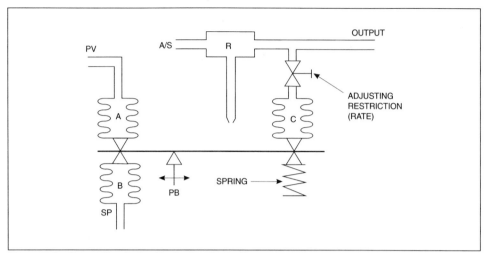

Figure 7-4. Proportional-plus-Derivative Controller

in the reset bellows changes slowly, as air is passed through the restriction, to gradually counteract the pressure in the feedback bellows. This has the effect of removing the feedback action and raising the gain of the controller to the value that could result without any feedback action. The effect of this higher gain is to reduce the offset between the set point and the measured value of the variable (PV).

When the output stops changing, pressure in both bellows is equal. The time required for equalization of pressure in the two bellows (C and D) depends strictly on the amount that the adjustable restriction has been opened.

Proportional-plus-Integral-plus-Derivative Controller

This type of controller is sometimes called a three-mode controller or PID controller. This controller has all three control modes or algorithms incorporated in one device, but their individual basic characteristics remain the same. Each control algorithm makes its contribution to the reduction or elimination of error, but the response is more difficult to identify because the PV curve, when recorded,

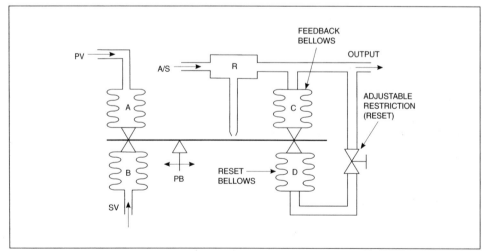

Figure 7-5. Proportional-plus-Integral Controller

may show all three simultaneously. Figure 7-6 illustrates a schematic of the PID controller.

There is an ideal setting for each of the three adjustable features: proportional band (%), integral (repeats per minute), and derivative (minutes).

 The time constant of the derivative restrictor must be shorter than that of the integral restrictor to avoid regenerative action. That is why the two restrictions are installed in series. A proportional band setting alone cannot keep the process variable in line with the set point after a load change has occurred.

Integral action automatically brings the value of the process variable back to the set point by continuing to change controller output as long as deviation or error exists. In a pneumatic controller, the PV input-to-output pressure relationship is automatically moved up or down scale until the error is eliminated.

Derivative action adds to the effects of proportional and integral control when the error is increasing so that all three actions try to prevent further increase of error and bring the process variable to the set point. Derivative action subtracts from the effects of proportional and integral actions when the error is decreasing. Thus, it counteracts tendencies of the proportional and integral actions that cause overshoot or slow reaction.

Often, derivative action before integral action is included in the hardware of a controller to prevent overshoot on start-up and shutdown. In fact, it makes the derivative contribution to control a function of the rate of change of the process variable rather than a function of the rate of change of the error. This derivative-before-integral action is usually not selected where there are transmission lags around a control loop (including the process and control system).

Manual Control

Manual control is usually required during start-up and shutdown activities and when there are process disturbances or when the automatic part of the controller malfunctions. Most controller hardware is equipped with a bypass circuit and an adjustable restriction to manipulate the output pressure.

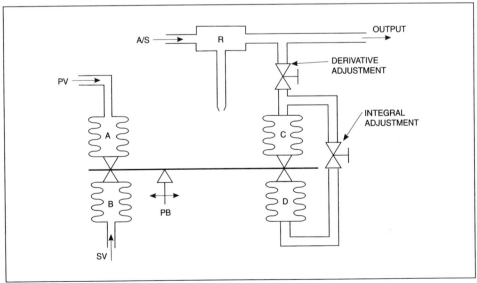

Figure 7-6. Proportional-plus-Integral-plus-Derivative Controller

Some controllers are dedicated for manual control; they are called manual loaders. Figure 7-7 shows a diagram of a manual bypass circuit in an automatic controller.

The orifice in the bypass line reduces the pressure from 20 to 15 psi, the three-way valve shuts the automatic controller off, and the output pressure can be regulated by hand using the three-way valve. Obviously, the transfer from automatic to manual and vice versa is not "bumpless" in the example. More circuits and regulators are required to make the transfer bumpless, and most controllers today have bumpless transfer capabilities built in.

Indicating Controllers

Basically three categories of pneumatic PID controllers are available:

(1) Indicating controllers

(2) Recording controllers

(3) Blind controllers

The examples here are actually blind controllers, but the majority of controllers have an indication of the process variable value, set point value, and valve position. The indicator may be a dial or a scale (either vertical or horizontal), usually graduated in percentage, square root or in engineering units. The recording controllers record the process variable, and pointers indicate the set point and valve position.

The pneumatic controllers discussed here are analog controllers. That means that the process variable value is analog to the input signal value and the output signal value is analog to the valve position.

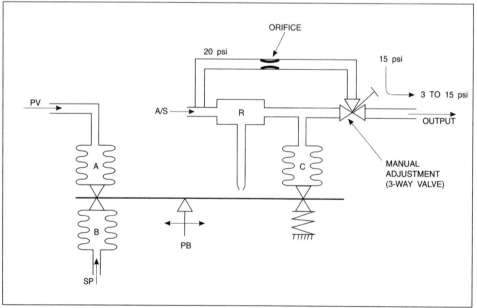

Figure 7-7. Automatic/Manual Controller

Pneumatic Auxiliary Devices

Pneumatic Logic Control Devices

A complete pneumatic logic circuit can be built by employing miniaturized pneumatic valves called elements. These devices can perform specific functions as OR, AND, NOT, memory, delay, timing, pulse, and other specialized functions (see Figure 7-8).

These elements are diaphragm-operated poppet-type valves and are totally air actuated. Ordinary plant air from 30 to 150 psig can be used for their operation.

Some of the elements will be presented here. The OR element combines two air signals so that either can produce the output. The output port will be pressurized (c = ON) when either one or both of the input ports is pressurized (a = ON or b = ON).

The AND element combines two air signals so that both must be ON to create an output. The output port will be pressurized (c = ON) only when both of the inputs are pressurized (a and b are ON). It can also by used with timer elements to produce time delay functions.

Pneumatic Computing Elements

A computing element is a relay that receives information in the form of one or more instrument signals and modifies the information (or its form) before transmitting one or more signals. This type of relay is not a controller or a switch; it performs mathematical calculations to solve an equation for an unknown process variable. Some of the computing functions are:

(1) adding and subtracting,

(2) multiplying and dividing,

(3) square root extraction (linearizing a nonlinear signal),

(4) ratio and bias operation,

(5) averaging several signals,

(6) high- or low-signal selection, and

(7) integration.

An example of the square root extractor will be presented here. Figure 7-9 depicts a diagram of a pneumatic square root extractor mechanism. This type of computing relay is necessary to linearize the signal of a differential pressure transmitter, which transmits a flow measurement signal, especially when the flow is being measured with an orifice flange. Chapter 1 stated that the measured differential pressure and corresponding flow rate have a square root relationship.

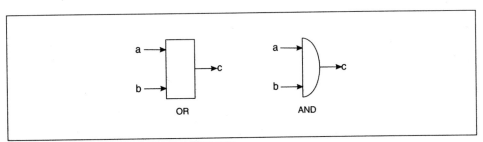

Figure 7-8. Pneumatic Logic Control Elements

The output pressure of the dP cell transmitter activates the receiving bellows (X). The bellows pushes against the differential beam, which is balanced by a calibrating spring. On an increase of differential pressure, the differential beam will always be forced in a clockwise direction about its fulcrum. Then the cam will bear down on the follower to make the flapper approach the nozzle. (Figure 7-10 shows more details.) As it does so, pressure in the nozzle will increase and the output from the booster relay (R) will increase. This output loading pressure will expand the square root bellows (Y) and change the position of the square root beam about its fulcrum.

Figure 7-10 shows that as differential pressure increases due to increased flow the cam is depressed a distance, d, that is proportional to the differential (not proportional to the square root of the differential). The flapper being forced against the nozzle causes an increase of pressure in, and an extension of, the square root bellows, rotating the square root beam through the angle α.

Values of α less than 20° are exactly proportional to the square root of cam drop d. But α is proportional to transmitter output. Thus, output in psi represents the square root of the differential pressure caused by the orifice flowmeter being inserted in the pipe.

> **For every value of the measured flow, there is only one position of the differential beam and one position of the square root beam; therefore, there is only one value of the square root output.**

Electric and Electronic Controllers

Electronic analog controller functions are basically the same as those of the pneumatic controllers. They can be classified by control mode or algorithm.

On-Off Controller

On-off or two-position control may be used satisfactorily if some cyclic fluctuation of the controlled variable is permissible. On-off action can be achieved by any device that opens or closes an electrical circuit when the measured variable

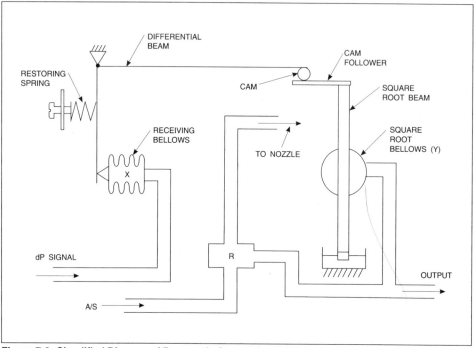

Figure 7-9. Simplified Diagram of Pneumatic Square Root Extractor

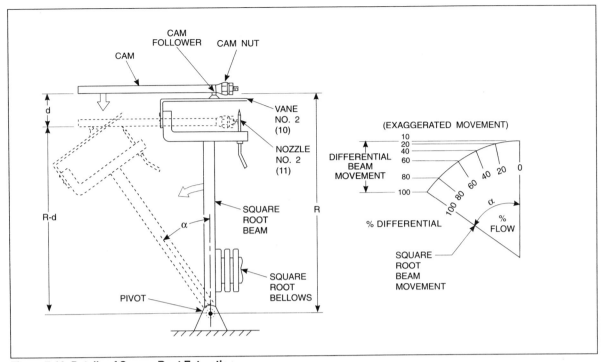

Figure 7-10. Details of Square Root Extraction

departs from the set point. Typical examples are: level switches, flow switches, pressure switches, and thermostats (refer to Chapter 1). In fact, the controlling means (final element) is actuated by the measuring or primary element. The set point has to be adjusted in the primary element.

Three-Mode or PID Controller

The three-mode controller provides three control functions — proportional action, integral (reset) action, and derivative (rate) action — for the regulation of a wide variety of processes. The three-mode controller in its entire configuration will be discussed here, since most electronic controllers are supplied with proportional and integral action with optional derivative action. The described controller has indicators for set point, measured variable value, and valve position (output signal).

In the analog controller circuit shown in Figure 7-11, the deviation amplifier receives a process variable signal and a set point signal. The amplifier compares the two signals to produce a signal equal to the difference between the signals. The polarity of the deviation amplifier output depends on which input is larger. If the noninverting (+) input is larger, the output is positive. If the inverting (–) input is larger, the output is negative.

The deviation meter displays the output of the deviation amplifier. If the process variable signal deviates above or below the set point signal, the deviation meter reads above or below the index line.

With the automatic/manual switch set to auto, the deviation amplifier output signal is inverted in the mode amplifier and converted into controller outputs in the current driver. The proportional band and rate (derivative) settings control the mode amplifier response to the deviation signal. The reset (integral) setting controls the rate at which an unchanging deviation signal changes the output of the mode amplifier.

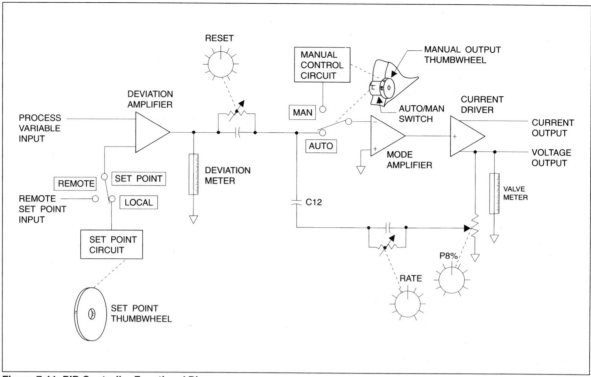

Figure 7-11. PID Controller Functional Diagram

When set to manual, the automatic/manual switch connects the manual control circuit to the mode amplifier. The manual control circuit generates pulses as the manual output thumbwheel is rotated to increase or decrease the controller output, depending on the direction of thumbwheel rotation.

Mode Amplifier

The mode amplifier (see Figure 7-12), consists of operational amplifier U3 and its associated circuits. Potentiometer R60, an internal adjustment, compensates for any accumulated error in the deviation amplifier and mode amplifier circuits. The input to the mode amplifier is zero when the set point and the process variable are equal in the automatic mode. In the manual mode, the input to the mode amplifier is zero as long as no pulses are coming from the manual control circuits. Capacitor C17 in manual mode or C12 in automatic mode provides capacitive feedback around the mode amplifier to maintain the mode amplifier output at a constant level when its input is zero. When the mode amplifier input signal deviates from zero, its output changes until the input returns to zero. Capacitor C17 in manual mode or C12 in automatic mode charges to this new level of mode amplifier output; then it maintains that output.

When set to manual, switch S101 connects the manual control circuit to the input of the mode amplifier. Switching to manual mode is bumpless because capacitor C16 holds the mode amplifier output constant during transfer. The manual control circuit injects precise pulses into the mode amplifier as the manual output thumbwheel is rotated. Depending on the polarity of the pulses, the mode amplifier output will increase or decrease for the duration of each pulse.

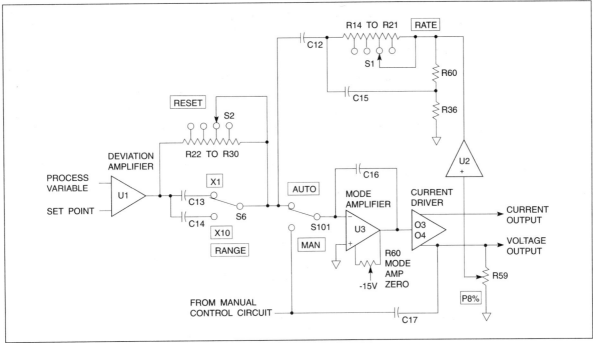

Figure 7-12. Mode Controls Schematic

Mode Controls

As shown in Figure 7-12, potentiometer R59 (PB %) feeds back a portion of the voltage output to the mode amplifier input. This feedback controls the electrical gain of the controller. Because electrical gain is an inverse function of proportional band, decreasing the PB % control setting (decreasing the percent of proportional band) increases controller gain. Buffer amplifier U2, a unity gain amplifier, prevents loading of the proportional band adjustment to keep the adjustment linear.

The rate (derivative) circuit of the controller is an RC network connected between the output of U2 and capacitor C12. The RC network delays the feedback according to the RC time constant and the rate of change in controller output. A delay in feedback momentarily increases controller gain and enables the controller to correct for a process deviation before it becomes too large.

The controller reset (integral) circuit, which consists of switch S2 and resistors R22 to R30, determines how fast the controller output changes to correct for a steady state process deviation signal from the deviation amplifier. The process deviation signal passes through capacitor C13 or C14 only when the signal is changing; therefore, the only path for the steady state signal is through the reset circuit. As the reset control setting is increased (in repeats per minute), the reset resistor decreases, increasing the rate at which the mode amplifier corrects for the steady-state process deviation signal.

Range switch S6 places either capacitor C13 or C14 into the mode amplifier input circuit, and this affects the ratio of feedback impedance over input impedance. Switching from the × 1 position to the × 10 position decreases the mode amplifier gain by a factor of ten. This increases the PB% units and the reset units by a factor of 10.

The controller input and output are either a voltage signal that ranges between 1 to 5 V DC, or a current signal (4 – 20 mA DC) that is converted by a range resistor jumper (250 Ω) to a 1 to 5 V DC signal.

Electronic Auxiliary Devices

Electronic Analog Computing Elements

Electronic analog computing elements provide functions similar to those of the pneumatic types. They have, however, the versatility needed by the systems designer to implement more sophisticated control schemes. Besides the basic computing functions, electronic computing elements offer special-purpose computing functions.

The algebraic signal conditioning functions are reviewed briefly in the following paragraphs.

The square root extractor solves the basic equation:

$$Y = \sqrt{A}$$

The adder/subtractor solves the basic equation:

$$Y = \pm A \pm B \pm C \pm D \pm E$$

It is commonly used to add or subtract flow signals or to calculate differential temperature. It may also be used as an inverter.

The multiplier/divider solves the following basic equations:

$$Y = \frac{AB}{C} \qquad Y = A^2$$
$$Y = \frac{A}{C} \qquad Y = \frac{A^2}{C}$$
$$Y = AB \qquad Y = \sqrt{AB}$$

The multiplier/divider is most commonly used to compute a process variable that cannot be measured. Some typical examples follow.

Partial computation of the gas mass flow rate equation:

$$\sqrt{\frac{(P)\,(\Delta P)}{T}}$$

(to temperature and pressure compensate a differential pressure signal.)
where:

P = process pressure
P = differential pressure
T = process temperature
Output = temperature and pressure compensated dP signal

This output would be sent to a square root extractor to derive gas mass flow rate.

Computation of mass flow rate of solids in a slurry:

$$\text{Output} = (P_s)\,(Q)\,(D)$$

where:

P_s = % by weight of solids (density gage)
Q = linear volumetric flow
D = average density of solids and carrier (This must be approximated and is a constant.)
Output = mass flow rate of solids

The most common application for a square root extractor is linearization of a differential pressure signal to derive linear volumetric flow.

Other applications are:

(1) Btu rate computation used for calculating the rate of hear transfer,

(2) internal reflux computation for reflux control in a distillation column,

(3) computation of mass flow rate of solids on a conveyer,

(4) computation of liquid mass flow rate with changes in density due to changes in the mix or composition of the liquid,

(5) ratio station with remotely adjustable ratio value, and

(6) ratio computation of two process variables.

The ratio station solves the basic equation:

$$Y = (\text{Ratio})\, A$$

The integrator solves the equation:

$$Y = \int A$$

The integrator outputs are not 4 to 20 mA DC but rather a series of pulses that are used to step digital counters. The output pulses of the linear integrator occur at a rate that is linearly proportional to the input magnitude. The integrator is most commonly employed where totalization of flow is required. For example, each output pulse represents some number of gallons or liters of flow. Since flow is proportional to the square root of differential pressure, the square root of a differential pressure signal must be extracted either by a square root extractor and fed to a linear integrator or by a square root integrator. This computing element can also be used to totalize Btu of heat transfer by linearly integrating the output of a Btu rate computer (multiplier).

The selector/limiter can select the highest or lowest of up to four input variables. Signal selection is necessary when more than one controller operates a signal valve, as in an override control system, or when a process input receives redundant transmitter signals. The selector/limiter also has high and low signal-limiting capability, which is typically used to maintain a remote set point within predetermined limits, such as in a cascade control loop to maintain the slave controller's set point within predetermined limits.

Special-purpose analog computing elements are customized analog computers that perform a necessary function that cannot readily be performed with standard instruments. Each computing element of this type is designed and built to individual customer needs and specifications. The following are some typical applications:

(1) Three-element boiler feedwater computer
$$Y = K_1\, \sqrt{A} - K_2\, \sqrt{B} - K$$

(2) Net positive suction head computers
$$Y = A - [f_B\, B + f_A\, C + K]$$

(3) Derivative summing boiler computer
$$Y = A \pm K_1\, \frac{dB}{dT} + K_2$$

(4) Lead-lag computer (feed forward)

The most common application for ratio stations is to set a controlled flow equal to a specific percentage of a wild flow.

$$Y = K_1 A \frac{T_1 S + 1}{T_2 S + 1}$$

(5) Three-halves power extractor

$$Y = K_1 A^{3/2} \pm K_2$$

where:

$$
\begin{aligned}
A,\, B,\, C, &= \text{input variables} \\
K &= \textit{constants} \\
Y &= \text{output} \\
f &= \text{function of}
\end{aligned}
$$

The ratio station maintains a predetermined ratio between the input and output signals. Input bias and output bias adjustments modify the ratio to fit process loop requirements. Operation of the ratio station is explained mathematically as follows:

$$V_o = \pm R\,(V_i - B_i) + B_o$$

where:

$$
\begin{aligned}
V_o &= \text{output signal (1 to 5 V DC or 0 to 100\%)} \\
R &= \text{ratio (adjustable from 0.3 to 3.0 on linear scale or 0.55 to 1.732 on} \\
&\quad \text{square root scale)} \\
B_i &= \text{input bias (1 to 5 VDC or 0 to 100\%, internal adjustment)} \\
B_o &= \text{output bias (1 to 5 VDC or 0 to 100\%, external adjustment)} \\
V_i &= \text{input signal (1 to 5 VDC or 0 to 100\%)}
\end{aligned}
$$

Input and output signals may be converted into current values.

To develop the equation, the ratio station circuitry is connected as shown in the simplified schematic shown in Figure 7-13. Detailed schematics are shown in Figures 7-14, 7-15, 7-16, and 7-17.

Buffer Amplifier

The buffer amplifier (see Figure 7-14) provides a high input impedance for the input signal. The amplifier also provides the subtraction of the $(V_i - B_i)$ term of the output equation. Amplifier gain is unity since R19 equals R13 in value. For an 8-Hz corner frequency, C2 is added to the amplifier AR1A feedback circuit. The control action switch determines the algebraic sign of the equation. Changing the switch affects the output signal. The input bias is adjusted by R8. Changing the input bias affects the output signal.

Summing Amplifier

The summing amplifier (see Figure 7-15) receives its input signal from the buffer amplifier through thumbwheel potentiometer R34. R34 adjusts the signal between 0.1 and 1.0 volt. Since the summing amplifier gain for the signal is three (determined by R6, R30, R18), the ratio becomes 0.3 to 3.0. The output bias portion of the amplifier circuitry is connected at Pin 5 of AR1B to form a summing junction, hence the plus (+) sign in the equation. The amplifier gain for the output bias is one since R16, R17, and R18 are all equal in value.

Current Driver

The current driver (see Figure 7-16) uses transistors Q1 and Q2 in the Darlington configuration for high current gain. Zener diode CR3 limits the current driver input voltage signal and, therefore, limits the output of the ratio station. This prevents possible transducer damage due to excessive current or voltage output.

The output current range depends on the value of resistor R_{out}. The output voltage is indicated on the ratio station valve meter. Potentiometers R2 and R7 are internal adjustments that set the valve meter midscale and span, respectively. This current driver is similar in operation to the one in the PID controller (Figure 7-12).

Bias Supply

The bias supply (see Figure 7-17) modifies the oncoming +15 V DC and –15 V DC power for use by the ratio station. Additional filtering is provided to the 15 volt supplies by capacitors C8 and C9. Resistors R20 and R21 are current-limiting resistors for the two Zener diodes, CR1 and CR2, which provide a regulated +6.2 and –6.2 volts DC, respectively.

Digital Controllers

Digital process controllers have become more commonplace. Especially with the arrival of modern and lower-cost digital electronic technology, it has become economically and technically feasible to provide new hardware solutions for process control. In the last 15 years the following digital control systems have evolved:

(1) Direct digital control

(2) Supervisory control

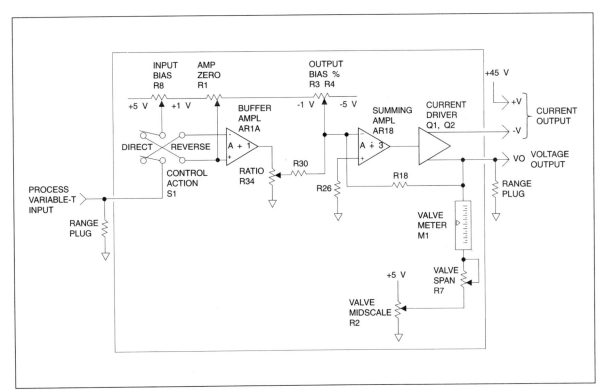

Figure 7-13. Ratio Station Functional Diagram

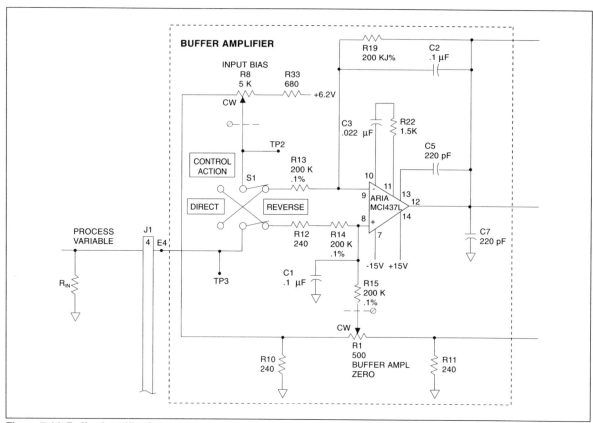

Figure 7-14. Buffer Amplifier Schematic

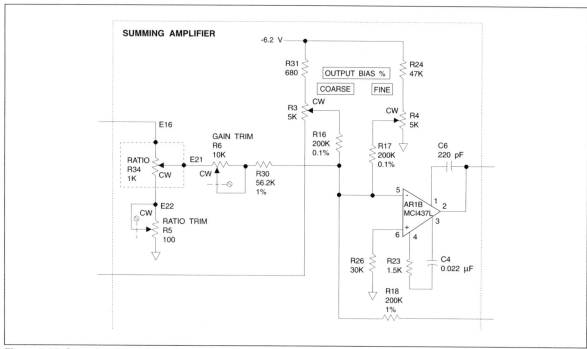

Figure 7-15. Summing Amplifier Schematic

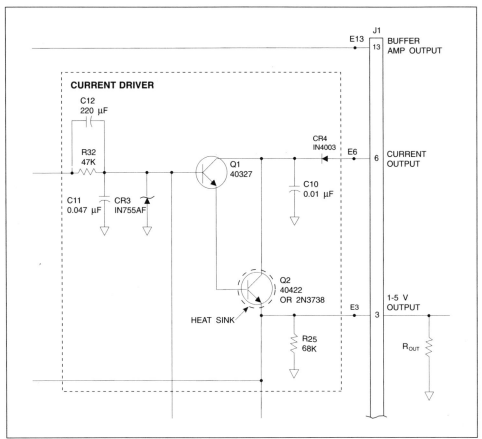

Figure 7-16. Current Driver Schematic

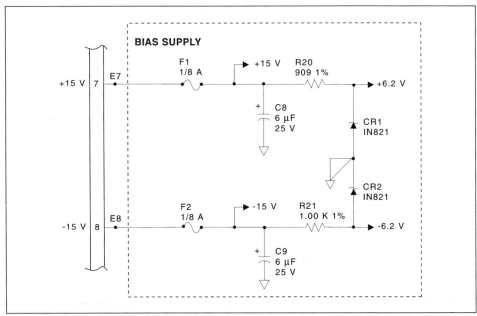

Figure 7-17. Bias Supply Schematic

(3) Hierarchical computer control

(4) Distributed control

The distinctions between the various types of digital control systems have been mainly historical in nature. With the introduction of low-cost microprocessor chips, these areas have overlapped and merged.

Some digital electronic concepts will be examined before those of digital control systems. Since the microprocessor-based controller is the most recent development in digital control and also the most applied system in process control, this section will concentrate on its theory and operation.

The difference between analog and digital systems is their mode of operation. The circuits used to handle analog information electronically are usually considered to be linear circuits. Digital systems, on the other hand, handle the information in digital form. The system quantities or system information is made up of a combination of separate bits.

A digital function is an operation (such as adding two numbers, or selecting a code when a switch is thrown, or storing a combination of bits that represent a number) that is performed with digital circuits using digital information. All the digital functions required to perform complete system tasks can be put in a space as small as about 1/4-in. (6.35 mm) square. This is done on a small piece of silicon semiconductor material. This component is known as a digital integrated circuit chip or microprocessor chip.

The development of microprocessors is based on LSI (large-scale integration) technology. There are several hundred circuits on such chips. Two types of logic chips have been developed: CMOS (complementary metal oxide semiconductor) and TTL (transistor to transistor logic). "Logic" is a shorter term for digital integrated circuit.

The following are the building blocks of any digital control system, mini- and microcomputers:

(1) Processor or control (CPU)

(2) Memory

(3) Communication (input/output)

These system building blocks are shown in Figure 7-18. They are not independent of the outside world; they require inputs from the outside world to sense and react to. They sense inputs and act on these inputs to provide outputs. Even the memory of the system may receive inputs or provide outputs to the outside world.

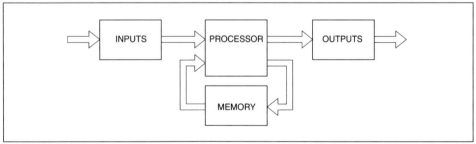

Figure 7-18. Computer Building Blocks

Processor

This building block is referred to as a CPU (central processing unit), which is the brain of a processing system. It performs the actual execution of instructions. The CPU also controls sequencing of operations and controls communications with the memory and I/O (input/output) interface building blocks. The CPU is supported by RAM and ROM memory, a clock, an interconnecting bus system, and I/O devices. Microcomputers have all these devices on a single chip.

Within the processor are also a number of devices called registers that are used for temporary storage and the manipulation of numbers and instructions. Registers are made up of circuits called flip-flops, which are two-state devices; that is, they operate in either an ON or an OFF condition, and they can be changed from one state to the other at any time on command from control circuits within the CPU.

One type of register is the clock, which consists of circuits called clocked flip-flops. Microcomputers use clock signals to time the operation of all their circuits. Clocked flip-flops and other memory elements allow this timing to be precise. The clock's function is to generate all timing signal pulses for the microprocessor, memory, and the I/O devices. This function is important because it synchronizes the microprocessor to the memory.

Another type of register is the program counter, which is a function block that contains and sends out the memory addresses that locate the next instruction.

Usually data (information) must be used by the processor when executing an instruction. This data must be located by an address either in memory or from an input. The microprocessor must keep track of this data address. The functional block that saves the data address is called a data address register. The microprocessor architecture schematic is illustrated by Figure 7-19.

An address is a pattern of characters that identifies a unique storage location.

Another important subsystem is the arithmetic logic unit (ALU). If the processor is to be able to do more than merely transfer information around the computer, it must have circuits that will perform arithmetic and logic operations. This group of circuits is the ALU, which typically provides addition, subtraction, the

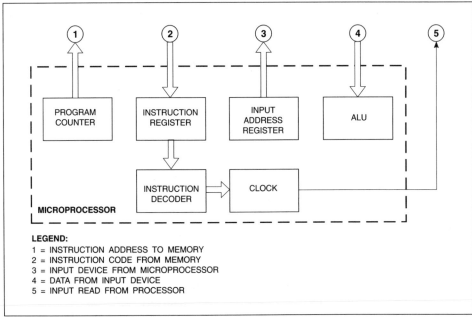

LEGEND:
1 = INSTRUCTION ADDRESS TO MEMORY
2 = INSTRUCTION CODE FROM MEMORY
3 = INPUT DEVICE FROM MICROPROCESSOR
4 = DATA FROM INPUT DEVICE
5 = INPUT READ FROM PROCESSOR

Figure 7-19. Microprocessor Subsystems

basic logical operations, multiplication, and division. Logic circuits such as AND, OR, and NOT allow the processor to make decisions such as >, <, =, +, –, etc. Data is brought from memory or from I/O units. The ALU has storage devices (registers) that provide temporary storage for the often-used data. It is from these internal ALU data registers that information flows to the other functional blocks within the computer.

Memory

Numbers, words, or characters that a computer system must use in performing its tasks, its data, are stored in its memory. It stores this information as bits. The arrangement of bits form codes to identify each number or character. The memory is also used to store or remember the instructions or sequence of operations the processor is to perform. Such a memory is called a program memory. The instructions are also coded with bits. The sequence of instructions, stored in order, one word after another, is called the computer program.

A word has 16 or more bits, a byte has 8 bits, and a nibble has only 4 bits. Word length should not be confused with memory size. Memory sizes are usually specified as 4K, 16K, 32K, 64K, and 128K words of memory, with word lengths of either 4 bits, 8 bits, 16 bits, or 32 bits. To identify memory size, storage capacity and word size have to be combined.

For example, 4K by 16-bit memory means that it can store 4,096 words with a 16-bit word length. The bit words represent ASCII characters. All memories are classified as volatile and nonvolatile. A volatile memory cannot retain data after its power has been removed. Nonvolatile memory, once programmed, will retain its information after the power is removed.

Conceptually there are two types of memory: main memory and extended memory. Main memory is usually of the semiconductor type. An extended memory can be magnetic tapes, discs, cassettes, etc.

RAM (random access memory) is a read/write type of memory that can be randomly accessed. These memory types store multiple-bit digital codes that represent instructions, data, and control signals. A memory for this purpose can be provided by using many registers, or many lines of multiple flip-flop rows, for the multiple-bit storage location. A control read/write signal, 1 for write and 0 for read, tells the memory to store the inputs or read the outputs. RAM may be either static or dynamic. However, in both cases a backup power supply is required to prevent loss of data, since RAM is volatile.

ROM (read-only memory) means that the memory has a program of bits that is fixed in its array during the time of manufacturing. The connections are made by the photographic masks while the material is processed. When it is addressed, it will always read out the same information from the addressed location.

PROM (programmable read-only memory) means that the memory bits can be programmed into the array by the user. Usually, the full array is made with ones in each location; then the user burns away the connection between the crossing wires by a pulse of current to make a zero and get the required code in the array.

EPROM (erasable programmable read-only memory) means that the code stored in the memory array can be programmed by the user, erased, and then reprogrammed to a different code. Special equipment such as ultraviolet light fixtures are required to erase the units. They must be removed from the system to be erased and reprogrammed.

EAROM (electrically alterable read-only memory) means that the memory array can be programmed and erased while still in the circuit.

Magnetic Bubble Memory

Bubble memories have very large storage capacities and are nonvolatile. They are relatively slower than the semiconductor memory types. Bubble memory is grown in a substrate material when magnetic bubble domains are formed.

A magnetic bubble is a minuscule cylindrical magnetized region (0.2 mil or 5 micrometer) that can drift around in a thin film of certain magnetic crystalline materials such as "yttrium-iron garnet" (YIG). The YIG film is grown epitaxially (from a hot gas) on a slice of a nonmagnetic crystalline material called "gadolinium-gallium garnet" (GGG). This material is then cut into memory chips. Each chip is sandwiched between two flat permanent magnets that provide a field pointing the opposite way from that of the bubbles. The bubbles are generated by pulses of current, they can also by destroyed by a pulse in the opposite direction. The presence of a bubble stores a one, and the absence of a bubble, where one might be, stores a zero.

Bus

A bus is required in a microprocessor to route all data traffic to and from devices. It usually consists of two or more conductors that run parallel. When data is to leave the microprocessor, it is usually done through an ACIA (asynchronous communication interface adapter) and/or a modem (modulator/demodulator).

Input/Output (I/O) Interface

In order to be able to communicate with the outside world, a CPU needs an additional subsystem called the I/O interface. Depending on the application of the digital computer, two basic types of I/O devices are available: operator I/O and process I/O.

OPERATOR I/O

Operator I/O allows human beings access to the process. Process operators manipulate such devices as push buttons, toggle switches, and keyboards to input data or commands to the computer, and they receive information from the computer via such devices as a CRT (cathode-ray tube) screens, LED numerical displays, and pilot lights.

PROCESS I/O

Process I/O communicates directly between the CPU and many types of process devices. Typical devices in a real-time process system include sensors, limit switches, tachometers for input; control valves, motor starters, stepping motors, and variable speed controllers for output. Since most process devices are analog in their operation, signal converters are required, such as A/D (analog-to-digital) converters for input and D/A (digital-to-analog) converters for output.

THE DIGITAL I/O SUBSYSTEM

The digital I/O subsystem communicates with process devices that have only two possible states: ON or OFF. Digital inputs are created by contact closures such as level switches, MCC contacts, limit switches, etc. Digital outputs are activated by ON/OFF commands from the CPU. I/O devices also interface with the extended or bulk memory hardware such as disk drives, cassette recorders, and drums.

Summary of Digital Devices

To clarify some of the digital technology terminology, some concepts will be summarized. A digital computer consists of at least one CPU (together with input, output, and memory units), in which information is represented in discrete coded form such as ones and zeros, or ON and OFF voltage levels. A minicomputer is a computer in a certain range of size and speed, generally smaller, slower, and less sophisticated than a "full-size computer." A microcomputer is a computer in a certain range of size and speed, generally smaller, slower, and less sophisticated than a "minicomputer." A microprocessor is an integrated circuit or set of a few ICs that can be programmed with stored instructions to perform a wide variety of functions and consists at least of a controller, some registers, and some sort of ALU — that is, the basic parts of a simple CPU.

More advanced microprocessors have additional devices on the same chip, such as a decoder/encoder (hexadecimal to machine language conversion), a clock to handle timing and control functions, a serial I/O control for interfacing other devices, and the interrupt control (which interrupts the data process when a high-priority task is to be performed), a register array (which also includes stack pointers), a program counter, and incrementer/decrementer registers. These registers are in conjunction with the program counter register. Figure 7-20 depicts a schematic of a single-chip microprocessor with an 8-bit internal data bus.

GATE

A gate is a solid-state logic circuit that has one output channel and one or more input channels, such that the output signal is completely determined by the state of the input signal(s). Such a "logic gate" is an AND, OR, NOT, NAND, NOR or "Exclusive OR" gate.

AND

The AND gate is a circuit with two or more inputs of binary digital information and one output, whose output is 1 only when all the inputs are 1; the output is 0 when any one or more inputs are 0.

OR

An OR gate is a circuit with two or more inputs of binary digital information and one output, whose output is 1 only when any one or more inputs are 1; the output is 0 only when all inputs are 0.

NOT

A NOT gate is occasionally used to represent an "inverter." An inverter is a binary digital building block with one input and one output. The output state is the inverse or opposite of the inputs state.

NAND AND NOR

A NAND gate is a circuit that acts as an AND gate followed by an inverter, and a NOR gate acts as an OR gate followed by an inverter.

EXCLUSIVE OR

An Exclusive OR gate acts as a certain combination of other gates.

Figure 7-21 illustrates some of the logic gate symbols and the truth tables for an OR gate and an Exclusive OR gate. The truth table for a logic gate shows, for each information output, the logic state that results from each combination of logic states at the information inputs. The logic states are 1 (yes, true) and 0 (no, false).

Flip-flop circuits are made up of gate circuits, and gates consist of transistor circuits or diode circuits. The flip-flop, which forms the basis for registers and

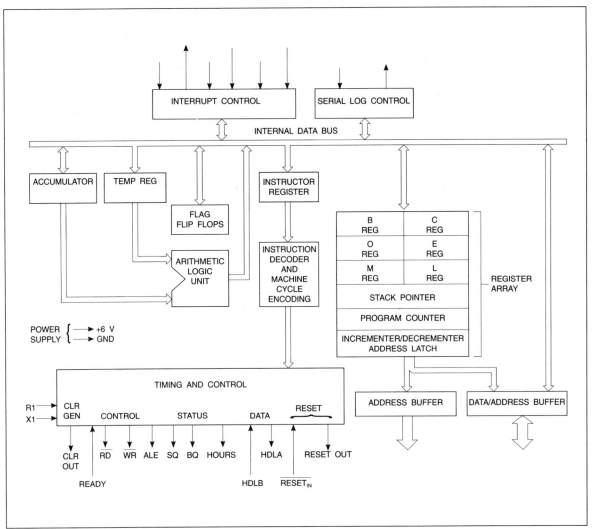

Figure 7-20. Single-Chip Microprocessor

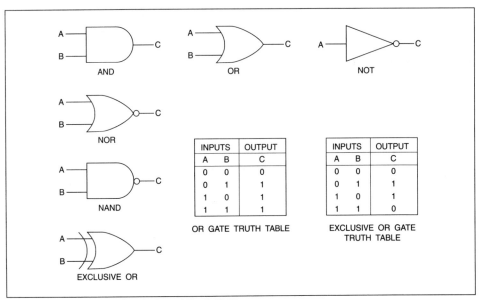

INPUTS		OUTPUT
A	B	C
0	0	0
0	1	1
1	0	1
1	1	1

OR GATE TRUTH TABLE

INPUTS		OUTPUT
A	B	C
0	0	0
0	1	1
1	0	1
1	1	0

EXCLUSIVE OR GATE
TRUTH TABLE

Figure 7-21. Logic Gate Symbols and Truth Tables

the storage capability, is constructed out of combinational logic elements, as shown in Figure 7-22. This circuit is referred to as an R-S flip-flop.

The term R-S is derived from Reset and Set. Two cross-coupled OR gates and NOT gates are combined to make a flip-flop and store one bit of data. NOR and NAND gates are also used for this purpose. Another type of flip-flop is the D flip-flop: an additional gate circuit that carries a data input is coupled with an R-S flip-flop. The D flip-flop is one type of clocked flip-flop that is used in a timing device. Still another flip-flop is the JK flip-flop, which is also clocked out and can be used to build a counter. Figure 7-23 shows one way to build a JK flip-flop. J and K are called control inputs because they determine what the flip-flop does when a positive clock edge arrives. The RC circuit of the clock input has a short time constant (differentiator circuit), thus converting rectangular clock pulses into narrow spikes. Because of the AND gates, the circuit is positive-edge triggered.

Programming Concepts

All the digital hardware that has been discussed is not able to perform any control function without software or programming. The digital or computer system must be programmed with the process information, control algorithms (PID), and operator interface instructions that are necessary for proper operation.

Some control functions can be preprogrammed by the system supplier, other functions must be configured by the user's control system engineer. Most microprocessor and computer-based control systems come with standard fill-in-

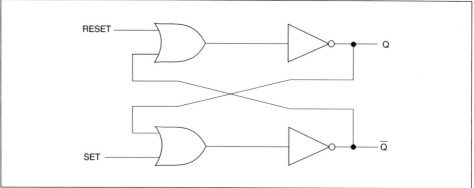

Figure 7-22. R-S Flip-Flop Circuit

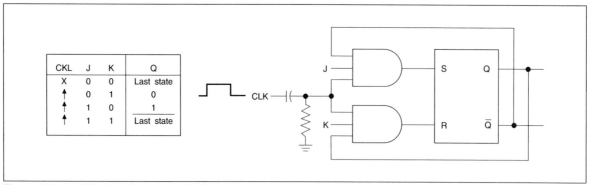

CKL	J	K	Q
X	0	0	Last state
↑	0	1	0
↑	1	0	1
↑	1	1	Last state

Figure 7-23. J-K Flip-Flop Circuit and Truth Table

the blanks software for data acquisition, process control, alarming, and operator displays. Software can be classified into three categories: application, system support, and executive.

APPLICATION SOFTWARE

Application software consists of programs for tasks that are directly related to the primary functions of the system, such as a program to read analog inputs into memory, a program to compute control outputs based on input and set point values, and a program to convert this data into engineering units.

SYSTEM SUPPORT SOFTWARE

System support software consists of programs that aid the user in the development of application programs. These programs are generally vendor-supplied. Such programs are: computer language processors that translate high-level language (BASIC, FORTRAN, Pascal) programs into machine-language programs; editors to facilitate the creation or modification of user-written programs; and debugging aids (to find program errors).

EXECUTIVE SOFTWARE

Executive software, the operating system of the computer, consists of programs that oversee or supervise the actual operation of the system while it is running and performing such functions as scheduling and starting the execution of system-application programs; allocating main memory and loading programs into main memory from bulk memory (cassettes, drums, floppy discs, etc.); and supervision of I/O operations.

Programming Language

A program is a series of actions proposed in order to achieve a certain result. Activities such as locate, read, interpret, and execute the instruction are repeated for each instruction in sequence throughout the program.

A programming language is a set of representations, conventions, and rules used to convey information. The language of the digital electronic circuits inside the computer is one composed of the binary codes that represent the numbers, letters, symbols, and commands used by humans to give instructions to the computer. A conversion is required from the human language to the digital codes that the machine understands. This digital code is called machine language.

To facilitate communications between humans and machine, another type of language is required: high-level language. An intermediate language is required to bridge the high-level language with the machine language.

A high-level language programmer writes the program in a powerful, general-purpose language (FORTRAN, BASIC, Pascal, ADA) that is organized in a way directly related to the way humans solve problems. This is converted to the assembly language program with a compiler.

An assembly language programmer writes the program using instruction abbreviations called mnemonics. This is converted to a machine language program with an assembler.

A machine language programmer writes the program directly in terms of the binary codes the computer can understand or decode and execute. No further conversion is required.

When the digital control system is purchased, the computer is usually preprogrammed to communicate in a high-level language. Some systems are also equipped with configurable prewritten programs for specific applications. These systems are referred to as a configurable digital system. They are easily configured by engineers with no specialized computer background. Their total capability includes all conventional control system functions plus many capabilities not easily implemented with analog hardware, including feedforward, dead time compensation, and multivariable control.

Dedicated Single-Loop Controller

Figure 7-24 shows a dedicated single-loop controller using a microprocessor. The microprocessor is factory preprogrammed, except for the tuning constants. This controller operates as any conventional analog controller. It can also communicate with a main frame (master computer) through its serial data transceiver.

Single-loop controllers are available that are designed to solve mass flow, energy, and volume calculations, including a versatile executive program for control of the various subroutines such as math, input scaling, calculations, etc.

Distributed Control

This system consists of several single-loop and multi-loop controllers in a file or nest. Many independent microprocessor systems monitor and control an entire plant. The controllers also have the ability to communicate with each other via a data bus or highway. This bus is a serial digital transmission line (coaxial cable) that links the controllers and allows the operator interface console or main frame (support computer) to communicate with them. The operator interface is usually a CRT console driven by a microprocessor. In many cases no main frame is required, since the CRT console is a computer in itself and can perform computer-like functions such as present tabular and graphic displays, drive printers, and store historical data for logging and trending. In many distributed systems, the console is capable of scan and alarm functions, thus eliminating the need for a hard-wired annunciator system. Push button and status indication functions can be performed by the CRT console. A second or third console rounds out the system's reliability and flexibility. Failure of the CRT console does not affect operation of the controller units that provide automatic PID control (see Figure 7-25).

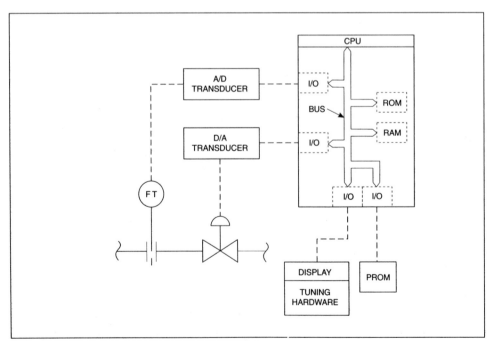

Figure 7-24. Microprocessor-Based Flow Controller

Chapter 8 discusses distributed control systems in more detail.

Programmable Logic Controllers

Not long ago, industrial processes and control systems used relays, timers, and counters, and their control logic was represented in the language of "ladder diagrams." Relay ladder diagrams are universally understood in the industrial world. The programmable controller was originally developed as a replacement for standard hard-wired relay control logic (electromechanical relay controls). It can be easily programmed and modified without actual rewiring. The system usually employs a microprocessor.

A programmable logic controller, according to NEMA standards, is a digitally operating electronic apparatus that uses a programmable memory for the internal storage or instructions that implement specific functions such as logic, on/off control, timing, counting, complex logic, arithmetic calculations, PID (simulated analog) control, data logging, and CRT displays.

There is a tendency to confuse PLCs with computers, minicomputers, and programmable process controllers that are used for numeric control and for position control. PLCs are used for sequence control. Unlike computer control, the PLC does not require very sophisticated programming, debugging, and maintenance techniques. Programming is accomplished via symbolic relay logic ladder diagrams.

As with any other digital control device, the heart of a PLC is the CPU. Input commands, device status, and instructions are converted to logic signals: one for input present, and zero for no input signal in positive logic. These logic signals are then processed by the CPU. As in traditional ladder diagrams where the

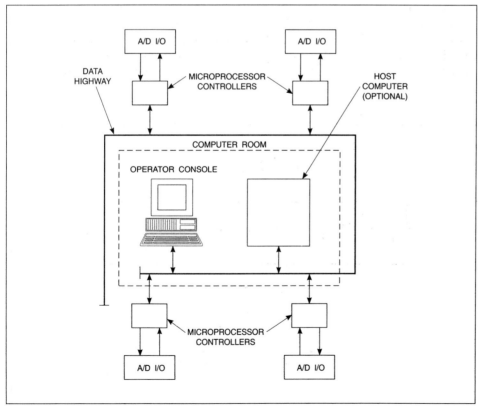

Figure 7-25. Microprocessor-Based Distributed Control System

NO/NC (normally open/normally closed) contacts of the field devices activate relays and timers, PLCs process logic signals and activate output triacs that can be normally energized or de-energized.

In a relay control system, NO/NC contacts from relays are available for use in the control scheme. Similarly, in a PLC, internal and output coils have NO/NC contacts that can be used in the control scheme. No wiring is needed for implementing the control logic in a PLC. All sequence control logic is internal to the PLC and is processed by the CPU.

Many PLCs have various microprocessor chips that are preprogrammed with a main "executive program." This program enables the CPU to understand input command instructions and status signals and provides logic processing capability. Figure 7-26 illustrates the basic architecture of a PLC system.

MEMORY

The memory hardware is similar to the ones discussed previously. The memory size furnished in the PLC varies with the size of the control functions to be performed. Various PLCs have different limitations on the number of horizontal and vertical contacts that can be programmed into each step. This affects net memory usage for a given ladder schematic. Also, the number of words of memory used per contact varies from model to model.

An important concept in PLC operation is "memory scan." For example, all user programs are entered in the multi-node format. The multi-node format allows for up to ten elements of the program in each horizontal rung of the ladder diagram. Up to seven of these rungs can combined into a network of relay contacts and other programming elements (timers, counters, etc.). Each network can have up to seven coils placed at the extreme right of the network. The network becomes one basic parameter of the ladder diagram. The controller will solve each network of interconnected logic elements in their numerical sequence, the order in which they were programmed. The first network is scanned from the time power is applied, first from top left to bottom left and then continuing to the next vertical column to the right. Within a network, the logic elements are solved during the scan; then the coils are appropriately energized or de-energized to com-

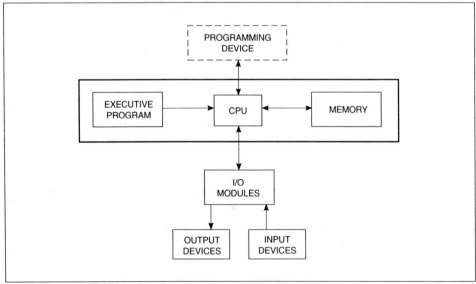

Figure 7-26. PLC System Basic Architecture

plete the scan. Since the scanning rate is very fast (some milliseconds), it appears that all logic is solved simultaneously.

INPUT/OUTPUT MODULES

PLC input/output modules are designed to interface directly with industrial equipment. Input modules can interface with a variety of signal levels, such as 120 V AC, 48 and 24 V DC, 4-20 mA DC, and 5 V DC (TTL). Most manufacturers offer optically isolated inputs, which permits mixing of discreet and analog inputs. Output modules are also available in the same wide ranges of signal levels as input modules. Field devices such as motor contactors, valves, solenoids, MCC interlocking, and lights can be directly operated from the output modules.

PROGRAMMING

All PLCs are provided with the ability to program or simulate the functions of relays, timers, counters, and other functions. Most programming is done in the basic format discussed in the memory section. Other PLC manufacturers use Boolean algebra equations. An example of a partial relay logic ladder diagram is shown in Figure 7-27.

Chapter 9 discusses programmable logic controllers in more detail.

When programming a relay contact into the general format, any horizontal arrangement of contacts can be used.

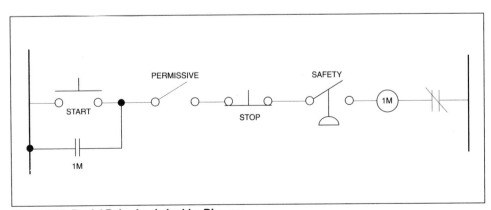

Figure 7-27. Partial Relay Logic Ladder Diagram

About the Author

Donald A. Coggan is owner/engineer of his own consulting firm in Quebec, Canada, which is involved in all areas of control, automation, and training including computer-aided training. A graduate of McGill University in Montreal, Mr. Coggan's early interest in the computer as an instrument for control and training led to his developing Computer*Ease, a computer literacy program, and CHINEASE, which is software that teaches writing and pronunciation of the Chinese language. His expertise in HVAC instrumentation and controls resulted in his writing and presenting short courses and training manuals in the field.

Mr. Coggan has written more than 60 technical papers and articles. He is a member of the Order of Engineers of Quebec, the Instrument Society of America, the Association of Professional Engineers of Ontario, and the American Society of Heating, Refrigerating, and Air-conditioning Engineers.

8

Distributed Control Systems

In broad terms, instrumentation can be divided into two groups — primary and secondary. Primary instrumentation consists of the sensors and final control elements that are located near the process being controlled. This is an area in which quiet improvements occur continuously. Secondary instrumentation is essentially what one sees in a control room. It consists of the equipment used to indicate, alarm, record, and control. It is in this area that revolutionary changes are made and distributed control systems are playing a major role.

Indicating, alarming, and recording support the principal function, which is controlling. In a continuous process environment, control in its simplest terms means keeping the process variable equal to the set point. In a batch process environment, control means keeping the process variable equal to the set point as well as keeping all physical events synchronized with the sequence of events of the process recipe. Advanced control means determining correct set points or an ideal process recipe. If the control system design results in the right set points being effectively maintained, the result is consistent and efficient process performance.

Introduction

A number of technological advances in instrumentation have improved the performance of conventional control applications:

(1) Pneumatic telemetry permitted the beginning of centralized control rooms.

(2) Electronic analog controls improved accuracy and allowed more compact arrangements of control panels.

(3) Digital technology (minicomputers) introduced sophisticated and advanced controls, which allowed logical alarming and indication through CRTs.

(4) Distributed control systems (DCS) have allowed lower installation costs due to the interconnecting and grouping of control modules, lower maintenance cost, better system reliability, ease of configuring the process concept, and ease of expandability.

Other key developments are shown in Figure 8-1, which illustrates the general progression of industrial control from its mechanical-regulator beginnings to state-of-the-art distributed control systems.

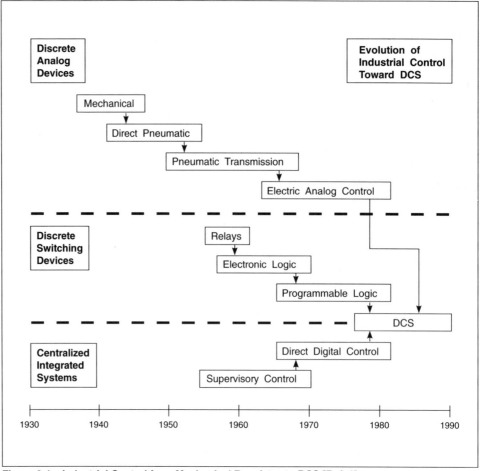

Figure 8-1. Industrial Control from Mechanical Regulator to DCS [Ref. 1]

Definition of a DCS

As implied by its name, a distributed control system is one whose functions are distributed rather than centralized. A DCS consists of a number of microprocessor-based modules that work together to control and monitor a plant's operations. The modules are distributed geographically. This reduces field wiring and installation costs. It also reduces risk by distributing the control function throughout a number of small modules rather than concentrating it in one large module.

A DCS is a computer network. It differs from an office or personal computer network: a DCS does real-time computer processing as opposed to the batch-processing done by data processing office computers. The difference between real-time and data processing computers is the way they execute their programs.

Data processing computers typically do a single program operation at a time. The program will start with some fixed data, perform a complex set of calculations, and provide a set of results. Once the program has done its job, it stops until it is instructed to run again with new data. An example would be the monthly processing of invoices by a utility company.

As in batch processing, the real-time computer also executes its program by using fixed data, performing calculations, and providing a set of results. The difference, however, is that it runs the same program repeatedly with updated data, sometimes several times a second.

A simple example of a real-time operation is the computerized cruise control on a car. For example, assume the speed set point is 100 mph (60 kph). The real-time computer continuously scans the car's actual speed. If the actual speed is lower, say 90 mph (55 kph), the computer will increase the speed by calculating and increasing the amount of fuel to the engine. Similarly, if the measured speed were higher, say 110 mph (65 kph), the computer would decrease the fuel intake. It continuously makes tiny incremental adjustments several times a second by scanning the actual speed, comparing it to the set speed, and recalculating the fuel requirements. A DCS does exactly the same repetitive scanning and recalculating for hundreds or even thousands of devices throughout a plant.

Basic DCS Functions

The DCS, like the programmable logic controller, is connected to primary control elements such as temperature and pressure transmitters, flowmeters, gas analyzers, pH and conductivity sensors, weigh scales, contact switches, valves and motors, and so on. From these field devices it receives electrical signals, for example 4-20 mA, 1-5 V DC, 24 V AC, and 120 V AC. The DCS converts these signals (digitizes them). Once converted, they can be used by the computer to:

(1) control loops,

(2) execute special programmed logic,

(3) monitor inputs,

(4) alarm the plant operations,

(5) trend, log, and report data, and

(6) perform many other functions.

Field signals are divided into two basic categories — analog and discrete. Analog signals are continuously variable; they act like the dining room dimmer, which changes the lighting intensity in a gradual manner. Discrete signals can have only two values or positions and are called two-position or on-off or snap-acting. They are often associated with contact devices, such as the light switch in a home. There is no "in between" with discrete devices — they are either open or closed, true or false, on or off, etc.

Analog loop control often involves simply maintaining a process variable (such as temperature or pressure) equal to a set point. It is like the cruise control maintaining its set speed. Of course, many different types of control loops (feed-forward, lead-lag, cascade, etc.) are being executed in a DCS, but simple, set point-maintaining loops often account for the bulk of them.

Discrete control very often consists of simple logic statements coupled with field sensors to provide logic interlocks or process sequences. For example, consider a tank to be filled with a liquid and then heated. To protect the product and/or equipment one could use a logic interlock that says:

(1) IF the level is below a minimum point,

(2) THEN the heater coil cannot be turned on (or must shut off).

The process might also call for the liquid to be stirred with an agitator. The previous logic interlock could be coupled with sequencing logic that says:

(1) First, fill tank.

(2) Second, turn on heater.

(3) Third, start agitator.

(4) Fourth, empty tank.

In the sequence, the second step cannot take place until the first is completed. Likewise, the third step cannot start until the second step is completed and so on. By adding the IF-THEN logic interlock, if the level should ever drop below the minimum level, the heater would still trip off.

Role of the Computer in DCS

Because a DCS is computer-based and all its information is in digital form, it can easily combine analog control loops with discrete logic (interlocks and sequences). The above example of using sequencing and logic to control the heating and stirring of a liquid in a tank could also incorporate an analog loop to maintain a constant temperature in the liquid. As illustrated in Figure 8-2, all functions would execute simultaneously.

A DCS can involve as little as a few hundred inputs, outputs, control loops, and logic interlocks or tens of thousands of them. It can scan all the primary elements or sensors, characterize the input signals and alarm them, recalculate loop parameters and execute logic, and then send the results to motors and valves throughout the plant. It constantly reevaluates the status of the plant and makes thousands of incremental decisions in fractions of a second. It is capable of all this and more for two main reasons:

(1) A DCS is made up of many independent control modules that can operate simultaneously and independently.

(2) It has the ability to carry out rapid communications between these and other modules by means of a communications link called a real-time data highway.

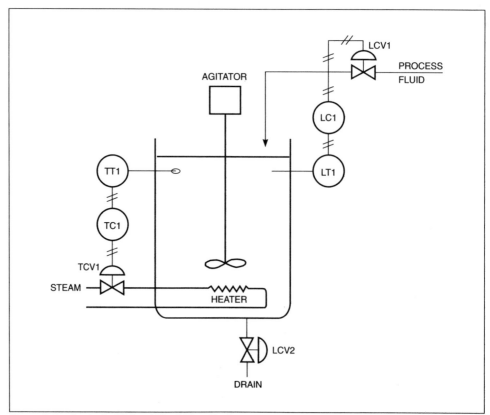

Figure 8-2. Control of Stirred Liquid in a Tank

Multiple, high-speed, incremental control adjustments permit close and coordinated plant control that results in more consistent production. This close control provides a plant with the means to fine tune the process (for example, to accommodate variations in feedstock). It also provides the means to maintain a more consistent product within process areas, thereby minimizing the amount of compensating done downstream. A more consistent product means fewer rejects and a more efficient operation.

Close control is only the first step to efficient production. Many plants find that their process units need to make adjustments not only for varying feedstock characteristics but also for varying end product requirements and varying operating techniques. To keep track of and coordinate all these fluctuating circumstances, a DCS incorporates extensive capacity for communications and data storage and retrieval. This, then, is another key DCS function, because it enables plant personnel to make the right decisions by supplying information that is both accurate and timely.

Most DCSs are capable of rapidly displaying process information and storing it to be retrieved, reviewed, and analyzed at a later date. Typically, this information would be used by all the departments in the plant, from process engineering to maintenance to production to plant management. A good DCS provides quick and easy access by the appropriate personnel to the appropriate information.

Being computer-based, the DCS also offers intelligent alarm management. It can force the operator to focus on the most important alarm, thus allowing him or her to respond more appropriately to the situation. Some alarm functions include the ability to:

(1) filter out nuisance alarms,

(2) recalculate alarm limits,

(3) re-alarm lingering alarms, and

(4) prioritize alarms.

DCS and Expert Systems

To reduce costs and improve performance, DCS manufacturers are incorporating many enhancements introduced by the computer industry:

(1) Higher resolution CRTs

(2) Better networking between the control areas of complex processes

(3) Megabyte memory chips and 32-bit microprocessors

(4) Enhanced algorithms to continuously tune loops and assure that every loop performs optimally

(5) Many more state-of-the-art technologies applied to solve classic instrumentation problems

The most important DCS enhancement, however, is due to the great strides made by the computer industry in artificial intelligence (AI), particularly expert systems. Expert systems have already shown tremendous potential not only as a diagnostic tool but also as a development aid for the control engineer.

Expert systems are attractive because they clone the knowledge of a small number of experts and then make it usable by a large number of nonexperts. This is particularly applicable in the control industry. Control systems are usually operated in the automatic mode, because efficient operation in the manual mode depends on the skills of a particular operator. Some operators are more effective than others.

Expert systems that capture the expertise of the most skilled operators can allow less-skilled operators to perform their tasks with considerably increased proficiency. Such a technique can be used to optimize start-ups, optimize grade changes in a process, and execute emergency shutdowns.

Overview

Because of its mandatory dependence on computer technology, the DCS is clearly software intensive, and practitioners cannot neglect this particular aspect. Nevertheless, advanced computer-related topics such as artificial intelligence, expert systems, management information systems, optimization, simulation, and modeling will not be discussed in this chapter on DCS. There are two reasons for this. The first is that this book has a chapter devoted exclusively to the role of computers in industrial control. The second reason is that implementing advanced computer capabilities varies widely not only in the manner in which such capabilities are used but also where and why they are used. The choices and decisions associated with them depend on the philosophy and operating needs of an individual plant. For a book on basics, such topics are a little too advanced to be included.

The goal of this chapter is to present basic information about DCSs in as straightforward and complete a manner as possible. Although it covers the fundamentals of DCSs in terms of the tangible aspects of the equipment, it also addresses the intangibles, such as implementation. This is done in the light of a DCS being a long-term, living investment rather than simply a one-time computer purchase.

Most DCSs provide similar capabilities in a comparatively cost-effective way for a given project. However, they may vary greatly in cost of ownership. Simple things such as ease of system start-up (and restarting after a shutdown) may have profound effects on the overall cost of a DCS project. The addition of a seemingly small number of extra modules or even a software upgrade can double project costs and delay start-up. A DCS purchase requires careful evaluation and future planning.

Other factors, such as site preparation costs, ease of expansion, product obsolescence, upgradability of hardware and software, backward and forward systems compatibility, maintenance, training, and integration with other computers, are important issues in the proper evaluation of any DCS.

DCSs can also vary widely in terms of reliability and availability. Some suppliers offer proven, off-the-shelf products; some products are still in the developmental stage. Software that is continually promised but never delivered has come to be known as "vaporware." "Vendor support requirements" is an expression related to how much an owner alone can do versus having to return to the vendor to have it done. Any of these factors could make the difference between an easily implemented, cost-effective computer system and a financial "black hole." Often, apparently low-cost systems turn into overbudget problems because the right questions were not asked in the beginning.

Thus, this chapter presents DCS basics in a larger scope from hardware and software to start-up, expansion, maintenance, upgrades, and purchasing strategies.

DCS Architecture

Overall Structure

The structure of a DCS is often referred to as its architecture. In terms of functional modules, DCSs from the various vendors have a lot in common . This section therefore examines these functional modules from the point of view of a generic system that is representative of all manufacturers. Figure 8-3 illustrates the architecture of a such a generic system in terms of functional modules. The key word is "functional." The modules do not necessarily represent physical components; some manufacturers may combine two or more functions in one physical component.

In addition to the process instruments (such as temperature transmitters, flowmeters, pH sensors, valves, and so forth), which are common to any process control approach, there are six generic functional modules:

(1) Input/output or I/O modules scan and digitize process instrument input/output data. Some may perform elementary simple logic.

(2) The local I/O bus links I/O modules to controller modules.

(3) Controller modules read and update field data and perform control calculations and logic to make process changes.

(4) User interfaces include operator interfaces and engineering workstations.

(5) The data highway is a plant-wide communications network.

(6) Communication modules provide a link between the data highway and other modules, typically controller modules and user interfaces.

Each DCS vendor has a proprietary approach, and it is possible, for example, for the functions of control and I/O to be combined in the same physical component. Nevertheless, it is still possible, even preferable, for a DCS to be described by means of the generic functional modules.

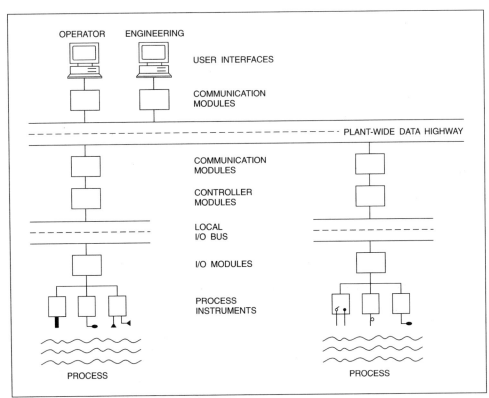

Figure 8-3. Architecture of a Generic DCS

Input/Output Modules

Input/output modules provide the main interface between the DCS and the process being controlled. They convert the information provided by the process instruments into digital form. They also provide signal filtering, contact debouncing, and in some instances they can also do alarming, signal characterizing, and low-level logic. Four basic types of signals connect to I/O modules:

(1) Analog inputs, also called analog ins or AIs

(2) Analog outputs, also called analog outs or AOs

(3) Digital inputs, also called digital ins or DIs

(4) Digital outputs, also called digital outs or DOs

Analog inputs are gradually varying signals (as opposed to two-position), typically connected to sources such as 4-20 mA and 1-5 V DC transmitters, thermocouples, and RTDs (resistance temperature detectors). Analog outputs are gradually varying signals, usually 4-20 mA, typically connected to devices, such as valves, dampers, and variable speed motors.

Digital inputs are typically connected to two-position devices such as limit switches, relays, and pulse contacts. Digital outputs are contact openings and closings that operate controlled devices (such as valves, dampers, and motors) in a two-position manner.

I/O modules are typically designed for varying levels of input/output loading, for example:

(1) A single board connected to a single field device providing single-point integrity

(2) A single board connected to a single input device and a single output device providing single-loop integrity

(3) A single board connected to multiple (4, 8, 12, 16, 32) inputs

(4) A single board connected to multiple (4, 8, 16) outputs

(5) A single board connected to multiple inputs and multiple outputs (for example, eight in and four out)

I/O modules may have separate, individual circuits, or they may share components such as analog-to-digital and digital-to-analog converters and multiplexers. Typical features to look for in I/O modules are:

(1) Isolated or nonisolated grounding on a per-point or per-board basis

(2) Level of fusing protection on a per-point, per-circuit, or per-board basis

(3) Accuracy and linearity of the sampling frequency

(4) Protection from electromotive force (emf) and transients

(5) Immunity to radio frequency (rf) interference

(6) Fail-safe positioning

(7) Overload and surge protection

(8) Impedance matching with field devices

(9) Loop feedback sensing

(10) Manual override of loop control

(11) Mean time between failure (MTBF) and mean time to repair (MTTR) (field values, not theoretical)

(12) Criticality — that is, if the board fails, what else will be affected

With these criteria in mind, one should be able to evaluate the level of reliability of I/O modules when comparing various vendors' systems. This will indicate when and where to apply redundancy at this level.

Systems from different vendors have different redundancy needs based on criticality and reliability.

Local I/O Bus

The local I/O bus provides a bridge between the I/O and controller modules and, by definition, is restricted in terms of geographical area and data loading. It typically operates at a slower speed than the plant-wide data highway, although communication rates can range from 9,600 to 250,000 to 1 million bits per second.

I/O buses can connect varying numbers of I/O and controller modules. The manner in which they provide communications can also vary, from polling or scanning of the I/O by the controller modules to serial communications between I/O and controller modules. They can also be arranged for serial or parallel communications or a combination of both.

While I/O buses are seldom a bottleneck or a limitation, they become a critical component if they fail. The loss of a single I/O bus can affect the control of many end devices.

When evaluating a system design, one is well advised to consider redundant I/O modules as a key requirement.

Controller Modules

Controller modules are the true brains of a DCS. Their primary function is to use continuously updated information from I/O modules and then perform the complex logic and analog loop calculations needed to produce the controller output signals that keep process variables at the desired values. It is at the controller modules that many DCS functions, such as the following, are performed:

(1) I/O signal characterization

(2) Signal filtering

(3) Alarming I/O modules

(4) Ranging and engineering units

(5) Control logic

(6) Control interlocks

(7) Sequencing

(8) Batch control

(9) Passing on of trending information

(10) Passing on of report information

Controller modules are microcomputers and, as such, have similar limitations. Although the various numbers associated with the various types of controller modules can have a mesmerizing effect, not all of these numbers are important in one's evaluation of controller module performance. The key ones are:

(1) available memory for configuration,

(2) available idle time (based on a given scan rate),

(3) I/O loading or criticality,

(4) number of available software addresses for input/output blocks, and

(5) number of available software addresses for control blocks.

In the sizing and selecting of a DCS, it is vitally important to ensure that there is enough processing power not only to serve the active I/O and control functions but also to provide some spare capacity for future I/O expansion, additional logic, and extra things such as totalizers. This is an important consideration, because adding this processing power after the fact doubly penalizes the owner. First there is the added cost of the extra modules and other associated equipment, such as communication modules, power supplies, and cabinets. This added cost is often determined on a noncompetitive basis and is, therefore, higher than it would have been if purchased as part of the initial contract.

The second penalty is inferior performance due to the extra loading put on the original and the new controller modules, the communication modules, and the data highway. This extra loading is the result of controller modules doing link communications instead of simple control. Link communications are those that pass high volumes of information between control processors. Such communications consume large amounts of memory and scan time in the associated controller and communication modules and load the data highway. A simple way to avoid this potentially reduced performance is to specify suitable values of I/O loading, memory usage, and idle time for controller modules. For example, for a given scan cycle (1/4, 1/2, or 1 s on average), one can specify the amount of spare memory and idle time to be available in the controller module after execution of the I/O and control functions. Spare memory and idle time should normally range from 20% to 60%, depending on the application. Limiting the number of I/O and control functions executed in a controller module is a good idea for three reasons:

(1) It ensures the availability of the microprocessor power needed to carry out the specified functions and thereby simplifies configuration engineering.

(2) It allows for easier, more flexible future expansions and reduces the risk of link communications.

(3) It reduces the criticality of any given controller module by limiting the number of I/Os and loops controlled, thus limiting the damage caused by failure of the module.

Communication Modules

Communication modules are also microcomputers, but they differ from controller modules in function. Rather than execute control strategies, communication modules manage the flow of information between the data highway and controller modules, user interfaces, and gateways to host computers and PLCs. Although there is always a physical limit to the amount of data that communication modules can handle, they are not often a bottleneck.

If problems do occur, the communications rate and memory capacity should be checked. Performance improves if one either decreases the number of communication modules or decreases the number of devices served by single modules. Again, there should always be room for expansion. Communication modules are critical to proper operation of a DCS; without them, the operator may be blind to the process.

Specifying redundant communication modules is almost always a good idea.

Real-Time Data Highway

Real-time data highways come in many variations. Topologies can be linear, loop, or star, and they may or may not include "traffic controllers." Since a data highway is a microprocessor-based module, it should be viewed as considerably more than one or two cables strung out across the plant.

If controller modules are the brains of a DCS, then the data highway is its backbone. It is an active component through which pass the system's messages and file transfers, all in real time. It constantly updates the consoles, gateways, and other modules connected throughout the system countless times each second. It is probably one of the most critical DCS modules, because it is common to all other plant-wide components. If the data highway should fail, operators are cut off from the process, link communications are lost, and process control is affected. The data highway is the one DCS component that should almost always be made redundant. In this case, redundant does NOT mean one highway is active and one is a hot standby; it means that both highways are active, permitting a bumpless transfer between highways without need for human intervention. If traffic directors are part of a data highway, they should also be made redundant.

The following are principal issues to be addressed in the evaluation of a DCS data highway:

(1) Synchronized versus nonsynchronized

(2) Deterministic versus nondeterministic

(3) Token passing versus report by exception

(4) Variation in protocol types (all are proprietary)

(5) Peer-to-peer versus collision detection-based communications

(6) Speed of data transmission

(7) Maximum transmission distance

The evaluation of the security and reliability of a data highway is not straightforward because many factors are involved. Most importantly, speed isn't everything. Other key factors are module highway access, message buffering and prioritizing, and efficiency. For example, highways based on collision detection and report by exception can lose 70-80% of their rated capacity when message loading increases due to alarm burst and process upset conditions. Unfortunately, it is under such conditions that it is most important for the data highway to perform efficiently. Generally, one should evaluate a data highway design based on a worst-case scenario. Consideration should be given to:

(1) the number of tags (I/Os and control loops) that are connected to the highway,

(2) how much trending and reporting information is being transferred,

(3) the volume of link communications, and

(4) the number of alarm points.

Once the required data highway capacity is known, the size, number, and configuration of highways (and traffic directors) can then be specified.

Repeaters or gateways are an integral part of real-time data highways. When one data highway is fully loaded and more capacity is still needed, additional highways can be used. Two common approaches are used to permit communications between highways. The first is to link the highways together via a higher level or so-called super highway. Each real-time data highway is joined to the super highway by means of gateway modules, which are usually redundant. This would mean that connecting two redundant real-time data highways together would require four gateway modules. The second approach is a straightforward highway-to-highway connection via highway interface modules. In this second approach, there is no super highway acting as a go-between.

Whichever approach is used, if one ends up with a requirement for multiple highways, extra costs should be expected. If the requirement happens to be "unplanned," the extra cost could be substantial, considering the gateways, other interface hardware, software, engineering, and possibly re-engineering — all added "after the fact." Sizing a real-time data highway means looking as far as possible into the future and planning for maximum loading.

Host Computer Interfaces and PLC Gateways

A requirement in many DCS applications is the transfer of information to and from other types of computers. This can be required for a variety of reasons, such as:

(1) integration with management information systems (MIS) computers,

(2) integration with optimizing or modeling computers,

(3) integration with production and maintenance computers or computer networks already in place (or to come), and

(4) integration with other process control computers (such as PLCs).

Whatever the situation, the distinctly different computer systems must be able to communicate with one another. That is, the real-time computer system may have to talk to MS-DOS-, PS/2-, or UNIX-based computers (see Chapter 5). As there is no universal agreement on operating systems, all DCS vendors have taken the approach of a "translator box" or "host gateway." Typically, this gateway is a passive device in that it does not initiate communications but merely translates and transports information. Typically, it does this in a method similar in concept to that used in a post office box as illustrated in Figure 8-4.

This method is often explained in terms of a data transfer table and is generally an efficient means of communication. It is faster and accommodates more data than an approach that uses a direct question and answer on a point-to-point basis. Gateways can also accommodate file transfers of large quantities of data, such as trend or report files, although not all gateways have these abilities.

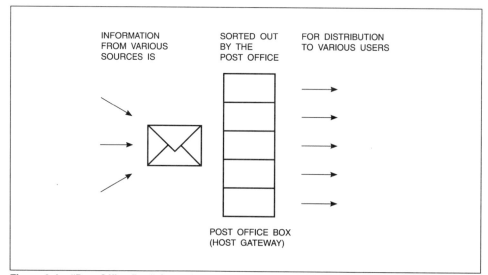

INFORMATION FROM VARIOUS SOURCES IS

SORTED OUT BY THE POST OFFICE

FOR DISTRIBUTION TO VARIOUS USERS

POST OFFICE BOX (HOST GATEWAY)

Figure 8-4. "Post Office Box" Operation of a Host Gateway

Since a host gateway module is normally a passive device that simply translates, it needs to be told what information to translate and when to read and write to the various system registers. In short, it requires a driver device with driver software to take charge of the communications. This setup is often a master-slave relationship between the DCS and the host computer.

In communications with a PLC, it is usually the DCS that is the master handling the driver software. The reverse is normally true when a DCS communicates with a host computer. It is essential to know if a vendor includes the driver software with the interface or gateway. Proven, off-the-shelf driver software is highly preferred to software that must be custom developed. In the latter case, a user must be prepared to pay a high premium and, in addition, suffer the frustration of on-the-job debugging. Custom software development is very expensive in both the short and long terms.

While a host gateway module is passive in terms of communications, it is an active computer device. It therefore has memory and scan time limitations to be aware of in terms of:

(1) size of data base,

(2) speed of communications,

(3) rate of data base refresh, and

(4) types of data accessible (for example, trend files, report files, types of live data, and so on).

Power Distribution System

This is the part of a DCS that is most often overlooked and, like the real-time data highway, it is a system component common to all others. It is the DCS component that takes raw electrical power, converts it, conditions it, and regulates it for the various other computer modules in the system.

The typical power distribution system can be split into two parts — bulk power and power regulation. With bulk power, the key issue is to make sure that variations in the main AC source do not exceed the capabilities of bulk power supplies. Battery backup is usually mentioned in the same breath as bulk power supplies and may appear in various forms: uninterruptible power supply, separate battery packs, or integral battery packs. Whichever approach is used, the batteries should be able to take over instantaneously if power fails or dips. Loss of power to the microprocessor modules could erase some sections of memory and also require a reboot of the system. Battery backup is sized to keep the system energized long enough to meet essential needs. Typical backup times may range from two or three minutes to two hours.

Power regulation is also vital to the operation of a DCS but is almost never lacking in capacity. However, redundant power regulation is recommended for most system modules and most applications.

The power distribution system is not a high cost DCS component, but it is important and should not be skimped on. Planning for future needs and partial loading of 50 to 75% of the rated capacity is highly recommended. Redundancy in bulk power and power regulation is a wise investment.

User Interfaces

Introduction

The user interface has undergone quite a revolution throughout many industries over the last 50 years. With today's complex processes, user interfaces are needed for engineering personnel as well as for process operators.

In the days of smaller and simpler industrial plants, manual control was the first form of operator interface. An operator would walk a tour, check tank levels and pressure gages, and adjust valves. As industrial operations grew in size and sophistication, tours became longer and more operators were required. Adjustments became less straightforward as processes became more interrelated. Manually collected data became less useful in terms of providing a true picture of what was going on.

With the advent of relay logic panels and pneumatic transmission came the concept of a centralized control or monitoring room. For the first time, information was brought to the operator. However, the centralized information was incomplete, and what was there tended to be unreliable. Consequently, operators still made their tours, reading strip chart recorders and local panel indicators.

Electronic analog controllers and PLCs soon made for more precise, cost-effective, and reliable control. It also made for a better centralized control room. Fewer operators were required for ever larger and more complicated plants. However, since much of the old pneumatic control equipment and relay panels still existed, manual intervention was still required and operators continued to do their tours. If an operator went out for an hour or two and a process upset occurred just after he or she left, it could be some time before it was corrected. Off-spec products and rejects usually resulted.

In the late 1960s and early 1970s the first digital computers emerged, quickly followed by the first CRT-based display stations. For the first time virtually any instrumented information about the process was available at the touch of a button.

Since then, CRT-based consoles have increased in power, speed, and reliability. They permit almost instantaneous access to measured variables throughout even the largest production plant. Single control centers that operate entire paper mills, steel mills, and refineries are increasingly replacing the multiple operator rooms distributed throughout a site. For the first time, operators of the individual process units are finding it easy to work together. Soon, multiple computer systems will use common operator interface stations, with multiple displays on a single CRT, reducing the number and types of consoles required.

The operator or man-machine interface, is the "razzle-dazzle" module in a DCS. It is the one device that strongly influences people's perceptions of the entire DCS. An impressive operator interface translates into an impressive DCS. The question is, "Is this just a pretty face or is the beauty more than skin deep?" The answer to this question depends on the extent to which the following features are present:

(1)   accessibility to and size of the data base,

(2)   integrity of information,

(3)   screen build (or image) speed (static and dynamic), and

(4)   reliability and redundancy.

Being a computer, operator interfaces come in many different configurations, with very different abilities and methods of operation. Most operator interfaces are "tag limited," that is, they are restricted to the size of data base they can access. DCS vendors offer PC-based interfaces in consoles that range in capacity from 500 to 10,000 tags. These interfaces usually have a live RAM-resident data base that duplicates the data base resident in the I/O and controller modules.

This RAM data base is continually updated from the field, and screen display information is continually refreshed from it. One should be cautioned here that the operator sees information from the interface data base and not directly from the field measurement. If the system operates on a report-by-exception basis and communication is lost, the operator could be unaware that the process is doing something completely different.

So-called display-based consoles do not use a RAM-resident data base. These interfaces store various display pages, which, when loaded onto the screen, will have their dynamic points refreshed directly from field measurements.

The time it takes to refresh a screen image (screen build) is an important factor in the evaluation of an operator interface. A screen display is usually divided into areas of static and dynamic data. Static data is associated with objects (such as boxes, tanks, pipes, and valves) that do not change (except for color, flashing, or reverse video) with the process. They are a fixed part of the display.

Dynamic data consists of live field data such as process variable values, loop set points, controller outputs, and contact status. It is information that is constantly updated, or refreshed, on the CRT.

Screen builds can be as fast as 1/4 second and as slow as two to 15 seconds. The speed can be affected by things such as:

(1) size of display page (amount of static information),

(2) number of dynamic points on display pages,

(3) location of resident display pages (hard or floppy disk), and

(4) use of display pages from another highway module.

Typically, RAM-resident display pages with few dynamic points will build very quickly. Screen build times could be much longer if an interface has to display many dynamic points that are scattered throughout a plant and have to travel over a busy data highway.

Some operator interfaces have a multitasking ability; that is, they can carry out several functions at the same time. Thus, while an operator is using the interface, things such as trending, reporting, and alarm and event logging are going on in the background. An operator interface can sometimes be configured to act as an engineering workstation as well. These extra tasks should not affect the fundamental purpose of the interface, which is to give the operator an efficient window on the process being controlled. This should not be a problem if the added functions are treated as enhancements to the interface.

Multitasking operator interfaces cost more than the single-tasking units but can be more cost-effective than buying a number of separate single-tasking units.

Another important factor in evaluating an operator interface is how many the job needs, including the associated electronics, hard disks, and so on. The following are typical choices:

(1) Number of CRTs

- one per operator

- three CRTs for two operators

- two or more per operator

(2) Number of sets of electronics

- one per CRT

- one per two or more CRTs

(3) Number of hard disks

- per interface

- per electronics

In evaluating needs, it is a good idea to imagine what happens to the operator if any one component fails. For example, if there are four CRTs and two sets of electronics and two hard disks, one would ordinarily expect to be secure. How-

Interfaces tend to be expensive modules in a DCS. Adding them in an unplanned way often leads to a budgetary problem because of the added equipment and engineering required.

ever, if the electronics are each handling separate data bases and one fails, part of the plant may be running blind simply because it cannot access the other's data base. Running blind may not necessarily be fatal, but it is generally considered unsafe. Such a situation can be readily prevented with today's interfaces, which allow for appropriate levels of redundancy to be built in.

Operator Interface Hardware

The operator interface hardware consists of CRT displays, keyboards and other access devices, and hardcopy devices. It also includes power supplies, disk drive units, and card files.

The CRT video display unit is the main interface component. It is the vehicle by which system users operate, control, and manage the entire control process. CRT displays replace the conventional analog control panel, and provide the user with an easy view of the process by means of a hierarchical series of displays.

The layout of the various components depends on the manufacturer as well as the user's preferences. In principle, however, the interface should be designed to increase the operational capabilities of the people who use it. In other words, it should follow good ergonomic design principles (see Chapter 10). Examples of two such interfaces are shown in Figure 8-5.

CRT DISPLAY MONITORS

CRT monitors operate via a dedicated video module with its own processor, which may support alphanumeric keyboards, mice or trackballs, and an alarm horn. Figure 8-6 illustrates how the various components are typically connected.

Sizes of color monitors vary from 13 inches to 19 inches. Monitors may be mounted into a workstation or on a desktop. The video information displayed may include text, charts, and graphics. Ways to describe the display area are:

(1) 48 lines with 80 text characters per line,

(2) 640 horizontal by 480 vertical pixels for a high resolution graphics display, and

(3) 320 horizontal by 240 vertical pixels for a low resolution graphics display.

Monitors may include a touch-screen display as an optional feature. With this option the user selects display objects by touching them on the screen. This is accomplished by means of infrared LEDs and phototransistor detectors mounted opposite one another on the front surface and forming an invisible lattice of infrared light beams. Each time the beams are intercepted, either by a finger touching an object on the screen or by a pointing device, a signal is sent to the processor that indicates the position of the selection.

ALPHANUMERIC KEYBOARD

The alphanumeric keyboard is a regular computer terminal keyboard, with the standard QWERTY key arrangement and ASCII-formatted output. It may have additional keys to facilitate the work of the engineer in configuring the data base, building displays, and setting up system application packages such as historical trending. Figure 8-7 illustrates an alphanumeric keyboard.

OPERATOR KEYBOARD

The operator keyboard is a specialized device designed to make the operator's interaction with the process faster and easier. It can be a full-stroke keyboard or a spill-proof membrane style with tactile feedback. Various dedicated keys allow an operator to select important functions with a single keystroke. The actual

(a) THREE INTEGRATED CONTROL STRATEGIES
(Courtesy of ABB Kent Taylor)

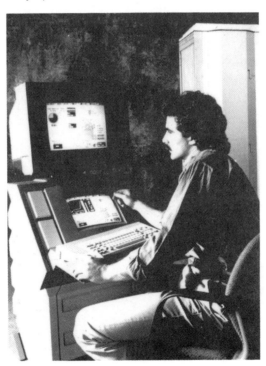

(b) INTELLIGENT AUTOMATION
(Courtesy of The Foxboro Company)

Figure 8-5.   Examples of Operator Interfaces [Refs. 2 and 3]

Figure 8-6. CRT Display Monitor and Related Components [Ref. 3]
(Courtesy of The Foxboro Company)

layout of the keyboard varies with the DCS supplier. Figures 8-8 illustrates two different types.

The keyboard show in Figure 8-8(b) contains an annunciator section with light emitting diodes (LED) and Mylar™ keyswitch. Each LED can be configured by the user as ON, OFF, or FLASHING according to process conditions. The numeric section contains numeric and data entry keys and cursor control.

TRACKBALL AND MOUSE

The trackball is a cursor control device. It allows users to control the cursor position by manually rotating a mechanical ball whose position is converted into data signals equivalent to those generated by the normal cursor control keys. It

Figure 8-7. Alphanumeric Keyboard [Ref. 2]
(Courtesy of ABB Kent Taylor)

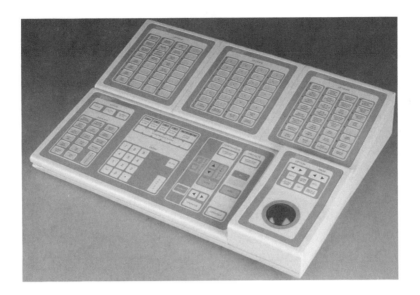

(a) OPERATOR KEYBOARD

(b) MODULAR KEYBOARD

Figure 8-8. Different Types of Operator Keyboard [Refs. 3 and 4]
(Courtesy of Rosemount, Inc.)

also has a push button to acknowledge alarms and messages and to select an action as one would with the "Enter" key on a regular keyboard.

The mouse is another table top cursor control device. When the user moves the mouse across a surface, an internal mechanical ball in contact with the surface rotates and generates cursor control signals similar to those of the trackball. Acknowledge and Enter push buttons are also provided. Figure 8-9 illustrates a typical mouse and trackball.

HARDCOPY DEVICES

Hardcopy devices include printers and video copiers. The printer is used for alarm and event logging, graphics, and reports. Printer output may be in color or in black and white. With a color printer, a user can print blocks, characters, and line graphics with different colors. Alarm conditions can print with a different color to distinguish them from normal operator actions.

A video copier reproduces an image of the CRT screen. The user may select image orientation, density, and colors. Figure 8-10 shows the hardcopy devices.

Operational Philosophy

The role of the operator can be described in terms of the four plant conditions given in Table 8-1.

Table 8-1. Plant Conditions and Role of Operator

Level	Plant Condition Description	Operator Mode Description
1	Steady-state operation	Overall process surveillance
2	Minor plant/unit upset	Unit and section monitoring
3	Planned start-up/shutdown or feed changes	Interactive
4	Major plant upset or emergency	Priority loop interaction

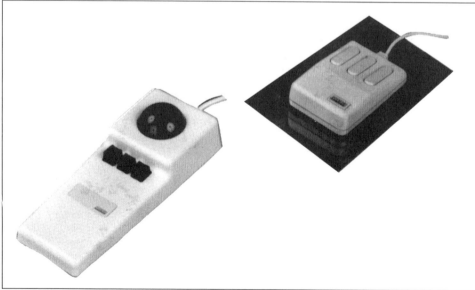

Figure 8-9. Typical Mouse and Trackball [Ref. 3]
(Courtesy of The Foxboro Company)

(a) COLOR DOT-MATRIX PRINTER
(Courtesy of The Foxboro Company)

Figure 8-10. Printer and Video Hardcopy Devices [Refs. 2 and 3]

Level 1 - Steady-state operation. In a steady-state plant condition, the operator's role is normally limited to an overview of the whole or part of the process plant and monitoring the trend records of a number of key variables or variables in an acknowledged alarm condition.

Level 2 - Minor plant/unit upset. During the condition of minor upsets on a particular section of the plant or in anticipation of an upset caused by changing conditions in an interrelated process area, the operator needs to take a closer look at the plant area in question. In this mode of operation, it is essential to have easy access to a summary of the loops involved as well as the related trends. Usually two CRT consoles are required to maintain sound process control; while the operator looks at the individual loops, a standby overview display should be on all the time.

Level 3 - Planned start-up/shutdown. In this operational mode, the information needs are similar to those of the previous unit monitoring mode; however, the operator is now interacting with the process by his or her own need to change and manipulate the control loops.

Level 4 - Emergency/major upset. A major upset condition causes the operator interaction to be most intense. It is essential to have speed and simplicity in accessing the required information and in manipulating the available parameters of the priority loops. Special page displays that give status and multitrend reports should be formatted in advance for use under upset conditions, recognizing full well that there can always be an unpredictable major upset. In this situation, the judgment of the operator, along with parallel safety interlock and shutdown systems, should override any predefined procedures.

Interface Displays

The configuration of each operator console and the number required depend upon the type of plant involved. Regardless of the configuration, however, when the system is powered up the first screen displayed is usually the main menu, which can be retrieved at any time during operation. Figure 8-11 shows a typical display.

Interface displays are described in terms of:

(1) display hierarchy,

(2) plant overview displays,

(3) trend displays,

(4) alarm displays, and

(5) graphic displays.

The display hierarchy is a sequence of different displays that take the operator from the general to the specific, for example, from a plant overview display to a group display showing one group or unit or individual loop. Figure 8-12 illustrates a typical display hierarchy.

The plant overview display includes the process area name, bar graphs of normalized deviation, color alarm blocks, list of alarms, and the date and time. The actual display configuration would depend on the manufacturer's specifications. From the overview display, no direct operator action can be taken; this would occur only upon selection of a more specific display, such as a loop in alarm. Other overview displays are obtained by a page-forward and page-back process. Figure 8-13 shows a typical plant overview display.

Group displays may contain individual control loops with tag name, bar graph of measured variable, output value, set point, process value in engineering units, set point source (remote, local, or tracking), output mode (auto, manual, track-

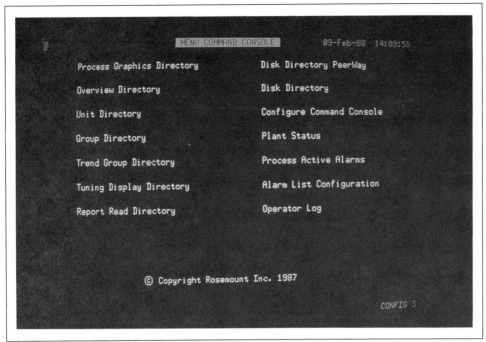

Figure 8-11. Typical Main Menu Display [Ref. 4]
(Courtesy of Rosemount, Inc.)

ing), and also alarm information. A page-forward and page-back technique can be used to select other groups. Figure 8-14 shows a typical group display.

Trend displays indicate the rate of change of key variables in a process and are important indicators of plant status. This information can help the operator identify or anticipate process upset conditions. The user can select a real-time or historical mode (from one second to one month, typically), and scale in both percent and engineering units for all variables and for the time base. Trends are often recorded on a hardcopy printer. Figure 8-15 illustrates a typical trend display. Table 8-2 lists the principal features to look for in trend displays.

Table 8-2.   Principal Features of Trend Displays

(1)	Sample frequencies available (one per sec, one per five sec, one per minute)
(2)	Number of variables that can be trended
(3)	Duration of trending periods available (one hour, one shift, one day, one week, one month, one year)
(4)	Amount of RAM and amount of hard disk space available for trend files
(5)	How sampling and duration are related and how they affect each other

Trending and reporting are methods of archiving information and require a method of backing up hard disk files. It is important for the operator to understand how this backup is accomplished and how the data can be retrieved (and reviewed) later on.

Alarm displays provide a list of alarms with their tags, types, descriptions, priorities, and acknowledgment status. Alarms are listed chronologically according to the time of detection of the alarm. The size of an alarm list is defined by the user and usually contains approximately 200 alarms. As the number of alarms exceeds this figure, the oldest alarms drop off. Dedicated keys on the keyboard

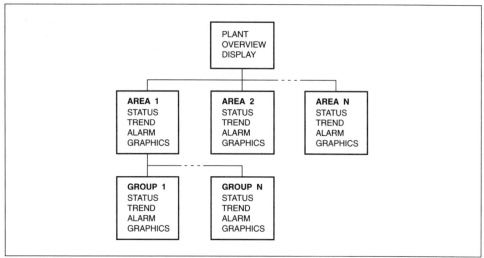

Figure 8-12. Typical Display Hierarchy

permit the operator to quickly identify active alarm conditions, to scroll up and down the alarm listings, and to acknowledge alarms.

The graphic display is a schematic of the process being controlled. The display is dynamic, giving the operator real-time data concerning the condition of the process. The display is user-definable and is constructed with a variety of geometric shapes, texts, and process control symbols. Graphic displays indicate loop tag names, measured values, output values, output modes, and engineering units. The operator can control directly from the graphic representation by means of animated symbols and color changes of things such as tank levels and process temperatures. Figure 8-16 shows a typical graphic display.

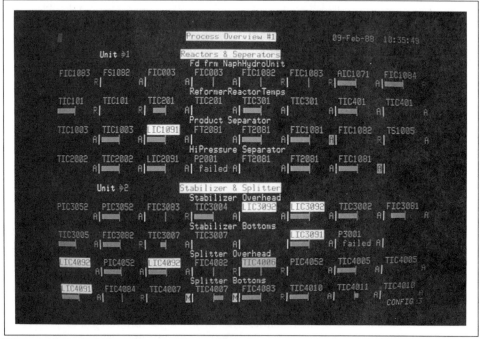

Figure 8-13. Typical Plant Overview Display [Ref. 4]
(Courtesy of Rosemount, Inc.)

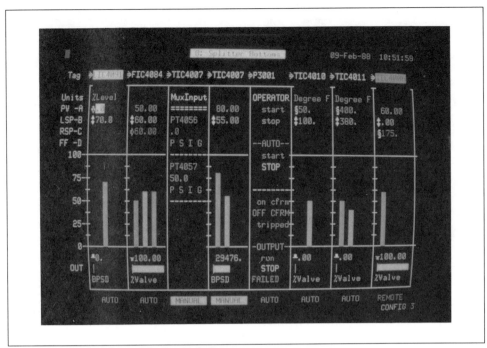

Figure 8-14. Typical Group Display [Ref. 4]
(Courtesy of Rosemount, Inc.)

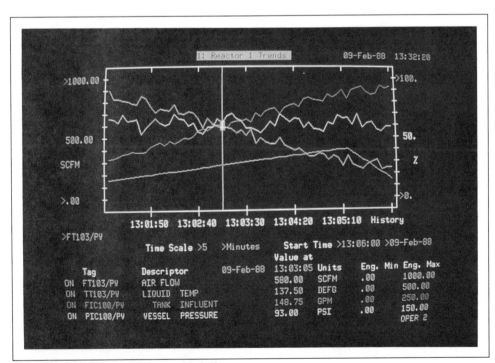

Figure 8-15. Typical Trend Display [Ref. 4]
(Courtesy of Rosemount, Inc.)

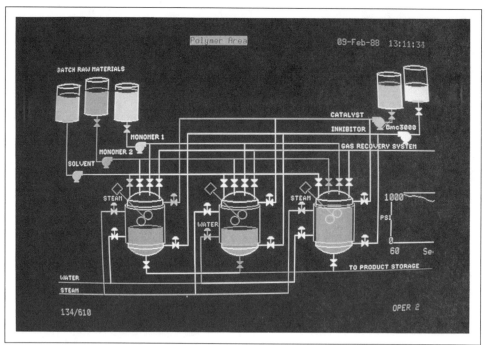

Figure 8-16. Typical Graphic Display [Ref. 4]
(Courtesy of Rosemount, Inc.)

Engineering Workstations

The engineering workstation is used principally to:

(1) configure the data base and console,

(2) update and decompile the data base, and

(3) implement application software.

Engineering workstations are usually physically separated from the location of the operator interface, which permits engineers to work independently, before, during, and after installation of the DCS.

Engineering workstations are often put together with the same hardware as the operator interface. They include CRTs, electronics, keyboards, hard disks, floppy disks, and/or tape drives. The key difference is the software. Thus, an engineering workstation could be used as an operator interface if it had the proper operator interface software.

The engineering workstation is typically used offline; however, with the right software it can be used online as a diagnostic tool. In general, however, the engineering and operator functions are separated, even by different physical locations of the workstations. This is to minimize any potential interference in the day-to-day running of the plant.

Basic DCS Software Modules

Programming Concepts

All computer systems need software (programming) to execute their assigned tasks. The DCS, which is highly computer-based, must be programmed with process information, control algorithms, and the operator interface instructions that are necessary for proper operation.

Some control functions can be preprogrammed by the system supplier, while other functions must be configured by the user's control system engineer. Computer-based control systems come with standard fill-in-the-blank software for data acquisition, process control, alarming, and operator displays.

Software can be classified as executive, system support, and application, even though descriptions from the manufacturers do not always seem to fit neatly into these categories.

Executive Software

Executive software is the operating system of the computer. It consist of programs that oversee or supervise the actual operation of the system while it is running. It performs such functions as:

(1) scheduling and starting the execution of system-application programs;

(2) allocating main memory and loading programs into main memory from bulk memory such as cassettes, drums, and floppy disks; and

(3) supervising I/O operations.

System Support Software

System support software consists of programs that aid the user in the development of application programs. These programs are generally vendor supplied. They include:

(1) computer language processors that translate high-level language programs such as BASIC, FORTRAN, and PASCAL into machine-language programs;

(2) editors, to facilitate the creation or modification of user written programs; and

(3) debugging aids (to find program errors).

Application Software

Application software consists of programs for tasks that are directly related to the primary functions of a system. Examples are reading analog or digital inputs into memory, computing control outputs based on input and set point values, and converting this information into engineering units.

Communications Software

The communications system facilitates the exchange of information between process control and information devices. Communications software is proprietary, even though there is a standard based on the International Standards Organization (ISO) Open System Interconnection (OSI). This standard provides for connection between one communication system and another using a standard protocol. A protocol is a set of conventions that govern the way in which devices communicate with each other.

The OSI reference model of seven layers of communications networks is an industry standard for linking intelligent devices in a distributed applications environment. The seven layers are described in Table 8-3.

Table 8-3.   Seven Layers of the OSI Reference Model

Layer 7	Application — Provides the interface for application to access the OSI environment
Layer 6	Presentation — Provides for data conversion to preserve the meaning of the data
Layer 5	Session — Provides user-to-user connections
Layer 4	Transport — Provides end-to-end reliability
Layer 3	Network — Provides routing of data through the network
Layer 2	Data link — Provides link access control and reliability
Layer 1	Physical — Provides an interface to the physical medium

System Configuration

System configuration, which includes data base and console configuration, is the main engineering function. Data base configuration is used to create, maintain, and document the system data base. It involves building, compiling, installing, and downloading the data base, and then updating it at runtime.

Console configuration is used to specify the contents and to define the portion of the data base assigned to each console. It involves:

(1) defining the console environments or scope,

(2) specifying the loops to appear on the operational displays,

(3) specifying the graphics to be associated with the operational displays, and

(4) specifying the users that are allowed access to each console.

The configuration process is organized in a hierarchical manner as exemplified in Figure 8-17.

Configuration techniques can vary widely from manufacturer to manufacturer. They can range from syntax-laden, line-by-line programming to small interactive logic blocks or large multifunction blocks. A block for one DCS vendor does not mean the same thing as a block for another. In the same way, a display page means different things to different suppliers. Some are prestructured while others are customized graphics.

Ideally, the vendor will the system configuration. The complexity of different configuration techniques can easily lead to extra engineering hours. Long-term costs may be incurred if one considers troubleshooting, modifications, and expansions. Examples of the specific tasks involved in configuration (for which an owner is well advised to enlist the help of the vendor) are indicated in Table 8-4.

Finally, note that an upgrade to the system software will likely involve modifications to the configuration. This can be almost as expensive as buying the original system. This topic will be discussed in more detail later in the chapter.

An owner should exercise caution if planning to do system configuration entirely on his or her own or through a third party.

Supplier Software Examples

As provided by a supplier, software does not always fit the neat categories given above. Examples from two industry suppliers, The Foxboro Company and ABB Kent-Taylor, are given below to illustrate this point.

No matter what the supplier's approach to software, it can be safely stated that there is no guarantee that it will be bug-free. As a rule, it will be more trouble-free the more it has been field-tested in actual installations. Generally speaking,

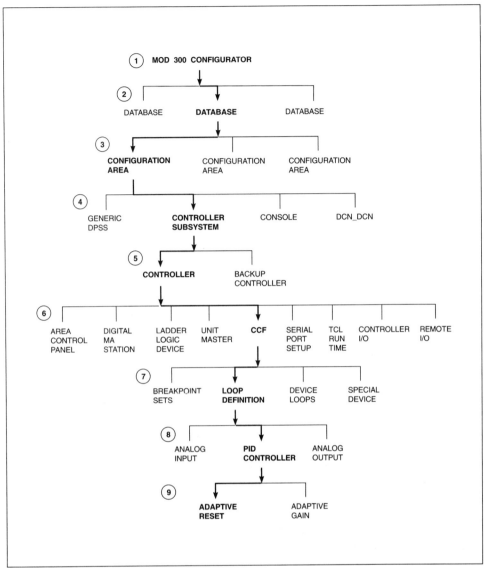

Figure 8-17. The Configuration Process [Ref. 2]
(Courtesy of ABB Kent-Taylor)

Table 8-4. Typical Configuration Tasks

Configure an input block.	Develop faceplates.
Do signal characterization.	Define groups (normally four or eight faceplates).
Define alarming parameters.	Define multitag displays (16, 32, 64...).
Set range/scale and engineering units.	Design dynamic graphics.
Define links to control blocks.	Define trend reports.
Configure control blocks with PID or other algorithms.	Define alarm and event logs.
Configure displays.	Define other modules.
Develop alarm banners.	

a single integrated package is better than a number of little packages. Another consideration is the consistency of operation in a software package. That is, are the same commands and techniques used from one application package to another. As always, the simpler the better.

Foxboro I/A Series

SYSTEM SUPPORT SOFTWARE

System support software is that the collection of programs necessary for the system to perform its basic functions.

Real-time display software enables a workstation to interact with any and all of the necessary real-time plant and process data that is available in the system.

Default display software allows immediate and complete access to an installed process control scheme through a series of automatically generated displays. The default display subsystem, without user configuration, constructs a series of interactive real-time displays that allow access to any control compound, block, or block parameter available to the system.

Trend subsystem software is used to preformat real-time and historical trend displays.

Alarm subsystem software manages the initialization, configuration, generation, display, and actions of alarm messages within a workstation.

System management software obtains current and historical information about the system, displays it, and allows a user to intervene in system operation and perform diagnostics.

Electronic documentation software permits self-documenting in electronic format for the for the I/A series of systems.

System configurator software is used to specify the system network and packaging and to document the system.

Real-time display configurator software is used for the displays builder and display configurator, which are used to build display templates and configure dynamic display objects.

Graphics utilities software contains a set of general-purpose editors and tools for the construction and editing of Foxboro-format fonts and markers used in the real-time display Configurator.

APPLICATION SOFTWARE

Application software consists of the various modules of integrated control software that can be used independently or combined.

Continuous control software uses blocks with predefined algorithms from which the user can select, organize, and configure in order to perform a specific control task.

Sequential control software complements the continuous and ladder logic domains by providing sequencing capabilities.

Ladder logic software provides on-off control, counting, and timing capabilities.

Taylor MOD 300

CONFIGURABLE CONTROL FUNCTIONS

This software consists of modules for the functions of continuous process control, discrete device control, sequential process control, standard operational displays, and alarm detection. There are three main classes of modules.

Loop class modules perform a specific control function and can be used to define control, indication, and calculation loops.

Function class modules are really a subset of the loop class modules. They are used to perform a specific function, such as combining analog input s, PID algorithms, and analog outputs, to form a control loop.

Device loop modules are used to control discrete devices such as solenoid valves and motors.

TAYLOR CONTROL LANGUAGE (TCL)

TCL is a real-time, high-level language used to write process control programs, including:

(1)   sequential/batch control,

(2)   complex arithmetic and logic functions, and

(3)   supervisory tasks such as start-ups and shutdowns.

TCL is intended especially for translating process requirements in to a structured executable program. Table 8-5 indicates the usual computer programming language abilities featured in TCL.

Table 8-5.   Computer Programming Operations of TCL

Type	Functions
Arithmetic	ADD, SUBTRACT, MULTIPLY, DIVIDE
Relational	EQUAL TO, NOT EQUAL TO, LESS THAN, GREATER THAN
Logical	NOT, AND, OR
Conditional	IF, THEN, ELSE
Looping	FOR, TO, REPEAT, GOTO
Others	MATHEMATICAL, TRIGONOMETRIC, STRING MANIPULATION, DATE/TIME

TCL also has a number of capabilities that are related specifically to process control:

(1)   Control block parameters associated with dynamic program control (for example, state, status, and mode)

(2)   Program flow statements that include conditional statements and looping statements

(3)   Peripheral I/O statements that provide a means of data exchange between the TCL program and I/O ports

(4)   Data base manipulation and process control used to read and write data base items

(5)   Taylor Ladder Logic (TLL), which is used for ladder logic functions such as relay, timer, and counter manipulation

Installation

Installation of a DCS involves physical location of components, environmental conditioning, power distribution, and wiring.

Physical Location

The components of a DCS may or may not be widely distributed geographically throughout a plant. Variations from one project to another are due to the type of plant involved and the philosophies of plant operations. One possibility for the distribution of DCS equipment is indicated in Table 8-6.

Table 8-6. Typical Distribution of DCS Equipment

Control Room	Engineering Room	Computer Room
Operator consoles	Engineering workstations	Cabinets containing: - process I/O cards - controllers - communications modules
Operator keyboards	Printers	
Printers		
Video copiers		

The basic intent of control room design is to provide optimal equipment operation as well as personal comfort for the operator. Future expansion is also an important consideration in the layout of a control room. Figure 8-18 shows a typical control room and computer room layout.

Vendors provide the equipment dimensions and clearances that are essential in planning equipment layouts or the movement of equipment through doorways and halls. Figure 8-19 shows typical examples of the kind of information supplied by manufacturers.

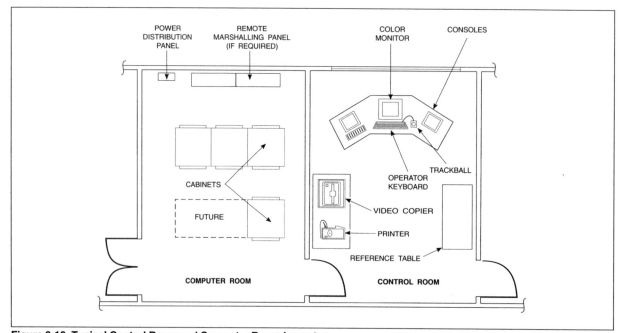

Figure 8-18. Typical Control Room and Computer Room Layout

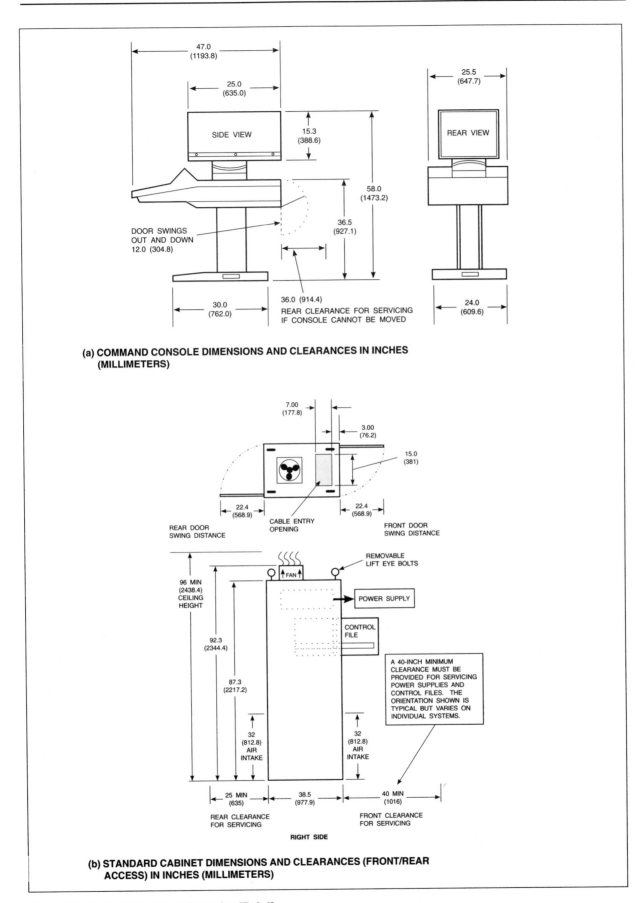

(a) COMMAND CONSOLE DIMENSIONS AND CLEARANCES IN INCHES (MILLIMETERS)

(b) STANDARD CABINET DIMENSIONS AND CLEARANCES (FRONT/REAR ACCESS) IN INCHES (MILLIMETERS)

Figure 8-19. Typical Clearance Information [Ref. 4]
(Courtesy of Rosemount, Inc.)

Lighting is an important aspect of control room design. Recommendations from manufacturers include the following:

(1)   Use indirect or recessed incandescent lighting fixtures with diffusion lenses to prevent glare and to ensure uniform illumination.

(2)   Provide illumination levels of approximately 420 lumens per square meter where CRTs are operated.

(3)   Provide illumination levels of approximately 650 lumens per square meter where CRTs are not operated.

(4)   Where variable lighting intensity is desirable or required, consider task lighting and/or lighting controls, such as dimmers.

Emergency lighting is needed so that during a power failure essential functions are not shut down due to a lack of sufficient illumination. Emergency lighting can be powered from a battery backup system or an emergency generator.

Take vibration into account if the control room is adjacent to large machinery such as shakers or presses. To protect sensitive DCS equipment, shock absorbers or isolation pads must be installed on the offending machinery. Th e operational and the static vibration limits are specified by the DCS equipment manufacturer.

Environmental Conditioning

Environmental controls maintain both optimal equipment operation a nd personal comfort. Temperature, humidity, and air filtration must be considered. Temperature is generally maintained between 65°F (18°C) and 75°F (24°C). Relative humidity should be around 50%. In addition to contributing to optimal equipment operation and personal comfort, proper environmental conditioning also helps reduce the buildup of static electricity charges. See Table 8-7.

Table 8-7.   Typical Equipment Environmental Characteristics

Characteristics	Controller Subsystem Signal Conditioning Card File and Terminations	Data Processor Subsystem (with Disk Drive and Printers)	All Other Subsystems and Interfaces (without Printers or Disk Drives)
Operating temperature	32 - 122°F (0 - 50°C)	50 - 104°F (10 - 40°C)	32 - 104°F (0 - 40°C)
Storage temperature	−40 - 140°F (−20 - 60°C)	−40 - 140°F (−20 - 60°C)	−40 - 140°F (−20 - 60°C)
Relative Humidity	10 - 90%	20 - 80%	10 - 90%
Maximum Wet Bulb Temperature	90% RH at 90°F (32°C)	90% RH at 80°F (26.7°C)	90% RH at 90°F (32°C)

To determine the amount of air conditioning in the control room, one must consider the internal and external factors that contribute to heat gains and losses in the control room area. Internal factors include equipment, lighting, and people — all of which generate heat. External factors include outside temperature, exposure to sun, and wind. If the control room is located inside a building, generally only internal factors will be taken into consideration.

To reduce the risk of exterior airborne contaminants causing damage to electrical components or disk and tape drives, the control room is maintained under a positive pressure of approximately 25 pascals with respect to the control room exterior.

Different levels of corrosion protection are given in ANSI/ISA-S71.04, Environmental Conditions for Process Measurement and Control Systems: Airborne Contaminants.

Power Source

POWER SOURCE QUALITY

Depending on the country in which a DCS is installed, system equipment will operate on 115-230 V AC, 50 to 60 Hz, single phase. Dedicated power supplies mounted in the system cabinetry supply DC power for the actual system components. Manufacturers specify the quality of power needed for acceptable operation of their equipment by means of parameters such as:

(1) range of voltage variation from the nominal value,

(2) range of frequency variation from the nominal value, and

(3) harmonic distortion.

The total AC power requirements can be calculated by adding up the individual power consumption values for each device. These values are specified by the manufacturer.

Table 8-8 shows the power consumption and cooling requirements for a typical system.

As an option, some maunfacturers offer a backup power arrangement. When the primary AC power fails or drops below approximately 15% of the nominal voltage level, internal batteries supply DC power to the system for a predetermined period of time. The AC power is reconnected when the level returns to its nominal value. A simple battery backup does not regulate either frequency or voltage when the system works from primary AC power. This would be done if the backup power is provided by an uninterruptible power supply.

Since the power supply is such a critical element, installation planning may include uninterruptible power supplies (UPS) to ensure that vendor's equipment specifications are met.

GROUNDING SYSTEM

To minimize the effects of electrical noise (steady-state or transient) caused by large electrical equipment, large relays, and motor contactors, the primary AC power source must be stable and noise-free and possibly even completely independent. In addition, an adequate grounding system is needed to avoid conditions that lead to equipment operability problems. Such conditions include electrical noise caused by ground currents circulating through the system and by static electricity discharges

The grounding system should terminate at common grounding electrodes (ground rods) and have a resistance to earth of one ohm or less when the system grounding is used in combination with power generation equipment. When it is not, five ohms or less is recommended. Grounding systems require periodic inspection and testing. Figure 8-20 illustrates a typical grounding installation.

The grounding system will also minimize the hazard of electrical shock to personnel.

Wiring

CABLING AND TERMINATIONS

Wiring coming from field equipment may terminate directly at cabinets that contain the I/O cards or at conveniently located remote marshalling panels. The remote marshalling panel is simply an electrical box that contains terminal blocks

Table 8-8.   Typical AC Power Consumption and Cooling Load

Equipment Description	AC Power Draw (V A, Typical)	Cooling Load (Btu/h, Typical)
Console subsystem with 2 color monitors and keyboards	900	3200
Data processor or gateway subsystems	1000	3200
Controller subsystem (3 fully loaded card files)	900	3105
SC controller subsystem (3 fully loaded card files)	1464	4850
Multi-bus I/O system (fully loaded)	1600	4000
19-in. Color monitor without touch screen	80	276
19-in. Color monitor with touch screen	95	300
13-in. Color monitor	70	250
System maintenance terminal	60	110
Black and white or multi-color printer		
operating	138	431
standby	55	170
Video copier		
operating	400	1050
standby	150	470
Video processor	120	410
Video multiplexer (operating)	17	52
Systems terminal printer	120	410
Workstation processor	820	2830
Local control panel	42	150
Remote hardened console	95	300

and is designed to facilitate wiring between field junction boxes and I/O cabinets. Wiring cabinets have cable entry openings at top and bottom. This facilitates computer room design when there is a raised floor or an elevated cable tray.

Figure 8-21 illustrates typical cable routing inside a cabinet, and Figure 8-22 illustrates analog inputs and outputs and field wire termination.

TYPE OF CABLE

Cables recommended for 120 V AC service would have the following characteristics:

(1) Conductors: seven-strand, class B, 14 AWG

(2) Ground conductor: single, seven-strand, class B, concentric bare

(3) Inner jacket: high-temperature PVC

(4) Outer jacket: flame-retardant PVC

(5) Armor: interlocking aluminum alloy or galvanized steel

Single-pair or triad cables recommended for 24 V DC and 4-20 mA service would have the following characteristics:

(1) Conductors: seven-strand concentric bare copper, class B, 16 AWG

(2) Shield: aluminum/Mylar tape shield with tinned copper drain wire

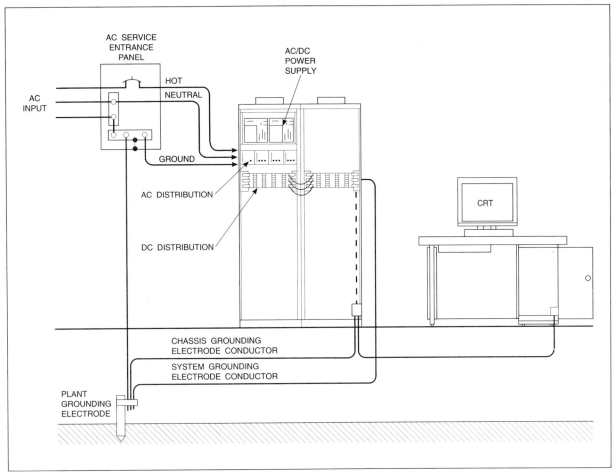

Figure 8-20. Typical Grounding Installation [Ref. 4]
(Courtesy of Rosemount, Inc.)

(3) Lay of twist: two inches nominal

(4) Jacket: FR PVC

Multiple-pair or multiple-triad cables recommended for 24 V DC and 4-20 mA service would have the following characteristics:

(1) Conductors: seven-strand concentric bare copper, class B, 20 AWG

(2) Shield: aluminum/Mylar tape shield with tinned copper drain wire, individual and overall shield (4-20 mA analog signals), overall shield (24 V DC and on-off signals)

(3) Lay of twist: two inches nominal

(4) Jacket: FR PVC

The choice of armored or nonarmored cable is up to the user, but generally armored cable is the safe choice.

Intrinsically Safe Barriers

An intrinsically safe barrier has terminals for connecting field and control room wiring, thus minimizing the number of panel terminals. Also, in the barrier system, a sharp line of demarcation between the hazardous area and the safe area is provided. The barrier is a completely passive device that requires no power

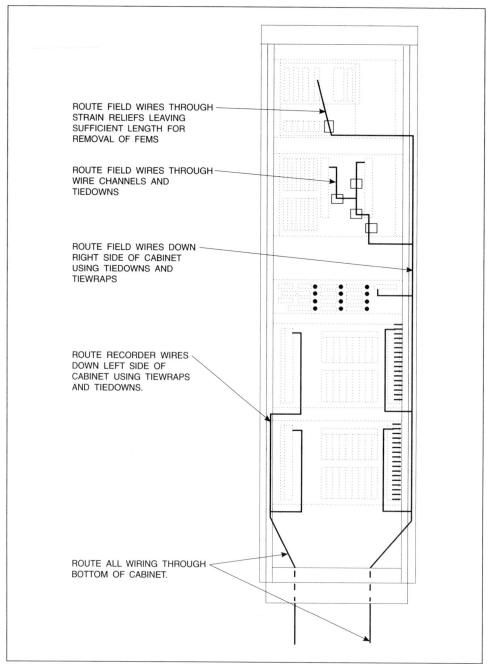

ROUTE FIELD WIRES THROUGH
STRAIN RELIEFS LEAVING
SUFFICIENT LENGTH FOR
REMOVAL OF FEMS

ROUTE FIELD WIRES THROUGH
WIRE CHANNELS AND
TIEDOWNS

ROUTE FIELD WIRES DOWN
RIGHT SIDE OF CABINET
USING TIEDOWNS AND
TIEWRAPS

ROUTE RECORDER WIRES
DOWN LEFT SIDE OF
CABINET USING TIEWRAPS
AND TIEDOWNS.

ROUTE ALL WIRING THROUGH
BOTTOM OF CABINET.

Figure 8-21. Typical Cable Routing in a Wiring Cabinet [Ref. 4]
(Courtesy of Rosemount, Inc.)

source and passes a 4-20 mA DC signal at a nominal 24 V DC power rating with virtually no degradation (less than 0.1%).

In the barrier are wire-wound resistors for current limiting and redundant Zener diodes for limiting the voltage. A fuse is in series with the resistors. The question often arises why a fuse alone would not do the job. If a high voltage were placed on the input terminals with only a fuse and resistors present (without diode), an explosion could occur in the hazardous area before the fuse could blow. A fuse can pass sufficient energy to cause the explosion before it opens. Only microseconds are needed, under the right conditions, for an explosion to take place. Without some method of voltage limiting, a fault voltage appearing on

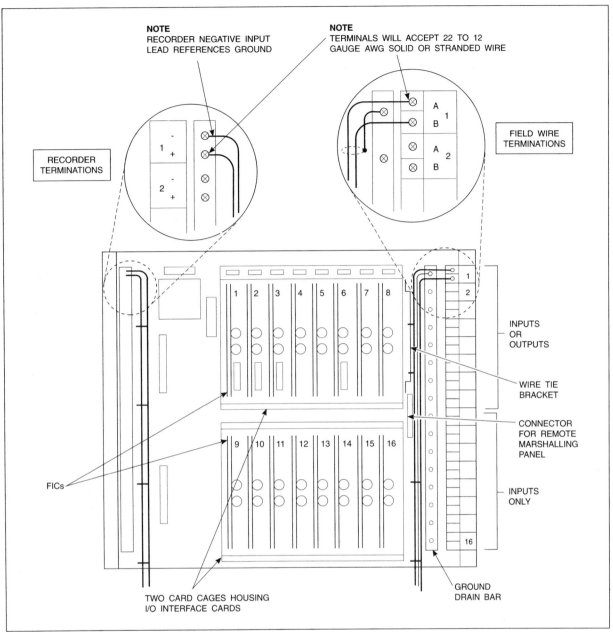

NOTE
RECORDER NEGATIVE INPUT
LEAD REFERENCES GROUND

NOTE
TERMINALS WILL ACCEPT 22 TO 12
GAUGE AWG SOLID OR STRANDED WIRE

FIELD WIRE
TERMINATIONS

RECORDER
TERMINATIONS

INPUTS
OR
OUTPUTS

WIRE TIE
BRACKET

CONNECTOR
FOR REMOTE
MARSHALLING
PANEL

INPUTS
ONLY

FICs

TWO CARD CAGES HOUSING
I/O INTERFACE CARDS

GROUND
DRAIN BAR

Figure 8-22. Typical Field Termination of AIs and AOs [Ref. 4]
(Courtesy of Rosemount, Inc.)

the barrier input terminals, together with increased current, could be passed along
to the hazardous area.

System Checkout and Site Power-Up

This activity is almost always the exclusive domain of the DCS vendor because it is associated with the warranty of the system. An owner should always
be sure that these two functions are included in the vendor's price. The owner
may not have the expertise to do it and certainly does not want to pay the exhor-
bitant price asked by some vendors after the fact. It is clearly a good idea for the
system checkout and power-up to be included in the vendor's original price and
quoted under competitive conditions. There should also be a payment retention
(ranging from 5 to 20%) payable upon system acceptance; this ensures that the

system is up and running before the final payment is made. Experienced owners know that "money talks."

Typical System Layouts

Although the approach of this chapter is to look at distributed control sytems in a generic way, each vendor tends to have a unique and proprietary approach. This has already been illustrated with regard to software. Figures 8-23, 8-24, 8-25 and 8-26 show how how the overall structure can vary among four different manufacturers.

It would be of little consequence to analyze each structure for its various strengths and weaknesses because each is generally the result of the vendor's proprietary approach to DCS. For the purposes of this chapter, it can be simply stated that there are more similarities than there are differences.

Start-Up Services

DCS start-up services may be easily overlooked in the evaluation of a vendor's system because the buyer doesn't readily see the importance of them at the time of purchase. Usually, it is only after installation that the tremendous importance of things such as staging, functional testing, and training become apparent.

STAGING

This is not a well-known service, especially to first-time buyers. Depending on an owner's experience, it could be seen either as a waste of time or as a valuable tool that is instrumental to an efficient, on-time start-up. Staging may be defined as a type of stress or operational test aimed primarily at DCS hardware.

It is an activity in which the entire DCS is completely assembled and cable connected to duplicate the eventual site installation. After powering up with only the

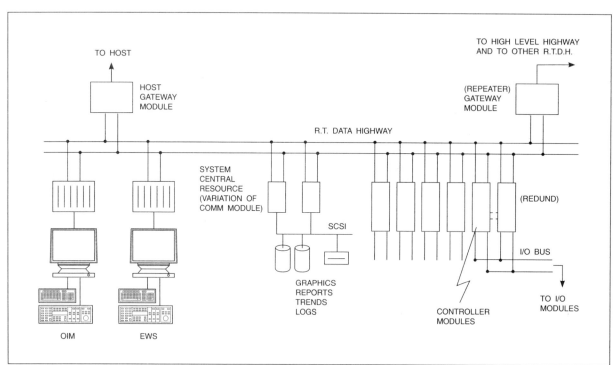

Figure 8-23. DCS Structure Typical of Foxboro

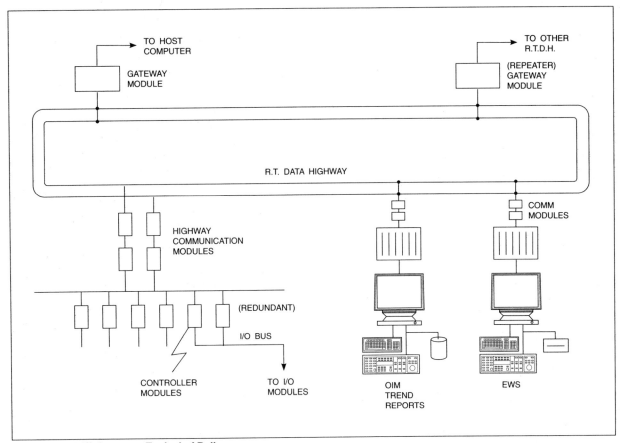

Figure 8-24. DCS Structure Typical of Bailey

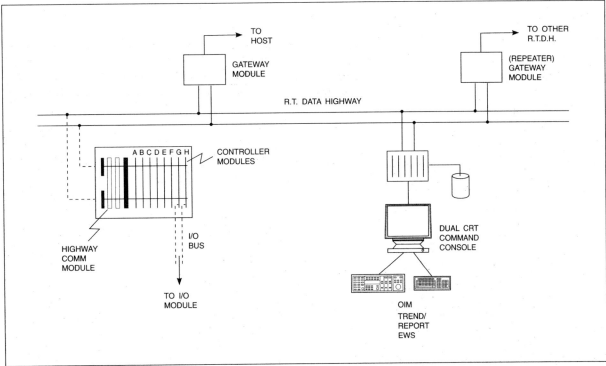

Figure 8-25. DCS Structure Typical of Rosemount

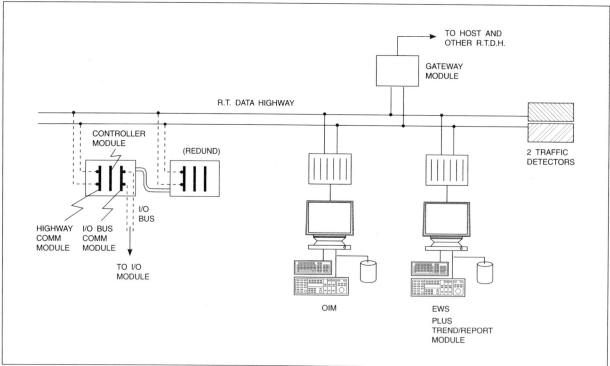

Figure 8-26. DCS Structure

system software programs running, the system is expected to operate without problems for a given period of time (usually 48 hours or more). Some DCS vendors will further stress test the system in an overheated environment, cycling the AC power supply and increasing the surrounding temperature to 122°F (50°C) or more.

Stress testing is done to eliminate the infant mortality associated with computer components. Infant mortality is the expression used to describe the tendency for a component that is destined to fail to do so very early in its life. The rate of component failure will drop to a low level for a long period of time and then increase again as the aging process takes over. As shown in Figure 8-27, the curve expressing this process graphically takes on the shape of a bathtub.

The time required for staging is considered time very well spent, even (especially) when a project is on a tight schedule. On-site component failures during start-up can cause frustrating delays that can be so readily avoided by the staging process. In addition, it ensures that there are no missing cables, connections, or system components when the DCS arrives at site.

SYSTEM FUNCTIONAL TEST

The system functional test is similar to staging except that its focus is on the configuration rather than the DCS hardware. This service involves assembling and powering up the system with the system software programs and also includes the configuration software. The purpose of this test is to eliminate problems in the loops and logic of the particular application. Inputs and outputs are verified and simulated and logic interlocks and loops are flexed.

Generally speaking, this service is best performed offsite, away from the distractions of the plant. The time required for a functional test should be two or three weeks for everything from assembly through testing, recrating, and reshipping. The advantage of the system functional test is that it allows the vendor's technicians to identify and correct problems under conditions that are as close as

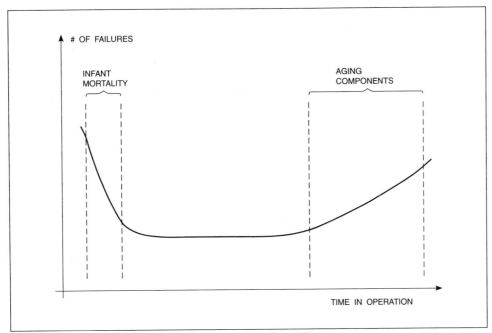

Figure 8-27. Mortality of DCS Components as a Function of Time

possible to ideal. Troubleshooting on the job is difficult at best. Besides the distractions and the pressure to perform, it is often more complicated to isolate the source of a problem. In other words, is it the DCS configuration or is it the field equipment?

TRAINING

Every manager wants his staff properly trained in running the computer system as well as diagnosing and repairing it. Most project specifications call for some combination of off- and on-site training. Off-site training is usually done in a formal classroom setting by instructors who are often, but not always, full-time professionals. Ideally, the classroom instruction is supplemented by hands-on laboratory work.

On-site training has the advantage that it is done with the actual equipment that will be operated. On the other hand, the instructor is often an installation technician who, although knowledgable about the system, can be greatly lacking in teaching ability. In addition, on-site training is frequently a cram session in which too much information is passed on in too short a time for the operator to really absorb it effectively.

Because the importance of training is often undervalued, what little is done is often not of much use. A proper training program can involve the development of quite an elaborate master plan. A good one takes into account the actual and required skill levels of the people who will have to deal with the DCS. It will include all the training programs necessary to take individuals from their actual to their future skill levels. This learning is best done over an extended period of time so that it is effective and, therefore, useful. Training courses that are part of the master plan may come from any number of sources that include vendors, local colleges, owner in-house programs, and so on.

Finally, training should be seen as an ongoing process with refresher courses a part of the overall plan. The world of DCS is closely connected to computers, and therefore, is evolving at breakneck speed.

System Documentation

System documentation consists of the collection of written (and sometimes video) materials that describe the DCS. Ideally, it must be complete enough to tell an owner everything that needs to be known about the DCS, and, at the same time, it must be as easy to use as picking up a phone and calling the vendor. Anyone who has used a personal computer and any assortment of available software knows how difficult it is to end up with "ideal" documentation. On the other hand, a DCS is a complex thing, so complete documentation will necessarily be complex to use. Figure 8-28 provides an overview of one supplier's set of DCS documentation.

Maintenance

Placing an order for a DCS signals the start of an owner's concern with a number of factors involved in the satisfactory operation of the system. These include: site preparation, mounting of hardware, installation time, staging, documentation, commissioning, training, and finally, maintenance.

Although the design of today's modern digital circuits allows for minimal maintenance and a high degree of self-monitoring, an owner still needs to know how they work and how they should be maintained.

Flexible Level of Support Programs

Historically, the level of maintenance provided is determined by what amount will be most effective in providing satisfactory operation at a reasonable cost. The strategy for supplying effective maintenance has to be worked out by both the user and the manufacturer. Manufacturers are generally quite willing to work closely with customers to help out with things that are unfamiliar.

For smaller systems, many owners are interested in doing as much as possible with their own resources. This is to be encouraged as long as it does not result in an operational hazard or equipment damage. When it comes to complex maintenance activities (those that require specially trained service personnel), the owner should consider a higher level of maintenance support. Choosing a combination of support services that best fits the needs of the owner will result in lower product costs and higher quality. Since a DCS can be considered a production tool, its real value depends on how effectively it is used. A well-designed maintenance program adds to the effectiveness.

As described in Figure 8-29, a reactive maintenance philosophy of replacement and repair shortens a systems life. A more proactive approach of preventive and predictive maintenance extends the life of the system. However, the payoff occurs when various operational services and system upgrades are provided. System life is extended almost indefinitely and the DCS provides greater functional value.

Categories of Maintenance

Every organization has its own maintenance philosophy, and it is up to management to determine what overall approach to maintenance is needed for acceptable reliability of the DCS.

Maintenance can be seen as falling into three categories:

(1) Enhancement maintenance

(2) Preventive maintenance

(3) Corrective maintenance

TDC 3000 BOOKSET DIRECTORY
Section 2

The dates in the Revision Date column are the publication's revision date at the time this Information Directory was printed. Publications with the same publication number but with the same or a later date are valid, those with an earlier date are invalid. *When you receive revised versions of these publications or Document Change Notices for them, we strongly urge you to mark the new revision date in this section, and replace the publication or changed pages.*

BINDER TDC 1010: SYSTEM SUMMARY

Tab Publication Title	Publication Number	Revision Date
Read Me First Read Me First	SW01-310	8/90
Information Directory Information Directory	SW01-300	7/90
System Overview System Overview	SW70-300	*
Specification and Technical Data 1 System Technical Data 2 Application Module Specification and Technical Data 3 Computer Gateway Specification and Technical Data 4 CM50S Specification and Technical Data 5 Computing Module 60 Specification and Technical Data 6 History Module Specification and Technical Data 7 Hiway Gateway Specification and Technical Data 8 Local Control Network Specification and Technical Data 9 Processor Gateway Technical Data 10 Universal Station Specification and Technical Data 11 Universal Work Station Specification & Technical Data	SW03-300 AM03-300 CG03-300 CM03-150 CM03-301 HM03-300 HG03-300 LC03-300 PG03-300 US03-300 UW03-300	* * * 9/88 * * * * 5/90 * *

BINDER TDC 1020: LCN SITE PLANNING & INSTALLATION

Tab Publication Title	Publication Number	Revision Date
Planning 1 LCN Site Planning Manual 2 LCN Guidelines - Implem., Troubleshooting, Service	SW02-300 LC09-310	7/90 6/90
Installation LCN System Installation Manual	SW20-300	1/90
Checkout LCN System Checkout Manual	SW20-310	2/90

* Not yet available as of the date of this publication

Figure 8-28. Typical DCS Documentation [Ref. 5]
(Courtesy of Honeywell, Inc.)

BINDER TDC 1030: IMPLEMENTATION/START-UP & RECONFIGURATION-1

Tab Publication Title	Publication Number	Revision Date
Overview Implementation Overview	SW09-360	7/90
Startup 1 System Startup Guide—Floppy Drives 2 System Startup Guide—Cartridge Drive	SW11-303 SW11-304	6/90 6/90
Messages Messages Directory	SW09-307	4/90
Network 1 Network Form Instructions 2 Network Data Entry	SW12-305 SW11-3-5	4/90 5/90
[In Pocket] Engineer's Digest	SW09-306	1/90

BINDER TDC 1030: IMPLEMENTATION/START-UP & RECONFIGURATION-2

Tab Publication Title	Publication Number	Revision Date
Reference 1 Engineer's Reference Manual 2 Configuration Data Collection Guide 3 System Control Functions	SW09-305 SW12-300 SW09-301	6/90 * 6/90

BINDER TDC 1031: IMPLEMENTATION/CONFIGURATION FORMS

Tab Publication Title	Publication Number	Revision Date
Forms 1 Network Forms 2 Data Hiway, Box/Slot, and Data Point Forms 3 Logic Block Forms 4 Hiway Gateway Library Forms 5 History Module History Group Forms 6 Area Forms 7 Picture Editor Forms 8 Button Configuration Forms 9 Free Format Log Forms 10 Application Module Forms 11 Computer Gateway Forms	SW88-305 HG88-300 MC88-300 PC88-345 HM88-300 SW88-380 SW88-350 SW88-370 HM88-360 AM88-300 CG88-300	4/90 3/90 5/90 6/90 5/90 3/90 11/89 1/90 1/90 6/90 8/90

* Not yet available as of the date of this publication

Figure 8-28 (continued).Typical DCS Documentation [Ref. 5]
(Courtesy of Honeywell, Inc.)

BINDER TDC 1032: IMPLEMENTATION/ENGINEERING OPERATIONS-1

Tab Publication Title	Publication Number	Revision Date
Text Editor Text Editor Operation	SW11-306	1/90
Utilities Utilities Operation	SW11-307	2/90
Data Entity Building Data Entity Builder Manual	SW11-311	5/90
HM History Groups HM History Group Form Instructions	HM12-300	5/90

BINDER TDC 1033: IMPLEMENTATION/ENGINEERING OPERATIONS-2

Tab Publication Title	Publication Number	Revision Date
Area Area Form Instructions	SW12-380	5/90
Picture Building 1 Picture Editor Form Instructions 2 Picture Editor Data Entry 3 Picture Editor Reference Manual	SW12-350 SW12-350 SW12-350	1/90 1/90 12/89
Button Configuration 1 Button Configuration Form Instructions 2 Button Configuration Data Entry	SW12-370 SW12-370	1/90 1/90
Free Format Logs 1 Free Format Log Form Instructions 2 Free Format Log Data Entry	HM12-360 HM11-360	1/90 1/90

Figure 8-28 (continued). Typical DCS Documentation [Ref. 5]
(Courtesy of Honeywell, Inc.)

523

BINDER TDC 1034: IMPLEMENTATION/HIWAY GATEWAY-1

Tab Publication Title	Publication Number	Revision Date
Control Functions 1 HG Implementation Guidelines 2 HG Control Functions	HG12-310 HG09-301	5/90 3/90
Data Points Data Hiway, Box/Slot, and Data Point Form Instructions	HG12-300	5/90
Parameters HG Parameter Reference Dictionary	HG09-340	3/90
Logic Blocks 1 Logic Block Form Instructions 2 Logic Block Data Entry	MC12-300 MC11-300	5/90 4/90

BINDER TDC 1034: IMPLEMENTATION/HIWAY GATEWAY-2

Tab Publication Title	Publication Number	Revision Date
Control Language/MC 1 Control Language/MC Data Entry 2 Control Language/MC Reference Manual	PC11-385 PC27-310	1/90 7/90

BINDER TDC 1035: IMPLEMENTATION/AM-1

Tab Publication Title	Publication Number	Revision Date
Control Functions AM Control Functions	AM09-302	2/90
Data Points AM Form Instructions	AM12-300	6/90
Parameters 1 AM Parameter Reference Dictionary 2 AM Implementation Guidelines	AM09-340 AM12-310	3/90 5/90
Algorithms AM Algorithm Engineering Data	AM09-301	5/90

Figure 8-28 (continued). Typical DCS Documentation [Ref. 5]
(Courtesy of Honeywell, Inc.)

BINDER TDC 1035: IMPLEMENTATION/AM-2

Tab Publication Title	Publication Number	Revision Date
Control Language/AM		
1 Control Language/AM Overview	SW27-300	1/90
2 Control Language/AM Reference Manual	AM27-310	1/90
3 Control Language/AM Data Entry	AM11-385	1/90

BINDER TDC 1036: IMPLEMENTATION/CM60

Tab Publication Title	Publication Number	Revision Date
CM60		
1 CG Form Instructions	CG12-300	3/90
2 CG Parameter Reference Dictionary	CG09-340	3/90
3 CM60 User Manual	CM27-310	6/90
Pascal		
CM60 Pascal Language Manual	CM27-330	1/89

BINDER TDC 1037: IMPLEMENTATION/COMPUTER GATEWAY

Tab Publication Title	Publication Number	Revision Date
Computer Gateway		
1 CG Form Instructions	CG12-300	3/90
2 CG Parameter Reference Dictionary	CG09-340	3/90
3 CG User Manual[1]	CG11-310	6/90

BINDER TDC 1038: IMPLEMENTATION/PROCESSOR GATEWAY

Tab Publication Title	Publication Number	Revision Date
Processor Gateway		
1 CG Form Instructions	CG12-300	3/90
2 CG Parameter Reference Dictionary	CG09-340	3/90
3 Processor Gateway User Manual	PG11-310	*

* Not yet available as of the date of this publication

[1] If interface is with a Bull DPS 6, Honeywell 45000, or DEC VAX computer, refer to the *CM60 User Manual, Processor Gateway User Manual,* or *CM50S User Manual,* respectively.

Figure 8-28 (continued). Typical DCS Documentation [Ref. 5]
(Courtesy of Honeywell, Inc.)

BINDER TDC 1039: IMPLEMENTATION/CM50S

Tab Publication Title	Publication Number	Revision Date
CM50S		
1 CG Form Instructions	CG12-300	3/90
2 CG Parameter Reference Dictionary	CG09-340	3/90
3 CM50S User Manual	CM11-310	6/90

BINDER TDC 1040: IMPLEMENTATION/PROCESS MANAGER-1

Tab Publication Title	Publication Number	Revision Date
Overview		
Process Manager Specification and Technical Data	UC03-300	3/90
Planning		
1 Process Manager Site Planning	HN02-300	5/90
2 Process Manager Implementation Guidelines	PM12-300	6/90
Installation		
1 Process Manager Installation	HN20-300	4/90
2 Process Manager Checkout	HN20-310	1/90
[In Pocket]		
Process Manager Parameters Pocket Guide	MG09-341	5/90

BINDER TDC 1040: IMPLEMENTATION/PROCESS MANAGER-2

Tab Publication Title	Publication Number	Revision Date
Control Functions & Algorithms		
Process Manager Control Functions and Algorithms	UC09-300	3/90
Forms		
Process Manager Configuration Forms	UC88-300	4/90
Parameters		
Process Manager Parameter Reference Dictionary	MG09-340	2/90
Control Language/PM		
1 Control Language/Process Manager Overview	UC27-300	*
2 Control Language/Process Manager Reference	UC27-310	6/90
3 Control Language/Process Manager Data Entry	UC11-300	1/90

* Not yet available as of the date of this publication

Figure 8-28 (continued). Typical DCS Documentation [Ref. 5]
(Courtesy of Honeywell, Inc.)

BINDER TDC 1041: UCN SITE PLANNING & INSTALLATION

Tab Publication Title	Publication Number	Revision Date
Overview UCN Specification and Technical Data	UN03-300	6/90
Planning & Installation UCN Site Planning and Installation	UN02-300	2/90

BINDER TDC 1050: PROCESS OPERATIONS

Tab Publication Title	Publication Number	Revision Date
Process Operations Process Operations Manual	SW11-301	8/90
[In Pocket] Operator's Digest	SW11-315	*

BINDER TDC 1060: SERVICE

Tab Publication Title	Publication Number	Revision Date
System Service 1 System Maintenance Guide 2 Maintenance Test Operations	SW13-300 SW11-302	7/90 7/90
Module Service 1 Universal Station Service 2 History Module Service 3 Five/Ten-Slot Module Service	US13-300 HM13-300 LC13-300	6/90 6/90 6/90
LCN Test Programs 1 Test System Executive 2 Hardware Verification Test System 3 Core Module Test System 4 LCNI Network Communications Test	SW13-210 SW13-211 SW13-212 SW13-208	2/90 7/90 7/90 7/90

* Not yet available as of the date of this publication

Figure 8-28 (continued). Typical DCS Documentation [Ref. 5]
(Courtesy of Honeywell, Inc.)

BINDER TDC 1061: PM SERVICE

Tab Publication Title	Publication Number	Revision Date
PM Service Process Manager Service	HN13-300	6/90
PM Test Programs 1 Manager Module Test System (MMTS) 2 Process Manager Test System (PMTS) 3 Process Manager Test Executive (PMEX)	UC13-200 UC13-210 UC13-220	7/90 3/90 7/90

BINDER TDC 1070: IMPLEMENTATION/LOGIC MANAGER

Tab Publication Title	Publication Number	Revision Date
Overview Logic Manager Specification and Technical Data	LM03-300	6/90
Planning & Installation 1 Logic Manager Site Planning 2 Logic Manager Installation 3 Logic Manager Implementation Guidelines	LM02-300 LM20-300 LM12-300	7/90 7/90 6/90
Control Functions 1 Logic Manager Control Functions 2 Logic Manager Parameter Reference Dictionary 3 Logic Manager Forms	LM09-300 LM09-340 LM88-300	6/90 6/90 5/90
Process I/O 1 621 I/O Specifications User Manual 2 621 Isolated Analog Input Module User Manual	620-8995 621-8988	4/90 4/89
Programming 1 623 MS-DOS User Manual 2 623-51 Loader/Terminal User Manual	623-8986 623-8999	6/89 10/87
Service Logic Manager Service	LM13-300	7/90

Figure 8-28 (continued). Typical DCS Documentation [Ref. 5]
(Courtesy of Honeywell, Inc.)

BINDER TDC 1080: IMPLEMENTATION/PLC GATEWAY

Tab Publication Title	Publication Number	Revision Date
Overview PLC Gateway Specification and Technical Data	PL03-300	1/90
Planning & Installation PLC Gateway Planning, Installation, and Service	PL02-300	5/90
Control Functions 1 PLC Gateway Control Functions 2 PLC Gateway Parameter Reference Dictionary 3 PLC Gateway Implementation Guidelines 4 PLC Gateway Forms	PL09-300 PL09-340 PL12-300 PL88-300	4/90 4/90 * 4/90

BINDER TDC 1090: IMPLEMENTATION/MICRO TDC 3000

Tab Publication Title	Publication Number	Revision Date
Overview 1 Micro TDC 3000 Specification and Technical Data 2 Micro TDC 3000 User's Manual	MT03-300 MT11-300	7/90 3/90
Service Multinode Module Service	MT13-300	10/89

STAND-ALONE PUBLICATION

Publication Title	Publication Number	Revision Date
Universal Work Station Installation, Operation, and Service	UW02-100	10/88

* Not yet available as of the date of this publication

Figure 8-28 (continued). Typical DCS Documentation [Ref. 5]
(Courtesy of Honeywell, Inc.)

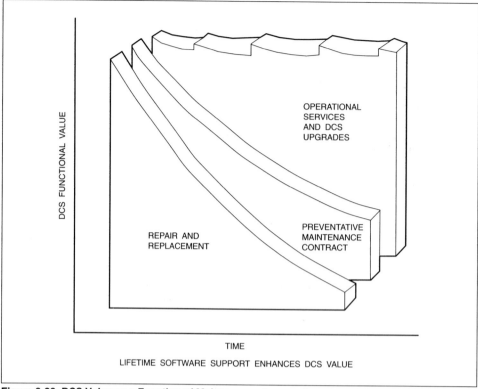

Figure 8-29. DCS Value as a Function of Maintenance Approach

Manufacturers and designers will customize their programs to meet the requirements in the three areas.

ENHANCEMENT MAINTENANCE

This involves using the most recent releases of hardware and software, that have been redesigned for improved maintenance and functionality. Keeping a DCS at current revision levels provides a cost-effective way to operate at peak performance.

The DCS manufacturer can implement software enhancement and support programs in three ways:

(1) Periodic software enhancements issued to the customer

(2) Technical engineering assistance made available to the customer at any time

(3) Management by the manufacturer of the client's DCS documentation

PREVENTIVE MAINTENANCE

Preventive maintenance is intended to keep a system from breaking down by providing equipment inspections on a regular basis. Since there is no obvious immediate benefit from this type of maintenance, it can vary widely according to individual preferences (and perhaps to the skill of the salesman selling the idea). It is up to management to provide a structure that will result in a balanced program using in-house resources together with the support programs offered by the systems manufacturer.

Experience indicates that the ratio of the cost of preventive maintenance to the cost of repair and replacement is about 1:2.

CORRECTIVE MAINTENANCE

Corrective maintenance consists of performing qualified repairs to a system that has failed and, thereby, returning the system to its original usable condition.

Providing this type of service on such a complex integrated system as a DCS can be a formidable challenge to both plant maintenance departments and equipment vendors.

Owners must find an efficient means to supplement their in-house capabilities with outside industrial support groups. It is very important that such outside support services be provided by an experienced team and with the same promptness that the owner would provide from in-house resources, had they been available.

Service Contracts

Each organization has its own maintenance philosophy and must, therefore, determine what service program best suits its needs, so that the DCS remains available, capable, and dependable. The service activities needed to support a proper program of maintenance are shown in Figure 8-30.

A variety of services are available from outside firms:

(1) On-call service

(2) Equipment maintenance

(3) Equipment or system audits

(4) Maintenance service retainers

(5) System utilization services

(6) Software support services

(7) Resident field engineers

On-call service is service done by a support group that maintains well-trained service engineers who are capable of troubleshooting almost any DCS.

Equipment maintenance is done according to a contract. Routine preventive and corrective maintenance are performed by a qualified engineer who is located in the general area of the client.

Equipment or system audits are done according to an agreement that includes periodic review of system performance by a qualified engineer who is familiar with the process that the DCS system is controlling. This sometimes includes added features, such as extended support journals, semiannual performance reviews, technical support with PC computer-based bulletin boards, and so on.

Maintenance service retainers are agreements whereby demand service is made available at a level of so many days, at an agreed cost per day, for a specified term. Emergency demand service is not the same thing; it means delivering adequate service support in time to avert possible downtime. This would be an extra to the retainer.

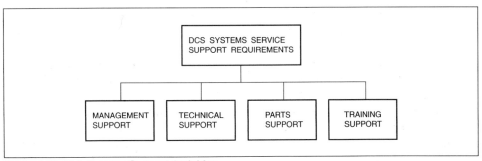

Figure 8-30. Maintenance Support Activities

System utilization services are provided according to an agreement whereby a process specialist is scheduled to actually operate the process in order to optimize it for the customer.

Software support services are provided via telephone system and modem, usually by qualified systems and application engineers. This can include periodic system update letters, software updates, and remote diagnostics.

Resident field engineer services involve an agreement whereby an engineer resides in the immediate area of the client's plant and reports to work only at that plant.

Note that the organization supplying service support may be the original equipment manufacturer or a sophisticated repair center established by a third party. Third parties can provide maintenance and repair for laboratory, industrial process control, and electronic test equipment. This support can include warranty and nonwarranty service as well as installation and start-up assistance.

Purchasing Strategies

Long-Term Buying

The purchase of a DCS must be made in the context of a long-term management activity. Planning in advance will make sure that answers are more readily available when problems occur. The following are some of the factors to be considered in such long-term purchases:

(1) System effectiveness

(2) Technical performance

(3) Capability

(4) Availability

(5) Support effectiveness

(6) Reliability

(7) Maintainability

(8) Safety

(9) Accessibility

(10) Software configuration

(11) Quality

(12) Software enhancement

System effectiveness is the ability of the DCS system to sustain successfully the overall system demand within a given time when operated under specific conditions. It also can be characterized as capability, function availability, and support effectiveness.

Technical performance is similar to technical capability but is often differentiated as a measure of what total technical support the manufacturer can offer.

Capability is a measure of how well a system performs.

Availability is the probability that a system will operate satisfactorily and effectively when needed or when used to control a process under specific conditions.

Support effectiveness is the service support required to sustain the equipment and keep it operating effectively by furnishing it with whatever enhancement it needs.

Reliability of a system is the probability that a system will perform as intended without failure for a specific time period under specific design conditions.

Maintainability of a system such as a DCS is the inherent ability of applying technical knowledge and management skills to the developed system so that it can be effectively operated and efficiently maintained.

Safety of an overall system is an assessment of the equipment to assure that both system and human safety levels are attained at all times during utilization.

 Safety design is of great importance to protect the system against failure and breakage and to eliminate the hazardous conditions that can cause operating accidents and injury.

Accessibility is the ease with which a control system can be accessed in order to contact part of an assembly or subassembly, as well as its ability to give answers to specific questions quickly.

Software configuration is what an organization must consider in the makeup of a DCS in terms of physical properties and functional characteristics. The design of the hardware/software must be intelligently engineered, and the specific DCS configuration installed at the plant must satisfy that customer's needs.

Quality of equipment and service is especially important in reducing system downtime. Quality means that the system should be at least 99% fault-free. DCS suppliers must have the capability of providing quality in both products and services.

Software enhancements must be considered so that the system's functionality can be expanded or improved with every new product released by the manufacturer. Software enhancements provide the platform for new system applications by adding new system functions, improving system performance, and introducing new technology innovations.

Expanding and Upgrading a DCS

When it comes to the expandability of a DCS, modularity is often the only thing considered. Unfortunately, in most cases, there is little discernible difference between one DCS and another; all DCSs are modular by nature and, therefore, expandable. The only difference is where the break points are in terms of add-ons. The term "break point" refers to that number of inputs and/or outputs beyond which there is a need for additional hardware — sometimes associated with a dramatic increase in cost. It is possible for the addition of a single point to require extra cabinetry, power and communication modules, and data highways.

The unseen difference in system expandability often lies in its backwards compatibility and its upgradability.

All DCS designs are changing constantly and at a pace unfathomable to most laymen. Computer chips can become obsolete within months. DCS vendors have to keep up with such changes to stay competitive. In addition, customers are constantly demanding added functionality. As a result, software is continuously being written and integrated into vendors' systems.

Dealing with Obsolescence

This constant evolutionary change has both good and bad points. The good news is that DCS costs will drop while the systems become easier to use, more powerful, and less human-labor intensive. This makes for more efficient and competitive plant operations. The bad news is that system components will become obsolete (and therefore costly to support, repair, and replace) and will require a program of upgrading to stay current.

Many DCS owners have found that there are new, more powerful modules or even whole new systems less than a year after they made the initial purchase. The question then becomes whether to buy the "slightly out of date" version and make it compatible or update to the "new" technology. It is a complex decision and involves potentially skyrocketing costs.

Costs can be excessive in either case. Older components can become very costly after being discontinued or pronounced obsolete. They are more difficult to get repaired. Spares and technical support are harder to obtain and are more expensive. Upgrading existing equipment can also be an expensive proposition depending on the design of the system. This is one area of real differentiation between DCS vendors. Every vendor will swear to "backwards compatibility" and, to some extent, will have it. There are systems with good compatibility and others with very poor compatibility. The authors are aware of an instance in which only a software upgrade cost over half a million dollars — almost the cost of an entire system.

It is often a good idea to review the research and development (R&D) program of DCS vendors being considered. This will indicate what the outlook is for the future and will put a company's product in a better perspective. It is also a good barometer of the company's stability and commitment to the product. Having one or two years of system upgrades built into the project is a good insurance policy.

Upgrading a DCS

Avoiding obsolescence can be done by identifying an "upgrade path." In other words, find out exactly how upgrading will be done. The following are some of the alternatives:

> **Most reputable companies will guarantee the availability of a product for about 10 years after it has been discontinued. Get it in writing!! Read it and understand it.**

(1) Upgrade the system software only.

(2) Upgrade both the system and configuration software.

(3) Upgrade hardware modules modified with system software.

(4) Use all new hardware and software.

Understanding the relationship between a vendor's hardware and software will make it easier to decide on the "what and when" of upgrades. For example, one ought to consider upgrades to a DCS that will not be expanded. Keeping a DCS up to date is a good way of protecting one's initial investment.

With a proper program of upgrades, one can help prevent the wholesale replacement of an older system that has become obsolete.

This sort of upgrading means that, after a couple of years, when significant (to the owner) functional improvements have been made, one brings the system up to the current technology. This spreads out the costs and implementation time involved in DCS changes. If it seems that few functional changes are taking place in a system, chances are little vendor R&D is going on. If so, it would be wise to hold off on an upgrade and perhaps even look at alternative DCS vendors.

Upgrade paths will vary from vendor to vendor. Some will be more costly and difficult than others. Some require downtime; others do not. As much as possible, find out what is likely to be involved before making a final selection. Talk to other owners to get a feel for what they have done. This will often provide a good indicator of things to come.

References

1. Lukas, Michael P., *Distributed Control Systems*, New York: Van Nostrand Reinhold Company, 1986.

2. Taylor, *MOD 300 — Overview*, Rochester, NY: ABB Kent-Taylor.

3. Foxboro, *Intelligent Automation Series: Hardware Overview*, Foxboro, MA: The Foxboro Company.

4. Rosemount, *System 3*, Eden Prairie, MN: Rosemount, Inc.

5. Honeywell, *TDC 3000 Bookset Directory*, Minneapolis, MN: Honeywell Inc.

6. ANSI/ISA-S71.04, Environmental Conditions for Process Measurement and Control Systems: Airborne Contaminants, Research Triangle Park, NC: Instrument Society of America, 1986.

7. The RPGO Series of Recommended Practices for Control Centers, Research Triangle Park, NC: Instrument Society of America.

8. Wade, H. L., eds., *Distributed Control Systems Manual*, Research Triangle Park, NC: Instrument Society of America, 1991.

9. Herb, S. M., and Moore, J. A., *Understanding Distributed Process Control*, Research Triangle Park, NC: Instrument Society of America, 1987.

About the Authors

Dan Bellefontaine has been a DCS specialist for Rosemount since 1987. A graduate of McGill University in engineering, he was a sales engineer for Ingersoll Rand and Koppers Engineering before joining Rosemount.

Maurice L. Pyndus is a graduate of the Montreal Institute of Technology and has attended McGill University and Sir George Williams University. He evaluates existing control systems in various government facilities for Public Works, Canada. He is developing specifications for the replacement of equipment with DCS controls using microcomputer-based technology. Mr. Pyndus is an active Senior Member of ISA, who spent the last ten years of his 35 years with Bailey Controls as Manager, Service and Sales.

Alberto Dufau has, for the past eighteen years, worked as an engineering consultant in chemical and petroleum industries in Canada and Argentina. A graduate of the Universidad Technologica Nacional in Mechanical Engineering, he is employed by SNC-Lavalin Inc., in Montreal.

9

Programmable Logic Controllers

Any discussion of programmable controllers should first define the beast. A programmable controller is an industrial, real-time, modular, solid-state control system that operates under a stored program written in ladder logic, complete with monitoring and program development facilities. Originally, programmable controllers were referred to as PCs, but with the introduction of the IBM personal computer (IBM PC™) came confusion, so the acronym for the programmable "logic" controller became PLC. Conforming to common usage, and for the sake of clarity, this chapter uses the term PLC exclusively throughout. To better understand the multi-worded PLC definition above, let us now examine some of the more important terms.

"System" means that the PLC is a complete package with all parts designed and tested to work together. It includes software, hardware, and documentation. Unlike computer systems that include components and software from different suppliers, the PLC comes from one source that makes sure it all works together.

"Modular" indicates that capacity and functionality can be added as required, and then removed to be reused elsewhere.

"Industrial" denotes suitability for the temperature, humidity, vibration, and voltage variations found in an industrial plant, and that the design is for 24-hour operation, seven days a week. It also implies having the appropriate electrical approvals and inputs and outputs suitable for the range of voltages used in industry.

"Real-time" signifies that the PLC is always working , always in a state of readiness, in order to react without unusual prompting when the operator presses a button or a sensor is activated. There are levels of real time, and some PLCs may not be able to suit all needs. Some high-speed machines exceed the capabilities of any PLC.

"Solid-state" infers no vacuum tubes, pneumatics, or hydraulics. Even disk drives are not used in a PLC.

"Stored programs" in memory are used to run a programmable logic controller. Gone is the hard-wired logic that operates according to the way it is connected together. Furthermore, the PLC runs programs in ladder logic, a high-level Boolean-type language that is readily understood by plant electricians. Many PLCs offer extensions to the basic ladder logic or other languages that can be used in addition to the ladder logic language. Finally, the PLC includes some facilities for its programs to be developed and its operation to be monitored while it is controlling a machine or process.

 The PLC is a general-purpose device. It is not specific to a class of machines such as machine tools or packaging equipment, nor is it limited to specific functions such as batch control, sequence control, and so forth. Nevertheless, the PLC

is not a general-purpose computer. It is not designed for data processing, or handling large quantities of text or files, or operating video terminals.

Introduction

History

The first PLC was specified in 1968 by the Hydramatic division of General Motors to replace the inflexible relay-controlled systems then used on their assembly lines. The first PLC was delivered in 1969 and by 1971 was spreading to other industries. In the early 1970s, arithmetic and data-manipulation functions were added to the basic relay-logic functions. Not long after, the CRT display was implemented for programming and monitoring. As electronic components became smaller and more reliable, PLCs became faster, more powerful, and capable of controlling more inputs and outputs. In the late 1970s high-speed local communications networks were developed that allowed a large process or machine to be controlled by several smaller controllers. This also allowed the integration of materials-handling machines with the production equipment.

The 1980s brought ever faster and more powerful PLCs, with expanded instruction sets and 16-bit capabilities instead of 8-bit, and even floating-point calculations instead of integer ones. The size of the PLC shrank to the point that a single card can do what it took an entire rack to do in the past. Small and very small PLCs have also appeared for applications that would normally have required as little as half a dozen relays.

Recent developments have been in the area of communications between different brands of PLCs and between PLCs and personal and mainframe computers. There has also been an ever-increasing selection of devices for operator interfacing that can be added to PLCs without having to be controlled directly by the PLC. Some PLCs now have computer modules that plug into their rack and allow direct communications. There has been a constant improvement in the range and capabilities of modules available for analog and special input/output modules.

Principles of Operation

PLCs operate very much like computers. Figure 9-1 shows the parts of a typical PLC, which is remarkably similar to the computer in its basic arrangement. It might be worthwhile for the reader to review the pertinent section of Chapter 5, since this chapter will not repeat the description of the basic operation of a computer. Basically, the PLC reads the input devices, executes its program using the status of the input devices, writes the appropriate values to the output devices, and then starts over again. This is one of the major differences between a general-purpose computer and a PLC. The general-purpose computer reads and executes the instructions typed on the keyboard, including those that tell it to read more information from the disk drive or to execute other programs. In addition, such general-purpose computers do not usually keep executing the same program all the time, although they can be programmed to do so.

The number of input and output devices can be anywhere from six to several thousand, with programs varying from one thousand bytes to one hundred thousand bytes. The time required to execute the program once is called the "scan time," which varies from less than a millisecond to a few seconds.

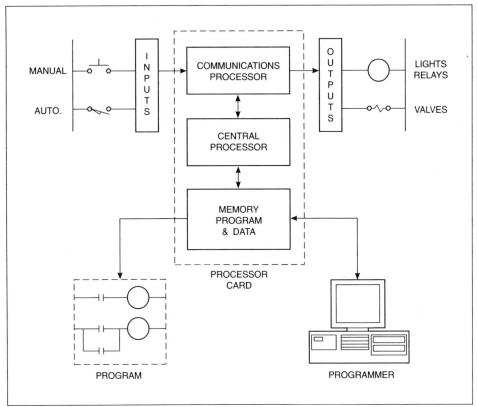

Figure 9-1. Parts of a Typical PLC

Typical Areas of PLC Application

Not long ago authors would give lists of the areas where PLCs were used or could be used. Such lists are now irrelevant because PLCs are used everywhere. They are found in dock levelers and stretch wrappers, packaging machines and assembly lines, power distribution systems and process plants. With the advent of small self-contained PLCs, it is now very easy for the average electrician to design, install, and program small applications that previously would have been done with relay logic.

PLCs vs. Other Types of Controllers

RELAY LOGIC

PLCs are quickly taking over relay-logic applications. Manufacturers are finding that even applications requiring only a few relays can be replaced on a cost-effective basis, especially if there are several variations that would require individual wiring diagrams. Nevertheless, there is resistance to conversion to PLCs for some applications that require just a few relays or that require relays for interfacing, isolation, or power handling.

The difficulty of modifying hard-wired relay panels will not disappear. The expandability of PLCs is often one of the major reasons for their being selected over hard-wired relay logic. Even a minor expansion will give the PLC an advantage over the hard-wired relay-logic system. Add the ease of troubleshooting a PLC-based system and it's hard to understand why anyone would still want a relay-based system. The answer probably lies in the training and programming equipment necessary to get started. Also, some companies (especially smaller ones in more remote areas) may be uncomfortable with new technology, and they may want to stay with the old standby.

PLCS VS. COMPUTER CONTROLS

The similarities of the PLC to the general-purpose computer have already been mentioned; let us now look at some of the important differences.

Computers are not generally designed for the industrial environment. They require conditioned control rooms and very clean power, while PLCs are, by definition (and when placed in an appropriate enclosure), designed to run in the usual factory environment.

Everybody knows that computers need programmers to program them and operators to run them. PLCs need neither. They are designed and built to be programmed by electricians, technicians, and engineers, and they do not require operators.

PLCs are single-task machines. They run one task and do it continually. Computers, on the other hand, are usually doing several things at once.

Computers and PLCs each have their place, and it is not difficult to obtain a good understanding of the role of each.

PLCS VS. PERSONAL COMPUTERS

The comments listed above for computers also apply to personal computers, but there are other factors involved.

Personal computers now come in industrial versions that can run in the same environment as PLCs. The weak point is still the disk drives required for an application of any size.

Although personal computers have lots of input/output (I/O) hardware available for them, complete systems from a single vendor are not common. This potential lack of responsibility to make the whole thing work makes some people favor PLCs over personal computers.

The personal computer is well accepted as a partner in automation with the PLC, the most widespread application being as a programming device. The next most common is probably as an operator interface device. These applications make good use of the personal computer's text handling and communicating abilities—abilities that the PLC does not have. As time goes on, there may be more blurring of the differences between personal computers and PLCs.

PLCS VS. DEDICATED CONTROLLERS

With the advent of inexpensive controllers on a chip that can be configured to match the user's need, the industry is seeing more electronic controllers designed for a particular application. These generally fall into two categories—generic controllers and specific ones. A specific controller would be made by an equipment builder to control his own equipment. For example, a compressor manufacturer could have a controller designed for a whole family of compressors. One piece of hardware would handle the entire product line; only the program, or perhaps only the operating parameters, would have to be changed. A generic controller, on the other hand, is available to anyone who needs to perform the operation it was designed to handle. A good example of the generic controller is the multi-zone temperature controller that also includes some sequencing functions.

In single-loop or zone control, the PLC is not competitive, and the dedicated controller has the advantage. As the number of zones or loops increases, or as the logic required to handle the application gets more involved, the PLC becomes more interesting. This is especially so when the application requires a PLC for other parts of the process or machine anyway. Often, the generic, dedicated controller is created to suit the common needs of a number of applications or to serve a large number of possible users for one application. A good example is the use of multi-zone temperature controllers on machinery for the plastics industry.

The choice between dedicated controllers and PLCs is often difficult, and it is clouded by other issues such as a machine builder's marketing strategy or the cost of developing something in-house versus buying the hardware outside and just doing the software. Often the end user is not in a position to make the decision.

PLC Sizes

One measure of the size of a PLC is the number of I/Os it can handle. Generally, the more I/Os a PLC can handle, the larger and more complex it is. There is no established list of sizes and features, so there is overlap among the categories. This overlap allows for a smooth change as applications get larger and more demanding. Without overlap there would probably be a large jump in the cost of a system as break points are reached.

Figure 9-2 is a graphical representation of the five sizes of PLC. The smallest, or micro-PLC, is also often called a "relay replacer" and is for those applications that require only a minimal amount of I/Os. The small, medium, and large sizes are self-explanatory. There is considerable overlap in these sizes and in the features offered. The super PLC is really a different animal from the others. It is designed for the very large or complex operations found in the process industries and is often used with other PLCs as slaves. The sheer size of the super PLC is daunting, and designing is usually a team effort by people specialized in the process concerned. With the advent of networked PLCs, the super PLC may be less commonly used.

Benefits of Using PLCs

Table 9-1 gives some of the benefits of using PLCs. The list cannot be complete, and some may argue that many of the benefits can be obtained by other means, but it's clear that relay or hard-wired systems cannot claim these benefits as advantages over PLCs.

Table 9-1. Benefits of Using PLCs

High Reliability
Flexible Control
Easily-Modified Program
Easy Troubleshooting
Reduced Space Requirements
Reusable
Lower Cost
Modularity

Ladder Logic Concepts

Introduction

Because PLCs were created to replace relays and their ladder logic, a brief description of relay logic is now in order. The description that follows will not make one an expert in relay logic; that would require considerable study and practice. Relay logic was developed from components available early in this century

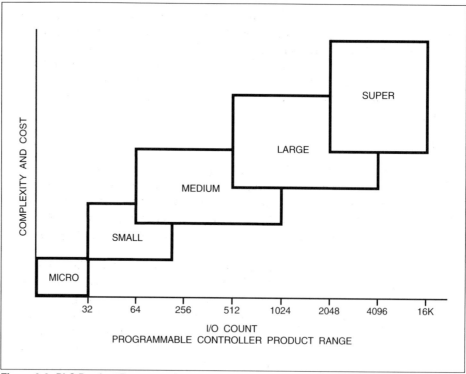

Figure 9-2. PLC Product Range

and had its roots in the relays used in the telephone industry. It is a visual language based on the way the switching elements in the actual devices look, so it was easy for early electricians to learn and use. Most electrical control equipment manufacturers have extensive libraries of relay circuits available in the form of wiring diagrams. Once understood, these basic circuits can be adapted to a wide range of control problems.

Devices

RELAYS

A relay is an electromechanical device that consists of a fixed coil that moves a set of contacts that make or break the path for the electric current. Figure 9-3 shows examples of typical relays. When power is applied to the coil, the armature is attracted towards the coil and the contacts close, allowing current to flow. When the power is removed from the coil, a spring (or gravity) moves the armature away from the coil and the contacts close. Many mechanical arrangements are available to operate the relay contacts, the choice being determined by the current the contacts must carry, the voltage they are expected to handle safely, speed of operation, size constraints, mounting position, the operating life of the relay, and cost.

Relay contacts come in two types—normally open (NO) and normally closed (NC). The word "normally" applies to the state of the contacts when no power is applied to the relay. Thus, when no power is applied to the relay, the NO contact is open and the NC contact is closed. Once power is applied to the relay, the NO contact closes and the NC contact opens. Figure 9-4 shows the symbols used for the relay and its contacts. Relays can have both NO and NC contacts, in varying quantities from one to 16.

The inflexibility of relays was reduced in three ways: overdesign, convertible contacts, and modular relays. It was standard practice to install relays with an assortment of contacts so that many future changes could be accommodated. To

Since the number and type of contacts (or poles, as they are also called) is determined at the time of manufacturing, changes to the initial configuration cannot be made afterwards. If a circuit design requires a relay with one NC and three NO contacts, that's what is ordered and installed. Later discovery that the design really needed four NO contacts means replacing the relay.

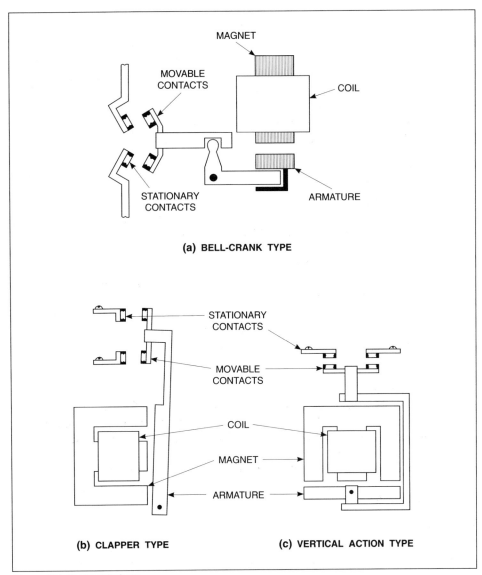

Figure 9-3. Typical Relays
(Courtesy of Square D Company)

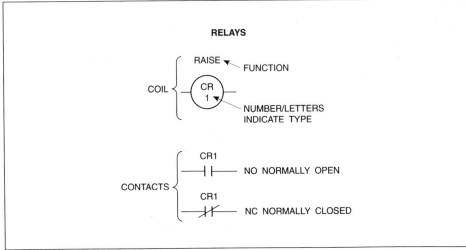

Figure 9-4. Relay Coil and Contact Symbols

some extent, this also reduced the number of relays that had to be stocked. Over-design made control panels larger and more expensive than needed. Convertible contacts allowed the installer to change from NO to NC or vice versa. This reduced even further the number of items to be stocked, as well as the space required for them. The final improvement was modular relays that could be expanded from two to 12 poles, as required, from standard modules. The user could build them up as needed for the application. Space still had to be left for expansion, however, and the modular relays were more expensive that the fixed ones.

PUSH BUTTONS

The ubiquitous push button is an industrialized version of the doorbell button. Push buttons are momentary action devices; as long as they are held down, the circuit remains closed, or open, and it returns to its normal state when the button is released. Push buttons come with NO and NC contacts like relays. For momentary action devices, the normal condition is the inactivated position, just as it comes out of the box. Figure 9-5 shows the symbols for NO and NC push buttons. Push buttons are relatively modular, consisting of an operator and contact blocks that can be stacked together as required. The difficulty with push buttons used in pre-PLC times is the wiring that had to be done between the relay panel and the push buttons out on the machine. It was common to have hundreds of wires going out to push buttons, and any change in the relay logic often required changing wiring to the push buttons.

SELECTOR SWITCHES

Selector switches are like push buttons except that they remain in whatever position they are moved to. They are also called "maintained devices," because they maintain their last position. There is no convention for describing the maintained contacts of selector switches as there is for NO and NC contacts for relays or push buttons. The selector switch symbol has an arrow or truth table to indicate the state of its contacts for each position it can take. Figure 9-6 shows some symbols for selector switches.

OTHER DEVICES

Another common device, the limit switch, is a snap-acting device that is operated by a cam or some other mechanical device. The snap-acting limit switch will change the state of its contact when the operating rod is moved a very small amount from one position to another. Typically, the rod has to move only a few thousandths of an inch to cause the contact to transfer. Figure 9-7 shows the contact representation of a limit switch; Figure 9-8 shows the operating characteristics of a typical switch. Since limit switches are momentary devices, they have NO and NC contacts, as do relays and push buttons.

Limit switches are used in many other sensing devices such as level, flow, pressure, and temperature switches, in which a mechanical sensing system (such as a float or Bourdon tube or capillary tube) moves a small amount to activate the

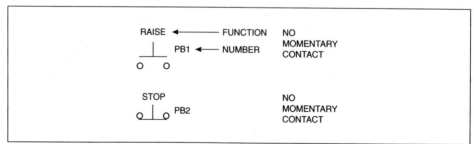

Figure 9-5. Push Button Contact Symbols

limit switch. Adjustment is made by moving the limit switch closer or farther away.

Some Basic Circuits

Before describing some of the basic relay-logic circuits, we should look at the basic conventions used in drawing these diagrams. Momentary devices are shown in their normal state (NO push buttons are shown open, relays are shown in the de-energized condition, and so on). Maintained devices are drawn with an indication that details the state of the device. Intersecting lines are connected only if a dot is placed at the intersection.

The most elementary circuit is one to convert a momentary action into a maintained one. Figure 9-9 shows this basic circuit, which consists of two push buttons and a relay. If the start push button is depressed, power is applied to the relay coil, and the NO relay contact closes. Once the relay has been energized, the start push button may be released, because the NO relay contact will allow current to flow to the relay to keep it energized. This NO contact is called a "seal-in contact" because it seals the circuit to keep the relay energized after the push button has been released. Since relays generally operate within a few cycles, the push button contact needs to remain closed for only a short time (e.g., a tenth of a second). When the relay is energized, it is also said to have been "picked up."

There is another way the relay may be turned off. If the power supplying the relay circuit is removed, the relay will drop out and will remain in that state even if power is restored. This feature is called "low voltage release" and is regarded as an important safety feature. A power loss will, therefore, shut everything down, and a positive operator action will be required to start it up again.

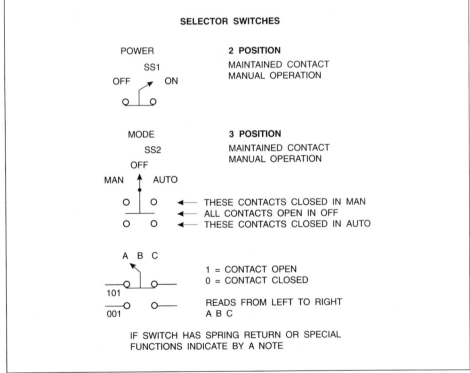

Figure 9-6. Selector Switch Contacts

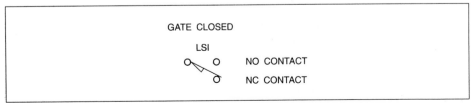

Figure 9-7. Limit Switch Contacts

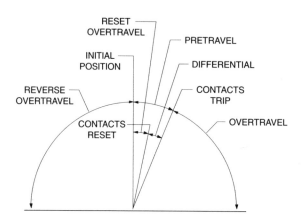

OPERATING DATA

	Type of Switch					
	Universal Type T					
	No. 1		No. 2		No. 3	
	Single Pole Double Throw Spring Return	Three Point Double Throw Spring Return	Single Pole Double Throw Spring Return	Three Point Double Throw Spring Return	Single Pole Double Throw Maintained Contact	Three Point Double Throw Maintained Contact
Operating Data	Initial Position and Counter-clockwise	Initial Position and Counter-clockwise	Initial Position and Counter-clockwise	Initial Position and Counter-clockwise	Spring return of arm to initial position, contact position maintained until operated in reverse direction	Spring return of arm to initial position, contact position maintained until operated in reverse direction
Pre-travel‡	14°		Int. Pos. –9°, Final Pos. –16°		7°	
Over-travel	74°		73°		74°	
Reverse Over-travel	74°		73°		—	
Differential	12°		5°		—	
Oper. Force	10 in.-lbs.		10 in.-lbs.		10 in.-lbs.	
Repeat Accuracy w/1-1/2″ arm	±.002″		±.002″		±.002″	
To convert sequences, remove base plate, pos. plate and latches. Reassemble pos. plate and latches as shown.	POSITIONING PLATE — LATCHES		POSITIONING PLATE — LATCHES		POSITIONING PLATE — LATCHES	

‡ Pre-travel listed may vary up to 5° additional for universal switches or up to 2° additional for standard switches due to free travel of lever arm at initial position.

Figure 9-8. Heavy Duty Limit Switch Operating Characteristics
(Courtesy of Square D Company)

This term comes from the early days of relay logic (when relays energized, a small flag was raised).

Once the relay is turned on, how is it shut off? Pushing the NC stop push button will break the circuit to the relay, causing the NO contact to return to its open state. Once the relay has dropped out (become de-energized), the stop push button may be released, and the relay will remain in the de-energized state until the start button is pushed again.

This basic circuit does more than simply operate a relay. One of the relay contacts in the figure is shown operating a small motor that is connected to a different power supply. Note that the relay operates on a relatively safe 24-volt supply, whereas the motor runs on 220 volts in this case. For larger motors, a device called a contactor, or starter, is used. It works just like the relay, but it is designed to handle large amounts of power, and it usually has one to four power poles and up to four auxiliary contacts for relay use.

Figure 9-10 shows the same circuit but with several extra push buttons to allow control of the relay from more than one place. Notice how the NC stop buttons are connected in series, and the NO start buttons are connected in parallel. Push-

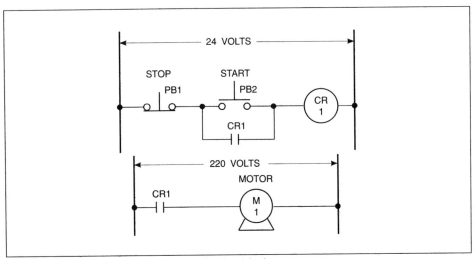

Figure 9-9. Converting Momentary to Maintained Action

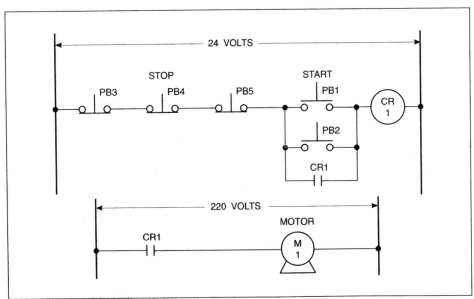

Figure 9-10. Controlling Relay from More Than One Point

For a concrete example, consider a circuit that controls a garage door. Figure 9-9 or Figure 9-10 would work nicely to get the door started, but how can it be stopped when it is all the way up? The answer is to add a limit switch that opens, and thereby stops the motor, when the door is up. Figure 9-11 shows such a circuit. Notice that the limit switch contact is shown as NC so as to open when the door actuates it.

So far there's a circuit that opens the door, but there is no way to close it. Figure 9-12 shows a second relay added for the down direction, with a second limit switch that opens when the door is all the way down.

Note that the same stop button serves both up and down directions. Putting separate stop buttons for each direction would be confusing and even unsafe. Other features could be added to this basic circuit.

There is often a need to avoid engaging two or more mutually exclusive functions at the same time. This is called interlocking, and a good example would be the raise and lower function of the door. The idea is to avoid running both the up and down motors at once, even if some smart aleck purposely pushes both the up and the down buttons at the same time. Figure 9-13 shows that this can be done by using a normally closed contact of the up relay in series with the down relay and vice versa. Now the relay that energizes first will prevent the other relay from energizing.

Another common circuit in door operators is an emergency reverse that operates if the door hits something on the way down. In this case, a switch on the bottom of the door is activated, thus sealing in a relay that causes the door to stop and then rise to the fully open position, after which the auto-reverse relay is de-energized. Figure 9-14 shows one way this can be done.

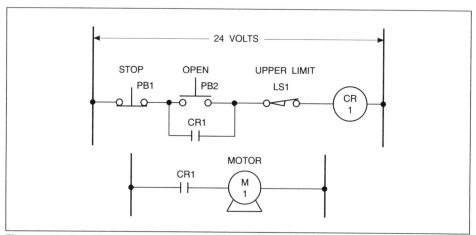

Figure 9-11. Door Opener Relay Circuit with Limit Switch

This rather brief introduction to ladder logic should provide some feeling for it, but a real appreciation comes only after designing some circuits and working with one that goes on for several pages and has extensive interlocking.

ing any of the start buttons will provide power to the relay and cause it to seal in. Pushing any one of the NC stop buttons will break the circuit, drop out the relay, and turn off the motor.

By now, it's clear that ladder logic uses several basic building block circuits, over and over again, to provide some desired operation. This convenience has led to some abuses of ladder logic. To describe how something functions, some designers have used ladder logic rather than a written description that may take several paragraphs. Some designers even work directly in ladder logic without first describing what the circuit is supposed to do. This can lead to circuits that are hard to understand or that do not really do what they were expected to do.

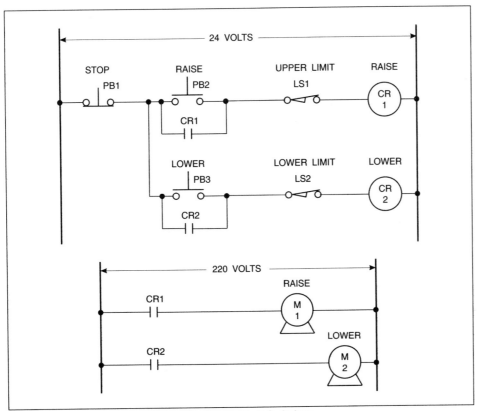

Figure 9-12. Door Open and Close Circuit with Limit Switches

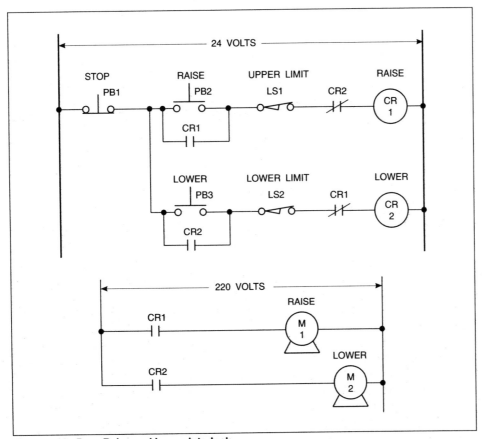

Figure 9-13. Door Raise and Lower Interlock

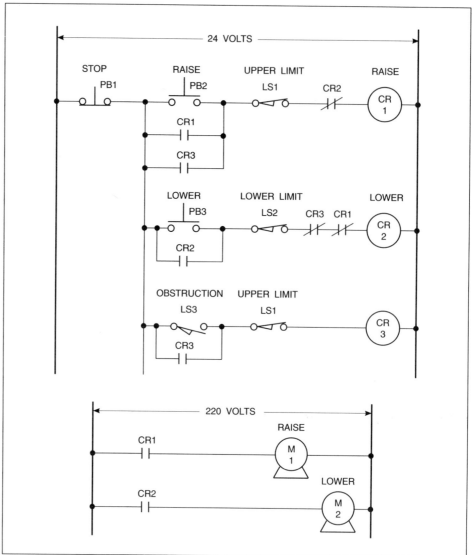

Figure 9-14. Door Opener Auto-Reverse Circuit

Processors

Introduction

The modern PLC is a very specialized computer, so it shares much of the computer architecture described in Chapter 5. Figure 9-15 shows a generalized block diagram of the PLC. Since computer technology is continually changing, the way the various parts of the process work and communicate is also changing. Products designed three years ago will be different from those designed today or next year. The processor is the brain of the PLC, and it consists of five basic parts: the *CPU or central processing unit*, which controls the operation of the PLC; the *memory*, which contains the instructions and data that tell the CPU what to do; the *communications processor*, which handles communications with the outside world; a *battery*, which keeps time and maintains the program; and a *power supply*, which provides the energy necessary to run the processor.

The processor can vary in size, from a part inside the one-piece PLC commonly known as a "shoe-box," to a single card in an input/output rack in some

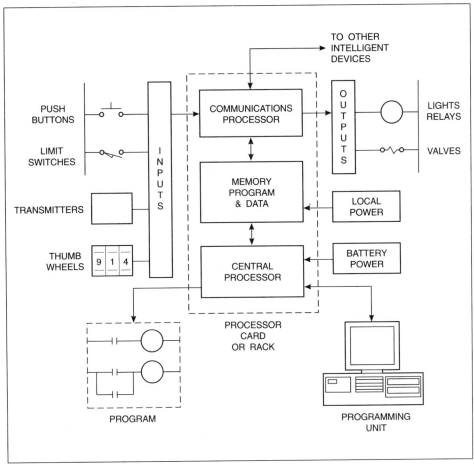

Figure 9-15. Generalized PLC Block Diagram

medium-sized PLCs, to an entire rack with one or more cards for each part of the processor in large PLCs. The central processing unit (CPU) can be a 4-, 8-, 16-, or 32-bit microprocessor. The larger the number of bits the microprocessor can handle at a time, the more work it can do; and, hence, the faster it will execute a program. With the advent of specialized microprocessors, the communications processor may very well be a microprocessor that is designed only to handle the communications between the real world and the memory used by the CPU. In other designs, this function may be handled by several standard or custom chips. Even the CPU itself may have some specialized functions (such as mathematical operations and PID control done by microprocessors that are designed just for these operations). This use of several microprocessors within the same processor is called multi-processing; it provides a faster operation than would result from doing everything in one microprocessor.

See Chapter 5 for a more complete discussion of the operation of computers and microprocessors.

Processor Scan

Unlike the office or general-purpose computer, the PLC executes only one program, and it keeps on doing it over and over again. The time the PLC takes to execute this program once is called the scan time. Not only is the scan time itself of interest, but also how much it varies under differing operating conditions. Figure 9-16 is a block diagram of the steps involved in a complete cycle.

Examine the PLC's execution time for instructions that accomplish the same function and then choose the faster one. It may be faster to multiply by 0.5 than to divide by 2. Similarly, a multipurpose function may be slower than a dedicated function. The only way to find out is to study the published information.

In small PLCs, program execution is often expressed as y milliseconds per k of program. Medium and large ones give the time it takes for each instruction, often with an infinite number of details to handle all the variations of each command, and the user is left to figure it all out. Many PLCs have an internal function that records the scan time and even gives the maximum and most recent values.

Note the distinction between "program scan time" and "complete scan time." The difference can be as much as 100 percent.

During the input scan, the processor updates the information on all the inputs. This may take a fixed amount of time, say, one millisecond, or it may be expressed as x milliseconds per k of inputs. The output update time is usually expressed the same way, and sometimes the two are combined into one figure.

Although a fast scan time is often desired, a repeatable scan time can be even more important, especially where reliable timing of events is required. The experienced designer can choose the logic functions that will result in the shortest execution time. Many PLCs will reduce their program scan time by stopping the solution of a line of logic once it has determined it is false, regardless of the rest of the line. Figure 9-17 shows how this saves scan time. The price one pays for this extra speed is often a scan time that is more variable. By the judicious ordering of contacts, the designer can tune a program to suit the need.

Once the PLC has read its inputs, executed its program, and set its outputs, one would expect it to be able to start over again right away. Unfortunately, a few other things must be taken care of first. The programming unit, if it is connected, needs some time to communicate with the processor, especially if the program logic is being changed. In some older PLCs, there was a noticeable effect on scan time with the programming unit connected. Even more recent PLCs can have problems if the same communication network is used for both programming and operation. Once again, only the manufacturer can say for sure what the limitations are.

The PLC also does some housekeeping things such as checking battery status, verifying memory integrity, checking power supply voltages, and resetting the watchdog timer. Generally, these functions do not take much time, and they are often buried in the input/output scan or update.

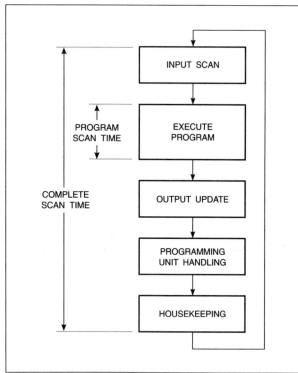

Figure 9-16. Steps in PLC Scan Cycle

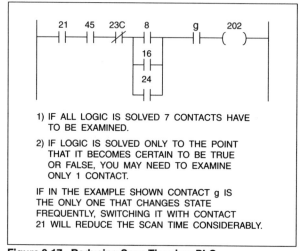

1) IF ALL LOGIC IS SOLVED 7 CONTACTS HAVE TO BE EXAMINED.

2) IF LOGIC IS SOLVED ONLY TO THE POINT THAT IT BECOMES CERTAIN TO BE TRUE OR FALSE, YOU MAY NEED TO EXAMINE ONLY 1 CONTACT.

IF IN THE EXAMPLE SHOWN CONTACT g IS THE ONLY ONE THAT CHANGES STATE FREQUENTLY, SWITCHING IT WITH CONTACT 21 WILL REDUCE THE SCAN TIME CONSIDERABLY.

Figure 9-17. Reducing Scan Time in a PLC

Input/Output Scan

The processor can use two methods to communicate with its input and output equipment. In the parallel mode, all the bits of a word are transferred at one time in parallel. This method is fast but limited to short distances due to timing changes caused by cable capacitance. In the serial mode, the bits of a word are transmitted one at a time; this method is slower but it can be made to work very well over almost any distance. Generally, the processor can update all its parallel inputs in one scan but requires several scans to update all its serial inputs. One can say that parallel I/Os are synchronous with the PLC scan, but that serial I/Os are asynchronous. PLC manufacturers vary in the way they handle asynchronous I/Os. Some will access one rack each scan, while others will provide data only on how long it takes to update the I/O table.

The asynchronous nature of serial I/Os has serious implications for scan time. The scan time is no longer a good indication of how fast the PLC is or how fast an input change can be seen by the PLC. For example, a PLC with a 40 millisecond scan time could easily handle events that lasted for 100 milliseconds if it is using parallel I/Os. If, on the other hand, it is using serial I/Os and the update time is 100 milliseconds, the PLC could still have a 40 millisecond scan time and not be sure of seeing a 100 millisecond event.

To get around some of these I/O update limitations, most PLCs have a feature that will do an immediate update of the I/Os during the scan instead of waiting for the end of the scan. This works very well with parallel I/Os but is useless with serial I/Os. The immediate update also slows down the scan time, since the processor has to stop executing the user's program while it updates the I/Os. In all cases, the I/O immediate update works only on a limited number of bits, usually one word. This is still a very useful tool, however, when only a small amount of high-speed logic is required.

Power Supply

Needless to say, PLCs require power to operate. Two sources of power are used in PLCs: battery power and local power from the plant or utility. In many cases, the battery is used to maintain the user's program, keep the time-of-day clock running, and retain the status of certain bits and registers. Some small PLCs do not use a battery and so cannot maintain a real-time clock. Whenever a battery is part of the PLC, it should be monitored in the user's program, with adequate warning given when it is getting low. Memory and program backups are covered in the next section, and power supply wiring and installation are in the section on installation.

Power quality will affect the operation of a PLC. There are two places to provide this quality: in the power supply itself or in extra equipment added to the "upstream" side of the power supply. There is no free lunch, because skimping on the power supply too often requires additional equipment that can wipe out all the savings. A well-designed PLC will include circuitry in the power supply or elsewhere that continually monitors all necessary voltages and performs an orderly shutdown when the voltage drops below the level required to maintain reliable operation. Similarly, there will be circuitry to allow the processor to run only after determining that all the voltages necessary for operation are present and stable. It is important to have a dead band between shutdown and start-up to avoid oscillations or other problems.

Look at both the range above and below nominal when considering power supply capabilities. General practice for utilities is ±10 percent of nominal, but manufacturers of PLCs may give narrower limits such as ±5 percent or +5 percent -10 percent. Although low voltages have been the most common problem, high voltages can cause as many, if not more, problems.

Programming Devices

Unlike computers, PLCs do not come with the software and hardware necessary to write, document, and debug the programs that run in them. With a personal computer, one needs to buy only a compiler or an interpreter (software) to write programs; the hardware is already there. PLCs require the purchase of hardware (and often software) to create, debug, and document programs. The most basic programming device is a hand-held single instruction device that allows one to program one instruction at a time and to see the result on an LCD screen. This device is inexpensive and easy to use. It is also very limited and is generally used only on the small shoe-box PLCs. A step up is the hand-held line-of-logic device. This device has a larger LCD display that shows an entire line of logic in relay ladder symbols instead of only one instruction. It is clearly much easier to use and more flexible, but also more expensive. The line-of-logic device is available for the small shoe-box PLCs and some medium-sized controllers. Some hand-held units have the capability to save a program to a separate tape unit and then reload the program from the same tape. Some PLCs can use an ordinary audio tape unit, while some require a much more expensive digital tape unit. Some hand-held devices will also program memory chips.

The dedicated programming terminal is designed solely for programming the PLCs of a single manufacturer and, often, only one family or type of that manufacturer's product line. As with the hand-held units, ladder logic is displayed on a screen, often only one line at a time, but other features such as documentation, searching, and copying are offered. The program can be saved, usually on tape either in an integral unit or in a separate device. These programming devices are rugged industrial units, but they are also heavy, expensive, and not useful for anything else. Due to such limitations, this type of programming device is fast disappearing.

The most powerful programming device is the personal computer with a suitable adapter card and software. This equipment is available from the PLC manufacturer or a third-party vendor; both sources have pros and cons. The software usually provides all the features found in the dedicated programming device, plus saving and loading from diskette, formatted printing, more extensive documentation, and cross referencing. In addition, the computer can be used for other tasks.

 A few points must be considered when choosing a computer-based programming system. Many PLC manufacturers offer an industrialized computer with the necessary hardware and software to program their PLCs. In creating this package, the PLC manufacturer may use a nonstandard basic input-output system (BIOS), special cards or boards, and even a different operating system. This can result in problems for users who want to run word-processing, database, or other programs. Installing software from another PLC manufacturer in the same computer can cause even more problems. Fortunately, most PLC manufacturers and all third-party software manufacturers offer software that will run on any standard computer with the specified memory, drives, and monitor. Almost all options require buying the adapter card that allows the computer to communicate with the PLC.

The selected programming device must permit one to:

(1)   copy the program to some secure media (copying the screen display of a hand-held device by hand onto a sheet of paper is an option),

(2)   reload a program from the storage media,

(3)   monitor the program while it is running,

(4)   make changes to the program, and

(5)   develop new programs.

The ease, flexibility, and cost can vary considerably and are often major considerations in the choice of a PLC.

The Memory System

Overview

As with any computer, a PLC needs memory to store its operating system, data, and program. Unlike personal computers, the PLC must not lose its program when power fails and must restart itself automatically after a power failure. These requirements can be met in many different ways depending on the size of PLC, its speed, and cost targets. Bear these differences in mind when comparing different PLCs for the same job. Sometimes the PLC manufacturers will bring them to a buyer's attention, especially if they feel their design has an advantage over a competitor, but one should not rely on this.

Structure

The two basic parts of PLC memory are the system memory and the user memory. The system memory contains the programs that tell the microprocessor how to operate as a PLC. There is also some temporary memory for the microprocessor to use in executing its programs. The system memory is programmed by the PLC builder and is clearly not under user control. The system memory has three parts: the executive programs, the scratch pad, and the system status areas. The scratch pad, system status areas, and read/write memory are constantly being changed while the PLC is running. The executive programs do not change and are thus located in programmable read-only memory (PROM). Some manufacturers may even put the executive in a custom-made read-only memory (ROM), but this requires extremely high volumes to be economically viable.

Under some circumstances, the executive memory may be changed. If the PLC manufacturer adds features, enhancements, or bug corrections in the executive program, these changes may be made available to users who have the old version. Sometimes these upgrades are free, usually if it involves corrections of bugs or features that were promised but not available at the original delivery time. On other occasions there is a charge for such upgrades. Some manufacturers simply send the chips with instructions and new labels for the PLC and the user changes the chips himself, while others require sending the old PLC back for an upgrade. Some users prefer changing the chips themselves, because such an approach involves work that is easier to schedule and requires only minimal downtime; however, many users do not have the tools, or desire to get into chip changing, especially where static-sensitive complimentary metal-oxide silicon (CMOS) chips are concerned. In many cases where the board or PLC has to be returned to the manufacturer for upgrade, it is because the changes involve more than merely changing a plug-in chip, and the manufacturer judges it best to make these changes for the user.

This brings us to the really interesting part—the user memory. User memory contains the following items: input data table, output data table, internal bits, internal registers, and the user program. Each PLC manufacturer will have proprietary terms for these items; they differ slightly from the ones used here, but they should still be easily recognized. There are three ways to allocate the memory among these five items: fixed allocation, user-configurable hardware, and user-configurable software.

In fixed allocation, the PLC manufacturer determines all the allocations beforehand for a particular model and leaves the user to choose only what suits the need. For user-configurable hardware, the user can choose from a limited list of hardware options to suit the application. In the user-configurable software version, all the choices are made by the user in the memory, subject to a maximum memory limitation. Table 9-2 summarizes these choices.

Table 9-2. Memory Types and Configurations

ITEM	USE	MEMORY TYPE	HARDWARE CONFIGURABLE	MEMORY CONFIGURABLE	SIZE 1 WORD = 16 BITS	COMMENTS
EXECUTIVE	Operating system for the PLC	Read-only	System	System		
SCRATCH PAD	Temporary data for the system	Read/write	System	System		
SYSTEM STATUS	Clock error conditions, scan time, etc.	Read/write	I/O system	System	64 words	
INPUT DATA TABLE	Status of inputs, force table	Read/write	I/O system	User memory	32 to 256 words	At least 2 bits per input
OUTPUT DATA TABLE	Status of outputs, force table	Read/write	I/O system	User memory	32 to 256 words	At least 2 bits per output
INTERNAL BITS	Single bits use for program control	Read/write	Data memory	User memory	16 to 128 words	
REGISTERS	Data and valves used in the user program	Read/write	Data memory	User memory	512 to 4096 words	At least 2 registers per timer/counter
USER PROGRAM	Controls the machine	Read/write	Program memory	User memory	2048 to 32768 words	

The fixed allocation is usually found in the small shoe-box PLCs. Here, there may be 10 inputs, 20 outputs, 64 timers/counters, 200 internal bits, and so forth, and the only way to change this is to buy a different model. The fixed allocation makes it easier for the system programmer, who has complete control of everything. It may also make the PLC faster, since the hardware and software can be optimized for one configuration.

With user-configurable hardware, the user gets to choose a configuration to suit the specific application, for example, from among 256, 512, 1024, 2048, or 4096 I/Os and 1K, 2K, or 4K registers. Some manufacturers may allow the selection of any values for I/Os and registers, while others limit the choice to a smaller range (such as 1K I/Os and 1K registers, or 2K I/Os and 2K registers, or 4K I/Os and 4K registers). Since the hardware is usually on a card, the penalty for not getting enough is buying a new I/O memory card (sometimes a credit is available for the old one). This is because the cards are different, and one cannot simply plug in a few more chips and upgrade from 1K to 2K. The user program memory is on another card and is usually field expandable, either by adding chips or by adding another card, or both. Care must be taken in reading the PLC manufacturer's literature and manuals to understand the various upgrade possibilities and their costs.

User-configurable memory occupies a single block of memory to provide all the functions required by the system. Some partitions, such as I/O tables, may be of fixed length based on the maximum I/Os the PLC can handle; others, such as internal bits and registers, are user-definable to suit the needs of the program. This overhead reduces the memory available for the program itself. The original 8K of memory could become 6K or even 4K after all the necessary bits and registers have been given the space they need. Configurable memory provides greater flexibility, but the user has to manage it and understand the limitations.

It's easy to see the user's dilemma—how much memory is need for I/Os? Internal bits and registers? Program space? How does this change if a different PLC is used? All these questions must be answered before the program is written. One can estimate the needs reasonably well, given a well defined application with a complete I/O list, a completed description of what the PLC should be doing, and a good knowledge of the PLCs concerned. If any one of these three things is missing, and it usually is, the decision is not so easy. Most designers will choose to buy more memory, feeling that the cost of trying to program around a memory limitation is going to be far more than the cost of the extra memory. In the case of unique applications or applications that involve only a few copies, this is probably justified.

> If many copies are to be made of the same application, a better approach is to do a pilot project with whatever it takes; then once the application is well developed, see what can be done to optimize the memory requirements.

Memory Types

The first consideration in the choice of memory types is the method of accessing the data—random or sequential. In random access memory (RAM for short), any bit or word can be accessed directly without having to read the bits or words that come before the desired bit. Sequential memory requires accessing the memory locations one after another in exact physical sequence, like reading a tape. All PLC memory is RAM of one type or another. Sequential memory is found only in some types of charge-coupled devices (CCDs for short) that are used in vision systems to read the camera image.

The next consideration has to do with the operations permitted on the memory—reading and writing. Memory can be read-only, read and write, or write-only. Memory operations are also carried out on diskettes. The only standard acronym is ROM for read-only memory. Some examples of write-only memory are printer and screen memory.

The last consideration is how long the memory will retain the data stored in it. Volatile memory loses the data as soon as power is removed. Nonvolatile memory will retain the data intact when power is removed.

Of the nonvolatile read-only memory types, the most stable is the ROM chip. This chip is made like a memory chip except that the data values are designed right into the chip and cannot be changed. This process is used only for high-volume applications such as the executive program. Changing the program requires designing a new chip.

Programmable read-only memory (PROM) can be programmed once and only once. The chip is made with fusible elements for each bit and thus comes with all bits set to 1. To program the chip, a high current that melts the fusible link is passed through those bits that are to become zero. PROMs are often used for smaller volume applications that do not justify the time and expense of ROMs.

A further development is the erasable programmable read-only memory (EPROM). This chip has a window over the storage area, allowing the chip to be erased by exposing it to an ultraviolet (UV) light source. Complete erasing usually takes several minutes, but several chips can be done at the same time. The programming process does not physically destroy parts of the circuit, but instead places a charge at the desired location in the memory in order to indicate a 1. Since the memory cells are made with no discharge path, the charge remains until it is erased. There is a limit to the number of times an EPROM can be reprogrammed, but it is not usually considered a problem. Inadvertent erasing of the chip can happen if the window over the memory area is not covered and a source of UV light is present, as from a welding torch or a fluorescent light. To avoid this problem, the window is usually covered with a suitable sticker after programming. There have also been questions about the long-term memory retention of EPROMs, considering that such things as cosmic rays could cause a single bit to change. These considerations are probably irrelevant for most PLC users.

At the other extreme is the volatile read/write memory in which a solid-state memory chip uses a collection of transistors to form a bistable memory cell that can be read without destroying the data in the cell. For details, consult an electronics textbook on the subject. Various technologies are used, each having its own costs and benefits, and all this is constantly changing as existing technologies are improved and new ones are developed. The chip has four sets of connections—power, control, address, and data. The address lines are used to specify the exact memory location to be read or written, and the control lines specify the type of operation to be performed. The data lines contain the data to be written or the values found in the memory cell. Proper manipulation of these lines permits the reading and writing of data. The major drawback is that power is required to maintain the data. Remove the power and the memory is lost. When the power is restored, the individual memory cells may take any value, so one of the processor's start-up tasks is often to set all memory to 1 or 0 before going on.

Consider now the nonvolatile RAM. One of the older types is EEPROM, sometimes called E2PROM, for electrically erasable programmable read-only memory. These chips are similar to the UV erasable ones, except that the erasing can be done electrically. This makes them ideal for storing programs and other data that must not be lost. Although these chips can be programmed in the PLC, the programming does take a long time compared to the PLC scan time. In addition, there are more stringent limits on the number of times the chip can be erased. These limitations are more than offset by the advantages of having a non-volatile memory. Another option is battery power for the memory. In this case, a regular volatile RAM chip is used for the memory, and power is supplied to it from a battery in the PLC. Since the battery is powering only the memory, it can maintain the data for at least a year, and some of the newer lithium batteries are said to be good for five years.

 Note that although battery-backed memory is nonvolatile, it is not as long-lived as the other types of nonvolatile memory. Realize also that removing the battery with power off clears the memory.

Some manufacturers put the battery right on the memory card so that the memory card can be removed without losing the program. Others put the battery in the power supply card with the result that removing either the memory card or the power supply card will clear the memory. Some users have been caught by this! So-called NOVRAM chips combine a regular RAM, EEPROM, and control circuitry, all in one chip. The PLC sees only the RAM, and it activates a control line in the NOVRAM to copy the RAM to EEPROM, or vice versa. In some applications, this copying is automatic; in others, it is up to the user to issue the save instruction. Some systems have been made with enough energy storage in them to allow the NOVRAM to make a copy of the RAM to EEPROM when it senses that the power is failing.

What does all this mean for the average PLC user? First, one must decide how permanent the programs must be. EPROM or EEPROM would be used for critical systems or for systems or machines shipped to places where field service is not justified simply for the loss of a program. In such cases, the program would be developed in-house, debugged, and then the final version copied into EPROM or EEPROM. On the other hand, if one develops all programs in-house, makes constant changes, and has good backup and control systems, battery-backed memory is recommended because of its convenience.

Manufacturers do not offer all the above options in every PLC they make. One common practice is to provide for a memory expansion module that can be either additional RAM or nonvolatile memory. The implications of this practice can be serious for users who need lots of memory and EEPROM. For example, one manufacturer offers a module that allows 8K of RAM and 8K EEPROM as back-up, or 8K RAM plus 6K of additional RAM. If the program runs in less than 8K, there is no problem, but if, say, 9K is needed, EEPROM backup is not possible. If the backup is needed, the size of processor must be increased by one.

Memory Maps

The arrangement of PLC memory is known as a memory map. Computers also have memory maps, as described in Chapter 5. As mentioned earlier, there are many ways to design the memory of a PLC, and in some of them the user can also configure the memory to suit his needs.

Figure 9-18 shows a memory map for a controller that uses a fixed memory map. A more interesting example is Figure 9-19 which shows the memory map for a user-configurable hardware system. Note how the manufacturer has assigned the memory addresses so that programs can be expanded easily and

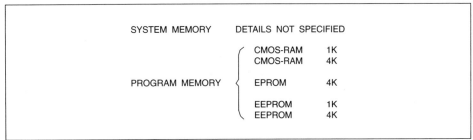

Figure 9-18. Fixed Memory Map

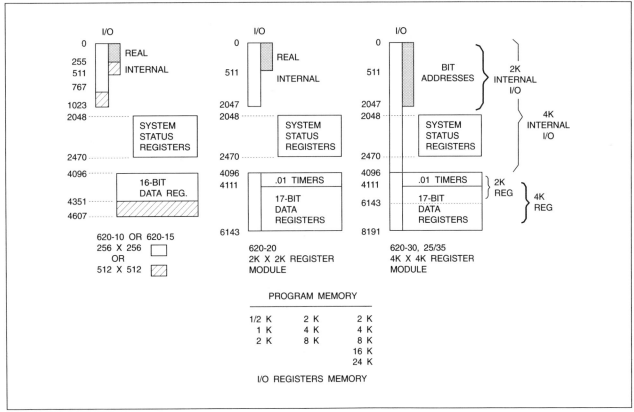

Figure 9-19. Memory Map—User-Configurable Hardware
(Courtesy of Honeywell, Inc.)

upgraded without having to recode the entire program because the addresses are now different. Not all designers are as considerate. With a limited number of resources it is up to the user to keep track of his usage and manage it carefully. Some manufacturers will include forms to allow the use of the available memory to be mapped out as the programmer or designer goes along.

Figure 9-20 shows the basic memory map for a system that allows the user to configure the memory himself. Note that only the input and output data tables are mandatory; all the rest is optional. Figure 9-21 shows the same type of memory but with some values filled in. Note how the number of bytes taken per element varies with the type of data being stored (from one to three bytes per element). With user-configurable memory, some consideration must be given to future expansion, especially if it is to be done on-line. Most PLCs of this type put the data tables at the beginning of memory, followed by the user program. This allows the user program to expand and contract in the memory not used by the data tables. Changing the data table size requires stopping the processor and rearranging the memory. This precludes making on-line data table size changes. If development is being done in the lab, this is not a problem; but in the case of an operating machine, it may not be easy to get the downtime required, especially if the shutdown and start-up processes are long, costly, or both. The usual practice is to allow some extra space in each data table data type so that future on-line modifications can be done easily. Care still must be taken to keep this reserve memory from using up too much of the available memory to the detriment of the program. User-configurable hardware does not have this problem, because the only change one can make is to add more hardware and that has to be done with the PLC stopped.

DATA FILES

WORDS	NAME	
32	OUTPUT IMAGE	
32	INPUT IMAGE	
32	STATUS	
3-1002	BIT	
5-3002	TIMER	
5-3002	COUNTER	**MAY BE REPEATED AS REQUIRED**
5-3002	CONTROL	
3-1002	INTEGER	
4-2002	FLOATING POINT	

PROGRAM FILES

WORDS	TYPE	
4 UP	PASS WORD AND IDENTIFICATION	
4 UP	SEQUENTIAL FUNCTION CHART	
4 UP	MAIN PROGRAM	**ONE ONLY**
4 UP	LADDER LOGIC SUBROUTINES	**REPEAT AS REQUIRED**

Figure 9-20. User-Configurable Memory

DATA TABLE		
FILE NO.	TYPE	WORDS
0	OUTPUT	32
1	INPUT	32
2	STATUS	32
3	BIT	30
4	TIMER	8
5	COUNTER	11
6	CONTROL	8
7	INTEGER	3
8	FLOATING POINT	4
9	INTEGER	8
10	BIT	64
TOTAL		232

PROGRAM FILE		
FILE NO.	DESCRIPTION	WORDS
0	SYSTEM DATA	10
2	MAIN	384
3	SUBROUTINE 1	28
4	FAULT HANDLER	34
5	MESSAGES	210
TOTAL		835
GRAND TOTAL		1067

Figure 9-21. Completed Example—User-Configured Memory

It should be possible to determine the memory required for the operating system, I/O tables, and minimal data tables. The problem is in determining the memory required for the program and realistic data tables. Since the complexity of the application is not usually known beforehand, experience has led to the development of some rules for sizing. Hughes (see the References and Bibliography) says to take 10 times the I/O count, then suggests leaving a further 25-50 percent for future growth. Others suggest leaving 20-40 percent available for future growth without saying how to calculate the base amount. Most also say that extensive use of mathematical or data manipulation functions, or special functions such as PID loops, require additional memory.

Memory Size Considerations

Memory size is a slippery subject. The first thing to know is the number of bits in a byte or word. The most common value is eight bits to the byte and two bytes (or 16 bits) to the word. Some manufacturers use a 16-bit byte as standard. Others, although they use an 8-bit byte, operate on two bytes at a time and so require all operations to use 16 bits even if they don't need them. Why do they do this? The 8-bit byte allows 256 possible values, which is not really enough for most mathematical operations and would limit the number of I/Os in a PLC to 256. The 16-bit word allows 65,536 combinations, commonly referred to as 64K. This certainly gives enough I/O space and is reasonable for mathematical operations. To be able to express signed numbers, the maximum value is cut in half because one bit has to be used for the sign. Thus $\pm 32,767$ is often seen as the largest integer expressible in a PLC. To complicate matters, at least one manufacturer uses a 17-bit register that permits the expression of values up to $\pm65,535$. This is possible because of the use of dedicated memory for each type of data and a design that allows for 17-bit registers.

Consider a PLC that uses only 8-bit bytes and needs, say, 2K to do its job. Another PLC that uses a 16-bit word and the same number of words will take 2K words or 4K bytes (because there are two 8-bit bytes to the word) to do the same job! In practice, things are usually not this straightforward, because the larger PLCs can usually do more with one word than the smaller ones can do with one byte.

One must also consider any overhead needed for the operating system, status tables, and so forth that may require user memory. Some manufacturers quote the total memory in the processor (say, 64K) and then state somewhere in the documentation that the operating system requires 34K (leaving only 30K for programs).

The Discrete Input/Output System

Introduction

The input/output (I/0) system provides the physical connection between the process equipment, or machinery, and the processor. As can be seen from Table - 9-3, many different types of devices can be connected to the processor. This section will discuss the most common class of I/O devices in a PLC system—discrete devices. By definition discrete devices have only two states—on and off. Some other names for these states are closed and open, true and false, high and low, 1 and 0. Discrete devices are easily handled by the processor, because its memory also has two states only.

Table 9-3. Typical Discrete I/O Field Devices

Input Field Devices	Output Field Devices
Selector switches	Annunciators
Push buttons	Electric control relays
Photoelectric cells	Electric fans
Limit switches	Lights
Logic gates	Logic gates
Proximity switches	Alarm horns
Process switches (level, flow, etc.)	Motor starters
Motor starter contacts	Electric valves
Control relay contacts	Alarm lights

One of the major advantages of PLCs is their modular construction. By using an input/output system that is based on a relatively small number of modules, the user can build a system to suit the needs, and then later, expand or contract it as needs change. Troubleshooting and repair are also simplified, as only the suspect part needs to be changed, which can usually be done without removing wiring.

Some smaller PLCs come with the I/Os in the same package with the processor, then allow additional expansion blocks with various combinations of I/Os to be added on. Medium and large PLCs use a rack system. The rack (also called chassis or cage) is an equipment housing that receives the various modules that make up the system. Figure 9-22 shows a typical rack arrangement. In most systems, the location of the power supply, processor, and communications cards is fixed, and the remaining slots can be filled by any type of module. The I/O modules generally have an edge connector on one end that plugs into the communications backplane and a connector on the front that receives the terminal block or wiring arm to which the physical connections are made.

I/O Rack Enclosures and I/O Table Mapping

The I/O rack provides a foundation for the various modules in the system. It offers physical support and restraint of the modules against vibrational forces, usually without screws or other cumbersome hold-down arrangements. It provides a power distribution system for the modules that supply the voltages necessary for the proper operation of the modules in the rack. The backplane also includes the parallel communications channel that allows the modules to communicate with the local processor or communications card. Most manufacturers also include a keying function that, if used, ensures the correct module in the correct slot. There are also provisions for grounding and labeling and, sometimes, programming ports. With all these features many manufacturers refer to the rack as a universal I/O rack. The racks are often available in several sizes: full, half, quarter, and eighth to suit the user's needs.

The power supply may be mounted in a slot in the rack or outside the rack, with a cable to the rack. Most manufacturers offer several sizes of power supplies to suit the power requirements of the modules to be installed in the rack. Power supplies are often dual voltage and dual frequency in order to reduce the number of parts that have to be manufactured. The user has to set the switches properly, or wire differently, depending on the type of supply. There is usually a power-on light, and sometimes several lights, to indicate input power, output power, and fault conditions. An on/off switch is very useful but is not always provided. Without a switch, one has to shut off all power just to change a fuse or to make some configuration switch changes.

The next module in the rack is the processor or communications module. If the processor is in the I/O rack, the processor controls all communications with the I/Os in the rack. If there is no processor in the rack, a communications module is required to communicate with the processor and the I/O modules in the rack. This communications module communicates constantly with the modules in its rack via the rack backplane and, on command from the processor, communicates via another cable with the processor. This communications module also performs some control functions such as shutting down the I/O modules if communication is lost with the processor or if too many errors occur.

The I/O rack is a passive device but it does have some configuration options that are determined by the rack configuration settings. These switches are usually located on the backplane and can be accessed by removing some of the modules. They determine such things as the type of rack, the starting address, the module type or size for each slot, the shutdown option (turn everything off or leave outputs in their last state), the location of the power supply, and so on. Some manufacturers use switches in the communications module for some of these items. Since there are no standards as to where these settings should be made, each manufacturer makes that decision, and the designer has to carefully read the documentation to understand how to configure the rack.

Take care in counting the number of slots available in a rack, especially when making comparisons between manufacturers. Although most manufacturers count only the slots available for I/O modules, there may be times when one of the slots will be pre-empted. When a rack that normally uses an external power supply uses a rack-mounted power supply, one slot is lost. In some cases, power supplies may be paralleled for greater capacity, and so another slot may also be lost. Furthermore, some modules take more than one slot, or have other requirements that may reduce the number of I/O points that can be installed in a rack.

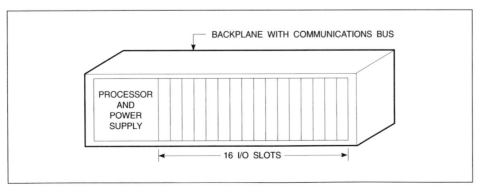

Figure 9-22. A Typical I/O Modular Housing [Ref. 14]

Record configuration details of each and every rack and make them available to the operating and maintenance personnel. When communications modules, processors, or racks are changed, it is essential to properly configure the new ones. Many hours of downtime can be caused by improperly configured modules or interchanged modules. Pay particular attention to the switch designations in the manual and in the hardware. The manual may indicate the switches should be open or closed, whereas the switch in the module may have its positions indicated as on and off. Although we all know that "on" equals "closed" and "off" equals "open," it is easy to end up with the switches set exactly backwards to what they should be. Assume nothing! Check even the factory settings to see that they really are as they should be.

Some newer rack designs have eliminated the configuration switches; instead, they use an EEPROM or other memory device in the rack or communications module to contain the configuration information. The configuration is then done with the programming unit, thus making it easier to verify the configuration and correct it as required.

ADDRESSING

Before discussing the I/O table mapping one has to understand something about I/O addressing. With varying numbers of I/O modules with different numbers of points in various racks located just about anywhere, some method is required to uniquely identify every input and output. The manufacturer has to answer three basic questions in designing an addressing scheme: Will the addresses be fixed or user-determined? Will the addresses be based on inputs and outputs or insensitive to whether the point is an input or output? Will the addresses be decimal or word-and-bit? Fixed addressing is used only by the smaller, shoebox PLCs. The lowest address is in the processor unit, and it increases for each expansion module added on (see Table 9-4). Medium and large PLCs require the

Table 9-4. Fixed-Addressing Example

Name	Allocation No.	No. of Points
Input	0-7, 10-17, 20-27, 30-37, 40-47, 50-57, 60-67, 70-77, 80-87, 90-97, 100-107, 110-117, 120-127, 130-137, 140-147, 150-157	128
Output	200-207, 210-217, 220-227, 230-237, 240-247, 250-257, 260-267, 270-277, 280-287, 290-297, 300-307, 310-317, 320-327, 330-337, 340-347, 350-357	128
Internal Relay	400-407, 410-417, 420-427, 430-437, 440-447, 450-457, 460-467, 470-477, 480-487, 490-497, 500-507, 510-517, 520-527, 530-537, 540-547, 550-557, 560-567, 570-577, 580-587, 590-597, 600-607, 610-617, 620-627, 630-637, 640-647, 650-657, 660-667, 670-677, 680-687, 690-697	240
Special Relay	700-707, 710-717	16
Timer	0-63	64
10-msec Timer	64-71	8
External Timer	72-79	8
Counter	0-44	45
Reversible Counter	45 (dual pulse), 46 (up/down selection), 47 (up/down selection & high-speed counter stage selection)	1 each
Shift Register	0-127 (Bidirectional)	128
Shingle Output	0-95	96
Coincidence Output of Counter	FUN100-FUN147	48
≥ Output of Counter	FUN200-FUN247	48

user to determine the addresses within established design limits. If the addressing is I/O-insensitive, the user can establish any mix of inputs and outputs up to the system maximum and place the modules anywhere in the rack. When the addresses are I/O-sensitive, there are distinct maximums for inputs and outputs, and the user must place his modules in specific racks or positions in the rack, depending on whether they are inputs or outputs. The final choice is based on whether the address is a decimal value that does not take into account the physical location of the module or whether it's a word-and-bit combination that is determined by the module's physical location in the rack. Figures 9-23 and 9-24 give examples of two possibilities currently used.

The I/O-sensitive, word-and-bit addressing scheme is the most common, probably because it easy to use and understand. On the other hand, it is more restrictive because it requires more care to design a system and can require an extra rack if the quantities of inputs and outputs are substantially unequal.

I/O TABLE MAPPING

The I/O table map is a tabular or graphical representation of the inputs and outputs connected to the PLC. There are three parts to the I/O map: the list of addresses, the location and type of all modules actually installed, and the details of the field device connected to each input or output point. Figure 9-25 shows a map for a processor using a word-and-bit addressing scheme, while Figure 9-26 shows a

ADDRESS	FUNCTION	
0	INPUT1	
1	INPUT2	
2	INPUT3	
3	INPUT4	ONE 8 POINT INPUT CARD
4	INPUT5	
5	INPUT6	
6	INPUT7	
7	INPUT8	
8	OUTPUT1	
9	OUTPUT2	
10	OUTPUT3	
11	OUTPUT4	ONE 8 POINT OUTPUT CARD
12	OUTPUT5	
13	OUTPUT6	
14	OUTPUT7	
15	OUTPUT8	
512	INTERNAL BIT	
513	INTERNAL BIT	
		TO PROCESSOR LIMIT
4096	REGISTER1	
4097	REGISTER2	
4098	REGISTER3	
	ETC.	TO PROCESSOR LIMIT

Figure 9-23. Decimal Non-I/O-Sensitive Addressing Example

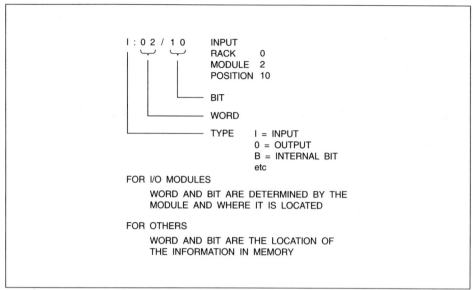

Figure 9-24. Word-and-Bit I/O-Sensitive Addressing

map for a decimal addressing scheme. Manufacturers often have forms available for making an I/O table map with the allowable addresses already listed, thus making it easier for the user to complete them. Figure 9-27 is an example of such a form. Some users prefer to make their own forms or even build a database for this information. Figure 9-28 is an extract from a custom database for a PLC using a decimal-based addressing scheme that is I/O-insensitive. Notice how the type of module (input or output) has been added along with the rack, module, and terminal positions, to allow the user to know exactly what is installed where and what field devices are connected. Some documentation packages provide extensive mapping functions and will be discussed further later in this chapter. Figure 9-29 shows the complete hardware-to-software interface for a word- and bit-addressed system that puts into perspective much of what has been already discussed here.

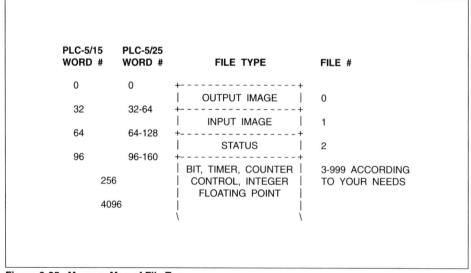

Figure 9-25. Memory Map of File Types
(Courtesy of Allen-Bradley, a Rockwell International Company)

Remote I/O Systems

One should really speak of serial and parallel I/Os, as opposed to remote and local I/Os. Many manufacturers call parallel I/Os local and serial remote, even if they are all in the same enclosure! To avoid further confusion, this discussion will refer to them as parallel and serial I/Os. Parallel I/Os use a cable that has one wire for each bit transmitted. All eight or sixteen bits are transmitted simultaneously. This method is fast and relatively easy to use. Its limitations are sensitivity to noise, distortion caused by cable length and capacitance, and restriction to about 100 feet of cable per system. Serial I/Os send the bits, one after the other, over one or two pairs of wires. Serial communications are slower than parallel communications and require more hardware and programming, but they are less affected by noise and cable length. In addition, they can go up to 8000 feet without repeaters and even beyond this distance if repeaters or fiber optic cables are used.

Some manufacturers offer both parallel and serial I/Os for all I/Os, while others offer only one rack of parallel and require the rest to be serial. Again, there is no standardization, and the market is continually changing. When comparing systems, check the arrangements the manufacturers offer and also the prices; it is not always as clear and easy as one would think.

Discrete Inputs

Because the microprocessor chip in the processor works only with binary logic levels of 0 and 5 volts, some level conversion and threshold detection are required. The voltage levels commonly used are shown in Table 9-5. In addition to the AC And DC values, some manufacturers offer modules that will accept either AC or DC. The differing voltage levels come from the variety of devices that can be connected and from the practices of various industries. The voltage sensed by the input module is provided from some source outside the module, often from the field devices but in many cases from the same supply as the PLC.

Table 9-5. Typical Discrete I/O Signal Voltage Levels

I/O Signal Types	I/O Signal Types
5 volt DC	120 volt AC/DC
12 volt DC	230 volt AC/DC
24 volt AC/DC	Relay contacts
48 volt AC/DC	100 volt DC

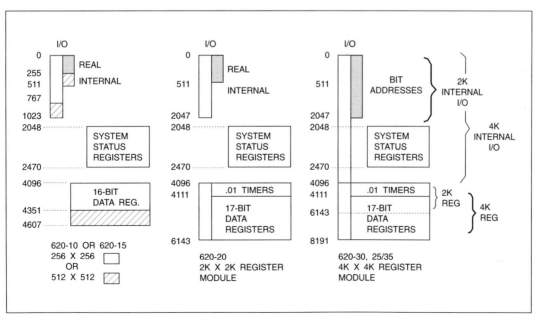

Figure 9-26. System Status, I/O, and Register Table Configuration
(Courtesy of Honeywell in U.S.A.)

TYPE **PLE-30R ALLOCATION TABLE**

Sheet No. _____ of _____

Part No. Symbol & Name		Part No. Symbol & Name		Part No. Symbol & Name		
Input	0	Internal Relay	40	Internal Keep Relay	80	
	1		41		81	
	2		42		82	
	3		43		83	
	4		44		84	
	5		45		85	
	6		46		86	
	7		47		87	
	8		48		88	
	9		49		89	
	10		50		90	
	11		51		91	
	12		52		92	
	13		53		93	
	14		54		94	
	15		55		95	
	16		56		96	
	17		57		97	
Output	18		58		98	
	19		59		99	
	20		60	Timer	0	
	21		61		1	
	22		62		2	
	23		63		3	
	24		64		4	
	25		65		5	
	26		66		6	
	27		67		7	
	28		68		8	
	29		69		9	
Internal Relay	30		70	Counter	0	
	31		71		1	
	32		72		2	
	33		73		3	
	34		74		4	
	35		75		5	
	36		76		6	
	37		77		7	
	38		78		8	
	39		79		9	

Title		Approved by	Checked by	Designed by
Name of Program				
Date	Dwg No.			

Figure 9-27. Sample I/O Table Map Form
(Courtesy of IDEC Systems & Controls Corporation)

SEQ NO	TRK NO	MOD	CCT	ADDR	CODE	PART NO	WIRE	LINE	OLDWIRE	OLDLINE	MACHINE FUNCTION	DESCRIPTION	DEVICE NUMBER
314	2	E	01	224	IN	621-1100	117	3			PRINT 2	BINDER LEVEL 1 LOW	LL3
315	2	E	02	225	IN	621-1100	118	3			PRINT 2	BINDER LEVEL 1 MED	LL3
316	2	E	03	226	IN	621-1100	119	3			PRINT 2	BINDER LEVEL 1 HIGH	LL3
317	2	E	04	227	IN	621-1100	120	3			PRINT 2	BINDER LEVEL 2 LOW	LL4 FUTURE
318	2	E	05	228	IN	621-1100	121	3			PRINT 2	BINDER LEVEL 2 MED	LL4 FUTURE
319	2	E	06	229	IN	621-1100	122	3			PRINT 2	BINDER LEVEL 2 HIGH	LL4 FUTURE
320	2	E	07	230	IN	621-1100	124	3			PRINT 2	BINDER LEVEL ON	SS21
321	2	E	08	231	IN	621-1100	125	3			PRINT 2	BINDER LEVEL PUMP 1	SS37 FUTURE
322	2	E	T1		IN	621-1100							
323	2	E	T2		IN	621-1100							
324	2	E	B1		IN	621-1100	X2						
325	2	E	B2		IN	621-1100	X2						
326	2	F	01	232	IN	621-1100	126	3			PRINT 1	DOCTOR BLADE IN	PB34
327	2	F	02	233	IN	621-1100	127	3			PRINT 1	DOCTOR BLADE OUT	PB35
328	2	F	03	234	IN	621-1100	128	3			PRINT 1	NIP CLOSE	PB36
329	2	F	04	235	IN	621-1100	135	3			PRINT 1	NIP OPEN	PB37
330	2	F	05	236	IN	621-1100	138	3			PRINT 2	DOCTOR BLADE IN	PB48
331	2	F	06	237	IN	621-1100	144	3			PRINT 2	DOCTOR BLADE OUT	PB49
332	2	F	07	238	IN	621-1100	145	3			PRINT 2	NIP CLOSE	PB50
333	2	F	08	239	IN	621-1100	146	3			PRINT 2	NIP OPEN	PB51
334	2	F	T1		IN	621-1100							
335	2	F	T2		IN	621-1100							
336	2	F	B1		IN	621-1100	X2						
337	2	F	B2		IN	621-1100	X2						

Figure 9-28. Custom Database Extract of I/O-Insensitive Decimal-Based Addressing Scheme

To avoid problems caused by the loss of the supply voltage, connect it directly to an input, and then use this input to provide an appropriate action in the program. This can be especially useful if several different supply voltages all have to be present for the proper operation of the system. It can also help prevent misleading, or false, alarms caused by the loss of one of the supply voltages.

When in operation, the input module senses the voltage supplied to its terminal and converts it to a logic level signal that the processor can use. A logic 1 indicates the presence of a voltage at the input module's terminal and corresponds to the switch connected to it being ON or CLOSED. A logic 0 indicates the absence of voltage at the input module's terminal, and corresponds to the switch being OFF or OPEN. Figure 9-30 shows a typical block diagram for an AC input. The details will vary widely among PLC manufacturers as new components come on the market and as prices change. The power section converts the AC voltage to DC in the bridge rectifier, filters the resulting DC to remove the AC ripple and some of the noise that comes in, and then passes the voltage to a threshold detection circuit. The noise filter introduces a signal delay of 10 to 25 milliseconds. The threshold detector determines the minimum level required for a logic 1 and the maximum level for a logic 0. There is a dead band to eliminate chatter if the signal is near one of the threshold values. The input indicator light is usually placed after the threshold detection but is sometimes placed on the logic side.

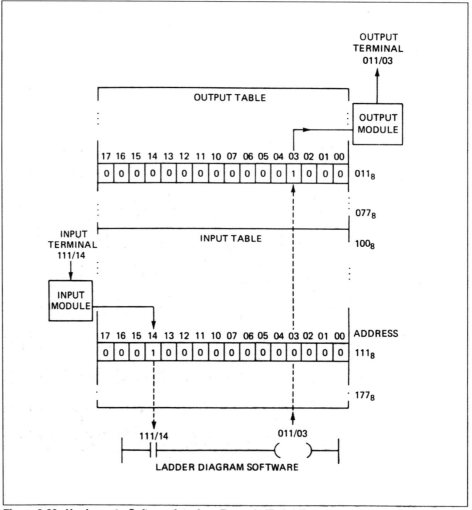

Figure 9-29. Hardware-to-Software Interface Example [Ref. 14]

Once the threshold detector has processed the input signal, it sends the result to the signal isolator. Since microprocessors are very sensitive to high voltages, isolation is needed between the high voltages present on the input cards and on the processor data bus. Generally, optocouplers are used to provide 1000 to 2500 volts isolation between the input and output sides of the coupler. These values may sound high, but spikes on the AC line can easily reach several hundred volts, and even higher, as a result of arcing contacts or welding flash. After passing through the signal isolator, the signal reaches the logic circuitry, which sends it via the data bus to the processor when it is requested to do so. Input modules that accept both AC and DC values will have the same general arrangement as shown in Figure 9-30, because the DC voltage will pass unaffected through the bridge rectifier. The values for the threshold detector will be higher for DC than for AC, because the AC values are usually measured as root mean square (RMS) values, and the detector looks only at peak values. The specification sheet should give the exact values.

For DC voltages only, there is an arrangement that is similar except for the elimination of the bridge rectifier. Also, the noise filter is often different since it does not have to filter out the 50- or 60-cycle ripple from the AC line. This is why the DC inputs often have a faster response time. DC input modules also come in "sink" and "source" versions, depending on the type of input device that will be used with them. The terms "sink" and "source" refer to the electrical configuration of the electronic circuit in a device. A sink device receives current when it is on and is said to be sinking current. A source device supplies current when it is on and is said to be sourcing current. To work properly, a source field device needs to be connected to a sink input module, and a sink field device needs to be connected to a source input device. Some manufacturers make modules that can be configured as source or sink, as required for the application. Others offer different modules, while some may only offer one type of module. Similarly, field devices may be offered as sink, source, or convertible.

> *When using sink and source modules, examine the module connection diagrams carefully to be sure that the manufacturer's terminology is understood. It's possible for the manufacturer's name for the module to be confusing, because it will often refer to the type of device to be used with the module and NOT the configuration of the module. Check also the field device data sheet, especially if devices need to be substituted. Much time can be lost tracking down mismatched devices and modules.*

Figure 9-31 shows how to convert field devices by adding a resistor in the appropriate place. The resistor is usually 1200 ohms but could be different depending on the internal circuit of the input module.

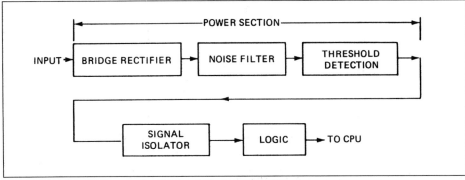

Figure 9-30. Block Diagram for AC Input Module [Ref. 14]

HIGH-SPEED INPUTS

Most manufacturers also offer high-speed DC input modules. These modules are usually restricted to 24 volts or less and have less filtering but faster circuits inside. They can be very useful in monitoring high-speed events. They are more sensitive to noise, so care has to be taken with the input wiring.

ISOLATED INPUTS

All the modules discussed so far have at least one common terminal and work well for most applications. Some applications require separate commons for each input. In such cases, there is an isolated input module for which each input has two terminals that are isolated from all other inputs. Since more terminals are now needed for the same number of inputs, there are usually fewer isolated inputs per module than for the ordinary non-isolated ones. This is due to the extra circuit board space required and also because the manufacturer will probably use the same wiring connector for both isolated and non-isolated inputs.

Some I/O capacity is lost in using isolated inputs. For example, a system that uses 8-point modules may only get six isolated points in the same module. Some systems that offer both 8- and 16-point inputs may only offer a six-point isolated input module, further reducing the number of I/Os available. This is no problem for only a few isolated inputs, but for a large number, a larger capacity system is needed.

CONNECTIONS

Figure 9-32 shows a generalized input module connection diagram. There are three sets of connections: power hot or plus, power neutral or negative, and the individual point connections. Manufacturers usually make only one connection block (or wiring arm) for both inputs and outputs, so some connections may not be used for all applications. For inputs, the power hot or plus is often not required, but the neutral or negative is always required, as are the individual point connections. Some manufacturers allow the individual circuits to be split and have separate commons or power. In this case, the jumpers between the power and common terminals are removed, and the individual power or common wires are connected. Manufacturers will often identify the terminals with the bit positions if they use a word-and-bit identification scheme; otherwise, they will number them 1 through *n* so the installer does not have to keep counting the terminals to see where he is. Labels are often provided to allow the installer to identify the inputs with their address, and, if space permits, their function.

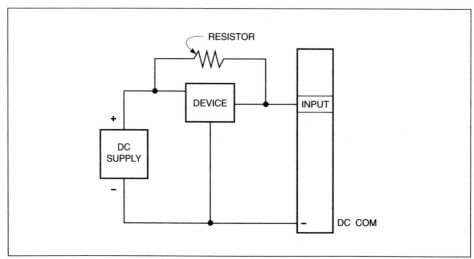

Figure 9-31. Converting Field Device with Resistor

Discrete Outputs

Outputs are the complement of inputs, so the following discussion starts at the processor end and goes toward the field device. Figure 9-33 is the block diagram of a typical AC output module. The logic is updated by the processor and latches the value sent by the processor. This latching is necessary because the processor is not in continuous communication with any module but, instead, sends signals to each module as its turn comes during the I/O scan. These latches allow the selection of default values for the outputs should the module lose communication with the processor. For the output to be turned off, all the latches and the outputs are set to OFF. For the outputs to remain in their last state, the latches are left as is and the outputs will remain as they were until power is removed. The logic circuit sends a signal to a signal isolator that is just like the ones used on inputs and that serves the same purpose—protecting the processor from damage. From the isolator, the signal goes to an AC switch, usually a triac, and then to some filtering circuits that protect the triac from noise and from transients generated when

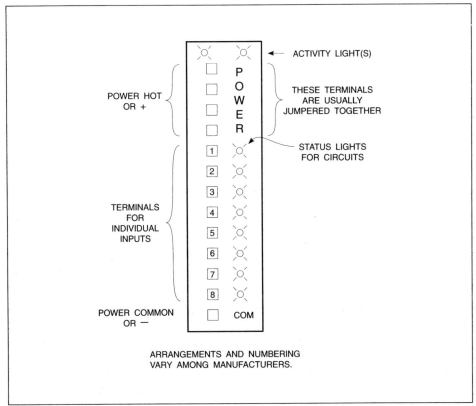

Figure 9-32. Generalized Input Module Connection Diagram

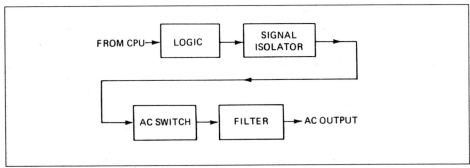

Figure 9-33. Block Diagram for AC Output Module [Ref. 14]

inductive loads (such as relay coils and solenoid valves) are switched off. Finally, there are fuses, either one per output, or a common fuse for all the outputs. Modules with a fuse usually incorporate a light to indicate that the fuse is blown.

The direct current output module has a similar block diagram, except that the AC switch is replaced by a DC switch—usually a power transistor. In addition, DC outputs, like inputs, can be either source or sink. Most DC outputs are sources that require loads to sink current. Sinking loads is the usual case, since most loads are not solid-state devices. Interfacing DC outputs to other solid-state devices (such as inputs or computers) requires care in design.

ISOLATED OUTPUTS

Like inputs, outputs come in isolated versions, for the same reasons and with the same advantages and disadvantages. Isolated outputs are also available as "dry contacts" or "reed relays." Such a module has a small relay with one (or sometimes two) contacts per output. The contacts can be normally open, normally closed, or one of each, depending on the manufacturer and the module selected. The contacts are designed for low-power, low-voltage signals that require contacts (as opposed to a voltage source). They are often used with electronic devices such as adjustable speed drives or with interfaces to computers or other specialized systems. One can always obtain a relay output by wiring a relay to any appropriate output. Now that small track-mounted relays of the DIN type are readily available and inexpensive, the reed relay module will probably be used less often than it once was.

COMBINATION MODULES

Some manufacturers offer I/O modules that have both inputs and outputs in the same module. Some even offer stand-alone modules that can be programmed for any mix of inputs and outputs. This type of I/O can be very useful for a distributed system that has a small number of I/O points in many places. In general, they work in the same manner as regular I/Os and can be used where appropriate.

Forcing

"Forcing" is a PLC function that allows the user to require that a particular input, output, internal coil, instruction, and the like, take some predetermined value irrespective of the value determined by the program running in the PLC. Forcing takes control away from the program and gives it to the user. In most cases, the user is a technician or an engineer involved in designing or troubleshooting the machine or process; only rarely would it be the process or machine operator.

The force function is available from the programming terminal or computer; it is not accessible from within the PLC program itself. Generally, it is manually entered by the operator, but it could be done automatically by a computer that has access to the programming port.

One would use the force function during:

(1) development to check out operation on a mockup or simulator,

(2) checkout and start-up to verify the operation of outputs and the program,

(3) debugging to verify the operation of the PLC program or to test various conditions,

(4) troubleshooting to isolate the problem and even to provide temporary fixes, and

(5) future development work to temporarily change the logic of the program to test some new process, material, or operation.

How does one know when there are forces in the program? If the manufacturer does not provide a way to find all forces in the program, be prepared for trouble, especially if he also allows removing the loader with forces active. Many manufacturers have a light on the PLC that indicates that it is running with forces active. Many also have a search function in the programming unit that allows a search through the program to find the forces, one by one. Some documentation packages also provide a list of forces and their location. Armed with this information, one can then decide what needs to be done.

To include the force function in a PLC, the manufacturer must make some design decisions for both software and hardware. Since all manufacturers do not have the same outlook, they all do not offer the same scope for the force instruction. The scope of the force function can also vary within the product line of a single manufacturer, especially in going from the shoe-box PLC to the giant that can control an entire factory. Generally, there is a force table somewhere that contains the information about the forced elements. Since the force status can take any one of three values—forced on, forced off, and not forced—the table is not just a binary one. Clearly, some memory has to be set aside for this function.

Manufacturers can offer a wide scope of choices for forcing inputs and outputs. For example, some might limit the user to forcing all occurrences of the inputs and outputs in the program. Some may offer the option of forcing individual contacts, one by one, as they occur in the program in addition to forcing all occurrences. This individual forcing can be very useful for debugging or troubleshooting. Some systems permit activating and deactivating the forces without removing them from the program. This would be useful for setting up a particular pattern of forced elements to test some function. Activating them all at once ensures knowing just how the sequence was activated.

The I/O table and the force table can be separate tables that are combined by the processor as it solves the logic, or the force table data can be applied to the I/O table, directly overriding any action of the processor. If the I/O and force tables are combined by the processor, examining the I/O table will not necessarily give one the true state of the inputs and outputs; one would also have to look at the force table and then check to see if forces were enabled. This is often done for the designer by the programming package, but, examining the PLC I/O table directly—say, from a computer—will require the designer to make all these checks himself. Again, manufacturers vary, so check this out for each system used.

Some users believe that forces should be used for testing and debugging, not as program fixes. In actual production applications, PLCs with forces are used for various reasons and for varying lengths of time. For critical processes, where people should not be able to force anything, a PLC for which the force function can be disabled can be used. Some PLC manufacturers also allow password protection for program changes and forces, or else they have switch settings in the processor to disable the force function as well as on-line program changes.

Interpreting I/O Specifications

Making the most effective use of discrete I/Os means understanding the manufacturer's specifications. Some manufacturers make this easy, while with others it is more difficult. Often it is necessary to make additional calculations or comparisons to make a decision. The I/O specifications appear in at least two places: in the general system specifications and in the module specifications themselves. Some manufacturers put the specifications of all their modules together in one booklet; others have a booklet for each module that comes with the module; some have a booklet for all the common or simple modules and individual booklets for each of the more complex ones. There is also the question of which revisions apply. With the constant evolution of electronic equipment, the specifications of today's module may not be the same as those for the one bought last year. This can be a problem when modifications have to be made to an existing system for which the documentation has been lost.

Once collected, how are the appropriate documents interpreted? There are four groups of specifications: general items, specific items applying to both discrete inputs and outputs, input specifics, and output specifics. Each will now be ex-

While on the subject of forces, let us not forget the original forcing tool—a jumper wire, (with or without alligator clips!). Even in this electronic age, jumpers are still used to defeat program logic and avoid nuisance trips. This type of force is usually done by the technician or electrician, often as a quick fix, but usually for a good reason. This type of force can be hard to find in a large system, so remember to look for it if problems arise.

amined in turn. For the sake of completeness, self-explanatory items are listed without discussion.

GENERAL ITEMS

"A chain is only as strong as its weakest link" certainly holds true for PLCs. The specifications of the rack power supply, the communications module, and the processor affect the performance of the I/O modules. Also, most manufacturers design all their components to the same specification, at least in one family.

Voltage. This usually applies to the power supplies in the system.

Frequency. Again, this applies to the power supplies. Most manufacturers design for the ranges normally encountered, but one should check carefully if operating from locally generated power.

Electromagnetic Interference. There are various tests and standards for the level of noise that the system will tolerate. Discussion of these tests is beyond the scope of this book.

Operating vs. Storage.

The following are straightforward but share an important distinction. Two values are often quoted for them: one for operation and another, usually more generous, for storage.

Temperature. This is the temperature inside the panel, with the doors closed. It might be a good idea to measure it at various times of day or under different operating conditions, if the system appears to be operating at too high a temperature.

Humidity.

Vibration.

SPECIFIC ITEMS

The following apply to both discrete inputs and outputs.

Points per Module.

Wire size. This may affect the ability to connect the field wiring directly to the module terminal blocks. Also, there is usually a minimum size , although it is often not given.

Isolation. This is the isolation between the field devices and the processor electronics provided by the module.

Location of Indicators. This identifies where the indicator lights are connected—on the logic side or the power side.

Wiring Diagram.

Factory Settings.

Accessories. This describes items such as wiring arms or terminal blocks, keying pins or bands, and so forth, which may or may not be supplied with the module but are usually necessary for proper operation of the system. It is annoying to discover at the time of installation that all the terminal blocks are missing because they were supposed to have been ordered separately at an extra cost!

Voltage. This is the range cf voltages the module will accept without damage and still function correctly. It is expressed as a nominal value plus a tolerance or, alternatively, as a range.

INPUT MODULES

The following apply to input modules.

Current. This is the current required to operate the module in the ON state. The input device must be able to supply this current continuously in the ON state.

Surge Current. This is the maximum current drawn by the module from the field device as it turns on. This is a transient condition, but the field device must be able to handle it.

An inexpensive max/min thermometer can be obtained at most hardware stores to record the temperature inside the cabinet without opening the doors. To observe the temperature inside without opening the door, use an indoor/outdoor thermometer. These thermometers read to at least 50°C and so will be okay for a maximum below that value. They may have to be checked for accuracy before any drastic action is taken, but they can be a very cost-effective tool.

Minimum Turn-on Current. This is the maximum leakage current that a solid-state device can have before it is likely to turn the input on. Some solid-state devices will pass a small amount of current even when in the OFF state. If this leakage is too large, it may cause a false operation of the input module.

Input Threshold Voltage. This is the voltage at which the module recognizes the field device as being on. Some manufacturers will also give the voltage level, below which the field device is considered to be off. This is also called switching level by some manufacturers.

Input Delay. This is the time the input must remain at the ON or OFF level for the module to recognize it as valid. It is due to the filtering circuits and the response time of the detection circuits. Two values may be quoted: one for the off-to-on transition and another for the on-to-off transition. Usually the on-to-off transition is longer. AC/DC inputs are nine to 25 milliseconds, and DC inputs are one to three milliseconds.

Operating Current. This is the current the module draws from the rack power supply. It is also called load or loading. When designing a system, one must ensure that the total current required by all the modules located in a rack can be handled by the rack power supply.

OUTPUT MODULES

The following are for output modules.

Current Rating. This is the maximum current that a single output can safely carry under load. Many manufacturers also limit the current each common can carry and the total current the module can carry.

Power Rating. This defines the total power an output module can handle safely. Some manufacturers quote figures for all outputs, while others stipulate total current. Many also provide further information on how these values are affected by ambient temperature. Generally, the lower the ambient temperature, the more power the module can handle. Since there is no standardization, one must carefully examine the manufacturer's specifications and maybe even call his application engineer.

Surge Current. The surge current, also called inrush current, defines the absolute maximum current the output can carry for a very short time. The specification usually includes the time for which this current can be supported, along with a further warning that it is non-repetitive. Non-repetitive means the output has to be turned off and allowed to cool before the output can safely carry the maximum surge current. This surge current is usually caused by inductive devices such as starter coils, solenoids, and motors that draw a large amount of current (up to six times normal) when they are first turned on.

Output Delay. Like the input delay, this is the length of time the module takes to change state once it has received a command from the processor. Some manufacturers do not give figures for this because the timing can be adversely affected by the field devices connected to the output.

Off-State Leakage. This is the other half of the minimum turn-on current discussed above for inputs. The off-state leakage current is the current that flows through the solid-state devices when they are in the OFF state. It can be up to a few milliamperes, and it may cause trouble when interfacing to other solid-state devices such as inputs. Older PLCs had problems with this, but that is pretty much a thing of the past now.

Voltage Drop. Solid-state devices do use some of the voltage they are switching. Typical values are two volts. This means that a 115 volt power level on the input of the module will be only 113 volts when it appears at the output terminal. In general, this is an issue only at lower voltage levels; a two-volt drop on five volts can be a problem! Also, if the line voltage is already low, the voltage drop could cause trouble.

To avoid power problems on output modules, use relays for large loads, either the electromechanical type for loads up to 10 amps or solid-state ones for higher loads, or where there is a lot of cycling.

Field Power Requirement. Modules that switch power from the field need a certain amount of power above and beyond the current drawn by the field devices. This is usually stated on a per-output, energized basis and is not usually a problem except where the power supply is marginal. This is a concern only for DC-output modules.

Fusing. This describes what type and size of fuse is used in the module for each output or for the common fuse, as the case may be. Where a fuse is used for each output, it may be important to check how the fuse acts under inrush conditions. Some fuses will blow when an inductive load is energized, even though they will safely carry the load once the coil has passed its inrush condition.

Operating Current. This is the same as the operating current for input modules except that usually an additional amount for each output is energized. Many people miss this part in doing their load calculations.

The input and output delays discussed above are for the module only and do not include the processor scan time, the time required to update the I/O rack over the parallel or serial communications network, or the response time of the field devices. These factors are very important and must be considered for any time-sensitive applications.

> Fast-blow fuses are the worst for inrush circuits; in some cases, it may be necessary to put a relay between the output and a starter coil greater than size 3 in order to avoid blowing the fuse.

The Analog I/O System

Introduction

The last section covered discrete I/Os that were limited to the binary values of 1 and 0. This section will cover analog I/Os (ones that vary over a continuous range and can take an infinite number of values). Figure 9-34 shows the variation in analog values. The wide variation and the speed at which the variation takes place both impose a substantial burden on digital devices, but PLCs are up to the task. Table 9-6 gives a brief overview of the types of devices that can be used with analog I/Os, along with some of the variables that they measure.

Table 9-6. Typical Analog I/O Field Devices

Input Field Devices	Output Field Devices
Flow transmitters	Analog meters
Pressure transmitters	Electric motor drives
Level transmitters	Chart recorders
Temperature transducers	Current-to-pressure units
Analytical instruments	Electric valves
Potentiometers	

Analog I/Os are an important part of the general field of data acquisition, and all the theory, practices, and skill involved in data acquisition should be observed when using analog I/Os. This brief survey of analog I/Os cannot hope to cover all the data acquisition practices that should be observed; it will simply concentrate on the parts that are unique to PLCs.

Generally, the analog I/O system must take a value that varies continually over a wide range and convert it to a series of 1s and 0s that a processor can understand. One should already be familiar with the binary number system and the basic idea that any numerical value can be converted to a binary equivalent. The analog input converts an input signal into a binary number of n digits; the analog output does the reverse: it converts a binary number of n digits to an analog

Consider the following example, which assesses the resolution and accuracy of 12-bit, 10-bit, and 8-bit devices. Most common resolution is 12 binary digits (12 bits), which gives 4096 possible discrete values. A quick calculation shows that one bit is 0.025 percent of the maximum value (4096) that can be displayed. Similarly, 10 bits or 1024 discrete values gives a resolution of 0.098 percent. Eight-bit devices with 256 discrete values are available, but the low resolution of 0.39 percent may prove to be a handicap. Now, the question is do these resolutions equate to the accuracy of the analog-to-digital converter? Clearly, the answer is no; a manufacturer could offer a 12-bit device with an overall accuracy of only 0.4 percent of full scale. Others might not even give the accuracy explicitly, but leave it for the user to find. Accuracy is determined not only by the resolution but also by the design and quality of the components in the analog circuit, as well as the quality of the analog-to-digital converter.

value. The greater the number of binary digits used, the greater the resolution but not necessarily the accuracy. Also, the more digits, the greater the cost.

Another concept in analog I/O systems is multiplexing. Since the components that are needed to convert signals from analog-to-digital representation are expensive and operate very fast, most manufacturers use one converter for many inputs. This is done by coupling the converter with an electronic multipole switch that connects the inputs one at a time to the converter. The control circuitry on the card then places the binary value that was read into the memory location that corresponds to the input. Since the converter chip may represent 25 percent of the cost of an 8-input module, it's easy to understand why designers try to make good use of it. Multiplexing, however, requires some care on the part of the user. The converter sees the input for a short time only and gives an instantaneous reading of the voltage. Unfortunately, most analog signals have some noise and ripple. As a result, a series of instantaneous readings will probably vary more than is desirable and thus require the designer to provide some filtering, either in the analog circuit or in the digital circuit.

Analog Input Hardware

One of the most basic parameters of an analog input module is the number of channels, but the definition of a channel needs some clarification. Voltage can be measured with respect to a common by running a single wire to the common (called single-ended) or by running two wires to the measurement point and measuring the difference between the two points, irrespective of the common (called differential). Many cards can do both measurements but provide only half as many channels for the differential measurement; for example, a card may be quoted as eight channels differential and 16 channels single-ended. When making comparisons, be careful to verify the type of channel.

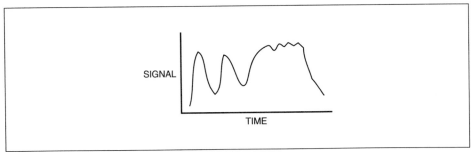

Figure 9-34. A Continuous Analog Signal

Like discrete modules, analog modules may be isolated or not. Since analog modules have to transmit an accurate representation of the analog signal across the isolation device, the isolation equipment is more difficult to design and is more costly. It is essential, however, in cases where the signals come from different power supplies or have common mode voltages that exceed the capabilities of the regular non-isolated card.

Analog modules need calibration. Some have a single set of span and offset adjustments for all channels, while others have individual adjustments. Making one adjustment is probably faster and more convenient, but it may reduce the accuracy of the system. Even if there is only a single adjustment, it is possible to adjust the span and zero of each channel in the software if desired. Execution time and memory will suffer, of course, but not usually very much.

The ranges typically offered are shown in Table 9-7. Note that the resolution in bits remains the same for all ranges, but the resolution in volts changes. The exercises one goes through in choosing transmitter ranges are applicable here also.

Table 9-7. Standard Analog Signal I/O Ratings

Analog Signal	Analog Signal
1 to 5 volt DC	4 to 20 mA
0 to 5 volt DC	0 to 20 mA
-5 to +5 volt DC	-20 to +20 mA
-10 to +10 volt DC	0 to 10 volt DC

CONFIGURATION

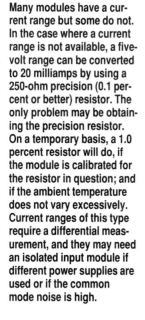

Many modules have a current range but some do not. In the case where a current range is not available, a five-volt range can be converted to 20 milliamps by using a 250-ohm precision (0.1 percent or better) resistor. The only problem may be obtaining the precision resistor. On a temporary basis, a 1.0 percent resistor will do, if the module is calibrated for the resistor in question; and if the ambient temperature does not vary excessively. Current ranges of this type require a differential measurement, and they may need an isolated input module if different power supplies are used or if the common mode noise is high.

Having discussed some of the features of analog modules, let us see how to choose among these possibilities. Figure 9-35 is a graphical representation of the types of analog modules available. The first selection decision is whether the module will be general-purpose or dedicated, within the class of inputs or outputs. Since the hardware required for inputs is quite different from that of outputs, there would be little advantage to putting both inputs and outputs on the same board. This is contrary to the practice in personal computers where it is common to see a universal I/O board with both analog and discrete inputs and outputs on the same board. Justification for the difference is that slots in a personal computer are very limited and that the I/O mix of the universal card suits any needs. Among input modules, the most common dedicated modules are the thermocouple and the RTD (resistance temperature detector) inputs. The particular signal processing needs of these types of inputs are not very well handled

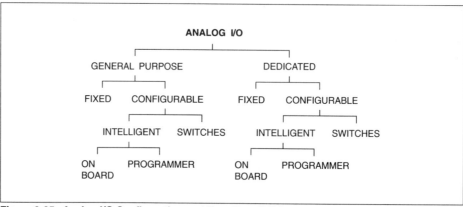

Figure 9-35. Analog I/O Configuration and Type Possibilities

in a general-purpose module, and such inputs are plentiful. Output modules tend to be general-purpose by their very nature.

The next selective decision is whether the modules will be fixed or configurable. This distinction is fast disappearing with the advances in electronics and packaging. It is hard now to find a module that is not configurable in some way. Configuration can be done by setting a series of switches on the module or by issuing instructions from the processor or programming terminal. Switch-configured modules are the simplest and probably the most reliable. The cost of the simplicity and reliability is their inflexibility for reconfiguration. Changing the configuration requires powering down the rack (and maybe even the processor), removing the module, changing the switches, replacing the module, and then restarting the system. This clearly requires some coordination with the process under control. It may even be an intolerable limitation. Another difficulty is knowing how the module is configured without removing it from the rack. If the configuration is unclear or it needs to be checked, most systems require that the module be removed and examined firsthand. Switch-configured modules usually have some error-indicating bits or words, but they are not extensive and will not support user-provided limits. Of course, the programmer can provide these features in the software as required.

Intelligent modules allow the configuration to be done via programming in the processor or from the programming terminal. There is one further variation for intelligent modules regarding the location of the configuration data. It can be in the processor program or in a nonvolatile memory chip in the module itself. Modules with an onboard memory are less likely to be reconfigured by accident or by data transfer problems. Modules that are configured from the processor require that the configuration information be sent each time the module is powered up. Because this information is part of the program, it can be altered at any time, either by design or by accident. The effects will be seen only the next time the module is configured, which may be hours, days, or even weeks later.

This reconfigurability is a great advantage when the process demands different calibrations for different products or operating conditions. A grade or product change can be made completely under program control with the ease and reliability that provides. Most intelligent modules allow the range, output format, and sample time to be set for all channels, and they allow individual values for the high and low alarms and calibration data. One cannot usually select range and format on a per-channel basis. Some systems allow one to read back the configuration parameters in the module to verify they are the desired ones, while others do not provide any means of verifying the parameters, other than the successful transfer of the data to the module.

Analog Input Data Representation

The data that an analog input module places in its register can take several forms depending on the design of the module and the use to be made of the data. Obviously, manufacturers do not necessarily offer all the possibilities, so the designer's task is often to determine how the available options can be used with programming to give the desired result. Generally, the output from the analog-to-digital converter will be a signed binary number from which the module can create a variety of other formats. One may have to convert the format for a number of reasons: to work in engineering units, (or 10 times the engineering unit for better accuracy); to convert from integer units to real values in order to make certain calculations; or to convert the value in engineering units to a bit pattern that is suitable for driving a display or printer. The conversions are not always reversible. For example, conversion to ASCII representation is not usually reversible, and rounding and sign removal are not reversible. Some operations require a specific data format (or type). Mathematical operations usually require either integer or real data types.

The data returned from the analog input module has three parts: the value bits, the sign bit, and the status bits. These bits may all be parts of one word, as shown in Figure 9-36, or they may be in different words of a multi-word format, as shown in Figure 9-37. The single-word format is common in switch-configured and fixed-configuration modules, while the multi-word format is common for intelligent modules. The single-word format cannot be used directly as is. At best, the status bits have to be removed before doing any other operations. There are, of course, variations on these practices.

Analog Data Handling

Analog data, by its very nature, is more complex to process and use, but it also does much more than discrete data. There are many issues, all of them having an impact on program size, complexity, and speed. Are the analog modules read every scan, every second, or every minute? Are all the channels read simultaneously, or one-by-one? Are all analog modules read every scan? Often it is not only the reading of the analog module that has to be examined, but also the other

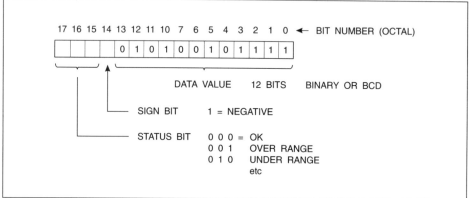

Figure 9-36. Single-Word Analog Data Representation

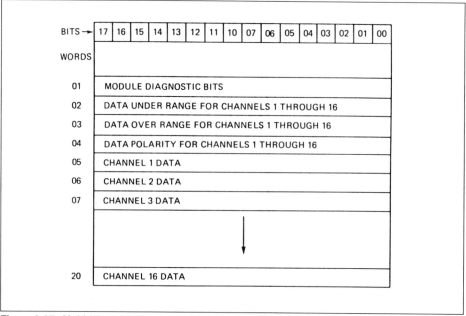

Figure 9-37. Multi-Word Analog Data Representation [Ref. 14]

calculations that follow (e.g., scaling, alarming, PID, and output operations). In many cases, the process does not require high-speed operation of the analog part of the program, but it may require a faster scan time for the discrete part of the program. A common trade-off is to perform part of the analog processing in each scan. The system goes from one point per scan to half the points per scan. The needs of the process and the capabilities of the processor will determine the appropriate value. When processing is done in more that one scan, care must be taken to synchronize the various parts of the program and to resynchronize them should they go out of synchronization. This extra work must be taken into account when planning the effort required to program and commission a system.

Processors may have limitations on the communications with analog modules, either directly (x module calls per scan) or indirectly if they use message or data queues in the communication process. The direct limitation is easy to understand, but the indirect ones are more difficult. Some processors have a separate communications system for intelligent modules or for modules that transmit multiple words of data. This communications processor receives requests from the program processor, does the communication with the intelligent module, puts the desired data in the program processor's memory, and then signals the program processor that it is finished. This communication process is not synchronized with the program scan, so a queue is established to contain the requests. The program processor continually adds its requests to the queue, and the communications processor deletes them as soon as they are done. If the requests come in faster than the communications processor can handle them, the queue eventually fills up, and further requests are blocked. The blocked requests are usually lost forever, so there has to be some way for the program to handle this situation. Sharing the work over several scans or on a time basis is one way. Another is to have a program that monitors the queue and then takes action to avoid problems as the queue fills up.

Some processors do not use a queue system; they simply make the scan longer as the load goes up. In such instances, the programmer may not realize that there is a problem until the program is run and the scan time is found to be excessive. Again, the same remedies apply, but here the programmer has to watch the scan time and reduce the load until the scan time is reasonable and the process requirements are met.

To further complicate matters, processors that use serial and parallel I/Os may have different rules and response times for the parallel and serial I/Os. Generally, parallel I/Os will be faster and more deterministic than serial I/Os. Often the delay for serial I/Os will be given in milliseconds and the programmer will have to ensure that the program waits the required time after a request before reading the data. Furthermore, the serial I/O system may be able to handle only a limited number of requests at a time; it may have a request queue as described above, or it may simply ignore surplus requests. In many of these cases, there will be no indication in the module data that the data is invalid or has not changed since the last time it was read! The same data may be appearing over and over again. Eventually the process will go out of control, or it will become evident that values do not change; but this discovery could take a while, especially if such a problem is not expected.

The programming considerations just mentioned may actually restrict the size and speed of processor required to certain of the larger models. This serious question should be investigated before the hardware is chosen. If several brands of equipment are being examined, it may be necessary to go into some detail to be sure all the constraints are considered. In some cases, it may be necessary to simulate the programming or application to see if it is feasible in the hardware under consideration. One final caution—leave some room for future expansion.

Analog Outputs

To a very large extent, analog outputs have considerations similar to analog inputs. The main difference is that the data flows in the opposite direction, so it is not necessary to repeat what has already been covered for inputs.

Forcing Analog I/Os

Most manufacturers will not allow forcing analog I/O values. If values cannot be forced, how is it possible to check the operation of the program or calibrate modules? Programming is the answer. Arrange the program so that the analog module data are not used directly but rather are allowed to pass through one or more intermediate registers that can be monitored or even manipulated, irrespective of the real values, to allow one to calibrate or test the program. If calibration is going to be frequent, it is desirable to write a subroutine, just for calibration, that can be called either from the programming terminal or the operator's console. If required, partial operation of the process in this manner may be allowed.

Interpreting Analog I/O Specifications

Many of the points mentioned for discrete I/O specifications apply also to analog ones. The general items and those common to both input and outputs apply. There are some other unique points to consider.

RESOLUTION, ACCURACY, AND REPEATABILITY

These are all defined in ISA-S51.1, Process Instrumentation Terminology, but check the manufacturer's literature carefully anyway and ensure that all needs are met, especially since the digital resolution may be very different from the accuracy.

TEMPERATURE COEFFICIENT

This is the rate of change of the accuracy with temperature (also called drift). It is usually quoted as percent of full-scale deflection per degree Celsius. This may be important for an application that requires high accuracy or is in an ambient that varies widely.

STABILITY

This is similar to drift caused by temperature but is due to changes in the power supply and to aging of the components.

CALIBRATION

In addition to the calibration possibilities, the recommended calibration interval is also given. The time between calibrations should be respected in order to maintain quoted accuracy.

INPUT SAMPLING RATE

This indicates how often the module can read the inputs. If it is given in channels per second, divide by the number of channels to determine the time between successive readings of a channel. The reciprocal of this is sometimes used instead (it is called conversion speed). Again, if it is given on a per-channel basis, multiply (not divide) by the number of channels to get the time between successive readings. Note that analog cards process all the channels all the time, so it's impossible to get a shorter time between successive readings by leaving some of the channels unused.

SETTLING TIME

This is the time the output of an analog module takes to reach desired value. It is usually specified in microseconds for a given step change.

SLEW RATE

For analog output modules this is a measure of how fast the output can change with respect to time.

Special Function Interfacing

Introduction

This section discusses all the other modules found in PLCs. Some manufacturers will have modules, not discussed here, to perform functions that are not available in their processors. Remember also that manufacturers may choose not to offer all the possible types of modules because the volume would not justify the development cost. Often a manufacturer will develop a module to suit an existing market niche or one to be entered; others follow suit if they feel it is worthwhile.

Since these specialized modules vary substantially from manufacturer to manufacturer, no attempt is made here to go into any detail on their use and programming and only a brief description will be given of the function they perform and how they could be useful.

Sensors

This classification includes a wide range of modules designed to interface with various types of sensors used in industry.

TTL INTERFACES

These modules are designed to operate with the five-volt logic levels associated with transistor-transistor logic (TTL). They generally require an external five-volt supply and have much shorter filtering times. They are often used to interface with display devices, multiplexers, digital drives, and other electronic devices that require this type of signal.

THUMBWHEEL AND DISPLAY DRIVERS

These modules have circuits to interface with BCD (binary-coded decimal) thumbwheel switches and display units. Many also have multiplexing capabilities that allow one module to drive many modules. Similar results can often be obtained with discrete DC I/O modules, but they may not be as clean. Many third-party vendors offer compatible devices.

MULTIPLEXERS

These modules allow a single analog input or output to handle many more channels. The speed drops, of course, but the cost of a multiplexer is much less than the cost of another analog input or output card. These modules do take rack space, but all the engineering has been done, so they can be relatively easier to apply compared to a homemade solution. Several third-party manufacturers offer external multiplexers with engineered solutions for this problem.

PULSE INPUTS

These high-speed modules allow PLCs to read high-speed inputs such as pulse tachometers, rate meters, or similar devices. They are started and stopped under program control and retain their counts for later processing. They are useful for rate calculations or sometimes for encoders. They can also be used to catch narrow pulses that would not be detected by even high-speed input modules.

ENCODER INPUTS

Many different modules are available to interface with encoders. Some come with their own encoders; others will adapt to many of the standards available from other suppliers. The basic encoder cards convert the encoder's absolute position information into a value the processor can use. More advanced ones also provide on-board comparison outputs that turn on or off according to presets supplied by the processor. This allows a basic cam switch function that is still under processor control.

HIGH-SPEED COUNTERS

These are similar to encoder inputs except that they receive a pulse train and provide outputs when predetermined processor-established limits are met. A module may include several channels; these are useful for applications that require frequent preset changes and exceed the processor's speed capabilities.

SIMULATOR MODULES

This is essentially a regular module with switches and lights incorporated. This allows the user to input information directly to the processor without the expense of wiring switches up to a module. Although suggested as a testing aid, it can be used to configure systems or perform diagnostic tests.

SUPERVISORY MODULES

These modules are designed to supervise the circuits attached to them, report the failure to the processor, and, in some products, disconnect the output by blowing a fuse. They are used where it is essential to be certain of the input or output status. They are sometimes called wire-break or fail-safe modules.

REAL-TIME CLOCK

For those processors that do not have a built-in clock, this card is just the thing. Whether a clock is needed or there is simply concern about the clock in the processor being changed inadvertently, an extra one in the I/O rack may be quite reassuring.

PID Controls

Some systems have a processor option for PID control that includes additional memory, dedicated hardware, and other items to enhance the processor's performance in PID loops. This was often a cost-effective way to provide the higher speed performance required for some processes. The content of the PID processor or package varies widely, but it can be very useful indeed, whatever the form it takes.

Another option for PID control is the dedicated PID module that has inputs, outputs, and control capabilities all on the module. The module, once configured by the processor, can run independently of the processor or be adaptively controlled for improved performance. The major benefit is reliable operation, and sometimes lower cost, if only one or two loops need to be controlled.

Some manufacturers also offer an interface to a range of process control equipment that operates on a distributed network, thus offering the best of all worlds.

Positioning Controls

These controls are designed to allow the processor to control applications that need precise positioning or speed control and require frequent changes in the parameters. These modules are all capable of running by themselves once they have been programmed and have received the necessary data from the processor. The hardware for even a simple application may be more expensive than the PLC, and a good application may require substantial knowledge of motion con-

Note that a simulator is not needed for outputs; just use the least expensive output module and look at the status lights.

The programming for a successful PID operation may be complicated, but it can be worth the effort if it improves the process control, reduces variability, or allows faster and more reliable grade or product changes.

trol and mechanical engineering. The major benefit is in those applications that require the PLC to change the motion control parameters to suit production needs. Some applications are done using a personal computer or a link to some other plant computer.

Third-Party Devices

Some companies have specialized in making devices that interface to PLCs made by others. In some cases, the devices simply connect to existing modules and follow the manufacturer's published specifications. In other cases, the third-party supplier has entered into an agreement with the PLC manufacturer in order to obtain the proprietary information necessary to allow him to do the interfacing. The bulk of these are data display and entry devices, but other types are also offered. Some third-party vendors offer features not available from the PLC manufacturer, but some compete directly.

Third-party devices can be very useful and cost-effective if they meet the need; however, some cautions apply. There may be a lead-time problem when the PLC manufacturer changes the interfacing arrangements. The third party may not find out about it until after the change reaches the market; as a result, the development work may be late in starting.

Communications Interfaces

Everybody is into communications these days, and PLCs are no exception. There is a wide range of devices for communicating with just about anything, but they may not be compatible with the brand one needs. Many vendors will say they can communicate with a certain manufacturer's equipment using a certain model of module. Often it works, but sometimes it doesn't. In some instances, development work is needed to get it going. Ask for specific cabling diagrams and programs that have been tested for the application—and ask for a written guarantee. The following descriptions may appear a bit confusing because of the wide variation among manufacturers and how they chose to present their products.

SERIAL COMMUNICATION FOR I/OS

Although already discussed in several places, it is worth mentioning again here. The offerings in this area are wide and even include fiber optic communications to allow long-distance or disturbance-free communications. Some manufacturers include serial communications as standard fare in their processors, while others offer it as an option at extra cost. Choose the option that suits the application.

INTER-PROCESSOR COMMUNICATION MODULES

These modules allow processors from the same manufacturer to share data. Some are dedicated high-speed modules that link individual pairs of processors. Others are networks that allow several processors to communicate at a slower pace. Some manufacturers include network communications in the processor at no extra charge. Data communications among several processors can be difficult, especially if the problems are intermittent. Solving them can require specialized communications skills or training.

PROPRIETARY LINK MODULES

These modules provide a gateway to another computer system or to another PLC system. The idea is to allow two different PLCs to talk to each other or to allow a PLC to talk to a distributed control system or industrial minicomputer. Usually these modules exist because a particular network has become so widely used that the manufacturer has to offer the capability. There is still the programming problem since the module only provides a way to talk—it does not provide simultaneous translation. If the information is well structured and the application well understood, there may be no problems.

ASCII MODULES

These modules provide memory for canned messages and a serial communications port for serial devices such as printers, display units, and even computers. They usually operate in one direction only and are still used for driving serial devices. The memory and port capabilities vary extensively. The processor selects a message from those in the module memory, then selects the port to send it on, and the module does the rest. Some allow the inclusion of variable data and time and date information. Display units with built-in memory and port controllers have replaced ASCII modules in many cases.

DATA COMMUNICATIONS

These modules are designed for bidirectional serial communications with other intelligent devices such as scales, gages, and computer systems. They do require some programming and a good knowledge of the systems concerned. For some applications, they are the only way to communicate.

Computer Modules

Many manufacturers offer a module that allows the user to write programs in BASIC or other computer languages while accessing the processor's data table. These modules are usually used to generate reports or other text-type information that a PLC handles poorly. In many cases, the module communicates over the backplane with the processor, while in others it requires a data communications card. If speed and timing are important, then the backplane is the only way to go. Even with the backplane, there are limitations on the quantity of data that can be transferred without increasing the scan time. Some programming is usually required in the PLC to provide the data required by the computer module. The computer module also requires programming to provide the functions required. Many of these computer modules have the hardware and software necessary to permit communications with so-called dumb terminals that allow an operator to input data to the process.

Peripherals

A growing range of peripheral devices can be connected to PLCs. Some are offered by the PLC manufacturer, others by third parties. Many available I/O networks allow the user to install operator interface terminals of various types on just about any system. Four avenues used to provide this type of interface: the regular I/O system, a dedicated communications module, the programming port, and the manufacturer's data highway.

REGULAR I/O SYSTEM

These devices are connected to the regular I/O connections, parallel or serial as required, and they communicate with the processor in the same way as the other I/Os. By sending particular codes, they can be made to do much more than the regular I/O modules. They do take up I/O data table space, however, and they require processor time. Some examples of these easy-to-use devices are canned-message display units and intelligent push button stations.

DEDICATED COMMUNICATIONS MODULES

In this case, the peripherals are connected to a module that plugs into the I/O rack, and that does the communications. Usually, all the programming and interfacing are provided, thus making the application easier. Many systems use some of the I/O data table; this makes it easy to interface to the PLC's program. The response time of these systems varies from milliseconds to seconds, so it is necessary to examine the application carefully before adopting any particular solution. Operators are often confused (or annoyed) when there is a two-second delay between the time the button is pushed and the light comes on or the action starts. With screens, the delay may not be as much a problem.

PROGRAMMING PORT DEVICES

These handy little devices plug into a PLC's programming port and allow the operator to access timer and counter presets and enter data into the processor's data table. Since they do not require any additional communications cards, they are relatively inexpensive. Some suppliers even have port multiplier cards that allow the installation of several devices on the same programming port. Since the programming port is not a high-speed device, there can be timing problems when using these devices to control the process. They are really meant for occasional use where a more complete system would not be justified.

MANUFACTURER'S DATA NETWORK

Many PLC manufacturers now have a local area network that allows a user to program the controllers as well as send information to them and receive information from them. In addition to their own hardware and software for these networks, the manufacturer often provides a card and driver software to allow personal computers to operate on the network. Several third parties have taken advantage of this to offer software and hardware packages that provide everything from simple data gathering to complete supervisory control. Such packages can be expensive and require a good bit of time to develop, but they are powerful, and they provide great advantages for distributed process control. There are also many experienced consultants who are willing to help get a project designed and running with a local area network.

Programming Languages

Introduction

This section provides an overview of most of the languages available for programming PLCs. There is not enough space to provide a course in any of these languages, so discussion is limited to a description of the language, how it is used, and the applications that make good use of it. Remember that there are no industry standards for programming languages for PLCs. Although the diagrams may look alike, the actual programs are specific to the manufacturer and even to the PLC model. It is not generally possible to convert a program from one manufacturer's PLC directly to that of another. One must enter the program manually in the development system of the target controller. Some third-party vendors offer conversion facilities or programs that can be used for more than one manufacturer, but these represent only a very few of the possible applications.

Ladder Logic

Ladder logic programming was described briefly in an earlier section. It is the most common programming language for PLCs. The PLC has instructions that allow it to duplicate all the common relay functions. Programming is done in a format that closely resembles hard-wired ladder logic. This language is very good for replacing hard-wired relay logic and for those applications that involve many logic operations. The ladder logic instruction set is shown in Table 9-8, but manufacturers may offer some extras or even delete some functions.

Table 9-8. Ladder Logic Instruction Set

Function	Symbol	Function	Symbol
Normally open contact	—| |—	Off delay timer	—(TOF)—
Normally closed contact	—|/|—	Counter	—(CTR)—
Output	—( )—	Skip	
Transition	(One Shot)	End of skip	
Retentive output	—(R)—	Subroutine call	
Latch output	—(L)—	Begin subroutine	
Unlatch output	—(U)—	End subroutine	
On delay timer	—(TON)—		

Extended Ladder Logic

The advent of the PLC allowed ladder logic to be extended to areas that were never possible when it had to be done using relays and hard-wired connections. Table 9-9 lists some of the extensions to basic ladder logic. These functions are especially useful in the process industries where mathematical and data operations are required for good control of the process. Some of them are also very useful in assembly line operations. Most PLCs now offer at least some of the extended capabilities.

Table 9-9. Ladder Logic Extensions

LADDER LOGIC EXTENSIONS

SEQUENCER
DATA MANIPULATION
SHIFT REGISTOR
ARITHMETIC INSTRUCTIONS
MATRIX INSTRUCTIONS
DIAGNOSTICS
I/O UPDATE
COMMUNICATIONS AND HANDLING
MESSAGES
PID

Boolean Logic

Boolean logic is based on the principles of Boolean algebra with the addition of some symbols that make it easier to understand. Any ladder logic diagram can be converted to Boolean and vice versa, so why use one or the other? The Boolean form is more compact and closer to the actual computer instructions. Boolean is, however, harder for many people to visualize, especially for complicated logic. Boolean is seen in the smaller shoe-box PLCs, where the cost of a CRT programmer would be prohibitive. Since Boolean is simpler, it can be accomplished with a relatively small, hand-held programmer with a one-line display. Some small PLCs have a personal computer-based programming system using ladder logic that is later compiled into Boolean for use in the PLC. This allows having the best of both worlds. Table 9-10 shows some of the Boolean mnemonics and their ladder logic equivalents.

Table 9-10. Typical Boolean Set and Ladder Diagram Equivalents [Ref. 14]

Mnemonic	Function	Ladder Equivalent
LD	Load input	┤ ├
LD NOT	Load NC input	┤ / ├
AND	Logical AND	──
OR	Logical OR	⌐⌐
OUT	Energize output	— ( ) —
TIM	Timer	— (TON) —
CNT	Counter	— (CTN) —
ADD	Addition	— (+) —
SUB	Subtraction	— (−) —
MUL	Multiplication	— (x) —

Sequential Function Charts

The sequential function chart language was developed from the structured programming techniques used in high-level computer languages. It is designed to produce a modular program that is self-documenting and easy to read as well as being faster to execute. By breaking a process down into a number of steps and defining the conditions necessary to enter and exit each step, the task of programming a complex system is greatly reduced. Figure 9-38 shows a typical sequential function chart program. Each of the numbered process steps contains instructions in sequential function chart language (or in ladder logic or some other convenient

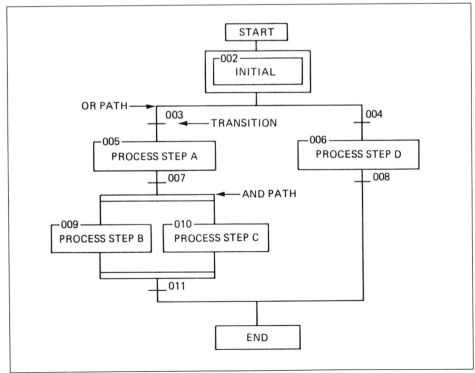

Figure 9-38. Typical Sequential Function Chart Program [Ref. 14]

language). When displayed on a CRT, each step can be expanded to see the logic and to monitor the operation of the process. The execution rules for a sequential function chart require that one, and only one, process be executed at a time. This leads to some problems in real-world applications that have several things happening at once. One approach is to add interrupt and jump statements, but these break the modularity of the program and make it difficult to understand. The French Grafcet standard avoids this by allowing simultaneous control sequences. In addition, Grafcet also allows a final control block in ladder logic that facilitates the inclusion of emergency stops and safeties. Several manufacturers now offer either the sequential function chart language or Grafcet as an option in their PLCs.

High-Level Languages

PLCs have to accept ladder logic programming to be called PLCs, but there are times when it is convenient to have other languages available inside the PLC for certain tasks. Some manufacturers offer languages such as BASIC or PASCAL as an extension to their ladder logic language. This is not the same thing as having an external module that executes a BASIC or PASCAL program; it is actually part of the ladder logic program and is executed within the ladder logic scan time instead of being done outside the scan. For process applications that require extensive mathematical operations or data handling, this programming approach may be a big timesaver.

Fill in the Blanks

Some specific applications are programmed by supplying the appropriate data to the PLC. This approach is often used for configuring intelligent modules (e.g., PID and motion control applications). The concept is that the application is well enough understood that one need supply only the name of the application, along with the predetermined parameters, to program it completely. This approach usually requires several pages of instructions in the manual regarding the parameters of the application. The manufacturer must provide sufficient information for the user to be able to determine the correct values to use, and he must allow the user to tailor the application to suit his needs. Figure 9-39 is a example of the choices available for a PID control loop. This approach is used as an extension to the ladder logic programming language; it cannot replace it.

Custom Function Blocks

Many manufacturers offer function blocks for predefined operations such as PID control, data manipulation, communication with intelligent modules, and so on. In addition, some manufacturers offer users the option to "roll their own," and create function blocks that perform operations they find useful. The user must design and test the block and document it, but once this is done the block can be used over and over again, with the assurance that it will work as expected all the time. This can be used to enforce modularity or uniformity, or even to ensure that good design practices are followed. It can also be very useful where a small number of routines are used frequently. The only disadvantages are the time to test and develop the custom function blocks and getting locked into one supplier for the hardware.

A similar result is possible using subroutines along with some of the advanced editing capabilities of many of the programming packages. The ease of use and modularity would be there, but there would be no assurance that the module had not been modified by the programmer except by manual or semiautomatic com-

```
           equation:  1 (O:AB/1:ISA)          feed forward:     0
               mode:  0 (O:auto/1:manual)  max scaled input:  100
              error:  0 (O:SP-PV/1:PV-SP)  min scaled input:    0
    output limiting:  0 (O:NO/1:YES)              deadband:     3
    set output mode:  0 (O:NO/1:YES)      set output value %:   0
  set point scaling:  0 (O:YES/1:NO)        upper CV limit %:   0
   derivative input:  0 (O:PV/1:error)      lower CV limit %:   0
    deadband status:  0                     scaled PV value:   30
upper CV limit alarm:  1                        scaled error:  25
lower CV limit alarm:  0                       current CV %:   56
setpoint out of range: 0
           PID done:  1
        PID enabled:  1                           set point:    55
                              proportional gain (Kc)   [.01]:  100
                              res. time (Ti) [.01 mins/repeat]: 30
                              derivative rate (Td)  [.01 mins]:  0
                              loop update time      [.01 secs]: 20


Enter value or press <ESC> to exit monitor.
N10:13 = █
remote program   no forces        decimal data  decimal addr  PLC-5/15 Addr 32
```

Figure 9-39. Fill-in-the-Blanks Display of PID Control Block
(Courtesy of Allen-Bradley, a Rockwell International Company)

parison. The custom function block guarantees that the logic is identical; the only questionable item is the selection of the I/Os and parameters.

Other Languages

Despite what has been said in earlier sections about the importance of ladder logic to the identity of a PLC, some systems do not use ladder logic. Some manufacturers of high-speed controllers call their equipment PLCs even though they are programmed in BASIC, FORTRAN, or some proprietary language that is similar to one of the standards. This equipment is often specialized for some type of application such as motion control, speed, small size, or low cost. This specialization detracts from the universality of the PLC, and the use of a language that requires a programmer rather than an electrician or technician further restricts its use on the shop floor. Although such a specialized controller may be used, just remember it is not a PLC in the generally accepted sense.

PLC System Documentation

Introduction

Documentation is the road map others use to understand, debug, and modify a system. How well they can do their job depends on the quality and completeness of the documentation. Since most systems of any size involve several people, and often several companies, it is essential to document what and how things are done. Once the system is installed and operating, the documentation must be updated to reflect any changes made along the way, and any troubleshooting and maintenance information must be added. As a last step, the appropriate number of copies of the documentation have to be made and distributed to the appropriate individuals or organizations. As changes are made, the documentation must be revised, copied, and distributed.

The documentation for even a small shoe-box PLC-based system will be 30 to 50 pages, in addition to the manual for the controller and other devices. The documentation package for a large system can easily make a stack six feet high, and, if the system is complex, it can be quite a bit more. An organized approach is required to maintain control, to make sure nothing is missing or lost, and to make it easy for the users to find what they need.

What Should Be in the Documentation?

TABLE OF CONTENTS
The first item in any documentation should be a table of contents that gives a complete list of what is in the documentation package. The list is best done in a computer database program so changes can be made easily. It also allows some sorting of the information to be done easily at any time. The following information should be included in the table of contents:

(1) Name of document

(2) Source

(3) Identification or part number

(4) Revision

(5) Number of pages

(6) Location in manual

(7) Other information that may be useful

SYSTEM DESCRIPTION
The next item is a description of the system and how it is supposed to function. This is the foundation for all the programming and other work that follows, so it should be reasonably complete; however, it need not be an operating or training manual. There should be a clear statement of the control problem or task, as well as a description of the design strategy used in designing the PLC system.

SYSTEM CONFIGURATION
This is a system arrangement diagram and should include sufficient information to allow the reader to know what hardware is in the system and where it is located. On large, spread out systems, this may require several drawings, but it may be only a single sheet for small systems.

HARDWARE LIST
This is so easy to forget, yet is so easy to do and so indispensable, especially when processor model number information is needed and the system has to be shut down and the processor removed in order to get the information.

WIRING DIAGRAM
At the very least, the user needs to know how all the I/Os are wired. It is also very helpful to have information on power wiring for the PLC and other wiring that may affect it (e.g., emergency stops).

DATA TABLE AND MEMORY MAP
The following refer to the PLC data table and memory map:

(1) Input/Output Address Assignments

(2) Internal Storage Address Assignments

(3) Register Assignments

MANUFACTURER'S MANUALS

There should be a copy of the manual for every piece of hardware supplied with the system. The input and output devices that are connected to the PLC should be included, or, if they are part of another documentation package, then that package should be cited in the table of contents.

CONFIGURATION OR SETUP DATA

This refers to configuration data, switch settings, jumper positions, and the like, for all the modules in the PLC. The same information for any devices that interface with the PLC should be included.

ANNOTATED PROGRAM

A printout (hardcopy) of the program, with the descriptive annotation and program cross references, is part of the documentation. For a description of the annotation, see Figure 9-42.

SPECIALIZED DEVICE PROGRAMS

Any modules or devices included with the PLC should have their programs and documentation included with that of the PLC. This would include modules such as ASCII and PID or devices such as canned-message display units.

MACHINE-READABLE PROGRAM COPIES

These are also known as backup copies. They should be ready to load into the PLC.

SOFT COPIES OF DOCUMENTATION

There should be a backup diskette or tape of the documentation for the PLC program as well as any other parts of the documentation that were done on a personal computer.

SPARE PARTS LIST

This list should also include a list of replacement fuses, preferably on a per-module basis.

PROGRAMMING AND CALIBRATION EQUIPMENT REQUIRED

The list should include software and hardware and accessories such as cables. If some equipment was not provided with the system, it should be so noted.

CALIBRATION PROCEDURE AND PREVENTIVE MAINTENANCE

This should include any information specific to the application and not be merely a copy of the manufacturer's calibration procedure.

The preceding list is fairly general and could be shortened for small systems, but the full list should be kept in mind at all times. Some of the information may be combined with other items. For example, the system configuration and device configuration can be combined with the wiring diagram, and the memory map information may be done in the program annotation package.

Program Annotation

The original PLC programs were done in ladder logic with reference numbers only. Figure 9-40 is an example of such a bare program. It was quickly realized that programs of any complexity would be impossible to follow without more documentation. The advent of minicomputers and personal computers allowed for a more complete and useful documentation system. The basic feature of all these systems is element documentation.

For easy reference, organize the documentation into a set of binders in the same order listed in the table of contents. A little care in the organization of what is in each binder can make the documentation much easier to use. Putting the program cross reference in a separate binder from the program makes it easy to flip through the cross reference while studying the program.

PLC Development Systems

This book uses the word "development" instead of "documentation," because these systems do more than merely provide a way for the designer to document what he is doing or has done. They provide tools to speed up the writing and development of programs. They are to programming what word processors are to writing. Unfortunately, there is not as much choice in development systems as there is in word processors. The reason is that the cost to develop these programs is very high and the number of copies sold is low compared to word processor programs.

Figure 9-43 gives an overview of the three basic types of systems available: dedicated hardware, IBM-compatible systems running MS-DOS™, and others. The dedicated hardware systems are fast disappearing except for the small shoe-box systems. The trend is definitely toward systems that operate on any 100-per-cent-IBM-compatible clone under MS-DOS™, but there are still exceptions that one should know about.

The other category includes systems that run on IBM clones but use an operating system other than MS-DOS™. Many European systems were first developed under CPM and are still offered this way. These systems may have problems co-existing with MS-DOS™, and will take up much more memory, and will require users to be familiar with yet another operating system. It is also more difficult to import or export data between two different operating systems.

Another variation in the other category is the modified compatible. Some manufacturers have a clone that is not 100 percent compatible and so may not run all software, particularly that necessary to program other PLCs. In many cases, these modifications were made in good faith to allow the equipment to be suitable for the industrial environment.

> **To see how much easier it is to use a well-documented, consider the following example. Figure 9-41 shows the types of information that can be displayed for a single element. Documentation for each line and each page is also available. This allows the designer to provide comments on what is being done and why and, if desired, even the troubleshooting hints. Figure 9-42 shows the same line of programming, but fully annotated. Notice how much easier it is to use.**

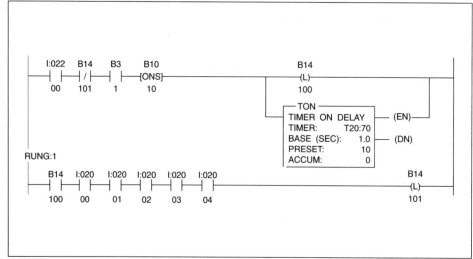

Figure 9-40. A Program with No Documentation

If a user is going to have only one brand and only one model of that brand of PLCs and doesn't want to use the equipment for anything else, then he or she can choose just about any development package that does the job. To be able to support several brands of PLCs, or in some cases several models from the same manufacturer, the 100-percent-IBM-compatible clone is a better route. It is quite feasible to have the software and even the hardware for several different PLCs in the same computer, along with copies of the programs used in the plant. This can substantially reduce the cost of supporting several brands in one plant.

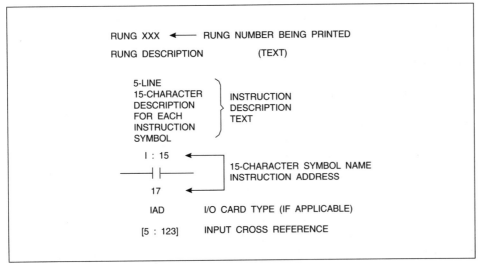

Figure 9-41. Element Annotation

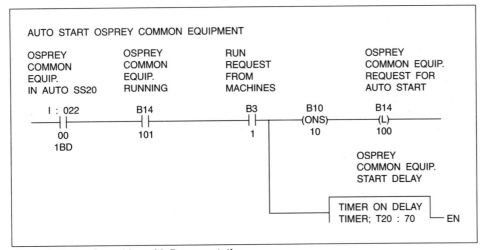

Figure 9-42. The Same Line with Documentation

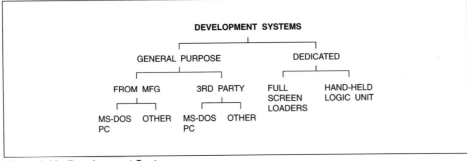

Figure 9-43. Development Systems

Drawing Practice

Now is the time to address a fundamental issue in PLC installation—what numbering system should be used for the wires and devices connected to it and to other pieces of equipment. The ISA standards for symbols and identification (S5.1, S5.3,) work well for process control loops and process industry applications, but they were not designed for automated systems such as robot cells, packaging equipment, material-handling equipment, and specialized machinery. These industries have developed standards, but many users and manufacturers have struck out on their own and ignored the published standards.

There are three basic numbering methods: sequential numbers from 1 on up (the JIC method), numbering based on the drawing sheet and line number, and numbering based on the PLC addresses. Each method is discussed in turn, with a detailed description including the pros and cons.

SEQUENTIAL NUMBERING

Sequential numbering dates from the early days of relay logic. It assigns a unique wire or device number to every wire or device. If one follows the JIC convention, the wire numbers are sequential numbers, while the device numbers are made up of a sequential number, a device letter code, and a contact number. There are many variations, the most popular being to make the device number a letter code and a sequential number, with no contact number.

The major advantage of the sequential method is its open-endedness. It is also independent of hardware configurations, and it can be easily modified to provide a systematic numbering scheme. Its disadvantage is the record keeping required. The designer must keep a list of all numbers used, with a cross reference to where they are used on the drawing.

SHEET AND LINE NUMBERS

This system uses the sheet and line number, along with a sequence number on that line. A typical number would be 227604 where 22 is the sheet, 76 indicates the line on the sheet, and 04 indicates the fourth item on the line. Wires would have only the six-digit number, and devices would add the device letter code to the six digits to form a complete number.

The major advantage of the sheet and line number system is that it does not require a list of numbers used or an index. Since one knows exactly where the wire or device is on the drawing, the device and its number are easy to find. The final two digits of a new number may be found by inspection. This system is also open ended (within the numbers of digits chosen, that is, a three-digit system allows nine sheets with nine lines; four digits allows 99 sheets with 99 lines, and so on) and is independent of hardware configuration. However, this method is forever tied to the physical appearance of the drawing. The original sheet and line location of every element must be preserved in all revisions, unless one is prepared to change the numbering on the machine or in the field. In addition, wires or devices that appear on more than one line or sheet present special problems that require specific rules for the numbering of these items. Another disadvantage is the size of the numbers: five- and six-digit numbers are common compared to three-digit ones for the sequential system.

PLC ADDRESS

The final system uses the PLC address as the number to associate with the wiring and devices for that address. Where more than one wire or device is required per address, a letter suffix or decimal suffix is used (e.g., 112A or 112.2).

The major advantages of this system are the ease of knowing where something is connected to the PLC and the minimal record keeping required. Its disadvantages are serious. There is no provision for numbering anything that is not connected to the PLC. This system is also dependent on the hardware used. Changing the brand of PLC can change the numbering required. Changing the wiring from one output to another requires renumbering the wires and devices. A cross reference is still required to know where the PLC addresses are located on the drawing. In addition, using one of the other two methods for the non-PLC wiring requires care to avoid using the same number twice.

The sheet and line method is often used for mass-produced systems that are very stable, because it does provide troubleshooting advantages that outweigh its rigidity in this case.

The PLC address method has been used by panel builders who make only the panel and may not even know where the rest of the equipment is going. It is often combined with the sheet and line method, because this requires the minimum amount of design and drafting time to set up.

The sequential number method is widely used but is becoming less popular among machine builders because of the time required to keep the necessary records, especially in large projects where many people will be working on the job at the same time.

The moral is to specify the system that meets the need. Provide the necessary documentation complete with examples and preferred codes for devices. This is especially important if the machine is being built in Europe or Japan where there may also be language and cultural issues to deal with. If the numbering method is specified in the very beginning, the manufacturer can build it into his pricing, often at a very low cost, if it is a custom or built-to-order system.

Implementing and Programming the PLC

Control Definition

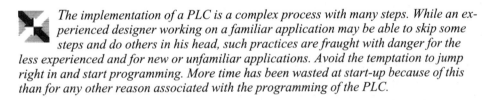

The implementation of a PLC is a complex process with many steps. While an experienced designer working on a familiar application may be able to skip some steps and do others in his head, such practices are fraught with danger for the less experienced and for new or unfamiliar applications. Avoid the temptation to jump right in and start programming. More time has been wasted at start-up because of this than for any other reason associated with the programming of the PLC.

The only place to begin the design process is with a definition of the task to be accomplished. Depending on the organization and the process, this step can be anything from a one-page memo to a 100-page book. Everyone involved in the process should be consulted—manufacturing, research, maintenance, operators, engineering, marketing, shipping, and so on.

The more people consulted beforehand, the less explaining there is to do afterwards, when their needs are overlooked.

DEFINING THE TASK
The type of process will greatly influence the amount of time and consultation needed to make a complete control definition. Duplicating an existing process should be the easiest, since all the information is probably readily available.

Do not rely on the original control definition unless it has been carefully checked for correctness. Too often, the process is updated or changed, and the control definition is left untouched.

Probably the next easiest job to describe is a modification of an existing process. All the comments above apply. In addition, one has to really understand the modifications required and how and if they will be changed during the implementation process.

The most difficult definition is for a new process. While it may be possible to make use of some existing information to develop the definition, there can be substantial uncertainty that must be resolved in order to arrive at a good definition.

 One of the most important aspects of the control definition is how it may change with time, both during the implementation and afterwards. Overlooking this could have disastrous results.

Control Strategy

Once the goal has been set, a way has to be found to get there. Control strategy is just another term for the method used to solve the problem. There is nothing fancy about selecting a strategy—just a lot of hard work that many people find boring. Make a list of all known ways to reach the goal, analyze their strong and weak points, and synthesize new ones. Test the various strategies and determine which one best suits the need. It sounds easy, but the temptation is to choose a familiar or common strategy without looking at very many possibilities. The evaluation of strategies can be long and complex, especially for large projects. Also, there may be some interaction between the hardware and software and the strategy, especially if cost, lead time, and other such considerations are involved. Ideally, these concerns should be left for later, but in practice, they are often factored into the strategy decision.

The most fundamental rule in defining a strategy is to think first and program later. Too many projects have come to grief because this rule was neglected, due to ignorance or to time or other pressures. A program will be only as good as one's understanding of the possible solutions to the problem. It's not unusual to spend two days working out a strategy and then one day programming it. Starting the programming as the first step could leave one with two days of programming and a week of trying to debug the mess. Those who want to live dangerously may choose to ignore the "think first, program later" rule; if so, it should be only on a small and not too important project.

Large complex processes will have to be broken down into manageable chunks that can be handled by different teams. This approach requires careful monitoring to ensure the parts all fit together afterwards. A structured approach is often used in these cases to ensure all groups cover all points of concern. The development of such structured approaches is beyond the scope of this book. In cases where the success of the project may be affected, one should treat the team approach as a control problem by itself, to be solved well before the project is started.

Once the control strategy is developed, the user should have most of the information necessary to design the PLC application required to implement it.

Implementation Guidelines

The two preceding sections apply to any control problem—not just to PLC applications. The rest of this chapter centers on using PLCs to implement the control strategy that was chosen to meet the control definition. Table 9-11 gives eleven steps that should be followed in implementing a PLC application. Earlier sections have already covered the first two items in general terms; the following is a bit more specific about the subsequent nine steps.

FLOWCHARTING

Many writings have touted the benefits of flowcharting, but too few of us really practice what the experts preach. The control definition and strategy give general guidelines but usually do not go into any detail. The program flowchart should be more detailed and should completely describe the relationships between the various parts of the process. The usual practice is to start with an overall flowchart and then flesh out the details. In the making of the flowchart, structure and modularity can be given to the program. Resist the temptation to just jump in and start programming. Think about the process, and spend some time

Table 9-11. Steps in Developing a PLC Application

NAME	ACTION
Control definition	Understand the desired functional description of the system.
Control strategy	Review possible control methods and optimize.
Flow chart	Describe in detail the program steps.
Input output list	List all the input and output devices required.
Select	Choose the programmable controller that best suits the need of the process.
Configure	Assign I/O addresses and memory map.
Program	Convert flow chart into program code and document.
Check	Ascertain that the program does what is required.
Install	Install the hardware.
Commission	Put the system into operation.
Finalize documentation	Update the documentation to reflect all changes.

working out all the details that have to be considered: power-on initialization, error recovery, error checking, start-up and shutdown sequences, fault handling, operator interface handling, alarm handling, and so on. These are all the things included in instructions such as "use good engineering practice" or "follow standard industry practice" related to control definition. The work represented by these innocuous phrases can soak up a lot of engineering time, if it is not included in the flowcharting process and has to be added after the program is all written.

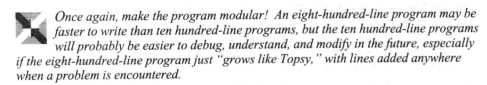 *Once again, make the program modular! An eight-hundred-line program may be faster to write than ten hundred-line programs, but the ten hundred-line programs will probably be easier to debug, understand, and modify in the future, especially if the eight-hundred-line program just "grows like Topsy," with lines added anywhere when a problem is encountered.*

I/O LIST

Creation of an I/O list is at this point for two reasons—it is needed for the next step, and by now there should be enough information to make a good I/O list. Some designers may prefer to do the I/O list before starting the flowchart in order not to forget some points or parts of the process. During the flowcharting process, one should pick up some of the I/Os that were forgotten or unknown, and after the flowchart is done, make, or revise, the list. It should contain the name and tag of each point, along with the type of signal (discrete, analog, input or output), the voltage and current rating, and any other information that may be desired. The I/O list will be the starting point for the hardware selection, so take the time needed to check it carefully.

PLC SELECTION

To this point in the design, it is not known which brand or type of PLC is going to be used. Everything done so far should have been independent of the hardware chosen. Now a specification for the required hardware can be written and sent to suppliers for them to prepare proposals. In many projects, the designer has the freedom to choose from all the PLCs available in the market, but in an increasing number of cases the choice may be restricted by company policy or by a user's preferences. Where the choice is limited, some of the following will not be necessary.

In developing the specification for a PLC for a particular application, one needs to know the scope of the application. Table 9-12 gives the four possible scopes for a PLC application. This discussion focuses on the scope for hardware only, since the other possibilities are more in the project management area. The steps required to implement a PLC are the same regardless of the scope of the purchase order; generally, the purchase order scope determines only who does what—not what needs to be done. In the end, the designer is responsible for all the details required to make the process work.

Table 9-12. Four Possible Scopes for PLC Specification

TYPE	DESCRIPTION	SUPPLIER
Hardware Only	Hardware only, user does programming.	Manufacturer, Wholesaler, Panel Builder
Hardware and Pgm	Hardware and software, user does installation.	Manufacturer, Panel Builder, System House
Hardware, Pgm, and Install	User does definition and management, vendor does the rest.	System House, Consulting Firm, Constructor
Turnkey	User does definition, vendor makes decisions and makes it work.	Consulting Firm, Constructor

A PLC specification for hardware only should contain all the information listed in Table 9-13. This information is necessary for the supplier to make a well-informed proposal. If some of the information is missing or incomplete, suppliers will make their own assumptions, which may or may not be to the advantage of the designer. Also there is no guarantee that all suppliers will make the same assumptions for the missing information, with the result that the proposals may not be comparable.

PLC SPECIFICATION ITEMS

The I/O layout allows the supplier to determine the most economical arrangement for serial versus parallel I/O racks and the amount of modules required for each rack. It should be referenced in the I/O list so the supplier can easily determine which I/Os are where.

Peripheral equipment includes anything that is not already in the I/O list and has to be supplied or interfaced with. It includes operator interfaces from thumbwheels to screens, communications interfaces to computers or other devices, programming equipment, and so on. Although the flowchart should include the logic to operate these devices, it is best to be explicit about what is to be included.

Any special environmental constraints should be spelled out. Among these are corrosive environments, high, or low temperatures, humidity, elevation above sea level, shock and vibration, and magnetic fields from welding or arc furnaces.

Table 9-13. PLC Specification Contents

Input output list
Flow chart (may be summarized)
I/O layout
Peripheral equipment required
Environmental requirements
Operating voltages, frequencies, and tolerances
Extra capacity requirements
Loading levels for outputs, power supplies, etc.
Accessories
Everything necessary to make it work clause
Warranty requirements
Commercial clauses
Training and start-up assistance
Spare parts

Operating voltages, frequency, and their tolerances can be very important and must be included. If line voltage varies by ±10 percent, say so. Also, watch systems that are going to be exported to other countries because power quality may be very different from that at home. Also, beware of isolated plants out in the country or at sea; they can have problems all their own.

How much room is needed for expansion? Many designers like to have 25 percent extra I/O space installed, with the possibility of adding another 25 percent by installing additional racks. Individual applications may need less or more; it's up to the user to decide. Make similar allowances for memory, registers, and so forth. The suppliers may say it is not necessary, so it's best to be well prepared before talking to them.

What loading should be figured on for power supplies and other modules such as outputs? Will the system be loaded right up to its limit as specified, or will there be some capacity left for expansion possibilities? It is not very useful to have 25 percent empty rack space if modules cannot be installed in it because the power supplies are already fully loaded. The same goes for output modules that have a per-output and per-module rating. Once the module limit is reached, the remaining outputs are useless. A safe practice is to require that the power supplies be able to handle a fully loaded rack.

Accessories can be contentious items. Some accessories are essential to the operation of the system, whereas others are just nice to have. Cables and connectors are not always included in the cost of the module, and they may have to be ordered separately. Some installation and test equipment is also included as accessories. Any known accessories should be described here.

Documentation is often disappointing. Every manufacturer includes a manual with each module, but sometimes a user ends up with 25 copies of the input module manual and only one processor manual, when he would really like three of each. Spell out the requirements here. If it's three copies of everything, say so. Some manufacturers will provide them free; others charge, or else take forever to send them. Consider also the language of the documentation. With the world market, documentation is sometimes available in several languages, which can be a plus or a minus. Some translations are not very good, and this can make learning difficult.

Some users like to include a clause that requires the supplier to furnish everything that is necessary to make the system perform the operations requested or, alternatively, furnish a list of what the purchaser needs to add to make the system work as requested. This will make the supplier think twice before signing off on the proposal. The supplier will make sure nothing has been forgotten or at least be explicit about what is not included. This is especially important if one is not familiar with the supplier's equipment or if the application is very involved and requires a large amount of detail work. If the supplier's literature does not spell out what is necessary to make it work, there will be problems. This situation can arise where the supplier is an agent or a wholesaler and not the actual manufacturer, or where the product line is new and people have not yet had the time to become completely familiar with it.

Many warranties start when the goods are received, some when the goods are manufactured, and others when the system is commissioned. Specify what is needed, along with the type of service desired (immediate exchange, repair exchange, or only repair), and specifically who pays the transportation, brokerage, and duty charges, if applicable. If the system is to be shipped outside the country in which it was sold, one may have to request or negotiate special warranty conditions.

Some spare parts are necessary. Fuses and batteries are the first item to look into. Ask for a suggested list to see what is needed to ensure reliable operation.

Training and start-up assistance should be included in the specification if they are important for the application. The services to be provided and the schedule for their delivery should be described. If the next training course is in March and the job has to be completed in February, there will be a problem. Also check where the training will be offered. Some manufacturers offer free training at the factory, but if the user must pay for travel and living expenses, these could be more than the cost of a local course.

Technical people often neglect commercial clauses, but they must be included. See the purchasing or legal department for the necessary information, especially if different countries are involved.

List the documentation expected with the proposal. If a complete parts list and a copy of all the manuals is needed, spell it out. Without this type of information, it may be difficult to properly evaluate the proposal. If the documentation is extensive, there might be an added cost for it, or it may be available only on loan, which may be good enough.

The specification items listed above are for a system of reasonable size and complexity. For a single shoe-box PLC that sells for less that $1000, it would be overkill. On the other hand, for a machine builder who wants to choose a shoe-box PLC to use in all machines and expects to use several hundred a year, the complete specification would probably be justified.

SPECIFICATION PROBLEMS

What problems can one expect to have in developing a PLC specification? The first two items to follow will probably be the most difficult, so some of the issues related to them will be addressed in detail.

The first problem is confidentiality. A user may not want to tell others about a new control strategy, process, or manufacturing line. Without this information, however, it would be difficult to determine the PLC functions required. The I/O list and layout can be rendered nonconfidential by removing only the information that needs to be kept confidential. A summary of I/Os by type is often enough information for the supplier to do the job. The process description and flowchart are more difficult. One approach is to start from the flowchart and make a list of the operations required, along with the quantities of functions (such as timers, counter, shift registers, PID loops, and so forth) that are needed by the application. This is exactly the work the supplier will have to do anyway, so if it is done, the process description and flowchart can be omitted. The risk here is missing

---

Warranties are often negotiable.

Sound judgment is required to tailor the list to the needs of the project, but remember—too much information is better than too little.

some of the functions required or not making the best use of the capabilities of each PLC, thereby ending up with an application that is less than optimal. An experienced designer working with suppliers who are well known and familiar can do a good job. Lack of familiarity with the equipment or lack of experience in using PLCs can lead to problems.

The other type of problem is insufficient information to develop the process description or flowchart, or even the I/O list or layout. This usually happens when the process is new or still under development and is not fully understood. This situation requires using past experience to provide the information necessary, probably in the same form as used for confidential applications. Here the risk is greater, so the requirements for extra capacity and future expandability are very important. Care must still be taken not to over specify, because this could drive the cost up to the point that the project becomes uneconomic.

PROPOSAL EVALUATION

Using the information supplied in the specification, suppliers should be able to determine the hardware required for the application, estimate the amount of memory, and choose the type of processor from their product line. Once the suppliers' proposals are in hand, they need to be evaluated. There are various methods for evaluating bids based on point systems. A good one is described in the Kepner-Tregoe book [Ref. 17]. All these systems require that one evaluate how well each fulfills a list of requirements. This list would include many items not mentioned here (price, delivery, financial resources, and so forth), which have varying degrees of importance depending on the user and the application.

Each proposal should be carefully examined to see that it really does conform to all the requirements of the specification. Errors can be made, and clauses can be overlooked or even ignored. It's not surprising to receive quotes that have omitted items that were explicitly described in the specification. Check the data sheets for the I/O modules to see that they meet the needs. See if the voltage, frequency, and temperature tolerances are what were specified. Check the power supply and output module loadings. Verify the expansion capabilities. In short, double-check the proposal. This will lead to a good understanding of the supplier's capabilities and to a more informed decision.

When finally placing the order for the PLC, base it on the specification, not the supplier's quotation. This makes certain that what is supplied is what is wanted, and not simply what the supplier listed on the quote sheet. If some parts do not perform as specified but are within the manufacturer's specification, there is a good basis for a claim; otherwise one must rely on the supplier's goodwill. A supplier who has made a careful study of the specification and included all the requirements in the bid should not object to this method of ordering.

CONFIGURATION

The selected hardware must be configured. Configuration refers to choosing the values for the various switches and configuration tables. This is not as easy as it seems. The difficulty is in finding the optimal configuration for both present and future needs. Since the configuration requirements vary widely, there are really no general rules to follow, only a few broad guidelines.

Is it possible to continue to add I/Os and racks up to the maximum the system will address? Some configurations will reduce the maximum number of I/Os that can be addressed because of limited rack space. In some cases, the maximum number of I/Os might be reduced to half its value simply by choosing the wrong configuration.

Can the configuration be changed later without having to make massive changes to drawings and programs? If the answer is no, think twice before proceeding.

Provide rack space for modules that require high-speed communications. In some systems, both high-speed and analog modules need to be placed in certain locations to get the best performance. This is usually the case for systems that have both parallel and serial communications. If there is only one parallel rack, leave space in it for future high-speed modules, even if this means having to add an extra rack.

Although no system is perfect, try to place I/Os in a systematic manner. It will make troubleshooting much easier. There are many systems; only a few are described here.

Assign I/Os by function (e.g., all start buttons, all stop buttons, all photoswitches, and so forth). To be effective, leave some space in the assignment for future additions so they can remain in sequence. Some people also like to assign the addresses in sequence by type (e.g., by device number, or by physical location on the machine or process). This gives a lot of problems if changes are made and is usually workable only in a completely tested and proven machine.

Another variation is to assign addresses so that only the same type of device is in a particular module. This gives a module of start buttons, followed by a module of stop buttons, followed by a module of photoswitches, and so on. If the design requires several modules for a type of device, they should be in consecutive slots. Consider leaving some empty slots for future use.

> **Another possibility is to assign the addresses by process block. Consider the following example. An application has 10 motors, each with three push buttons, two limit switches, and a pressure switch. This gives six inputs per motor. One could assign them in blocks of six, always in the same order. For eight-point modules, one could assign each motor to a different module and leave the two extra inputs for future use. For sixteen-point modules, one could assign the first six points in a module to one motor, skip the next two, and then assign the next six points to the next motor. Another option would be to use one module for each of the functions for all the motors. This would require a sixteen-point module or two eight-point modules and would leave six points unused per function. These unused points may be useful for expansion.**

Configuring also includes the assignment of internal memory. Here, there's more freedom to leave holes for future expansion or to make the program more readable. With a flowchart already done, the logic should be well enough understood for a determination of the memory required for internal bits, data registers, timers, counters, and so on. The patterns mentioned above for I/Os can be applied to memory (to the limitations imposed by the processor's programming conventions). In those processors that allow multiple files for timers, counters, and so forth, it may be useful to set aside files for specific uses. For example, one might want several timer files (one for confirmation delay, one for alarm damping, and one for general purposes). This may even permit the timer number to be the same as the associated device, thus making the result even more convenient for troubleshooting personnel.

PROGRAMMING

This brings us almost to the nitty-gritty, but not quite. Complete all the necessary documentation before starting the programming. Document all the I/Os and internal locations. This has several benefits, the most important of which is eliminating errors in programming and transcription. A systematic approach to I/Os and internal bits means that the documentation too can be very systematic. One can give it to a clerk or secretary to enter all the required information, then

---

Leave free rack space where it will do the most good. If I/Os exist in several physical locations, you probably need free space at each location for both inputs and outputs. A good knowledge of the application may suggest leaving more room in certain areas than others. It's asking for trouble to leave a full and unexpandable rack in a remote location. Murphy will ensure that all expansion takes place there!

simply check it afterwards. Since it is systematic, errors will usually stick out like a sore thumb.

With the documentation completed, programming can now start. First, look for the modular items to see what parts should be done first and then copied. Do the common stuff first. When unsure of the logic, do a mockup, then test it on a simulator or on a real PLC to see that it works as expected. Be modular in the approach. Test at each step, then make the copies as needed. Document on the fly, since it is much easier to do this than to leave it all for later. Have somebody else read over the program for errors or use the cross reference to find unusual occurrences that may actually be errors.

It is beyond the scope of this chapter to go into programming examples and the details of converting ladder logic or flowcharts into finished programs. Several of the references have chapters or sections on programming with examples and case histories; see them for more information.

CHECK

This is a step that some people love to hate. Before the finished program is installed, it must be checked for both logic and documentation. The best person to check the program is someone other than the programmer, but that is sometimes difficult to arrange. A simulator is useful for this, but it does take a certain amount of work to set up the test conditions. If the program is large or the start-up time is short, this may be the only option. The flowchart, control strategy, and control definition should be used to set up the test conditions. In some cases, it may be necessary to test only the basic modules, while in others, the entire program will have to be tested. If a test program is developed, it should be documented and kept for future use. Part of the problem in testing is determining all the combinations of I/Os and internal bits that need to be tested. Entire books have been written on software testing, and the principles that apply to computer programs also apply to PLC programs.

FINALIZE DOCUMENTATION

Installation and commissioning will be covered later in the book, so this chapter deals only with finalizing the documentation. During the installation and start-up processes, changes will be made. These changes will require changes to the process description, control strategy, flowchart, and so on. The time to do it is the time the changes are made, while things are still fresh. The best procedure is probably to keep revising these documents as things occur, even if it is only in the form of notes on the most recent copy. This will act as a memory jogger and facilitate the task of making complete changes later. This step is often left out—usually because the project is late and over budget by then. This oversight is costly when the process or machine has to be modified at a later date. When documentation is not accurate, everything has to be checked out after the fact—sometimes at a significant added cost.

Guidelines for Installation, Start-up, and Maintenance

System Layout

Although PLCs are industrial equipment, they usually require some protection from the contaminants found in the industrial environment. In addition, people must be protected from the voltages involved. There is also a certain amount of other equipment that may be controlled by the PLC.

System layout includes integrating all the above into a system that satisfies the needs of the designer, the operator, and the maintenance personnel. The design of control panels is covered elsewhere, so this section is limited to what concerns

> Talk to the people who will maintain the equipment and see what they think about the assignment. They will be using it long after the project was first started, so listen to them and do whatever is possible to meet their needs.

PLCs. Table 9-14 lists the types of equipment that will have to be mounted and wired. In large systems, there may be further constraints due to maximum shipping dimensions and weight.

Table 9-14. PLC Equipment to Be Mounted and Wired

Programmable Controller

Processor
Power supplies
I/O racks
Cables
Accessories

Support Equipment

Power transformer or conditioning equipment
Power distribution
Emergency stop relay(s)
Interface cards or relays
Terminal blocks

The various enclosure types usually hide the PLC from view under normal operating conditions. In the panel doors, provide windows located to allow visual inspection of all the indicator lights on the I/O racks and on the processor, as well as any key switches or other devices that may be of interest. Quite often, it is necessary only to observe the status of the I/O lights for a point or series of points to know what is wrong with the equipment. This also makes it easy to keep the panel doors closed and to maintain the PLC in the desired environment.

The PLC manufacturer will usually provide a recommended layout for various models and configurations and even drilling templates for the racks and other devices. Remember, these are minimum requirements, and one may want to be more generous. Figure 9-44 is a typical minimum spacing drawing. This drawing is for equipment mounted on a flat plate in an industrial cabinet. Another type of mounting is in racks used originally in the electronics industry but now common to the process industries, especially for computer or control room use. In rack mounting, the equipment is mounted one above the other in standard racks, usually a nominal 19 inches wide. Wiring is then run to terminal blocks elsewhere.

Heat dissipation is a concern in panel layout. Generally, the heat-producing components are mounted near the top, the other components farther down. In many cases, there is considerable heat produced in the racks themselves due either to rack-mounted power supplies or to the power handled by the output modules. The space occupied by wireways must be considered in allowing adequate room for ventilation; minimum spacings may have to be increased as a result. In some cases, fans may be required to circulate air inside the cabinet. Air conditioners may be necessary, especially in areas with a high ambient temperature.

Proper grounding is an essential requirement for both safety and good operation. Figure 9-45 shows the grounding requirements for one PLC; others are similar. Be careful to observe the manufacturer's instructions about which terminals should be grounded and which should not. The routing of the ground

In planning the layout, consider wire routing for various signal levels and types. While 120-volt AC signals are not usually affected by noise, they themselves can generate noise that affects other low-level signals (e.g., analog inputs and outputs, high-speed, low-level DC signals, and so on). These constraints will affect the I/O configuration, so give them due consideration when configuring the system.

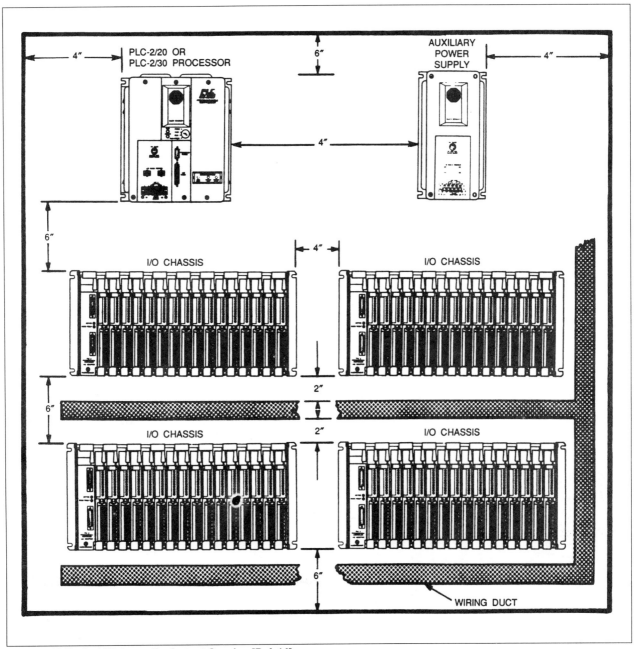

Figure 9-44. Typical Minimum Equipment Spacing [Ref. 14]
(Courtesy of Allen-Bradley, a Rockwell International Company)

wires and the connection points is very important. For a more complete description of the problems, causes, and cures, see IEEE Standard 518.

Remember to leave room for cables and accessories. Cables are often a problem, especially if there is not enough room left to plug and unplug them. Some manufacturers give good information on the space required for this, while others are remiss or make the information hard to find. In any case, always leave adequate space for the cables connecting the various parts of the system, and route them away from sources of interference or noise.

Leave space for future additions of racks, power supplies, terminal blocks, and so forth. Do not forget the space for wiring these items as well as the distribution equipment they will require. Consider drilling and tapping the mounting holes for these future devices to avoid a possible shutdown in the future. It is also a good way to remind others that the space is already reserved for equipment.

Power Distribution

All electronic equipment needs clean, reliable power. PLCs, because of their industrial design, should be less demanding than other electronic equipment, but there are limits to everything. Since most industrial equipment operates on 220-, 440- or 575-volt three-phase power, and PLCs tend to operate on 120-volt, single-phase power, some transformation is necessary. There are many ideas on how to distribute power to the PLC and its I/Os. Some prefer to use a single transformer for everything, saying the same quality of power will be distributed everywhere.

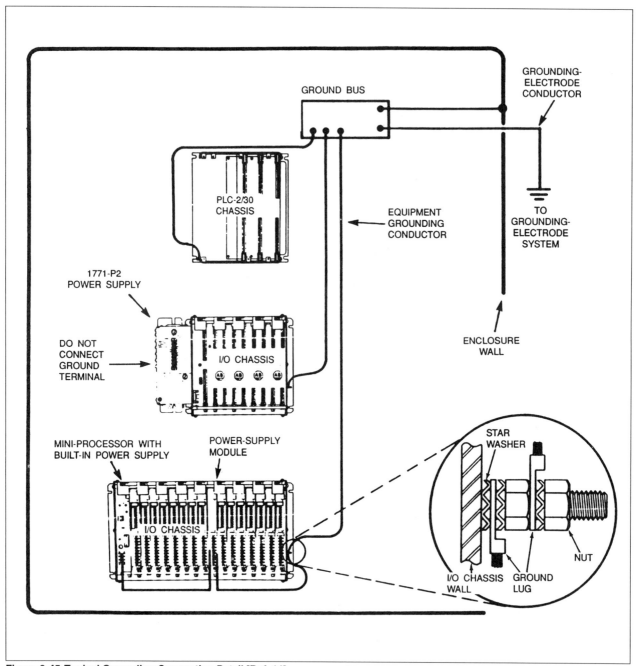

Figure 9-45. Typical Grounding Connection Detail [Ref. 14]
(Courtesy of Allen-Bradley, a Rockwell International Company)

Others suggest using separate transformers for the PLC power supplies and the I/Os to avoid noise and load problems being communicated from the output devices to the processor electronics. When using multiple voltage levels and both AC and DC power, it is no longer necessary to have a common supply. Problems could occur in one power supply but not in the others.

Figure 9-46 is a typical power distribution diagram. In this case, the same transformer provides power for everything. Although only one fuse is shown in the transformer secondary, it is better to provide separate fuses for the PLC, the inputs, and the outputs. This way, a fault in the wiring or an overload takes out only part of the equipment, making it easier to find the source of the problem. Another possibility is to use circuit breakers rather than fuses. They cost a little more but are much more convenient, and they can be used to remove power selectively when performing maintenance or testing.

In addition to the controller and its I/Os, power is also needed for lights and tools that may be used in servicing the equipment. Provide at least one 15-amp circuit from a separate source for servicing. The separate source avoids polluting the supply for the PLC. Hand tools, such as drills and vacuums, can generate a lot of electrical noise, especially if they are a bit worn. Also, while doing cleaning or modifications, consider turning off the power to the controller.

Since PLCs need to be programmed and since the programming equipment requires power, provide an outlet next to the PLC that is powered from the same circuit for the programming equipment. Label the outlet "to be used only for programming equipment." The programming equipment is best fed from the same circuit as the PLC in order to avoid any problems with differences in grounding or neutrals. It is not unusual for problems to be caused by missing or defective grounding connections on the circuits feeding programming equipment. When the ground is missing, the data cable becomes a grounding conductor, and problems will occur because it is definitely not designed for this use.

On large, spread out systems, provide a communications system to link the various parts of the machine and control equipment. Two wires, some phone jacks, a set of headphones, and a battery will provide a workable system for start-up and troubleshooting. Otherwise, consider using walkie-talkies or runners, bearing in mind that this alternative may not work too well in an industrial environment.

If several circuits provide power to I/Os and if problems are likely to occur with shorts or grounds, it may be in order to consider monitoring the power supply circuits with the PLC. DC power supplies can be especially annoying if they have automatic protection circuits that reduce the output voltage to zero under adverse conditions. The really bad ones restore the voltage as soon as the problem goes away. Try to find an intermittent short with that type of supply! In this case, a relay makes a good voltage detector, and its contact can be easily brought into the PLC and an appropriate alarm generated.

> Install a power outlet fed from the PLC circuit, along with a data circuit plug, on the outside of the cabinet, protected with a cover or warning saying "this outlet is to be used only for programming equipment." This allows one to monitor or even program the controller with all the cabinet doors closed. In dusty, hot, wet, or other problem areas, this is a real advantage. Consider additional outlets with the same warning at remote rack locations, even on the process machinery, if the manufacturer will allow this extra wire and connections.

Safety Circuitry

There was a bit of relay logic in Figure 9-46. The "master control relay" is designed to allow shutting the system down in the event of a serious problem, such as the PLC going berserk or an output becoming stuck in the ON position. In this case, the relay disconnects power to the outputs, regardless of the actions of the PLC. This is also called an emergency stop relay, and the push buttons are usually made with large red mushroom-type buttons that are easy to spot and actuate in case of trouble. This independent emergency stop circuit is essential and is required by most codes. Some considerations in its design may not be obvious.

One should not interrupt the power to signaling devices such as message displays or pilot lights. To do so blinds the operator to the status of the machine and requires that he reset the emergency stop to see what happened. Consider also what will happen when de-energizing outputs that control solenoid valves, brakes, valve operators, and so forth. Depending on the pneumatic and other considerations, one could actually create a more hazardous condition by removing power from some outputs.

The status of the master control relay should be brought into the PLC and used to provide an orderly shutdown of the logic and a graceful restoration after the master control relay has been reset. The process will dictate the arrangements required, but they can be complex.

Terminals and Wiring Space

It is a common practice to bring all field wiring to a set of terminal blocks and from these terminal blocks to the I/Os of the PLC. This is often done because the field wiring is several sizes larger than can be accommodated by the terminal blocks on the PLC. In addition, the common wiring can be more easily handled. There is a price for this convenience: the field wiring terminal blocks can occupy

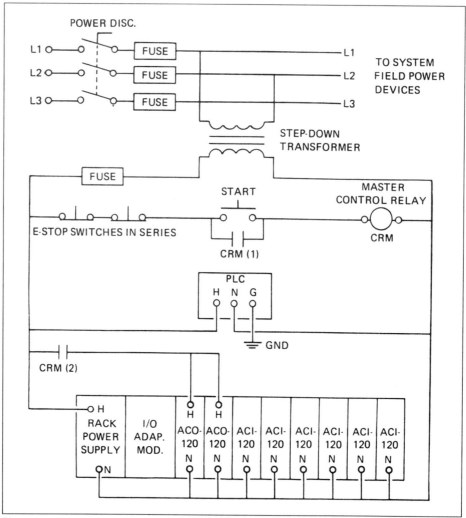

Figure 9-46. Typical AC Power Distribution Drawing [Ref. 14]

as much space as the PLC! Carefully examine the wire sizes that can be handled by the terminals on a PLC. Eight-point modules will often handle 12-and 14-gage wire. Sixteen-point modules will have trouble with 12-gage wire, and the 32-point modules will have trouble even with 16-gage wiring. In addition to the wire size, be sure to consider the insulation thickness. The insulation for 16-gage, 1000-volt wire is considerably larger than for 300-volt wire.

The wiring duct used with terminal blocks is always too small! Even designers with a lot of experience in this area always seem to be faced with bulging ducts and no place for spares or the extra wire that was left uncut—just in case. A good practice is to mount the terminal blocks on standoffs and use four-inch deep ducts. Since the panels are usually eight to 18 inches deep, this is no problem. What is really needed, however, is a six-inch deep duct! Be sure to leave enough room between the duct and the terminal block. Two inches on each side of the terminal is a minimum. For wire numbers that are long, more room may be needed.

Make room for spare wires. Do not bury them in the bottom of the wiring duct where they can never be found again. Provide an empty duct just for spares, put them there without cutting them short, and identify them so they may be easily found later. The maintenance people will love this!

Table 9-15 is a typical control panel specification. It can be modified to include the extra items useful for a particular job. Note that it refers in several places to "the enclosed drawings." Provide a list of the enclosed drawings at the end of the specification so there is no problem due to misunderstandings or to lost drawings.

I/O Assembly

If this chapter's advice on I/Os has been followed, assembly should be a piece of cake. The configuration information should include all the locations of the I/O modules, along with the switch and jumper settings required. The installer just has to go by the numbers. If this was not done, the installer will have to work it all out alone, will probably not get it right, and somebody (guess who?) will have to figure it all out on start-up, when time is at a premium.

Checkout

It's a good idea to separate checkout from start-up because they are two very different procedures. Checkout verifies the proper installation and operation of all the devices connected to the PLC, whereas start-up gets the machine or process running. A good checkout can make for an easy start-up; a bad or nonexistent checkout can make the start-up a nightmare.

Double check all dip switch settings. Many manufacturers use a variety of switches with different markings. The most common problem is with "on" and "off" compared to "closed" and "open." It's not uncommon to encounter modules with their switches set to just the opposite of what they should be.

Table 9-16 shows a typical checkout procedure for the PLC. To make this work, the drawings and other information must be available to check against. Also it is essential to have a written checklist and to check off the items as they are done. If necessary, add dates and initials so that there's a record of what was done and when, just in case some time elapses between checkout and start-up. Figure 9-47 is an example of a checkout sheet, but note that it has been combined with a start-up verification that will be discussed later.

Checkout takes at least two people—one to actuate the devices and one to observe the PLC's response and record the results. One of the best people to be involved in the checkout is the future operator of the machine or process. This person has a vested interest in knowing everything about the machine and how it works. What better way is there than to do the checkout? Be sure the people doing the checkout understand what they are doing and why and are not just working by rote. Devices should be operated several times to ensure they are not stuck. All positions of a device should be tested, not just one. All pilot lights should be tested to see that they light up on command.

Devices such as message display units can be tested by activating one control line at a time instead of all the possible combinations. In the case of a 256-mes-

Table 9-15. Typical Control Panel Specification [Ref. 14]

The control panel furnished under this specification shall be supplied complete with the instruments and equipment listed on the enclosed drawings, installed and electrically wired, and ready for wiring to field instruments and equipment.

The control panel shall conform to the following:
1. The control panel shall be fabricated with cold-rolled steel plate.
2. All miscellaneous items, such as wire raceways, terminal strips, electrical wire, etc., where possible, shall be made of fire-resistant materials.
3. The grounding bus bar and studs shall be pure copper metal.
4. A minimum of two 120-V ac, 60-Hz utility outlets shall be installed in panel.
5. An internal fluorescent light with a conveniently located ON/OFF switch shall be installed in the panel.
6. A circuit breaker panel with circuit breakers to accommodate all panel lighting, power supplies, instruments, I/O modules, processors, and any other loads listed on the system drawings shall be provided.
7. Each electric wire over 4 inches in length shall be identified at each end with a wire number per the electrical drawings for the system.
8. Terminal strips with screw-type connectors shall be used and no more than two wires shall be terminated on any single terminal.
9. Nameplates shall be made of laminated plastic with white letters engraved on a black background for both front and rear panel-mounted instruments and components, such as power supplies, transformers, I/O racks.
10. The ac wiring shall be separated from 4 to 20-mA dc current and digital signals by a minimum of 24 inches, and they must be wired to separate terminal strips.
11. All electrical wires and cables shall enter the control panel through the top. Sufficient space shall be provided to allow ac power cables entering the top of the panel to continue directly to the circuit breaker panel.
12. All wires and cables shall be routed and tied to provide clear access to all instruments and components for maintenance and removal of defective components.
13. All ac wires shall have a minimum insulation voltage rating of 600 volts ac, and all ac wires shall have a minimum insulation voltage rating of 300 volts dc.
14. All electrical conductors shall be copper of the correct wire size for the current carried with 98% conductivity, referenced to pure copper.
15. Wireways shall be attached securely to the control panel.
16. Metal surfaces with wires or cables passing through them shall be furnished cables.
17. All wire bundles or cables shall be clamped to the panel at all right angle turns.
18. All wires entering or leaving a wire bundle shall be tied to the bundle at the point of entering or leaving the main wire bundle.

sage unit, this would require eight tests instead of 256, which represents a good saving in time.

Do not attempt to correct any problems encountered during checkout itself. Instead, note the problems or anomalies on the report sheet and give it to another crew for correction. Once the correction has been made, retest at least the questionable items. There are three reasons for this: the people doing the checkout may not be qualified to correct the problems; interrupting the checkout will probably make it take much longer; and a complete list of problems will give a better idea of what action to take. For example, crossed connections, defective modules, and so on, may become very apparent when the entire results are viewed. Systematic errors become apparent and may be corrected altogether instead of one at a time.

ELECTRICAL CHECKOUT

MACHINE #:	SECTION #:	DATE:	PAGE:

# EQUIPMENT	# DWG REF	DESCRIPTION	# PLC INPUT	PILOT LIGHT	PROGRAM	# PLC OUTPUT	PROGRAM	PILOT LIGHT	FUNCTIONAL TEST	SUCCESSFUL	CHECKED BY
	8001	CF. VACUUM SECTION 1	I2000	√	√						
	8002	CF. CHOPPER INS. 1	I2001	√	√						
	8003	CF. VACUUM CHOP. 1	I2002	√	√						
	8004	CF. VACUUM ANVIL 1	I2003	√	√						
	8005	CF. VACUUM SEC. 2	I2004	√	√						
	8006	CF. VACUUM SECT. 3	I2005	√	√						
	8007	CF VACUUM CUT AND PLAC	I2006	√	√						
	8008	CF. VACUUM SECT. 6	I2007	√	√						
	8009	CF. FINAL CHOPPER	I2010	√	√						
	8010	CF. VACUUM FINAL CHOPPER	I2011	√	√						
	8011	CF. VACUUM ANVIL FINAL	I2012	√	√						
	8012	CF. VACUUM FOR DUST.	I2013	√	√						
	8013	FREE	I2014								
	8014	FREE	I2015								
	8015	FREE	I2016								
	8016	MAIN DRIVE TEMP. OK.	I2017								
	8101	PULL CORD	I2100	√	√						
	8102	EM STOP IN PBS1	I2101	√	√						
	8103	EM STOP IN PBS2	I2102	√	√						
	8104	EM STOP IN PBS3	I2103	√	√						
	8105	EM STOP IN PBS4	I2104	√	√						
	8106	EM STOP IN PBS5	I2105	√	√						
	8107	EM STOP IN PBS5	I2106	√	√						
	8108	EM STOP IN PBS7	I2107	√	√						
	8109	EM STOP IN PBS8	I2110	√	√						

Figure 9-47. Typical Checkout Sheet

Table 9-16. Typical Checkout Procedure [Ref. 14]

I. Visual Inspection

1. Verify that all system components are installed per the system drawing.
2. Check I/O module location in equipment racks per I/O drawings.
3. Inspect switch settings on all intelligent modules per drawings.
4. Verify that all the system communication cables are correctly installed.
5. Check that all input wires are correctly marked with wire numbers and terminated at correct points on input module.
6. Check that all output wires are correctly marked with wire numbers and terminated at correct terminals on output module.
7. Verify the power wiring is installed per the ac distribution drawing.

II. Continuity Check

1. Use ohmmeter to verify that no ac wire is shorted to ground.
2. Verify continuity of ac hot wiring.
3. Check ac neutral wiring system.
4. Check continuity of system ground.

III. Input Wiring Check

1. Place the programmable controller in test mode.
2. Disable all output signals.
3. Turn "on" ac system power and power to input modules.
4. Verify the E-stop switch removes power from the system.
5. Activate each input device, observe the corresponding address on the programming terminal, and verify that the indicator light on the input module is energized.

IV. Output Wiring Check

1. Disconnect all output devices that might create a safety problem, such as motors, heaters, or ccontrol valves, etc.
2. Place programmable controller in test mode.
3. Apply power to the programmable controller and the output modules.
4. Depress the E-stop push button and verify that all output signals are de-energized.
5. Restart system and use the forcing function in the programming terminal to energize each output individually. Measure the signal at the output devices and verify that the output light on the module is energized for each output tested.

FUNCTIONAL CHECKOUT

Once the module and wiring checkout has been done, it is time to do the functional checkout. For example, it may have already been verified that the starter for motor 12 picks up when commanded to do so; now it must be verified that motor 12 runs when the starter is energized and that it turns in the correct direction. Similarly, pneumatic and hydraulic equipment must be checked to see that the motors and cylinders that move are the right ones and that the direction of movement is correct.

Here, the purpose is to see that the desired action really takes place. Again, the future operator is the best candidate for the job, supported by the appropriate technical people. A test sheet should be made up and checked off as the tests are done, and corrections should be left till after the testing is done.

At this time, any calibration or setup work that is required should also be done to ensure that all is ready for the start-up.

Start-Up

After the successful installation and operation of all the devices and equipment connected to the PLC, the program can finally be installed. The program should be reasonably bug-free if it was tested on a simulator or by some other means.

Now comes the acid test of all the hard work to date. Table 9-17 gives the start-up test procedure for the PLC. Item four is time-consuming but essential to having a good process. This should pick up most of the remaining bugs. By separately testing each major component of the system, the start-up is a lot easier. Again, documentation is necessary to be sure nothing has been overlooked. A printout of the program can be used as the log sheet, with the results recorded as work progresses. The procedures used to test the program by itself may be adequate for testing the completed assembly.

Table 9-17. Typical Start-Up Procedure [Ref. 14]

Operation Test

1. Place processor in "Program mode" and turn on main power switch.
2. Load the pre-tested control program into the programmable controller.
3. Disable all outputs, select the run mode on the processor, and verify that the run light on the processor is activated.
4. Check each rung of logic for proper operation by simulating the inputs and verifying on the programming terminal that the correct output is energized at the proper time or sequence in the program.
5. Make any required changes to the control program.
6. Enable output modules and place processor in "Run" mode.
7. Test control system per process operating procedure.

Testing the live controller with the outputs disabled requires some extra work in order to simulate the confirmations and field signals that would follow the activation of the logic. For complex sequences, this may be almost impossible to do. In such cases, it may be necessary to dry run the process or machine to verify the operation before putting material in the system. In other cases, a safe substitute may be used, or critical devices can be simulated by custom-made hardware. Eventually, the real process is started, its behavior is observed against the original description, and changes are made where necessary. Some of these changes will be due to programming or installation problems, others to problems in the description, and some to unexpected behavior of the process. Finally, improvements will be suggested based on the experience gained in operating the process.

Maintenance

PLCs are very reliable. Unfortunately, that very reliability leads people to neglect them, the feeling being that no maintenance is needed. The biggest enemy of electronic equipment is heat, and the most common causes of excessive heat are loose connections, accumulations of dirt and dust, and failed ventilation. The most common cause of erratic operation of PLCs is loose or dirty connections. A good preventive maintenance program will control all of these problems rather well.

Table 9-18 lists the most common preventive maintenance points along with a rough idea of how often they should be done. This has to be adjusted for the specific situation and hardware. The weekly maintenance is important in providing a continuing presence at the equipment and a history of the application, which can provide useful diagnostic information. If the frequency of diagnostic messages suddenly goes up, one may suspect a different problem than if it goes up slowly over a period of time.

Keep a log of everything done on, to, or around the PLC. This should also include program modifications. If the budget does not allow a sophisticated computer-based system, or even an inexpensive computer-based system, at least

Perform any maintenance inspections long enough before any scheduled shutdown to allow time to take corrective action, if required. This is especially important if shutdowns are few and far between.

Table 9-18. PLC Preventive Maintenance

Routine Maintenance Perform weekly

Change filters.

Check error log and scan time.

Check diagnostic lights.

Back up program.

Visual Inspection Perform quarterly

Check for proper operation of ventilation system.

Check for buildup of dust or dirt.

Is any extraneous material in the cabinets?

Look for loose connections.

Look for signs of heating, burning, or water infiltration.

Are there signs of temporary modifications, jumpers, etc.?

Is all the documentation present and up to date?

Maintenance Perform yearly

Reseat all cards, cables, and connectors.

Tighten all connections.

Replace batteries.

spend $10 for a bound log book that resides near the PLC. Use the book to record all the happenings on the PLC. Then periodically review the log book and look for patterns or chronic problems. Since a lot of repairs to PLCs are done by changing modules, track these repairs to see if there really is a common problem, and then take corrective action even if it is only to advise the manufacturer.

Get on the manufacturer's list for information on problems, updates, and new products. Then read and, if necessary, study the information that arrives. It could help identify repairs or changes that can be made before they evolve into problems, or it may provide the answer to a current problem. This information is available from most manufacturers. Sometimes it can be obtained through a distributor.

Keep software up to date. Both the programming software and the processor software should be current. As time goes on, it will become more and more difficult to get older versions of software, and support people may not be familiar with it. It is much better to upgrade a processor during a scheduled shutdown when there is still a functioning processor of the old design that can just be put back in the slot.

Programming software is continually evolving. Each new version has its minor bugs to discover and correct. On the other hand, to make one more productive, there are also improvements such as added features, increased speed, and reduced memory requirements. One must also consider that the new and improved computers that come out may require a new version of the programming software to fully exploit their capabilities.

 Diskettes are fragile and easily mislaid. Provide a central location for backups with a fireproof safe, and control the access to it. If labor unrest is a problem, be extra careful with backups. When choosing a fireproof safe for diskettes, make sure it is designed for diskettes and not paper. Diskettes deform above 140°F and become unusable. Paper, on the other hand, does not burn until 451°F. A data safe must have at least twice as much insulating material as a regular fireproof safe. This is often accomplished by adding more insulation to a regular fireproof safe. This results in less storage space and it costs more, but the peace of mind is worth it. Another option is to

store a set of diskettes off-site at a company that specializes in the storage of computer tapes and magnetic material. For an inexpensive solution that provides at least minimal coverage, take a set of diskettes home. The chances of both one's home AND the factory going up in smoke at the same time are almost nil. Someone living near the plant may have to be more careful about making this choice. Confidential programs should not be stored just anywhere.

Check backups from time to time. It's entirely possible for them to be unreadable. Although this is not common, one should never rely on a single unverified backup. At the least, try to read the files immediately after making them, or use the verify option in DOS. The best check is to load the program back into a processor and see that it works, or at least runs.

For plants with a number of PLCs, someone should be responsible for seeing that backups are made and maintained and that everything is kept up to date.

Troubleshooting

One can only troubleshoot a system that worked properly at some time. If the system never worked properly, then it's still in the start-up phase. The rest of this discussion assumes that the system worked correctly at some time in the past. The Kepner-Tregoe book [Ref. 17] describes some basic tools for problem solving and decision analysis that should be understood and used by anyone who attempts to troubleshoot a PLC system, or any system for that matter. Although the title may be misleading, make no mistake about it—this book is not just for managers, it's for everybody. In fact, the examples are mostly down-to-earth production problems.

A good troubleshooter needs the orderly approach described in Kepner-Tregoe as well as a good understanding and knowledge of the system: the process, equipment, PLC, program, and operators; but other specialized knowledge may also be required. There are several ways to approach these requirements. Perhaps there is an individual with all the expertise necessary, or a team can be formed to do the job. Kepner-Tregoe would probably recommend that the manager help the other members use the method to arrive at a solution, but anyone can learn the method and use it alone or with others to solve the problem. Table 9-19 lists, in decreasing order of occurrence, the cause of problems with PLCs. A quick examination of this figure will give an idea of how to approach the problem. Check the field devices and I/Os first! Once it is established that all field and I/O devices are working properly, then, and only then, look into the programming and other areas.

A number of steps should be followed in the troubleshooting process. Some of these are almost obvious, but they bear repeating as done in Table 9-20. Get the facts straight from the operator (or operators if the problem occurred on more than one shift). It is always amazing to see the distortion that can occur when information passes through many hands before it gets to the person in charge. Check the log books, not just the one for the PLC, but also the machine log or any others that may bear on the problem. Get as much information as possible on the problem before trying to do anything.

Be sure to use the most up-to-date information, program, drawings, and process documentation. Be careful about the program in the PLC. Most monitoring software uses the copy on the program on the hard disk, not the one actually in the PLC. If there are any doubts, do an upload and regenerate cross references before proceeding. When the problem has been found, share it with others, at least by entering it in the log book.

What are some of the tools available for troubleshooting PLCs? Many on-line monitoring programs have data-gathering possibilities to make troubleshooting easier. Generally, these options allow one to record the status of a number of bits

Table 9-19. Problem Sources in PLC Applications

Field devices
Wiring
I/O modules
Program
Processor, racks, etc.

Table 9-20. Steps in Solving PLC Problems

Talk to the operator(s).
Check the log book(s).
Make sure you have the current program and documentation.
Verify the operation of the I/O.
Change only one thing at a time.
Write the results in the log book.

or registers over time. Some will give a continuous representation of the status, while others will show transitions only. Some will record the values at specified time intervals. In working with this type of tool, it is important to understand its limitations. The response time is limited to the communication delay between the processor and the loader. Generally, this is no more than once a scan, so events that happen several times in one scan cannot be monitored.

In addition, the communications delay between the PLC and the monitoring computer is a significant consideration. Communicating over an RS-232 port at 9600 baud really slows things down; it may take several hundred milliseconds to update a display. Some manufacturers have dedicated cards that provide higher speed access to the PLC, but some figures limit the resolution to 100 milliseconds. If the program has a scan time of 40 milliseconds, that really means seeing the update only every third scan! Special monitoring cards can be added to the processor in the PLC to allow much faster monitoring, but they are costly and require modifications to the processor. It's not simply a matter of walking up and plugging one in while things are running.

These monitoring programs create histograms (timing charts) of the logic as it happens, and then display it on the screen or store it in memory for later playback. This technique is especially useful when one is concerned with the sequence of events that precedes a problem or occurs during an upset. Some systems will trigger the recording from the process to get only the parts of interest. Of course, given enough disk space, one can let the recording go on for a long time, but then it has to be looked at thoroughly to find the parts of interest.

Another useful tool is the multi-rung display. Many monitoring programs will display several nonconsecutive rungs on the screen at once. There is still a limitation as to what will fit on the screen, but sometimes seeing two or three rungs at the same time will answer a lot of questions. In most cases, documentation can be suppressed to get more on the screen at one time.

All systems allow the monitoring of large blocks of I/Os, bits, or registers at once but without any logic or descriptions. This can be useful if one's eyes are quick and the problem is not complex. For example, one can often observe 16 bits at once and see that one or several change state at the same time or some unexpected change occurred that would have been missed if only a couple of lines of logic were being looked at.

Some systems provide custom display screens that monitor selected registers or bits in the processor, along with some documentation. These screens are developed and stored as files on the hard disk and can be recalled at any time. They can be particularly useful for monitoring PID loops, shift registers, and so on, that may need periodic observation. A judicious choice of variables can result in all the information needed to monitor a process. This can make it easy for an electrician or technician who may not be totally familiar with the process to monitor what is going on. It is also a time saver since it requires setting up the screens only once. If the design process has been orderly, it may even be possible to design only one screen and then simply change the register numbers for different parts of the process.

One very useful tool is the array of diagnostic lights on the modules, processor, and other cards. One of the most elementary checks is to see if the module light agrees with the field device position and the internal bit in the processor. Differences will show if the problem is the field device or the module. If the module is suspect, check again after replacing it; some voltage problems can simulate module problems.

> If several modules are bad in the same position, suspect the connector, either for the field wiring or the rack backplane or the field wiring.

The lights on the processor and other modules will generally indicate only major faults (such as no power or a dead battery), but check them if a large part of the system is dead. Sometimes one bad communications module can bring the entire system down.

Troubleshooting is a learned skill. One has to do it, and make mistakes, and learn from them as well as from the mistakes of others. Repeated efforts bring improved performance. To develop troubleshooting skills, be prepared to tolerate the downtime required for people to practice, or, alternatively, build a simulator for them to practice on. Remember that when people are learning they will take more time. Do not rush them too much. If there's already one good troubleshooter on the team and others need to be developed, put up with the extra time the trainees will need to diagnose the problem while they are learning. Plan for it in the project and do everybody a big favor.

> **Another useful tool is to add programming to trap errors or particular events. The PLC can even monitor itself sometimes. By connecting a suspect output back to a spare input and adding a little logic, one can verify that the output goes on when it should. This trick was used to prove there was a problem with a one-revolution clutch. By counting the number of times the program activated the output, the number of times the input tied to the questionable output was activated, and the number of times the stacker actually moved, it was possible to show that the problem was the clutch (not the program or the output). In that case, the clutch missed about once every 300 cycles and caused a jam. With this information the clutch manufacturer was able to rebuild the clutch. Other problems require different methods, but the idea is simple—write a program to isolate the problem, especially if it is intermittent.**

References and Bibliography

Magazines

There are no longer any magazines dedicated exclusively to programmable controllers. The magazines listed here have a regular feature on PLCs at least once a year.

One magazine was dedicated to PLCs and copies may still be found in libraries. It started out in 1982 as *The PC User*, became *Programmable Controls, The User Magazine*, in 1984, and became an ISA publication in 1987. It has been included in *Industrial Computing* since 1989.

1. *Automation*, 1100 Superior Avenue, Cleveland, OH 44197-8065.

2. *Automation Products and Technology*, 395 Matheson Blvd. East, Mississauga, ON, Canada, L4Z 2H2.

3. *Control Engineering*, Paid Subscription Service Center, 44 Cook Street, Denver, CO 80206-5191.

4. *ESD Technology*, 100 Farnsworth, Detroit, MI 48202.

5. *I&CS, (Instrumentation & Controls Systems)* P.O. Box 2026, Radnor, PA 19089-9985.

6. *Industrial Computing plus Programmable Controls*, ISA Services, Inc., P.O. Box 12277, Research Triangle Park, NC 27709.

7. *INTECH*, ISA Services, Inc., P.O. Box 12277, Research Triangle Park, NC 27709.

8. *Machine Design*, 1100 Superior Avenue, Cleveland, OH 44114.

9. *Manufacturing Systems*, P.O. Box 3008, Wheaton, IL 60189-9972.

10. *The PLC Insider's Newsletter*, Carefree Communications, Box 5268, Carefree AZ 85377.

Books

11. Bryan, L. A., and Bryan, E. A., *Programmable Controllers Selected Applications*, Volume 1, Chicago, IL: Industrial Text Co., 1988.

12. Bryan, L. A., and Bryan, E. A., *Programmable Controllers Workbook and Study Guide*, Chicago, IL: Industrial Text Co., 1988.

13. Gupta, Sanjay, and Gupta, Jai P., *PC Interfacing for Laboratory Data Acquisition and Process Control*, Research Triangle Park, NC: Instrument Society of America, 1989.

14. Hughes, Thomas A., *Programmable Controllers*, Research Triangle Park, NC: Instrument Society of America, 1989.

15. Johnson, David G., *Programmable Controllers for Factory Automation*, New York, NY: Marcel Dekker, Inc., 1987.

16. Jones, Clarence T., and Bryan, Luis A., *Programmable Controllers: Concepts and Applications*, Atlanta, GA: International Programmable Controls Inc., 1983.

17. Kepner, Charles H., and Tregoe, Benjamin B., *The New Rational Manager*, Princeton, NJ: Princeton Research Press, 1981.

18. Kissell, Thomas E., *Understanding and Using Programmable Controllers*, Englewood Cliffs, NJ: Prentice Hall, Inc., 1986.

19. Whilhelm, Robert E., *Programmable Controller Handbook*, Hasbrouck Heights NJ: Hayden Book Company, 1986.

Standards and Practices

20. FIPS PUB 94, "Guidelines on Electrical Power for ADP Installations."

21. IEEE Std. 518-1977, "IEEE Guide for the Installation of Electrical Equipment to Minimize Noise Inputs to Controllers from External Sources."

22. ISA-RP60.8, Electrical Guide for Control Centers.

23. ANSI/ISA-S5.1, Instrumentation Symbols and Identification (Formerly ANSI Y32.20).

24. JIC EMP-1-67, Electrical Standards for Mass Production Equipment.

About The Author

James E. Bouchard is Senior Project Leader for Johnson & Johnson, Inc. of Montreal and a graduate of McGill and Sir George Williams Universities. He has been responsible for the technical quality of the work and for developing standards. He provided technical guidance, support, and training for others of the electrical engineering group and was responsible for the programming equipment and software used. A Senior Member of ISA, Mr. Bouchard is editor and author of the Programmable Controller/Computer column in the "News Meter." He now provides all project-related electrical engineering for ultra-thin feminine hygiene protection production equipment.

10

Ergonomics and Occupational Safety

The term "ergonomics" is derived from the Greek *ergon*, meaning "work," and *nomos*, whose most apt meaning is a "system," or "set of laws." Since most people are normally employed in gainful work in order to earn a living, this expression is applied to the ways in which humans best need to interface with all forms of machines, control systems, and management structures they are involved with in their daily activities. Thus, ergonomics is concerned with defining the basic principles that provide all persons with the maximum of safety and comfort in the workplace, while at the same time it attempts to improve both worker and manager productivity for the greater benefit of all sectors of society.

Initial efforts to organize these principles were begun as early as 1915 from studies conducted in England, but these studies ended without meaningful result. Later, the growing need for improved human factors definitions and guidelines in industry received great impetus in the United States during World War II, when the previous studies were revived and, combined with new studies and further research, resulted in considerable development in this area for both military and industrial purposes. So it was that ergonomics had its formal beginnings as an applied science in 1949, following several international meetings of academics, researchers, and engineers.

The Aims of Ergonomics

Practice

Ergonomics is an applied technology, i.e., it is a method by which our physical surroundings can be altered in an organized and purposeful way. Therefore, its practice can be applied to many branches of engineering and science: psychology, medicine, physiology, sociology, environmental control, industrial design, business administration, and so on. In real-life situations, the application of ergonomic rules to *any* machine or system *always* needs a multidisciplinary approach.

The primary aim is to optimize the functioning of external systems and machines by adapting them to human needs and capabilities. In other words, the rules governing the practical application of ergonomics must be tailored to an average of the human biological makeup. Machines must be made to fit man, not vice versa. The same guidelines should also apply to management systems with respect to their hierarchical work groups.

Inherently tied to these basic aims are a number of related human factors activities such as technical skills training, industrial hygiene and occupational safety. More on this later.

Planning

In terms of performing actual work, this is the most essential part of any ergonomic activity, and it is carried on in four major branches:

(1) Technical Planning

(2) Time Planning

(3) Personnel Planning

(4) Planning the Workplace

Technical planning concerns itself with total system/machine operations and performance, with defining criteria for the allocation of man/machine tasks. Very efficient methods now exist, using charts, logic layouts, and other aids, that provide a systematic and organized approach to this planning process.

Personnel planning, increasingly a more important and more complex unit of industrial activity, works in three main areas: selection, training, and resource allocation. A few of the many and varied facets of this activity are: recruiting, defining physiological and psychological criteria for job functions, writing job descriptions, calculating staffing requirements, developing and completing training programs, and so on. A brief discussion of technical training and training simulators is included in a later part of this chapter.

Time planning involves coordinating the activities of some or many specialists working in different disciplines to achieve an ordered system (or project) completion. Among the several methods of time planning that have been developed are: PERT (Program Evaluation and Review Technique) and CPM (Critical Path Method). Time planning can also be used to define the time relationships in the system being developed. In this way it is possible to determine the optimum task allocation between man and machine both in occupational safety terms and in economic terms.

Planning the workplace means that first consideration must be given to environmental and occupational safety concerns for workers, Before optimizing machine environment and performance.

A good ergonomic design is an investment in increased employee safety and comfort, in better system performance, and in improved productivity. Although it should not be assumed that optimum results are easily obtained, the quality of the results is immutably linked to the time and effort required to produce them.

Ergonomics as a science is divided into two major branches:

(1) Biomechanical ergonomics, which covers human tasks where muscular power is involved, i.e., loading and unloading trucks, felling trees, or any other work involving physical labor; and

(2) Information ergonomics, where humans receive many types of data and stimuli from their surroundings in order to control the operation of functional systems, i.e., machine tools, industrial parts assembly, office machines, etc.

It is the second area, as it applies to industrial process control, that will be reviewed here.

Information Ergonomics in Industrial Control

Man/Machine Interface

In the course of a day's work the average worker, whether in an office or in a production plant, is exposed to a huge mass of visual and audible information. Some of this data is recognized by the worker as being meaningful to the work at hand, and some is not. Thus, in order for this information to have a useful meaning it must be received in an ordered fashion and correctly interpreted; otherwise, it is disregarded by the worker and becomes simply audible and visual "noise."

The interaction between man and functional system (or man and machine) is a procedure by which man receives this external data generated and emitted by the system. The data is impressed on the human senses, processed by the brain, and normally results in a series of physical actions by the person on the control elements of the machine or system. For this data to provide rapid recognition and correct response, it must be appropriately coded. This coding process needs to respond to what is called the *perceptual organization* of the human brain, and it is one of the cornerstones on which information ergonomics is founded.

A simple system showing the relationship between man and machine appears in Figure 10-1.

The man is the controller of the system. He receives information about the operation of the machine through his senses. After processing the information he then performs appropriate control actions to keep the machine in proper working order. The area of contact between man and machine is the interface, which is the main area of interest for the practice of information ergonomics.

Surrounding environmental conditions affect both man and machine and the overall quality of their performance. Often, for reasons of personal safety and comfort, the environment surrounding the man must be totally different from that surrounding the machine. The contact area in this case may be a grouping of instruments and control devices on a control panel placed in an area adjacent to the machine. It may also consist of a separate control room in which all operating functions are located.

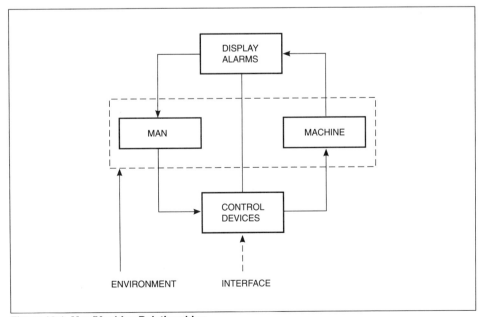

Figure 10-1. Man/Machine Relationships

Task Allocation

The division of tasks between man and machine is called *allocation.* A great deal of early work in ergonomics was done in attempting to determine what type of work is most suitable to each, based on both physical performance and economic returns. The summary results of this work still largely hold true, although the effects of increasing computerization are causing some shifts in traditional allocation patterns. Table 10-1 contains these summary results:

Advantages of Man	Advantages of Machines
Better ability to detect faults.	Fast and accurate responses.
Better at recognizing and differentiating patterns.	Ability to carry out repetitive work.
Flexibility to change to different activities.	Good short-term memory.
Good long-term memory.	Ability to do several tasks simultaneously.
Ability to make decisions from small or incomplete amounts of information (induction).	Ability to make decisions from very large amounts of data (deduction).

Signal Reception

In order to be able to accept system signals, man needs some form of *measuring equipment.* This equipment is the collection of sense organs that recognize the essential codes provided by the system and, once decoded by man, will result in the proper control responses. In the workplace environment the human senses most often used for the direct reception of ergonomic data are vision and hearing. The other senses are only little used for this purpose. Luckily, these signals are also the easiest to organize and codify.

The sense of touch as an information receptor is limited to recognizable physical phenomena within the range of human capabilities: hot-cold, soft-hard, smooth-rough, and so on. However, these perceptions are useful only when related to data being received by the other senses. The sense of smell is more limited still and normally involves only reactions to "safe-unsafe" conditions. The sense of taste is not normally used in an industrial context except as a quality control factor in the food industry.

Signal Codes

Identification of signals made possible through the *perceptual organization* of the brain means that certain types of signals, or codes, are more easily interpreted than others, probably because of the physical design of the brain and the nervous system. It is not known whether the perceptual organization has a specific organic center located at any particular point in the brain; but it is known that the work it performs has a most definite result.

The function of coding is to rationalize the information to suit the perceptual organization as naturally as possible. Presentation of the information should be designed in such a way as to be clearly associated with correct meaning. The coding process is akin to creating an *artificial language*—one possessing its own clearly defined set of rules.

This artificial language must have its own *semantics*, that is to say, it must have a set of meaningful and easily recognizable symbols. The meaning of the different symbols used must be agreed upon and understood by all. It must also have its own *grammar*, which defines the different ways that the symbols can be combined.

In industrial applications both figurative and nonfigurative codes are used. Figurative codes are simple sketches that resemble the apparatus or machines they represent. Nonfigurative codes are normally acronyms (a grouping of letters) or abstract figures, with meanings that can be easily explained and readily understood.

In terms of an artificial language, figurative symbols are suitable for use as "nouns" (machinery, things, etc.), nonfigurative symbols for "verbs" (actions). Numerals are also frequently used and may represent both identification and quantification.

The visual and audible codes as used in industrial control vary to some extent in different countries and also vary with different types of industry. However, by a sort of loose agreement there exists a very general set of guidelines for signal codes that, although not definitive, are helpful to the system designer. These guidelines are summarized in Table 10-2.

The degree of use of standardized coded information is by no means uniform in the industrialized world.

Table 10-2. **Table of Recommended Signal Codes**

Code	Max.	Recommended
Colors		
Lamps	10	3
Surfaces	50	9
Design		
Letters/Numbers	no limit	no limit
Geometrical	15	5
Figures/Diagrams	30	10
Size		
Surfaces	6	3
Length	6	3
Lightness	4	2
Frequency (flashing)	4	2
Sound		
Frequency	(Large)	5
Loudness	(Large)	2

It is important to note that the number of codes recommended for use is quite low. For instance, in the color codes only three are recommended for foreground use (lamps), and only nine for background (surfaces), out of a maximum range of perhaps 50 different colors. In the audible codes as well, the recommended number is low: five for frequency and two for loudness.

For the practical use of colors in the design of signal codes in normal environments, the first rule is to use no more than the six basic colors: red, yellow, green, blue, black, and white. For systems that must take into account persons with color blindness, this range must then be limited to red, blue, black, and white.

The size of figures, be they symbols, letters, or numbers, is critical to a good design. Established standards for obtaining the best possible reliability in reading letters/numbers describe the following three characteristics as being the most important:

One rule of thumb for setting the minimum height of a letter, number, or figure in millimeters is to multiply the reading distance in meters by ten and divide by three.

(1) The ratio between height and thickness of stroke

(2) The relation between height and width

(3) The appearance (light/dark contrast)

At a reading distance of three meters the minimum symbol height would be 10 mm. The recommended width is 70 to 80 percent of the height.

Instrument scales are monitored by operators on an almost continuous basis, so it is important that the scales provide good readability. A frequently used formula for determining good scale size is:

$$D = 14.4L \qquad (10\text{-}1)$$

where D is the reading distance, L is the length of the scale, and D and L are in the same units. Thus, a scale length of 70 mm is acceptable for reading at a distance of one meter. At a normal reading distance of three meters, a scale should be about 210 mm long.

The grouping of symbols and letters can also have an appreciable effect on speed of recognition. The easiest to recognize is a grouping of three (3). Table 10-3 shows some suggestions for suitable ways of grouping figures.

Table 10-3. Suggestions for Grouping Figures

No. of Digits/Symbols	Alternatives	
7	3,2,2	3,4
8	3,3,2	2,2,2,2
9	3,3,3	

Visibility

Good visibility means that coded visual symbols should be of the correct size, be sufficiently well lit, and have a good contrast level. The human eye is able to sense electromagnetic radiation in the visible spectrum in wavelengths from 300 nm to 1500 nm. This represents a relatively wide bandwidth of 10^9 and makes the eye a reasonably efficient sensor. However, for visibility to be effective, it is subject to some important contributory factors:

(1) Luminance discrimination (the ability to see differences and variations in brightness)

(2) Sharpness (the ability to differentiate figures)

(3) Temporal visual ability (the ability to distinguish changes and movements over time)

(4) Depth discrimination

(5) Color discrimination

Hearing

The human ear can distinguish pressure variations in air of frequencies between 20 Hz and 20,000 Hz. This capability is not linear, so that the ability of the ear to sense change varies with both the frequency and intensity of the sound. Changes in frequency are most readily distinguished at the low end, while changes in sound level are most easily detected at high levels and high frequencies. Also, in order for a sound to be heard in a noisy background, it needs to be louder than the background noise (called masking noise).

It is not obvious that any generally accepted standards, or even any practical guidelines, have been developed for using sound codes. At present sound codes are almost exclusively used for alarm annunciation. Most installations use a "loud" alert for noisy locations and a "soft" alert for control rooms, often with siren-like howls in varying frequencies. A most effective technique, universally used in computer systems, uses tone bursts or "beeps" of differing pitches and modulations for warnings or alarms.

Decision and Motor Action

The time taken for a person to react to an external stimulus by carrying out a certain action in response is called "reaction time." The greatest part is taken up by the brain in its work to arrive at a decision. Actual values of reaction time response differ with different sense organs; for hearing and touch it is about 140 ms, while it is about 180 ms for sight. For smell and taste it is in the range of 500-1000 ms. This time difference can vary widely—it can decrease with proper signal coding or with increases in signal strength. It can also decrease with higher levels of operator training. At the same time, the response time increases in a logarithmic manner with the number of signals that need action.

 The average person cannot handle signals more frequent than one every 300 ms, or about three per second. Therefore, for reliable regulation of events that are faster than three per second, man cannot be used as the controller.

For controlled movements, where each movement is a response to a signal, the human limit of speed is also about three responses per second but can vary as well with the person's training and age.

From the foregoing it can be seen that to have effective control of continuous or discontinuous industrial processes where many events occur (for all practical purposes) simultaneously and at high speed *automatic controllers* must be used. In turn, the controllers must be monitored by operators, who must also perform all associated nonautomatic tasks.

In industrial process control work, operator actions built into the control system design are usually limited to hand/finger motions. By far the greater number of these are either push button or rotary selector switch actions, or they are computer keyboard actions. Most of the components obtainable for this purpose today are of ergonomic design and require only that a judicious selection be made of suitable types for the desired application.

Planning Control Centers

Fundamentals

Today's industrial process plant, with its complex array of machinery, requires very large numbers of display and control devices, which are normally grouped together on instrument panels. This, of course, holds true only when computer-based control systems are not used. Traditionally, an instrument panel is mounted into a larger control panel. Control panels are relatively large metallic structures of rugged design that provide for mounting displays and instruments on the front, with cable, compressed air, and other connections in the rear. Current practice is to install these panels in environmentally conditioned control rooms. A number of different styles are common and are reviewed below.

Physical design of a control panel begins with the determination of certain fundamental measurements. The first of these is the natural sightline, i.e., the operator's eye height is determined, both in standing and sitting positions, so that the optimum viewpoints of the proposed instrument grouping or console can be set. Most other measurements can then be fixed from this first position: chair seat height; arm's reach distances; heights of indicators, manual switches, push buttons, and levers; heights of pedals and footrests; and so on.

Standard measurements for both men and women have been established for a number of years (anthropometrics), and virtually all panel design today is based on these measurements or similar ones. Figure 10-2 contains much of this basic information.

Computers or Unit Controllers?

Rapid technological evolution in the computer field is now an everyday thing. One result of this is that the choices now open to the control system designer have been radically changed for all time. Today's choices lie first and foremost in determining whether to use individual control instruments and displays mounted on control panels or to select a computer system using video control consoles.

Microprocessor-based distributed control systems (DCS) have become the accepted standards of the industry. For all but the smallest projects the choice today is invariably DCS. Despite this, it is certain that for particular requirements conventional instruments mounted in control panels will continue in use for many years and in many applications.

Computer systems are thoroughly covered in other chapters of this book, so this chapter's review will be limited to control panels and video consoles.

Instrument Control Panels

Control systems used in the process industries normally do not contain control devices manipulated by an operator that are in direct contact with machinery or moving parts. Rather, all operating machinery is remotely controlled, be it in manual mode or in automatic mode. Instruments used for remote control in the main perform the following functions:

(1) Quantitative monitoring

(2) Qualitative monitoring

(3) Process control monitoring and alarming

(4) Comparisons of monitored data

(5) General overview of process operations

(6) Starting and stopping process operations

In designing an instrument display, the purpose and functions of the proposed instrument grouping or console should be clearly described. The positioning of instruments in a group is critical; those that provide the operator with the most relevant information should be placed so that it is obvious they are the most important. When an instrument panel is designed as a model of the process, the relative importance of individual instruments must become clearly apparent. This design principle is relatively simple and should be in common use. In practical terms, instruments and controls need to be positioned according to the following criteria:

(1) Frequency of use

(2) Sequence of use

(3) Degree of importance

(4) Operating function

Depending on the size and complexity of the process to be controlled, instrument groupings can at times become so large and complex that their relationship with the various units of process equipment becomes too vague, and rapid operator recognition becomes quite difficult. In these cases a pictorial presenta-

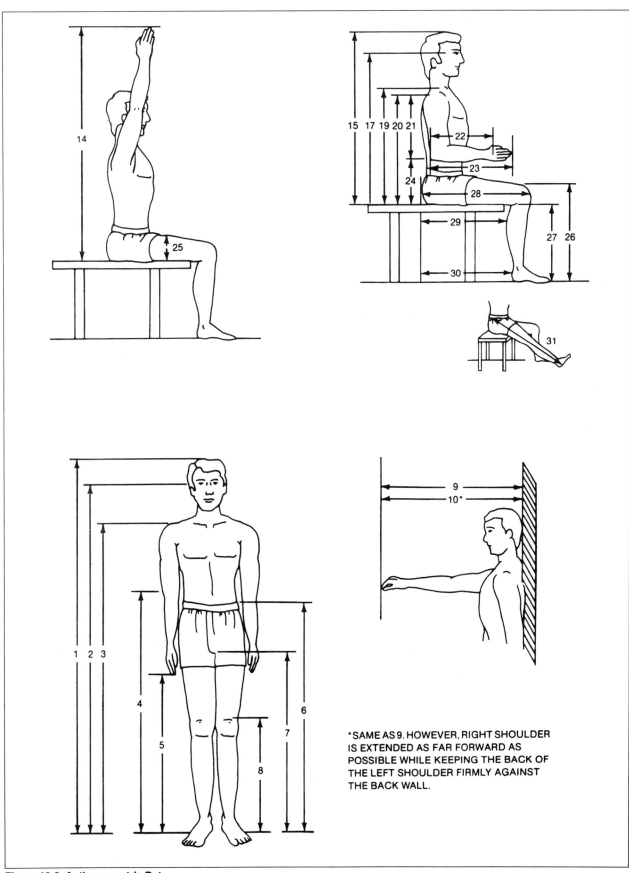

Figure 10-2. Anthropometric Data

*SAME AS 9. HOWEVER, RIGHT SHOULDER IS EXTENDED AS FAR FORWARD AS POSSIBLE WHILE KEEPING THE BACK OF THE LEFT SHOULDER FIRMLY AGAINST THE BACK WALL.

tion of the process itself, called a process graphic panel, is included as part of the instrument display. This process graphic panel can currently be obtained in two popular styles called the full graphic and the semigraphic.

Panel Design Considerations

Because the measure of efficiency of a control system is to such a large extent dependent on an operator's ability to absorb visual process data and to respond quickly, control panels of suitable ergonometric and anthropometric design should satisfy the following considerations:

(1) Provide the most rapid operator access to the critical controlling elements: instruments, push buttons, selector switches, emergency stop push button, and so on. Visual indicators, regulators, panel meters, and displays grouped together at operator eye-level; hand-control devices grouped separately below the visual groups, usually at about waist level, and more accessible to hand/arm movements.

(2) Allow for rapid operator action to particular control situations that may periodically occur: process start-up, normal process shutdown, emergency shutdown, and repair or maintenance of equipment.

(3) Provide an amount of spare capacity for future expansion and modernization of the control system.

(4) Permit convenience of installation and ease of maintenance for all panel-mounted equipment.

(5) Provide an aesthetically pleasing appearance.

(6) Provide a functional arrangement for all other auxiliary devices, including alarm annunciators and other accessories.

Panel design is influenced to a degree by the type of process involved and the location of the panel in the plant. The size of individual panel sections will often be governed by the route that must be taken in transporting them to their final location. Panel weight must be considered in relation to the bearing capabilities of the floor. Normally, panels and their housings are mounted on a concrete pad or curb.

The material most often used for panel construction is mild carbon steel in thicknesses of 1/8, 3/16, and 1/4 inch. Stainless steel and fiberglass-reinforced resin are usually used for more difficult locations but are finding increased popularity for general use.

Surface finishes for panels should be relatively light in color but somewhat darker than individual instrument housings. Edges and frames of instrument panels should not give a strong contrast.

COMPONENT GROUPING

Two basic options are open to the panel designer: (1) conventional control instrument groupings without process graphics or (2) graphic display panels. The choice depends on a number of factors, including the size and complexity of the system, the degree of operator skills available, the requirements for training of new operators, etc. In cases where a graphic display panel is the logical choice, the designer has the option between full graphic and semigraphic panels.

Full graphic panels represent pictorially on the panel face the entire flow diagram, in large scale, of the process to be controlled. Regulators, recorders, and indicators are located in panel cutouts inside the appropriately pictured process vessels. This system has several advantages: in allowing rapid recognition of particular components, in the training of new operators, and in improving the under-standing of complex systems.

While well-designed full graphic panels readily fulfil these functions and are impressive in general appearance, they are also relatively expensive, wasteful of panel space, and present cost problems of major magnitude when, for any reason, the flow diagram needs modification.

Semigraphic panels allow more efficient high-density grouping of instruments on the panel face. A reduced-scale flow diagram of the process covers the upper part of the panel, with the instrument grouping located immediately below.

HIGH-DENSITY LAYOUT

Recent trends in the design of electronic instruments have resulted increasingly in component miniaturization. This, of course, allows for more economical high-density panel layouts, where most of the usable panel space is devoted to the process instruments. Here, two pitfalls lie in wait for the designer: the first is overcrowding, resulting in the loss of identity of individual components both at the front and the rear of the panel; the second is a marked reduction in rapid recognition of specific visual data required for optimum operator response. This type of layout, then, must be planned with great care. Figure 10-3 shows a variety of instruments grouped on a panel.

CONTROL PANEL TYPES

The following are the most common panel styles in current use.

(1) Vertical flat panel. This is the most popular and certainly the most economical because of its simplicity of design. Mounting control components is simple due to the regular shape, and the components are readily accessible for maintenance. Normally mounted on a rear-access cabinet, the panel can be modified to mount against a wall, on legs, or on channels.

(2) Slant top section. This is a variation of the vertical flat panel. The top section is tilted forward, usually about 15 degrees, and is most often used for mounting alarm annunciators, process indicators, and semigraphic displays. The tilt provides improved visibility with reduced glare or reflection and a degree of aesthetic relief from the stark lines of the flat panel.

(3) Standing console. This can be provided either as a self-contained unit or as part of a vertical flat panel. It effectively moves the lower section of the flat panel into a position that improves operator accessibility and vision. It is normally used for mounting manually operated push buttons, motor controls, and other switchgear as well as operational indicators and signal

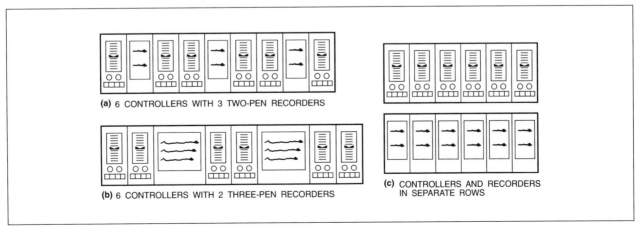

(a) 6 CONTROLLERS WITH 3 TWO-PEN RECORDERS

(b) 6 CONTROLLERS WITH 2 THREE-PEN RECORDERS

(c) CONTROLLERS AND RECORDERS IN SEPARATE ROWS

Figure 10-3. Typical Panel Instrument Grouping

lights. When used as part of a vertical flat panel, it increases the available mounting space.

(4) Desk console. The version illustrated in Figure 10-4 is shown in combination with a vertical "look-over" panel. This version is normally used where observation of a second panel is necessary or to keep a critical process in view. The "look-over" option is intended for use by a standing operator. The desk function provides writing space and can be used in either the standing or seated position. The desk may be equipped with a drawer for storage of records, charts, pads, etc.

(5) Breakfront panel. This increasingly popular construction improves the visual quality of both upper and lower panel sections without seriously degrading serviceability. It has a highly aesthetic appeal, which may justify its somewhat higher cost by better looks and improved operator acceptance. It finds increased use in larger control rooms, particularly in oil refineries, chemical plants, and nuclear power stations.

Figure 10-4 shows a cross-sectional view of the panel front shapes described above.

WIRING CONSIDERATIONS

The rear of the control panel is usually allocated for cable entry, wire ducts, terminal blocks, and other connections. Control systems should so designed that most of the active components that connect to each other are located within the same control panel unit. Adequate space between cable ducts must be provided, because crowding of instrument wiring can add heavily to both initial assembly and later maintenance costs.

PANEL LIGHTING

Lighting should be planned to reduce operator fatigue, stress, and errors in reading. Instrument panels should be illuminated with diffused lighting, directed so that no reflections occur from instrument glass or from the panel itself. Many modern instruments contain either internal lighting or have digital displays that provide their own illumination.

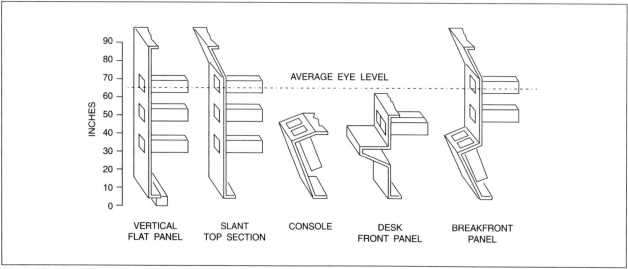

Figure 10-4. Panel Front Shapes

Alarm Annunciators

In addition to control and indicating devices, instrument control panels must contain warning devices, connected to the process equipment, that serve to notify operating personnel of abnormal operating conditions. This is most conveniently accomplished with an alarm annunciator system. The standard alarm annunciator is normally a housing or box, the front of which is fitted with one or more rows of plastic, backlit windows. The windows are lit by lamps whose circuitry is connected into the process with various types of limit switches. The switches are intended to turn on when an abnormal process condition is reached, which in turn lights the appropriate window and alerts the operator of an alarm condition.

Where control panels with conventional instrument group displays are used, one or more alarm annunciators are mounted on the control panel, usually located above the main instrument grouping.

Annunciator circuitry can be either electromechanical (relays), solid-state, or microprocessor-based. The design varies depending on the manufacturer, but certainly the microprocessor type is already the most popular and probably the most cost-effective.

A terminology that reflects the nature of the instruments and how they work has developed around these devices. ANSI/ISA-S18.l, Annunciator Sequences and Specifications, covers the terminology used for operating and specifying annunciator systems. The following terms are typical:

Abnormal operating conditions—the monitored variable is not within the specified operating limits (also known as off normal).

Acknowledge—recognition by an operator that an alarm condition exists, done by pressing the "ACK" push button.

Alert—the state of the system when an alarm point is activated.

Flasher—a circuit in the annunciator that makes the alarm lamp flash on and off during an alarm condition.

High-low—an annunciator that indicates "abnormally high," "normal," and "abnormally low" process conditions.

Memory or *lock-in*—a feature of the annunciator that keeps the alarm lamp flashing until it is "acknowledged," even though the original alarm condition may have turned off.

Normally open contacts—during normal operation of the process, the contacts of the alarm limit switch are in the open position.

Normally closed contacts—during normal operation of the process, the contacts of the alarm limit switch are in the closed position.

Reset—a push button that returns the annunciator window to its normal state after the alarm condition has returned to normal.

Test—a push button that turns on all the alarm lamps together to check if they work.

Sequences—this denotes several different methods of annunciator operation, more or less complex, available from suppliers. These are divided into four classes:

Class 1, the simplest standard operational sequence.

Class 2, same as Class 1, but includes an "alert step."

Class 3, same as Class 1, but with "high-low" option.

Class 4, same as Class 1, but can do multiple simultaneous alerts.

The simplest operational sequence goes like this:

(1) The process limit switch goes off normal; the alarm horn sounds; the alarm window starts flashing.

(2) The operator responds by pressing the "acknowledge" push button; the horn goes silent; the alarm window stays lit, but stops flashing.

(3) The operator clears up the alarm condition; the limit switch goes back to normal.

(4) The operator pushes the "reset" push button to put the alarm window back into normal operation; the alarm window turns off.

First out sequence—allows the identification of the first alarm point to turn on. In many situations one alarm condition may trigger a whole series of additional alarms that may be a result of only the first abnormal condition. The ability to identify the first alarm can save considerable downtime in getting the system back to normal.

Annunciators with backlit windows are by far the most commonly used, although other types exist for special applications. The window type is manufactured in various sizes. Figure 10-5 is a typical example. The plastic windows are available in several translucent colors, the most widely used being white with black letters.

Video Control Consoles

Distributed control systems and computer control systems provide one or more video consoles for use as the operator's window into process operations. This normally consists of desk-mounted or tabletop video screens with specially designed operator keyboards. Typical video consoles located in a central control room may have several screens and keyboards.

The introduction of the computer video screen in industrial applications brings to process control a totally new information tool. Not only does it provide a great deal more information than conventional instrument displays, it also enables control functions to be performed in infinitely better and different ways. By their very nature and because they present information to the operator in such small sizes compared to conventional instrument panels, video screens possess their own set of ergonometric design rules.

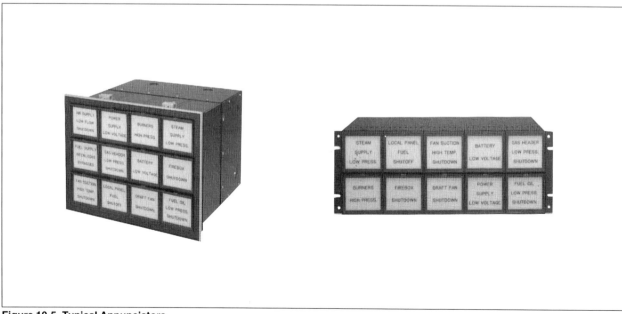

Figure 10-5. Typical Annunciators
(Courtesy of Ronan Engineering Ltd./Industrial Projects Ltd.))

The most usual screen configuration has a dark background with light foreground/text. It requires a very careful consideration of lighting conditions; surrounding light levels should be considerably reduced, and any risk of glare must be avoided.

The following are some recommended basic rules for optimum legibility and coding for the design of video screens:

(1) Character size: Height should be between 16 and 27 minutes of arc. Ratio of width to height equal to 0.75. Ratio of stroke width to height between 1:10 and 1:6.

(2) Character shape: A matrix of 9×7 points is best (for low-resolution screen).

(3) Viewing position: About 70 cm is best viewing distance with text character size about 3.2 mm; directly in front, with a maximum angle of 30 degrees.

(4) Colors: Those in the yellow-red-green range are good. Avoid dark blues.

(5) Flashing markers: A rectangular shape with a 3 Hz flashing rate is best.

(6) Code labels: Alphanumeric is best, color second best for identification. No more than seven letters and/or numbers should be used.

There are no set rules as to the density of information contained on a given page. It has been suggested that a person's search time (of screens) in seconds is, in general, one fifth of the number of alternatives (symbols) presented on the screen. What is certain is that increased concentration of information in a given area increases both search time and the number of errors made.

Figure 10-6 illustrates a typical distributed control system console complete with video screen, keyboard, memory cassettes, and a hard copy printer.

A video screen is a modern tool that provides the operator greater flexibility and controllability for the process than ever before, particularly when the information it presents is prepared with thoughtfulness and care. It will be around for a

Figure 10-6. DCS Operator Console
(Courtesy of ABB Kent Taylor)

long time to come. It is essential, then, that we understand the ways in which it differs from the conventional control panel and the changes that these differences will likely make in the future in improved control room work methods and improved process control techniques.

Among the major differences, three are fundamental:

(1) A given screen page presents only a limited amount of *effective* information, so many pages may be needed to represent a process unit.

(2) The screen presents information in *time sequence*, one page after another, while a panel presents all process information in parallel at the same time.

(3) The video console is able to provide an enormously greater amount of process and production information than the operator has ever had before.

Control rooms must always contain a minimum of two (or better, three) video screens for each process unit to be controlled.

Control Rooms

The purpose of a control room is not only to group all control functions associated with a process in a central location, but also to provide an adequate level of comfort for process operators who spend most of their working time there. In addition, it is not only the operators who need a comfortable environment; many computer control suppliers insist on an equal level of comfort for their machines.

A control room of modern design is sufficiently large to prevent crowding when all operating equipment, chairs, tables, and auxiliary devices are installed. The actual size is, of course, related to the application, but in no event should it be smaller than 25 square meters. The room should be fitted with a computer-type raised floor, tilted windows to avoid glare, controlled incandescent (not fluorescent) lighting, air filtration, and climate control.

The design of the room must meet a number of specific environmental conditions:

(1) Climate control. Temperature and humidity should conform to ANSI/ISA-S71.01, Environmental Conditions for Process Measurement and Control Systems: Temperature and Humidity. Air filtration should conform to ANSI/ISA-S71.04, Environmental Conditions for Process Measurement and Control Systems: Airborne Contaminants. Air intake ducts for the room must be carefully designed so as not to draw in contaminants from the process area. Also, it is preferable to maintain the room at a slight positive pressure (0.08 inches of H_2O), as heavy traffic may cancel out the effects of environmental control.

(2) Audible noise reduction. Acoustic treatment should be provided to reduce noise. A level of 55 dBa is desirable.

(3) Electrical interference (noise) reduction. The room should be located away from high-voltage electrical rooms, high-voltage cables, and other possible sources of electromagnetic interference (emi). Where this is not possible, suitable protection using high-permeability metallic screening should be provided. Radio frequency interference (rfi) from radiotelephones (walkie-talkies) and similar devices must be avoided.

The control room is in most cases not one single room but two. Located adjacent to the control room proper is an equipment room that contains process control cabinets, cable termination cabinets, computers and their associated devices, and any other equipment that is related to process control but does not need to be accessed on a continuous basis. Normally, this room is substantially the same size as the control room, or somewhat larger, and designed to the same specifications.

Adequate lighting is essential to good operator comfort. Generally, three lighting levels are needed:

(1) For video display areas. A controllable level of 100-450 lux of indirect, nonglare, incandescent lights.

(2) Control panel areas, printers, recorders. A level of about 650 lux.

(3) For equipment maintenance, room cleaning, etc. A level of 1200 lux is necessary.

Personnel Training

The Industrial Training Program

One ability, very characteristic of man, and the overriding factor that distinguishes man from machines, is the ability to adapt to different situations. This adaptation often takes a certain time. When it occurs systematically, or according to a certain method, it is referred to as education or training.

The increasing use of computers in industrial production brings to the production floor, and consequently to process operators, a vastly greater amount of process and production information than was ever before available. This information is now available instantaneously, is able to spread horizontally through an entire production area, and is often available to operators at an earlier point in time than it is to management.

This situation is, of course, in opposition to the very concepts of traditional management, which operate in a vertical line. The result is that it forces operating personnel, consciously or not, to have a much greater influence than ever before in day-to-day decision-making functions in the factory at the production level. As a rule, operators have little or no training in this function, and, indeed, normal management practices would prevent operators from taking part in the decision-making process. This is a phenomenon not yet generally recognized at the management level, but it is bound to have an impact on future management strategies.

What is universally recognized, however, is the need for an increasing level of skills training in all industrial workers. Skills training is a tool for improving not only the productivity of machines and processes but also the well-being of the workers who operate them.

Virtually all industrial enterprises committed to their future success and to the welfare of the employees on whom this success depends participate in, or operate directly, an industrial skills training program. This usually involves all plant technical and operating personnel and, very often, management personnel as well. It is normally an ongoing plan in which new employee training, as well as upgrading skills of existing employees, is a nonstop effort.

Naturally, a complete training program is organized as a multidisciplinary plan that contains a number of course structures suited to the different job functions concerned.

An effective course structure consists of three essential parts: objectives, course outlines, and lesson plans.

Clearly defined objectives should be the primary basis of any training activity. They determine the scope of the training and help guide the selection and preparation of materials to be used.

After the objectives of the plan have been set, the next step is to develop an outline of what is to be covered. Topics and subtopics should be arranged to show the order in which instructions will be given.

The lesson plan is the blueprint for presenting, section by section, the material contained in the course outline. Lesson plans standardize training. They are the instructor's guides, which help him to:

(1) present material in proper order,

(2) place proper emphasis where it belongs,

(3) avoid omissions,

(4) conduct classes according to a schedule,

(5) provide for trainee participation, and

(6) gain confidence, especially for new instructors.

In ergonomic terms, a number of ground rules should be observed in the design of training programs if good results are to be obtained:

(1) People can absorb only a limited amount of information at one time. Training must be divided into stages to avoid overloading.

(2) Demands for improved speed and accuracy of trainees as a result of the program must be determined very accurately for the different stages of the learning process. People need to be told what is expected of them.

(3) Students should be given continuous feedback on their performance during the entire training process.

(4) Students should be prompted to use correct methods at every step of the training cycle. They should be prevented from making errors if possible.

(5) Students should be able to set their own rate of learning speed by controlling the teaching speed.

Training programs need to utilize all available training aids, such as films, videotapes, teaching machines, simulators, and other audiovisual products. They must also incorporate in their teaching methods modern personnel management techniques: work analysis, physiological and psychological evaluations, and the like.

In industrial process control, as in many other industrial activities, one training aid that responds well to the various ergonomic requirements of skills training is the computer-based process simulator.

Process Training Simulators

Process training simulators are a combination of: a mathematical model (a computer program that represents a plant's operating equipment); the control strategy and control system configuration (application software that massages the program); and computer hardware (which can be a copy of the hardware used in the actual process operations). The hardware may, in fact, be any computer that is capable of running the program and performing the training functions. The simulator video screens and keyboards, however, must resemble the actual operating ones as closely as possible.

In a simulator training session, an operator-trainee is seated in front of the simulator control console. With the keyboard he or she attempts to control the operation of what looks like the real process, as shown by the data seen on the screen.

The computer runs a real-time process model and changes the flows, temperatures, levels, pressures, fluid concentrations, etc., in response to control actions taken by the trainee.

An instructor sits at another console and, through predefined commands, monitors the results of the actions taken by the trainee during the session. The instructor is able to insert faults into the operation to simulate real problems that the operator may face in the actual plant: equipment failure, erratic conditions in product feed, control valve malfunctions, etc.

By combining judicious fault insertion, cold-start simulations (process starts from empty), and the like, an instructor is able to train operators much more rapidly and realistically than can be done with classroom lectures and site visits.

Training simulators offer numerous benefits. Among them are:

(1) familiarization and training on the use of the operating console itself,

(2) familiarization with the console graphic displays and the relationships between the keyboard and the graphic elements,

(3) familiarization with the alarm-handling facilities of the console,

(4) teaching of process operations and principles,

(5) training to recognize problems and to solve them,

(6) improving teamwork among operators,

(7) getting feedback on the efficiency of the graphic displays, and

(8) the possibility of creating an unlimited number of teaching scenarios for ongoing training to upgrade operator skills.

TRAINING SIMULATORS HARDWARE

A training simulator consists, physically, of: a computer that can range in size from a small personal computer to a large minicomputer on which simulation software runs; a process control station with one or more graphic display screens and keyboards on which the operators are trained; and an instructor monitoring station, as well as related peripherals such as printers, data storage devices, and so on. The process control station the trainee uses closely resembles the actual control room console that will be used after the training is complete.

TRAINING SIMULATORS SOFTWARE

Simulators are custom built by specialized companies that employ experienced engineers and technicians in a multitude of disciplines, such as process design, process control, electrical engineering, instrumentation and control engineering, and computer software and hardware development. Their role is to convert the client's P&ID diagrams, control configuration, SAMA drawings, and logic drawings into software that simulates the dynamics of the process.

The simulation software consists usually of a proprietary executive program and a library of subprograms that represent a host of unit operations and control modules. Figure 10-7 illustrates the interaction of the various software components.

Occupational Safety

Definition

In all industrialized countries, employers are held accountable for the safety and security of their workers on the job. As a result of this, a very large amount of legislation and regulation exists, at all governmental levels and in all jurisdictions, that attempts to define what safety and security are; what industrial hygiene is; and what an industrial accident is, how to prevent it, and what is likely to happen as a result of one.

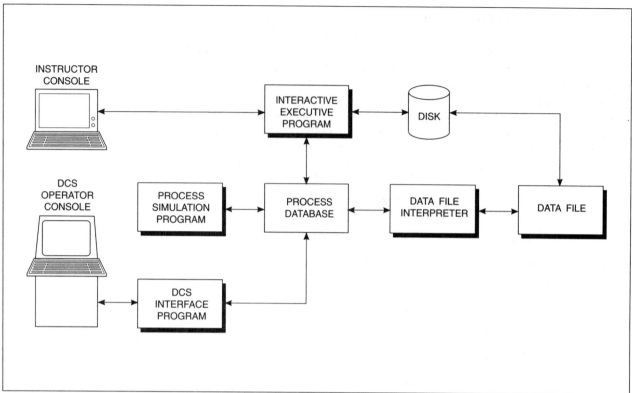

Figure 10-7. Training Simulator Software Schematic

Employers are required to provide a workplace that is equipped with suitable protection from all hazards on the job: adequate heating and ventilation, sanitary and washing facilities, sufficient illumination, and no overcrowding. Employers must provide protection against occupational diseases, fire and explosion hazards, toxic chemicals, electrical shocks, excessive noise, open tanks and exposed areas, and protection against accidents that may be caused by the operation of machinery.

Fire and other types of alarm systems, sprinklers, extinguishers, protective clothing, first aid and survival gear, emergency showers and eyewashers, equipment and piping identification codes, and signs and warnings are some of the safety features that must be installed on the premises and/or be available at all times. Finally, a number of automatic safety devices must be inserted in the process equipment to protect against accidental injury in cases of abnormal process conditions.

Employees are required to conform to all laws, codes, and safety regulations and to avoid accidents.

This very complex interrelation between protective legislation, safety devices, security protection, first aid, emergency rescue, safety and security training, and inspection and surveillance must of necessity use ergonomic methods of planning in order to achieve any success. However, their primary aim in this case is to prevent injury rather than to optimize system functions.

Safety Standards and Codes

Accident prevention and job safety have long been a preoccupation of industry, labor, and government. The very large body of legislation that exists to regulate industrial activity in all countries attests to that fact. In North America, the first practicable worker's compensation act was passed in Wisconsin in 1911.

A number of organizations, made up of representatives from government, labor, and industry, have come into being for the purpose of defining those guidelines that later become legislation. Principal among these is The National Safety Council, founded in Chicago, Illinois, in 1913, after several years of preparatory work.

Both the United States and Canada have their respective versions of an Occupational Safety and Health Agency (OSHA), with inspection and regulatory powers supported by appropriate legislation. States, provinces, and other jurisdictions enact refinements to these laws, made to suit local conditions. Parallel organizations exist in most European countries, Japan, and elsewhere in the world.

In the United States a number of standards-generating organizations, grouped around ANSI (the American National Standards Institute), are very active. Among these are the Instrument Society of America (ISA), the American Petroleum Institute (API), the National Fire Prevention Association (NFPA), and others, as well as many engineering associations: mechanical, chemical, automotive, electrical, etc. These groups perform a most important role in establishing technical standards of quality for industrial material and in setting up uniform safety standards for their use. The body of work generated by these groups is willingly accepted by all industries because of their excellent and well thought out technical content. In Canada, where the C.S.A. (the Canadian Standards Association) performs a similar function, ANSI standards are also universally accepted.

There are, in addition, private organizations, such as insurance underwriters (Underwriter's Laboratories, Factory Mutual), who operate testing laboratories for verifying the fire-resisting qualities and other features of industrial materials and who are able to impose a number of regulations of their own by charging lower insurance rates to individual plants that comply with their recommendations and higher rates to those that don't.

One of the pillars of occupational safety regulations, which exists in most countries, states, provinces, and other jurisdictions, is a National Electrical Code (NEC) or equivalent. This defines a number of plants and plant areas in which there is a greater or lesser risk of fire and explosion from electrical sources, be it from normal operations or from malfunctions. It also defines the types of electrical material to be used in the separate locations and the standards for their installation, thereby reducing the risk of injury to persons and damage to materials.

Protection against Hazards

HAZARDOUS LOCATIONS

The National Electrical Code "area classification" system is a standard accepted universally in the United States and Canada and has its equivalents in all industrialized countries. It is recommended by the National Fire Protection Association (NFPA) and adopted by most insurance underwriters. Article 500 of the NEC provides a definition of three classes of hazardous locations, which is also given in API-RP 500:

Class I:	Locations made hazardous by flammable gases or vapors.
Class II:	Locations made hazardous by combustible dusts.
Class III:	Locations made hazardous by ignitable fibers or filings.

These classes are further divided into Division classification:

Division I:	Locations that may contain hazardous mixtures under normal operating conditions.
Division II:	Locations in which the atmosphere is normally nonhazardous but may become hazardous under abnormal circumstances such as equipment failure, failure of ventilating systems, etc.

What this classification procedure means is that, depending on the hazards presented by the materials and products processed in a plant, different styles of housings for electrical switchgear must be used in those areas designated as hazardous. The type of switchgear classified as "explosionproof" is, of course, constructed much differently from that which is not "explosionproof."

In both Class I and Class II hazardous areas, the NEC requires that enclosures for electrical devices and connection boxes must be constructed in such a manner that an explosion would be contained should it occur within the enclosure. Because these explosionproof enclosures are both very bulky and expensive, in certain circumstances the NFPA allows for the use of purged and pressurized nonhazardous enclosures in hazardous areas in lieu of explosionproof ones.

This means using a standard enclosure into which a small amount of compressed air is released on a continuous basis, thereby dissipating any gas concentrations inside. Specifications are given in NFPA Standard 496, "Standard for Purged and Pressurized Equipment in Hazardous Locations," or in ISA-S12.4, Instrument Purging for Reduction of Hazardous Area Classification.

Nonexplosionproof electrical equipment can cause fire or explosions in areas that contain flammable liquids, gases, or dusts. An ignition (and consequent explosion) takes place when the following three conditions exist at the same time:

(1)   The presence of a fuel

(2)   The presence of oxygen

(3)   A source of heat (high temperature)

Flammable gases are the most hazardous, and each gas or vapor has a range of concentrations in air in which explosions can occur. When concentrations are outside this range, the mixture is either too rich or too lean to burn. One method of protecting against explosions is to provide sufficient ventilation to dilute the hazardous vapors or gases to a concentration below their lower explosive limit (LEL).

The enrichment of air with oxygen extends the explosive range of flammable mixtures. Gases and vapors that are lighter than air diffuse quite readily into the atmosphere and become diluted. Heavier-than-air gases are less likely to disperse quickly. Flammable liquids are not as hazardous as flammable gases. Liquids do not ignite until a sufficient amount has evaporated and the vapor reaches the LEL.

Liquids are classified by the National Fire Protection Association (NFPA) according to their volatility: the more volatile the liquid, the sooner its vapor concentration reaches the LEL. The flash point of a liquid is the temperature at which it gives off enough vapors to form an ignitable mixture. Volatile liquids with low flash points present the same danger as flammable gases.

The NFPA has three classifications of flammable liquids:

Class I: Very volatile liquids whose flash points are below 100°F. Class I has three categories that define the relationship between flash point and boiling point.

Class II: Liquids with boiling points between 100° and 200°F.

Class III: Liquids with flash points above 200°F. These liquids are not considered very hazardous.

The auto ignition temperature is that degree of temperature to which a fuel-air mixture must be heated in order for it to ignite spontaneously.

Dust explosions can occur when flammable dust particles are suspended in air and are ignited by flames, sparks, or hot surfaces. The properties of flammable dusts that contribute to ignition are their particle size, concentration, minimum explosive concentration, ignition temperature, and minimum cloud ignition energy. The smaller particle sizes are more susceptible to explosions. Metallic dusts present special problems because their conductivity causes shorting and arcing of ordinary electrical devices and, in effect, creates their own source of ignition.

INTRINSIC SAFETY

Another method of eliminating the risk of ignition or explosion from electrical equipment operating in hazardous locations is to apply the principle of intrinsic safety. This is a method of ensuring that electrical control wiring leading from a nonhazardous area into a hazardous area is incapable of carrying sufficient electrical current to cause an explosion in case of malfunction. ISA-S12.1, Definitions and Information Pertaining to Electrical Instruments in Hazardous Locations, states: "Intrinsically safe equipment and wiring are incapable of releasing sufficient electrical or thermal energy under normal or abnormal conditions to cause ignition of a specific hazardous atmospheric mixture in its most ignitable concentration." NFPA 493 also establishes standards and guidelines for its use and installation, as well as test apparatus and procedures for testing installed intrinsically safe systems.

Certification for the use of intrinsic safety equipment is granted on a system basis rather than for individual devices, because the effective safety value of any one component depends on the characteristics of all other elements in the system. Certifying agencies include Underwriter's Laboratories, Factory Mutual, and the Canadian Standards Association (C.S.A.).

Designing an intrinsically safe system for electronic instrumentation means that energy limiting is accomplished by controlling the voltage and current carried by the signal cables between the hazardous and nonhazardous areas. In addition, the quantity of stored electrical energy in field instruments must be limited to levels that cannot cause ignition.

One method requires that the power supply transformer within each control room instrument have total isolation between primary and secondary windings. Current-limiting resistors are then added in series with the signal wiring, which is then routed into the hazardous area.

A second method, illustrated in Figure 10-8, places an energy-limiting barrier in each signal pair before it enters the hazardous area. Control room instruments need not be intrinsically safe and can be connected to any device, as long as the voltage present is not greater than 250 V rms. The barrier system is most often selected because it is more versatile and safer than other methods.

The barrier system provides a sharp line of demarcation between the hazardous area and the safe area. The barrier itself is a completely passive device that requires no power source and passes a 4 to 20 mA signal at a nominal 24 V DC power rating with virtually no degradation (less than 0.1 percent). Inside the bar-

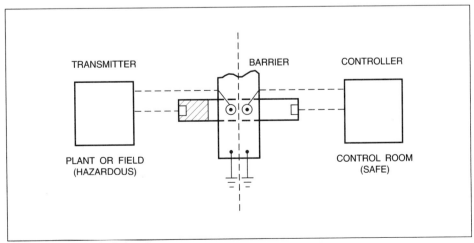

Figure 10-8. Intrinsic Safety System

rier are wire-wound resistors for limiting current and redundant Zener diodes for limiting the voltage. A fuse of suitable capacity is in series with the resistors. Figure 10-9 shows a typical construction.

Examining Figure 10-9 may prompt the question of why a fuse alone would not do the job. If a high voltage is impressed on the circuit terminals with only a fuse and resistors present (without the Zener diodes), an explosion could result in the hazardous area because a relatively large amount of current needs to flow in the circuit before the fuse can blow. A fuse is inherently a slow device and can pass considerable energy before it opens. Only microseconds are needed under the right conditions for an explosion to take place. Without some method of voltage limiting, a fault voltage appearing at the barrier input terminals would be passed along the cables.

With Zener diodes across the two signal wires, any voltage outside the range of -0.7 to +30 V DC causes the Zeners to conduct. If a fault voltage of high enough potential and of sufficient duration appears at the safe area terminals, the current through diode D1 increases until the fuse blows. If the fuse does not blow immediately and the power dissipation of D1 is exceeded, the Zener is designed to short circuit before it opens. The shorting D1 current will rise rapidly, thus blowing the fuse. An excess current is always diverted to ground by the barrier and is never allowed to reach the hazardous area. The time lapse difference between the fuse blowing and the Zener diode opening is about three orders of magnitude.

The barrier shown in Figure 10-9 is a "positive" barrier and is used in systems where the negative side of the signal line is grounded. If the Zener diodes are inverted, the barrier is a "negative" barrier and the positive side of the line is grounded. The barrier itself must be grounded, because grounding eliminates any common mode voltages from creating a potentially hazardous field condition. The residual resistance between the grounding electrode and the barrier must be less than one ohm.

VENTING AND OVERPRESSURE

An important consideration of process design is the protection of workers from the risk of accident resulting from equipment failure due to overpressure in pressurized vessels or equipment. A number of specially designed pressure-relief devices are available for this purpose.

Equally important is the venting of storage tanks that contain corrosive, flammable, or other hazardous liquids, flammable or toxic gases, and the like. Whether pressurized or not, they are most often located outdoors. Unpressurized vessels usually need to be vented to atmosphere for process requirements, personal safety, or other reasons. A number of techniques exist to prevent damage or personal injury from spills, fire, explosion, overpressure, or tank-wall failure.

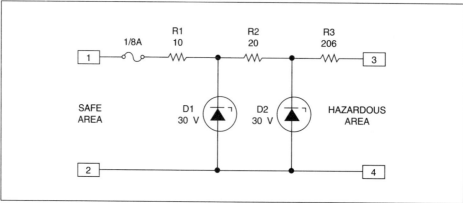

Figure 10-9. Intrinsic Safety Barrier Circuit

As an example, tanks that contain dangerous acids are required to be located outdoors inside a ditch or sump of special construction, complete with retaining wall, large enough to accept the entire contents of the tank in case of a spill, thereby reducing the risk of injury from overfilling or equipment failure.

Tanks that contain flammable or explosive liquids are often vented to the outside through specially designed flame arresters, which are metallic louvers of different styles made up of several layers of fine wire mesh inside a frame, through which flames cannot pass, but which prevent the tank from becoming pressurized.

Overpressure in process equipment may develop from:

(1) malfunctions or operating errors,

(2) thermal expansion,

(3) chemical reactions,

(4) explosions,

(5) external fires,

(6) material fatigue,

(7) power failure,

(8) automatic control failure,

(9) cooling apparatus failure, or

(10) combinations of the above.

Protecting people is the uppermost consideration in the use of pressure-relief devices; the next is protecting process equipment from damage. One factor that must not be overlooked is provision for the safe disposal of the material released by an overpressure condition.

The pressure-relieving devices in common use are safety relief valves and rupture discs. Codes and recommended practices pertaining to pressure-relieving devices have been prepared by the American Society of Mechanical Engineers (ASME) and by The American Petroleum Institute (refer to Section VIII of the ASME Boiler and Pressure Vessel Code, and API RP52O, and RP521).

To resume, pressure-relief devices are installed:

(1) to provide safety for operating personnel,

(2) to protect the environment,

(3) to reduce material loss, and

(4) to prevent downtime.

A safety relief valve is an automatic device actuated by the static pressure upstream of the valve and is characterized by rapid full opening or pop action. It is used for gas or vapor service and also for steam and air. The action is accomplished by a force-balance system acting on the closure of the relieving area. The orifice size of the pressure-relief valve is selected to pass the required flow at specific conditions. This opening is closed with a disc held in place by a spring until the preset amount of overpressure is reached. The contained system pressure acts on one side of the disc and is opposed by a spring force on the opposite side. Figure 10-10 shows a typical arrangement.

Formulas, specially designed computer software programs, and other design data are readily available for calculating safety valve types and sizes for liquid, gas, and steam services.

The following formula is an example for steam service. This formula is contained in API-RP 520, Appendix D:

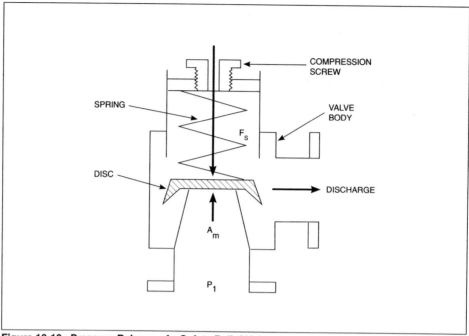

Figure 10-10. Pressure Balance of a Safety Relief Valve

$$W = 50AP_1K_{sh} \tag{10-2}$$

or

$$A = \frac{W}{50P_1K_{sh}} \tag{10-3}$$

where:

W = flow rate (lb/hr)

A = effective discharge area (in.^{2})

P_1 = upstream pressure (psia) (set pressure times 1.03, or 1.10 plus the atmospheric pressure

K_{sh} = correction factor for the amount of superheat (1.0 for saturated steam at any pressure)

50 = constant used in API equations (Section VIII of ASME Boiler Code uses 51.1)

A rupture disc consists of a thin metal diaphragm held between flanges. Its purpose is to fail at a predetermined pressure, serving essentially the same purpose as a pressure-relief valve. Rupture discs are fabricated from carefully selected sheets of metal. They have defined limitations inherent in their ultimate tensile or compressive strengths, as well as in creep, fatigue, or corrosion resistance.

Rupture discs may be grouped into four different design types:

(1) Solid metal

(2) Composite

(3) Reverse-buckling

(4) Shear type

The construction materials used for rupture discs are usually stainless steel, copper, nickel, aluminum, Monel™, Inconel™, and sometimes titanium and tantalum. These metals are used in foils and sheets in soft annealed condition.

Bibliography

1. Ivergard, T., *Information Ergonomics*, Chartwell-Bratt, 1982.

2. Diffirient, N.; Tilley, Alvin R.; and Bardagjy, Joan C., *Humanscale 1/2/3*, MIT Press.

3. Grandjean, E., *Fitting the Task to the Man*, London, 1980.

4. Stewart, T. F. M., "Displays and the Software Interface," *Applied Ergonomics*, Vol. 7, 1976.

5. Wellford, A. T., Skilled Performance, Dallas, TX: Scott, Foresman and Co., 1976.

6. Peters, N., "Commissioning a Training Simulator: Preparation and Follow-up," *Pulp and Paper Canada*, 91(4); T127-130, April 1990.

7. Province of Québec: *Santé et Sécurité au Travail*, Québec, 1978.

8. National Safety Council: *Accident Prevention Manual for Industrial Operations*, Chicago, Ill.

9. ISA-RP60.3, *Human Engineering for Control Centers*, Research Triangle Park, NC: ISA, 1985.

10. ISA-RP60.8, *Electrical Guide for Control Centers*, Research Triangle Park, NC: ISA, 1991.

About the Author

Eddie Marquis is a senior Automation/Instrumentation Specialist at Bechtel Canada and has over 30 years of experience in pulp and paper, chemicals, and metallurgy.

11

Project Management Strategies

This chapter does not go into the technical aspect of design, which is covered in detail in other chapters. It deals with the aspect of project management, from start to finish. Although written for those who are managing a project for the first time, some points may offer a fresh perspective to more experienced instrumentation project lead engineers. It does not intend to review all the possibilities that can arise in the business of project management; however, it does present a goodly number of observations and experiences that could be useful to the practitioner but are not easily found in conventional textbooks.

Standards

The instrumentation and process control industries use three broad categories of standards: design, documentation, and safety.

Design standards, which have been agreed upon by various regulating bodies, refer to all aspects of defining quality and methods for the fabrication or installation of materials and equipment. Standards can be mandatory requirements or simply recommended practices.

Documentation standards refer to both the content and the presentation of documents. The presentation includes symbol definitions as well as the way these symbols are to be linked together. They are a gathering of different conventions that can be set by regulating bodies, by trade groups, or by any group with a common interest.

Safety standards are usually set, or at least are enforced, by regulating bodies such as state or national governments. Their purpose is to provide for safe equipment and installation with the protection of the worker and the public uppermost.

Mandatory standards are usually grouped within codes, supplemented by rules and regulations. They are revised on a regular basis, along with technical development or accident historical data. They tend to vary from country to country.

Recommendations are not limited by borders and can, therefore, be applied anywhere as long as they do not counter any local code, rule, or regulation. When they are followed, communications are greatly enhanced because everybody uses the same dictionary!

At the beginning of each project, great care must be exercised to determine which standards apply. Depending on where the installation will be done and depending on the client, who may have his own standards and practices, the applicable standards may be more stringent than those of a government or regulating agency. A list should be established and circulated to all concerned and a project library established for easy reference if the necessary standards are not already part of the company's library.

The following are some major standards-developing organizations in North America:

ANSI	American National Standards Institute
API	American Petroleum Institute
ASHRAE	American Society of Heating, Refrigerating and Air-Conditioning Engineers
ASME	American Society of Mechanical Engineers
ASTM	American Society for Testing Materials
CSA	Canadian Standards Association
EIA	Electronic Institute of America
FM	Factory Mutual
IEEE	Institute of Electrical and Electronic Engineers
ISA	Instrument Society of America
NEMA	National Electrical Manufacturers Association
NFPA	National Fire Protection Association
OSHA	Occupational Safety and Health Association
SAE	Society of Automotive Engineers
UL	Underwriters' Laboratory
ULC	Underwriters' Laboratory of Canada

Design Standards

Design standards tend to be grouped together under a general umbrella within a country, such as ANSI in the USA, CSA in Canada, or DIN in Germany, or over a wider area, such as IEC in Europe. Much work is still to be done in order to have "standard" standards. For example, in North America, ISA's "standard reference temperature" is both 0°C and 15°C, while ASHRAE calls for 18°C, 20°C, and 24°C. This produces confusion on projects, even when definitions are given in the project specifications. A compressor supplier's attractive bid may have to be rejected because the same "standard temperature" was not used to calculate the capacity of the equipment. Imagine the results if standards did not exist!

With the growing internationalization of engineering, standards such as those written by ISO or IEC are gaining increased international acceptance. Such progress will eventually correct existing aberrations, whereby, for example, an adapting spool piece is needed to match an ISO flanged valve to an ANSI standard flanged pipe.

Recommended practices, such as ISA-RP60.8 and ISA-RP60.9, suggest practical methods for the design of wiring and tubing in control panels. If no standards exist in an organization, it is preferable to follow such recommended practices rather than to reinvent the wheel.

Documentation Standards

Documentation standards are the visual dictionary of communication in engineering. Symbols are the most important; they constitute the alphabet. Symbols have evolved within different trades, and, even today, different symbols are used to illustrate the same device or equipment within different trades.

An outstanding feature of ANSI/ISA-S5.1, Instrumentation Symbols and Identification, is that it defines a set of standard symbols for use on process (or piping) and instrumentation diagrams (P&IDs) and establishes a nomenclature to define the functions attached to the symbols. These have been internationally adopted in the petroleum and chemical industries. Their use in other industries, while spreading, is still somewhat limited. As a matter of fact, ISA is nearly unknown in the material-handling industry, which tends to adhere to JIC (Joint Industry Council) standards in North America, which apply to electrical equipment manufacturers. JIC standards are also well used for control schematics in North America, but JIC's nomenclature sometimes conflicts with ANSI/ISA-S5.1 nomenclature for the functional representation. In other parts of the world, IEC standards (with many adaptations) are used to represent control schematics.

None of these nomenclatures lends itself easily to integration with modern engineering techniques that use computer databases and programmable controller programs; this can lead to as many as three correspondence tables on the same project! Generally, there is one for interfaces between P&ID tag number and programmable controller bit or internal reference number and a second one with the computer or distributed system's own point or tag number. This proves cumbersome and is a source of errors and potential danger, especially during maintenance.

Other standards are more oriented toward the overall document aspect, for example, ISA-S5.4, Instrument Loop Diagrams, which suggests the content and presentation for loop diagrams.

Safety Standards

Safety standards are directed toward the safety or protection of installations, workers, and the environment. Standards such as those from OSHA, NFPA, FM, and UL are written mainly for that purpose. They set clear rules about the design of equipment and installations. Equipment is tested according to the set standards and bears a seal to that effect. These standards set rules according to local resources, environment, and practices. This is why their application seems, for the time being, to be restricted to the country in which they have been developed.

A clear example is the set of rules and standards for intrinsically safe installations. The American standard has definitions and levels different from the IEC standard; the Canadian standard is in between, and the British standard is still different. Any equipment to be used for such installations must bear the stamp of the country where it is to be used and installed according to the regulations in effect in that country.

> Standards are necessary, but there is room for improvement in the standards jungle. Whatever the application, however, they should be well known and applied properly.

Documentation

The documentation involving the instrumentation and control (I&C) group during the course of a project is quite varied in its content and presentation. Some documents are produced for detailed engineering purposes and rarely leave the project files. Others are produced solely to circulate information and provide a record of what has been done. Still other documents are produced for the installation contractor to be able to do the work. Finally, some are produced as guides for operating and maintenance people. Some well-used types will be described with reference to related standards for details.

Process (or Piping) and Instrumentation Diagram

The process (or piping) and instrumentation diagram (P&ID) is the first document that involves the I&C designer. It follows the process flow sheet prepared by the process, the mechanical, or the chemical group. When there is no process group, the I&C group prepares the whole P&ID, as it is the best format for presenting the general operating strategy of an installation while readily identifying the critical points, from the standpoint of both operation and safety. The P&ID shows all process fluid lines, all equipment, and all valves that influence operation, using ANSI/ISA-S5.1 symbols and nomenclature. From the preliminary engineering to the final detail design phases, it can go through three degrees of detailing, as shown in the standard. An example is shown in Figure 11-1.

Operating Description

Process descriptions are prepared by process people. They usually identify important control points and safety considerations. The operating description describes the actions required from the operator and the information presented to him, whether locally or in a control room. Operating descriptions initially prepared by the process group are normally best completed by the I&C group.

The operation description should be well structured and should account for every process unit and process line. It is a dynamic document that eventually includes detailed references to equipment and instrument numbers and to all related drawings. It will be complete in every respect when it reaches its final destination—the operating manual. The operating description must have a balanced structure and presentation. If the text is cluttered with references, it will be hard to read and follow. If there are not enough references, it may be difficult to establish the proper relation between the description and the plant.

Data Sheets

Data sheets are used to convey all useful information about an instrument. They contain the necessary process data, physical mounting requirements, environmental conditions (normal, maximum, design), electrical requirements (voltage, current, classification), and expected operating behavior for each instrument on a project before the instrument is selected.

A data sheet constitutes a short form of specification for an individual instrument or instrument type. It is invariably a single sheet of paper that can be circulated to the process or mechanical engineer for coordination purposes and transmitted to vendors for bids. When the instrument is selected, the full catalog number and all necessary options are added to provide a complete record of the particular instrument. It is then passed to the installation contractor and others for site coordination and calibration. The final document is passed over to the client's maintenance group.

A good collection of sample data sheets can be found in ISA-S20, Specification Forms for Process Measurement and Control Instruments, Primary Elements, and Control Valves (see Figure 11-2).

Calculations

Calculations are the basis for the selection of control valves, safety valves, and flow elements. Like all engineering calculations they must be well documented, clearly legible, dated, and signed. Any subsequent revision must be systematically identified: what, by whom, and when.

Instrument List

The instrument list has traditionally been the document on which all instruments are shown with their principal technical characteristics, usually complemented by the requisition and complete model numbers as well as the number of the loop where each instrument appears.

With the increased use of the personal computer, the instrument list has become a real working tool that saves a lot of valuable time. For example, once the basic process information has been entered in an overall database, it can later be extracted by the process and piping groups using advanced computer-aided

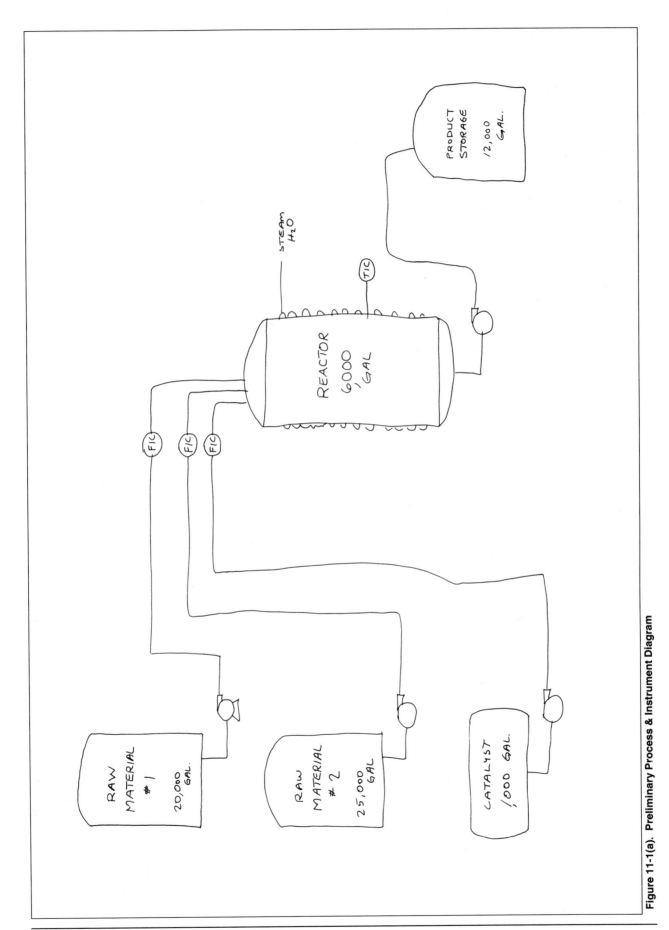

Figure 11-1(a). Preliminary Process & Instrument Diagram

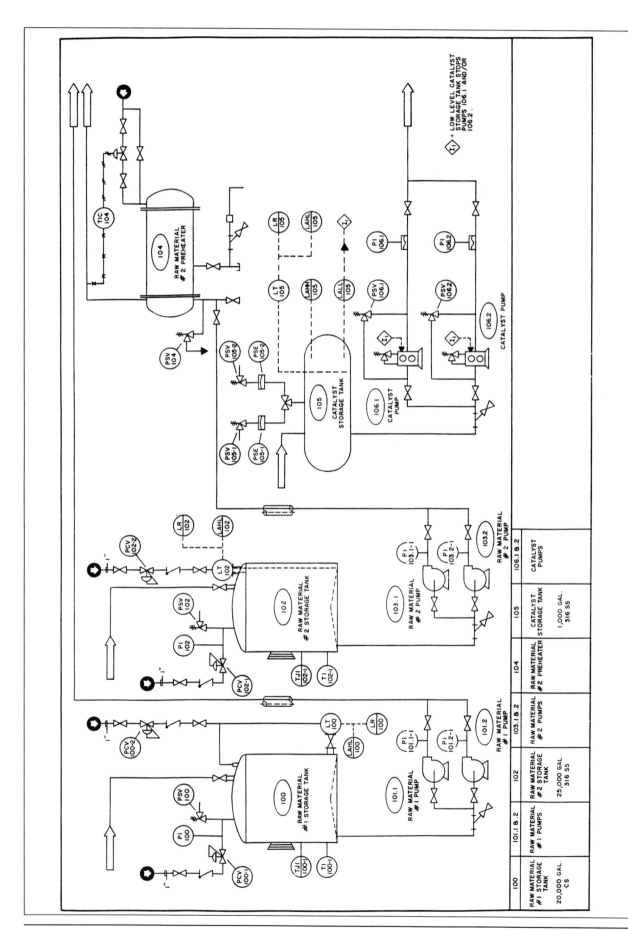

Figure 11-1(b). Preliminary Process & Instrument Diagram (continued)

© ISA S20.50, Rev. 1   **CONTROL VALVE DATA SHEET**   **Second Printing**

PROJECT _____	DATA SHEET _____ of _____
UNIT _____	SPEC _____
P.O. _____	TAG _____
ITEM _____	DWG _____
CONTRACT _____	SERVICE _____
*MFR. SERIAL	

#			Units	Max Flow	Norm Flow	Min Flow	Shut-Off	
1	Fluid					Crit Press PC		
2		Flow Rate					—	
3		Inlet Pressure						
4	SERVICE CONDITIONS	Outlet Pressure						
5		Inlet Temperature						
6		Spec Wt/Spec Grav/Mol Wt					—	
7		Viscosity/Spec Heats Ratio					—	
8		Vapor Pressure P_V					—	
9		*Required C_V					—	
10		*Travel	%				0	
11		Allowable/*Predicted SPL	dBA	/	/	/	—	
12								

#				#		
13	LINE	Pipe Line Size	In ____	53		*Type ____
14		& Schedule	Out ____	54		*Mfr & Model ____
15		Pipe Line Insulation		55		*Size ____ Eff Area ____
16		*Type ____		56		On/Off ____ Modulating ____
17		*Size ____ ANSI Class ____		57		Spring Action Open/Close ____
18		Max Press/Temp ____		58	ACTUATOR	*Max Allowable Pressure ____
19		*Mfr & Model ____		59		*Min Required Pressure ____
20	VALVE BODY/BONNET	*Body/Bonnet Matl ____		60		Available Air Supply Pressure:
21		*Liner Material/ID ____		61		Max ____ Min ____
22		End	In ____	62		*Bench Range ____ / ____
23		Connection	Out ____	63		Actuator Orientation ____
24		Flg Face Finish ____		64		Handwheel Type ____
25		End Ext/Matl ____		65		Air Failure Valve ____ Set at ____
26		*Flow Direction ____		66		
27		*Type of Bonnet ____		67		Input Signal
28		Lub & Iso Valve ____ Lube ____		68		*Type ____
29		*Packing Material ____		69	POSITIONER	*Mfr & Model ____
30		*Packing Type ____		70		*On Incr Signal Output Incr/Decr ____
31				71		Gauges ____ By-pass ____
32		*Type ____		72		*Cam Characteristic ____
33		*Size ____ Rated Travel ____		73		
34		*Characteristic ____		74		Type ____ Quantity ____
35	TRIM	*Balanced/Unbalanced ____		75	SWITCHES	*Mfr & Model ____
36		*Rated C_V ____ F_L ____ X_T ____		76		Contacts/Rating ____
37		*Plug/Ball/Disk Material ____		77		Actuation Points ____
38		*Seat Material ____		78		
39		*Cage/Guide Material ____		79		*Mfr & Model ____
40		*Stem Material ____		80	AIR SET	*Set Pressure ____
41				81		Filter ____ Gauge ____
42				82		
43		NEC Class ____ Group ____ Div ____		83		*Hydro Pressure ____
44		____		84	TESTS	ANSI/FCI Leakage Class ____
45		____		85		
46		____		86		
47	SPECIALS/ACCESSORIES	____				
48		____				
49		____				
50		____				
51		____				
52						

Rev	Date	Revision	Orig	App

*Information supplied by manufacturer unless already specified.   ISA FORM S20.50, Rev. 1

Figure 11-2. Control Valve Data Sheet

design technology. This avoids copying mistakes and improves tracking of changes. The instrument list data, after addition of I&C information, can be used to generate a good part of the data sheets through custom programs that do batch processing of the data. The rest of the information needs only to be added, thus saving a great deal of time.

A further step is to define a standard connection reference for each instrument. As an example, a pressure, flow, or level switch with one single-pole, double-throw contact can be assigned the attribute "CONN1." In another database, CONN1 can be further defined as a device with three terminals identified "C," "NO," "NC." An application program can then, for each such instrument, automatically assign one three-conductor cable from the instrument to a junction box and, once the junction box is identified, produce a connection list on which, for one instrument, "C" would be connected to terminal 1, "NO" to terminal 2, and "NC" to terminal 3, with the grounding being taken care of automatically. There are applications in which the automation is taken a step further; that is, approximately 70 percent of the loop drawing is drawn automatically by defining how instruments having the same loop number are connected together.

Document integration and automation can be a big help in speeding up the detail design process. The document production can be started late in the project when the rate of change has somewhat slowed down. Extensive integration improves coordination, but there are always changes. Taking care of these changes in a coordinated manner, as well as introducing special items that are not according to the predefined typicals, has always been the challenge of all these semi-automated design tools. Sometimes, handling these special cases can consume as much time as producing the rest of the documentation.

Despite their drawbacks, these systems have allowed man-hours to be saved compared to the traditional drafting approach, but their greatest merit has been an appreciable shortening of the engineering schedule, especially on grass roots projects where large investments are committed.

The instrument database can also be used to define the scope of work when several contractors will share the work under different contracts. As an example, it can be used to define which party (with the associated contract number) supplies the instrument, which physically mounts it, which performs the electrical connections, and which does the testing and calibration of it. This information is easily entered at the time the detail design is done; it may take a lot more time to recall and enter two or three months later when the pressure to issue field contracts arises. Even if a few mistakes seep through, a well detailed scope of work such as this allows for much better pricing by bidders because the "fuzz factor" is greatly diminished.

Entering purchase order numbers and delivery dates with the cooperation of the purchasing group can also speed up the planning work of the contractors on site as well as provide a good base from which to evaluate the work load for receiving, checking, and calibrating the instruments.

The instrument list, through computerization, can be used to obtain specific ongoing coordination and exception reports, such as special lists to coordinate ranges and set points. An overly detailed instrument list, circulated for coordination purposes, may be too intimidating to read and may simply end up accumulating dust on a shelf. Circulating a list with only the pertinent information that will fit on one line per instrument is more likely to be read and, therefore, useful.

The instrument list and the P&IDs constitute the key documents of a project and require care, appropriate access, and regular backups.

If you expect feedback, ask for one or two specific categories of information at a time. Do not bury your colleagues with too many requests, which will likely be ignored or discarded.

Logic Diagrams

Logic diagrams can take many different forms. They are used to convey, in a simple form, the sequence of operations that takes place in an automated installation. Logic diagrams are usually presented using a set of symbols defined in ANSI Y32.14 or ISA-S5.2, Binary Logic Diagrams for Process Operations. A relatively new method of representation is the state and transition diagram, introduced from Europe. It has the advantage of being more easily understood by people who do not have an I&C background. Another important advantage is that a growing number of programmable controller manufacturers provide direct programming using this form. The only drawback seems to be that the Boolean equations required to program transitions do not accommodate the ISA standard tagging very well; either the sheer number of characters required is too much for the allowed memory or it is just too complex to display on a screen. A correspondence table is then needed.

Logic diagrams can be used for shutdown or emergency sequences, but a cause and effect chart, which is more compact, can serve the same purpose. It is an X-Y table on which instruments are identified along columns and situations appear as rows. For each row or situation, a symbol such as a 1 or a 0 is marked to identify the position or status of an instrument. Such a tool can be used to simulate dangerous situations, examine the status of each instrument as designed, and modify the logic if dangerous or potentially hazardous situations are discovered. The cause and effect chart is a systematic technique that reduces dependence on the experience of an individual. These documents are relatively easy to coordinate with process, chemical, or mechanical engineers as well as operators, which is not always the case for conventional logic diagrams.

Installation Detail Drawings

Installation detail drawings (see Figure 11-3) are prepared, for the most part, as typical drawings for the installation of each type of instrument. Special instruments and analyzers require specific installation details. Usually, most of them can be drawn from an internal library or the client's library. Petrochemical companies have well-established standard installation details for their own use, and they expect engineering firms to follow them for their projects.

When the relative position of hardware components is significant, this information is communicated by detail drawings, which are usually shown as dimensionless isometrics. Important dimensions that must be respected are clearly indicated. Installation detail drawings may include other additional information as needed for requests for bids on a turnkey contract. In instances where the bidder is expected to use his/her standards, it should be requested that these be submitted for evaluation with the bid. Otherwise, the detail drawings would be more complete, showing all fittings, block valves, and materials to be used.

Good installation details are important for all aspects of field work and contractual matters. They allow for easier quality control and provide a ready yardstick for progress payments, additional work, or unit pricing.

Control Schematic Drawings

Control schematic drawings or elementary diagrams are used to represent the control logic associated with equipment. They follow the JIC standards in North America; that is, relay, push button and switch contacts are connected to active devices such as motor starters, solenoids, and motorized valves through horizontal lines running between two power supply lines. This presentation resembles a ladder and is well known as the "ladder diagram." Usually drawn on large draw-

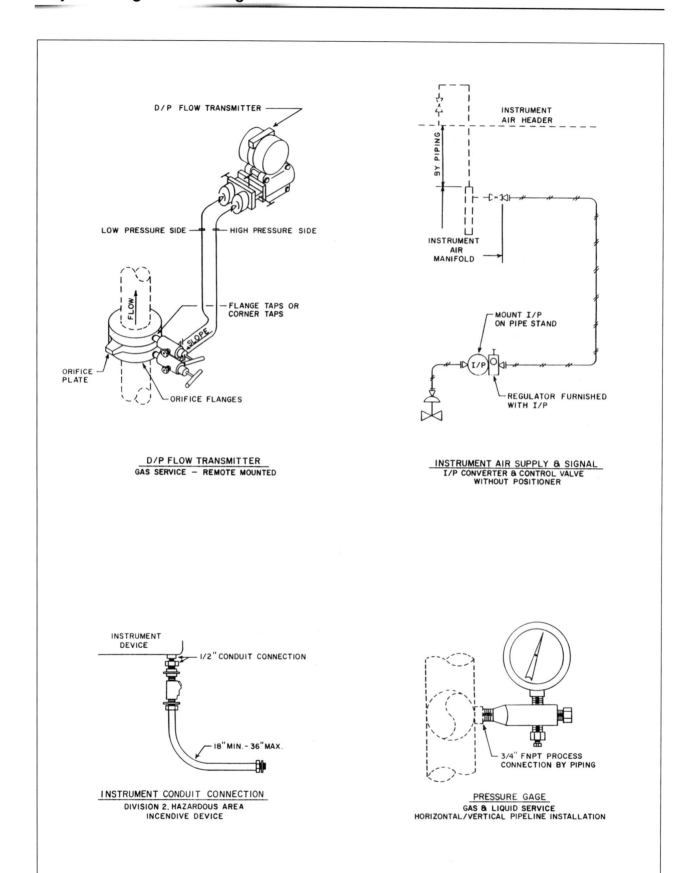

Figure 11-3. Installation Details Drawing

ings where several ladders can be shown side by side, the ladder diagram is well adapted to simple machine tool drawings because all logic can fit on one drawing.

The IEC presentation format, used outside North America, shows the logic as a horizontal ladder. All references are given by column and page number. It has the advantage of being compact; a complete control schematic can be as little as twenty pages long, easy to reproduce, and able to fit nicely into a binder.

With the advent of the programmable controller and the distributed control system (DCS), one finds fewer conventional control schematic diagrams than in the past. Nevertheless, motor starters still exist, and some form of manual intervention is always provided apart from the more sophisticated control equipment for maintenance or emergency. Standardizing this simple logic is a great time-saver. Fewer than a dozen diagrams can cover most projects, from the simple fractional horsepower motor to the hundred-horsepower reversing motor. If they are standardized early enough, they can be imposed on all equipment suppliers on the project as soon as a motor control center (MCC) vendor has been chosen. While it may seem at first glance to add to the cost of the equipment, it is largely compensated for by the savings in coordination and engineering and by the avoidance of costly interface problems and reworks.

Input/Output (I/O) Drawings

Input/output drawings show all the input and output devices and signals connected to a PLC or a DCS. They are equipment-oriented: they show all connections per physical I/O block, block type, reference tag, I/O number, and terminal number on each block. They are well adapted to computerized drawing due to their repetitive graphical representation. The presentation of these drawings generally follows the control schematic format. They are also well suited to small-size pages, accommodating one block per page.

Configuration Drawings

Configuration drawings are used for PLC and distributed control systems. One PLC or DCS unit is represented per drawing, which shows the constituent components and their physical arrangement. It includes racks, power supplies, memory modules, local and remote I/O modules, and other required hardware. Their goal is to show the complete physical arrangement of components as well as the spare space required. They can be a great help not only for ordering PLCs and DCS units but also for their mounting. They also provide for a fixed allocation of modules that will be used for address coding and programming.

Panel Layout and Detail Drawings

Panel layout drawings can be as simple as showing the relative position of components, such as the push buttons, lights, meters, and controllers, on a control panel. The control panel supplier will then produce detail drawings for review and approval.

The overall schedule can be improved, although at a greater engineering cost, if the drawings show the exact layout desired, with dimensions and catalog numbers for each component. The panel-layout approach is usually followed on "fast track" projects. Detail drawings that show the layout of terminal blocks and auxiliary components are then produced. Detail wiring diagrams can be forwarded later, as the sheer fabrication and physical mounting can take ten to fifteen weeks after award of a panel fabrication and assembly contract. This eliminates the several cycles of drawing submittal and review found in the first approach. The panel builder fabricates and assembles as per the supplied drawings. The engineer is then fully responsible for the quality of the documents and all necessary rework. This is, in practice, balanced by the saving of at least two months of schedule time.

Location Drawings

The purpose of location drawings is to show where instruments are located relative to equipment, vessels, or piping. While some users in the petrochemical industry have developed the practice of showing these drawings with great detail using isometric views, it is more common to use piping or equipment drawings as a washed-down background on which to show the approximate location of the instruments and field panels. The specifications then describe the location shown as approximate, with the exact location to be determined on site. Defining the location within a few meters and indicating the elevation for instruments to be mounted above floor level avoids minor contract disputes with the installing contractor, and the instrument location can be determined easily on site when the equipment is actually seen in place. It also accounts for the constraints of site conditions and allows for the determination of the best place for access, since most equipment drawings fail to show details of field-run conduits and minor pipes that can be in the way.

Major cable trays and conduit runs can also be shown, but these would usually appear on electrical drawings produced by the electrical specialty. This makes it much easier to coordinate the layout, the space allocation, and the spacing of trays between different voltage levels, especially when entering electrical equipment rooms where PLCs will be installed.

Wiring Diagrams

Wiring diagrams show the wiring and connections between control components (see Figure 11-4). They are produced to show connections within a panel or cubicle, connections of field devices to a panel, or connections between panels. All components are shown simply by a square or round box or a characteristic symbol such as a push button or a meter. All cables, conductors, and terminals are identified with their respective number name or color code.

Connections to field components or between panels can also be produced in the form of a connections list. Each conductor is shown with a wire number and the originating terminal and device as well as the destination terminal and device. When the design is automated, it is a document that is easily and quickly produced. It is also easy for the field electrician to use in connecting to terminals. The connections list is standard practice in some large organizations and is widely used in Europe. In the author's experience, however, it has proved to be the type of document that, if produced manually, is quite prone to error, especially after a few modifications have been made since they are not always easy to track. Even extensive ongoing checking will result in the discovery of most mistakes only at the last minute, when time is scarce. It therefore ends up being an expensive approach. On the other hand, it can be cost-effective for an application such as the production of several pieces of equipment or machines of the same model; once the first one is debugged, the others are built the same way.

Loop Drawings

Loop drawings (see Figure 11-5) are a form of wiring diagram showing all instruments and components related to a control loop, including hardware and software components such as a loop controller and an indicating or alarming point. All cabling, tubing, and connections are shown and identified. This is the drawing that will be used during the lifetime of the plant by the maintenance

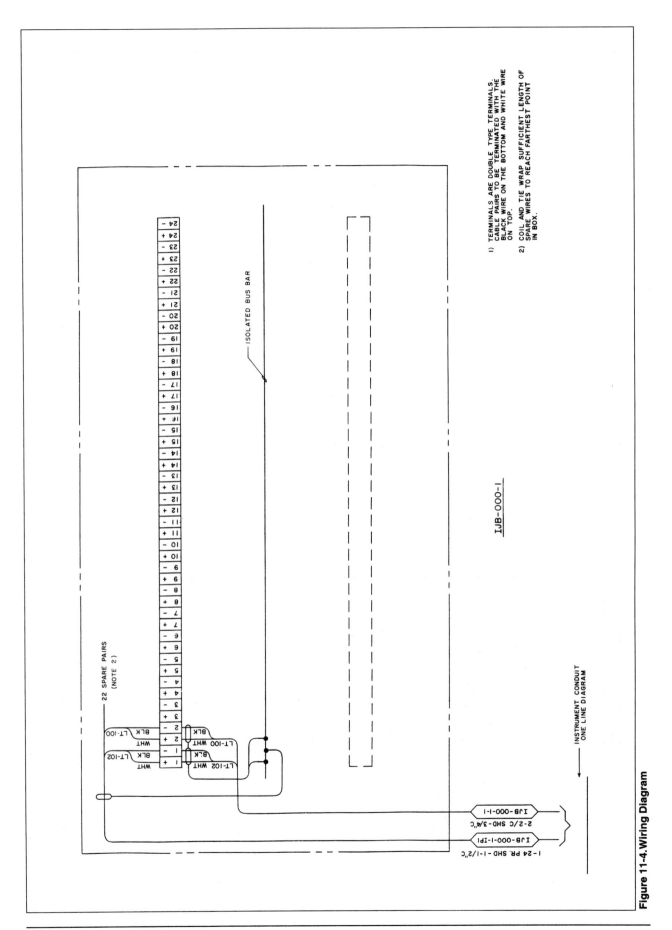

Figure 11-4. Wiring Diagram

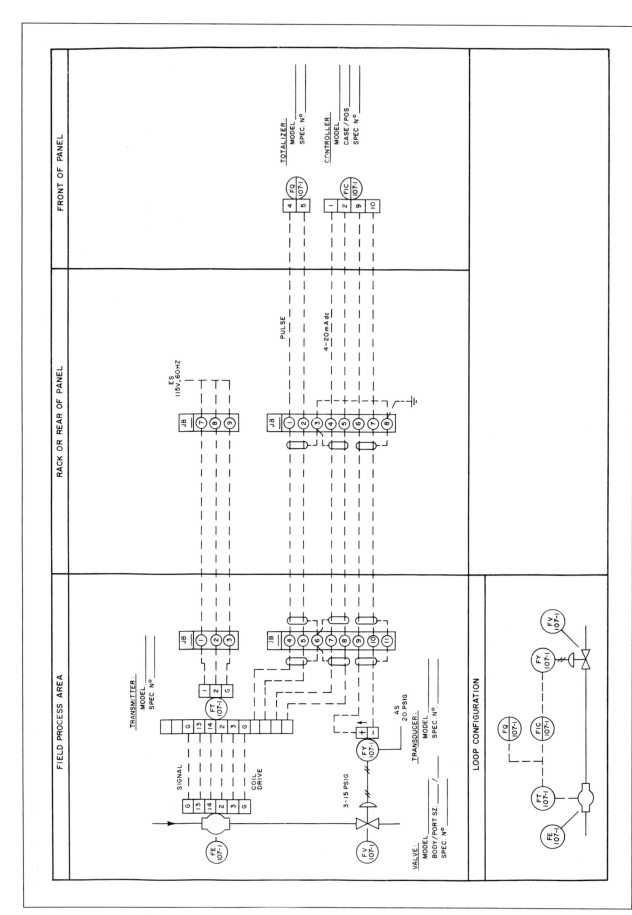

Figure 11-5. Typical Loop Diagram

group. ISA-S5.4, Instrument Loop Diagrams, defines the typical presentation for loop drawings.

As has been seen, preparing loop drawings can follow a certain degree of automation. When not equipped with highly sophisticated design tools, one can nevertheless use graphical representation techniques to an advantage. Examining project requirements reveals that many loops are similar. Masters can be produced for each type of loop, omitting tag numbers, cable numbers, and addresses. Then it is a simple matter to fill in the blanks to produce a great quantity of loop drawings.

Loop diagrams are usually restricted to analog signal devices. There is more and more interrelation between analog signals and on/off devices. For instance, a control valve can be fitted with limit switches or have a local high-pressure override. It is advantageous for a maintenance person to find all these devices on one drawing when that person is in front of the valve. To avoid having two types of overlapping documents (namely, loop drawings and wiring drawings for on/off devices), some organizations have standardized the loop drawing approach to all field devices.

Graphic Display Drawings

Graphic display drawings require attention and consume engineering time. They will provide the interface between operators and the process. They have to be laid out carefully, with due consideration given to what information will be provided and how it will be provided. Emergency conditions must be addressed. If a process usually runs automatically, will the operator be sufficiently alert and find the required manual procedures within the allowed time if an emergency condition requires manual intervention by the operator?

Graphic displays need ample coordination with plant operating personnel. Often, more time is spent on the colors to be assigned to process equipment than on the real function of the display—the operation. Quick communication is essential. ISA-S5.3, Graphic Symbols for Distributed Control/Shared Display Instrumentation, Logic, and Computer Systems, and ISA-S5.5, Graphic Symbols for Process Displays, address the symbols to be used and provide a general guide to display presentation.

Program Documents

The programming of PLCs and DCS has become more attractive to many I&C designers who see the possibility of exercising their genius. Beware! The program carries the instructions for the control strategy, sequences, recipes, and secure operations. Sometimes it involves shutdown and safety sequences that cannot be easily solved. One may also have to address questions of redundancy and complex strategies.

Great care and a systematic approach must be applied to programming. A complete knowledge of the basic software as well as the behavior of the hardware under different conditions is necessary. Strict standards should be developed and enforced on a project where several persons will be programming different parts of the process. This will help preserve uniformity in addressing similar control problems and ensure uniform operator interface throughout the plant regarding actions, feedback, color codes, emergency or exception procedures, identical start-up and maintenance procedures, and identical alarm treatment. It is therefore highly recommended that the designer define standards, such as motor control blocks or macros, as allowed by the chosen equipment in order to enhance the quality of the end product with the ultimate goal of safety in mind. A good operation description should be completed before any programming is attempted. It should be complemented with some form of sequence or logic diagram that shows the overall control strategy for each process line or unit.

Programs should carry comments. Some equipment software allows for the insertion of comments; other software is best served by third-party software packages that allow an off-line treatment of the program for more comprehensive documentation. Good documentation can save time and also provide for added safety with the inclusion of warning notes. It is best done by the programmer, who is usually the one most aware of the operating sequence. Three months later, when there are questions that are not answered by appropriate comments, the ingenious ideas of the programmer may not be obvious to the person doing the field check and start-up. Parts of the software may therefore get reprogrammed, with the result that important safety features may be defeated. Systematic quality control is a "must" for safe and clear programming.

Specifications

Specifications are written to convey, in words, the desired product, how it should perform, and how it should be installed. Such specifications are best conveyed by a long list of data in the form of data sheets or tables. As such, it is recommended that a detailed scope of work be presented in a subset of the instrument list.

Specifications are produced to purchase instruments, systems, and control panels and to define the broad scope of work to contractors. Specifications should contain at least the following information:

(1) Purpose

(2) Location of work or where to deliver

(3) Codes and regulations that apply

(4) Inspection quality level expected that will be subject to controls

(5) Process data or background

(6) Function to be met or goals to be achieved

(7) Methods to use, with appropriate reference to standards

(8) Materials to use, with appropriate reference to standards

(9) Checking or testing methods

(10) Quantity expected

(11) Description of the documentation expected with a submittal schedule

(12) Definition of the submittal and review procedure with a maximum turn-around time

(13) Delivery date or schedule

(14) Date or schedule for input to be provided later to the vendor, supplier, or contractor

(15) List of documents that are to be considered as part of the specifications

(16) Rule(s) to interpret eventual conflict between the text of the specifications and other documents

(17) Ruling on the acceptance of equivalence and alternative proposals

(18) Detailed definition of payment items

(19) Guarantee and payment clauses if not included in the commercial documents

(20) Penalty clauses for delays and defaults

The language should be clear and direct. If the document is multilingual, state which language will prevail in the case of misunderstanding, dispute, or arbitration. Remember: when bidding, most suppliers and contractors have lawyers review the specifications to detect loopholes! These can prove to be costly to the owner.

A good specification is more than just a good technical document.

Instrumentation Projects

The engineering approach to an instrumentation project usually depends on the application, established tradition, or, sometimes, the lack of tradition existing in the milieu.

An instrumentation project in the municipal field is generally engineered with the performance specification approach, wherein the supplier-contractor is chosen for his/her experience in the field and is required to do much of the detailed engineering. The instrumentation engineering team responsible for the project concentrates its efforts on defining the goals, the system architecture, the quality of equipment to be used, and the documentation to be supplied.

A petrochemical project tends to be fully engineered by a consulting firm that is familiar with the application and the client's intricate standards, which have been developed over the years mainly for safety reasons. Its documentation is usually extensive, and good engineering quality control is required. In the industrial domain it is surpassed only by nuclear and military or space technology projects in terms of documentation and systematic quality control.

At the other end of the spectrum there is the "turnkey" approach, in which a complete unit or plant is to be delivered based on production performance specifications.

These approaches will be reviewed briefly to show how they affect the instrumentation engineer's task.

Performance Specification Approach

USE
The performance specification approach is used primarily when:

(1) engineering contracts are awarded solely on a project percentage basis, such as contracts awarded by different government agencies;

(2) the process is public technology;

(3) some suppliers of instrumentation and control systems have developed a specialty and have recognized expertise in a given application;

(4) there are enough specialized suppliers to offer competitive bidding; and

(5) the instrumentation and control systems engineering is part of a turnkey package.

DOCUMENTATION
The documentation required to assemble a performance specification package typically consists of the following:

(1) A functional description of the process and the operator interface

(2) A Level 1 P&ID, as defined by ISA-S5.1

(3) A control system hierarchy diagram

(4) A typical specification for each type of instrument or a list of specific makes for some applications, if:

    (a) the engineers feel that the possible bidder's own line may not be suitable for the application or

    (b) there is a requirement for standardization with existing equipment to keep the maintenance and spare parts inventory low

(5) A list of all standards that apply to the specific project, avoiding being too general (Listing nonapplicable standards opens the door to arguments about the applicability of standards that the engineer wants to be applied.)

(6) A sample of each type of document to be produced or specific reference to recognized standards whenever possible (If necessary, indicate the content of the document; always be specific as to the symbols to be used. The size of the documents with the corresponding lettering size should also be specified if they are to be microfilmed. This is generally specified in the general contract conditions but, as the size of some documentation such as loop drawings may not be used by other disciplines on the project, they may be overlooked. Specify which document must be submitted for review, approval, information, and final documentation.)

(7) A good description of the presentation and the contents of both the operator manual and the maintenance manual

(8) A good installation specification with typical installation detail drawings to convey clearly to the supplier the quality level of the final installation required (Define clearly the materials to be used, the type and size of cables, trays, tubing, manifolds, block valves, the minimum and maximum allowed mounting height, etc., if not already covered by an applicable specification, rule, or regulation.)

(9) A detailed description, including drawings, control points list, and layouts of the interface with existing installation, if applicable

(10) A checking and calibration procedure to be followed, with the supporting documentation required (Most large suppliers have their own; however, unless it is requested that it be submitted with the bid, they may be tempted to use shortcuts to increase their competitiveness when faced with some less well-known supplier who may slip in with an attractive price.)

(11) A checking and start-up schedule when interfacing with existing installations

(12) In cooperation with the contract team, well-defined interface points between the various disciplines involved, the document submittal schedule, the payment schedule definition, and, if possible, a sizable progress payment for the delivery of approval documents (This encourages general contractors to put pressure on their subcontractors; otherwise, they tend to delay I&C work because it hurts their cash flow. Payment on the production of documents helps to overcome the cash flow issue and has a favorable impact on the project's schedule.)

RESPONSIBILITY

It is the lead I&C engineer's role to see that a good package is assembled, from both the technical and contractual perspectives. This is especially true when the I&C package is only a subcontract to a general contractor or even a subcontract to an electrical or a mechanical subcontractor.

The scope of work definition is then of utmost importance. Work not clearly defined, especially an ill-defined interface involving several trades, will most certainly pop up at some field meeting as a pretext to make claims for extra payments. Whether or not such claims are accepted, they will nevertheless delay the work and, if numerous, will overshadow more important items. Be clear: do not rely on "We usually do it this way..." Because of the national and international movement of contractors and manpower, there will always be a contractor who is accustomed to doing it another way.

Detail Design Approach

USE

The detail design approach is generally used for specific process control projects, although it is not unusual to find all approaches within the same project. That is, some specific process units may involve turnkey contracts, while others, such as utilities, may be awarded as performance specifications supplied and installed by a third party. The detail design approach may be done in house by the owner's engineering group or, more generally, it may be done by an engineering consultant or an engineering consultant-contractor firm. In both cases, it is important to have a clear understanding of the project's requirements, which include the following:

(1) The budget

(2) The schedules, for both the overall project and the detail engineering

(3) The scope of work

(4) The standards agreed upon

(5) The methodology

(6) The engineering review phases and related documents—the definition of owner input as well as the committed dates to supply the information (This input may be process data, information about a similar plant, drawings of existing equipment to be reused, or vendor drawings from equipment directly purchased by the client)

(7) The local standards, rules, and regulations

DOCUMENTATION

The starting point for the I&C engineering group may be:

(1) a brief process description, a rough process flow sheet, or both;

(2) a detailed process flow sheet and a well developed P&ID, as frequently seen in the petrochemical industry; or

(3) a plant layout or an elementary process flow diagram with reference to existing older installations and a primitive list of what should not be done the same way.

Detail engineering requires that a complete array of I&C activities take place. Even more work may be required, however, if only superficial preliminary engineering has been completed or, as sometimes happens, a major process change occurs just as the owner decides to go ahead with the project and some added basic engineering has to be done.

Depending on the industry, the I&C engineer may have access to a process engineer to supply the basic process information in the form of detailed flow sheets or process descriptions or both. The resource for the I&C engineer may also be mechanical engineers who have good knowledge of the equipment and its operation and, if lucky, experienced plant operators.

From the input gathered from these various sources, the I&C engineer prepares documents that define and fix the requirements for the control systems. This documentation should be as short as possible, easy to edit or modify, and with a minimum of technical terms since they should be easily understood by project leaders who may be mechanical or civil engineers or even non-engineers. It should be very clear what the I&C team plans to design and build and how it will operate in order to be in line with the human relations policy of the owner, who can then plan his personnel recruitment accordingly.

Although all activities may take place, it is not always necessary to produce all documentation in its final form. Some clients require keeping only a draft form of background and working documentation, such as logic diagrams or purchase requisitions. Clients who do their own purchasing will require only specifications and data sheets from the engineering group.

Again, it is important to have a well-defined scope of work prior to commencing the work.

RESPONSIBILITY

I&C represents about three to five percent of the total capital cost of a project. I&C engineering is the last to leave the design group, and the I&C contractor is the last to leave the site. As a result, the owner's capital investment cannot be fully realized until the I&C group has done its job completely. On a major project (say with a one billion dollar total investment), a 10% interest rate means that the owner bears an interest burden of approximately two million dollars a week, not counting contractors' charges and production losses.

It should now be easier to see why so much pressure is put on the I&C group to produce documents on time. This group must be go-getters of information. They can't rely on engineers of other disciplines to forward it to them; the other engineers are too busy with their own schedules and priorities. The cooperation of the purchasing group is sought to follow up with vendors for the data and the shop drawings that are needed to complete the work.

The success of the project depends greatly on the lead engineer whose role is more that of planner and organizer than technically oriented person.

Turnkey Approach

USE

Turnkey projects are carried out by engineering-contractors or specialized vendor companies. A fixed price is committed against an agreed-upon production performance and a final delivery date of a complete plant, process unit, or intricate machinery.

DOCUMENTATION

The documentation varies greatly, depending of the type of project. If it is custom-built, pretty much the same documents have to be produced as for detailed-design projects. If it is only an adaptation of a standard supply, much of the documentation can be drawn from previous projects and modified to suit. Document reproduction services are best put to use to produce new tracings that can be modified, if modern computer technology has not been used previously, to maintain low engineering cost. It may be effective to digitize previous documents for increased competitiveness on future projects. All cases should be analyzed on their own merits.

Often, most of the detail data sheet preparation and instrument calls for bids can be skipped because of satisfaction with the cost, performance, and small number of problems encountered with specific instruments. This results in a reduced cost of engineering, purchasing, and follow-up because of less dependence on vendor data.

Important pieces of documentation on this type of project are the complete operating and maintenance manuals, which are the responsibility of the entire engineering group.

RESPONSIBILITY

The engineering group is faced with the same responsibilities found in the detail-design approach. Moreover, most of the time the same people are involved with the testing and start-up. They therefore require the flexibility needed to work with construction contractors, as well as an awareness of all field safety procedures, local rules, and labor union contracts. They should also be able to work under the pressure of the construction manager, who absolutely wants to meet the target date and even improve upon it since that will mean added profit and benefits for the construction company.

Engineering Phases

Preliminary or Basic Engineering

The preliminary engineering consists of: (1) preparing sufficient documentation to ensure that the client's needs and goals are well understood and (2) clearly describing methods, with possible alternatives, for meeting these goals to the team that will do the detail engineering. That team may be one's own or it may be a vendor, an engineer-contractor, or a different group within one's organization. Basic engineering goes through several different steps.

INFORMATION GATHERING

Preliminary meetings with the representatives of the client will give an overview of what the client would like to see. It may be a straightforward, down-to-earth, exact copy of an existing plant process or an idealized view of a process of the future.

In the first case, one should request a visit of this plant as soon as possible with the following questions in mind:

(1) Where is the plant located?

(2) How old is the plant?

(3) Is the same equipment still available?

(4) Can newer technology be applied to advantage?

(5) What is the organizational work and responsibility division?

(6) Is it a one-, two-, or three-shift operation?

(7) Does it run seven days a week?

(8) Are there planned shutdowns? How often?

(9) Are there specific problems or clues to possible problems that could arise if not addressed?

After this information has been put together, a brief report containing the findings and suggestions of possible alternatives, supported with advantages and rough cost comparison figures, can be presented to the client for evaluation and discussion. The client may reconsider some of the initial ideas expressed, which will narrow down the practical ideas to develop.

CONTROL PHILOSOPHY

The expression "control philosophy" often brings smiles to some people who think, "Here come the dreamers..." Philosophy really is the right word, however, because the design exercise and the resulting documents lead to a translation into hardware and software of the management policy that the owner wishes to maintain in the new plant:

(1) Who will be responsible for starting a process unit? The area engineer? The foreman? The chief mechanic or electrician?

(2) Who can stop a piece of equipment? The area engineer? The foreman? The chief mechanic or electrician?

(3) Who really needs what category of equipment operation information? Process information? Performance or throughput information? Most importantly, what will they do with and how will they use this information?

When these questions are answered to the I&C engineer's satisfaction, it is time to prepare a small document describing how and where each category of equipment should be controlled in this new plant. Is a pump normally started from local push buttons, from a local panel grouping different pieces of equipment, or from a central control room with local emergency stop only? Many variations must be addressed because they will have a strong impact on the control system design and cost. This document will be reviewed and modified a few times until it is found acceptable by the client.

The next step is to sketch some possible control system configurations, then evaluate their cost, the possible suppliers, and the impact of the choice of approach on the delivery and the overall project schedule. Usually two schemes may be retained for future detailed review at this stage.

A global cost evaluation of each scheme can be made by comparison with the actual cost of similar projects, taking into account local factors and regulations. Then all these documents can be summarized in a report, which is normally sufficient to establish what is called a "feasibility" study. When the go-ahead is given, the basic engineering can proceed.

The basic engineering requires the addition of process and instrumentation diagrams, generally defined as P&ID Level 1 by ISA. It is normally a joint effort of the process and I&C groups, but when a process group is not available, the responsibility rests with the I&C group.

Basic design criteria should be drafted (or standard ones adapted) to permit an instrument and input/output count and to ensure that the detail engineering will be done using the predetermined approach and within the established budget. It is a good idea to support the basic design criteria with: (1) drawings or sketches illustrating typical motor starting circuits with their appropriate control function and location, (2) sketches of typical control loops to support the I/O count, and (3) drawings of typical control and operator panels, using previous project experience.

BUDGET

The important remaining step is to establish the budget. Individual instruments can be counted from the P&IDs and their budget cost verified by telephone with current vendors. Installation costs can be estimated, taking into account the location of the plant, the type of material required, and productivity factors adjusted for local conditions.

The I/O count can be made by totalling the number of process motors, assuming five to seven I/Os per motor (depending on the typical drawings established) and adding 20 percent for nonprocess motors and I/Os that will pop up at the detail-design stage.

Costs can be attached to control panels and consoles at a global value based on experience or at an estimated cost per unit if a modular approach is retained or at a unit price per unit of length. These guidelines for an overall cost estimate are used when little detailed information is known. Various factors are then applied, according to specific project conditions.

A good source of information, at least for checking purposes when one has no previous experience with a given type of industry, is specialized magazines. For example, *Chemical Engineering* (McGraw-Hill Publications) regularly publishes construction cost indices for the chemical industry in North America. These figures will allow one to verify that the ratio of the I&C to the total project estimate is reasonable. One must be careful; however, because the scope of work assumed by the magazine may not be the same as the project under consideration. The cost of control cables, their installation and connections may be included with the electrical discipline by some; others might include these costs with I&C. In this discussion it will be considered as part of the I&C work with the estimation made accordingly.

There are different ways to estimate the engineering effort. Here, the hours per P&ID, the cost per I/O, and the total project ratio have been used. Typically, all three are used and compared before a final figure is set. It is difficult to give a rigid guideline because these figures vary widely from one type of industry to another and from project to project; they may even vary by as much as 20 percent for two identical projects with two clients. Table 11-1 gives the range used here as a general guide for a wide range of projects, that is, from 2,000 to 150,000 hours of I&C detail engineering.

Table 11-1. Typical Engineering Estimate Hours

	MIN	MAX
Per P&ID (hours)	1200	2000
Per I/O (hours)	5	10
Percent of Total Engineering	12	20

The higher figures tend to apply to the petrochemical industry while the lower are for general material-handling industries where a good part of the detail work is done by suppliers. These figures include vendor data checking and some field assistance. Field checking and start-up are extra.

> Using this method, an I&C engineer estimated a project bid at 4,200 man-hours. A vice-president said, "This is too high— we have to do it for under 2,000." The project was sold at the lower figure and given to another senior I&C senior engineer to design. There were many animated discussions with the client as a result of efforts to keep costs down and petition for extras; however, when the project was finally completed, the actual time was 4,300 hours!

With the geopolitical shrinking of the planet, engineering is becoming more international, with the result that documentation must often be generated in more than one language. For each additional language 10 percent should be added to the total I&C engineering budget.

The basic engineering itself, with P&IDs, design criteria, system architecture, typical motor schematics, typical loop drawings, datasheets, preliminary instrument lists and estimates, may represent 15 to 20 percent of the detail engineering budget—another explanation for the wide range of estimating factors for the detail engineering, which depends greatly upon the quality of the basic engineering.

Detail Engineering

Detail engineering is the heart of a project. It should be approached and planned with great care in order to respect schedules and budgets. To a large extent, the satisfaction of the total engineering effort of a project is measured by the quality of the I&C detail engineering: Is the plant starting at the schedule date? Is the normal production rate achieved at the expected date? Is the plant running well? In the long term, are there more or fewer than expected unscheduled shutdowns due to instruments and control systems?

PLANNING

The planning of the work is crucial. It is more than simply trying to establish a detailed list of documents, especially when the input is scarce. Planning requires a close examination of the total situation, achieved with both technical and project management representatives of the client.

SCOPE OF WORK REVIEW

The scope of work review should be approached methodically by obtaining satisfactory answers to a list of basic questions, by examining project documentation and the contract, and by holding both technical and managerial meetings with the client's representatives. The following are some of the important questions to be answered:

(1) What information is there to start with?

    (a) How good is the basic engineering?

    (b) Are there alternatives to explore?

    (c) Is the process well defined?

    (d) What influence does the choice of equipment or process vendors have on the choice of instruments? Systems? Detail control design?

    (e) What standards apply? How familiar are they?

(2) What are the expectations?

    (a) Review the basic engineering with the client and confirm that the goals are the same in terms of approaches to system and plant operation philosophy. Note carefully all deviations or suggestions in order to introduce alternatives. Make sure that complete minutes of meetings are taken in order to obtain management approval. Many points raised at a technical review are rejected at the management level, so be careful not to spend any design time on unapproved work!

    (b) Is a formal call for bids needed for every type of instrument? Does the client have preferred suppliers? (This should be asked, because sometimes the vendor used routinely by the client may not offer the best ratio of cost to performance, which can be mentioned. Perhaps some equipment is preferred due to factors that do not appear in the literature. Without asking questions, one is in a poor position to make suggestions.)

(3) What is the definite project approach?

    (a) Complete detail design?

    (b) Partly detail design and partly vendor designs and vendor data review?

    (c) What is the designer's responsibility with respect to the vendor's engineering? Legally, the vendor may be responsible, but who is responsible for the total project's schedule, regardless of the vendors' performances?

    (d) What quality assurance level is required?

(4) What is the up-to-date schedule to work with?

Do not be shy when it comes to asking about the schedule—not only the overall schedule but also the detail schedule. Be careful to check the forecast dates for major process equipment purchases in relation to P&ID approval dates and for approval by the client of major decisions such as what control system architecture is to be used. Changing the specifications for the control system supplied by a process or mechanical equipment vendor after the contract has been awarded can have a disastrous effect on cost, schedule, and project management.

(5) What documents are to be produced?

Review the required documents with the client, and, for each category, decide what the purpose of each should be:

    (a) For approval. If so, agree on when they should be submitted and define a maximum duration for the client's review, comments, and approval. A maximum turnaround time frame from the time of the first submittal should be agreed upon (some clients tend to comment on rather than approve each submittal and therefore formal approval of drawings occurs only at the end of the project). Usually, P&IDs, system architecture, main control panel and control room layout drawings, and general purchase and construction specifications should be approved by the client (insist on this point).

    (b) For information. Once the format has been agreed upon, the client wants only an idea of how well the principles are adhered to but demands the ability to assess the progress of the work reported!

    (c) For maintenance. Detailed information should be included with the maintenance manual.

    (d) For the record. Documents and studies kept with the engineering files will be produced upon request.

Once the type of project and scope of work have been assessed, the original budget is ready for review and the detail engineering can be planned.

ENGINEERING BUDGET REVIEW

The original engineering budget is now ready for reassessment. The same methods referred to previously are used, taking into consideration the latest information. It should be easier at this point to account for contingencies. Additionally, a preliminary activity schedule, as opposed to a document schedule, can be established.

This activity schedule should consider the following:

(1) Technical review meetings with the client

(2) Preselection meetings with I&C vendors

(3) Preselection meetings with mechanical and process equipment suppliers

(4) Regular engineering progress review meetings

(5) Regular interdisciplinary coordination meetings

(6) Meetings with authorities or regulating bodies such as insurers and electrical and environmental authorities

(7) Visits to similar installations, whether belonging to client, vendor, or others

(8) Meetings with the client's representatives to discuss comments on documents submitted for approval

(9) Specifications, data sheets, drawings, and requisition lists

(10) A list of the mechanical or process equipment requisitions where I&C input and follow-up are required

(11) A list of the field trips that will be necessary for proper coordination

(12) A list of the forecast field interventions or support activities that will be charged to the engineering budget

(13) A list of any other items that can affect the total hours, the rate of progress of the engineering, and the overall schedule

Care should be taken to list all activities because they reflect one's understanding of the scope of work, they allow one to communicate that understanding to others, and they provide a basis for defense of the budget during the course of the project.

Once this list is complete, the time to complete each item can be established. Many will be specific to the client and some will be specific to the project. As an example, if the project engineering phase has a forecast duration of 14 months and a weekly coordination meeting is scheduled, then the total meeting time allowed should be 180 hours (60 weeks times three hours per week). If it is the practice of the chief engineer to require meeting attendees to take turns preparing minutes of meetings, perhaps two hours per meeting should be added. In this manner, one can estimate the time required for coordination meetings. This may seem like a small item, but many hours will be consumed by meetings over the course of the project. If this time is not planned for in the original engineering budget, the end of the project might be delayed.

Most project managers prefer an "hours per drawing" approach, which may apply well to civil, structural, or even architectural disciplines but is very difficult to apply to I&C. Many activities take place at different times in the engineering process. Being unequally distributed and very project-dependent, they can lead to disproportionate hours per unit for some types of documents, resulting in an unrealistic measure of the progress of the work. How can progress of field trips and technical reviews with the client, which will affect the whole project, be reflected? On the other hand, if there is a specific activity entitled, "Define motor control strategy with the client," the activity can be followed and progress measured. If the project manager is not sufficiently convinced of the value of the activity approach to adopt it, a complete list of documents may be used to show a correlation between real progress and actual hours spent.

MANPOWER PLANNING

Establishing a manpower plan is an important step. The basic structure may vary from project to project. A pyramidal structure, where the lead discipline engineer looks after everything, has been customary. This is still valid for small projects, but it would likely create a bottleneck for larger projects where more than five or six people are involved in I&C design. Larger projects tend to use a matrix approach in which different persons are made responsible for the detail design of specific areas of the project. They are usually referred to as "discipline area" engineers as opposed to "project area" engineers who are responsible for the project coordination, equipment purchases, and installation contract awards within their areas. Depending on the particular structure, the discipline area engineer may report either to the project area engineer or to the lead I&C engineer for administrative, budget, and progress purposes. In the first case the I&C lead discipline engineer is normally responsible for establishing project standards, ensuring design uniformity, providing for quality control, and supplying appropriate personnel.

In all cases, a manpower projection must be established. Typical project document production schedules as they apply generally to fast track projects are presented in Figures 11-6 and 11-7. Figure 11-8 presents the optimum distribution of the engineering man-hours when a low level of basic engineering is provided. This distribution covers projects where good coordination is achieved and when the activities are performed at their due time during the course of the project. It has proven to be correct within about five percent on all types of projects. On fast track projects the installation contract is awarded at about the 65 to 70 percent point of the engineering design activity. Whenever budget overruns have been encountered, they can usually be traced to activities that have been done before they should have (for instance, having to produce cable lists or connection lists before vendor data was obtained). They can also be traced to activities started late due to missing information that required too much time to obtain or required considerable overtime to complete.

It is apparent from Figure 11-8 that the hour distribution is highly nonlinear. This translates into a basic project team plus temporary personnel for a part of the project. This temporary manpower can come from within the organization, where people are familiar with the working methods and standards, or it can come from outside, with the hiring of contractual people on both long-term and short-term contract bases. It is an essential part of management's responsibility to be aware of the resources pool within the area, considering both the quality and the quantity available. It takes approximately one month to bring a good engineer or technician to a 100 percent performance level in a new environment and up to three months, on average, if the position also includes project management responsibilities. This knowledge of human resources is essential to establishing a valid and realistic engineering budget.

Productivity factor estimates must now be added to the budget and the manpower curve reprocessed. Then the budget and personnel requirements are ready to be presented to the engineering manager and the recruiting department.

GETTING THE WORK DONE

Once the planning fits within the project schedule and the budget receives approval, the job must be done.

TEAM BUILDUP

In a moderate-sized organization a basic team can be built with inside resources and supplemented with outside resources. Interpersonal relationships must be monitored when people must work closely under the pressure of deadlines.

The lead engineer's biggest challenge is to build and maintain team spirit. A harmonious group is an efficient group. A project that requires a special effort, for whatever reason, makes its own contribution to team spirit, and people seem

A team must be built with a good balance of people who can get the work done (and can do some of it themselves...) and technical people who are content to do things well without being disturbed by meetings or administrative work.

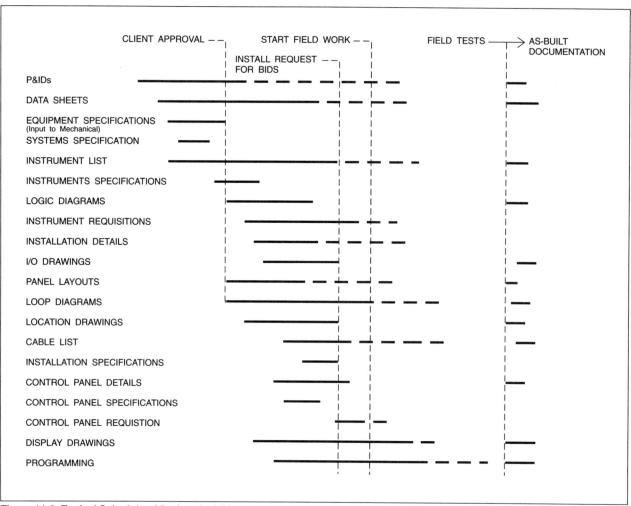

Figure 11-6. Typical Schedule of Project Activities

to rally and become more cooperative. A good lead engineer will maintain a high level of enthusiasm throughout the project.

Some team leaders have apparent success by applying a more strict, less human approach, but it is the author's experience that well-qualified personnel do not respond favorably to dictatorial treatment. Without cooperation from all members of a team, deadlines are sure to be missed and budgets exceeded.

MONITORING PROGRESS

Success in meeting schedules and budgets can be achieved by building up the team along the forecast curve and monitoring the status of the work. It requires reasonably good follow-up because I&C work is highly dependent on the basic work of the process, piping, and mechanical groups. That means regular adjustments must be made. It is necessary to maintain a good file of any changes (with a description of the work or the delay), an estimated man-hour impact, and a record of who actually does the work. Most organizations have a regular budget update review, and normally all client-requested changes are well monitored. But management tends to underplay internal changes and squeeze them into the project budget. Of course, an allowance for unforeseen work should be included, but, if such events are not tracked from the beginning, they can add up quickly and result in headaches when additions to the budget must be requested. If well

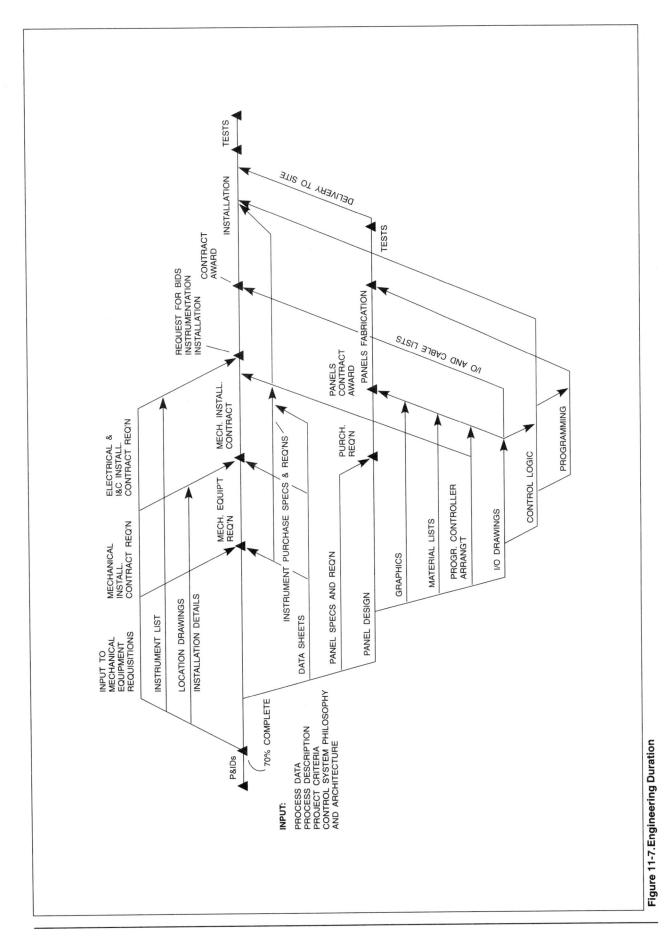

Figure 11-7. Engineering Duration

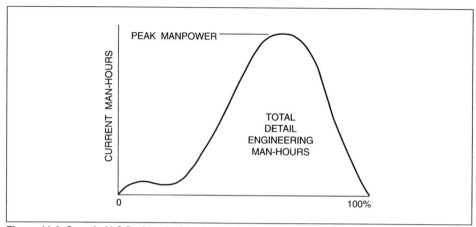

Figure 11-8. Sample I&C Project Activity Network—Fast Track Project

documented, such internal changes may justify a budget transfer from another discipline, but budget additions that are hard to identify are never welcome.

As previously advised, a good way to monitor progress is to monitor activities. An improvement on this is to evaluate the effort expended to date, along with what is needed to finish according to an estimate from the people who do the work, and then compare these with the budgeted values. The lead engineer must then interpret the figures and rationalize them. Some people will state the engineering design hours to complete documents; others will give the overall hours to complete an activity, including "overhead hours" related to meetings, visits to vendors, and other activities not strictly related to document production. Theoretically, the hours used to date (actual progress, or *AP*) plus the hours of work remaining (*WR*) should equal the hours in the original budget (*OB*) plus any approved extra (*AE*) hours or:

$$AP + WR = OB + AE \qquad (11\text{-}1)$$

In practice, what really exists is:

$$AP + WR = \frac{1}{E}(OB + AE) \qquad (11\text{-}2)$$

where E is the relative team efficiency factor, 1.00 being the planned value. A team working at an efficiency of 0.67 would therefore use up 1.50 times the total budgeted and authorized extra hours. Management would not be pleased.

ANALYZING THE RESULTS

The engineering group efficiency follows a learning curve as people join the team and become familiar with the project standards and routine. Efficiency factors can be typically as in Table 11-2 but vary on each project.

Table 11-2. Typical Efficiency Variations along a Project

Progress %	Efficiency
30	0.70
50	0.95
60	1.00
70	1.15
95	0.90

The overall project efficiency should be 1.00 if the hours spent are exactly on budget. This is almost never the case; they are more likely to be over or under budget. A good analysis is extremely important; if the difference is large, good explanations must be found. If the budget has been exceeded for no good reason, the project administrator will be considered incompetent and will probably not be in charge of the next project. Being under budget earns a good rating if the administrator can support the use of the methods employed on the project to achieve the under-budget performance. If not, credibility is lost with respect to establishing budgets, which may result in automatic, and unjustified, budget cuts on future projects.

Variations in efficiency are also caused by the current state of the economy. When resources are scarce and work is abundant, people are relaxed and tend to take it easy. During a recession, people are more careful and the efficiency factor is better. Project estimates will consider these and other things, such as general trends due to changes in technology (for example, computer-aided design and drafting and the use of word processors for spec writing).

Quality Control

Quality control is a project engineering activity that is often underrated. On a small project it is frequently performed as a matter of course by the lead engineer, who checks and approves the work of the assigned design engineer, ideally before the work leaves the department. On a large project, several engineers are responsible for different areas of the project. This, along with the overall number of persons working on the same project, makes it impossible for the lead engineer to check all the aspects of the design and documentation.

PLANNING

Planning for quality control is the right way to start a project. Planning is a synonym for classifying thoughts; it calls for the establishment of a strategy and the implementation of some means to ensure that the strategy is followed. Planning for quality control then leads to:

(1) establishing the list of *Objectives*;

(2) choosing or establishing *Standards*; and

(3) assigning someone or a small group of individuals who will be responsible for:

- answering questions relating to these standards,

- amending them as necessary,

- proposing new standards during the course of the project, and

- seeing to their *Implementation*.

OBJECTIVES

Quality control starts by determining the goals to be achieved. As a minimum, these goals should include the uniform presentation of drawings and documentation and the uniform implementation of solutions for similar control and measurement problems throughout the plant. A good number of hours should be spent on:

(1) the importance of the problem,

(2) the choice of solutions,

(3) spotting and reporting hard-to-solve problems,

(4) minimizing the rework required to fix inadequate or nonworking solutions, and

(5) using proven technology without excluding new technical advances.

These points represent nothing less than good and efficient engineering combined with some degree of systematic approach.

STANDARDS

Standards are very convenient for a small project. They are mandatory for any medium-to-large project if one wants to respect an engineering budget. The standards should set the goals for quality but may also be used to minimize wasted time. For example, on a larger project, when many temporary people are added to the team for a certain period, a good way to introduce them to the project is to give them half a day to look at the project summary and standards. Half a day may not seem like much, but it is usually sufficient for someone with some experience to glance through the standards, see their contents, and take note of what is similar to what they are used to and what is different. Then they can refer to it when required. The purpose is not to learn these standards by heart but to know what topics are covered.

These standards may be those of the client or the project management company. If none are specifically mentioned in the engineering contract, those of the project management company should be approved by the client before the design criteria are defined. If none of these exists, the use of recognized standards should be recommended, but care should be taken that a copy is presented to the client for approval, because they may be quite different from what is used in the client's country if this is an international project. The client may prefer to use some of his or her country's standards, especially regarding symbols, to ensure that maintenance personnel will be familiar with the documentation. The same reasoning holds for the choice of units to be used. Nearly every country has its own units, or at least its own flavor, in the use of the "English" system as well as of the "international," "SI," or "metric" system (each of the latter three is a variant of the same, depending on the country of application).

IMPLEMENTATION

At the beginning of the project, when the workload is light, the lead I&C engineer may handle the gathering of the standards and seek the client's approval. When the project starts picking up steam, one person should be assigned the responsibility of looking after the standards. This responsibility should consist mainly of:

(1) answering any questions and listening to suggestions,

(2) circulating proposals,

(3) organizing meetings of senior I&C personnel working on the project to discuss any amendments or additions, and

(4) coordinating with other disciplines such as electrical, piping, or mechanical as appropriate for symbols addition or modification.

If there is any overlapping with respect to standards, it may be necessary to ask the authority to amend or add to the project standards, FOR THE PROJECT ONLY. Authority may not exist to amend a company's standards. Another person or group responsible for company standards might seek approval or at least check that such amendments do not encounter major objections.

As a minimum, standards should cover:

(1) symbols to be used for all aspects of the work;

(2) the style and structure of each type of document to be produced on the project, with a sample outlining the degree of detail expected;

(3) the procedure for the routing, checking, and approval of each document (Part of this procedure, or at least for some of the documents, is generally covered in the global project standards. Include a photocopy of this procedure.); and

(4) the style and structure of each type of drawing, with a sample set of typical drawings and instructions for repetitive work such as motor control circuits, loop drawings, program modules (DCS, PLC, computer), displays, and control panels.

For a motor control circuit, there should be a standard for each group of starter size and type: single-speed, reversing, and variable speed. This will ease any interface with the electrical group and greatly facilitate document coordination with vendors. Loop drawing standards should include typical control valves, as well as temperature, level, pressure, and differential pressure measurement loops for different applications that will be encountered many times on the project. Of course, program modules depend on the application and function. There may be a typical pump-starting module, including a time delay on start-up, a pressure or flow device confirming the operation, and local and remote control and display points. Another one may be a typical bucket conveyer with all the required safeties and interlocks with the main motor and the maintenance motor, if so equipped.

Displays on conventional graphic panels, DCS screens, or PLC operator panels should also be standardized early with regard to:

(1) the shape of the symbols;

(2) the colors, both static and dynamic;

(3) the operator interface approach, that is, what controls the operator has and what information is presented as well as how the information is presented: digital values, analog meter-type representation, or graph bar for measurements, or a symbol change of color or written message for equipment status;

(4) the alarm strategy; and

(5) the safety requirements and shutdown principles.

 Great care should be exercised for all these choices, taking into account ergonomic principles as well as safety consequences.

Documents produced by the I&C group should be reviewed on a regular basis by the quality control person. A bottleneck would be created by delaying the quality control review until just before going out for bids. True quality control is an ongoing procedure that increases in momentum as the project progresses. It is difficult to implement a strict procedure, however, because some people tend naturally to know the standards and follow them pretty well, while others are "too busy" to check them. Ideally, there should be a closer or more frequent check in such cases; a more common procedure is to establish a minimum schedule of check points according to the degree of advancement of each area or part of a project. A simple way to help catch deviations from the standards is to ensure that the quality control person be on the interdisciplinary coordination distribution, which occurs at well-defined steps during the engineering phase.

The same person or a subordinate should follow the same quality control procedure when it comes to the people assigned to the programming when it is done in house. It is usually more difficult to run a close check on programming done by vendors or mechanical or process equipment suppliers; however, a special effort should be made to carry out functional tests at the vendor's premises before the equipment is shipped to the site. Insist on a valid simulation procedure, which should be specified or at least submitted for approval before the contract is signed.

Call for Bids

The engineering work is part of the preparation of contractual documents. Even if the project is described by performance specifications, they contain at least a functional description of the work expected and some sort of yardstick to measure its attainment in order to authorize payment. Such a document would be poor from many aspects if it did not define in some way the quality of material expected from contractors or suppliers. If the vendor's equipment works only for a year or two and then becomes a maintenance nightmare for the client, complaints to the design engineer and possible legal action could ensue.

The legal aspect is important and, in most organizations, there are people in the purchasing or contract department who should be consulted by the design engineers during the preparation of the project specifications. The engineer has to define exactly what is expected in terms of scope of supply, scope of work, types and quality of instruments, equipment, material, installation, and performance. The contract person, who can also be an engineer, is generally more concerned with schedules and administrative matters.

Nevertheless, the I&C engineer, through experience and contact with several vendors, is expected to know the standard lead time for different types of instruments or equipment. As an example, control valves can be expected to be delivered within 26 to 28 weeks after award of contract. That means that, taking the various internal delays into account, the I&C engineer should forward his control valve requisition to the purchasing group at least eight months before the site-required date. This knowledge has a direct impact on the establishment of the engineering schedule. The following discussion describes the different types of calls for bids that the I&C engineer may encounter on a project.

INSTRUMENT PURCHASE

For this relatively simple procedure, the work expected from the design engineer may take two forms. One is to indicate a complete catalog number of an instrument on a document that could be, in the simplest form, a memo to the purchasing department. This is customary in an operating plant when an exact replacement is required for a failed instrument. It is simple, quick, and... dangerous. Instrument vendors have a quick turnover of available sizes and materials, and their models change frequently. Unless the engineer's company has regular contact with a trusted vendor who knows the company's exact application, too little care given to an order may invite disaster. The purchasing agent in this case has little recourse for an exchange, or even less for a refund, if the instrument does not fit or fails to meet the intended purpose.

Clearly, the more information provided, the better. Even if the engineer is well aware of what is needed, the vendor should still be provided with information about: the process application, working ranges, operating limits, normal environment (temperature, pressure, humidity, atmospheric severity), physical mounting interface or constraints, supply or control voltages, and electrical classification. With all this information provided, a vendor can bear some responsibility regarding the performance and durability of the instrument. This information is best conveyed through the use of a data sheet.

If the make of the instrument is not important and, as it so often happens on projects, prices will be obtained from a number of vendors, be sure to consult with the end user or the client. All makes may not be available in the area where it will be installed, or the user may have a bad history of complaints with the local representative.

On the other hand, the product that the engineer is familiar with may exhibit specific problems in an unfamiliar installation. In these cases, the data sheet should be supplemented with a short specification that describes the application, the environment, and any specific information that will ensure that, whatever the choice, it will at least be an acceptable one. The quote of any respectable vendor will list areas of the specification and data sheet that cannot be met. In such instances, the vendor will often supply alternative proposals that mention where the specification is not met or, in some cases, where it is exceeded, especially if the alternative is more expensive than what would be required to meet the specification. At least the engineer is in a position to examine the offers technically and to make recommendations to the client or the end user.

DCS AND PLC PURCHASE

Distributed control systems and programmable logic controllers have been discussed in other chapters. With respect to the selection and purchasing process, they typically require the same approach used for instrument purchases, in applications where the programming is not done by the vendor. These systems must be selected at the beginning of the project because no standards cover such items as physical size, hardware and software configurations, communication networks and protocols, and programming.

Systems become "long delivery" items when the project requires more than half a dozen PLCs or two operator consoles. The vendor has to schedule their fabrication into the yearly plant production, which implies between 18 and 24 months lead time. It also implies that the vendor is relied upon to lend or rent sufficient equipment to allow the programming of the system to start early during the project phase. All such nontechnical points must be precisely defined in the bid document. It is the responsibility of the assigned I&C engineer to define all these requirements since they will play an important role during the bid evaluation and may influence the purely monetary evaluation when comparing two equally acceptable systems from the technical aspect.

On a project with several facilities it is customary for programming and installation of DCS units or PLCs to occur at different times in the project schedule. If this is the case, the bid documents should include a very realistic expected delivery schedule, carefully spelling out who will store the equipment (and support the cost) in case of schedule changes. If it is not the vendor, he should be required to list the storage conditions. It is surprising to note the differences in environmental operating conditions listed in the published brochures compared to the conditions expressed by the vendors when they are tied to strict warranty clauses.

These points are summarized in Table 11-3.

PLANT EQUIPMENT PURCHASE

The purchase of plant equipment is the responsibility of process and mechanical engineers; other equipment, such as emergency power plants or laboratory equipment may be the responsibility of other disciplines. Whatever the application, the I&C discipline should be involved. While this is obvious for process equipment, it may not be so for the rest. Nevertheless, even if it is not required immediately, the tendency is that at one time or another all equipment will be connected to some form of data gathering equipment. The client should be advised that provisions be made to do that at a later date. It may take the form of available

Remember: an instrument is only as good as the service that is provided when it is most needed.

Table 11-3. Some Points to Remember When Purchasing Systems

Technical requirements

Hardware
   Spare parts
   Technical support

Realistic delivery schedule

Storage
   Conditions
   Responsibility

Programming
   Hardware resources
   Training courses for engineering group, client's representatives,
     maintenance personnel
   Support

Field service
   Support
   Conditions
   Cost

Warranty
   What
   When
   For how long
   Exchange/return policy

Terms of payment
   On hardware
   On documentation
   On training session

options or a built-in alternative that can be specified at the time of purchase. The other aspect is instrument and control equipment standardization, which results in a smaller inventory of spare parts and in ease of maintenance and translates into reduced operating costs.

The minimum that should be specified for plant equipment with integrated controls is listed on Table 11-4. Participation of an I&C group representative in the bid analysis is a must. Too often there are quite expensive adjustments to the originally submitted price for requirements that had not been established at contract award because of "lack of time for coordination." Be vigilant and look at the equipment purchase schedule to be able to alert concerned parties early enough for proper coordination.

SITE CONTRACTS

Site contracts fall into several different categories:

(1) Supply and install instruments

(2) Install only (instruments and control equipment supplied by owner)

(3) Check and calibrate

The first two categories may include checking and calibration; however, it is a growing practice, at least on the larger projects, to separate them. This practice has several goals:

(1) To have instruments checked by qualified personnel when they are received on site (Nonconforming or damaged instruments can then be returned for replacement or repair without delays to the schedule.)

(2) To provide quality control for the installation

Table 11-4. Minimum Specifications for Plant Equipment Purchase

Necessary	Optional But Recommended
Functional description	Drawings symbols and standards
Control interface list and characteristics for input signals and output signals	Programs to be used for instrument list and other data
Local indication for operator	Programming standards

Physical interface
  Flange sizes for instruments and
    control valves
  Units of measure
	List of approved vendors for the project

Standards and regulations	
Electrical classification	

(3) To provide calibration in a well-documented and orderly manner before start-up

INSTALLATION

The first two types of site contracts will be covered together. If the contract also includes the supply of instruments, all aspects of instrument supply should be as described for the performance-specification project. In general, there are two categories of contracts: fixed price and unit price. A third category is the cost-plus contract, which is generally found on smaller projects, usually involving modification to an existing installation. The cost-plus contract is not often used on larger projects due to the inherent need for intensive monitoring of material and labor costs. Most of the recommendations mentioned for unit price contracts apply equally well to cost-plus contracts.

FIXED PRICE CONTRACTS

The quoted price for a contract is inversely proportional to the quality of the documentation supplied at bid request time. If the bid documents are vague, the bidder has two choices: give a low price with an extensive list of exceptions, or give a price that should cover the worst case. What will be obtained is closely related to the general economic conditions. In prosperous times, contractors can be more selective. They may decline to bid on projects for which the documentation is poorly done because they have the alternative of bidding on projects for which the documentation is well done. When the economy is gloomy, they may decline to bid on a poorly written specification because they estimate their risk as too high for their own economic situation. Alternatively, they may present a bid with an endless list of exceptions.

Ideally, a contractor bidding on a fixed price contract should possess all the documentation necessary to do the work. In practice, on a large project, bids are invited when the I&C engineering is between 60 and 70 percent complete; that is, a fixed price contract is obtained for what is indicated on drawings and described in contract documents with some form of either unit price or cost-plus definition for extra work that will be defined later. Time means money, so a realistic schedule is needed to indicate when further documentation will be supplied. Typical information to be provided at bid time is shown in Table 11-5.

UNIT PRICE CONTRACTS

Unit price contracts are the only choice when there is not enough information to obtain a reasonable price for the installation with a fixed price contract. Contractors may refuse to bid on projects when the specifications lack all the informa-

Table 11-5. Typical Bid Information for Fixed Price Contract

Necessary	Optional But Recommended at BID TIME
Scope of work	Panel wiring drawings
Instrument list	Cable routing lists
Location drawings	As much information as possible
Typical installation drawings	
Electrical room drawings	
Control room drawings	
Piping and vessels drawings	
Cable tray drawings	
Cable lists	
Typical connection diagrams or lists with estimated quantities of connections	
Control panel arrangements	
PLC or DCS units drawings	

Installation specifications:
   Materials to be used
   Standards, codes, regulations
   Tagging methods
   Checking methods	

Description of interfaces with other contracts	
Description of units of work subject to unit prices or method of payment for extra work	

Schedule of delivery on site for:
   Instruments
   Systems
   Panels
   Other owner-supplied equipment	

Schedule of delivery of final updated documents by the engineer to complete the work	
Schedule of access to different areas at site, to commence installation	

tion necessary to enter a clear bid. When information is missing, both the designer and the contractor are forced to add costly safety margins to allow for the unforeseen. Also, specifications that are vague lead inevitably to claims for extra work with their own legal and cost considerations, not to mention the headache of simply facing and administering them.

The key to good unit price contracts is to completely identify and define all cost units. For instrument installations, this is accomplished with a complete list of typical installation details showing the sizes and materials of connectors, tubing, and supports and generally including everything required from the process interface to the cable tray, including supports and fittings. Cables can be priced per installed unit of length, including clamps. Unit prices for terminations can be provided for each type of cable, including cable identification, cable connectors, installation, conductor identification, connections, and continuity checks. These represent units that can be counted easily in the field and quantified by the project administration team for progress measurement and payment. A starting check list for this type of contract is given in Table 11-6.

Table 11-6. Typical Bid Information for Unit Price Contracts

Necessary	Optional But Recommended at Bid Time
Scope of work	Panel arrangement drawings
Instrument list	As much information as possible
Location drawings	
Typical installation drawings	
Electrical room drawings	
Piping drawings or P&IDs with sizes	
Cable tray drawings	

Estimated quantity of:
  Each type of cable
  Each type of tubing
  Connections
  Junction box with terminals
  Control panels with size
  PLC or DCS units	

Installation specifications:
  Materials to be used
  Standards, codes, regulations
  Tagging methods
  Checking methods	

Typical drawings for:
  Motor starter control
  Special equipment
  Junction box arrangement	

Description of interfaces with other contracts	
Full description of each unit of work	

Schedule of delivery on site for:
  Instruments
  Systems
  Panels
  Other owner supplied equipment	

Schedule of delivery of final updated documents by the engineer to complete the work	
Schedule of access to different areas at site, to commence installation	

CHECK AND CALIBRATE

The check and calibrate contract is a service contract: it does not involve installation or supply of materials but calls on independent firms with good teams of maintenance and calibration engineers and technicians. They are normally requested to set up a temporary facility on the job site to maintain good control of the instruments that are delivered. They are requested to provide lump-sum prices if the scope of work, the instrument list, and the delivery schedule are well defined and firm. If not, they are given approximate quantities and are requested to give a firm unit price for each type of instrument. Normally, the price is firm if the actual quantities are within 15 or 20 percent of the stated quantities, with provision for a unit price adjustment if they differ from this percentage. The activity of such a contractor is summarized in Table 11-7.

Table 11-7. Typical Activities of a Check and Calibrate Contractor

Verify instrument for transportation damages
Verify instruments when received against data sheet
Verify tagging and re-tag if necessary
Perform functional check and store
Calibrate instruments prior to installation

Check instruments delivered with equipment:
    Visual check for damages
    Proper engineering units, ranges

Take charge of the field tagging procedure, which has instruments given different color tags when:
    The instrument has been checked
    The instrument has been calibrated
    The instrument has been removed for repair or adjustment
    The instrument should not be touched
    The instrument is approved for start-up

Do final calibration of instruments supplied with equipment
Assist installation contractors in loop calibration, or at least witness these tests

Maintain a complete documentation file:
    Log for each instrument with settings, maintenance
    Calibration data for each loop

Generally provide support for the owner during start-up, but as a separate contract

Vendor Data Follow-up

Vendor data is needed to complete the project engineering, but it is provided by vendors or suppliers, and the outcome of the project is highly dependent on the quality of their work. One way to verify this quality before it is too late is to review, at given schedule milestones that are agreed upon before signing a contract, some key documents or drawings prepared by the vendor for the fabrication or assembly of the equipment.

It is extremely important to set fixed dates or a fixed duration after awarding of the contract for the submittal of such documents. Progress payments should be attached to these submittals as an incentive to produce them early. Too often documents are delivered late, often at a time when there is great pressure to deliver project components. Under such conditions it is difficult for the project managers to obtain documents that will assure that the equipment delivered meets the specifications. Hurried submittals may not precisely identify the interface details (such as wire numbers or terminal numbers in panels) that are needed to complete the overall project engineering.

 Vendor data should be adequately expedited and, when received, diligently returned with comments. Be careful not to stamp them as "approved" because, if the equipment does not meet the specifications during dry runs or at start-up, one might turn around and claim, "It has been approved by the engineer!" At this point there are seemingly endless discussions and even threats of legal action. Examples of better wording are "reviewed," "reviewed with comments," and "fabrication or delivery authorized."

Vendor data review may account for six to 10 percent of the vendor's engineering time if the quality is good. Some vendors have been known to return the same documents five or six times due to the poor quality of their engineering. In one instance the author's group spent as much as 60 percent of the design time in check-

> A tactic sometimes used by vendors is to ship roughly engineered equipment early in order to obtain payment on delivery and then be paid a daily fee for site assistance, during which time they will finalize the engineering, especially with regard to programming. As a result, the client may pay twice for the same work.

ing, commenting, and returning documents, because without this there would have been a delay of three months at start-up.

Project Engineering Wrap-up

When the project has been well organized and the files well kept, the wrap-up is easy to do with a relatively small commitment of time and personnel. This is not just because there is little time left in the engineering budget—it is usually because most of the original team players have been transferred to other projects or to the site or, as is often the case, have changed employers. The wrap-up is then a hard but still necessary task. The necessity is twofold. For one thing, engineers and designers, as individuals and as a group, have a legal responsibility for their designs. For another, the client wants and is entitled to valid documentation and backup data for his installation. The design files, assembled per responsibility group if possible, should contain:

(1) all preliminary data: scope, constraints (physical, environmental, regulatory, as well as those imposed by the client and the project engineering organization);

(2) all calculation notes signed with the name of the responsible individual, including revision dates, comments and approvals;

(3) all related minutes of meetings;

(4) all drawings that have been commented on by the client, authorities, or the project management organization; and

(5) a copy of each final document, updated as built or, if the construction supervision was not under the responsibility of the design organization, a copy of the last issue of each engineering document.

A good practice is to prepare additional notes during the course of the project regarding any special items that may be important to remember in case of a mishap, or just to provide a guide for the next project of a similar nature. These "notes to the file" should be dated and signed if ever they are to be considered a valid document in case of dispute.

Depending on the country and regardless of any civil responsibility, the engineer may also bear a professional responsibility for property damage or accidents. This responsibility is valid for an appreciable number of years. For protection, one should not simply rely on one's employer, who may, in case of dispute, have contrary interests. Engineers should keep copies of all documents that may be necessary for their defense in such cases.

The client will not need all this information, but the consulting firm should keep a copy (or microfilm) of all this data for the prescribed number of years. The client will require all technical data and vendor data, including operating manuals and maintenance manuals required from the vendors and suppliers by contract. This requires effort because some vendors are always reluctant to supply well-prepared manuals where one can readily identify what equipment has been supplied and especially what options apply. Many suppliers tend to send general manuals that make it difficult for anybody else to fathom the ocean of options available.

The client will also require a reproducible copy of all documentation certified "as-built." Attention should be paid to the legal aspect of signing "as built" documents. Even if the person who signs this documentation may never never set foot on the site, by signing, that person is considered to endorse all changes that have occurred. With a critical eye to all key documents, let the signer beware.

Although tedious indeed, project wrap-up is a necessary and important task.

Site Work

Site work is completely different from engineering at the office. One must be aware of the field practices specific to a job site. For example, working on a construction site may require taking a construction site safety course. For the most part, however, one needs to know how the trades are divided through their union contracts. On international projects in other countries, it is also important to know something about the interpersonal customs between laborers, foremen, and other degrees of authority. There must also be a good understanding of the lines of authority, both formal and informal, within one's own organization. Site work can be demanding and the hours long, especially when start-up approaches. One must deal with hard-nosed contractors clamoring for extras at job meetings. Deadlines must be met. Every lost day means lost money. These things have little to do with engineering but an understanding of them is essential if site work is to be productive. Site work is both a business and a technical challenge.

After being introduced, the first thing the I&C contract administrator should do on the site is read the contract documents from cover to cover until the content is familiar enough for a sustained conversation about any topic. Others working on the site are expected to be equally familiar with their own disciplines. Site personnel often have to field thought-provoking questions from visitors who are "just having a casual look around." Although this chapter does not present a detailed discussion of what site work consists of, the following is a summary of the various activities, not considering the levels of responsibility:

(1) Review the quality of materials, equipment, and installation according to the drawings, specifications, regulations, and good practice, including accessibility.

(2) Monitor the progress of work according to the schedule.

(3) Assess the progress of work for progress payments.

(4) Attend required progress review meetings with the contractor and make sure that minutes are taken and distributed the same day.

(5) Interpret the specifications and give a ruling if authorized to do so; otherwise, refer the issue to the proper authority.

(6) Receive, evaluate, and process change requests.

(7) Assure coordination with work done by other contractors, trades, and disciplines.

(8) Monitor the delivery of instruments, equipment, and all owner-supplied materials.

(9) Monitor the ordering of equipment supplied by contractors, who tend to delay ordering until the last minute in order to improve cash flow.

(10) Verify that all field changes are marked up on a master copy of the documents for later use in as-built documents provided to the client.

(11) Witness checking and calibrating procedures and tests.

(12) Maintain a well-organized filing system.

The following are some suggestions that can make the difference between a contract that runs smoothly and one that is hampered by delays and crushed under paperwork:

(1) Review design drawings before they are transmitted to the contractor.

(2) Offer to review the transmitted documents with the contractor, who will not be able to claim later that there are ambiguities or missing information. (If there is a question, it should be cleared up quickly through communication with the engineering office and confirmed soon afterward by the formal paperwork. A contractor is more likely to be cooperative and go ahead on the basis of a signed hand-drawn sketch if he's used to seeing the confirming change order follow quickly. When there are delays, the contractor may wait and possibly delay the work, all of which puts tremendous pressure on the project management team to get the paperwork out. The contractor may even submit a claim for having provided workers who were left idle due to a lack of information.)

(3) Make a daily round to assess the quality and quantity of work done by the contractor, to foresee delays due to progress of other trades, and to foresee interference problems.

(4) Know the contract well, but don't be a nitpicker. (As long as the quality level is respected, minor changes can be accepted and used as trade-offs when valid extras arise. Contractors are good at picking up small points that can translate into savings. Analyze them carefully.)

Not all of these pointers are part of formal contract administration, but they work. Study management's attitude at the site to see if there is the latitude to use this approach.

Conclusion

Instrumentation and controls project management has many aspects. Some are purely technical, such as the content of a data sheet; others deal more with project management and interpersonal relations. This includes everyone on the project management team, other disciplines, engineering management, the client, vendors, suppliers, and contractors. These aspects are not often covered in textbooks, and it is hoped that the discussion here has illustrated how they can contribute to the success of a project.

Bibliography

1. Bacon, John M., *Instrumentation Installation Project Management System*, Research Triangle Park, NC: ISA, 1989.

About the Author

Michel Spilmann, a 1967 graudate of École Polytechnique Universite de Montréal, has a degree in electrical engineering with automation specialization. He jointed Pratt & Whitney of Canada as a control engineer. After five years, he joined a major consultant firm, Lavalin, Inc., where from 1980 he held the position of Principal Engineer, Automation and Instrumentation, at the head office. He has worked as a consultant since 1986 and is now head of his own firm, SOMIS, Inc. He is a member of IEEE and a Senior Member of ISA. His main objective has been to improve quality in engineering and minimize overall project costs through standardization and organization.

Appendix 1
Laboratory Standards

This appendix has been written to help the teacher, student, and practitioner develop experimental exercises. These can be used to reinforce concepts learned in the book or to set up calibration facilities in an actual work setting. The material was originally prepared for use in technician level training programs at Monroe Community College in an Instrumentation Technology program.

The material has been prepared with a view to maximum hardware flexibility. Standard-measurement laboratory hardware should be suitable to accomplish most of the exercises.

The author would like to thank Professor Lawrence Skarin, Monroe Community College, and Mr. Otto Muller-Girard, Consulting Engineer, for their contributions and helpful references. Thanks also to Mr. Jack Moore, who allowed the use of some materials written by Mr. Austin A. Fribance in a manual published by Hickok Teaching Systems.

Introduction to Laboratory Standards

Pressure and Temperature

OBJECTIVES

(1) To study and use laboratory standards for the measurement of temperature and pressure.

(2) To become familiar with the terminology of metrology as it applies to the measurement of temperature and pressure.

(3) To compare measurements taken with common laboratory instrumentation against laboratory standards and determine instrument errors.

THEORY

This laboratory exercise introduces the important instrumentation concept of metrology, which includes laboratory standardization and calibration practices. All facilities where instruments are used or tested must rely upon certain special, high-quality, high-precision, accurate instruments that have been identified as *laboratory standards*. These standards become the basis of measurements for a specific laboratory. In the calibration process, standards are used as comparison devices. Instruments under test are compared to and adjusted to the standard. If the instrument's error cannot be removed by adjustment and is greater than the published tolerance limit for the instrument, chances are good that the instrument would not be used but returned to the manufacturer for factory recalibration and certification. Instruments that lose calibration frequently are generally taken out of service.

EQUIPMENT

Pressure standards:

> 0–30 psi standard gages
>
> 0–110 in. manometer or electronic manometer
>
> Portable pressure standard, (digital and/or analog)
>
> Dead weight tester

Temperature standards:

> Standard thermometers

Laboratory equipment:

> 0–5 psi gages
>
> 0–30 psi gages
>
> 0–100 psi gages

Industrial thermometers

> Pans
>
> Ice bath
>
> Pressure tubing and hardware

See Figures A1-1 and A1-2.

PROCEDURES

Low Pressure Standards (0–110 inches of water). Standard: 110-inch manometer or electronic manometer.

(1) Become familiar with the standard. Review instruction manual if available.

(2) Record: range, span, and accuracy of both the standard and the gage being tested.

(3) Set up the test circuit as indicated in Figure A1-1. Limit supply pressure to 3.0 psi. Use 0–5 psi gage as "gage under test."

(4) Prepare a table of values as follows: standard; gage under test; gage error.

(5) Check the 0–5 psi gage for the following pressure values: 0.50 psi; 1.00 psi; 1.50 psi; and 2.00 psi. Set the exact input pressure as determined by the standard. (Inches of water will have to be converted to psi.)

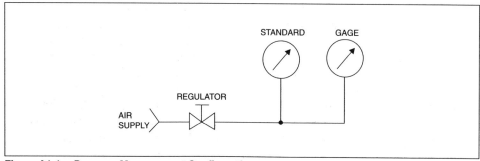

Figure A1-1.   Pressure Measurement Configuration

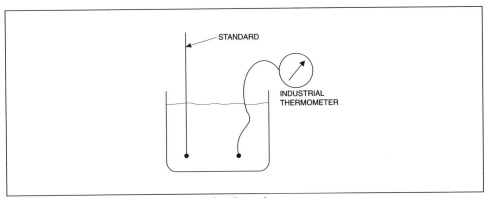

Figure A1-2. Temperature Measurement Configuration

Middle Pressure Standards (0–30 psi). Standards: 0–30 psi bench gages, portable pressure standards (analog and digital).

(1) Become familiar with the standard. Review available instruction manuals.

(2) Record: range, span, and accuracy of the standard and the gage being tested, 0–30 psi working gage.

(3) Set up the test circuit as shown in Figure A1-1.

(4) Check the "gage under test" against the standard for each major division indicated on the "gage under test." Set up data table, record data, and determine gage error.

(5) Repeat steps 1–4 above for each standard used in this pressure range.

High-Pressure Standards (30–100 psi). Standard: dead weight tester

(1) Become familiar with the dead weight tester pressure standard. Review instruction manual.

(2) Record: range, span, and accuracy of the standard and the 0–100 psi gage.

(3) Place the 0–100 psi gage in the proper position on the dead weight tester.

(4) Check the pressure at major gage division values. Tabulate the data as in previous procedures.

Temperature Standards (Range 32 to 200°F). Standard: Precision thermometers

(1) Become familiar with the standard thermometers and certification document.

(2) Set up three temperature conditions as follows: about 32°F; about 70°F; and about 200°F.

(3) Select an industrial thermometer. Record: range, span, and accuracy of both the standard and the industrial thermometer.

(4) Compare temperatures as indicated on the standard to those indicated on the industrial thermometer. Tabulate data indicating the error.

ANALYSIS AND QUESTIONS

(1) Define each of the following terms: range, span, precision, accuracy, repeatability, sensitivity, and calibration.

(2) Indicate some of the characteristics a laboratory standard should have.

(3) Indicate how a standard should be "cared for" in the laboratory.

(4) What process measuring instruments other than pressure instruments rely upon pressure standards for calibration?

(5) How is the accuracy of a transducer generally quoted in instrument specifications?

(6) What are "fixed points"? How are these points used in instrument calibration?

Low Voltage Sources

OBJECTIVES

(1) To introduce and learn to use accurate, precision sources for low-voltage (millivolt and microvolt) signals.

(2) To observe the operation of the mV potentiometer, a null-balance instrument.

(3) To draw correction plots for the instrument's two ranges using the digital multimeter (DMM) as a standard.

(4) To generate millivolt signals, to measure these signals, and to compare and evaluate results.

(5) Measure the output of a standard cell.

BACKGROUND

Accurate precise test equipment is used to generate and measure millivolt signals. VOMs (volt-ohm-milliammeters) are not good enough for DC millivolt measurements for two reasons: (1) they are not sensitive enough, and (2) they load measuring circuits too much. The potentiometer and the 4-digit or higher digital multimeter are the best instruments for the measurement of low-voltage signals.

Potentiometers use a precision linear voltage-divider resistor called a slidewire. Since the slidewire is very linear, one can fix a voltage accurately on the wire (a process called "standardization") and then accurately determine other voltages along the divider. This voltage is balanced against the unknown voltage through a null detector. The advantages of this system are twofold: (1) there is no loading, and (2) the accuracy of the null detector does not affect the accuracy of the measurement.

The digital multimeter has largely replaced the mV potentiometer as the instrument of choice in low-voltage measurements. It has extremely high input resistance and the convenience of direct reading. Its accuracy is as good as the potentiometer's and the cost is about the same.

EQUIPMENT

High-quality millivolt potentiometer
DMM (digital multimeter)
Calibrated voltage source

PROCEDURE

Familiarization with the High-Quality mV Potentiometer.

(1) Set the mV potentiometer. Optional: remove case of potentiometer.

(2) Optional: identify the detector, the working batteries, the standard cell, and the slidewires. Record the types of batteries in case cells need to be replaced.

(3) Consult vendor literature for the operation of the potentiometer.

Familiarization with Millivolt Sources.

(1) Obtain instruction manual for the mV source instrument to be used.

(2) Generate mV signals from 0–100 mV in 10-mV increments. Use the DMM to measure and verify the output of your millivolt source. Generate a table and tabulate this data.

Use of Millivolt Potentiometer as Source, and DMM.

(1) Generate mV signals from 0 to 64 mV in 4-mV increments using the calibrated voltage source.

(2) Measure these mV signals with the potentiometer and the DMM.

(3) Tabulate data.

(4) Using the source as a mV standard, plot source (input) vs. measured (output) for both the potentiometer and the DMM. (Note: Input is always on horizontal axis and output is on vertical axis).

Standard Cell.

(1) Measure voltage output of the standard cell with both potentiometer and the DMM.

QUESTIONS

(1) What is the specified accuracy of the mV potentiometer?

(2) What is the specified accuracy of the DMMs?

(3) If 50.0 mV was measured, between what limits is the "actual" voltage? (Based on mV potentiometer's specified accuracy, on the DMM specified accuracy).

(4) What is the maximum resolution of the mV potentiometer, the DMM?

(5) Which loads the circuit more — the mV potentiometer or the DMM? Explain.

(6) Comment on accuracy plots of the instrument.

(7) Comment on standard cell voltage measurement.

Static Resistance Measurements and Precision Resistance Sources

OBJECTIVES

(1) To make resistance measurements using the Wheatstone bridge, VOM, DMM, and precision Wheatstone bridge.

(2) To verify empirically the balanced bridge equation.

(3) To become familiar with the operation of standard decade resistance instruments.

BACKGROUND

Resistance measurements are important and occur frequently. For those measurements where the 5 percent accuracy of the VOM is not enough, the Wheatstone bridge instrument and the DMM are used.

Wheatstone bridges occur in the balanced form and the unbalanced form. The balanced bridge is used to measure resistance. In the unbalanced case, the battery voltage accuracy and the null detector accuracy have no effect on the measurement accuracy. In the balanced form, they are important.

The decade box is an instrument with a variety of resistors in a single source. It is used to provide exact values of electrical resistance. The decade box is often used to simulate a field-located sensor in the calibration of remote transmitter instruments. The decade box can provide resistance values of less than one ohm.

EQUIPMENT

DMM
Wheatstone bridge
VOM
Null detector
Bench power supply
Bssorted resistors
Decade box

PROCEDURES

Application of the Decade Resistance Box.

(1) Become familiar with the decade resistance box instrument, its measuring capability, and its operation.

(2) Record high and low range limits for this instrument.

(3) Measure resistance using the DMM and the decade box. The decade box is the standard for resistance input. Dial in and measure the following resistance values: 0.5 ohm, 1.0 ohm, 10.5 ohms, 995.3 ohms or representative values. Tabulate results.

(4) Record observations.

(5) Record the accuracy of the DMM for the specific ranges used.

Precision Resistor Measurements.

(1) Prepare a table of resistance measurements using three instruments for comparison (DMM, Wheatstone Bridge, and VOM). Measure the resistance of six unknown resistors.

(2) Tabulate resistance measurements for Wheatstone Bridge in column 1. Refer to instructions for the operation of this instrument.

(3) Tabulate resistance measurements for VOM in column 3.

(4) Tabulate resistance measurements for DMM in column 5.

(5) In columns 2 and 4, using the DMM readings as standard, tabulate percentage difference with actual values of the Wheatstone bridge and the VOM.

Breadboard Wheatstone Bridge.

(1) Breadboard a Wheatstone Bridge circuit as shown in Figure A1-3.

(2) Balance the bridge using R_x as 2.21 K ohms. Verify the points of equal potential around the bridge using the VOM. Tabulate.

(3) Place a 120-ohm resistor and a 500-ohm resistor, respectively, in the R_x position in the bridge. Verify the R_x resistance values by measuring the balancing resistor R_3 and comparing values with those of R_x.

Use of Commercial Wheatstone Bridge Standard.

(1) Select three resistors from those previously measured and determine resistance values using the Wheatstone Bridge. If there are a limited number of these instruments available, rotate the use of the instruments.

ANALYSIS AND RESULTS

(1) Determine the accuracy of the resistance measurements made with each instrument. Record.

(2) From the table of values obtained for the unknown resistors, give the best estimate of the exact values of these resistors.

(3) Explain the null balance method of measurement as used in this exercise.

(4) Upon what specific factors does the accuracy of a null balance instrument like the Wheatstone bridge depend?

(5) How do you account for the difference in resistance values obtained with the DMM and those dialed into the decade box standard?

Pressure

Manometer — Laboratory Standard

OBJECTIVES

(1) To make pressure measurements with a manometer.

(2) To construct a pneumatic circuit and to utilize a laboratory manometer as a standard for pressure determination.

(3) To calibrate a 0–5 psig pressure gage using oil-filled and mercury-filled manometers.

(4) To calibrate a 0–30 psig pressure gage using a commercial mercury manometer.

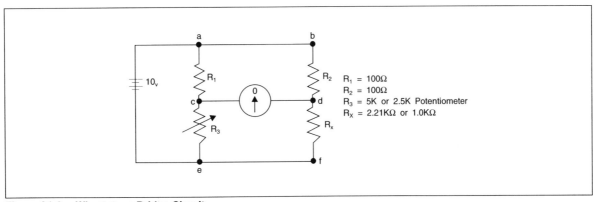

Figure A1-3.   Wheatstone Bridge Circuit

(5) To take several pressure measurements using commercial manometers of a variety of different types.

THEORY

One of the most accurate and dependable means of measuring pressures or pressure differentials employs the liquid manometer in which the pressure created by a column of liquid just balances that established by the pressure to be measured. Provided that the liquid is homogeneous and that its specific gravity is constant, the pressure established by a column of noncompressible liquid is directly proportional to its height. This same relationship between height of the material and the pressure created holds for many other conditions — the pressure created by the height of a stack of cans or bottles, for example, varies directly with the number of tiers of material. Because of the compressibility of gases, the pressure created by a very tall column is not a linear function of height. However, in many instances, the validity of the measurement depends upon the elimination of all frictional conditions. In the case of fluids, the frictional effects can be reduced to zero when there is no motion of the fluid.

Because of this absence of friction, the liquid manometer is used as a standard for pressure measurements to which almost any other low-pressure measurements can be referred.

The range of any given manometer, assuming adequate physical strength of the components varies with:

(1) The specific gravity of the liquid.

(2) The vertical differences (not the slant distance) in level of the surfaces of the liquid.

If a liquid metal, such as mercury, is used for the fill, then each two inches of difference in level represents about one pound per square inch pressure. If water is used in the manometer, it requires a column about 2.3 feet high to establish a pressure of one pound per square inch. If liquids with specific gravities significantly less than one are used, then one psi will require greater column heights.

For absolute pressure measurements using liquid-filled manometers, care must be exercised in selecting a fluid. The fluid should develop a negligible vapor pressure; otherwise, the accuracy of the measurement will be altered.

If absolute pressure measurements are to be made rather than gage pressure, which signifies pressures above atmospheric, the space above the fluid in one leg should be completely evacuated. Under these conditions, the pressure created by the difference in level of the two mercury columns represents the applied pressure. Atmospheric pressure is not adding its effects to those of the mercury in creating a condition of pressure balance.

EQUIPMENT

Mechanical breadboard unit (MBU)
Tubing
Assorted hardware
0–5 psi pressure gage
0–30 psi pressure gage
Commercial manometers (record instrument number)
Graph paper

PROCEDURE

(1) Secure a U-tube manometer mounted on a breadboard training stand. (Manometers contain different filling fluids, so be sure to note the specific gravity of the fluid in each manometer.)

(2) Construct the pneumatic test circuit as indicated in Figure A1-4. Adjust bench supply pressure to no more than 1 psi (for initial calibration).

(3) Use the red oil manometer specific gravity 1.0 to begin the calibration of the 0–5 psi gage from 0–1 psi in three steps. For each calibration step record the value of column height difference and pressure gage reading. Record information in a data table (see sample data table).

(4) Use the mercury-filled manometer to complete calibration at each major division on the gage (1, 2, 3, 4, 5 psig).

(5) Now select a 0–30 psi gage and calibrate this gage using a commercial mercury manometer. Make calibration in 5-psi increments. Record information in the data table.

(6) Plot on graph paper the following:

- For a specific manometer trainer — change in height vs. calculated pressure (three points)

- Calibration characteristic (curve) for the 0–5 psi pressure gage plot; calculated pressure (manometer) vs. gage indicated pressure

- Calibration characteristic for 0–30 psi gage — plot standard (calculated) pressure (manometer) vs. gage indicated pressure

Data Table

(1) Pressure values 0–5 psi gage.

- Gage indicated pressure (psi)   0 1 2 3 4 5

- Manometer height difference (ΔH)

- Calculated pressure ($P = K\Delta H$ in psi)

- Gage error

- Gage correction factor

(2) Same table structure for 0–30 psi gage.

(3) Comparison values using commercial manometers.

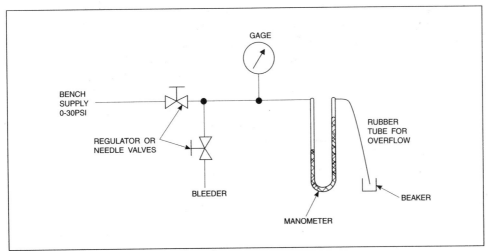

Figure A1-4.   Manometer Test Configuration

SAMPLE CALCULATIONS

Be sure to show sample calculations of pressure determination $P = K\Delta H$.

ANALYSIS OF RESULTS

(1) Describe the character of the plot of ΔH vs. pressure.

(2) Describe the calibration curve for each gage. Classify the type of error.

(3) What laboratory environmental conditions will affect the calculated pressure values obtained using a manometer?

(4) Tell how a manometer might be used for specific gravity determinations.

(5) Rate the manometer as a pressure measurement device for the following: range and span; precision; accuracy; linearity; speed of response; sensitivity.

(6) Briefly state why the manometer is accepted as a low-pressure standard.

(7) Comment on Table A1-1 and Figure A1-5.

Bourdon Tube Pressure Gage

OBJECTIVES

(1) To calibrate a bench standard gage against several laboratory master gages (standards): Heise gage, Wallace–Tiernan portable, or mercury manometer.

(2) To prepare an error deviation plot for the bench gage.

THEORY

Elastic materials undergo strains that are proportional to the stress to which they are subjected. Some require more stresses than do others to result in the same deformation. Young's modulus of 30,000,000 psi per inch of strain per inch of original length is virtually the same for all kinds of steel. It also tells us that at the original rate at which the steel started to deform under stress, a stress of 30,000,000 psi would have been required to strain a one-inch piece of the original sample so that its final length would be two inches, or each inch was strained one inch. Assuming linear responses, and the use of Young's modulus is based upon such an assumption, because most of the time the materials are used in the region where the strain is directly proportional to the stress, then 300,000 psi (which is 1/100th of the value of the modulus) would create 1/100th of the strain in each inch or 0.01 inch/inch. Similarly, 30,000 psi (being only 1/1000th of the value of the modulus) would strain each inch 1/1000th as much, or 0.001 inch/inch.

For most instrumentation work, every attempt is made to stay well within the region of linear response in order that there may be a direct relationship between stress and strain — the strain then being used as a measure of the stress.

If some of the elastic materials are formed into the shape of bellows, their large active area combined with the thin walls of the device produce relatively large deformations. In order to retain their linearity of response, care must be exercised to ensure that they do not acquire a set or permanent deformation.

The type of material used will vary with the number of cycles to which the bellows will be subjected. Operating conditions will dictate whether environmental considerations will dictate the use of materials such as stainless steel. The amount of the normal deflection or flexing will affect the life and, hence, the choice of material.

For higher pressures, the bellows is replaced by a noncircular elastic tube that often is formed in the shape of an oval. The tube in this form is known as a Bour-

Table A1-1. Conversion of Inches of Water to Pounds per Square Inch

Water		Water		Water		Water	
Inch	psi	Inch	psi	Inch	psi	Inch	psi
0.10	0.0036	5.10	0.1840	10.10	0.3644	15.10	0.5449
0.20	0.0072	5.20	0.1876	10.20	0.3680	15.20	0.5485
0.30	0.0108	5.30	0.1912	10.30	0.3717	15.30	0.5521
0.40	0.0144	5.40	0.1948	10.40	0.3753	15.40	0.5557
0.50	0.0180	5.50	0.1985	10.50	0.3789	15.50	0.5593
0.60	0.0216	5.60	0.2021	10.60	0.3825	15.60	0.5629
0.70	0.0253	5.70	0.2057	10.70	0.3861	15.70	0.5665
0.80	0.0289	5.80	0.2093	10.80	0.3897	15.80	0.5701
0.90	0.0325	5.90	0.2129	10.90	0.3933	15.90	0.5737
1.00	0.0361	6.00	0.2165	11.00	0.3969	16.00	0.5773
1.10	0.0397	6.10	0.2201	11.10	0.4005	16.10	0.5809
1.20	0.0433	6.20	0.2237	11.20	0.4041	16.20	0.5845
1.30	0.0469	6.30	0.2273	11.30	0.4077	16.30	0.5882
1.40	0.0505	6.40	0.2309	11.40	0.4113	16.40	0.5918
1.50	0.0541	6.50	0.2345	11.50	0.4150	16.50	0.5954
1.60	0.0577	6.60	0.2381	11.60	0.4186	16.60	0.5990
1.70	0.0613	6.70	0.2418	11.70	0.4222	16.70	0.6026
1.80	0.0649	6.80	0.2454	11.80	0.4258	16.80	0.6062
1.90	0.0686	6.90	0.2490	11.90	0.4294	16.90	0.6098
2.00	0.0722	7.00	0.2526	12.00	0.4330	17.00	0.6134
2.10	0.0758	7.10	0.2562	12.10	0.4366	17.10	0.6170
2.20	0.0794	7.20	0.2598	12.20	0.4402	17.20	0.6206
2.30	0.0830	7.30	0.2634	12.30	0.4438	17.30	0.6242
2.40	0.0866	7.40	0.2670	12.40	0.4474	17.40	0.6278
2.50	0.0902	7.50	0.2706	12.50	0.4510	17.50	0.6315
2.60	0.0938	7.60	0.2742	12.60	0.4546	17.60	0.6351
2.70	0.0974	7.70	0.2778	12.70	0.4583	17.70	0.6387
2.80	0.1010	7.80	0.2814	12.80	0.4619	17.80	0.6423
2.90	0.1046	7.90	0.2851	12.90	0.4655	17.90	0.6459
3.00	0.1082	8.00	0.2887	13.00	0.4691	18.00	0.6495
3.10	0.1119	8.10	0.2923	13.10	0.4727	18.10	0.6531
3.20	0.1155	8.20	0.2959	13.20	0.4763	18.20	0.6567
3.30	0.1191	8.30	0.2995	13.30	0.4799	18.30	0.6603
3.40	0.1227	8.40	0.3031	13.40	0.4835	18.40	0.6639
3.50	0.1263	8.50	0.3067	13.50	0.4871	18.50	0.6675
3.60	0.1299	8.60	0.3103	13.60	0.4907	18.60	0.6711
3.70	0.1335	8.70	0.3139	13.70	0.4943	18.70	0.6747
3.80	0.1371	8.80	0.3175	13.80	0.4979	18.80	0.6784
3.90	0.1407	8.90	0.3211	13.90	0.5016	18.90	0.6820
4.00	0.1443	9.00	0.3247	14.00	0.5052	19.00	0.6856
4.10	0.1479	9.10	0.3284	14.10	0.5088	19.10	0.6892
4.20	0.1515	9.20	0.3320	14.20	0.5124	19.20	0.6928
4.30	0.1552	9.30	0.3356	14.30	0.5160	19.30	0.6964
4.40	0.1588	9.40	0.3392	14.40	0.5196	19.40	0.7000
4.50	0.1624	9.50	0.3428	14.50	0.5232	19.50	0.7036
4.60	0.1660	9.60	0.3464	14.60	0.5268	19.60	0.7072
4.70	0.1696	9.70	0.3500	14.70	0.5304	19.70	0.7108
4.80	0.1732	9.80	0.3536	14.80	0.5340	19.80	0.7144
4.90	0.1768	9.90	0.3572	14.90	0.5376	19.90	0.7180
5.00	0.1804	10.00	0.3608	15.00	0.5412	20.00	0.7217

don tube and has gone unchanged for about 125 years. The theory of the tube is relatively complex if the behavior is examined in detail. For present purposes, one can consider that the forces acting on the interior curve exceed those on the exterior. This difference in forces then acts to "straighten out" the deformed tube. Because the difference in forces is small and virtually all of the force is "used" to change the shape of the tube, there is essentially NO force available to perform some function such as rotating a resistor, pushing some component, etc. The gear trains and their associated pointers must be almost frictionless if the response of the gage is to be correct. The small movement of the end of the Bourdon tube nor-

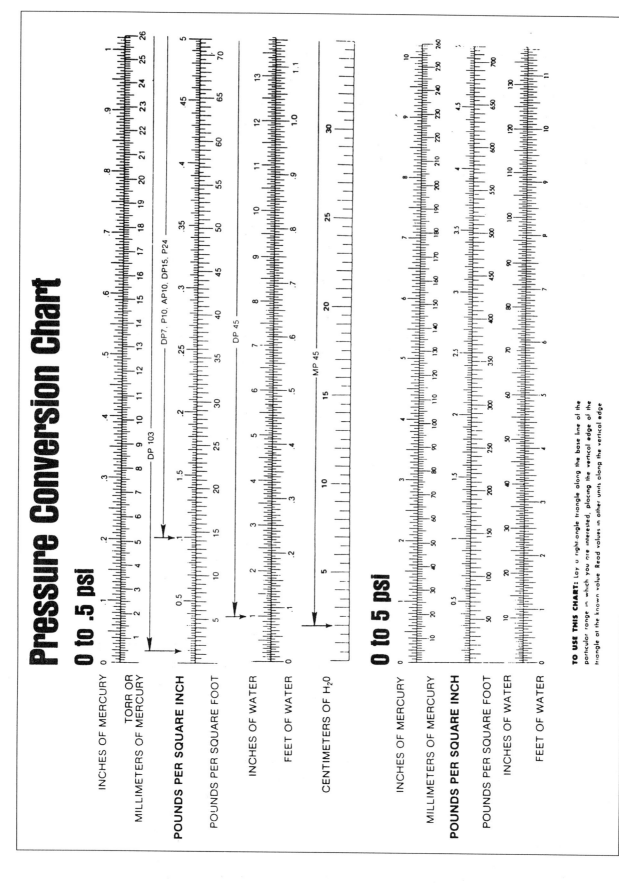

Figure A1-5.   Pressure Conversion

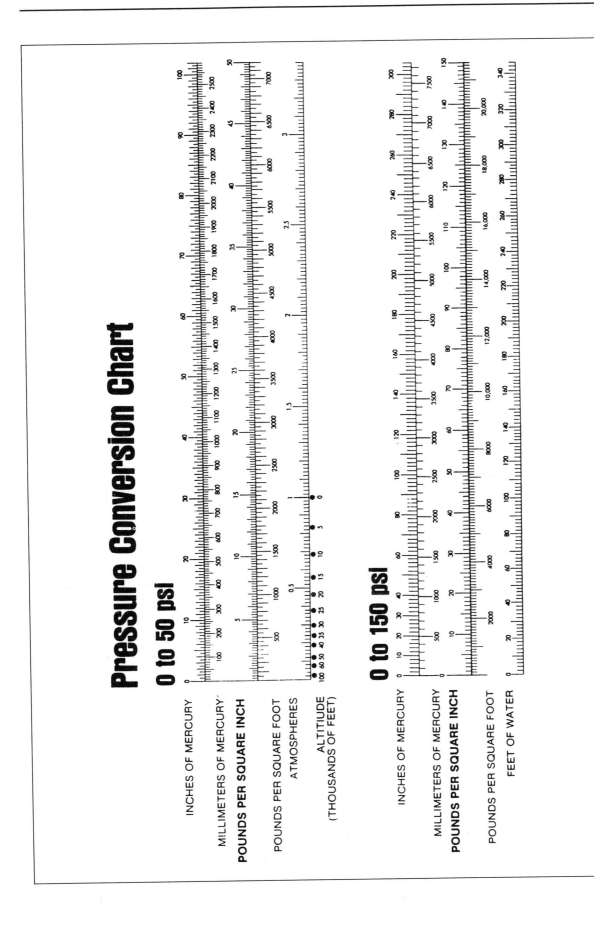

Figure A1-5 (continued). Pressure Conversion

mally is such that it must be amplified by gears (or other means) in order to be useful.

If the pressure gage can be made with a much longer Bourdon tube, shaped in the form of a helix, then the total movement of the free end of the tube may be sufficient to move the pointer through its required arc.

The source of the pressure is immaterial; whether caused by some process or by the expansion of some confined fluid, pressures create stresses that will always create strains. The resulting strain is then interpreted in terms of the pressure required to create it.

If sealing fluids or other means are used, contaminated process fluids can be prevented from entering the tube and plugging it or otherwise changing its response.

In order to calibrate pressure gages, several techniques may be used. For the low pressure bellows, the pressures that can be measured by the U-tube manometers can be used. For the higher pressure devices such as the Bourdon tube gages, master gages may be used. The standard master gage, in turn, should be calibrated using the deadweight type of tester.

There are three basic configurations of Bourdon tube pressure gage elements: the C-tube, the helix, and the spiral. Of these three, the most popular in use is the C-tube. However, the C-tube, although simple in basic design, has a major disadvantage when compared to spiral and helix configurations; it is the limited amount of travel of the elastic element. Since the tube movement is small, it is then necessary to attach the tube to a high-gain mechanical amplifier signal transmission system. In making calibration adjustments for Bourdon tube gage mechanisms, it is usually necessary to adjust various portions of this transmission system while being careful to preserve linearity and angularity of the gage output.

The amplification mechanism is like that of a jeweled watch, since little work (force) is available coming from the Bourdon tube itself. In repairing and calibrating Bourdon tube gage mechanisms, it is most important to minimize friction in the transmission mechanism.

EQUIPMENT

Commercial pressure gages: 0–30 psi
Commercial pressure standards
Assorted hardware and tubing

PROCEDURE

Examination of Bourdon tube gage mechanism.

(1) Remove the Bourdon tube from its case. Examine this mechanism.

(2) Apply 30-psi pressure to gage and observe the motion of the Bourdon spring.

(3) Measure the motion of the spring and compare it to the motion of the pointer. Determine approximate amplification produced by the mechanical linkage. Determine gage sensitivity.

(4) Diagram the Bourdon tube mechanism. Identify functional parts.

(5) Reassemble the gage; prepare for calibration and test.

Bench Gage Calibration.

(1) Select a 0–30 psi bench gage for a calibration check against laboratory master pressure gage standards.

(2) Use the standard pressure gage and/or mercury manometer for the calibration.

(3) Check the calibration at each major division indicated on the bench gage (both up and down scale).

(4) Prepare a calibration data table for the gage indicating: gage reading, standard (true) value, and gage error. From this data prepare an error plot for the gage.

(5) From the data table in (4) above prepare the following:

- Error plot for the bench gage (up and down scale).

- Gage accuracy graph for 0–30 psi bench gage. (Plot standard gage values as input axis and actual gage values as output axis.)

(6) Now draw tolerance limits on the graphs and rate this gage for accuracy from your experimental data.

ANALYSIS AND REPORT

(1) Calculate the amplification factor for the Bourdon tube pressure gage using Bourdon tube travel vs. pointer travel as the gain ratio.

(2) Be sure to label functional parts of the Bourdon tube gage on your diagram of the gage.

(3) Plot error deviation curve for the (bench standard).

(4) Rate the elastic pressure mechanisms for: accuracy, precision, sensitivity, responsiveness, linearity, range and span.

Pneumatic Transmitters and Recorders

OBJECTIVES

(1) To become thoroughly familiar with a commercial pressure transmitter. To examine receiver mechanism, mechanical amplification transmission linkages, pneumatic amplification linkages, and input vs. output characteristic curves for a pneumatic pressure transmission.

(2) To become thoroughly familiar with a commercial pressure recorder-indicator instrument. To examine receiver mechanism, mechanical transmission, amplification system, and plot input vs. output characteristic curves for a pneumatic pressure recording mechanism.

THEORY

Oftentimes, it is desirable to convert some measurement into a pneumatic signal that may be transmitted or used in some desirable manner. Temperature measurements or effects normally are not transmitted over 200 feet when using filled systems. However, the control room or display may be much farther from the measurement than that. Thus, in place of the original measurements, some suitable pneumatic replica is transmitted. Almost all pneumatic transmitters employ the clearance between a baffle and a nozzle to create the desired analog pressure.

If a flat baffle is brought close to a nozzle (see Figure A1-6), in which there is a restriction, the free flow of air is impeded. In turn, the pressure increases and, when the flow is almost totally restricted, will approach the pressure of the base system. Without the restriction, there would be no way to create a pressure change. A typical pressure-distance relationship is shown in Figure A1-7.

While the maximum pressure change occurs for a baffle movement of about 0.006 inch, the change in pressure or temperature required to move the baffle this much may be much more than desirable. Moreover, if appreciable amounts of air

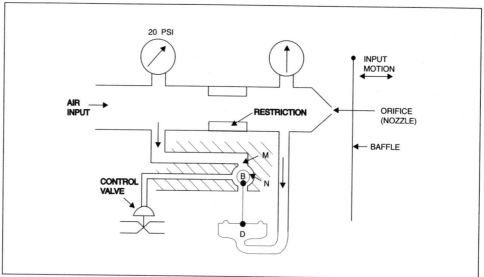

Figure A1-6. Pneumatic Relay Mechanism

are required for the actuation of a valve actuator or some other piece of equipment, then the small, long operating lines may be totally unacceptable.

In Figure A1-6, if the baffle is caused to move towards the nozzle, the back pressure will increase. This back pressure increase causes the diaphragm to move in an upward direction, thus causing the rod and ball to move upward also. The upward movement of the ball will cause a restricted air flow from the air input-to-the pilot relay output. As the baffle is moved away from the nozzle, the opposite effect occurs at the pilot relay output; i.e., back pressure will decrease, causing the diaphragm to move downward, thus causing the ball to also move downward and create an increase in the flow of air out from the pilot relay and an output pressure increase.

If a tangent is drawn to the pressure-clearance curve of Figure A1-8 for the values A and B selected, a net baffle-nozzle clearance change of 0.001 inch establishes a pressure change from 3 to 5 psi, or vice versa. In any event, a net change

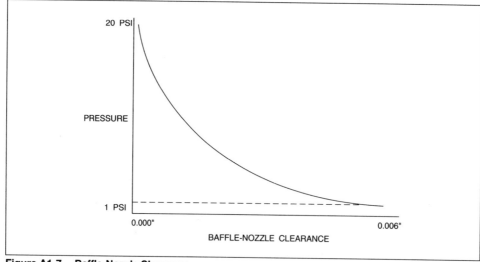

Figure A1-7. Baffle-Nozzle Clearance

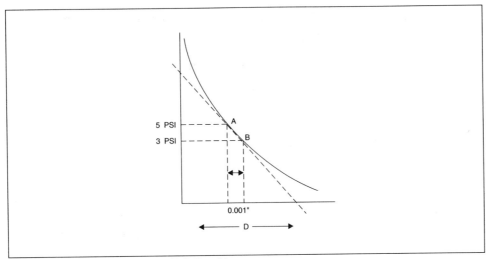

Figure A1-8.   Pressure-Distance Relationship

of 0.001 inch on the part of the baffle creates a variation in the pressure of 2 psi. If this pressure, which is created within the nozzle area, is transmitted to a flexible diaphragm D, as shown in Figure A1-6, an increase in the pressure will force the diaphragm D up. This will bring ball B closer to its seat at M, thus reducing the clearance at M while increasing it at N, and reducing the pressure that can be transmitted. Because an increase in pressure is created by the movement of the baffle with respect to the nozzle, which will create a decrease in pressure at the output, this is known as a reverse-acting relay. Some advantages of this device might at the moment not be apparent:

(1) A movement of the baffle of just 0.001 inch can force ball B through its total stroke. This increases the sensitivity and also gives a linear response. The pressure changes between points A and B are almost directly proportional to the amount of movement.

(2) Because the air for the controlled devices does not have to pass through restriction R, much greater amounts of air can be furnished in a given period of time. This speeds up the response of the associated equipment.

(3) The small change in the position of the baffle means that the monitored variable, i.e., pressure, temperature, etc., need change only by a small amount in order to move the baffle through its required maximum distance of 0.001 inch.

EQUIPMENT
Minimum three commercial pneumatic pressure transmitter instruments
0–30 psi standard gage
Pneumatic tubing
Pressure regulators
Pneumatic calibrators
Pneumatic recorders

PROCEDURES
Pressure Transmitters (Indicating and Blind).

(1) Read in the instruction manual specific sections on: principle of operation, transmission linkages, measurement linkages, and calibration of the transmission system.

(2) Remove front cover plate and thoroughly examine the mechanism.

(3) Using the manual and the instrument, identify the following:

- Pressure receiver element. (What type of elastic element is used?)

- Nozzle-flapper mechanism.

- Pneumatic relay.

- Mechanical transmission linkages.

- Range and span adjustments.

- Instrument specifications. (Record.)

(4) Instrument operation — determination of input vs. output characteristics:

- Find three pneumatic connection ports, identify function of each. Note: IN is supply in.

- Apply 20-psi supply signal to port labeled IN or SUPPLY.

- Connect 0–30 psi standard gage to port labeled OUT or OUTPUT.

- Now apply a varying pressure to the signal input. Note the response of the transmitter to this variable signal.

- Set up a data table. Vary input signal in suitable increments. Record pressure input signal, transmitter indicator signal, and transmitter output signal.

- Plot input vs. output characteristic curve for this instrument.

(5) Disassemble circuit and move to the next instrument.

Pressure recorder-indicators — three different models.

(1) Visually examine the instrument. Record manufacturer, serial number or model number, and range and span information as found on the instrument nameplate.

(2) Carefully remove the chart and front plate from the instrument so that the instrument mechanism can be readily observed.

(3) Identify the following functional parts of the mechanism:

- Elastic pressure receiver element. What type is used? Record.

- Link and lever mechanism for signal transmission.

- Draw a sketch of the mechanical transmission mechanism from receiver element to pen.

- Determine approximately the mechanical amplification from receiver output to pen output. This requires that receiver unit be supplied with a 0–full range signal. Compare pen travel with receiver output. Record amplification.

(4) Reassemble recorder mechanism and apply signal to the mechanism. Check input signal with 0–30 psi standard gage. Prepare data table and record input vs. output (indicated on recorder chart).

(5) Plot input vs. output response curve for this recorder.

ANALYSIS AND REPORT

(1) Prepare input vs. output characteristic curves for each instrument tested.

(2) Rate these instruments for the following:

- Accuracy

- Speed of response

- Responsiveness

- Sensitivity

- Linearity

(3) Comment on the functional units or subassemblies that are common to pneumatic transmitter mechanisms.

(4) Describe in the form of conclusions your observations in working with these commercial instruments.

Differential Pressure Transmitter

OBJECTIVES

(1) To study three differential pressure measuring systems.

(2) To plot energy characteristic curves for the three different transmitters studied.

(3) To test transmitters for repeatability.

(4) To determine the response time of the transmitters.

EQUIPMENT

Solid-state pressure transducer module
Commercial pneumatic and electronic differential pressure transmitters
Power supply
DMM
Pressure source
Standard pressure gage or manometer standard
X-T recorder

PROCEDURES

See Figure A1-9.
Solid-state Differential Pressure Transmitter Module.

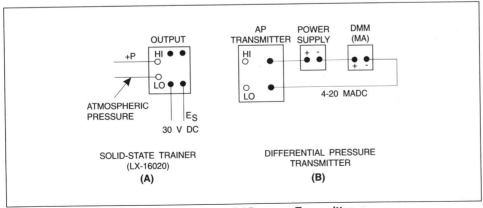

Figure A1-9.   Solid-State Trainer and Differential Pressure Transmitter

(1) Examine the electrical schematic of the chip.

(2) Prepare the energy transfer curve from a table of values. Input differential pressure 0–15 psi. From the energy transfer curve determine the transmitter sensitivity.

(3) Test transmitter for repeatability as follows: Test a couple of combinations of differential pressure where the differential pressure is kept constant. Example:

High 4.5 psi and low 2.5 psi $\Delta P = 2$ psi.

High 8.0 psi and low 6.0 psi $\Delta P = 2$ psi.

(4) Determine the response time for this transmitter by determining the time constant of the transmitter using the X-T recorder.

Differential Pressure Transmitter (Electronic).

(1) Select a specific commercial transmitter. Record model number, range, and span.

(2) Prepare energy transfer curve from range and span data.

(3) Take measurements at 0%, 25%, 50%, 75%, 100% of span. Plot.

(4) Test transmitter for repeatability.

(5) Determine response.

Pneumatic Differential Pressure Transmitter.

(1) Select a commercial pneumatic differential pressure transmitter. Record model number, range, and span.

(2) Repeat steps 2, 3, 4, 5 above with this transmitter.

ANALYSIS OF RESULTS

For each of the three transmitters:

(1) Comment on sensitivity.

(2) Comment on speed of response.

(3) Comment on transmitter repeatability.

(4) Indicate several possible applications.

Strain Gage

OBJECTIVES

(1) To study strain gages used in a static mode measurement of displacement (by monitoring strain in flexure).

(2) To use the same setup for a dynamic measurement.

(3) To calibrate a load cell unit.

THEORY

A solid wire has a resistance between its ends that is proportional to its length and inversely proportional to its cross-sectional area. If one knows the wire material, one can look up the resistivity, ρ, and compute the resistance as $R = \rho l/A$. This relationship is the basis for strain gage operation. If one stretches the wire, the l goes up and the A goes down, both slightly. These slight changes cause a slight change in the overall resistance, ΔR. Measuring this ΔR or using the strain gage in an unbalanced bridge arrangement will give an output indication that is proportional to strain.

Wire strain gages can be of two kinds: bonded, and unbonded. The gage glued to the steel rule in this experiment is purchased bonded to a strip of plastic. In application, the plastic is glued with epoxy or cyanoacrylate adhesive to the specimen whose strain is to be measured, and terminals are attached.

The unbonded strain gage has wires in free space. Usually, one cannot see the wires in unbonded applications as they are inside a protective enclosure. Statham force transducers are a commercial example of this kind of strain gage.

Load cells are applications of strain gage technology. The individual strain gages are usually configured in the form of a bridge. The output from the usually becomes the single leg of a Wheatstone Bridge measuring circuit.

EQUIPMENT

Strain gage trainer (strain gage mounted on a steel rule)
Digital multimeter
Storage oscilloscope
One-inch-depth micrometer assembly
Power supply
Optional: Tektronix TM 503 system: includes power supply DMM,
            differential amplifier, strain gage adapter, 6-conductor cable
Commercial load cell.

PROCEDURE

See Figures A1-10, A1-11, A1-12, A1-13, and A1-14.
Static Measurement Study

(1) Mount the depth micrometer on the cantilevered rule assembly. Have the spindle barely touching the rule.

(2) Prepare a table. Column 1 will be deflection, d, as caused by the depth micrometer pressing down on the rule. Column 2 is the calculated strain on the gage (see Figure A1-13). Column 3 will be the resistance measured (with as many places as your DMM will allow) at the specified deflection.

(3) Fill out column 1 with the following deflections: 0 to 0.400 in. in 0.050-in. increments. Calculate the strains for each deflection and enter in table column 2.

(4) Measure the resistance at each deflection with DMM. Note that it is recognized that this is an abuse of the micrometer subassembly.

(5) Using only the 0.400 in. data, calculate the gage factor, G. Typically, its value is between 1 and 2.

(6) Plot strain ($\Delta L/L$) vs. resistance change ($\Delta R/R$). This establishes the energy transfer curve for the strain gage system.

Dynamic Measurement Study
Using the Tektronix equipment assembly:

(1) Assemble the strain-gage adapter, power supply, differential amplifier setup. Set the adapter to 1 external arm. Set the power supply to 5.00 volts.

(2) The 6-conductor cable connecting the adapter to the power supply also brings the bridge output out to the two BNC jacks on the front of the power supply. These outputs should now be connected to the diff-amp inputs with two BNC-BNC cables.

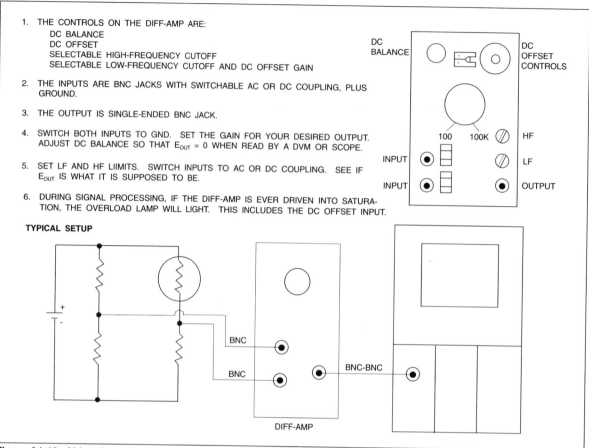

1. THE CONTROLS ON THE DIFF-AMP ARE:
     DC BALANCE
     DC OFFSET
     SELECTABLE HIGH-FREQUENCY CUTOFF
     SELECTABLE LOW-FREQUENCY CUTOFF AND DC OFFSET GAIN

2. THE INPUTS ARE BNC JACKS WITH SWITCHABLE AC OR DC COUPLING, PLUS GROUND.

3. THE OUTPUT IS SINGLE-ENDED BNC JACK.

4. SWITCH BOTH INPUTS TO GND. SET THE GAIN FOR YOUR DESIRED OUTPUT. ADJUST DC BALANCE SO THAT E_{OUT} = 0 WHEN READ BY A DVM OR SCOPE.

5. SET LF AND HF LIIMITS. SWITCH INPUTS TO AC OR DC COUPLING. SEE IF E_{OUT} IS WHAT IT IS SUPPOSED TO BE.

6. DURING SIGNAL PROCESSING, IF THE DIFF-AMP IS EVER DRIVEN INTO SATURA-TION, THE OVERLOAD LAMP WILL LIGHT. THIS INCLUDES THE DC OFFSET INPUT.

Figure A1-10. Abbreviated Instructions for AM-502 Diff-Amp Plug-In

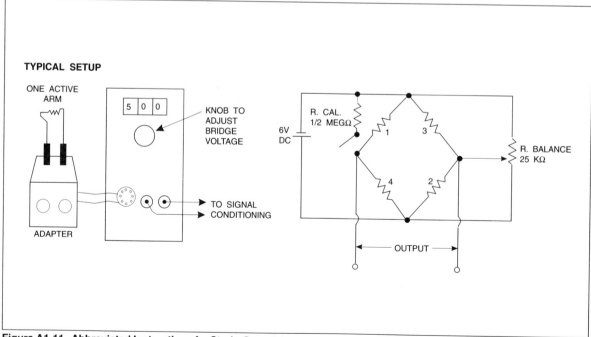

Figure A1-11. Abbreviated Instructions for Strain Gage Adapter and Transducer Power Supply

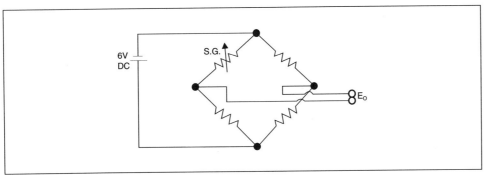

Figure A1-12. Bridge Output

(3) Set the diff-amp for a gain of 500. HF 3-dB point should be set to 1 kHz. LF 3-dB point should be set to DC OFFSET. This will allow correction of the residual bridge output at 0 deflection.

(4) Use the DC OFFSET coarse and fine controls to reduce the output voltage to 0 with no deflection.

(5) "Twang" the rule and note the decaying oscillation. Set the triggering controls for SINGLE SWEEP and NORM triggering.

(6) Record a twang on storage scope. Measure the frequency and the time it takes to decay to 63% of the initial amplitude.

(7) Calibration of commercial load cell. Review instruction manual for setup. Use standard weights for the generation of a calibrated force. Fill out calibration form and do an error plot of data obtained. How does your data compare to the specification?

ANALYSIS AND RESULTS

(1) Describe the nature of the curve of strain vs. resistance.

(2) Comment on the results using our training device vs. a commercial strain gage installation.

(3) Give some specific applications for strain gage transducers.

(4) What would be the effect of taping a weight to the end of the rule on the frequency of oscillation?

(5) Comment on the response of the strain gage transducer.

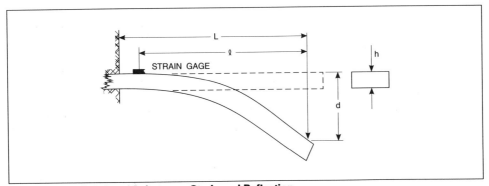

Figure A1-13. Relationship between Strain and Deflection

$$G = \frac{\Delta R/R}{\varepsilon}$$

WHERE:

ΔP = CHANGE IN RESISTANCE CAUSED BY STRAIN, E

P = UNSTRAINED RESISTANCE

E = STRAIN

G = GAGE FACTOR

EXAMPLE:

A STRAIN GAGE MEASURES 150.24 OHMS WHEN SUBJECTED TO 1200 MS. ITS UNSTRAINED RESISTANCE IS 150 OHMS. CALCULATE G.

$$G = \frac{.24/150}{(1200 \times 10^{-6})} = 1.33$$

EXAMPLE:

GAGE HAS A FACTOR OF 2.0. WHAT IS EXPECTED DR WHEN 1000 MΣ IS APPLIED? UNSTRAINED RESISTANCE IS 150 OHMS.

$$2.0 = \frac{\Delta R/150}{(1000 \times 10^{-6})}$$

ΔR = 0.3Ω

GAGE WILL READ 150.3 Ω

Figure A1-14. Strain Gage Factor

Electronic Pressure Transmitter

EQUIPMENT

Commercial electronic pressure transmitter
Standard pressure gage
DMM
Pressure switch

Power supply (24 volt DC)
Y-T recorder

PROCEDURE

(1) Carefully review technical literature or instruction manual on the transmitter.

(2) Examine instrument. Discern "how it works," i.e., theory of operation. What type of pressure transducer is used in the transmitter?

(3) From the manual, develop the following information:

- Block diagram of the various functions performed by this instrument, input to output.

- From the block diagram, draw a series of energy transfer curves that represent energy changes or transformations that occur within the instrument.

- Record input and output range, and span.

- Record accuracy statement.

- Record power requirements.

(4) Prepare a theoretical energy transfer curve, input to output. Assume the instrument is in perfect calibration.

(5) From the energy transfer data, determine instrument sensitivity.

(6) Now prepare the instrument for calibration and measurement.

(7) Check instrument calibration by sending in 10 pressure signals, 0 to full range, and record the output current or voltage values. Record in table form and plot on energy transfer curve.

(8) Verification of specification data:

- Record repeatability specification. Verify this spec with repeatability data (5 readings taken consecutively) taken at two different input pressures.

- Record accuracy data. Independent linearity — verify accuracy at 25% of span and 75% of span.

- Record time constant specification if available. Verify this spec using the Y-T plotter.

(9) Dynamic response. Apply a step change in pressure to the instrument while attached to the Y-T plotter and examine this instrument's response to the signal. Determine time constant.

(10) Return transmitter to original static condition.

(11) Repeat steps 1–10 with a second transmitter.

ANALYSIS AND RESULTS

(1) Show a measurement system block diagram with this transmitter as a part of the measurement system.

(2) Comment on the verification of instrument specification i.e., repeatability, accuracy, and speed of response.

(3) Comment on any other general observations.

Temperature

Liquid-In-Glass Thermometer Systems and Filled Thermal Systems Transmitters

OBJECTIVES

(1) To construct two volumetric water-filled thermometer systems.

(2) To plot column height vs. temperature for each system.

(3) To measure several known temperatures with these liquid-in-glass thermometers.

(4) To determine linearity, range, span, precision, accuracy, and time constant for each system.

(5) To compare sensitivity for each thermometer system.

(6) To determine time constant for commercial filled thermal system.

THEORY
Cubical Expansion Thermal Systems

Although almost everyone knows what is meant by the word "temperature" in everyday life, the measurement of temperature is not simple. As with so many other phenomena, it is not measured directly but rather must be interpreted in terms of how it affects other materials or their properties. While the temperature of an object is directly related to the thermal energy of the constituent particles, this fact is seldom used industrially. Rather, changes in dimension, pressure, resistance, color, or any one of a number of other relationships may be involved.

One of the very common results of changing the temperature of an object is to find that its dimensions and, hence, its volume have been changed. This is as true of liquids as of solids. When a solid changes into a liquid, there is a most pronounced change in volume. However, for many industrial temperature measurements, we use the change in volume of a liquid as a measure of its temperature change. By knowing the base or reference temperature, the volume and the temperature may be related at any desired point. For small temperature changes, the volume changes may be assumed to be linear. However, for extended ranges, there will be a divergence from linearity.

If the coefficient of expansion were strictly linear, which it is not in most cases, then the changes in volume at any given temperature might be expressed as:

$$L(1 + \alpha T) - L = \Delta V = \text{change in volume}$$

where:

L = original side of a cube
α = fractional change in dimension per degree temperature change
T = temperature change

Because α is a small quantity, when it is squared and cubed its values become so small that for most practical purposes they can be ignored. With these assumptions, if the above cubic expression is expanded and simplified by discarding the higher powers of α, then the change in volume = $V = (V)(3\alpha T)$ $V = V(1 + BT)$

The coefficient of volume expansion (B) nominally is equal to three times that for linear expansion (α).

Because the change in volume is obviously related to the original volume, the greater the amount of liquid involved, the greater will be the volume changes for any given temperature change.

Because a greater amount of fluid must be heated, and this will take more time because more heat must be transferred, greater volume changes and, thus, greater sensitivity in indicating temperature changes will be obtained at the expense of speed of response. Consequently, most designs become a series of compromises in which decreased sensitivity is accepted in order to decrease the response time.

In later parts of the experiment other factors governing the speed of response of a thermometer will be considered.

Gas Thermal Systems

When using all monatomic gases in the low-pressure ranges, it has been found that for constant volume conditions the pressure varies directly with the absolute temperature. Whether one uses the Kelvin or the Rankine scales for expressing the absolute temperature makes no difference. *If the absolute temperature is doubled while the volume is held constant, then the pressure of the confined gas will also be doubled.* Obviously, if the temperature is taken over any other range, the pressure will follow accordingly. This is in accordance with Charles' law for gaseous systems.

Because of this definite relationship between temperature and pressure, gas thermometers are often used. Moreover, they are exceedingly accurate while retaining the desirable characteristics of simplicity. As with other types of thermometers, actual commercial design is a compromise in which sensitivity is sacrificed for size and speed of response. Because of the temperatures involved, along with high pressures together with the need for chemically inert components, stainless steel may be used to hold the gas because of its mechanical strength and chemical stability. However, stainless steel is not a good conductor of heat and, hence, will impede the transfer of heat to and from the confined gas. Ideally, a thin silver container would have many of the desired characteristics, but practically it is too expensive or it may react undesirably with its surroundings.

While the type of container and its shape have no relationship to the temperature-pressure relationships, the speed of response is governed by the size and shape of the components used.

Vapor Thermal Systems

Vapor thermal systems follow Dalton's law of partial pressures. The relationship between temperature and pressure is nonlinear. In this respect, the vapor thermal system differs markedly from the gas-filled thermal system.

Vapor-pressure types of thermometers can be made more sensitive but are quite nonlinear. The pressure is a function of temperature only; it is independent of the amount of fill so long as the responsive fluid is present.

EQUIPMENT

Two extraction flasks: 100, 50 ml capacity
One standard lab thermometer
24-in. glass tubing (capillary)
One two-hole stopper, # 6
Constant temperature baths
One pan
Glass plug
Stop watch
Graph paper
Commercial filled thermal system thermometer
Set of standard thermometers
Meter stick

PROCEDURE

(1) Examine construction of simple liquid-in-glass thermometer (see Figure A1-15).

- Use 50-ml extraction flask, no. 6 two-hole stopper, 24-in. capillary glass tubing, glass plug, meter stick, and ice water. Prepare thermometer.

- Add enough ice water to the flask to thoroughly cover capillary tubing and fill flask to the brim. (Fluid should rise in capillary above stopper.)

- When water in the flask is at the same temperature as bath, add glass plug and depress plug so that fluid is visible above stopper in the capillary.

- Record liquid column height and bath temperature.

(2) Calibration of liquid-in-glass thermometer:

- Before beginning step 2, be sure that six bath temperature conditions have been established. (For example 30, 40, 50, 60, 70, and 80 degrees C).

- Now move the thermometer from bath to bath. Be sure to allow time for the thermometer to stabilize in each bath. Record bath temperature and capillary tube height.

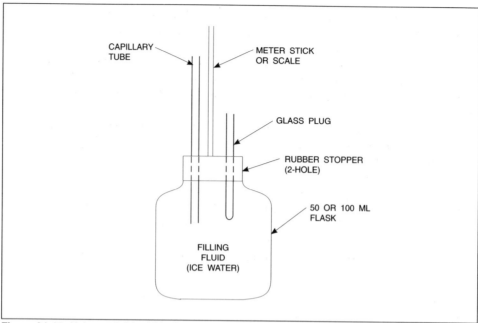

Figure A1-15. Volumetric Liquid-in-Glass Thermometer

- Prepare data table of column height vs. temperature. Plot for the thermometer.

(3) Determine room temperature with this thermometer. Record.

(4) Characteristics of this cubical expansion (liquid-filled) thermal system:

- Determine linearity of thermometer from data.

- Determine approximate range of thermometer. What are the limits? Why?

- Determine span of the thermometer.

- Take a second measurement of temperature in a constant temperature bath as an estimate of thermometer precision.

- Calculate thermometer sensitivity in inches/°C.

- Determine time constant of this thermometer. Use a step change of 40°C. Record time vs. temperature for the thermometer.

(5) Commercial vapor-filled and/or gas-filled thermal system.

- Thoroughly examine the commercial vapor-filled and/or gas-filled thermal system. Prepare a block diagram of this system, indicating functional parts.

- Determine time constant for this commercial vapor-filled and/or gas-filled thermal system.

- Prepare calibration form for this instrument. From calibration data prepare and error plot. Note instrument accuracy as compared to instrument specifications.

ANALYSIS AND RESULTS

(1) Calculate sensitivities for liquid-in-glass thermometer system.

724

(2) Determine time constant for each thermometer system.

(3) How does the commercial instrument compare to instrument specifications?

(4) Compute sensitivity for volumetric liquid-in-glass thermometers.

(5) Compare all systems for speed of response time constants.

(6) List several limitations to these types of temperature measurement systems.

Resistance Temperature Detectors (RTDs)

OBJECTIVES

(1) Plot the characteristic temperature vs. resistance curve for temperature values from 20°C to 100°C on graph paper.

(2) From this graph, determine the sensitivity of your bulb in ohms/°C.

(3) Determine temperatures of five different temperature conditions.

(4) Compare electrical temperature determination with those of lab standards.

(5) Determine repeatability of temperature bulb.

(6) Determine time constant of temperature bulb.

(7) Develop a measuring system using signal conditioning transmitter with the RTD, and record system output.

THEORY

The sensing element of the RTD is composed of a precision, noninductively wound, strain-free coil of reference-grade platinum wire. The resistance of this coil of wire changes with temperature in a highly reproducible manner. As such, these temperature elements are widely used by industry because of their accuracy and repeatability. Resistance temperature bulbs are generally attached to signal conditioning transmitters that convert resistance changes into proportional 4–20 mA current signals. Signal conditioning transmitters may be calibrated for a variety of temperature ranges.

EQUIPMENT
Bench power supply
DMM
Assorted electronic parts
Platinum type 311SS RTD
X-T plotter, temperature baths
Standard thermometers, graph paper
RTD temperature transmitter such as RIS
Model SC- 2374 signal conditioner or equal
Diwar flask
Recorder
Digital temperature indicator (used as a temperature standard)

PROCEDURES

(1) From the information in Table A1-2, prepare a plot of temperature vs. resistance for the platinum RTD bulb. This is the energy transfer characteristic for this detector.

(2) From this graph, determine the theoretical (average) sensitivity of the bulb in ohms/°C. Record.

Table A1-2. Resistance Temperature Table for Platinum Resistance Thermometer Elements (100 ohms at 0°C)

Temperature °C	Resistance of Element Ohms	Interchangeability		Temperature °C	Resistance of Element Ohms	Interchangeability	
		±Ω	±°C			±Ω	±°C
−200	16.99	0.22	0.50	230	188.00		
−190	21.43			240	191.68		
−180	25.83			250	195.35	0.09	0.25
−170	30.18			260	199.02		
−160	34.49			270	202.67		
−150	38.77	0.16	0.38	280	206.31		
−140	43.01			290	209.93		
−130	47.22			300	213.55	0.11	0.30
−120	51.40			310	217.15		
−110	55.56			320	220.74		
−100	59.69	0.10	0.25	330	224.32		
−90	63.80			340	227.89		
−80	67.89			350	231.45	0.12	0.35
−70	71.96			360	234.99		
−60	76.01			370	238.52		
−50	80.04	0.06	0.14	380	242.05		
−40	84.06			390	245.56		
−30	88.06			400	249.05	0.14	0.40
−20	92.06			410	252.54		
−10	96.03			420	256.01		
0	100.00	0.02	0.06	430	259.48		
10	103.96			440	262.93		
20	107.90			450	266.37	0.16	0.45
30	111.83			460	269.79		
40	115.75			470	273.21		
50	119.66	0.02	0.06	480	276.62		
60	123.55			490	280.01		
70	127.44			500	283.39	0.42	1.25
80	131.31			510	286.76		
90	135.17			520	290.11		
100	139.02	0.04	0.10	530	293.46		
110	142.86			540	296.79		
120	146.68			550	300.12	0.47	1.4
130	150.50			560	303.43		
140	154.30			570	306.73		
150	158.09	0.06	0.15	580	310.01		
160	161.87			590	313.29		
170	165.64			600	316.55	0.49	1.5
180	169.39			610	319.80		
190	173.14			620	323.04		
200	176.87	0.07	0.20	630	326.27		
210	180.59			640	329.49		
220	184.30			650	332.70	0.52	1.6

NOTE: For 200-ohm platinum elements multiply above resistance values by 2.
For 500-ohm platinum elements multiply above resistance values by 5.

Quadratic Approximation: $R = -0.00006\,T^2 + 0.3962\,T + 100.000$

(3) Attach bulb (white and red lead) to DMM and determine resistance equivalent of room temperature conditions. Establish best value of resistance and calculate room temperature using Table A1-2.

(4) Establish five (5) specific temperature conditions using water as a fluid medium as follows: Ice bath (Diwar flasks), room temperature pan, 50°C, 60°C, 70°C constant temperature baths, respectively. Immerse bulb in each bath and record measured resistance. Determine bath temperature from Table A1-2. Now measure bath temperature with temperature standard. Show comparison of values in a table.

(5) Take five successive temperature readings in the same bath to test bulbs for repeatability. Record results.

(6) Time constant determination using Y-T Plotter:

- Select a Y-T Plotter and several sheets of graph paper.

- Prepare two temperature baths using pans — hot and cold tap water in baths will be sufficient. Allow baths to stabilize in temperature (or two baths at two different temperatures). Make temperature difference at least 40°C.

- Determine the time constant by measuring millivolt change across a standard 250-ohm resistor in a bridge circuit. It is necessary to do this because of the low-level signal change involved and the desire not to introduce self-heating in the bulb due to high current values. Therefore, carefully construct the measuring circuit as shown in Figure A1-16. Leave battery disconnected until ready to take measurements.

- Now set up the Y-T Plotter to determine the time constant for the RTD.

- Energize the circuit; zero the plotter with the bulb in the cold water bath. (Be sure pen is just above zero of the x and y axes.

- Make a test run with the pen in the up position. Observe the plotter response as the bulb is transferred from the cold to the hot bath.

- When you are satisfied that the plotter is giving you the best representation possible, place the pen down on paper and draw the time constant curve.

- Calculate the time constant (63.2% value) in time for this temperature bulb and temperature difference.

(7) RTD temperature measurement system:

- Review the product data bulletin for the RIS Model 2374 or equal signal conditioning transmitter.

- Set up measuring circuit as shown in Figure A1-17

- Prepare energy transfer for the signal conditioning unit only.

- Using this curve and Table A1-2, determine room temperature, and two (2) other temperature conditions.

- Check temperatures obtained with lab standard.

ANALYSIS AND RESULTS

(1) Comment on the linearity of the temperature vs. resistance curves for this platinum resistance temperature bulb.

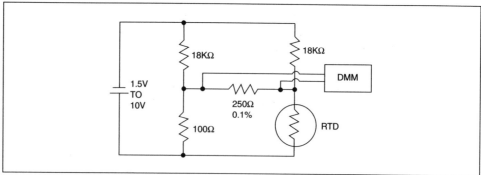

Figure A1-16. RTD Test Connections

(2) Explain the purpose of the two white lead wires on this bulb assembly.

(3) Comment on repeatability of this RTD.

(4) How does the value of the time constant compare to that published in the specifications? Why might the test sample be different from the published value?

(5) Comment on system temperature measurements.

(6) Name several instruments that could receive the output of the signal conditioning transmitter.

Thermistors

OBJECTIVES

(1) Plot energy transfer characteristic curves of temperature vs. resistance on semilog (three-decade) graph paper.

(2) To determine average sensitivity for the probe from the graphs.

(3) Measure actual resistance values for five different temperature conditions set up in the laboratory. Plot these values on the energy transfer curve.

(4) Compare resistance values obtained for actual temperature conditions with those indicated as specifications.

(5) Determine repeatability of temperature probe.

(6) Determine time constant for the probe.

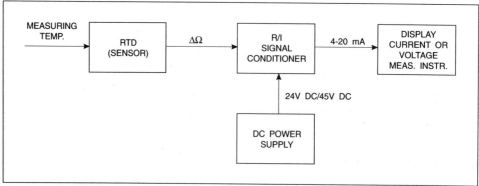

Figure A1-17. RTD Measuring System Block Diagram

(7) Be prepared to make a statement about thermistor accuracy from this investigation.

THEORY

Thermistors (thermally sensitive resistors) are mixtures of semiconductor oxides molded into a variety of physical configurations. Thermistors are very sensitive to changes in temperature; in fact, they are the most sensitive transducers for temperature measurement. Because of their high sensitivity, they are generally used in narrow span applications. The TCR (temperature coefficient of resistance) for the thermistor is generally negative and is a very high value when compared to the TCR of RTDs. Although repeatability has been a problem for some thermistor manufacturers, it will be noted in this laboratory that thermistor repeatability is comparable with RTDs.

Thermistors used in this laboratory are unshielded; therefore, it can be predicted that the time constant will be quite fast and is related solely to the mass of the thermistor probe.

EQUIPMENT

    DMM
    Electronics parts kit
    Keystone thermistor probe or equal
    HP plotter
    Standard thermometers
    Five temperature baths
    Three-decade graph paper

PROCEDURES

(1) Select thermistor probe — Keystone Series 25-73216 or equal — and specifications.

(2) From specifications, plot temperature vs. resistance values on three-decade graph paper.

(3) Determine average sensitivity for the probe unit. Keystone unit is _____ ohms/°C. See Table A1-3.

(4) Determine resistance values for each of the five temperature conditions established using DMM as the measuring instrument. Plot these values on the graphs previously drawn. How do the values compare to specifications? Conditions: 0°C, 30°C, 40°C, 50°C, 70°C, and 80°C.

(5) Determine repeatability of the thermistor by checking resistance under the same temperature conditions, at least five times. Allow time for device to stabilize.

(6) Time constant determination. Construct a simple voltage divider circuit using 250-ohm resistor or 100-ohm or 51.12, 1.5V battery and Keystone thermistor probe (see Figure A1-19). Connect the Y-T plotter across the terminals of the 250-ohm sampler resistor. Using room temperature water and water about 50°C, determine the time constant of this thermistor. Calculate the millivolt change that will occur across the 250-ohm resistor as the thermistor is moved from the initial temperature condition to the new temperature condition. If time permits, try the same determination for another thermistor probe.

(7) Examine other thermistor probe configurations.

Specifications

ELEMENT RESISTANCE:

Resistance at 32°F	Resistance Change at 32°F
100-ohm platinum	0.22 ohms per °F
200-ohm platinum	0.44 ohms per °F
500-ohm platinum	1.10 ohms per °F
120-ohm nickel	0.44 ohms per °F

ACCURACY:

Platinum Elements

±0.1°F or ±0.1% of the temperature being measured, whichever is greater, from 32°F to 900°F for ¼″ diameter standard length sensors.

±¼°F or ±¼% of the temperature being measured, whichever is greater, from −325°F to 1200°F.

(This accuracy applies to all sensors which are not ¼″ in diameter or are non-standard length.)

(Accuracy or interchangeability to the master resistance vs. temperature table expressed in degrees is given in the resistance vs. temperature tables on pages 8–9 for the 0.1% elements.)

Nickel Elements

±½% resistance tolerance.

Optional Higher Accuracies for Platinum Elements

±0.1°F over any specified span of 50°F or less from 32°F to 900°F. (For 500-ohm platinum elements only.)

±¼°F over any specified span of 75°F or less from 32°F to 900°F.

±½°F over any specified span of 100°F or less from 32°F to 900°F.

Optional Lower Accuracy for Platinum Elements

See separate catalogs and spec. sheets.

REPEATABILITY:

±0.1°F over range 32°F to 900°F.

Matched Pairs

Matched pairs for temperature difference applications have the following accuracies:

Platinum elements are matched to each other within 0.1°F or 0.1% of the temperature being measured, whichever is greater, from 32°F to 900°F.

Nickel elements are matched to each other within ½% of the resistance at the temperature being measured.

TIME CONSTANT

Two and one-half seconds for 63.2% response in water moving at 3 feet per second for the type "D" sheath (⅛″ diameter fast response tip).

Five seconds for the type "A" sheath (¼″ diameter).

SELF HEATING:

Typically 50 milliwatts power per °C error in 20°C water moving at 3 feet per second.

INSULATION RESISTANCE:

With dry external surfaces the insulation resistance between any lead wire and the metal sheath will be as follows:

Temperature of Sheath (°C)	Insulation Resistance (Minimum)
20	200 megohms at 100 Vdc
225	100 megohms at 100 Vdc
500	10 megohms at 100 Vdc
650	5 megohms at 100 Vdc

Figure A1-18. Typical RTD Specifications

TEMPERATURE
LIMITS:

Platinum Elements
Standard limits: –325° F to 900° F (element enclosed in type 316 stainless steel sheath).

Extended limits: –325° F to 1200° F (element enclosed in Inconel sheath) (higher temperature limits available).

Nickel Elements
–40° F to 500° F. (The life expectancy and performance of nickel elements improves if upper temperature limit is kept below 500° F.)

Hermetic Seal and Lead Wires
This portion of element assembly is rated to 400° F (higher ratings available).

Figure A1-18 (continued). Typical RTD Specifications

Table A1-3. Keystone Carbon NTC Thermistor Probes

Unit No.	0°C	25°C	37.8°C	50°C	75°C
1	147,743	44,451	25,410	15,479	6,208.3
2	151,579	44,495	25,407	15,478	6,178.9
3	147,307	44,489	25,426	15,482	6,187.1
4	151,344	44,441	25,388	15,450	6,197.5
5	151,762	44,490	25,388	15,454	6,174.8
6	149,956	44,499	25,424	15,479	6,190.1
7	151,000	44,494	25,506	15,513	6,261.1
8	146,968	44,456	25,413	15,483	6,197.0
9	151,588	44,503	25,419	15,459	6,178.1
10	144,685	44,465	25,431	15,500	6,212.5
11	151,200	44,453	25,406	15,480	6,195.5
12	151,161	44,485	25,440	15,504	6,245.3
13	151,234	44,436	25,369	15,458	6,170.9
14	147,000	44,481	25,424	15,473	6,183.3
15	151,436	44,444	25,384	15,474	6,174.4
16	151,474	44,453	25,376	15,441	6,168.1
17	151,361	44,465	25,397	15,450	6,177.4
18	151,222	44,485	25,418	15,488	6,209.0
19	151,517	44,480	25,408	15,589	6,210.1
20	151,453	44,500	25,426	15,506	6,201.2
21	151,464	44,434	25,366	15,450	6,165.0
22	151,402	44,473	25,419	15,456	6,182.5
23	151,635	44,495	25,388	15,549	6,183.9
24	150,282	44,436	25,375	15,440	6,172.4
25	151,370	44,502	25,432	15,522	6,201.9
26	149,213	44,476	25,394	15,451	6,171.8
27	151,355	44,479	25,422	15,485	6,199.4
28	150,800	44,487	25,434	15,515	6,231.3
29	151,245	44,511	25,443	15,503	6,211.5
30	151,306	44,446	25,389	15,456	6,182.9
31	151,213	44,481	25,416	15,499	6,208.5
32	150,717	44,480	25,420	15,488	6,199.4
33	144,566	44,507	25,419	15,464	6,186.5
34	151,337	44,476	25,425	15,502	6,255.6
35	151,412	44,444	25,372	15,444	6,168.8
36	151,586	44,498	25,407	15,475	6,186.5
37	148,777	44,460	25,402	15,442	6,166.9
38	151,168	44,437	25,386	15,483	6,184.7
39	151,399	44,471	25,390	15,472	6,179.7
40	151,162	44,438	25,402	15,479	6,195.4

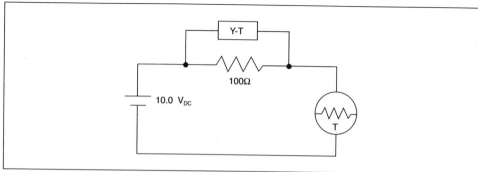

Figure A1-19. Thermistor Test Configuration

ANALYSIS AND RESULTS

(1) Comment on the linearity of the temperature vs. resistance curves for each of the thermistor probes tested.

(2) How do the sensitivity values of these probes compare to those of the RTDs?

(3) Comment on the limitations of each of the measurement instruments used to determine resistance of these probes. Which instrument is best? Worst? Why?

(4) Comment on repeatability as determined with available measuring systems.

(5) Compare time constant values with those of RTDs.

Thermocouple Transmitter

Objectives

(1) To plot the energy transfer characteristic curves for two thermocouple materials: Type J (iron-constantan) and Type K (Chromel-Alumel™) from the tables in the range of 0°C to 100°C.

(2) To determine thermocouple sensitivity in mV/°C for each of these materials from the above plots.

(3) To take specific measurements of constant temperature conditions using Type K and J thermocouples. To compare emf outputs with table values by plotting on the same graph.

(4) To determine repeatability of the thermocouple elements.

(5) To determine time constant for each of these thermocouple materials using Y-T plotter.

(6) To analyze a thermocouple measurement system using a commercial thermocouple transmitter and thermocouple element.

THEORY

For well over a hundred years, it has been known that when the junction of two dissimilar metals was heated while their other ends remained cool, a small electromotive force would be created. Because these potentials are measured in millivolts, great care must be exercised in measuring them, since the degrees per millivolt may be large, particularly with some of the thermocouples made of platinum and its alloys. Consequently, an error of but one millivolt may establish an error of 50 or 100 degrees. For temperatures below 2000 degrees, many combinations of thermocouple wires may be used. Their performance over the years has been carefully determined; so it is quite simple to go from temperature difference to emf or vice versa. Perhaps it is of greater significance to consider some of the factors that may influence the response of the thermocouple and modify the response, either in terms of magnitude or time.

A thermocouple produces an electrical output that is proportional to the actual difference between its heated junction and its cold one. If the hot junctions have a significant mass or if heat is transferred to them slowly because of some protective tube, the electrical responses may lag far behind the environment in which the thermocouple assembly has been placed.

For this reason, the hot junction should be made as small as possible, consistent with its performance.

When, for reasons of strength or protection against undesired chemical deterioration, a thermocouple must be protected by a sleeve or housing of some sort, each added element will both delay and attenuate the response of the thermocouple. To reduce the time constant of the combination to a minimum, means should be provided so as to provide the maximum rate of heat transfer to and from the thermocouple. The high-temperature thermocouples generally are made of platinum and its alloys. These require the protection of an impervious tube, which, in turn, requires mechanical protection. Thus, commercial thermocouple assemblies consist of a number of resistor-capacitor groups that both attenuate and delay the response.

EQUIPMENT
Thermocouple assemblies, type J and K
Diwar flask
Constant temperature bath
Standard thermometers
Y-T plotter
Graph paper
DMM
Thermocouple tables IPTS 1968: 20-110°C
Thermometers
Thermocouple transmitter (commercial)

PROCEDURE

(1) Determination of energy transfer characteristic. Plot temperature vs. emf output for K and J thermocouples. Plot approximately 10 points from 0°C to 100°C. Use manufacturer's tables.

(2) From the plots above, determine the sensitivity of each of the thermocouple (type J and K) in mV/ C.

(3) Set up the thermocouple measuring circuit as indicated in Figure A1-20. By using two ice baths or breakers and Diwar flask, verify that for the same temperatures at both junctions no output will be obtained from the circuit. Record observations.

(4) Measure the temperature at six (6) different but constant temperature conditions. Use both J and K type thermocouples. Plot values obtained on transfer characteristic curves. How do values compare to those plotted in No. 1?

(5) Determine repeatability of each thermocouple using room temperature water bath as the standard condition. Make five (5) successive temperature determinations.

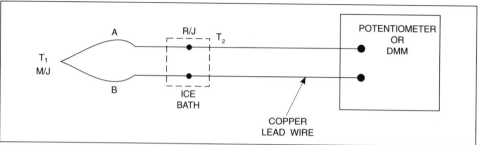

Figure A1-20. Keystone Carbon NTC Thermistor Probes

(6) Determine the time constant of one thermocouple sensor using the Y-T plotter. Obtain response curves. Calculate time constant.

(7) Build a thermocouple measuring system using the commercial transmitter. Using the calibration form, prepare calibration and generate and error plot graph. Compare results with published data for the transmitter. Now select two different temperature conditions. Compare measurements with standard thermometers. Determine system accuracy.

ANALYSIS AND RESULTS

(1) Describe the character of the energy transfer characteristic curve obtained in Step 1.

(2) Compare sensitivities of each thermocouple material. Which is most sensitive?

(3) State comparison of theoretical emf values (tables) as compared to actually measured values for specific temperatures.

(4) How do time constant values of thermocouples compare to those for resistance temperature bulbs, RTDs, and thermistors?

(5) From your results, rate thermocouples in accordance with span, range, accuracy, repeatability, sensitivity, and speed of response.

(6) Show energy transfer curve for the thermocouple measuring system. (Plot in engineering units.)

Displacement

Linear Variable Differential Transformer (LVDT)

Objective

(1) To study the operation and performance characteristics of the linear variable differential transformer (LVDT).

THEORY

The LVDT is a basic displacement-to-AC voltage transducer. The LVDT is a transformer device with a single primary winding and two secondary windings. The primary winding is placed exactly between two identical secondary windings. An iron core is positioned to couple flux to each of the windings. If the iron core is exactly centered, the flux in the two secondary windings is equal, thus producing no output from the two secondaries. If the core is moved away from the center, more flux is coupled into one and less into the other. The resultant is an output. This output is taken from the secondary windings connected in series opposition. Thus, when the core is centered, the output voltage is zero, and, if it is not centered, the output voltage is proportional to the amount of displacement.

EQUIPMENT

Schaevitz E 100 LVDT or equal (see Figure A1-23)
One-inch micrometer assembly
Oscilloscope
DMM
SK-10 circuit boards
Electronic parts:
One 0.1-μf capacitor
Two 100-μf capacitors
Two Ge diodes
One 470-ohm resistor
Two 2.2K resistors

PROCEDURE

(1) Record equipment numbers.

(2) Wire the LVDT according to the diagram in Figure A1-21.

(3) Plug in LVDT and measure primary voltage with the DMM. Record.

(4) Drop core in place and adjust micrometer for a null. Record micrometer reading. This is the null position.

(5) Make voltage readings on the DMM for 0.025-inch increments above the null and below the null. Note since the DMM is set to AC volts, it always reads positive regardless of the core position. Record data in a table.

(6) Plot the energy transfer characteristic of the LVDT, from the data in the table.

(7) Use the oscilloscope to observe the phase change that occurs at null. Record.

(8) Build a circuit that will give an output voltage (DC) whose polarity will indicate whether the core is above or below the null position. Circuit in Figure A1-22 will accomplish this.

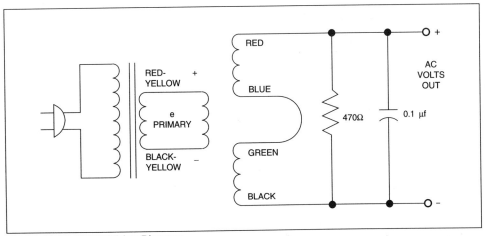

Figure A1-21. LVDT Wiring Diagram

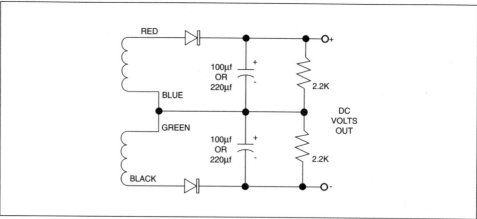

Figure A1-22. Null Measurement Circuit

(9) Use the circuit in Figure A1-22 to take data for the same 0.025-inch increments above and below the null. Use the 0.300 inch above and below null as limits. Record data in a table.

(10) Plot a curve relating the output voltage to the input displacement. Compare this energy transfer curve with the one plotted in Step 6.

SCHAEVITA "E"– LINE LINEAR VARIABLE DIFFERENTIAL TRANSFORMER DATA

A COMMON METHOD OF MOUNTING LVDT'S IS TO USE A SPLIT BLOCK, PREFERABLY ON NONMAGNETIC MATERIAL BORED TO FIT THE OUTSIDE DIAMETER OF THE UNIT. A CORE EXTENSION ROD HAVING A #4-40 THREAD AND MADE OF NONMAGNETIC MATERIAL IS THEN ATTACHED TO THE CORE. THIS CAN THEN BE ATTACHED TO A MICROMETER SHAFT OR ANY OTHER DISPLACEMENT MEASURING DEVICE.

ELECTRICAL CONNECTIONS AND CALIBRATION

TO OBTAIN DIFFERENTIAL OUTPUT FROM THE LVDT, CONNECT THE PRIMARY (YELLOW-RED AND YELLOW-BLACK LEADS) TO A 6-VOLT, 60-HZ REGULATED POWER SUPPLY, AND THE SECONDARIES (RED AND BLACK LEADS) TO THE OUTPUT MEASURING INSTRUMENT. THE SECONDARY MIDPOINTS (BLUE AND GREEN) SHOULD BE CONNECTED TO EACH OTHER.

ADJUST THE DISPLACEMENT MEASURING DEVICE, WHICH IN TURN WILL POSITION THE CORE IN THE LVDT, FOR MINIMUM READING ON THE OUTPUT MEASURING INSTRUMENT. THIS MINIMUM OUTPUT OCCURS WHEN THE CORE IS CENTERED IN THE LONGITUDINAL AXIS OF THE TRANSFOER AND IS THE NULL OUTPUT POSITION. DISPLACEMENT OF THE CORE FROM THIS POSITION RESULTS IN OUTPUT VOLTAGES PROPORTIONAL TO THE DISPLACEMENT (WITHIN THE LINEAR RANGE OF THE LVDT). WITHIN THIS SPECIFIED RANGE, THE LINEARITY WILL BE WITHIN 1 PERCENT.

SPECIFICATIONS

	E100
LINEAR RANGE	+0.100 INCH
OUTPUT IMPEDANCE	280 OHMS
OUTPUT VOLTAGE (FULL RANGE)	0.3 VOLTS WITH 500-OHM LOAD
INPUT POWER	0.52 WATTS

Figure A1-23. Schaevitz LVDT Specifications

ANALYSIS AND RESULTS

(1) What does infinite resolution mean? Give examples of infinite resolution.

(2) Why does the core get warm in the LVDT?

(3) Calculate the transducer constant for the first setup.

(4) Calculate the transducer constant (K) for the second setup. K = millivolts/inch displacement.

Flow

Positive Displacement Flowmeter

OBJECTIVES

(1) To disassemble, examine, identify functional parts, diagram, and then reassemble the nutating disc water meter mechanism.

(2) To calibrate a gallon container using volumetric flasks and buret tube.

(3) To check the accuracy of calibration of the nutating disc water meter on the flow trainer.

THEORY

For accurate flow measurements the positive displacement type of quantity flowmeter is a preferred measurement device. The most common of all meters of this type is the water meter. This meter employs a "nutating disc" in its design. The disc is inclined to the axis about which it revolves. A radial vane provides a sealing chamber for volumes both above and below and in this manner provides two separate chambers, one being filled while the other is being emptied. Each rotation of the disc is transmitted to a counter, and the counter records the passage of a definite volume of fluid. There is virtually no leakage from one chamber into the other when the meter is in good operating condition. Because of its efficiency and its excellent accuracy, it shall be used for a standard of reference in flow measurement experiments.

With mechanical types of readout devices, friction of stuffing boxes and related parts may constitute a major problem.

In order to determine the accuracy of a flowmeter, it is necessary to test the meter at several different flow rates. By measuring (1) the amount of fluid that passes through the meter and (2) the time required for this fluid to pass, the rate of flow can be determined. If one measures (volumetrically) or weighs the fluid that has been metered, its value can be compared with that registered by the meter. The accuracy of this calibration is dependent upon the accuracy of the weight or volume determination and the accuracy of the time determination instrument. Both of these variables are traceable to primary standards.

EQUIPMENT

    Flow trainer or equal
    Fluid flow source
    Nutating disc meter
    Gallon container
    Volumetric measuring glassware
    Large tank, 8–10 gal. capacity

Procedure

Calibration of a one-gallon container and large tank.

(1) Calibrate an ordinary one-gallon container using the volumetric flasks and buret tube. Using tape on the outside of the gallon container, mark the calibration point. 1 gal = 3785 ml

(2) Using the calibrated one-gallon container, now calibrate the large tank in one-gallon increments up to 8 gallons.

Determination of Accuracy of the Nutating Disc Meter Using Calibrated Tank as a Standard.

(1) Now discharge fluid into the calibrated tank from the flow trainer. Review the discharge procedure.

(2) Now discharge one gallon of fluid at a time into the tank, noting the meter reading after each gallon has been discharged.

(3) Prepare an accuracy statement for the meter on your trainer.

Nutating Disc Meter (Quantity Flowmeter).

(1) Record serial number and manufacturer of nutating disc meter.

(2) Completely disassemble the meter, being careful to put parts in a logical order so the meter may be reassembled.

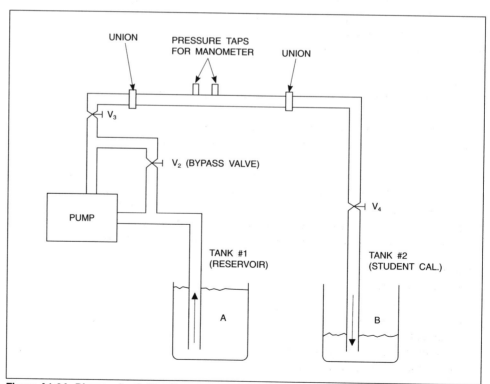

Figure A1-24. Diagram for Calibration

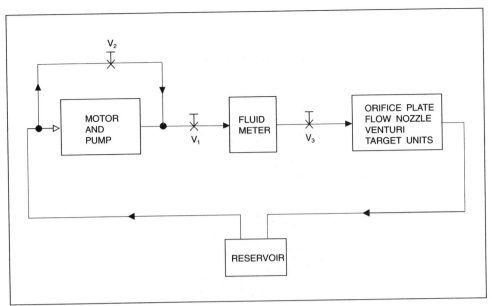

Figure A1-25. Block Diagram

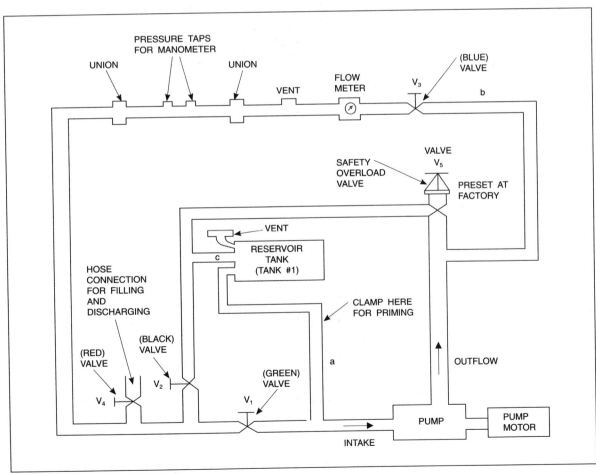

Figure A1-26. HTS LFL-1 Flow Unit

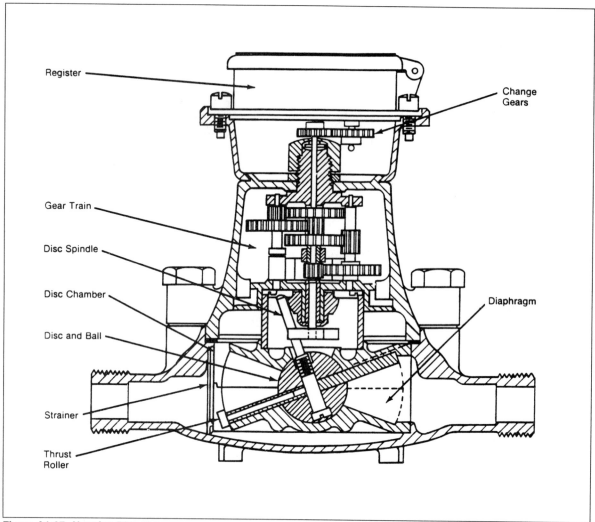

Figure A1-27. Nutating Disk Meter

(3) Diagram this meter mechanism and identify on the diagram all the functional parts of the mechanism.

(4) Determine the sensitivity of the nutating disc meter.

(5) Reassemble the mechanism (carefully).

ANALYSIS AND REPORT

(1) Comment on the accuracy of the scheme used for checking calibration of the nutating disc meter.

(2) Discuss accuracy of the meter. Published accuracy for this type of mechanism is +1.5 to 2.0%.

(3) Why is it necessary to remove as much air as possible from the flow trainer before taking measurements?

(4) What is the sensitivity of the nutating disc meter?

(5) Include labeled diagram of nutating disc meter in laboratory report.

(6) For each major part of the nutating disc meter identified in the diagram, give the function performed by the item.

Variable Area Meters

OBJECTIVES

(1) To examine the rotameter and identify functional parts of this meter.

(2) To check the calibration of the meter against the calibration of the nutating disc flowmeter on the flow trainer.

(3) To examine the purge meter and identify functional parts of this meter.

(4) To prepare a calibration scheme and check calibration of the purge meter.

(5) To examine the open channel flow installation.

(6) To check calibration of the Parshall flume using current meter as a standard.

(7) To measure flow using the rectangular weir configuration.

THEORY

Rotameters

As a fluid passes through a pipe, it possesses the tendency to carry objects with it. This ability varies with the speed of the fluid, the shape of the object, and the viscosity of the fluid. Thus, with suitable design, it is possible for any given rate of fluid flow to develop some definite force that will just equal the gravitational pull or some other desired force. For the minimum flow rate, the speed of the fluid would be such that even with the reduced area and the resulting increase in speed, the buoyant and drag forces would just equal the downward forces. Any increase in flow would increase the upward forces, with the other forces remaining constant.

Consequently, the "float" would move up until (the speed of the fluid at each higher section being reduced because of the greater cross-sectional area) once again it finds a position where the upward components equal the downward components. Thus, it is essentially correct to say that at each level where the indicator is at rest, the fluid flow rate is the same. Because the area of the tube increases with the square of the radius, it requires a ratio of radii of about 3 in order to provide for a flow rate change of 10 to 1 (which is the normal range of the device). Moreover, the flow rate is almost linear with the movement of the indicator. This is not true of most other designs. Also, devices that depend upon restrictions for their principle of operations are often limited to flow rate changes of 3 or 4 to 1. Rotameters have a rangeability of 1:10, where head flowmeters have rangeabilities of 1:3 or 1:4.

Parshall Flume

The Parshall flume developed by R. L. Parshall is an open channel variable area flow device. This device is a special type of Venturi flume. As fluid reaches the converging section of the flume, the level in the channel rises. The change in height of the flume is related to the volume of fluid flow through the device.

EQUIPMENT

Commercial rotameter
Flow trainer or fluid flow source about 10 gallons per minute rate
Purge meter
Stop watch
Volumetric flasks
Open channel flow trainer
Height gage
Velocity flow standard

PROCEDURE

Rotameter Study

(1) The commercial rotameter has been mounted on the flow trainer. Vary the flow rate and observe the action of the float mechanism.

(2) Using the stop watch and quantity flowmeter (nutating disc meter), check the calibration of the rotameter against the quantity flowmeter in five equal increments. Record data in a data table.

(3) Disassemble a rotameter and examine functional parts of the meter.

(4) Refer to Figures A1-28 and A1-29, exploded view and rotameter diagram. Complete the parts identification and attach to your report.

Purge Meter Study

(1) Examine the purge meter mechanism. Identify on a sectional diagram of this mechanism the following functional parts: needle valve, float, tapered tube, ball check valve, inlet, outlet.

(2) Construct apparatus for metering fluid flow with the purge meter. The purge meter is calibrated for water at flow rates up to 10 gallons per hour. This is a rather low flow rate.

(3) Check calibration of the purge meter using a known quantity of fluid flow and time. Catch a quantity of fluid in a volumetric flask. Record time and flow rate indicated on the purge meter. Give an accuracy statement for the purge meter based upon your results.

Open Channel Flow Installation — Rectangular Weir: Parshall Flume

(1) Diagram the open channel flow installation with the Parshall flume restrictor placed in the channel.

(2) Identify on diagram the following parts of the flume: converging section, diverging section, stilling well, throat area.

(3) Set up current meter or other flow velocity standard in the flow channel. With this instrument it is possible to make a determination of the velocity of flow of the fluid stream. To determine flow rate (Q) one also needs the area of the channel. Therefore, a height determination must be made for the channel at the point where the velocity is determined. Now apply the formula: $Q = $ (velocity $\times$ area) to determine the flow rate.

(4) Without changing flow conditions now determine channel height. Record height value.

(5) Repeat procedures 3 and 4 for two more different flow conditions. Be sure to measure height of the channel both in the stilling well and above the current meter.

(6) Plot flow rate against channel height for the rectangular weir. Compare results with those measured.

ANALYSIS AND RESULTS

(1) In your own words explain why the flow devices used in this laboratory exercise are classified as variable area flowmeters.

(2) Prepare accuracy statements for each of the devices tested.

(3) Which of the devices tested shows a nonlinear response to fluid flow? Why?

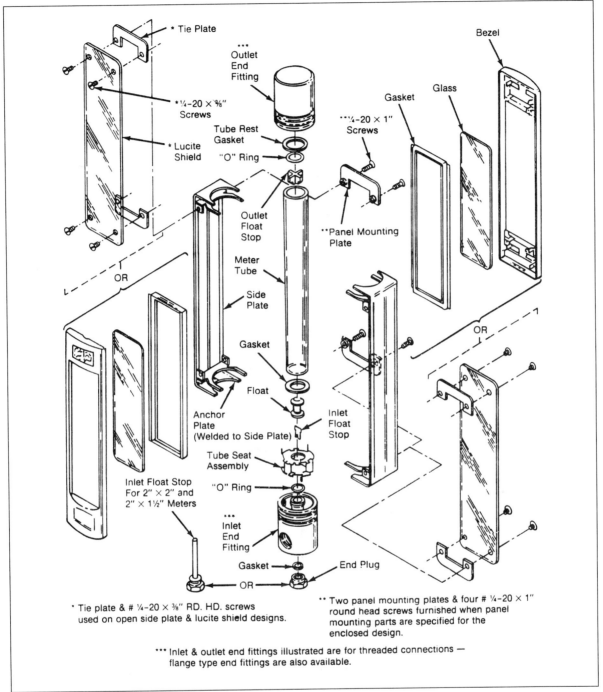

Figure A1-28. Rotameter — 3/4-in. and Larger Connections

(4) Make a statement concerning responsiveness of flow devices from your observations in this laboratory.

Head Flow and Differential Pressure

OBJECTIVES

(1) To construct a head flow measuring system.

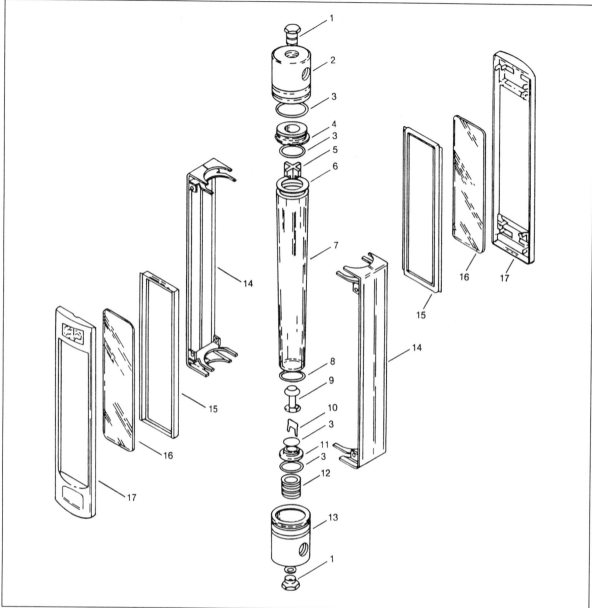

Figure A1-29. Rotameter — Exploded View

(2) To link the head flow primary to a commercial differential pressure transmitter.

(3) To prepare the energy transfer curve for the flow system.

(4) To interchange flow primaries and study the flow system.

(5) To observe the nonlinearity of output as flow rate is proportinally increased.

THEORY

Head flow primaries such as the orifice plate, the flow nozzle, and the Venturi develop differential pressures when these primaries are placed in a flow line. The differential pressures that develop are sensed by a differential pressure transmitter. The function of this transmitter is to amplify the pneumatic signal and then convert it to a standardize pressure output, generally a 3 psi to 15 psi signal. This

Directions: Match the numbered parts on the diagram with the part description at the right. Place the proper number in the space provided.

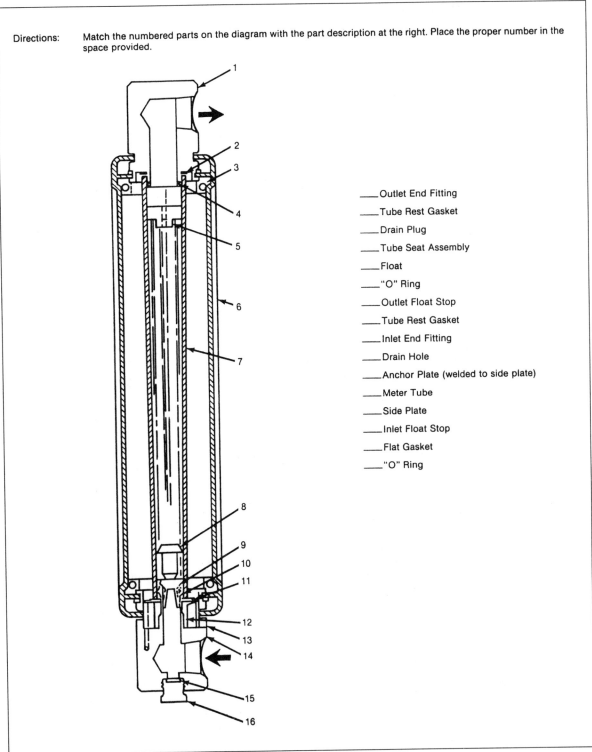

____Outlet End Fitting

____Tube Rest Gasket

____Drain Plug

____Tube Seat Assembly

____Float

____"O" Ring

____Outlet Float Stop

____Tube Rest Gasket

____Inlet End Fitting

____Drain Hole

____Anchor Plate (welded to side plate)

____Meter Tube

____Side Plate

____Inlet Float Stop

____Flat Gasket

____"O" Ring

Figure A1-30. Rotameter — Functional Parts Identification

standarized signal then may be used to drive any typical pneumatic display unit, such as a pneumatic recorder. This combination of flow primary, differential pressure transmitter, and display device then makes up the typical commercial pneumatic head flow system.

EQUIPMENT
Commercial differential pressure transmitter
Flow trainer
Flow tube primaries
Pressure display unit
Timing device

DIAGRAMS
Show a block diagram and schematic flow diagram of the system constructed.

PROCEDURE

(1) Construct the differential pressure flowmetering system consisting of a
flow primary, a commercial differential pressure transmitter, and a pres-

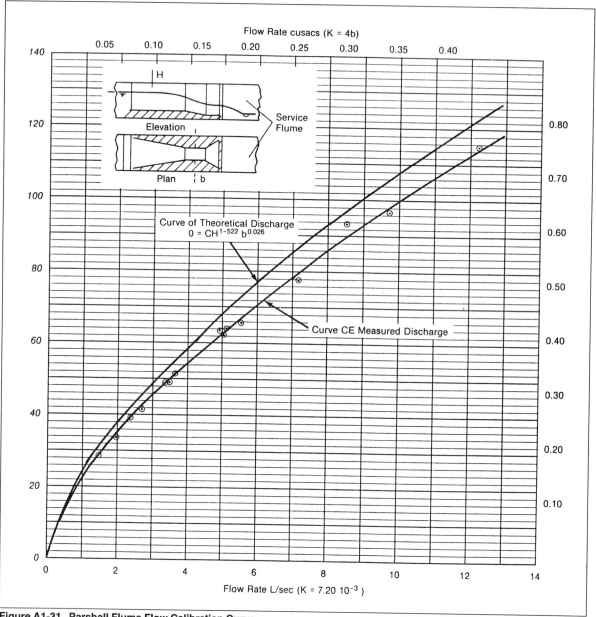

Figure A1-31. Parshall Flume Flow Calibration Curve

sure display unit. Be sure that all connections between flow tubes and differential pressure transmitter are airtight (no leaks). Be sure to remove all air from between the meter body and the flow primary.

(2) Test the system with several practice runs so that system potential is maximized and the output range is most closely matched to the differential pressure created by the specific flow tube used.

(3) Adjust flow rates in approximately ten steps. Record flow rate as indicated by nutating disc meter and stop watch, and output as indicated on the flow device. This information will be used to prepare the energy transfer characteristic curve input vs. output.

(4) Note that if the display device were a recorder, it would be possible to obtain a continuous record of flow rate changes made at the trainer. Connect a pneumatic recorder to the output of the flow transmitter and obtain a chart record of the flow trainer performance.

(5) Develop other experimental procedures as desired with this system.

(6) Move to another set up and repeat procedures above or change the primary element to a different primary element and repeat Steps 1-4 above.

ANALYSIS AND RESULTS

(1) Show a functional block diagram of the flow system developed in this laboratory.

(2) Explain the functional operation of each of the blocks listed above.

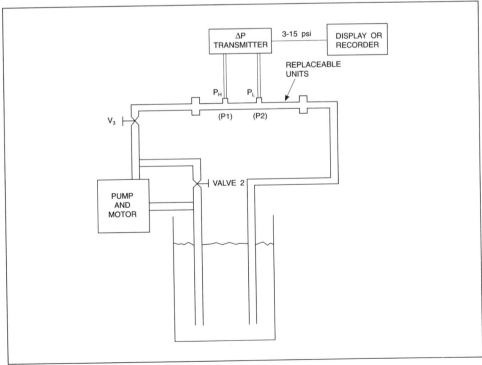

Figure A1-32. Head Flow Measuring System

(3) Show a pictorial diagram of the system developed with specific signal levels developed by specific components of the system.

(4) Show an energy transfer characteristic curve of Q (flow rate) vs. output for each system tested use rectangular graph paper.

Differential Pressure Transmitter — Electronic

OBJECTIVES

(1) To construct and test a specific flow system using a head flow-measuring element and electronic differential pressure transmitter (electronic).

(2) To observe and record the input/output characteristic curve.

(3) To step change the measuring system input and observe the time response.

(4) To check the system for repeatablity.

EQUIPMENT
Electronic transmitter
Flow trainer or equal, 0–10 gpm
Stop watch
Y-T plotter
Tools
Power supply

PROCEDURE

(1) Check out the flow trainer for operation. Check for type of flow primary element (orifice plate, Venturi, or flow nozzle).

(2) Differential pressure transmitter should have a span of about 0–50 in. or 0–80 in. water. Adjust if necessary.

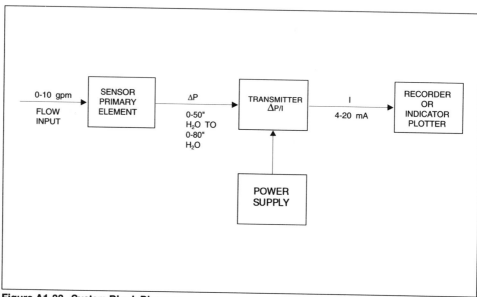

Figure A1-33. System Block Diagram

(3) When constructing the system, be sure that the meter body of the transmitter is completely filled with water. Open vents several times to remove air bubbles.

(4) Make several trial runs with the transmitter and flow trainer.

(5) Feed the transmitter output to a Y-T plotter. CAUTION: The Y- T plotter does not accept a current signal, so signal conditioning is necessary. (Use a 62.4-ohm resistor or a 250-ohm resistor to a develop voltage input signal.)

(6) Change the flow signal using the hand valve and observe the output signal obtained on the plotter.

(7) Record gallon/min vs. volts output, as determined by the recorder.

(8) Now sweep the flow from zero to full output. This will give the actual energy transfer characteristic curve of the system.

(9) Check for system repeatability. Establish a specific flow rate and check for system output repeatability.

(10) Vary the valve to create a step change in input on the trainer and note dynamic response of this flow measurement system.

ANALYSIS AND REPORT

(1) Cite any difficulties in setting up this system.

(2) Indicate the various energy conversions that took place in the system. Show energy transfer relationships.

(3) Is the output linear? If not, how may the signal be further conditioned to make it linear?

(4) Draw conclusions from data obtained in this laboratory problem.

(5) Comment on the dynamic response of most flow metering systems.

Liquid Level

Liquid Level Transmitters (Electronic and Pneumatic)

OBJECTIVES

(1) To set up a system that requires a level measurement (liquid column).

(2) To make the level measurement with an electrical pressure-measuring transmitter and with a pneumatic transmitter.

(3) To record level vs. analog electrical output and prepare an energy transfer curve for this system.

(4) Determine system response time.

(5) Determine system measurement repeatability.

THEORY

From the physical relationship $P = hd$ it is possible to measure the height of fluid (liquid level) in a container or column by a measurement of the pressure above a specific datum level. If this pressure is then sent to an electrical or pneumatic pressure transmitter, the level can be determined as an analog signal of the level. If the electrical or pneumatic analog signal is then converted to %, the level of the container will be known by % above datum level.

EQUIPMENT
Liquid column or tank
Commercial pneumatic and/or electronic pressure (level) transmitter
24-volt DC power supply
Electronic recorder (4–20 mA input)
DMM
Pneumatic hookup tubing
Standard pressure gage

PROCEDURES

(1) Select equipment for the pneumatic transmitter setup.

(2) Draw block diagram of this system with signals shown.

(3) Span the transmitter to the level of the tank or column. One inch of water column height equals one inch of water pressure above the datum.

(4) Prepare a data table; show column height, pressure, transmitter output, % column filled.

(5) Fill column or tank to the following levels and record data in the table: 0%, 20%, 40%, 50%, 60%, 80%, 100%.

(6) Prepare energy transfer curve of the pneumatic system performance.

(7) Repeat Steps 1-6 using an electronic pressure (level) transmitter.

(8) Attempt to make a step change in input and determine the time constant for the system. (Adding or subtracting a bucket of water will produce the step change in response.) Use Y-T plotter for time constant determination.

(9) Determine repeatability for this measurement system. For the same measured condition, measure this condition five times, determine average value and deviation of each value from the average value.

(10) For both the pneumatic and electronic transmitters, complete calibration forms. Prepare error plot for each transmitter. Record data and prepare graph.

(11) Connect electronic transmitter to commercial electronic recorder and vary level in the column. Note results.

ANALYSIS

(1) Comment on system accuracy as compared to transmitter specifications.

(2) What would be the effect of elevating or lowering the physical position of the transmitter relative to column datum?

(3) Discuss the theory of operation of each transmitter. How do these instruments work?

(4) Comment on the speed of response of this level measurement system.

(5) Comment on the repeatability of the electronic level measuring system.

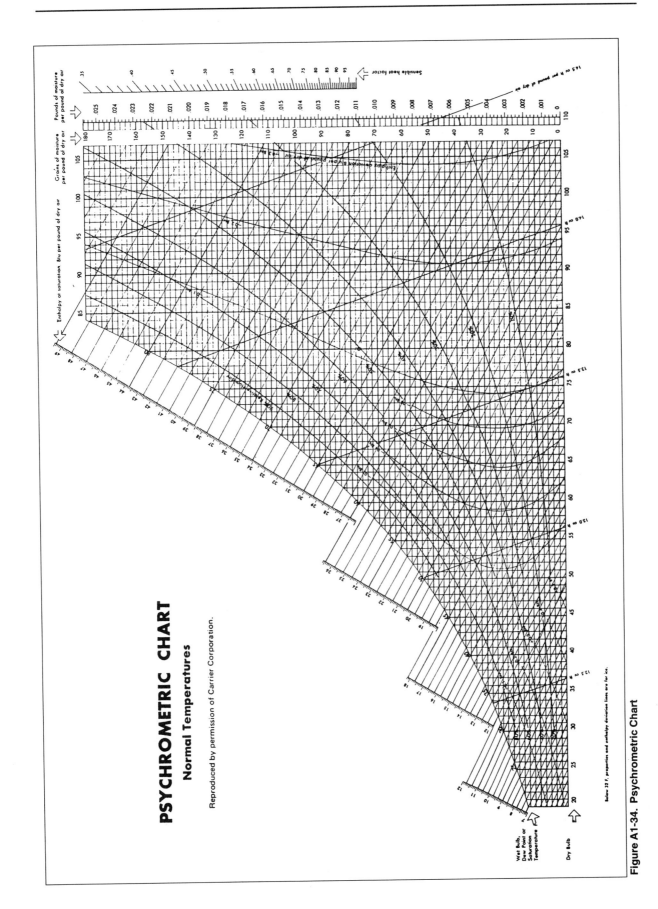

PSYCHROMETRIC CHART

Normal Temperatures

Reproduced by permission of Carrier Corporation.

Figure A1-34. Psychrometric Chart

Laboratory Standards

Table A1-4. Thermodynamic Properties of Moist Air at Sea Level

PB= 29.92 ALTITUDE= 0.

DB	WB	DP	RH	VP
70.0	70.0	70.0	100.0	18.77
70.0	69.0	68.6	95.2	17.87
70.0	68.0	67.1	90.5	16.99
70.0	67.0	65.6	85.9	16.12
70.0	66.0	64.0	81.3	15.27
70.0	65.0	62.4	76.9	14.44
70.0	64.0	60.8	72.6	13.63
70.0	63.0	59.1	68.3	12.83
70.0	62.0	57.3	64.2	12.05
70.0	61.0	55.5	60.1	11.28
70.0	60.0	53.6	56.1	10.53
70.0	59.0	51.7	52.2	9.79
70.0	58.0	49.6	48.3	9.07
70.0	57.0	47.4	44.5	8.36
70.0	56.0	45.2	40.9	7.67
70.0	55.0	42.7	37.2	6.99
70.0	54.0	40.1	33.7	6.32
70.0	53.0	37.3	30.2	5.76
70.0	52.0	34.3	26.8	5.02
70.0	51.0	31.1	23.4	4.39
70.0	50.0	27.8	20.1	3.77
72.0	72.0	72.0	100.0	20.09
72.0	71.0	70.6	95.3	19.15
72.0	70.0	69.1	90.7	18.23
72.0	69.0	67.7	86.2	17.33
72.0	68.0	66.1	81.8	16.44
72.0	67.0	64.6	77.5	15.58
72.0	66.0	63.0	73.3	14.73
72.0	65.0	61.3	69.2	13.90
72.0	64.0	59.6	65.1	13.08
72.0	63.0	57.9	61.1	12.29
72.0	62.0	56.1	57.2	11.50
72.0	61.0	54.2	53.4	10.74
72.0	60.0	52.2	49.7	9.99
72.0	59.0	50.1	46.0	9.25
72.0	58.0	48.0	42.4	8.53
72.0	57.0	45.7	38.9	7.82
72.0	56.0	43.2	35.5	7.13
72.0	55.0	40.6	32.1	6.45
72.0	54.0	37.8	28.8	5.78
72.0	53.0	34.8	25.5	5.12
72.0	52.0	31.5	22.3	44.48
72.0	51.0	28.2	19.2	3.85
72.0	50.0	24.5	16.1	3.23
74.0	74.0	74.0	100.0	21.49
74.0	73.0	72.6	95.4	20.51
74.0	72.0	71.2	91.0	19.55
74.0	71.0	69.7	86.6	18.61
74.0	70.0	68.3	82.3	17.69
74.0	69.0	66.7	78.1	16.78
74.0	68.0	65.2	74.0	15.90
74.0	67.0	63.6	69.9	15.03

PB= 29.92 ALTITUDE= 0.

DB	WB	DP	RH	VP
74.0	66.0	61.9	66.0	14.19
74.0	65.0	60.2	62.1	13.35
74.0	64.0	58.5	58.3	12.54
74.0	63.0	56.6	54.6	11.74
74.0	62.0	54.7	51.0	10.96
74.0	61.0	52.8	47.4	10.19
74.0	60.0	50.7	43.9	9.44
74.0	59.0	48.5	40.5	8.71
74.0	58.0	46.2	37.1	7.98
74.0	57.0	43.8	33.9	7.28
74.0	56.0	41.2	30.6	6.58
74.0	55.0	38.4	27.5	5.90
74.0	54.0	35.3	24.4	5.24
74.0	53.0	32.0	21.3	4.58
74.0	52.0	28.7	18.3	3.94
74.0	51.0	25.0	15.4	3.31
74.0	50.0	20.6	12.5	2.69
76.0	76.0	76.0	100.0	22.98
76.0	75.0	74.6	95.5	21.95
76.0	74.0	73.2	91.2	20.95
76.0	73.0	71.8	86.9	19.97
76.0	72.0	70.4	82.7	19.01
76.0	71.0	68.9	78.6	18.07
76.0	70.0	67.4	74.6	17.14
76.0	69.0	65.8	70.7	16.24
76.0	68.0	64.2	66.8	15.36
76.0	67.0	62.5	63.1	14.49
76.0	66.0	60.8	59.4	13.64
76.0	65.0	59.1	55.7	12.81
76.0	64.0	57.2	52.2	12.00
76.0	63.0	55.3	48.7	11.20
76.0	62.0	53.3	45.3	10.42
76.0	61.0	51.3	42.0	9.65
76.0	60.0	49.1	38.7	8.90
76.0	59.0	46.8	35.5	8.16
76.0	58.0	44.4	32.4	7.44
76.0	57.0	41.8	29.3	6.73
76.0	56.0	39.0	26.3	6.04
76.0	55.0	35.9	23.3	5.36
76.0	54.0	32.6	20.4	4.69
76.0	53.0	29.3	17.6	4.04
76.0	52.0	25.5	14.8	3.40
76.0	51.0	21.2	12.0	2.77
76.0	50.0	16.0	9.3	2.15
78.0	78.0	78.0	100.0	24.55
78.0	77.0	76.7	95.6	23.48
78.0	76.0	75.3	91.4	22.44
78.0	75.0	73.9	87.2	21.41
78.0	74.0	72.5	83.1	20.41
78.0	73.0	71.0	79.1	19.43
78.0	72.0	69.5	75.2	18.46
78.0	71.0	68.0	71.4	17.52

PB= 29.92 ALTITUDE= 0.

DB	WB	DP	RH	VP
78.0	70.0	66.4	67.6	16.60
78.0	69.0	64.8	63.9	15.70
78.0	68.0	63.2	60.3	14.81
78.0	67.0	61.4	56.8	13.95
78.0	66.0	59.7	53.3	13.10
78.0	65.0	57.9	50.0	12.27
78.0	64.0	55.9	46.6	11.45
78.0	63.0	54.0	43.4	10.66
78.0	62.0	51.9	40.2	9.87
78.0	61.0	49.7	37.1	9.11
78.0	60.0	47.4	34.0	8.36
78.0	59.0	45.0	31.0	7.62
78.0	58.0	42.4	28.1	6.90
78.0	57.0	39.6	25.2	6.19
78.0	56.0	36.6	22.4	5.50
78.0	55.0	33.2	19.6	4.82
78.0	54.0	29.9	16.9	4.15
78.0	53.0	26.2	14.2	3.50
78.0	52.0	21.9	11.6	2.85
78.0	51.0	16.7	9.1	2.22
78.0	50.0	10.1	6.5	1.61
80.0	80.0	80.0	100.0	26.22
80.0	79.0	78.7	95.7	25.10
80.0	78.0	77.3	91.6	24.01
80.0	77.0	75.9	87.5	22.94
80.0	76.0	74.5	83.5	21.89
80.0	75.0	73.1	79.6	20.87
80.0	74.0	71.7	75.8	19.87
80.0	73.0	70.2	72.0	18.88
80.0	72.0	68.6	68.4	17.92
80.0	71.0	67.1	64.8	16.98
80.0	70.0	65.5	61.2	16.06
80.0	69.0	63.8	57.8	15.15
80.0	68.0	62.1	54.4	14.27
80.0	67.0	60.3	51.1	13.40
80.0	66.0	58.5	47.9	12.56
80.0	65.0	56.6	44.7	11.72
80.0	64.0	54.6	41.6	10.91
80.0	63.0	52.5	38.6	10.11
80.0	62.0	50.4	35.6	9.33
80.0	61.0	48.1	32.7	8.56
80.0	60.0	45.6	29.8	7.81
80.0	59.0	43.0	27.0	7.08
80.0	58.0	40.3	24.2	6.36
80.0	57.0	37.3	21.5	5.65
80.0	56.0	34.0	18.9	4.96
80.0	55.0	30.5	16.3	4.27
80.0	54.0	26.8	13.8	3.61
80.0	53.0	22.6	11.3	2.95
80.0	52.0	17.5	8.8	2.31
80.0	51.0	11.0	6.4	1.68
80.0	50.0	2.0	4.1	1.06

Miscellaneous

Moisture Measurement—% RH and Dew Point

OBJECTIVES

(1) To study various methods for the measurement of % relative humidity and dew point.

(2) To learn to use psychrometric charts.

THEORY

The % relative humidity of homes, stores, and factories is becoming of increasing importance as we improve living conditions (air conditioning) and as we try to manufacture, fabricate, or construct devices made of increasingly delicate components whose dimensions are made more and more exact. The weaving of cloth and the color printing of papers require close control of the dimensions of the components.

Some materials undergo a marked change in dimension or shape when they absorb moisture, while others change their resistance when water is absorbed. Still other materials change certain equilibrium points as the % relative humidity changes. If the temperature of an object is lowered to the point at which moisture condenses on the object, this is called the dew point temperature. An important electrical moisture determination method uses an electro-optical sensor and a chilled mirror for accurate dew point determination. The common wet and dry bulb method for % relative humidity determination is also used. This is called psychrometry.

EQUIPMENT

Wet and dry bulb set
Hair hygrometer
Pneumatic humidity transmitter
Sling psychrometry
Matched platinum RTDs
DMMs
Fans
Psychrometric tables
Commercial % RH, dew point, and temperature sensor
Psychrometric chart

PROCEDURE

(1) Prepare a data table for recording of measurement of % relative humidity with the various instruments or dewpoint measuring systems. Determine %RH of your laboratory and one other condition. (Outside is another %RH condition.) The purpose of the table is to compare values and methods.

(2) Use of sling psychrometer (procedure):

- Wet wick with distilled water (room temperature).

- Sling for 30 seconds, moving around in the process.

- Read wet bulb first; record. Then read and record dry bulb.

- Repeat five times.

- From the average readings, determine %RH using psychrometric tables.

(3) Set up wet/dry bulb system. Determine %RH from this unit. Record in table.

(4) Record %RH from hygrometer. Examine these instruments and observe the method for %RH determination.

(5) Electrical psychrometric measurement. Design and then build an electrical wet/dry bulb %RH measuring system using two RTDs.

- Check the RTDs for matched output (same resistance value for same temperature condition).

- Place a piece of wet cloth (or wick) over an RTD. Saturate this wet RTD with distilled water solution.

- Using a fan, cause an air flow over both RTDs.

- Now determine and record temperature difference.

- Use psychrometric chart or tables to determine %RH.

- Enter this value in a table.

(6) Pneumatic %RH transmitter.

- Draw a block diagram of this measurement system using a 3–15-psi pneumatic recorder as the display unit.

- Review manual on this transmitter and determine theory of operation.

- Connect transmitter to supply air and estimate %RH from output obtained. Record value in table.

(7) Dew Point or a %RH Determination. Using the commercial %RH, dew point, and temperature measuring system, find the dew point and %RH of the laboratory area. How does this data compare to the data previously obtained? Record in table.

ANALYSIS AND RESULTS

(1) From the data obtained, comment on the true value of %RH for the conditions measured.

(2) Comment on accuracy and repeatability of instruments used to measure %RH.

(3) If we could create a step change in %RH, how would you rate these systems as regards their respective speed of response?

(4) Discuss the problem of getting commercial measurements that are accurate for the determination of %RH.

(5) What other methods are used for the determination of %RH electrically?

(6) Describe how the dew point sensor operates?

pH Measurements and Transmitters

OBJECTIVES

(1) To plot pH vs. voltage relationship as predicted by the Nernst equation for a combination electrode. Also to plot electrode curves for 40°C and 0°C, showing the effects of temperature on electrode output.

(2) To prepare solutions of known pH using buffer tablets. (This is a means for standardizing pH measuring systems.)

(3) To determine acidity or basicity of a solution using colormetric methods (Litmus papers).

(4) Use the portable meters, both analog and digital, for pH measurement.

(5) Use commercial pH meters.

(6) Determine pH of several unknown solutions.

(7) To examine a commercial industrial pH system and prepare a block diagram of the system.

THEORY

Whenever liquids dissociate into ions they become capable of carrying current. In addition, whenever metals or other materials are immersed in liquids, differences in potential often are generated. These potentials are, in turn, proportional to the amount of dissociation of the materials. The potential generated (E) created by the hydrogen ion concentration, at a constant temperature ($25°C$) may be expressed as follows:

$$E = 413 \text{ mV} - 59.120 \text{ mV (pH) @ } 25°C$$

Note that a completely neutral solution has a pH of 7 and that the output of the hydrogen cell then is exactly equal to zero.

Because the electrolyte creates the connecting path between the electrodes, any currents that are involved in the measurement must pass through the glass of the active electrode.

Because of its extremely high resistance (in megohms), the measuring current must be reduced as close to zero as possible; otherwise the entire potential created by the pH unit is "used up" in forcing current through the highly resistive portion of the circuit. Thus, there is virtually NO potential difference available at the terminals.

The secret of pH measurement with glass electrodes is to measure a current reduced to a minimum.

For the metered current to be in the microampere range, the amplifier must have a current amplification of the order of 1×10 to the sixth power or better if the current between electrodes is not to affect the results adversely.

EQUIPMENT

pH Hydron buffer tablets or capsules
Thermometer
Commercial buffer solutions
Distilled water
Portable pH meters (analog and or digital)
Distilled water
Commercial industrial pH measuring system
Commercial pH meters

PROCEDURE

(1) Prepare an energy transfer curve for electrical pH measuring system. Using the Nernst equation, plot a curve for mV vs. pH numbers for 25°C. Now using the table of values supplied (Table A1-5), ADD curves for temperatures of 0°C and 40°C to the curve of the Nernst equation data.

(2) From the above data, determine electrode sensitivity for each temperature condition.

(3) Prepare standard buffer solutions using tablets or capsules, pH 4.0, pH 7.0, pH 10.0. Buffer solutions will be used to standardize portable pH meters and bench-type pH meters.

(4) Use litmus papers to determine acidity or basicity of standard buffer solutions. Record results.

(5) Use these same papers to test the three unknown solutions.

(6) Procedure for use of portable pH meters and bench type pH meters.

- Check standardization of the pH meter. Be sure to check temperature of buffer solution and set temperature compensation dial accordingly.

- After standardization procedure is complete, determine exact pH of the three unknown solutions.

- Record data obtained from above.

(7) Use bench pH meters to determine the pH of the unknown solutions. Record values.

(8) Observe the commercial industrial pH-measuring system displayed. Prepare a block diagram of the measuring system.

(9) Calibrate the pH transmitter using the buffer solutions as standards. Prepare a three-point error plot from the data obtained. Compare accuracy to published data.

ANALYSIS AND REPORT

(1) Document observations made during this laboratory.

(2) Discuss several different methods for determining pH of a solution.

(3) What pH measurement device should give the most accurate results? Why?

(4) What are the practical problems involved with storage of electrodes?

(5) How is electrode sensitivity affected by temperature changes?

(6) How did the calibration of the pH transmitter compare to published specifications?

Table A1-5. Voltage to pH Conversions

Glass Electrode and Reference Electrode with Identical Internal Elements — Zero Volts at 7 pH*

pH						Voltage						
	0°C	10°C	20°C	25°C	30°C	40°C	50°C	60°C	70°C	80°C	90°C	100°C
0.0	-0.3794	-0.3933	-0.4072	-0.4141	-0.4211	-0.4350	-0.4489	-0.4627	-0.4766	-0.4905	-0.5044	-0.5183
1	0.3740	0.3877	0.4014	0.4082	0.4151	0.4288	0.4424	0.4561	0.4698	0.4835	0.4972	0.5109
2	0.3686	0.3821	0.3955	0.4023	0.4090	0.4225	0.5360	0.4495	0.4630	0.4765	0.4900	0.5035
3	0.3631	0.3764	0.3897	0.3964	0.4030	0.4163	0.4296	0.4429	0.4562	0.4695	0.4828	0.4961
4	0.3577	0.3708	0.3839	0.3905	0.3970	0.4101	0.4232	0.4363	0.4494	0.4625	0.4756	0.4887
5	-0.3523	3652	-0.3781	-0.3846	-0.3910	-0.4039	-0.4168	-0.4297	-0.4426	-0.4555	-0.4684	-0.4813
6	0.3469	0.3596	0.3723	0.3786	0.3850	0.3977	0.4104	0.4231	0.4358	0.4485	0.4612	0.4739
7	0.3415	0.3540	0.3665	0.3727	0.3790	0.3915	0.4040	0.4165	0.4290	0.4415	0.4540	0.4665
8	0.3360	3484	0.3606	0.3668	0.3730	0.3853	0.3976	0.4099	0.4222	0.4345	0.4468	0.4591
9	0.3306	0.3427	0.3548	0.3609	0.3669	0.3790	0.3911	0.4033	0.4154	0.4275	0.4396	0.4517
1.0	-0.3252	-0.3371	-0.3490	-0.3356	-0.3609	-0.3728	-0.3848	-0.3966	-0.4085	-0.4204	-0.4324	-0.4443
1	0.3198	0.3315	0.3432	0.3491	0.3549	0.3666	0.3783	0.3900	0.4017	0.4134	0.4251	0.4369
2	0.3144	0.3259	0.3374	0.3431	0.3489	0.3604	0.3719	0.3834	0.3949	0.4064	0.4179	0.4294
3	0.3089	0.3203	0.3316	0.3372	0.3429	0.3542	0.3655	0.3768	0.3881	0.3994	0.4107	0.4220
4	0.3035	0.3146	0.3257	0.3313	0.3369	0.3480	0.3591	0.3702	0.3813	0.3924	0.4035	0.4146
5	0.2981	0.3090	0.3199	0.3254	0.3308	0.3418	0.3527	0.3636	0.3745	0.3854	0.3963	0.4072
6	0.2927	0.3034	0.3141	0.3195	0.3248	0.3355	0.3563	0.3570	0.3677	0.3784	0.3891	0.3998
7	0.2873	0.2978	0.3083	0.3136	0.3188	0.3293	0.3398	0.3504	0.3609	0.3714	0.3819	0.3924
8	0.2818	0.2922	0.3025	0.3076	0.3128	0.3231	0.3334	0.3438	0.3541	0.3644	0.3747	0.3850
9	0.2764	0.2865	0.2967	0.3017	0.3068	0.3169	0.3270	0.3371	0.3473	0.3574	0.3675	0.3776
2.0	-0.2710	-0.2809	-0.2909	-0.2958	-0.3008	-0.3107	-0.3206	-0.3305	-0.3405	-0.3504	-0.3603	-0.3702
1	0.2656	0.2753	0.2850	0.2899	0.2948	0.3045	0.3142	0.3239	0.3336	0.3434	0.3531	0.3628
2	0.2602	0.2697	0.2792	0.2840	0.2887	0.2963	0.3078	0.3173	0.3268	0.3364	0.3459	0.3554
3	0.2547	0.2641	0.2734	0.2781	0.2827	0.2920	0.3014	0.3107	0.3200	0.3294	0.3387	0.3480
4	0.2493	0.2585	0.2676	0.2721	0.2767	0.2858	0.2950	0.3041	0.3132	0.3223	0.3315	0.3406
5	-0.2439	-0.2528	-0.2618	-0.2662	-0.2707	-0.2796	-0.2886	-0.2975	-0.3064	-0.3153	-0.3243	-0.3332
6	0.2385	0.2472	0.2559	0.2603	0.2647	0.2734	0.2821	0.2909	0.2996	0.3083	0.3171	0.3258
7	0.2331	0.2416	0.2501	0.2544	0.2587	0.2672	0.2757	0.2843	0.2928	0.3013	0.3099	0.3184
8	0.2276	0.2360	0.2443	0.2485	0.2526	0.2610	0.2693	0.2777	0.2860	0.2943	0.3026	0.3110
9	0.2222	0.2304	0.2385	0.2426	0.2466	0.2548	0.2629	0.2710	0.2792	0.2873	0.2954	0.3036
3.0	-0.2168	-0.2247	-0.2327	-0.2366	-0.2406	-0.2486	-0.2565	-0.2644	-0.2724	-0.2803	-0.2882	-0.2962
1	0.2144	0.2191	0.2269	0.2309	0.2346	0.2423	0.2501	0.2578	0.2656	0.2733	0.2810	0.2888
2	0.2060	0.2135	0.2210	0.2248	0.2286	0.2361	0.2437	0.2512	0.2587	0.2663	0.2738	0.2814
3	0.2005	0.2079	0.2152	0.2189	0.2226	0.2299	0.2373	0.2446	0.2519	0.2593	0.2666	0.2740
4	0.1951	0.2023	0.2094	0.2130	0.2166	0.2237	0.2308	0.2380	0.2451	0.2523	0.2594	0.2666
5	-0.1897	-0.1967	-0.2036	-0.2071	-0.2106	-0.2175	-0.2244	-0.2314	-0.2383	-0.2453	-0.2522	-0.2594
6	0.1843	0.1910	0.1978	0.2012	0.2045	0.2113	0.2180	0.2248	0.2315	0.2383	0.2450	0.2517
7	0.1789	0.1854	0.1920	0.1952	0.1985	0.2051	0.2116	0.2182	0.2247	0.2312	0.2378	0.2443
8	0.1734	0.1798	0.1861	0.1893	0.1925	0.1988	0.2052	0.2115	0.2179	0.2242	0.2306	0.2369
9	0.1680	0.1742	0.1803	0.1834	0.1865	0.1926	0.1988	0.2049	0.2111	0.2172	0.2234	0.2295
4.0	-0.1626	-0.1686	-0.1745	-0.1775	-0.1805	-0.1864	-0.1924	-0.1983	-0.2043	-0.2102	-0.2162	-0.2221
1	0.1572	0.1629	0.1687	0.1716	0.1744	0.1802	0.1860	0.1917	0.1975	0.2032	0.2090	0.2147
2	0.1518	0.1573	0.1629	0.1657	0.1684	0.1740	0.1795	0.1851	0.1907	0.1962	0.2018	0.2071
3	0.1463	0.1517	0.1571	0.1597	0.1624	0.1678	0.1731	0.1785	0.1838	0.1892	0.1946	0.1999
4	0.1409	0.1461	0.1512	0.1538	0.1564	0.1616	0.1667	0.1719	0.1770	0.1822	0.1874	0.1925
5	-0.1355	-0.1405	-0.1454	-0.1479	-0.1504	-0.1553	-0.1603	-0.1653	-0.1702	-0.1752	-0.1802	-0.1851
6	0.1301	0.1348	0.1396	0.1420	0.1444	0.1491	0.1539	0.1587	0.1634	0.1682	0.1729	0.1777
7	0.1247	0.1292	0.1338	0.1361	0.1384	0.1429	0.1475	0.1520	0.1566	0.1612	0.1657	0.1703
8	0.1192	0.1236	0.1280	0.1302	0.1323	0.1367	0.1411	0.1454	0.1498	0.1542	0.1585	0.1629
9	0.1138	0.1180	0.1222	0.1242	0.1263	0.1305	0.1347	0.1388	0.1430	0.1472	0.1513	0.1555

Table A1-5 (continued). Voltage to pH Conversions

Glass Electrode and Reference Electrode with Identical Internal Elements — Zero Volts at 7 pH*

pH	Voltage											
	0°C	10°C	20°C	25°C	30°C	40°C	50°C	60°C	70°C	80°C	90°C	100°C
5.0	-0.1084	-0.1124	-0.1163	-0.1183	-0.1203	-0.1243	-0.1282	-0.1322	-0.1362	-0.1402	-0.1441	-0.1481
1	0.1030	0.1068	0.1105	0.1124	0.1143	0.1181	0.1218	0.1256	0.1294	0.1331	0.1369	0.1407
2	0.0976	0.1011	0.1047	0.1065	0.1083	0.1119	0.1154	0.1190	0.1226	0.1261	0.1297	0.1333
3	0.0921	0.0955	0.0989	0.1006	0.1023	0.1056	0.1090	0.1124	0.1158	0.1191	0.1225	0.1259
4	0.0867	0.0899	0.0931	0.0947	0.0963	0.0994	0.1026	0.1058	0.1089	0.1121	0.1153	0.1185
5	-0.0813	-0.0843	-0.0873	-0.0887	-0.0902	-0.0932	-0.0962	-0.0992	-0.1021	-0.1051	-0.1081	-0.1111
6	0.0759	0.0787	0.0814	0.0828	0.0842	0.0870	0.0898	0.0926	0.0953	0.0981	0.1009	0.1037
7	0.0705	0.0730	0.0756	0.0769	0.0782	0.0808	0.0834	0.0859	0.0885	0.0911	0.0937	0.0963
8	0.0650	0.0674	0.0698	0.0710	0.0722	0.0746	0.0770	0.0793	0.0817	0.0841	0.0865	0.0889
9	0.0596	0.0618	0.0640	0.0651	0.0662	0.0684	0.0705	0.0727	0.0749	0.0771	0.0793	0.0815
6.0	-0.0542	-0.0562	-0.0582	-0.0592	-0.0602	-0.0621	-0.0641	-0.0661	-0.0681	-0.0701	-0.0721	-0.0740
1	0.0488	0.0506	0.0524	0.0533	0.0541	0.0559	0.0577	0.0595	0.0613	0.0631	0.0649	0.0666
2	0.0434	0.0450	0.0465	0.0473	0.0481	0.0497	0.0513	0.0529	0.0545	0.0561	0.0577	0.0592
3	0.0379	0.0393	0.0407	0.0414	0.0421	0.0435	0.0449	0.0463	0.0477	0.0491	0.0504	0.0518
4	0.0325	0.0337	0.0349	0.0355	0.0361	0.0373	0.0385	0.0397	0.0409	0.0420	0.0432	0.0444
5	-0.0271	-0.0281	-0.0291	-0.0296	-0.0301	-0.0311	-0.0321	-0.0331	-0.0341	-0.0350	-0.0360	-0.0370
6	-0.0217	-0.0225	-0.0233	-0.0237	-0.0241	-0.0249	-0.0257	-0.0264	-0.0272	-0.0280	-0.0288	-0.0296
7	-0.0163	-0.0169	-0.0175	-0.0178	-0.0181	-0.0186	-0.0192	-0.0198	-0.0204	-0.0210	-0.0216	-0.0222
8	-0.0108	-0.0112	-0.0116	-0.0118	-0.0120	-0.0124	-0.0128	-0.0132	-0.0136	-0.0140	-0.0144	-0.0148
9	-0.0054	-0.0056	-0.0058	-0.0059	-0.0060	0.0064	-0.0062	-0.0066	-0.0068	-0.0070	-0.0072	-0.0074
7.0	0.0000	0.0000	0.0000	0.0000	0.0000	0.0000	0.0000	0.0000	0.0000	0.0000	0.0000	0.0000
1	+0.0054	+0.0056	+0.0058	+0.0059	+0.0060	+0.0062	+0.0064	+0.0066	+0.0068	+0.0070	+0.0072	+0.0074
2	0.0108	0.0112	0.0116	0.0118	0.0120	0.0124	0.0128	0.0132	0.0136	0.0140	0.0144	0.0148
3	0.0163	0.0169	0.0175	0.0178	0.0181	0.0186	0.0192	0.0198	0.0204	0.0210	0.0216	0.0222
4	0.0217	0.0225	0.0233	0.0237	0.0241	0.0249	0.0257	0.0264	0.0272	0.0280	0.0288	0.0296
5	+0.027	+0.0281	+0.0291	+0.0296	+0.0301	+0.0311	+0.0321	+0.0331	+0.0341	+0.0350	+0.0360	+0.0370
6	0.0325	0.0337	0.0349	0.0355	0.0361	0.0373	0.0385	0.0397	0.0409	0.0420	0.0432	0.0444
7	0.0379	0.0393	0.0407	0.0414	0.0421	0.0435	0.0449	0.0463	0.0477	0.0491	0.0504	0.0518
8	0.0434	0.0450	0.0465	0.0473	0.0481	0.0497	0.0513	0.0529	0.0545	0.0561	0.0577	0.0592
9	0.0488	0.0506	0.0524	0.0533	0.0541	0.0559	0.0577	0.0595	0.0613	0.0631	0.0649	0.0666
8.0	+0.0542	+0.0562	+0.0582	+0.0592	+0.0602	+0.0621	+0.0641	+0.0661	+0.0681	+0.0701	+0.0721	+0.0740
1	0.0596	0.0618	0.0640	0.0651	0.0662	0.0684	0.0705	0.0727	0.0749	0.0771	0.0793	0.0815
2	0.0650	0.0674	0.0698	0.0710	0.0722	0.0746	0.0770	0.0793	0.0817	0.0841	0.0865	0.0889
3	0.0705	0.0730	0.0756	0.0769	0.0782	0.0808	0.0834	0.0859	0.0885	0.0911	0.0937	0.0963
4	0.0759	0.0787	0.0814	0.0828	0.0842	0.0870	0.0898	0.0926	0.0953	0.0981	0.1009	0.1037
5	+0.0813	+0.0843	+0.0873	+0.0887	+0.0902	+0.0932	+0.0962	+0.0992	+0.1021	+0.1051	+0.1081	+0.1111
6.	0.0867	0.0899	0.0931	0.0947	0.0963	0.0994	0.1026	0.1058	0.1089	0.1121	0.1153	0.1185
7	0.0921	0.0955	0.0989	0.1006	0.1023	0.1056	0.1090	0.1124	0.1158	0.1191	0.1225	0.1259
8	0.0976	0.1011	0.1047	0.1065	0.1083	0.1119	0.1154	0.1190	0.1226	0.1261	0.1297	0.1333
9	0.1030	0.1068	0.1105	0.1124	0.1143	0.1181	0.1218	0.1256	0.1294	0.1331	0.1369	0.1407
9.0	+0.1084	+0.1124	+0.1163	+0.1183	+0.1203	+0.1243	+0.1282	+0.1322	+0.1362	+0.1402	+0.1441	+0.1481
1	0.1138	0.1180	0.1222	0.1242	0.1263	0.1305	0.1347	0.1388	0.1430	0.1472	0.1513	0.1555
2	0.1192	0.1236	0.1280	0.1302	0.1323	0.1367	0.1411	0.1454	0.1498	0.1542	0.1585	0.1629
3	0.1247	0.1292	0.1338	0.1361	0.1384	0.1429	0.1475	0.1520	0.1566	0.1612	0.1657	0.1703
4	0.1301	0.1348	0.1396	0.1420	0.1444	0.1491	0.1539	0.1587	0.1634	0.1682	0.1729	0.1777
5	+0.1355	+0.1405	+0.1454	+0.1479	+0.1504	+0.1553	+0.1603	+0.1653	+0.1702	+0.1752	+0.1802	+0.1851
6	0.1409	0.1461	0.1512	0.1538	0.1564	0.1616	0.1667	0.1719	0.1770	0.1822	0.1874	0.1925
7	0.1463	0.1517	0.1571	0.1597	0.1624	0.1678	0.1731	0.1785	0.1838	0.1892	0.1946	0.1999
8	0.1518	0.1573	0.1629	0.1657	0.1684	0.1740	0.1795	0.1851	0.1907	0.1962	0.2018	0.2073
9	0.1572	0.1629	0.1687	0.1716	0.1744	0.1802	0.1860	0.1917	0.1975	0.2032	0.2090	0.2147

Table A1-5 (continued). Voltage to pH Conversions

Glass Electrode and Reference Electrode with Identical Internal Elements — Zero Volts at 7 pH*

| pH | \multicolumn{12}{c}{Voltage} |
	0°C	10°C	20°C	25°C	30°C	40°C	50°C	60°C	70°C	80°C	90°C	100°C
10.0	+0.1626	+0.1686	+0.1745	+0.1775	+0.1805	+0.1864	+0.1924	+0.1983	+0.2043	+0.2102	+0.2162	+0.2221
1	0.1680	0.1742	0.1803	0.1834	0.1865	0.1926	0.1988	0.2049	0.2111	0.2172	0.2234	0.2295
2	0.1734	0.1798	0.1861	0.1893	0.1925	0.1988	0.2052	0.2115	0.2179	0.2242	0.2306	0.2369
3	0.1789	0.1854	0.1920	0.1952	0.1985	0.2051	0.2116	0.2182	0.2247	0.2312	0.2378	0.2443
4	0.1843	0.1910	0.1978	0.2012	0.2045	0.2113	0.2180	0.2248	0.2315	0.2383	0.2450	0.2517
5	+0.1897	+0.1967	+0.2036	+0.2071	+0.2106	+0.2175	+0.2244	+0.2314	+0.2383	+0.2453	+0.2522	+0.2592
6	0.1951	0.2023	0.2094	0.2130	0.2166	0.2237	0.2308	0.2380	0.2451	0.2523	0.2594	0.2666
7	0.2005	0.2079	0.2152	0.2189	0.2226	0.2299	0.2373	0.2446	0.2519	0.2593	0.2666	0.2740
8	0.2060	0.2135	0.2210	0.2248	0.2286	0.2361	0.2437	0.2512	0.2587	0.2663	0.2738	0.2814
9	0.2114	0.2191	0.2269	0.2307	0.2346	0.2423	0.2501	0.2578	0.2656	0.2733	0.2810	0.2888
11.0	+0.2168	+0.2247	+0.2327	+0.2366	+0.2406	+0.2486	+0.2565	+0.2644	+0.2724	+0.2803	+0.2882	+0.2962
1	0.2222	0.2304	0.2385	0.2426	0.2466	0.2548	0.2629	0.2710	0.2792	0.2873	0.2954	0.3036
2	0.2276	0.2360	0.2443	0.2485	0.2526	0.2610	0.2693	0.2777	0.2860	0.2843	0.3026	0.3110
3	0.2331	0.2416	0.2501	0.2544	0.2587	0.2672	0.2757	0.2843	0.2928	0.3013	0.3099	0.3184
4	0.2385	0.2472	0.2559	0.2603	0.2647	0.2734	0.2821	0.2909	0.2996	0.3083	0.3171	0.3258
5	+0.2439	+0.2528	+0.2618	+0.2662	+0.2707	+0.2796	+0.2886	+0.2975	+0.3064	+0.3153	+0.3243	+0.3332
6	0.2493	0.2585	0.2676	0.2721	0.2767	0.2858	0.2950	0.3041	0.3132	0.3223	0.3315	0.3406
7	0.2547	0.2641	0.2734	0.2781	0.2827	0.2920	0.3014	0.3107	0.3200	0.3294	0.3387	0.3480
8	0.2602	0.2697	0.2792	0.2840	0.2887	0.2983	0.3078	0.3173	0.3268	0.3364	0.3459	0.3554
9	0.2656	0.2753	0.2850	0.2899	0.2948	0.3045	0.3142	0.3239	0.3336	0.3434	0.3531	0.3628
12.0	+0.2710	+0.2809	+0.2909	+0.2958	+0.3008	+0.3107	+0.3206	+0.3305	+0.3405	+0.3504	+0.3603	+0.3702
1	0.2764	0.2865	0.2967	0.3017	0.3068	0.3169	0.3270	0.3371	0.3473	0.3574	0.3675	0.3776
2	0.2818	0.2922	0.3025	0.3076	0.3128	0.3231	0.3334	0.3438	0.3541	0.3644	0.3747	0.3850
3	0.2873	0.2978	0.3083	0.3136	0.3188	0.3293	0.3398	0.3504	0.3609	0.3714	0.3819	0.3924
4	0.2927	0.3034	0.3141	0.3195	0.3248	0.3355	0.3463	0.3570	0.3677	0.3748	0.3891	0.3998
5	+0.2981	+0.3090	+0.3199	+0.3254	+0.3308	+0.3418	+0.3527	+0.3636	+0.3745	+0.3854	+0.3963	+0.4072
6	0.3035	0.3146	0.3257	0.3313	0.3369	0.3480	0.3591	0.3702	0.3813	0.3924	0.4035	0.4146
7	0.3081	0.3203	0.3316	0.3372	0.3429	0.3542	0.2655	0.3768	0.3881	0.3994	0.4107	0.4220
8	0.3144	0.3259	0.3374	0.3431	0.3489	0.3604	0.3719	0.3834	0.3949	0.4064	0.4179	0.4294
9	0.3198	0.3315	0.3432	0.3491	0.3549	0.3666	0.3783	0.3900	0.4017	0.4134	0.4251	0.4369
13.0	+0.3252	+0.3371	+0.3490	+0.3550	+0.3609	+0.3728	+0.3847	+0.3966	+0.4085	+0.4204	+0.4324	+0.4443
1	0.3306	0.3427	0.3548	0.3609	0.3669	0.3790	0.3911	0.4033	0.4154	0.4275	0.4396	0.4517
2	0.3360	0.3484	0.3606	0.3668	0.3730	0.3853	0.3976	0.4099	0.4222	0.4345	0.4468	0.4591
3	0.3415	0.3540	0.3665	0.3727	0.3790	0.3915	0.4040	0.4165	0.4290	0.4415	0.4540	0.4665
4	0.3469	0.3596	0.3723	0.3786	0.3850	0.3977	0.4104	0.4231	0.4358	0.4485	0.4612	0.4739
5	+0.3523	+0.3652	+0.3781	+0.3846	+0.3910	+0.4039	+0.4168	+0.4297	+0.4426	+0.4555	+0.4684	+0.4813
6	0.3577	0.3708	0.3839	0.3905	0.3970	0.4101	0.4232	0.4363	0.4494	0.4625	0.4756	0.4887
7	0.3631	0.3764	0.3897	0.3964	0.4030	0.4163	0.4296	0.4429	0.4562	0.4695	0.4828	0.4961
8	0.3686	0.3821	0.3955	0.4023	0.4090	0.4225	0.4360	0.4495	0.4630	0.4765	0.4900	0.5035
9	0.3740	0.3877	0.4014	0.4082	0.4151	0.4288	0.4424	0.4561	0.4698	0.4835	0.4972	0.5109
14.0	0.3794	0.3933	0.4072	0.4141	0.4211	0.4350	0.4489	0.4627	0.4766	0.4905	0.5044	0.5183

*Temperature correcting formula

$$pH = 7.0 + \frac{V}{0.00019842T}$$

where V = observed potential in volts

T = 273.16 + t

t = degrees centigrade

About the Author

Lowell E. McCaw is Professor Emeritus, Monroe Community College, Rochester, NY. He retired from active college teaching June 1988; since then he has been active in the community as a technical training consultant. Some of the area firms he has consulted for are Eastman Kodak, Xerox, E.I. Du Pont, ABB Kent Taylor, and Rochester Gas and Electric Company. He is currently on the Adjunct Teaching Staff of Rochester Institute of Technology and Monroe Community College. For the last eight years he has taught numerous courses for ISA, where he is a member of the Education Department, and is an Evaluator of Instrumentation Technology Programs for TAC/ABET.

A2

Basics of Electricity and Electronics

This appendix is meant as a brief review of electricity and electronics. No attempt is made to teach electricity, since there are excellent books available on this subject.

Electrons

Current is the flow of electrons, the electron being the negative particle of an atom. Electrons orbit around the central nucleus, which consists of protons and neutrons. The electron has a minute and definite amount of negative electrical charge. The proton is positively charged, while the neutron is neutral. The charge of an electron = 1.6×10^{-19} coulomb. The mass of an electron = 9.11×10^{-28} grams.

Direct Current

Direct current is defined as an electric current that flows in only one direction (see Figure A2-1). An important equation related to the simple direct current circuit is Ohm's law, which is stated as follows:

$$E = I R$$

where:

E	=	voltage across the circuit, volts
I	=	current through the circuit, amperes
R	=	resistance of the circuit, ohms

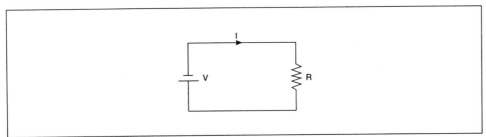

Figure A2-1. Simple Direct Current Circuit

Circuits in Series

When resistances are joined like the links of a chain, they are said to be in series (see Figure A2-2). The total resistance is equal to the sum of the separate resistances, as expressed in the following equation:

$$R_t = R_1 + R_2 + \dots + R_n$$

The sum of the voltage drops (*IR*) across each resistance in the circuit is equal to the applied voltage, according to Kirchoff's law of voltage, but the current through all the resistances is the same.

Circuits in Parallel

When resistances in a circuit are joined like the rungs of a ladder, they are said to be in parallel (see Figure A2-3). The reciprocal of the total resistance is equal to the sum of the reciprocals of the individual resistances, as expressed in the following equation:

$$\frac{1}{R_t} = \frac{1}{R_1} + \frac{1}{R_2} + \dots + \frac{1}{R_n}$$

According to Kirchoff's law of current, the sum of the currents entering a junction point in an electrical circuit is equal to the sum of the currents leaving the junction point. The ratio of the branch currents is the inverse of the resistance ratio.

Electrical circuits may have a combination of series and parallel arrangements and several sources of voltage. These circuits are called resistance networks. The resistance bridge is an application of resistances and Ohm's law.

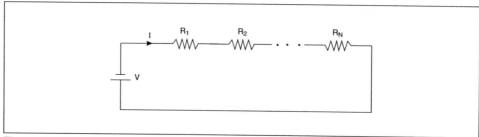

Figure A2-2. Resistances in Series

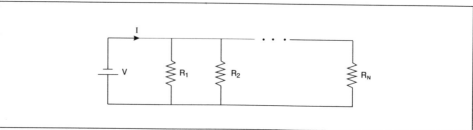

Figure A2-3. Resistances in Parallel

Bridge Circuits

RESISTANCE BRIDGES

An example of a simple bridge circuit is shown in Figure A2-4. It is a network that consists of four resistance arms that form a closing circuit. A current source is applied to two opposite junctions, and a galvanometer is connected to the other junctions. This circuit is often used to determine the value of an unknown resistance.

To achieve a balanced bridge circuit the null method measurement is applied. The bridge is balanced when $I_g = 0$. At this condition, points 2 and 4 are at the same potential. That is, the voltage drop across A is equal to that across X, or:

$$A\,I_{AB} = X\,I_{XS} \quad \text{also} \quad B\,I_{AB} = S\,I_{XS}$$

Therefore:

$$\frac{A}{B} = \frac{X}{S} \quad \text{or} \quad X = \frac{AS}{B}$$

The value of resistance X is unknown, but it can be determined if A, B, and S are known. This circuit, called the "Wheatstone bridge," is the heart of many measuring devices. This is because the resistance of materials changes with strain, temperature, chemical concentration, and so on. It may be used with alternating current (AC) circuits as well as with direct current (DC) circuits.

MISCELLANEOUS BRIDGE CIRCUITS

Many variations of the simple Wheatstone bridge circuit are used in instrumentation. These include: slide-wire bridges, fixed-ratio bridges, Kelvin bridges, megohm bridges, unbalanced bridges, Mueller bridges, self-balancing bridges, reactance bridges, simple capacitance bridges, Schering bridges, Maxwell bridges, Hay bridges, Owen bridges, Heaviside bridges, Wien bridges, and AC bridges.

Potentiometers

Careful voltage measurements are made in comparing the very low voltage generated by thermocouples with the temperature differences that created them. For this purpose a potentiometer is used. A potentiometer is defined as an instrument that measures an unknown potential difference by balancing that difference against a known potential difference. It consists of:

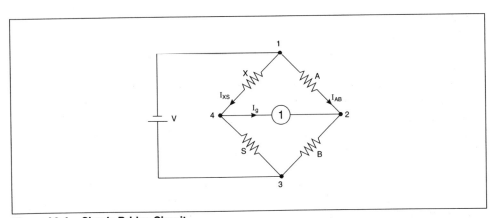

Figure A2-4. Simple Bridge Circuit

(1) a resistance element made of many turns of carefully spaced wire, and

(2) an arm that can contact any of the turns.

Effect of Temperature on Resistance

The resistance of most conductors and of all pure metals increases with temperature. Carbon (graphite) is an exception; it has a negative temperature coefficient of resistance; its resistance decreases as temperature increases. For engineering purposes, the resistance R at any temperature t can be expressed in terms of the resistance at some chosen reference temperature. This reference temperature is usually either 0°C or 20°C, and the proportionate increase in resistance with temperature for any material is called the *temperature coefficient of resistance*, designated α .

The relationship between resistance and temperature can be derived as follows. Let R_0 be the resistance of a conductor at the reference temperature 0°C and let the temperature coeffficient of resistance be ohm/ohm/°C at 0°C. From the definition of temperature coefficient, the change in resistance will be $R_0 \times \alpha \times t$, such that the resistance R_t at t degrees C is expressed by the equation:

$$R_t = R_0 + R_0 \alpha t$$

or

$$R_t = R_0 (1 + \alpha t)$$

Temperature coefficients are used in the design of RTD (resistance temperature detector) sensors.

Conductance

Conductance is the inverse of resistance. The mho (ohm reversed) is the unit of conductance, which is represented by the letter G. Conductance is a measure of the ease with which a conductor conducts electricity.

$$Conductance = \frac{1}{Resistance}$$

or

$$G\ (mhos) = \frac{1}{R\ (ohms)}$$

Magnetism

Everyone is familiar with the common magnet and with its two poles — north and south — located opposite each other. The law of magnetism states that like poles repel one another and unlike poles attract.

According to the molecular theory of magnetism, every molecule of a magnetic substance is itself a small magnet. In an unmagnetized material, the molecules lie at random, the tiny magnets neutralizing each other. When the material is magnetized, the molecules line up with the flux lines so that their north poles all point in one direction with their south poles in the opposite direction.

Magnetic flux is the total number of lines of force in a magnetic field. The unit of flux is the weber (Wb). A line of force is an imaginary line whose direction at any point gives the direction of the magnetic force at that point.

MAGNETISM AND ELECTRIC CURRENTS

Conductors of electricity set up around themselves magnetic fields that depend for their strength and direction upon the current flowing in the conductor. The magnetic field will cease to exist after the current has been interrupted. This principle is used in solenoid coils. If a conductor is placed in a magnetic field and then current is passed through it, the conductor experiences a force that tends to displace it in the field. All magnetic lines of force strive to take the shortest distance between poles. They can be said to resist distortion or stretching. This principle is made use of to produce mechanical motion from electrical energy in an electric motor. The lines of force that emanate from a magnet form closed loops. The path that the lines trace out around these loops is called the magnetic circuit.

The magnetic circuit passes through a series of different materials, including air. Each material offers some resistance to the passage of lines of force. This resistance is called the *reluctance* of the circuit. Reluctance is measured as the ratio of the magnetomotive force to the magnetic flux produced (an analogy with electrical resistance). The *magnetomotive force* is that force needed to establish the magnetic lines of force in a magnetic circuit. *Flux density* is the ratio of the total magnetic flux to the area through which it passes.

Permeability of a material is the measure of how much magnetic flux will be produced in a certain material by a given magnetizing force. The following units are used in magnetic circuit calculations:

H =   magnetomotive force (mmf), ampere-turns
Φ =   flux, webers
B =   flux density, webers/m^2
S =   reluctance
   =   permeabilty
A =   area, m^2

These quantities are related by the following equations:

$$B = \frac{\Phi}{A}$$

$$S = \frac{H}{\Phi}$$

$$\mu = \frac{B}{H}$$

INDUCTANCE

The process of converting electric energy into magnetic energy and back again into electric energy is called induction. *Inductance* is a measure of the ability of a circuit to induce energy into another circuit or back into itself. Also, inductance is the electromagnetic ability of a circuit or component.

There are two types of inductance: self-inductance and mutual inductance. Self-inductance is the ability of a circuit to induce electric energy from itself back into itself. Mutual inductance is the ability of a combination of two or more separate circuits or components to induce electric energy into each other. Inductances or coils may be joined in series or in parallel. When they are joined in series, the total inductance equals the sum of the individual inductances.

$$L_t = L_1 + L_2 + \ldots + L_n$$

When in parallel, the reciprocal of the total inductance equals the sum of the reciprocals of the individual inductances:

$$\frac{1}{L_t} = \frac{1}{L_1} + \frac{1}{L_2} + \ldots + \frac{1}{L_n}$$

Inductance is also the property of an electric circuit that resists a change in current.

Capacitance

Capacitance is the property that resists a change in voltage. The circuit component that possesses this property is called a capacitor. A simple capacitor consists of two parallel plates separated by air or other nonconductive material called the dielectric. An application of capacitance is for level detection and measurement. The symbol for capacitance is the letter C. The unit of capacitance is the farad. A circuit has a capacitance of one farad when a charge, Q, of one coulomb is required to raise the circuit voltage, V, by one volt, as expressed in the following equation:

$$C = \frac{Q}{V}$$

The coulomb is the unit of the electrical charge. Smaller units of capacitance are the microfarad (one millionth of a farad) and the picofarad (one millionth of a microfarad). Like resistors and inductors, capacitors can be connected in series or in parallel. When the capacitors are connected in series, the reciprocal of the total capacitance equals the sum of the reciprocals of the individual capacitances:

$$\frac{1}{C_t} = \frac{1}{C_1} + \frac{1}{C_2} + \ldots \frac{1}{C_n}$$

When the capacitors are connected in parallel, the total capacitance equals the sum of the individual capacitances:

$$C_t = C_1 + C_2 + \ldots + C_n$$

The buildup of voltage across a capacitor is called charging; the reduction of the voltage is called discharging. In actual circuits, a resistor is added to the circuit to limit the charging current (and thereby protect the power source). The effect of this current-limiting resistor is to increase the time required to charge the capacitor. When a capacitor is being charged through a resistor, the voltage will rise the instant the switch is closed (see Figure A2-5).

The time required for the capacitor voltage to reach 63. 2% of its final value is called the *time constant*. The time constant, τ, (in seconds) equals resistance, R, (in ohms) times capacitance, C, (in farads), as shown in the following equation:

$$\tau = RC$$

The capacitor discharge curve has the same shape as the charge curve except that it is inversed — virtually a mirror image. The time constant of discharge is the same as that for the charge to build up.

RC networks have many applications in instrumentation.

Alternating Current

An alternating current flows first in one direction, then in the opposite direction. The current starts at zero amplitude and builds up to a maximum value (peak value). It then drops back to zero, builds up to a peak value flowing in the opposite direction (negative value) and then rises to zero again (see Figure A2-6). The dotted line beginning at +10 represents a direct current of 10 amperes. The

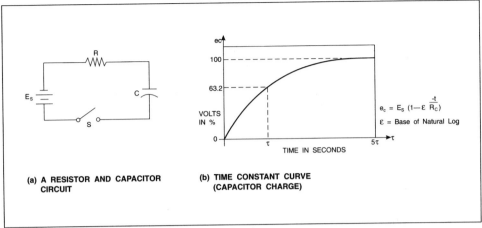

Figure A2-5. RC Circuit and Time Response Graph

sine wave represents an AC current flow. Each complete set of 360° is called a full wave or cycle. The number of complete cycles that occur in one second is referred to as the *frequency* of the circuit. The unit of frequency is the hertz (Hz).

AC Circuits

There are three basic AC circuits: resistive, inductive, and capacitive, depending upon the type of component in the circuit. All or any of these three types may be combined in one circuit.

RESISTIVE

If an alternating voltage is applied to a resistance or a number of resistances, the voltage and the current stay in phase with each other. The circuit calculations can be carried out in the same way as for DC. For example:

$$Power\ (\ in\ wats\ )\ =\ Volts \times Amperes$$

Examples of resistive-only loads are lighting and heating.

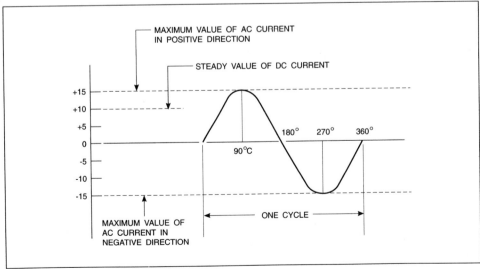

Figure A2-6. AC and DC Current Curves

INDUCTIVE

The principles of inductance were discussed in the section on DC circuits. Motion of either a conductor in a field or a field across a conductor produces an *electromotive force* (emf) and, according to Lenz's law, this emf will act to oppose the motion that caused it.

When an alternating current is applied to a coil, the same principle applies. Each time the current builds up, a changing magnetic field is produced. The buildup of this field cuts through the turns of the coil and produces an emf that opposes the current flow. The same effect, in reverse, takes place when the current reduces, and, thus, the field collapses. This effect is known as *inductive reactance*; it tends to restrict the current flow and also to cause the current to lag behind the applied voltage.

The unit of inductive reactance is the ohm, and the symbol is X_L. It is related to voltage, E, and current, I, by the following equation:

$$X_L = \frac{E_L}{I_L}$$

or

$$1 \; ohm \; of \; inductive \; reactance = \frac{1 \; volt \; across \; inductor}{1 \; ampere \; through \; inductor}$$

The magnitude of inductive reactance, X_L (in ohms), is directly proportional to the frequency, f (in hertz), and the inductance, L (in henrys), but not to the magnitude of the applied voltage. This is expressed as:

$$X_L = 2\pi f L$$

CAPACITIVE

When an AC supply voltage is applied to a capacitive circuit, the voltage across the capacitor changes continuously, following the pattern of the applied voltage. This means that the capacitor must be continuously charging and discharging, thus apparently causing an AC current to flow through it. The current is proportional to the rate at which the voltage across the capacitor is changing. When the voltage *is changing* at its maximum rate, the current *is at* its maximum (positive or negative) value. The current in this case leads the voltage by 90 degrees. The current and voltage maintain a fixed pattern; they have a constant ratio.

The ratio E/I represents resistance. Capacitance in an AC circuit behaves like resistance, except that it is not instantaneous. This resistance is called *capacitive reactance*, X_C, which is measured in ohms and is related to E_C and I_C by the following equation:

$$X_C = \frac{E_C}{I_C}$$

or

$$1 \; ohm \; of \; capacitive \; reactance = \frac{1 \; volt \; across \; capacitor}{1 \; ampere \; through \; capacitor}$$

The magnitude of capacitive reactance, X_C (in ohms), is inversely proportional to the frequency, f (in hertz), and the capacitance, C (in farads), by the following equation:

$$X_C = \frac{1}{2\pi f c}$$

Impedance

The total resistance to current flow in an AC circuit is called *impedance*, represented by the symbol Z, which is related to applied voltage, E, and circuit current, I, by the following equation:

$$Z = \frac{E}{I}$$

The value of impedance is obtained by a vector diagram (see Figure A2-7), using a rectangular graph with θ being the angle between the applied voltage and the resulting current. For a circuit that contains a resistance and an inductance in series, the magnitude of the impedance is:

$$Z = \sqrt{R^2 + X_L^2}$$

For a circuit that contains a resistance and a capacitance in series, the magnitude of the impedance is:

$$Z = \sqrt{R^2 + X_C^2}$$

For a circuit that contains a resistance, an inductance, and a capacitance in series, the magnitude of the impedance is:

$$Z = \sqrt{R^2 + \left(X_L - S_C\right)^2}$$

VALUES FOR AC POWER

Instantaneous value is the value at a particular instant of time. It will differ from one instant to the next.

Maximum or peak value is the greatest value reached during alternation, usually occurring once in each half cycle.

Average or mean value is the average value of the current or voltage. If an average value is taken over a full cycle, the positive and negative half-cycles will cancel out to give a zero result. The average value quoted is thus the average of one half-cycle.

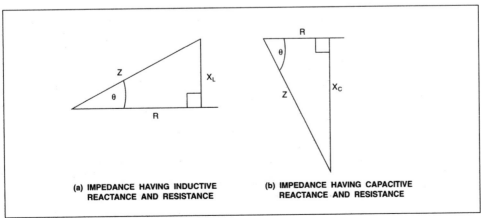

(a) IMPEDANCE HAVING INDUCTIVE REACTANCE AND RESISTANCE

(b) IMPEDANCE HAVING CAPACITIVE REACTANCE AND RESISTANCE

Figure A2-7. Impedance Triangles

Root-mean-square (RMS) or effective value of an AC current is that value which has the same heating effect as a DC current. The heating effect depends upon the square of the current flow.

The following is the relationship between the values when the shape of the wave form is the ideal sinusoidal:

Average value = $0.637 \times$ Maximum value

RMS value = $0.707 \times$ Maximum value

Maximum value = $1.41 \times$ RMS value

Solid-State Electronics

Diodes and transistors are made of materials called *semiconductors*, which are halfway between conductors and insulators. At one time, transistors were made of germanium; today, most are made of silicon. In their pure forms, neither silicon nor germanium are useful in semiconductor devices; impurities are added to alter their characteristics.

When a semiconductor material has been "doped" (impurity added) with arsenic, for example, the result is the formation of excess electrons and what is called an N-type semiconductor. (The N refers to the negative carriers — the free electrons.) Arsenic is called a donor impurity because it donates an easily freed electron.

To form a P-type semiconductor, a shortage of an electron is required. This refers to a "hole" in the crystal, and holes act as positive carriers. A material that will leave a hole in the covalent bond is indium. Indium is called an acceptor impurity because its atoms leave holes in the crystal structure that are free to accept free electrons. In addition to indium, boron and aluminum have been used as acceptor impurities.

There are five principal classifications of semiconductors:

(1) Diode

(2) Bipolar or junction transistor

(3) Silicon-controlled rectifier (SCR)

(4) Unijunction transistor

(5) Unipolar or field effect transistor

Diodes

A semiconductor diode is usually a single crystal of semiconductor material that is artificially created. One half of the crystal is made N-type and the other made P-type. Where the two sections meet is called a junction, giving rise to the term *junction diode*.

When a battery is connected across a junction diode, it is said to be biased. When the positive terminal of a battery is connected to the P-type material and the negative terminal to the N-type material, the junction is forward-biased. If the battery connection is reversed, the diode is reverse-biased. Figure A2-8 shows a forward-biased diode circuit.

Notice the symbol for a diode. The arrow part, called the anode, is the P-section of the diode. The flat part, called the cathode, is the N-section. The arrow of the diode symbol points in the direction of the "conventional" current flow (I_{cn}), which flows from positive to negative. The electron current flow (I_{el}), however, flows from negative to positive. Conventional current is used here in the discussion of applications of other circuit devices. In cases where the internal operation of semiconductors is explained, electron current flow is used. The junction diode

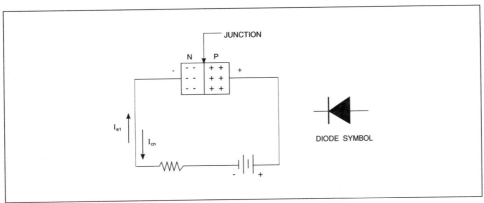

Figure A2-8.   Forward-Biased Diode Circuit

can be used as a rectifier to convert AC current to pulsating DC current. When a capacitor is connected across the load (resistance), this pulsating DC current is smoothed out into a more constant current. There are several other types of diodes in addition to the conventional junction type.

A *Zener diode* has a larger junction than the rectifier-type diode, and it works in the reverse direction from the conventional diode. The design is intended to protect the diode from excessive reverse voltages or breakdown voltages. This type of diode is used as a voltage reference and as a voltage regulator. Zener diodes can be made with breakdown or Zener voltages as low as one volt and as high as several hundred volts.

A *tunnel diode* is a highly doped junction diode made of either germanium or gallium arsenide. As a result of the high doping, the depletion region around the junction is extremely narrow. Because of the narrow depletion region, holes and electrons can cross the junction by more or less tunneling from one atom to another. If the voltage in a tunnel diode depletion region increases, the current decreases; if the voltage decreases, the current increases. This gives the opposite effect of resistance and is, therefore, called negative resistance. A circuit with a tunnel diode can oscillate.

The *positive-intrinsic-negative diode*, usually called a pin-diode, is a special diode that is doped so that it has very low resistance with a forward bias across it and a very high resistance with a reverse bias.

The *varactor diode* is one diode that is heavily doped and used as a variable capacitor. When the diode is reverse-biased, holes on the P-side and electrons on the N-side of the junction will move away from the junction. This is the same as moving the plates of a capacitor apart.

The *point-contact diode* is usually made with a small piece of N-type material. A large contact is fastened to one side of the crystal, and a thin wire is attached to the other side. A small region of P-type material is formed around the contact. This diode makes a better detector than the junction diode because it has a lower capacity.

The *optoelectronic diode* is simply a specially made diode that interacts with light to a useful extent. It is based on the principle that whenever light strikes semiconductor material it tends to knock bound electrons out of their sockets, so to speak, thus creating free electrons and holes. Conversely, when an electron falls into a hole, it tends to create a particle of light (photon). The photodiode is a light sensor and the light-emitting diode (LED) is a light source. LEDs are made of gallium arsenide; photodiodes are made of silicon.

Diodes may also be used in bridge circuits. For example, four diodes are used in a full-wave bridge rectifier circuit.

Semiconductor Triodes

Junction transistors (triodes) are made up of a single semiconductor crystal with three different regions: N-type, P-type, and N-type; or P-type, N-type, and P-type. The three sections are referred to as emitter, base, and collector. The base is located in the center region. Because there are two separate N-P junctions, these transistors are called bipolar transistors.

NPN TRANSISTORS

Figure A2-9 shows an NPN transistor connected to the external power source necessary to provide a current flow through the transistor. The negative terminal connection has the effect of repulsing the free electrons in the emitter to the emitter-base junction since this junction is forward-biased. At the same time, the positive terminal connection to the base repulses the holes to the emitter-base junction where the electrons and holes readily combine. The net effect is an electron flow across the junction, sustained as long as the battery is connected.

If the battery leads are reversed, the net effect is to prevent electrons from crossing the junction. The emitter-base circuit resembles a reverse-biased diode, and the transistor is said to be "cut off" here.

PNP TRANSISTORS

Figure A2-10 shows a common-base PNP transistor amplifier circuit. The N-type region is wired positive and the P-type region negative for the base-collector junction and is, therefore, reverse-biased. For the emitter-base junction, however, the N-type is connected negative and the P-type positive, and is, thus, forward-biased.

In a PNP transistor, holes are attracted from the emitter to the base because the base has been negatively biased by the battery. The collector is, in turn, negative with respect to the base. Since the base is relatively thin, most of the holes attracted to it from the emitter flow through it into the collector. In a given transistor, the percentage of holes that leave the emitter and get into the collector is relatively constant for a wide range of applied bias voltages. This value is around 98%. The remaining 2% of the holes are conducted out of the base lead.

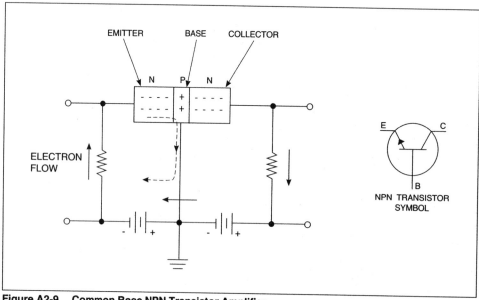

Figure A2-9.   Common Base NPN Transistor Amplifier

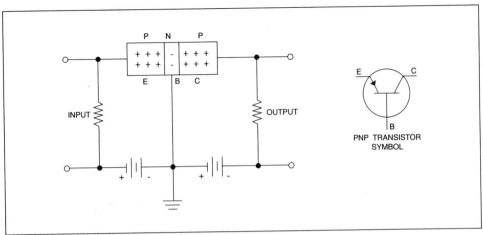

Figure A2-10. Common Base PNP Transistor Amplifier

The NPN transistor operates by having a large electron flow through the transistor from the emitter to the collector. The PNP transistor, however, operates on the flow of holes from emitter to collector. There is a flow of electrons through the transistor circuit in a direction opposite that of the flow of holes within the transistor.

Transistor Amplifiers

The function of a transistor is to amplify a small current flow between base and emitter to control a larger current flow between emitter and collector.

Figures A2-9 and A2-10 are examples of common base amplifiers. Even though the action of the transistors is different, the net result using the common base circuit is the same with both types of transistors. Two other amplifier circuits have been designed: the common emitter amplifier and the common collector amplifier; both use either an NPN or a PNP transistor. The most widely used circuit is the common emitter amplifier, because of its high gain. The ability of a transistor to amplify a signal current depends upon the fixed percentage of current that flows from the emitter to the collector. When a voltage is applied to the base in such a manner as to increase the forward biasing, or voltage, the emitter and collector currents are increased. The magnitude of amplification is a characteristic of the particular circuit. An example of a common base amplifier characteristic is the ratio, designated as , of the change in collector current to the change in emitter current and is expressed as:

$$\alpha = \frac{\Delta i_c}{\Delta i_c}$$

In the case where $\alpha < 1$, the circuit has a current gain of less than 1. Two other characteristics are the input and the output impedances. Input impedance is:

$$Z_1 = \frac{e_1}{i_1}$$

and output impedance is:

$$Z_2 = \frac{e_2}{i_2}$$

The ratio, β, of changing collector current to the changing base current is:

$$\beta = \frac{\Delta i_c}{\Delta i_c}$$

Equations relating the α and β of a transistor are:

$$\beta = \frac{\alpha}{(1 - \alpha)}$$

$$\alpha = \frac{\beta}{(1 + \beta)}$$

AMPLIFIER FEEDBACK

When part of the signal coming out of a transistor amplifier is fed back to the input, it is called *feedback. Negative feedback* results when the output signal being fed back opposes the input signal. This results in a lower gain for the amplifier, but it makes the amplifier more linear and more stable. If enough negative feedback is employed, the gain of the amplifier becomes almost independent of the transistor. This type of amplifier is called an *operational amplifier* (op amp).

If the voltage being fed back into the input tends to increase the input signal, the result is *positive feedback*. Positive feedback increases the gain of an amplifier but reduces its stability. If too much positive feedback is used, the output of the amplifier becomes independent of the input signal and the result is oscillation.

Transistor Oscillators

An *oscillator* is a circuit that is used to generate a varying voltage from a DC voltage source. There are two general types of oscillators: sinusoidal and nonsinusoidal. Sinusoidal oscillators generate a sinusoidal voltage (shaped like a sine wave). This is accomplished by applying the right amount of positive feedback to start and maintain oscillation. Nonsinusoidal oscillators generate a voltage that has a square or sawtooth shape. This is done by applying many times the amount of positive feedback that would be required for oscillation. Oscillators are used to generate audio frequency voltages for telemetering and remote control applications.

Other Electronic Devices

The *silicon-controlled rectifier* (SCR) is a type of diode with very low resistance when conducting and very high resistance in the off (or closed) state. Because of these characteristics SCRs are mainly used in power control (motor control) and switching circuits, and they may be used to control either AC or DC currents. The SCR is a four-layer semiconductor, classified as a thyristor. Its biasing is similar to that of a regular diode, but it will not conduct current until it is gated (switched). The formal name is "silicon reverse-blocking triode thyristor." An SCR can be considered as two transistors connected side by side.

The *triac* is another silicon-controlled device. It is the equivalent of two SCRs side by side. Its operation is similar to that of the SCR. The triac is used for "phase control" of a motor. Its formal name is "bidirectional triode thyristor."

The *unijunction transistor* (UJT) is a one-junction semiconductor. It is suitable as an oscillator. The UJT is connected with a capacitor in a relaxation oscillator circuit. The UJT can also be implemented to trigger an SCR at distinct time intervals to allow the SCR to turn on motor control circuits.

The *field effect transistor* (FET), or *unipolar transistor*, is made from a single silicon crystal (P-type or N-type) called a P-channel or an N-channel. The channel is mounted on a substrate, which is usually neutral, and has three terminals (called the source, the gate, and the drain) in two small regions on opposite sides of the crystal. When there are two embedded N-regions or two P-regions called the gate, it is called a dual-gate JFET. Its operation, called conductivity modulation, depends on either holes or electrons but not both. The FET is used as an amplifier. When the source-gate voltage is high enough, no more current carriers will reach the drain, and $I_D = 0$. This condition is called the pinchoff point. Different types of field effect transistors are available: the junction field effect transistor (JFET), the insulated gate field effect transistor (IGFET), and the metal-oxide semiconductor field effect transistor (MOSFET). The MOSFET is actually an IGFET.

An *integrated circuit* is a complete electronic circuit that contains transistors, diodes, resistors, and capacitors and their interconnecting electrical conductors. It is processed on and contained entirely within a single chip of silicon.

Integrated circuits are classified as either (1) switching or digital or (2) amplifying or linear. From integrated circuits, many other devices have evolved (TTL or transistor-transistor-logic circuits, MOS memory and calculator chips, microprocessor on a chip, and computer on a chip).

A3

Basics of Chemistry

A practical knowledge of chemistry is needed to understand the purpose of control systems in the process industry; however, it is beyond the scope of this book to teach basic chemistry. Therefore, this appendix simply reviews some of the basics and applications. An instrumentation designer must be aware that the process engineer deals with questions such as:

(1) What effects do large quantities of inert gases and monatomic gases have upon a system?

(2) Do chemical reactions always release heat? Do they release heat at different rates?

(3) Are electrical principles involved in all chemical reactions?

Process accuracy and output depend on the answers to questions such as these. Chemistry is the science that deals with the composition and properties of substances and the energy changes they undergo when they combine with one another.

All chemical changes are accompanied by energy changes — by either the absorption or emission of some form of energy — and such energy is predictable. In most chemical reactions the energy is released in the form of heat. Some chemical reactions require the absorption of heat. Chemical energy changes involve alteration of one or more of the molecules, the basic building blocks in any chemical reaction.

Basic Definitions

Atom. The smallest particle of an element that has all the properties of the element.

Atomic number. The positive charge on the nucleus of an atom and, therefore, the number of protons in the nucleus. The symbol Z is often used. Since atoms are electrically neutral, the atomic number also gives the number of electrons in orbit.

Atomic structure. Mentioned in the appendix on electricity. It is widely known that atoms consist of electrons, protons, and neutrons. The nuclei of atoms contain both protons and neutrons, except for hydrogen, which has only a single proton in its nucleus. The electrons orbit around the nucleus. The neutron is a particle that has no charge. In radioactive ray bombardment, however, neutrons may break down to form protons and electrons.

Atomic weight (of an element). The average of the weights of the naturally occurring isotopes, taking into account their relative abundance.

Compound. A substance that consists of two or more elements and can be decomposed chemically. Elements seldom occur freely in nature, but are found in compounds. Water is a well known compound that consists of oxygen and hydrogen, and by means of electrical energy it may be broken down into these elements. The proper ties of compounds are very different from those of their constituent elements. Obviously a change in energy of elements occurs when they combine to form compounds.

Diatomic element. A molecule of any element that contains only two atoms. Examples are: oxygen, nitrogen, and hydrogen.

Element. Any substance that cannot be further decomposed or broken down by chemical energy. Examples are: oxygen, silicon, aluminum, iron, and gold. Although 106 elements have been classified (1990), only 88 occur naturally; the others are synthetic.

Isotopes. Atoms of the same element that differ in mass because of a different number of neutrons.

Mass weight (of an atom). The sum of the protons and neutrons in the nucleus.

Mixture. The physical combination of two or more elements or compounds in no definite proportions by weight. The combining substances do not lose their original identity and may be separated from the mixture by mechanical means.

Molecule. The smallest quantity of any substance that has the physical and chemical properties of that substance.

Monatomic element. A molecule of any element that contains only one atom. Examples are: copper, iron, and helium.

Substance. Any variety of matter that has fixed and definite properties. Pure substances are classed as elements or compounds. Impure substances are mixtures.

Valence

The arrangement of the electrons of the planetary model of the atom is shown in Figure A3-1. The layers around the nucleus in which the electrons occur are called electron shells, or energy levels, and they are numbered or lettered.

The maximum capacity rule is: no shell may hold more electrons than 2 × the number of the shell. Electrons in the outermost shell are called valence electrons.

Active Elements

Elements whose outer electron shells are incomplete are called "active elements." There is a tendency for incomplete outer shells to be completed. This

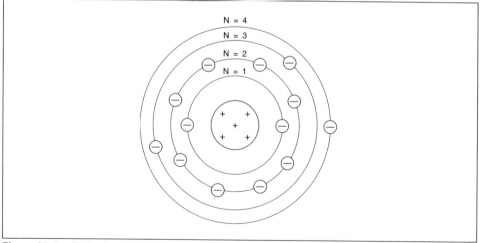

Figure A3-1.   Rutherford's Planetary Model of an Atom

happens when the spaces are filled with electrons from other atoms or when outermost electrons are lost, leaving the next inner orbit intact. The number of electrons so surrendered, acquired, or shared at the outermost orbit is known as the valence of an atom. Most of the information pertaining to the valence, atomic number, and atomic weight of an element can be found in the periodic table of the elements. Terms associated with valence are: ionization energy, electron affinity, and electronegativity.

Ionization energy is the energy absorbed by an atom to remove one outer electron from the attractive force of the nucleus to such a distance that the electron may be considered to be no longer attracted. Most metals have a low ionization energy level. The noble gas elements (such as helium, neon, krypton, and argon) have high ionization potentials.

Electron affinity is the energy released when an atom gains an electron to form a negatively charged ion (called an anion).

Electronegativity is a property of an element derived from ionization energy, electron affinity, and other properties. It may be defined as the measure of the attraction of an atom for the shared electron pair that holds the atom bonded to another. A chemical bond is simply any force of electrical attraction between electrons and positive nuclei of the atoms held together in a sample of matter. Bonds are broadly classified as covalent, ionic, or metallic, although intermediate types may occur.

Covalent Bonding

Covalent bonding occurs when atoms with the same or slightly different electronegativities form molecules by sharing pairs of electrons. The shared electrons hold the two atoms together.

A molecule is a definite unit of matter made up of atoms with relatively strong covalent bonds between them. In comparison, the forces of attraction between neighboring molecules are relatively weak (these are called van der Waal's forces). The chemical bond is considered ionic if the electronegativity difference is large.

The electrovalence of an atom is the charge on the ion formed when the atom gains or loses electrons in forming an ionic bond. Atoms become positive ions by losing electrons; they become negative ions by gaining electrons.

Symbols, Formulas, and Equations

Symbols and Formulas

A symbol represents one atom of an element; thus, there is H for hydrogen, Na for sodium, Fe for iron, and so on.

A formula represents a molecule if the compound exists as a definite molecular unit. Examples are: O_2 for oxygen, NaCl for salt, and CO_2 for carbon dioxide, and so on. A formula also shows the ratio of atoms in a crystal (such as SiO_2 for silica). Similarly, ionic compounds are given a formula that indicates the simple ratio of the ions in the compound.

Binary compounds have only two elements in the formula.

Equations

All chemical reactions may be represented by equations. The ability to balance simple equations is the key to understanding industrial chemical processes, to locating suitable methods and points of measurement, and to appreciating the results that may be anticipated.

A chemical equation is a short method of describing some aspects of a reaction. The formulas on the left of the equation show the reactants present before the change, and the products of the reaction are shown on the right. Here is an example:

$$\text{Sodium} \ + \ \text{Water} \rightarrow \text{Sodium hydroxide} \ + \ \text{Hydrogen}$$
$$\text{(Reactants)} \qquad\qquad\qquad \text{(Products)}$$

A chemical equation must be balanced. That is, the total of all atoms on the left must equal the total on the right. This follows from the law of conservation of matter: matter is neither created nor destroyed in a chemical change. Consider the following equation:

$$2H_2 + O_2 \qquad \rightarrow \qquad 2H_2O$$
$$\text{Hydrogen} \ + \ \text{Oxygen} \ \rightarrow \ \text{Water}$$

The above equation reads: two molecules of hydrogen react with one molecule of oxygen to produce two molecules of water. $H_2 + O \rightarrow H_2O$ is not correct because oxygen is a diatomic molecule (O_2). Sometimes one of the products escapes as a gas; this is shown by an upward pointing arrow. For example:

$$Zn + 2HCl \rightarrow ZnCl_2 + H_2 \uparrow$$
$$\text{Zinc} + \text{Hydrochloric acid} \ \rightarrow \ \text{Zinc chloride} \ + \ \text{Hydrogen} \uparrow$$

A downward arrow after a formula indicates that the substance is insoluble in water and is precipitated.

$$Pb\,(NO_3)_2 + KI \rightarrow \quad PbI_2 \downarrow + KNO_3$$
$$\text{Lead nitrate} + \text{Potassium iodide} \ \rightarrow \ \text{Lead iodide} \downarrow + \text{Potassium nitrate}$$

Thermochemical Reactions

The Law of Chatelier states that when a system in equilibrium is subjected to a change (in temperature, concentration, or pressure), there is a shift in the point of equilibrium that tends to restore the original condition or to relieve the strain.

When matter undergoes change, energy is always liberated or absorbed. In chemical changes, this reaction is often in the form of heat, known as the heat of reaction.

An exothermic reaction liberates heat:

$$C + O_2 \rightarrow CO_2 + 94{,}050 \text{ cal/mole}$$
$$\text{Carbon} + \text{Oxygen} \rightarrow \text{Carbon dioxide} + \text{Heat}$$

An endothermic reaction absorbs heat:

$$N_2 + O_2 + 43{,}000 \text{ cal/2 moles} \rightarrow 2NO$$
$$\text{Nitrogen} + \text{Oxygen} + \text{Heat} \rightarrow \text{Nitrous Oxide}$$

One mole of a substance is its molecular weight in grams.

Molecular Formulas and Percentage Composition

The molecular weight of an element or compound is the weight of one molecule relative to the weight of the C-12 atom, taken as 12.0000 atomic mass units, or amu (1 amu $= 1.66 \times 10^{-24}$ grams).

One mole of molecules of a compound contains the Avogadro number of molecules. Avogadro's number $= 6.024 \times 10^{23}$. Avogadro's law states that equal volumes of gases, when subjected to the same conditions of temperature and pressure, contain an equal number of molecules. Atoms and molecules are samples of matter and therefore have mass, which may be expressed in grams (amu).

The average molecular weight of a mixture of substances can be obtained by the following equation:

$$M_n = \frac{W_a + W_b + W_c + \dots}{W_a/M_a + W_b/M_b + W_c/M_c \dots}$$

where:

M_n = molecular weight of mixture
$W_{a,b,c}$ = weights of individual pure components
$M_{a,b,c}$ = molecular weight of pure components

The mole fraction is the ratio of the moles of one component to the total number of moles in the mixture.

The Mole Volume (Gas)

Many engineering calculations use the ideal gas law, which is expressed as:

$$PV = nRT$$

where:

P = pressure
V = volume
n = number of moles of gas
R = gas constant
T = absolute temperature

This equation shows that a mole of gas under conditions of standard temperature and pressure (stp) always occupies a definite volume regardless of the nature of the gas. This volume is called the mole volume. A gram mole of ideal gas at a temperature of 0°C and a pressure of 760 mm Hg occupies 22.41 liters. Other gas laws are given in the section on pressure in Chapter 1.

Gas Density

The mole volume is also called the gram molecular volume (gmv). If the formula of a gas is known, one may calculate its density, or the weight of any volume of the gas at any temperature and pressure, using the equation:

$$\text{Density of gas at stp} = \frac{\text{Mole mass}}{\text{Volume}}$$

Concentration

The reaction of solutions depends on their concentration. The molarity of a solute (the material that enters into and is dissolved in a solution) is the number of moles of solute per liter of solution. If the atomic weight and the valence of an element are known, its equivalent weight can easily be calculated, using the equation:

$$\text{Equivalent weight (grams)} = \frac{\text{Atomic weight}}{\text{Valence}}$$

The normality of a solute in a solution is the weight required to furnish or react with one mole of hydrogen ion, per liter of solution. When the atomic weight and valence of an element are known its equivalent weight can easily be calculated, using the equation:

Inorganic Chemistry

Solutions that conduct electrical current are called electrolytes; those that do not are called non-electrolytes. Electrical conductivity requires the presence of ions, which are produced when ionic solutes dissolve, or dissociate, in solution. Perfect dissociation and the number or ions present depend on several factors:

(1) The more weakly concentrated the solution, the higher the dissociation, and the greater the number of positive and negative ions. Water dissociates into positive and negative ions and is, in effect, a very weak electrolyte.

$$H_2O \rightarrow H^+ + OH^-$$

Water → Hydrogen ion + Hydroxide ion

(2) The more easily a solute dissociates in a solution, the greater the number of ions. Both the solute and the solvent must be considered. Hydrochloric acid dissociates 100% in water; in benzene it dissociates only 1%.

(3) Some substances dissociate easily at higher temperatures, others at lower temperatures. Temperature has a considerable effect on conductivity of any electrolyte.

Acids and Bases

Since the positive and negative ions in water are exactly in balance, many elements added to water can upset the balance of hydrogen and hydroxide ions.

ACIDS

Substances that increase the concentration of hydrogen ions are called acids.

$$HCl + H_2O \rightarrow H_3O^+ + Cl^-$$

Hydrogen chloride + Water → Hydrochloric acid

BASES

Substances that increase the hydroxide ion concentration are called bases.

$$NaOH \rightarrow Na^+ + OH^-$$

Sodium hydroxide → Sodium ion + Hydroxide ion

SALTS

Some salts are formed by neutralization and may be regarded as the compound formed when the positive ion from the base combines with the negative ion from the acid.

pH VALUES

One liter of pure water at 220°C contains 10^{-7} moles of hydrogen ion and 10^{-7} moles of hydroxyl ion. In water, the equilibrium product of the hydrogen and hydroxyl ion concentrations is a constant, 10^{-14} at 22°C. pH is mathematically expressed as:

$$pH = \log\left(\frac{1}{\text{Hydrogen ion concentration}}\right)$$

where the hydrogen ion concentration is expressed in moles/liter. Therefore, the pH of water, since it is neutral, equals:

$$pH = \log\frac{1}{10^{-7}} = 7.0$$

Acid solutions increase in strength as the pH values fall below 7. Basic or alkaline solutions increase in strength as the pH rises above 7.

Metals

An element acts like a metal if it readily surrenders its valence electrons. A metal does not share electrons — it loses them. Nonmetals tend to acquire or share valence electrons. There are more nonmetals than metals in nature; more than two-thirds of all elements are nonmetals. Metals are divided into groups according to the periodic table of the elements.

Corrosion is defined as the destructive alteration of a metal resulting from electrochemical or chemical action. There are several important kinds of corrosion.

Galvanic corrosion. Through a corrosive electrolyte, one metal is in contact with a more noble metal. The less noble metal is the "sacrifice metal." This is a pure electrochemical reaction that involves the transfer of ions accompanied by electric current. An example is a steel bolt in a copper plate.

Oxygen cell corrosion. This is similar in action to the ion transfer of galvanic action. A cell is formed by the current in a cycle from metal-low-oxygen to electrolyte to high-oxygen-metal.

Dezincification. This is the removal of less noble zinc from yellow brass or admiralty metal, leaving weak, spongy copper. The attack, which increases with temperature, may take place in either acid or alkaline solutions.

Blistering. The formation of blisters in steel, it results from acid concentration and corrosion due to the release of atomic hydrogen from such materials as H_2S and H_2SO_4.

Soil corrosion. This refers to the attack on metals buried in soil. Causes are:

(1) chemical composition of soil (its pH value),

(2) variations in soil composition,

(3) temperature of soil, and

(4) presence of bacteria leading to anerobic corrosion.

Amalgamation attack. This refers to the attachment of mercury to nonferrous metals.

Catalysis

A catalyst is any substance that alters the speed of a reaction without participating in it. A positive catalyst increases the reaction speed; a negative catalyst (inhibitor) decreases the speed. In neither case is the catalyst permanently changed, nor does it initiate chemical reaction.

Contact catalysts act by providing as much surface as possible, making more collisions between entering molecules. A mixture of iron oxide and potassium aluminate greatly accelerates the formation of ammonia from nitrogen and hydrogen.

$$N2 + 3H_2 \rightarrow 2NH_3$$

Nitrogen + Hydrogen $\rightarrow$ Ammonia

Catalysts are widely used in industrial chemical processes. Although they remain unchanged chemically at the conclusion of a reaction, there may be physical changes, such as size reduction.

Another example is the production of sulfur dioxyde, which uses finely divided particles of platinum as a catalyst.

$$S + O_2 \rightarrow SO_2$$

Sulfur + Oxygen → Sulfur dioxide

Many catalysts are sensitive to some substances, that become absorbed on the catalyst surfaces. If this happens, the catalyst is said to be "poisoned." Processes are available to regenerate poisoned catalysts.

Halogens, which belong to group VII A of the periodic table, are the most chemically active family of nonmetals. They are fluorine, chlorine, iodine, bromine, and astatine.

Organic Chemistry

Organic chemistry is a special branch of chemistry devoted to the study of carbon and its compounds. More than a million known compounds of carbon exist. Carbon is present in the atmosphere as carbon dioxide gas, and it is a constituent of living matter, coal, petroleum, natural gas, sugar, alcohol, and a host of other materials. It also occurs in rocks and diamonds.

Carbon is element number 6 (it has 6 protons, 6 neutrons and 6 electrons). Carbon has the particular ability to form long chains and rings in which carbon is linked to more carbon by sharing pairs of electrons. Coal is formed by the slow "carbonization" of vegetable matter, during which process there is an escape of hydrogen.

Hydrocarbons are compounds of carbon and hydrogen only. They are classified as open-chain (aliphatic) hydrocarbons and ring hydrocarbons (aromatics). Petroleum products are hydrocarbons.

Cracking is the thermal decomposition of large molecules into smaller ones. Cracking is used to get a higher yield of more desirable fuel elements.

Polymerization is the reverse of cracking. It is the union of two or more low-molecular-weight molecules to yield a molecule of higher molecular weight, the resulting constituents of which are in the same proportion. The molecular weight of such a polymer is always an exact multiple of the original substance. It is accompanied by heat and, usually, a catalyst.

The primary purpose of polymerization is to convert hydrocarbons, such as gasoline, into a more useful form by adding higher "fractions" of the original distillate to the gasoline. In other words, a chemical compound is transformed into a new chemical compound; for example, butane (C_4H_8) is polymerized to form isooctane (C_8H_{18}).

Aliphatic Hydrocarbons

Alkanes are saturated hydrocarbons with the general formula C_NH_{2N+2}. Methane gas (CH_4) is the simplest alkane. Adding another carbon atom results in ethane (C_2H_6). Refer to Figure A3-2.

Building up the chain with more carbon and hydrogen atoms results in propane (C_3H_8), butane (C_4H_{10}), pentane (C_5H_{12}), hexane (C_6H_{14}), and so on.

Alkenes are unsaturated hydrocarbons in which there is a double covalent bond between a pair of carbon atoms in each member of the group. The general formula is C_NH_{2N}. The simplest member is ethene (ethylene). See Figure A3-3.

Because a carbon atom has four valence electrons, four bonds must always be shown from each carbon atom in a correct structural formula. That is why the = symbol is shown between the carbon atoms, in Figure A3-3.

Alkynes have the general formula C_NH_{2N-2}; a triple covalent bond exists between two carbon atoms in the molecule. Acetylene (ethyne), C2H2, is the simplest member. See Figure A3-4.

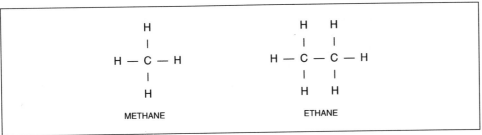

Figure A3-2. Structure of Methane and Ethane

The functional unit of the alcohols is an OH attached to a hydrocarbon chain. The most important alcohols are methanol (methyl alcohol), CH_3OH, and ethanol (ethyl alcohol), C_2H_5OH. The general formula for a monohydric alcohol (one OH group) is $C_NH_{2N+1}OH$. Methanol is produced from carbon monoxide and hydrogen heated with a nickel-copper catalyst.

$$CO + 2H_2 \rightarrow CH_3OH$$

Carbon monoxide + Hydrogen $\rightarrow$ Methanol

Ethyl alcohol (spirits) may be obtained from ethylene (ethene) but is mainly produced by the fermentation of sugar using yeast.

$$C_6H_{12}O_6 \rightarrow 2C_2H_5OH + 2CO_2$$

Sugar $\rightarrow$ Ethyl alcohol + Carbon dioxide

Alcohols may be oxidized to aldehydes and then to acids.

$$C_2H_5OH \rightarrow 2CH_3CHO \quad 2CH_3COOH$$

Alcohol $\rightarrow$ Acetaldehyde   Acetic acid

Structural isomerism is possible beginning with C_3H_7OH as shown in Figure A3-5.

Aromatic Hydrocarbons

Aromatics occur naturally in certain petroleum fractions or are produced in catalytic reforming operations. They can also be chemically manufactured from coal.

Aromatics with side chains may be converted into benzene by hydrodealkylation in a hydrogen atmosphere in the presence of a catalyst. The main aromatics for further processing are benzene, toluene, and the xylenes, of which there are three isomers (ortho, meta, and para). See Figure A3-6.

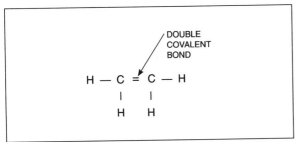

Figure A3-3. Structure of the Simplest Alkene — Ethene

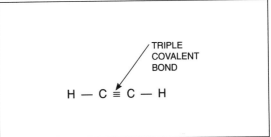

Figure A3-4. Structure of Simplest Alkyne — Acetylene

In general, hydrocarbons may be gaseous, liquid, or solid at normal temperature and pressure, depending on the number and arrangement of carbon atoms in their molecules. Those with up to four carbon atoms are gaseous; those with twenty or more are solid; those in between are liquid. Liquid mixtures, such as most crude oils, may contain either gaseous or solid compounds, or both, in solution.

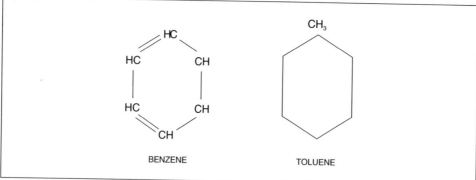

Figure A3-5. N-Propyl Alcohol to Isopropyl Alcohol

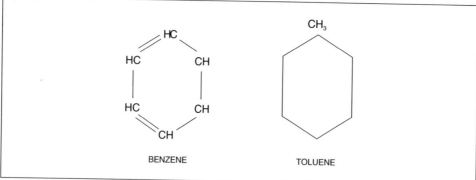

Figure A3-6. Aromatic Hydrocarbons — Benzene and Toluene

Index

Index

Index